Atsushi Uchida

Optical Communication with Chaotic Lasers

Related Titles

Paschotta, R.
Encyclopedia of Laser Physics and Technology
2008
Hardcover
ISBN: 978-3-527-40828-3

Moon, F. C.
Applied Dynamics
With Applications to Multibody and Mechatronic Systems
2008
Softcover
ISBN: 978-3-527-40751-4

Meschede, D.
Optics, Light and Lasers
The Practical Approach to Modern Aspects of Photonics and Laser Physics
2007
Softcover
ISBN: 978-3-527-40628-9

Kane, D., Shore, A. (eds.)
Unlocking Dynamical Diversity
Optical Feedback Effects on Semiconductor Lasers
2005
E-Book
ISBN: 978-0-470-85620-8

Beth, T., Leuchs, G. (eds.)
Quantum Information Processing
2005
Hardcover
ISBN: 978-3-527-40541-1

Bachor, H.-A., Ralph, T. C.
A Guide to Experiments in Quantum Optics
2004
Softcover
ISBN: 978-3-527-40393-6

Ghafouri-Shiraz, H.
Distributed Feedback Laser Diodes and Optical Tunable Filters
2003
E-Book
ISBN: 978-0-470-85622-2

Shapiro, M., Brumer, P.
Quantum Control of Molecular Processes
2011
Hardcover
ISBN: 978-3-527-40904-4

Atsushi Uchida

Optical Communication with Chaotic Lasers

Applications of Nonlinear Dynamics and Synchronization

WILEY-VCH Verlag GmbH & Co. KGaA

The Author

Dr. Atsushi Uchida
Saitama University
Department of Information and Computer Sciences
Saitama, Japan
auchida@mail.saitama-u.ac.jp

All books published by **Wiley-VCH** are carefully produced. Nevertheless, authors, editors, and publisher do not warrant the information contained in these books, including this book, to be free of errors. Readers are advised to keep in mind that statements, data, illustrations, procedural details or other items may inadvertently be inaccurate.

Library of Congress Card No.: applied for

British Library Cataloguing-in-Publication Data
A catalogue record for this book is available from the British Library.

Bibliographic information published by the Deutsche Nationalbibliothek
The Deutsche Nationalbibliothek lists this publication in the Deutsche Nationalbibliografie; detailed bibliographic data are available on the Internet at http://dnb.d-nb.de.

© 2012 Wiley-VCH Verlag & Co. KGaA, Boschstr. 12, 69469 Weinheim, Germany

All rights reserved (including those of translation into other languages). No part of this book may be reproduced in any form – by photoprinting, microfilm, or any other means – nor transmitted or translated into a machine language without written permission from the publishers. Registered names, trademarks, etc. used in this book, even when not specifically marked as such, are not to be considered unprotected by law.

Typesetting Thomson Digital, Noida, India
Printing and Binding Markono Print Media Pte Ltd, Singapore
Cover Design Grafik-Design Schulz, Fußgönheim

Printed in Singapore
Printed on acid-free paper

Print ISBN: 978-3-527-40869-6
ePDF ISBN: 978-3-527-64035-5
oBook ISBN: 978-3-527-64033-1
ePub ISBN: 978-3-527-64034-8
mobi ISBN: 978-3-527-64036-2

Persistence pays off.

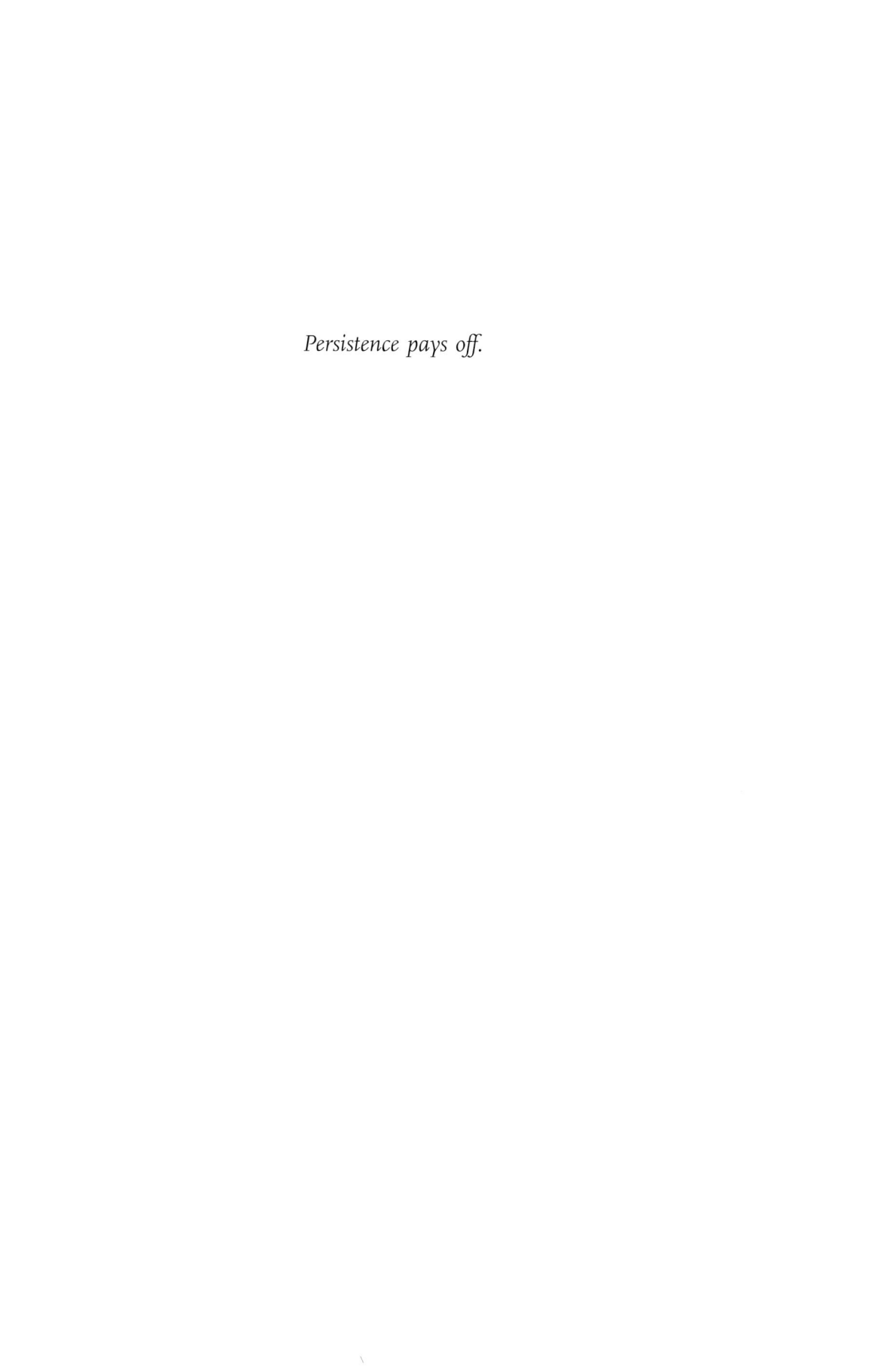

Foreword

In this book, Professor Atsushi Uchida lays the groundwork for many possible applications of nonlinear dynamical systems in optical science and engineering – including optical communications with chaotic lasers. The first part of the book introduces lasers and the phenomena of chaos and synchronization, while the latter part explores applications of chaotic lasers to communications, control and random number generation.

That lasers can exhibit irregular dynamics has been known since almost the first observations of laser operation. To actually understand and diagnose the origins of such behavior is an effort that still continues today. The realization that deterministic chaos, sources of random noise, and the interactions of multiple frequencies in a nonlinear optical medium could generate dynamics that range over many orders of magnitude in time scales has gradually grown and blossomed into a new research field. We know now that the development of accurate models to understand and predict the nonlinear dynamical behavior of lasers and optical systems requires new tools for observation, measurement and interpretation that involve both software and hardware, numerical models and stochastic simulations. The quantitative description of all the variables that describe the spatiotemporal dynamics of electromagnetic fields interacting with matter require new mathematical concepts, numerical simulation techniques, and considerable imagination to think of possible new applications that may prove useful in the future.

It is truly breathtaking to look back at the period in the 1950s and 1960s in which, almost simultaneously, the basic concepts of chaos, computers, information theory and lasers were discovered and developed. In the 1970s and 1980s, semiconductor lasers, integrated circuits, and fiber optics merged with and transformed the appearance, properties and applications of lasers. In the last two decades, miniaturized computer technology and digital imaging have begun to transform our world in ubiquitous ways that use lasers and electronics for everyday tasks and activities.

While almost all the mathematics and engineering needed to understand the vast majority of devices and instruments are "linear" in concept, we have learned that almost everything in nature and the laboratory behaves nonlinearly when one extends the ranges of parameters. As scientists engaged in exploring the unknown, we have glimpsed that in fact, nothing is truly linear in behavior, though linearity is a good and most useful approximation for many purposes. Chaotic and nonlinear dynamics appears everywhere we look, from the solar system to the motion of pendula, from turbulence in fluids to the fluctuations of light emitted by lasers. New paradigms have been established as the field of nonlinear dynamics of optical systems has been developed through the painstaking efforts of engineers, mathematicians and physicists.

As we begin to enlarge the boundaries of exploration, whether through new mathematical and computational tools or experimental instruments and measurements, we discover unexpected phenomena and begin to piece together parts of the infinite variety of patterns in space and time that constitute the dynamically evolving world around us. One of these new, unexpected phenomena is that of synchronization of chaotic nonlinear systems. While the concept of synchronization of periodic oscillators may seem natural to most people, the observation that two or more chaotic systems can synchronize is surprising and leads to novel possibilities for communication, control and sensing systems.

This book is a major step and a systematic effort to bring together those aspects of mathematics, physics and engineering that are basic in the emerging field of communication with chaotic dynamical systems. It describes many examples from recent literature that illustrate how one can communicate with lasers that are chaotic, using the phenomenon of synchronization of chaotic systems. The dynamics of laser and optoelectronic systems that can be used as transmitters and receivers for chaotic communication are often described by delay-differential equations. The complex and high-dimensional dynamics that can be generated by such systems, and their synchronization properties, are a rich and open field. The techniques developed to generate chaotic waveforms for communication at high bit rates, and the methods to encode and decode information, are active fields of research, with much work that remains to be done at the device and system levels. Communication networks and the privacy and security of information are important topics of exploration.

While I have only seen an overview of the text, it is clear that this book contains many interesting connections of the central topic to related areas in information science and communications theory. One of the novel features of this book is the chapter on random number generators using chaotic lasers, a newly emerging field where the author has made pioneering contributions. Another set of topics that one may not expect from the title are presented in the final chapter, on remote sensing, blind-source separation and fractal optics. These represent areas of the author's active research interests. Another great feature of the book are the many appendices to the chapters that provide detailed guidance to the student and researcher through the tricky landscape of computational algorithms. As a collaborator and colleague who has over a period of time spent many (many!) hours discussing research with Professor Uchida, and as a student of both optics and nonlinear dynamics, it is with much pleasure and anticipation that I look forward to seeing this book in print, and learning about many of the topics developed in it.

Rajarshi Roy
Institute for Physical Science and Technology
and Department of Physics
University of Maryland, College Park
USA

Contents

Foreword *VII*
Preface *XXVII*

Part One   Basic Physics of Chaos and Synchronization in Lasers

1   Introduction   *3*
  1.1   Lasers and Chaos   *6*
    1.1.1   Lasers   *6*
    1.1.2   Chaos   *6*
    1.1.3   Connection Between Lasers and Chaos   *7*
    1.1.4   Chaos and Noise   *8*
  1.2   Synchronization of Chaos and Optical Communication   *8*
    1.2.1   Synchronization of Chaos   *8*
    1.2.2   Optical Communication with Synchronized Chaotic Lasers   *9*
  1.3   Random Number Generation with Chaotic Lasers and Other Applications   *11*
    1.3.1   Random Number Generation   *11*
    1.3.2   Controlling Chaos and Other Applications   *12*
  1.4   Research Directions for Engineering Applications with Chaotic Lasers   *12*
  1.5   Outline of This Book   *13*
    1.5.1   Contents of Chapters: First Parts for the Basics   *14*
    1.5.2   Contents of Chapters: Second Part for the Applications   *16*

2   Basics of Chaos and Laser   *19*
  2.1   History of Instabilities of Laser Output   *19*
    2.1.1   Examples of Laser Instabilities   *19*
      2.1.1.1   First Observation of Laser Instabilities in a Ruby Laser   *19*
      2.1.1.2   Green Problem in a Solid-State Laser for Second-Harmonic Generation   *20*
      2.1.1.3   Feedback-Induced Instability in a Semiconductor Laser for Optical Disks and Optical Communication Systems   *21*
    2.1.2   Inherent Noise-Induced Instabilities   *23*
    2.1.3   Deterministic Chaos in Lasers and Lorenz-Haken Equations   *24*

2.2 Basic Chaos Theory 26
    2.2.1 Logistic Map in Discrete-Time System 26
        2.2.1.1 Recurrence Formula 26
        2.2.1.2 Chaotic Sequence 27
        2.2.1.3 Bifurcation Diagram 28
        2.2.1.4 Lyapunov Exponent 30
    2.2.2 Lorenz Model in a Continuous-Time System 31
        2.2.2.1 Lorenz Equations 31
        2.2.2.2 Temporal Waveform and Attractor 32
        2.2.2.3 Sensitive Dependence on Initial Conditions 32
        2.2.2.4 Bifurcation Diagram 33
        2.2.2.5 Lyapunov Exponent 33
    2.2.3 Reconstruction of Attractor in Time-Delayed Phase Space 35
    2.2.4 Types of Bifurcation and Route to Chaos 36
        2.2.4.1 Period-Doubling Route to Chaos 36
        2.2.4.2 Quasiperiodicity Route to Chaos 37
        2.2.4.3 Intermittency Route to Chaos 37
2.3 Basic Laser Theory 38
    2.3.1 Light–Matter Interaction for Laser Radiation 38
        2.3.1.1 Elements of a Laser 38
        2.3.1.2 Two-Atomic-Level Description and Mechanism of Laser Oscillation 39
    2.3.2 Radiative Recombination of Electron–Hole Pairs in Semiconductor Lasers 40
    2.3.3 Rate Equations for Laser Dynamics 41
    2.3.4 Relaxation Oscillation Frequency 44
    2.3.5 Mechanism of Chaotic Instability in Lasers 45
2.4 Connection Between Chaos and Lasers 46
    2.4.1 Single-Mode Laser Model Based on Maxwell–Bloch Equations 46
    2.4.2 Classification of Laser Models Based on Decay Rates 48
        2.4.2.1 Class C Lasers (Three Variables) 48
        2.4.2.2 Class B Lasers (Two Variables) 48
        2.4.2.3 Class A Lasers (One Variable) 49
Appendix
    2.A.1 General Formula of the Linearized Equations and the Derivation of the Linearized Equations in the Lorenz Model 51
    2.A.2 General Formula for the Calculation of the Lyapunov Exponent in Nonlinear Dynamical Systems Without Time Delay 53
    2.A.3 Rössler Model 54
    2.A.4 Derivation of the Relaxation Oscillation Frequency 55

3 Generation of Chaos in Lasers 59
3.1 Basics of Generation of Chaos in Lasers 59
    3.1.1 Classification of Generation Techniques of Chaos in Lasers 59

        3.1.1.1   Optical Feedback *60*
        3.1.1.2   Optical Coupling and Injection *62*
        3.1.1.3   External Modulation *62*
        3.1.1.4   Insertion of Nonlinear Element *62*
        3.1.1.5   Multimode Lasers and High-Dimensional Laser Systems *63*
        3.1.1.6   Class C Lasers Satisfying the Condition for Lorenz–Haken Chaos *63*
        3.1.1.7   Ikeda-Type Passive Optical Systems *63*
    3.1.2   Characteristics of Chaos in Lasers *64*
        3.1.2.1   High-Frequency Chaos *64*
        3.1.2.2   Existence of Two Oscillation Frequencies in Different Orders of Magnitude *64*
        3.1.2.3   Middle Degrees of Freedom *65*
        3.1.2.4   Physical Variables *65*
        3.1.2.5   Transmission of Chaos *65*
        3.1.2.6   Variety of Laser Models *65*
    3.1.3   How to Distinguish between Chaos and Noise from Experimental Data? *66*
        3.1.3.1   Experimental apparatus for Measurement of Instability in Laser Intensity *66*
        3.1.3.2   Examples of Chaos and Noise from Real Experimental Data *66*
        3.1.3.3   Observation of Bifurcation *67*
        3.1.3.4   Comparison with Laser Model *68*
3.2   Chaos in Semiconductor Lasers *69*
    3.2.1   Semiconductor Laser with Optical Feedback *69*
        3.2.1.1   Dynamical Regime and L–I Characteristics *69*
        3.2.1.2   Coherence Collapse and Chaos in Experiment *71*
        3.2.1.3   Numerical Results and Bifurcation Diagram *73*
        3.2.1.4   Low-Frequency Fluctuations (LFFs) *75*
        3.2.1.5   Short-Cavity Regime *78*
    3.2.2   Semiconductor Laser with Polarization-Rotated Optical Feedback *79*
    3.2.3   Semiconductor Laser with Optoelectronic Feedback *83*
    3.2.4   Semiconductor Laser with Optical Injection and Coupling *88*
        3.2.4.1   Temporal Dynamics and Bifurcation Diagram *88*
        3.2.4.2   Bandwidth Enhancement of Chaos by Optical Injection *91*
    3.2.5   Semiconductor Laser with Injection Current Modulation *92*
    3.2.6   Vertical-Cavity Surface-Emitting Laser (VCSEL) *93*
        3.2.6.1   Basic Characteristics *93*
        3.2.6.2   Optical Feedback *96*
        3.2.6.3   Optical Injection *97*

- 3.3 Chaos in Electro-Optic Systems 98
  - 3.3.1 Ikeda-Type Nonlinear Delay Dynamics 98
  - 3.3.2 Wavelength Chaos in Electro-Optic System 101
  - 3.3.3 Intensity Chaos in Electro-Optic System 103
  - 3.3.4 Phase Chaos in Electro-Optic System 105
- 3.4 Chaos in Fiber Lasers 107
  - 3.4.1 External Modulation 109
  - 3.4.2 Multimode (Dual-Wavelength) Dynamics 112
  - 3.4.3 Fast Polarization Dynamics 114
- 3.5 Chaos in Solid-State Lasers 115
  - 3.5.1 External Modulation 116
  - 3.5.2 Frequency-Shifted Optical Feedback 117
  - 3.5.3 Antiphase Dynamics in Multimode Laser 120
  - 3.5.4 Insertion of Nonlinear Crystal 122
- 3.6 Chaos in Gas Lasers 124
  - 3.6.1 Lorenz–Haken Chaos 124
  - 3.6.2 Insertion of Saturable Absorbers 128
  - 3.6.3 External Modulation 130
  - 3.6.4 Optical Injection 130
  - 3.6.5 Multimode Laser 132
  - 3.6.6 Low-Frequency Dynamics with Optical Feedback 134
- Appendix
  - 3.A.1 Numerical Model for Chaotic Dynamics in a Semiconductor Laser 135
    - 3.A.1.1 Coherent Optical Feedback 135
    - 3.A.1.2 Polarization-Rotated Optical Feedback 135
    - 3.A.1.3 Optoelectronic Feedback (Incoherent Feedback) 136
  - 3.A.2 Mechanism of Low-Frequency Fluctuations (LFFs) in Semiconductor Laser with Optical Feedback 137
  - 3.A.3 Numerical Model for Chaotic Dynamics in a Vertical-Cavity Surface-Emitting Laser (VCSEL) 139
  - 3.A.4 Numerical Model for Chaotic Dynamics in an Electro-Optic System 139
  - 3.A.5 Numerical Model for Chaotic Dynamics in a Fiber Laser 140
  - 3.A.6 Numerical Model for Chaotic Dynamics in a Solid-State Laser 140
    - 3.A.6.1 Single-Mode Laser Model 140
    - 3.A.6.2 Multimode Laser Model with Spatial Hole Burning (Tang–Statz–deMars Equations) 141
  - 3.A.7 Numerical Model for Chaotic Dynamics in a Gas Laser 142

4 Analysis of Chaotic Laser Dynamics: Example of Semiconductor Laser with Optical Feedback 145

- 4.1 Experimental Analysis of Semiconductor Lasers with Optical Feedback 145

- 4.1.1 Experimental Setup *146*
- 4.1.2 Light Power Versus Injection Current (L–I) Characteristics *147*
- 4.1.3 Temporal Waveforms, RF Spectra, Autocorrelations, and Optical Spectra *149*
- 4.1.4 Two-Dimensional Map *155*
- 4.2 Model for Semiconductor Laser with Optical Feedback *156*
  - 4.2.1 Lang–Kobayashi Equations *156*
  - 4.2.2 Derivation of the Electric-Field Amplitude and Phase of the Lang–Kobayashi Equations from the Complex Electric-Field Equations *158*
  - 4.2.3 Derivation of the Real and Imaginary Electric Fields of the Lang–Kobayashi Equations from the Complex Electric-Field Equations *160*
- 4.3 Analytical Approach of Semiconductor Laser with Optical Feedback *161*
  - 4.3.1 Steady-State Solutions without Optical Feedback *161*
  - 4.3.2 Linear Stability Analysis for Steady-State Solutions without Optical Feedback *162*
    - 4.3.2.1 Eigenvalues of Jacobian Matrix *162*
    - 4.3.2.2 Relaxation Oscillation Frequency *165*
  - 4.3.3 Steady-State Solutions with Optical Feedback *166*
  - 4.3.4 Linear Stability Analysis for Steady State Solutions with Optical Feedback *169*
- 4.4 Numerical Analysis of Semiconductor Laser with Optical Feedback *172*
  - 4.4.1 Numerical Results of Chaotic Dynamics *173*
    - 4.4.1.1 Temporal Waveforms, FFTs, and Attractors *173*
    - 4.4.1.2 Bifurcation Diagram and Two-Dimensional Dynamical Map *176*
  - 4.4.2 Linear Stability Analysis for Oscillatory Trajectory *179*
  - 4.4.3 Maximum Lyapunov Exponent *181*
- 4.5 Dimensionless Equations and Further Nonlinear Analysis *184*
  - 4.5.1 Dimensionless Equations *184*
  - 4.5.2 Numerical Results of Chaotic Dynamics *187*
  - 4.5.3 Linear Stability Analysis for Oscillatory Trajectory in Dimensionless Equations and Maximum Lyapunov Exponent *188*
  - 4.5.4 Lyapunov Spectrum *190*
  - 4.5.5 Kolmogorov–Sinai Entropy and Kaplan–Yorke Dimension *192*
- 4.6 Lang–Kobayashi Equations with Gain Saturation *194*
  - 4.6.1 Lang–Kobayashi Equations with Gain Saturation *194*
  - 4.6.2 Numerical Results and Histogram *195*
  - 4.6.3 Linear Stability Analysis for Oscillatory Trajectory and Measurement of Complexity *196*

Appendix
- 4.A.1 Derivation of the Rate Equations of Semiconductor Lasers *199*
- 4.A.2 Derivation of the Rate Equations of Semiconductor Lasers with Optical Feedback *205*
- 4.A.3 Analytical Approach of the Stability of Steady State Solutions for Semiconductor Laser with Optical Feedback Under the Limits of Weak and Short Feedback *206*
- 4.A.4 Runge–Kutta Method for the Integration of Ordinary Differential Equations *208*
- 4.A.5 Gram–Schmidt Orthogonalization *210*

5 Synchronization of Chaos in Lasers *211*

- 5.1 Concept of Synchronization of Chaos in Lasers *211*
  - 5.1.1 Introduction *211*
  - 5.1.2 What is Synchronization? *212*
  - 5.1.3 Why Should Chaos be Synchronized? *213*
  - 5.1.4 Characteristics of Synchronization of Chaos in Laser Systems *215*
  - 5.1.5 Synchronization of Chaos for Communication Applications *216*
- 5.2 History of Synchronization of Chaos in Lasers *217*
  - 5.2.1 Synchronization of Chaos in Electronic Circuits *217*
    - 5.2.1.1 Pecora–Carroll Method *217*
    - 5.2.1.2 Pyragas Method *218*
  - 5.2.2 Synchronization of Chaos in Lasers *219*
- 5.3 Coupling Schemes and Synchronization Types *223*
  - 5.3.1 Identical Synchronization *224*
  - 5.3.2 Generalized Synchronization (with High Correlation) *225*
  - 5.3.3 Synchronization of Chaos in Feedback Systems *225*
    - 5.3.3.1 Open-Loop Configuration *225*
    - 5.3.3.2 Closed-Loop Configuration *227*
  - 5.3.4 Mutual Coupling *228*
  - 5.3.5 Linear Stability Analysis and Conditional Lyapunov Exponent *229*
- 5.4 Examples of Synchronization of Chaos in Semiconductor Lasers *230*
  - 5.4.1 Semiconductor Lasers with Coherent Optical Feedback *231*
  - 5.4.2 Semiconductor Lasers with Polarization-Rotated Optical Feedback *236*
  - 5.4.3 Semiconductor Lasers with Optoelectronic Feedback *237*
  - 5.4.4 Semiconductor Lasers with Optical Injection *239*
  - 5.4.5 Semiconductor Lasers with Mutual Coupling *240*
    - 5.4.5.1 Symmetry Breaking and Leader–Laggard Relationship *240*
    - 5.4.5.2 Zero-Lag Synchronization *242*
  - 5.4.6 Vertical-Cavity Surface-Emitting Lasers (VCSELs) *243*
- 5.5 Examples of Synchronization of Chaos in Electro-Optic Systems and Other Lasers *245*
  - 5.5.1 Electro-Optic Systems *245*

- 5.5.2 Fiber Lasers *248*
- 5.5.3 Solid-State Lasers *250*
- 5.5.4 Gas Lasers *253*
- 5.6 Specific Types of Synchronization *254*
  - 5.6.1 Phase Synchronization *254*
  - 5.6.2 Generalized Synchronization (with Low Correlation) *258*
- 5.7 Consistency *263*
  - 5.7.1 What is Consistency? *263*
  - 5.7.2 Examples of Consistency in Laser Systems *265*
  - 5.7.3 Application of Consistency *269*
    - 5.7.3.1 Noninvasive Testing *269*
    - 5.7.3.2 Analysis of Brain Dynamics and Learning Process in the Brain *269*
    - 5.7.3.3 Physical One-Way Function *269*
    - 5.7.3.4 Teaching-Learning Methodologies *270*
    - 5.7.3.5 Design of Drug Delivery *270*
- Appendix
  - 5.A.1 Pecora–Carroll Method for Synchronization of Chaos *270*
  - 5.A.2 General Formula of Linearized Equations for Coupled Differential Equations *271*
  - 5.A.3 Procedure for the Calculation of the Conditional Lyapunov Exponent *273*
  - 5.A.4 Numerical Model for Synchronization of Chaos in Unidirectionally Coupled Semiconductor Lasers *274*
    - 5.A.4.1 Coherent Optical Feedback *274*
    - 5.A.4.2 Polarizaiton-Rotated Optical Feedback *275*
    - 5.A.4.3 Optoelectronic Feedback (Incoherent Feedback) *276*
  - 5.A.5 Numerical Model for Synchronization of Chaos in Unidirectionally Coupled Electro-Optic Systems *277*
  - 5.A.6 Numerical Model for Synchronization of Chaos in Unidirectionally Coupled Fiber Lasers *278*
  - 5.A.7 Numerical Model for Synchronization of Chaos in Unidirectionally Coupled Solid-State Lasers *279*
    - 5.A.7.1 Single-Mode Laser Model *279*
    - 5.A.7.2 Multimode Laser Model with Spatial Hole Burning (Tang–Statz–deMars Equations) *280*
  - 5.A.8 Numerical Model for Synchronization of Chaos in Unidirectionally Coupled Gas Lasers *281*
  - 5.A.9 Definition of Phase in Chaotic Temporal Waveform by the Hilbert Transform *282*

6 Analysis of Synchronization of Chaos: Example of Unidirectionally Coupled Semiconductor Lasers with Optical Feedback *285*

- 6.1 Experimental Analysis on Synchronization of Chaos in Two Semiconductor Lasers with Optical Feedback *285*

- 6.1.1 Experimental Setup for Synchronization of Chaos *285*
- 6.1.2 Experimental Results of Synchronization of Chaos *287*
- 6.1.3 Parameter Dependence of Synchronization of Chaos *290*
- 6.2 Model for Synchronization of Chaos in Two Coupled Semiconductor Lasers with Optical Feedback *293*
  - 6.2.1 Lang–Kobayashi Equations for Synchronization of Chaos in Unidirectionally Coupled Semiconductor Lasers with Optical Feedback *293*
    - 6.2.1.1 Coupled Lang–Kobayashi Equations *293*
    - 6.2.1.2 Identical Synchronous Solution *295*
  - 6.2.2 Derivation of the Electric-Field Amplitude and Phase of the Coupled Lang–Kobayashi Equations from the Complex Electric-Field Equations *296*
  - 6.2.3 Derivation of the Real and Imaginary Electric Fields of the Coupled Lang–Kobayashi Equations from the Complex Electric-Field Equations *299*
- 6.3 Numerical Analysis on Synchronization of Chaos in Unidirectionally Coupled Semiconductor Lasers with Optical Feedback *300*
  - 6.3.1 Measures for Synchronization of Chaos *300*
    - 6.3.1.1 Two Types of Synchronization and Cross-Correlation Values *300*
    - 6.3.1.2 Optical Frequency Detuning *302*
  - 6.3.2 Numerical Results of Temporal Waveforms and Correlation Plots *303*
  - 6.3.3 Parameter Dependence of the Two Types of Synchronization *305*
  - 6.3.4 Linear Stability Analysis of Synchronous Oscillatory Solutions for Identical Synchronization *308*
    - 6.3.4.1 Linearized Equations *308*
    - 6.3.4.2 Conditional Lyapunov Exponent *309*
    - 6.3.4.3 Numerical Results of Conditional Lyapunov Exponent *310*
  - 6.3.5 Open- Versus Closed-Loop Configurations *312*
- 6.4 Experimental Analysis on Generalized Synchronization with Low Correlation in Three Semiconductor Lasers in the Auxiliary System Approach *313*
  - 6.4.1 Experimental Setup for Generalized Synchronization with Low Correlation in the Auxiliary System Approach *314*
  - 6.4.2 Experimental Results of Generalized Synchronization *316*
  - 6.4.3 Parameter Dependence of Generalized Synchronization *320*
  - 6.4.4 Dependence of Generalized Synchronization on Optical Phase of Feedback Light *322*
- 6.5 Model for Generalized Synchronization with Low Correlation in Three Semiconductor Lasers in the Auxiliary System Approach *324*
  - 6.5.1 Coupled Lang–Kobayashi Equations for Generalized Synchronization in the Auxiliary System Approach *325*

6.5.2 Synchronous Solutions for Generalized Synchronization in the Auxiliary System Approach *327*

6.6 Numerical Analysis on Generalized Synchronization of Chaos in Three Semiconductor Lasers in the Auxiliary System Approach *327*

  6.6.1 Temporal Waveforms *329*
  6.6.2 Parameter Dependence of Generalized Synchronization *329*
  6.6.3 Linear Stability Analysis of Synchronous Oscillatory Solutions for Generalized Synchronization *331*
    6.6.3.1 Linearized Equations *331*
    6.6.3.2 Conditional Lyapunov Exponent *333*
    6.6.3.3 Numerical Results of Conditional Lyapunov Exponent *334*
  6.6.4 Two-Dimensional Map *334*
  6.6.5 Dependence of Synchronization Quality on Optical Phase of Feedback Light in the Closed-Loop Configuration *336*

Appendix
  6.A.1 Rate Equations for Identical Synchronization in Unidirectionally Coupled Semiconductor Lasers with Optical Feedback in the Closed-Loop Configuration *337*

Part Two   Application of Chaotic Lasers to Optical Communication and Information Technology

7 Basic Concept of Optical Communication with Chaotic Lasers *343*

7.1 History of Secret Communication *343*
  7.1.1 Cryptography *343*
  7.1.2 Steganography *344*
  7.1.3 Noise Communication *346*
7.2 Concept of Chaos Communication *346*
  7.2.1 Basic Idea of Chaos Communication *346*
  7.2.2 Features of Chaos Communication *348*
  7.2.3 Hardware Keys *348*
  7.2.4 Synchronization for Chaos Communication *349*
7.3 Characteristics of Chaos Communication *351*
  7.3.1 Hardware-Based Communication *351*
  7.3.2 Chaos-Synchronization-Based Communication *351*
  7.3.3 Privacy *352*
  7.3.4 Compatibility *352*
  7.3.5 Analog Communication *353*
  7.3.6 Subcarrier Communication *353*
  7.3.7 Coherent Communication *353*
  7.3.8 Multiplexing and Noise Tolerance *353*
7.4 Encoding and Decoding Techniques *354*
  7.4.1 Chaos Masking *354*
  7.4.2 Chaos Modulation *357*
  7.4.3 Chaos Shift Keying *360*

7.5 Tools for Quantitative Evaluation of Performance of Chaos Communication *362*
  7.5.1 Bit Error Rate, Q Factor, and Signal-to-Noise Ratio *362*
  7.5.2 Modulation Format and Eye Diagram *365*

8 Implementation of Optical Communication with Chaotic Lasers *369*
8.1 History of Chaos Communication *369*
  8.1.1 Chaos Communication in Electronic Circuits *369*
  8.1.2 Chaos Communication in Optical Systems *370*
  8.1.3 European Project for Chaos Communication *375*
8.2 Examples of Communication Systems with Various Chaotic Lasers *377*
  8.2.1 Semiconductor Lasers with Optical Feedback *377*
    8.2.1.1 Field Experiment of Chaos Communication *377*
    8.2.1.2 Transmission of TV Video Signal *381*
  8.2.2 Semiconductor Lasers with Optoelectronic Feedback *383*
  8.2.3 Electro-Optic Systems *384*
  8.2.4 Fiber Lasers *386*
8.3 Performance Evaluation of Optical Communication with Chaotic Lasers *390*
  8.3.1 Subcarrier Modulation *390*
  8.3.2 Photonic Integrated Circuit and Forward Error Correction for High-Performance Chaos Communication *394*
    8.3.2.1 Photonic Integrated Circuit *394*
    8.3.2.2 Chaos Communication Experiment *396*
    8.3.2.3 Forward Error Correction Technique *397*
    8.3.2.4 Bit-Error-Rate (BER) Performance *397*
    8.3.2.5 Analysis for Unauthorized Users *399*
  8.3.3 Optical Phase Chaos for 10-Gb/s Chaos Communication *400*
  8.3.4 Comparison of the Encoding and Decoding Schemes for Chaos Communication *403*
8.4 Privacy Issues in Optical Communication with Chaotic Lasers *405*
  8.4.1 Introduction *405*
  8.4.2 Reconstruction of Model and Parameter Settings by Time-Series Analysis *406*
  8.4.3 Parameter Estimation by Using Similar Hardware *406*
  8.4.4 Parameter Estimation by Time-Series Analysis *408*
  8.4.5 Direct Detection of Presence of Message from Time-Series Analysis *409*
  8.4.6 Summary of Privacy Issues *410*
8.5 Photonic Integrated Circuit for Optical Communication with Chaotic Lasers *410*
  8.5.1 Photonic Integrated Circuit for a Semiconductor Laser with all-Optical Feedback *411*
  8.5.2 Photonic Integrated Circuit for two Mutually Coupled Semiconductor Lasers *413*

        8.5.3   Photonic Integrated Circuit for Colliding-Pulse
                Mode-Locked Lasers   *415*
   8.6 Other Encoding and Decoding Techniques   *417*
        8.6.1   Spatiotemporal Encoding   *417*
        8.6.2   Polarization Encoding   *419*
        8.6.3   Multiplexing Communications   *421*
   8.7 New Perspective of Optical Communication with Chaotic Lasers   *423*
        8.7.1   Analogy to Biological Communication Systems   *423*
        8.7.2   Towards the World of Scientific Fiction   *424*

9 **Secure Key Distribution Based on Information-Theoretic Security with Chaotic Lasers**   *427*
   9.1 Introduction   *427*
        9.1.1   Secure Key Distribution   *427*
        9.1.2   Computational Security and Information-Theoretic Security   *428*
   9.2 Concept of Information-Theoretic Security   *429*
        9.2.1   History of Information-Theoretic Security and Maurer's Satellite
                Scenario   *429*
        9.2.2   Bounded Observability   *430*
   9.3 Implementation of Information-Theoretic Security with
        Chaotic Lasers   *431*
        9.3.1   Bounded Observability with Chaotic Semiconductor Lasers   *431*
        9.3.2   Public-Channel Cryptography with Coupled Chaotic
                Semiconductor Lasers   *435*
        9.3.3   Bidirectional Message Transmission with Mutually Coupled
                Semiconductor Lasers   *439*
   9.4 Information-Theoretic Security with Optical Noise   *441*
        9.4.1   Ultralong Fiber-Laser System   *441*

10 **Random Number Generation with Chaotic Lasers**   *445*
    10.1 Introduction   *445*
          10.1.1   Needs for Random Number Generation   *445*
          10.1.2   Extraction of Randomness from Chaotic Lasers   *447*
    10.2 Types of Random Number Generators   *447*
          10.2.1   What are Random Numbers?   *447*
                   10.2.1.1   Independence   *448*
                   10.2.1.2   Unpredictability   *449*
          10.2.2   Two Types of Random Number Generators   *449*
                   10.2.2.1   Physical Random Number Generators (RNG)   *449*
                   10.2.2.2   Pseudorandom Number Generators (PRNG)   *451*
          10.2.3   Issues of Conventional Random Number Generators   *451*
    10.3 Examples of Random Number Generators with Chaotic Lasers   *452*
          10.3.1   Monobit Generation with Two Lasers   *452*
                   10.3.1.1   Scheme   *452*
                   10.3.1.2   Parameter Dependence   *457*

10.3.2 Monobit Generation with One Laser *460*
10.3.3 Multibit Generation with One Laser *461*
10.3.4 Postprocessing for High-Speed Random Number Generation *463*
10.3.5 Bandwidth Enhancement of Chaotic Lasers for High-Speed Random Number Generation *465*
10.3.6 Photonic Integrated Circuit for Random Bit Generators *469*
10.4 Application of Chaotic-Laser-Based Random Number Generators to High-Speed Quantum Key Distribution *472*
10.5 Numerical Evaluation of Random Number Generator as Entropy Source *475*
10.5.1 Entropy Generation from Internal Noise by Chaotic Dynamics *475*
10.5.2 Estimation of Entropy *477*
10.5.3 Entropy Rate and Nondeterministic Bit Generation *478*
10.6 Conventional Methods for Physical Random Number Generators *480*
10.6.1 Thermal Noise *480*
10.6.1.1 Direct Amplification of Thermal Noise *480*
10.6.1.2 Metastability *481*
10.6.1.3 Two Oscillators with Frequency Jitter *482*
10.6.2 Quantum Noise *484*
10.6.3 Optical Noise (Spontaneous Emission Noise) *485*
10.6.4 Radiation from Radioactive Nuclide *486*
10.6.5 Chaotic Dynamics in Electronic Circuits *487*
10.6.6 Traditional Physical Devices *487*
10.6.7 Other Methods *488*
10.6.8 Commercial Physical Random Number Generators *488*
10.6.8.1 Intel Chip (Intel) *488*
10.6.8.2 Random Master (Toshiba) *489*
10.6.8.3 Random Streamer (FDK) *490*
10.6.8.4 Quantis (ID Quantique) *490*
10.7 Postprocessing Techniques for Improvement of Randomness *490*
10.7.1 von Neumann Method *491*
10.7.2 Exclusive-OR (XOR) Method *492*
10.8 Pseudorandom Number Generators *493*
10.8.1 Linear Congruential Method *493*
10.8.2 M sequence and Generalized Feedback Shift Register (GFSR) *494*
10.8.3 Combined Tausworthe Method *496*
10.8.4 Mersenne Twister *496*
10.9 Statistical Evaluation of Random Numbers with NIST Special Publication 800-22 Test Suite *497*
10.9.1 Strategies for Statistical Analysis of Random Number Generators *497*
10.9.2 Evaluation of p-Values *499*

- 10.9.3 Interpretation of Empirical Results *501*
  - 10.9.3.1 Proportion of p-Values *501*
  - 10.9.3.2 Uniformity of Distribution of p-Values *501*
- 10.9.4 Tendency of Passed/Failed NIST SP 800-22 Tests in Laser-Chaos-Based Random Number Generators *502*
- 10.9.5 Other Statistical Tests of Randomness *503*

Appendix
- 10.A.1 Recipe for High-Quality Random Number Generators *503*
- 10.A.2 Dichtl method for Postprocessing of Random Number Generators *504*
- 10.A.3 Algorithm of Mersenne Twister Pseudorandom Number Generator *505*
- 10.A.4 Detailed Description of NIST Special Publication 800-22 *506*
  - 10.A.4.1 Frequency (Monobit) Test *506*
  - 10.A.4.2 Frequency Test within a Block *506*
  - 10.A.4.3 Cumulative Sums (Cusums) Test *506*
  - 10.A.4.4 Runs Test *507*
  - 10.A.4.5 Tests for the Longest-Run-of-Ones in a Block *507*
  - 10.A.4.6 Binary Matrix Rank Test *507*
  - 10.A.4.7 Discrete Fourier Transform (Spectral) Test *507*
  - 10.A.4.8 Non-overlapping Template Matching Test *507*
  - 10.A.4.9 Overlapping Template Matching Test *507*
  - 10.A.4.10 Maurer's "Universal Statistical" Test *508*
  - 10.A.4.11 Approximate Entropy Test *508*
  - 10.A.4.12 Random Excursions Test *508*
  - 10.A.4.13 Random Excursions Variant Test *508*
  - 10.A.4.14 Serial Test *508*
  - 10.A.4.15 Linear Complexity Test *508*

11 Controlling Chaos in Lasers *511*
- 11.1 Classification of Controlling Chaos *511*
  - 11.1.1 Feedback Control Method *511*
    - 11.1.1.1 OGY Method *511*
    - 11.1.1.2 Occasional Proportional Feedback (OPF) Method *512*
    - 11.1.1.3 Continuous Feedback Control Method *513*
  - 11.1.2 Nonfeedback Control Method *514*
- 11.2 Examples of Controlling Chaos in Lasers *515*
  - 11.2.1 Feedback Control Method for Controlling Chaos in Lasers *515*
    - 11.2.1.1 Occasional Proportional Feedback (OPF) Method *515*
    - 11.2.1.2 Continuous Feedback Control Method *518*
  - 11.2.2 Nonfeedback Control Method for Controlling Chaos in Lasers *521*
    - 11.2.2.1 Loss Modulation *521*

- 11.2.2.2 High-Frequency Injection (HFI) Method for Semiconductor Lasers 523
- 11.2.2.3 Stabilization to High-Periodic Oscillations 525
- 11.3 Applications of Controlling Chaos in Lasers 525
  - 11.3.1 Suppression of Relative Intensity Noise (RIN) 525
  - 11.3.2 Chaotic Search and Adaptive Mode Selection 527
  - 11.3.3 Dynamical Memory 528
  - 11.3.4 Communication with Chaos by Controlling Chaos 529
- Appendix
  - 11.A.1 OGY Method for Controlling Chaos 530

12 Other Applications with Chaotic Lasers 533
- 12.1 Remote Sensing with Chaotic Lasers 533
  - 12.1.1 Chaotic Lidar 533
  - 12.1.2 Chaotic Radar 536
  - 12.1.3 Chaotic Correlation Optical Time-Domain Reflectometer 539
- 12.2 Blind Source Separation of Chaotic Signals by Using Independent Component Analysis 542
  - 12.2.1 Motivation for Blind Source Separation 542
  - 12.2.2 Principle of Independent Component Analysis 543
  - 12.2.3 Examples of Blind-Source Separation with Chaotic Lasers 544
- 12.3 Fractal Optics 547
  - 12.3.1 Chaos Mirror for Wireless Optical Communications 548
  - 12.3.2 Fractal Patterns in Regular Polyhedral Mirror-Ball Structures 550

References 557

Glossary 575
- G.1 List of Acronyms 575
  - G.1.1 Acronyms of Technical Terms 575
  - G.1.2 Acronyms of Units 578
- G.2 Source Codes of C Programming Language for Numerical Simulations 579
  - G.2.1 Logistic Map (Chapter 2) 579
    - G.2.1.1 C Source Code for Sequence of Logistic Map (Figure 2.5a) 579
    - G.2.1.2 C Source Code for Bifurcation Diagram of Logistic Map (Figure 2.8) 580
    - G.2.1.3 C Source Code for Lyapunov Exponent of Logistic Map (Figure 2.9) 581
  - G.2.2 Lorenz Euations (Chapter 2) 582
    - G.2.2.1 C Source Code for Time Series of Lorenz Equations (Figure 2.10) 582

G.2.2.2 C Source Code for Bifurcation Diagram of Lorenz Equations (Figure 2.12a) *584*

G.2.2.3 C Source Code for Lyapnov Spectrum (All the Lyapunov Exponents) of Lorenz Equations (Figure 2.12b) *586*

G.2.2.4 C Source Code for Synchronization of Chaos in Lorenz Equations (Diffusive Coupling, Section 5.2.1.2) *590*

G.2.3 Lang–Kobayashi Equations for a Semiconductor Laser with Time-Delayed Optical Feedback (Chapter 4) *592*

G.2.3.1 C Source Code for Time Series of Lang–Kobayashi Equations (Figure 4.13e) *592*

G.2.3.2 C Source Code for Bifurcation Diagram of Lang–Kobayashi Equations (Figure 4.16) *595*

G.2.3.3 C Source Code for Maximum Lyapunov Exponent of Lang-Kobayashi Equations (Figure 4.19) *599*

G.2.4 Synchronization of Chaos in Coupled Lang–Kobayashi Equations for Unidirectionally Coupled Semiconductor Lasers with Time-Delayed Optical Feedback (Chapter 6) *604*

G.2.4.1 C Source Code for Time Series of Synchronization of Chaos in Coupled Lang-Kobayashi Equations in Open-Loop Configuration (Figure 6.9) *604*

G.2.4.2 C Source Code for Time Series of Synchronization of Chaos in Coupled Lang–Kobayashi Equations in Closed-Loop Configuration (Appendix 6.A.1) *609*

G.2.4.3 C Source Code for Cross-Correlation Calculation of Synchronization of Chaos in Coupled Lang–Kobayashi Equations in Open-Loop Configuration (Figures 6.10a and c) *614*

G.2.4.4 C Source Code for Conditional Lyapunov Exponent of Synchronization of Chaos in Coupled Lang–Kobayashi Equations in Open-Loop Configuration (Figure 6.12) *620*

Index *627*

Preface

The aim of this book is to provide a comprehensive overview of research activities on both chaos (nonlinear dynamics) and lasers (photonics) that can be applied for engineering applications of optical communication and information technology. Many books and review articles have been published in the field of either chaos or lasers, however, the books related to both of these two major interdisciplinary fields are limited. This book covers the research fields of both chaos and lasers, and shows that the combination of chaos and lasers can result in novel applications to optical communications and information technologies.

This book is suitable for graduate students who are interested in learning about chaos and lasers and intend to start new research works in the interdisciplinary research fields. The comprehensive overview of experiments and numerical simulations for chaotic laser dynamics described in this book would also be very useful for professional researchers in the research fields of chaos or lasers and their engineering applications. Complicated mathematical formula are avoided in the main text in order to understand the essence of the topics treated in this book.

The author strongly acknowledges many researchers in the interdisciplinary fields of chaos and lasers. The author has been influenced by many enthusiastic discussions with these researchers in these research fields. It would be a great pleasure for the author if this book could contribute a small step by which these research fields of chaos and lasers would be widely recognized in other research communities. The author strongly feels that many researchers in these fields have contributed to accumulate a large amount of knowledge for both basic sciences and engineering applications.

The author strongly acknowledges Peter Davis, Ingo Fischer, Claudio R. Mirasso, Junji Ohtsubo, Rajarshi Roy, and Kazuyuki Yoshimura for reviewing some chapters in this book. Their comments and encouragements have been very valuable to improve the contents in this book.

The author acknowledges Peter Davis and Rajarshi Roy for long-term collaboration with the author. The author has been strongly influenced by their philosophies on scientific research.

The author thanks the following researchers for long-term collaboration with the author to produce fruitful research results, some of which are included in this book:

Tilmann Heil, Tohru Ikeguchi, Fumihiko Kannari, Yun Liu, Ryan McAllister, Toni Pérez, Shigeru Yoshimori, and Kazuyuki Yoshimura.

The author thanks the following researchers for helpful discussions and constructive comments at different occasions in the interdisciplinary research communities of chaos and lasers: Tahito Aida, Kazuyuki Aihara, F. T. Arecchi, Apostolos Argyris, Stefano Boccaletti, Adonis Bogris, Thomas Carroll, How-Foo Chen, Konstantinos E. Chlouverakis, Muhan Choi, Adam B. Cohen, Pere Colet, Ned J. Corron, Thomas Erneux, Ingo Fischer, Takehiro Fukushima, Jordi García-Ojalvo, Ignace Gatare, Daniel J. Gauthier, Athanasios Gavrielides, Shin-itiro Goto, Michael Hamacher, Takahisa Harayama, Toshimori Honjo, Yoshihiko Horio, Guillaume Huyet, Sheng-Kwang Hwang, Lucas Illing, Ido Kanter, Chil-Min Kim, Min-Young Kim, Song-Ju Kim, Takeshi Koshiba, Bernd Krauskopf, Jürgen Kurths, Fumiyoshi Kuwashima, Wing-Shun Lam, Laurent Larger, Tony Lawrance, Min Won Lee, Daan Lenstra, Fan-Yi Lin, Jia Ming Liu, Alexander Locquet, Paul Mandel, Christina Masoller, Riccardo Meucci, Claudio R. Mirasso, Linda Moniz, Jun Muramatsu, Thomas E. Murphy, Hiroya Nakao, Junji Ohtsubo, Toru Onodera, Kenju Otsuka, Edward Ott, Krassimir Panajotov, Jon Paul, Louis M. Pecora, Michael Peil, Luis Pesquera, Arkady Pikovsky, Will Ray, Elizabeth A. Rogers, Fabien Rogister, Damien Rontani, Ira B. Schwartz, Marc Sciamanna, K. Alan Shore, S. Sivaprakasam, Paul S. Spencer, Steven H. Strogatz, David W. Sukow, Satoshi Sunada, Dimitris Syvridis, Yoshiyasu Tamura, Shuo Tang, J. R. Tredicce, Ken Umeno, Gregory D. VanWiggeren, Raúl Vicente, Sebastian Wieczorek, H.-J. Wünsche, and M. Yousefi.

The author would strongly like to thank his graduate students, Hiroki Aida, Kazutaka Kanno, Takuya Mikami, Shinichiro Morikatsu, Haruka Okumura, and Taiki Yamazaki, for providing the original experimental and numerical data in Chapters 2, 4, 6, 10, and Glossary G.2 of this book. Without their contributions, this book would not have been completed.

The author thanks his graduate students who have enthusiastically worked all day long with the author: Yasuhiro Akizawa, Kazuya Amano, Kota Aoyama, Masaya Arahata, Kenichi Higa, Kunihito Hirano, Hidetoshi Iida, Masaki Inoue, Makito Kawano, Satoshi Kinugawa, Hayato Koizumi, Masahiko Kuraya, Takanori Matsuura, Isao Oowada, Mitsutoshi Ozaki, Satoshi Sano, Nagisa Shibasaki, Hiroyuki Someya, Junichi Takahashi, Toru Yamamoto, and Hoipang Yip.

The author would like to thank Valerie Molière and Ulrike Werner at Wiley-VCH Verlag for their encouragements to complete the book project. It was a long journey for this book project as a first experience for the author. The author thanks them for their warm patience.

Finally, the author thanks his family (Miho, Mikito, Tatsuya, Sadao, and Haruko Uchida) for their unconditional support for daily life.

Saitama, August 2011 *Atsushi Uchida*

Part One
Basic Physics of Chaos and Synchronization in Lasers

1
Introduction

The topics of this book widely cover both basic sciences and engineering applications by using lasers and chaos. The basic concepts of chaos, lasers, and synchronization are described in the first part of this book. The second part of this book deals with the engineering applications with chaotic lasers for information–communication technologies, such as optical chaos communication, secure key distribution, and random number generation. The bridge between basic scientific researches and their engineering applications to optical communications are treated in this book.

The history of research activities of laser and chaos is summarized in Table 1.1. Since the laser was invented in 1960 and the concept of chaos was found in 1963, these two major research fields were developing individually. In 1975, a milestone work was published on the findings of the connection between laser and chaos. In the 1980s, there were enormous research activities for the experimental observation of chaotic laser dynamics and the proposal of laser models that were used to explain the experimental results, from the fundamental physics point of view. Two important methodologies were proposed in 1990, that is, control and synchronization of chaos, which led to engineering applications of chaotic lasers such as stabilization of laser output and optical secure communication. There were many research reports on control and synchronization of chaos in various laser systems in the 1990s, and researchers tended to utilize chaotic lasers for engineering applications, rather than just for the studies on fundamental physics. Two important experimental demonstrations of optical communication with chaotic lasers were reported in 1998. Since then, many researchers have concentrated on working on the implementation of optical communication systems with chaotic lasers in the 2000s, including two major European projects, for the purpose of secure communication. High-performance demonstrations were reported in the field experiments of commercial optical-fiber networks, where bit rates of 10 Gb/s with low bit-error rate (BER) were achieved over 100 km fiber transmission. Meanwhile, other promising applications with chaotic lasers were demonstrated experimentally, that is, chaotic lidar and a random number generator, which could be research seeds for the next decade.

Optical Communication with Chaotic Lasers: Applications of Nonlinear Dynamics and Synchronization, First Edition. Atsushi Uchida.
© 2012 Wiley-VCH Verlag GmbH & Co. KGaA. Published 2012 by Wiley-VCH Verlag GmbH & Co. KGaA.

Table 1.1 History of research activities of laser and chaos (milestone studies).

Year	Research activity	References
1960	Invention of laser	(Maiman, 1960)
1963	Discovery of chaos	(Lorenz, 1963)
1975	Connection between laser and chaos	(Haken, 1975)
1979	Proposal of Ikeda chaos	(Ikeda, 1979)
1980	Proposal of Lang-Kobayashi equations for semiconductor laser with optical feedback	(Lang and Kobayashi, 1980)
1984	Classification of chaotic laser models	(Arecchi et al., 1984)
1980's~	Intensive works on the observation of chaos in lasers	(Gioggia and Abraham, 1983) (Weiss et al., 1983)
1990	Proposal of control of chaos	(Ott et al., 1990)
1990	Proposal of synchronization of chaos	(Pecora and Carroll, 1990)
1992	First experimental observation of controlling chaos in lasers	(Roy et al., 1992)
1993	First demonstration of chaos communication (circuit)	(Cuomo and Oppenheim, 1993)
1994	First experimental observation of synchronization of chaos in lasers	(Roy and Thornburg, 1994 (Sugawara et al., 1994)
1990's~	Intensive works on control and synchronization of chaos in lasers	(Meucci et al., 1994) (Fischer et al., 2000)
1994	Proposal of optical communication with chaotic lasers (numerical simulation)	(Colet and Roy, 1994) (Mirasso et al., 1996)
1998	First experimental demonstration of optical communication with chaotic lasers	(VanWiggeren and Roy, 1998a) (Goedgebuer et al., 1998)
2000's~	Intensive works on optical communication with chaotic lasers	(Tang and Liu, 2001c) (Kusumoto and Ohtsubo, 2002)
2004	First demonstration of chaotic lidar and radar	(Lin and Liu, 2004a, 2004b)
2005	First demonstration of field experiment on optical communication with chaotic lasers	(Argyris et al., 2005)
2008	Proposal of photonic integrated circuit for optical communication with chaotic laser	(Argyris et al., 2008)
2008	First experimental demonstration of random number generator with chaotic lasers	(Uchida et al., 2008a)
2010	High-performance demonstration of optical chaos communication with chaotic lasers (10 Gb/s)	(Lavrov et al., 2010) (Argyris et al., 2010c)
2010's~	What is next?	

This book overviews all the research activities related to laser and chaos in the recent half-century. The fundamental knowledge of chaotic laser dynamics is enormous thanks to the efforts of basic researchers in this field. This book describes chaotic temporal dynamics in various laser systems (Chapters 2–4), and techniques for synchronization of chaos (Chapters 5 and 6), which are key components for the implementation of optical chaos communication. The detailed descriptions of optical

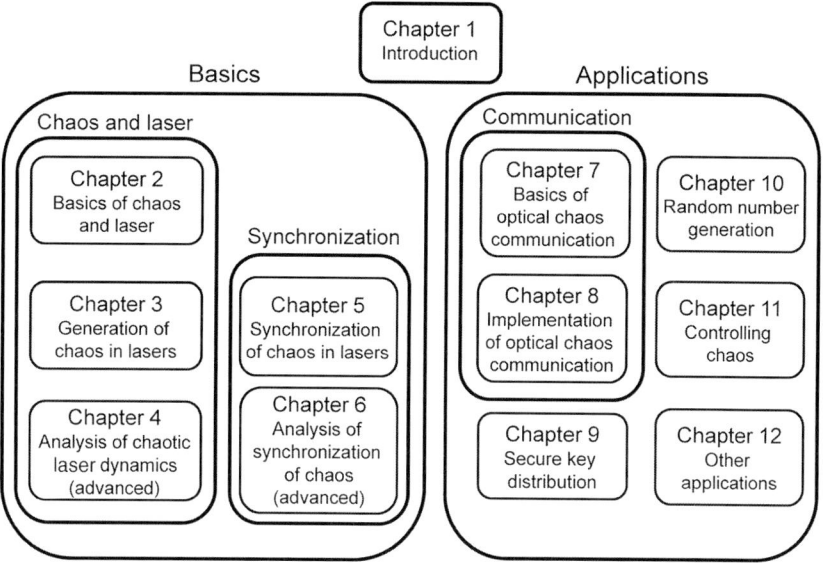

Figure 1.1 Contents of all the chapters in this book. It is recommended for readers to independently select and read the chapters that are interesting to them.

communication with chaotic lasers are found in this book (Chapters 7 and 8), which is one of the main goals of this book. Other possible and promising applications are described, such as secure key distribution, random number generation, dynamical memory, and chaotic lidar and radar (Chapters 9–12).

The contents of all the chapters in this book are summarized in Figure 1.1. The first part of the book corresponds to the description of the basic sciences of lasers, chaos, and synchronization (Chapters 2–6), and the second part describes their engineering applications (Chapters 7–12). All the chapters are written simply and comprehensively for beginners to avoid complicated mathematical formula. Mathematical formula is summarized in an Appendix of each chapter if necessary. For example, several mathematical models for various laser systems are described in the Appendices of Chapters 3 and 5. Chapters 4 and 6 are specially designed for advanced learners and are written with complicated mathematical formula and difficult technical terms, so beginners may skip Chapters 4 and 6. Readers may start reading any of the chapters based on their interests, since each chapter has been organized independently.

Source codes of the C programming language are listed in the Glossary for numerical simulations of laser models, and readers may try to do numerical simulations by themselves with chaotic laser models. This would be a good exercise for students and would be a good start of research work for researchers. The acronyms of technical terms and units are also listed in the Glossary.

1.1
Lasers and Chaos

1.1.1
Lasers

The term of "LASER" is an abbreviation of light amplification by stimulated emission of radiation. The laser is artificial light that has opened up a variety of applications in many scientific and engineering fields. The laser can be considered as one of the most important inventions in the twentieth century. The important characteristic of the laser is "coherence", which implies that the photons oscillate similarly in phase. Coherent light sources have outstanding characteristics compared with natural (incoherent) light: bright intensity output, high photon energy, good directionality, single wavelength, and narrow spectrum bandwidth in order to enable interference. This "clean" light has brought new technologies such as optical communication, compact disk systems, precise measurement, material processing, medical application, remote sensing, and so on. Most laser devices have nonlinear effects in their laser media, and inherent instabilities are sometimes unavoidable (see Chapter 2).

1.1.2
Chaos

From the beginning of laser history, instabilities of laser output are inevitable due to its inherent nonlinearity (Maiman, 1960; Maiman *et al.*, 1961), even though many efforts have been made to stabilize laser output for many engineering applications. Most lasers including semiconductor, fiber, solid-state, and gas lasers produce temporal and spatial instabilities of laser output at certain operating conditions or with an additional external perturbation. It has been known that these instabilities can be derived from a deterministic rule of laser dynamics, which can be described by using a set of differential rate equations. These types of instabilities have been known as "deterministic chaos" and can be distinguished from instabilities due to stochastic or quantum noise.

The term "chaos" is generally used to describe disturbance or turbulence in many situations. One of the most acceptable definitions of chaos in science is the instabilities derived from a "deterministic" rule. The term "chaos" has been used to describe fluctuations or time-varying (or space-varying) irregular phenomena that are governed by a deterministic rule, which can be described by using a set of mathematical equations (see Chapter 2). Chaos is a counterintuitive concept in the sense that one may find a mathematical rule in irregular fluctuations of complex dynamics.

One of the important characteristics of deterministic chaos is known as "sensitive dependence on initial conditions" (Lorenz, 1963). If two chaotic temporal sequences start from very close but slightly different initial conditions, the two sequences behave similarly at the beginning, however, they start to diverge exponentially in time and never show the same behavior again. This characteristic can be quantitatively measured by using the maximum Lyapunov exponent, and the existence of the positive maximum Lyapunov exponent is a proof of deterministic chaos. A tiny error of the initial conditions makes chaotic irregular sequences unpredictable. This fact implies that chaos is unpredictable for a long-term duration due to the sensitive dependence on initial conditions, although chaos is predictable for a short-term period due to the existence of deterministic rules.

1.1.3
Connection Between Lasers and Chaos

The connection between lasers and chaos was made in 1975 (Haken, 1975), where it was shown that a set of nonlinear differential equations from Maxwell–Bloch equations for a laser model resembles the Lorenz equations that are the basic model for deterministic chaos (see Chapter 2). Since then, many observations have been reported in experiments and numerical simulation in the 1980s (see Chapter 3). It was found that most lasers produce chaotic fluctuations of laser intensity. In the 1990s two major techniques for harnessing chaos were proposed: synchronization of chaos and control of chaos. Many attempts using these two techniques have been demonstrated in the literature. These research activities have led to engineering applications of optical communications with synchronized chaotic lasers.

One of the most well-known examples of chaotic laser output is a semiconductor laser with optical feedback. Semiconductor lasers have been widely used for optical compact disk and optical communication systems. A small fraction of the laser output is invariably fed back into the semiconductor laser because of the reflection occurring at the disk surface or the edge of optical fiber. Figure 1.2a shows a simple model of a semiconductor laser with optical self-feedback for the generation of chaos. A laser beam is reflected from an external mirror and fed back into the laser cavity. The optical intensity in the laser cavity is perturbed by the self-feedback light that affects the interaction between photons and the carrier density in the laser medium. A chaotic temporal waveform of laser output intensity can be observed experimentally, as shown in Figure 1.2b. Fast intensity fluctuation can be found of the order of

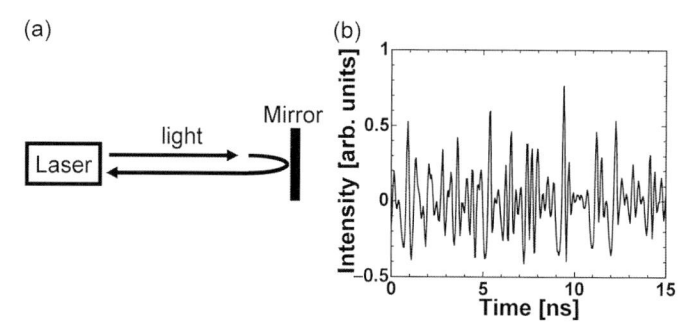

Figure 1.2 (a) Model of semiconductor laser with optical self-feedback. (b) Example of experimentally observed temporal waveform of laser output intensity in semiconductor laser with optical feedback.

GHz (the period of fluctuation is in the order of nanoseconds, 10^{-9} s), which corresponds to the relaxation oscillation frequency of the semiconductor laser (see Chapter 4).

Regardless of the observation of rich dynamics in chaotic laser output, many researchers did not pay attention to the instability of laser output. Most laser engineers avoided irregular fluctuations of laser output and considered that lasers were supposed to be stable, single mode (single wavelength), and have a narrow optical spectrum for use in engineering applications. Irregular pulsations of laser output need to be controlled and eliminated in order to generate stable output or periodic pulsations for large-intensity spikes, known as Q-switching pulses. On the other hand, the characteristics of rich temporal dynamics of laser output are suitable as a universal nonlinear dynamical model from the viewpoint of nonlinear dynamics in the interdisciplinary research fields. In addition, constructive combination between a laser and chaos could open up novel engineering applications, such as optical chaos communication, secure key distribution, and random number generation, which will be treated in detail in this book.

1.1.4
Chaos and Noise

One of the simplest questions is how to find chaos in laser intensity dynamics. What characteristics distinguish deterministic chaos from stochastic noise? The important feature of chaos is the fact that chaos is completely governed by a deterministic rule, which is described by a set of nonlinear equations without any stochastic terms. The determinism of chaos can distinguish irregular dynamics from noise. The instability of chaos results from inherent nonlinear interaction in dynamical systems, but not from stochastic noise.

By contrast, noise is defined as an irregular temporal waveform generated from a stochastic process that is based on a statistical law, but not on a deterministic rule. A simple example is throwing a dice. Each process of throwing a dice may not be deterministic, however, a statistical law may be evident after the repetition of many trials. Irregular temporal waveforms of noise are not completely deterministic, and cannot be described as a set of nonlinear equations. Instead, the behavior of noise can be sometimes described by a set of differential equations driven by a sequence of random numbers as a stochastic term (e.g., Langevin equations, see Chapter 2).

1.2
Synchronization of Chaos and Optical Communication

1.2.1
Synchronization of Chaos

"Synchronization" indicates temporal behaviors with a certain relationship. The same behavior of two temporal oscillations can be found for identical synchronization.

It has been known that two pendulums fixed on a common wall oscillate with the same frequency and phase (Pikovsky *et al.*, 2001). Synchronization of periodic oscillators is commonly used in many engineering applications of communications.

A question is whether chaotic temporal behaviors can be synchronized. Synchronization of chaos is a counterintuitive concept, since chaos has a strong dependence on initial conditions, which indicate that two nearby trajectories start to diverge exponentially in time and never reach the same state. The concept of synchronization of chaos thus contradicts the basic characteristics of chaos. It has been found that chaotic systems can be synchronized under certain conditions. A drive dynamical system itself may have unstable temporal waveforms, however, a response dynamical system with respect to the coupling signal from the drive system can become stable and the response system is able to follow the unstable temporal waveforms of the drive system. The change in the susceptibility to the coupling signal is the essence to obtain synchronization of chaos (see Chapter 5).

To achieve identical synchronization of chaos, similar hardware with similar parameter settings are required for coupled chaotic laser systems. This symmetry results in an identical synchronous solution of the coupled laser systems, and identical synchronization can be achieved. In addition, the identical synchronous solution must be stable. The stability of the synchronization depends on the particular dynamical system, but regions of stability in parameter space have been observed in various laser systems (see Chapter 6).

A remarkable demonstration of chaos synchronization was carried out and led to possible applications of secret communications (Pecora and Carroll, 1990). In optical systems, synchronization of chaos in coupled lasers has been investigated numerically and experimentally (Winful and Rahman, 1990; Roy and Thornburg, 1994; Sugawara *et al.*, 1994). Synchronization of chaos in one-way coupled optical systems and a potential application to optical secret communication have been studied (see Chapter 5).

1.2.2
Optical Communication with Synchronized Chaotic Lasers

Synchronization of chaos leads to an important application to optical chaos communications (VanWiggeren and Roy, 1998a; Goedgebuer *et al.*, 1998). Standard optical communication utilizes optical periodic carrier for encoding and decoding the message and there is no consideration for security in hardware level. Optical chaos communication leads to an additional layer of security or privacy in optical communication by using chaotic temporal carriers. These hardware-dependent optical communication systems have been developed and there have been international projects to implement optical chaos communication systems in real-world optical networks (OCCULT, 2001; PICASSO, 2006).

The basic concept of chaos communication is shown in Figure 1.3. A message is concealed in a chaos carrier in a transmitter, and the mixed signal consisting of the chaos and the message is sent to a receiver laser. The technique of synchronization of

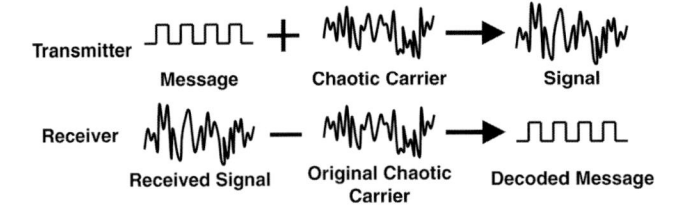

Figure 1.3 Schematics of the concept of optical communication with synchronized chaotic lasers.

chaos is used to reproduce the original chaos carrier in the receiver. The message can be recovered by subtracting the synchronized chaos carrier from the transmission signal. The quality of chaos synchronization strongly affects the degree of the recovered message signal (see Chapter 7).

Let us look at the procedure of the chaos communication in detail. The chaotic system in the transmitter produces a chaotic carrier to mask the message signal. The encoded signal is sent to the receiver and it is used for both decoding the message and achieving synchronization of chaos. In the receiver, a similar chaotic system can reproduce a nearly identical chaotic carrier by adjusting with a set of the static parameter values that are considered as static keys to be shared beforehand. If the receiver succeeds in synchronizing chaos by using the chaotic hardware and the static keys, it can reproduce a similar chaotic carrier and can succeed in decoding the original message.

The main purpose of chaos communication is to hide the "existence" of a message by a chaotic carrier waveform, which is known as steganography, compared with the technique of hiding the "meaning" of the message, known as cryptography (see Chapter 7). One of the most important techniques in chaos communication is to share the same chaotic carrier between the distant users by using synchronization of chaos. To achieve synchronization of chaos, similar hardware systems as well as similar parameter values are required in the transmitter and the receiver. The tolerance of synchronization against parameter mismatch is one of the measures for the level of privacy in chaos communication, that is, narrow parameter regions for achieving synchronization result in more privacy since it is difficult for eavesdroppers to achieve synchronization of chaos.

Recent advances on practical implementation of optical chaos communications will be described in this book (see Chapter 8), including the demonstration of chaos communication in commercial optical-fiber networks at 2.5-10 Gb/s over a 100-km distance with low BER (Argyris et al., 2005; Argyris et al., 2010c; Lavrov et al., 2010). Several advanced techniques including photonic integrated circuits and forward error correction have been implemented in chaos communications.

For chaos communications, it is necessary for two legitimate users to share a common secret key prior to the communication process. The system parameters provide a private key because the two communicating lasers must have nearly identical parameters, otherwise synchronization is failed. The distribution of a private key is the main weakness of any secure communication system, which is known as a secure key distribution problem (Maurer, 1993). Chaotic lasers could be useful for the physical implementation of secure key distribution based on information theory. The architecture based on chaotic lasers offers large key-establishing rates at long communication ranges. In addition, optical transmitters and receivers used for conventional optical communication systems, including erbium-doped fiber amplifiers (EDFAs) and dispersion compensation fibers (DCFs), can be used for chaos-based secure key distribution without using specially designed hardware. Several schemes for chaos-based secure key distribution systems have been proposed and demonstrated as a new way of secure key distribution (see Chapter 9).

1.3
Random Number Generation with Chaotic Lasers and Other Applications

1.3.1
Random Number Generation

Another promising application with chaotic lasers is random number generation (see Chapter 10). The output of chaotic lasers provides fast temporal dynamics of chaos with large spectral bandwidth. Typical bandwidth of semiconductor lasers is a few GHz, which is determined by the relaxation oscillation frequency. The speed of lasers is advantageous for the applications of physical random number generation. The combination of the characteristics of the complexity in chaos and large bandwidth in lasers could open up a new research field of fast physical random number generation. The output of chaotic devices could be both unpredictable as well as statistically random because they generate large-amplitude random signals from microscopic noise by nonlinear amplification and mixing mechanisms.

The concept of random number generation is shown in Figure 1.4. A chaotic signal of laser output is detected by a photodetector and converted to a binary digital signal

(a)

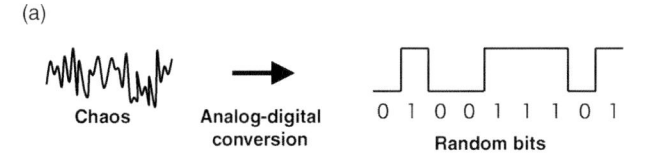

Chaos → Analog-digital conversion → 0 1 0 0 1 1 1 0 1 Random bits

(b)
```
000111100000011010111001000100101101101101111101101
100000100001010010101010101111111100100101100101000
010110100011010001111001101011010101011111001110101
110000000011011011001100001100001010010100001100011
100100010001110010011101101101000011010010000011101
010010101111100000001101110001101101110000111011100
```

Figure 1.4 (a) Schematics of the concept of random number generation with chaotic lasers and (b) the sequence of generated digital bits.

by an analog-to-digital converter (ADC). The ADC converts the input analog signal into a binary digital signal by comparing with the threshold voltage. The output binary random signal is a stream of bits as shown in Figure 1.4a (with the format of nonreturn to zero (NRZ)) and Figure 1.4b (as bits "0" or "1").

Random number generation with chaotic lasers has been intensively investigated since the first demonstration was published in 2008 (Uchida et al., 2008a). Many schemes have been demonstrated and random-bit generation rates from 1.7 to 400 Gb/s have been reported with verified randomness (see Chapter 10).

1.3.2
Controlling Chaos and Other Applications

Another direction of the research with chaotic lasers is controlling chaos, which is a natural way for engineering applications. Chaos needs to be avoided in many engineering systems, and the research on control and stabilization of chaos has strong motivation in nonlinear dynamical systems that are used for engineering applications. The research field on controlling chaos grew rapidly during the 1990s, where a chaotic temporal signal can be stabilized onto an unstable periodic orbit in a chaotic attractor based on control theory (Ott et al., 1990). The aim of chaos control is to obtain a stable or periodic temporal waveform by adding a small external perturbation. The techniques for controlling chaos have been applied to many interdisciplinary research fields, because there have been many dynamical systems that are required to stabilize chaotic instabilities and fluctuations (see Chapter 11).

Other promising applications with chaotic lasers have been proposed. Remote sensing applications with chaotic lasers have been reported as chaotic lider and radar systems. Blind signal separation using independent component analysis has been applied to chaotic temporal waveforms of laser output for the purpose of multiplexing communications. In addition, fractal optics have been used for wireless optical communication applications as a chaos mirror (see Chapter 12).

1.4
Research Directions for Engineering Applications with Chaotic Lasers

The research directions with chaotic lasers for engineering applications are summarized in Figure 1.5. There are three main research directions treated in this book, related to the concepts of how to harness chaos. For chaos communication applications, the characteristics of chaos are used in a straightforward way, that is, the determinism of chaos results in synchronization ability of chaos, and the middle degrees of complexity are suitable to hide a message signal. By contrast, for the applications of random number generation, the randomness of chaos needs to be maximized and determinism of chaos needs to be eliminated by converting analog chaos signals to binary signals. The important technique is how to extract and distill the randomness from deterministic chaos for this application. The research on

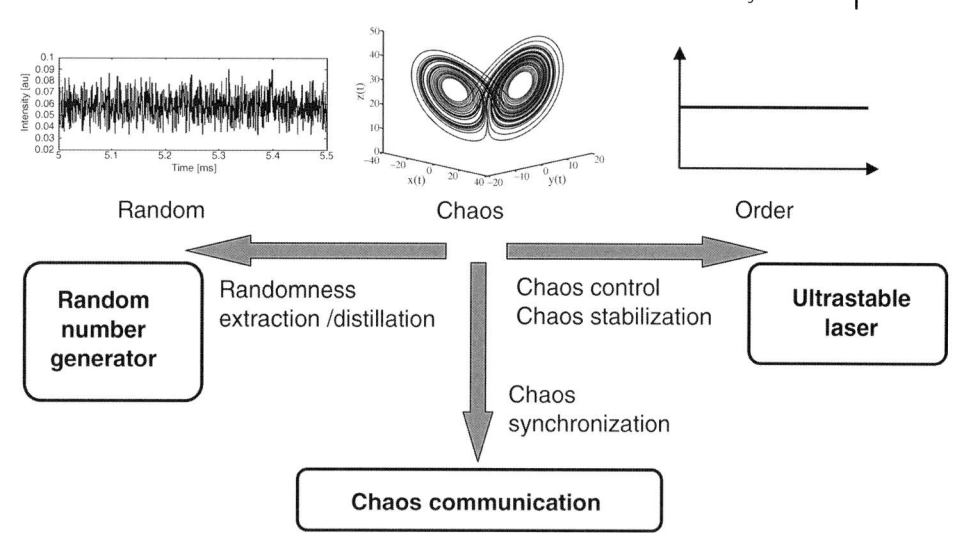

Figure 1.5 Three research directions for engineering applications with chaotic lasers.

random number generation requires a new engineering approach of chaos for maximizing the randomness of chaos.

By contrast, control and stabilization of chaos is a technique to completely avoid complexity and instability of chaos. To design and establish ultrastable lasers, chaos-control techniques may be useful for the suppression of chaotic instabilities. The features of deterministic chaos including unstable periodic orbits can be utilized for controlling chaos. The research on controlling chaos is the opposite direction of the research on random number generation, depending on minimizing or maximizing the complexity of chaos, respectively.

1.5
Outline of This Book

This book consists of two main parts: basics and applications. The first part corresponds to the description of basic physics of chaos and laser. Based on chaos and laser theories, the generation techniques of chaos in lasers are described in detail. Synchronization of chaos is also described and discussed, which is the basis for optical chaos communication. Chapters 2–6 cover the first part for the basics.

The second part of this book describes applications of chaotic lasers to optical communication and information technologies. The basic concept and recent advances of optical chaos communication systems are described in detail with experimental and numerical results from the literature. Two related topics to chaos communications are also described: secure key distribution based on information theory and random number generation with chaotic lasers. The use of chaotic lasers for fast physical random number generators has attracted increased interest, and this topic is overviewed in detail. Other possible applications of chaotic lasers are described, including controlling chaos, remote sensing, multiplexing communications with chaotic codes, and a chaos mirror for wireless communications. Chapters 7–12 cover the second part for the applications.

Readers do not need to read the text from Chapters 2–12 in sequence, instead, they can select and read chapters that are interesting to them. The contents of the chapters in this book are already summarized in Figure 1.1. The contents of each chapter will be described in more detail in the following.

1.5.1
Contents of Chapters: First Parts for the Basics

The first part of this book corresponds to the basics for chaos, laser, and synchronization, described from Chapters 2–6.

In Chapter 2, the basics of chaos and laser are introduced and explained. First the history of the observation of chaotic intensity fluctuations in lasers is described. The basic theories for chaos and laser are described separately. The connection between chaos and laser is made, and a basic single-mode laser model consisting of the variables of electric field, population inversion, and polarization of matter is described. The classification based on the decay rate of the laser model is explained with a rate-equation model.

In Chapter 3, the generation techniques of chaos in lasers are classified and overviewed with experimental and numerical examples from the literature. Most commercial lasers can be destabilized and produce chaotic output intensity by using the following techniques: feedback, coupling, modulation, and addition of nonlinear devices. These techniques are described in detail for various laser media, such as semiconductor lasers, fiber lasers, solid-state lasers, and gas lasers. The chaotic dynamics generated by using a passive optical system with time-delayed feedback is also described in electro-optic systems. The techniques for distinguishing between chaos and noise are discussed. Readers can learn a variety of chaotic temporal dynamics and bifurcations from many examples in various laser systems in Chapter 3.

Chapter 4 is specially designed for the readers who would like to learn analytical, numerical, and experimental techniques for the study on nonlinear dynamics in lasers. Standard techniques of nonlinear analysis in chaotic laser systems are introduced and explained in detail. A semiconductor laser with time-delayed optical feedback is used as an example of the analysis of chaotic laser dynamics. Experimental, analytical, and numerical results are shown. Experimental observations with temporal waveforms and RF spectra of chaotic laser intensities can be found in detail. For analytical techniques, Lang–Kobayashi equations are introduced and steady-state solutions are obtained. Linear stability analysis is also performed, which is a powerful tool to determine whether laser instability results from inherent nonlinearity of a deterministic model. For numerical simulation, temporal waveforms are generated from the Lang–Kobayashi equations. The maximum Lyapunov exponent, which is a reliable measure for the existence of chaos, is numerically calculated from the linearized equations. Other measures for complexity of chaos are introduced and calculated, such as the Lyapunov spectrum, Kolmogorov–Sinai entropy and Kaplan–Yorke dimension. Finally, Lang–Kobayashi equations with gain saturation are introduced to be consistent with experimental observations. The description in Chapter 4 may be very technical, but it would be useful for the readers who intend to start studying and working on chaotic laser dynamics.

In Chapter 5, synchronization of chaos in lasers is overviewed. The concept of synchronization of chaos is introduced and explained. The history of synchronization of chaos is described in electronic circuits and laser systems. Several coupling schemes for synchronization of chaos are explained, and synchronization is classified into two types: identical synchronization and generalized synchronization. These two types of synchronization can be distinguished by observing the delay time between two synchronized chaotic temporal waveforms in lasers with time-delayed feedback. Experimental and numerical examples of synchronization of chaos are also overviewed in various types of lasers, such as semiconductor lasers, electro-optic systems, fiber lasers, solid-state lasers, and gas lasers from the literature. Other types of synchronization such as phase synchronization and generalized synchronization are also described. Finally, a new concept of consistency is introduced and discussed, which indicates the reproducibility of a driven laser with respect to a repeated input signal.

In Chapter 6, standard techniques for the analysis of synchronization of chaos are introduced and explained in detail. Chapter 6 is specially designed for the readers who would like to learn experimental and numerical techniques for synchronization of chaos by using a realistic laser model. Synchronization of chaos is observed in two unidirectionally coupled semiconductor lasers with time-delayed optical feedback. Both experimental and numerical results are shown. For numerical simulation, coupled Lang–Kobayashi equations are introduced as a model for synchronization of chaos. Two types of synchronization (identical and generalized synchronizations) are observed and distinguished clearly in both the experiment and numerical simulation. The conditions for achieving synchronization of chaos are systematically investigated in wide parameter regions. Linear stability analysis is performed and the maximum conditional Lyapunov exponent is obtained, which is a good indicator of the stability of synchronous solutions. Generalized synchronization with low correlation is also investigated by using auxiliary system approach, where a chaotic output of a drive laser is injected into two response lasers that have identical parameter values but different initial conditions. High-quality synchronization of chaos between the two response lasers is obtained, while the correlation between the drive and each of the response lasers is relatively low for generalized synchronization. Chapter 6 may be very technical (as well as Chapter 4), but it would be very useful for the readers who intend to start studying and working on the analysis of synchronization of chaotic lasers.

1.5.2
Contents of Chapters: Second Part for the Applications

The second part of this book corresponds to engineering applications of chaotic lasers, described from Chapters 7–12.

In Chapter 7, the basic concept of optical chaos communication with chaotic lasers is introduced and summarized. The history of secret communication is described, and the technique of steganography leads to an invention of chaos communication. The concept and characteristics of chaos communication are discussed. Several encoding and decoding methods are introduced and characterized. The basic techniques for the evaluation of optical communication systems are also introduced.

Chapter 8 is considered as the main chapter for this book's main title, namely optical communication with chaotic lasers. The history of chaos communication is first described, and examples of optical chaos communication systems are introduced from the literature with different encoding/decoding techniques and laser configurations. Recent advances on optical chaos comunications are also described, including 10 Gb/s transmission over a 100-km distance with low BER. The privacy of chaos communication is also discussed in detail and several attempts of the attacks to chaos communication systems are described. Photonic integrated circuits have been invented for the implementation to real-world communication systems. Other encoding and decoding techniques are introduced, such as polarization encoding, spatiotemporal encoding, and multiplexing communications. Finally, new perspectives of optical chaos communication to biological communication systems are described. Readers will find a comprehensive overview in the field of optical chaos communication in Chapter 8.

In Chapter 9, secure key distribution based on information-theoretic security with chaotic lasers is described. Chaotic lasers can be useful for secure key distribution, and this topic results from the research activities in optical chaos communication. The importance of secure key distribution and the concept of information-theoretic security are described. Examples of the implementation of secure key distribution based on information-theoretic security with chaotic lasers or optical noise are described in detail from the literature.

In Chapter 10, a novel engineering application of random number generation with chaotic lasers is described and discussed. The use of chaotic lasers for random number generation was proposed in 2008 (Uchida *et al.*, 2008a), and intense research activities have been reported. The combination of chaos (complexity) and laser (fast oscillation with large bandwidth) results in a new research field of fast physical random number generation. The needs for fast physical random number generators are discussed and the types of random number generators are classified. Examples of random number generators with chaotic lasers are described in detail from the literature. Different methods for random-bit extraction from chaotic signals are proposed and discussed. The application of random number generation to quantum key distribution is also introduced. Numerical evaluation for nondeterminism of generated random bits is performed with a laser model. For comparison, conventional physical and pseudorandom number generators are introduced and discussed.

Finally, statistical evaluation of random numbers is described in detail. This chapter overviews the recent development of random number generation and would be informative for the readers who intend to start studying and working on random number generation.

In Chapter 11, controlling chaos is described. Unstable periodic orbits are used to control and stabilize chaos by using a perturbation with feedback. Nonfeedback control of chaos is also described and discussed. Several schemes of controlling chaos in lasers and their engineering applications are described from the literature.

In Chapter 12, other applications with chaotic lasers are introduced and described. Remote sensing with chaotic lasers is a promising application such as chaotic radar and lidar for precise detection. Blind-source separation of chaotic signals by using independent component analysis is described for mulplexing communication systems. Finally, fractal patterns in regular-polyhedral mirror–ball structures are observed and a chaos mirror is invented for the application of wireless optical communication.

Readers can learn the potential of the interdisciplinary research field between chaos and laser that would result in new findings in science and novel engineering applications.

2
Basics of Chaos and Laser

In this chapter, the history of chaotic instability in lasers is introduced and the basic theories of chaos and lasers are described. The connection between chaos and lasers is made and the rate equations for a basic single-mode laser model are shown for the classification of laser dynamics.

2.1
History of Instabilities of Laser Output

2.1.1
Examples of Laser Instabilities

2.1.1.1 First Observation of Laser Instabilities in a Ruby Laser

At the time when the laser was invented in the early 1960s, the output of laser light showed spontaneous irregular pulsations (Maiman et al., 1961). Much effort has been made to stabilize the fluctuations of both laser output intensity and optical-carrier frequency as a coherent light source that enables stationary interference. At the same time, many studies have been reported on the characteristics of such irregular pulsations and temporal fluctuations of laser output, and it has been clarified that unstable laser output results from inherent nonlinearity of a laser medium that causes complicated light–matter interaction. These types of instabilities are known as "deterministic chaos" in terms of nonlinear dynamics.

The first experimental demonstration of lasers was reported in a ruby (Cr^{3+} in Al_2O_3) solid-state laser (Maiman, 1960). Figure 2.1 shows the output pulse from the ruby laser measured by using an oscilloscope. In the report of the characteristics of emitted radiation (Maiman et al., 1961), it was described that the amplitude of the individual pulses making up the output are seen to vary in an "erratic manner." This description was the first historical observation of instabilities of laser output intensity, and it is probably considered as "chaos" nowadays.

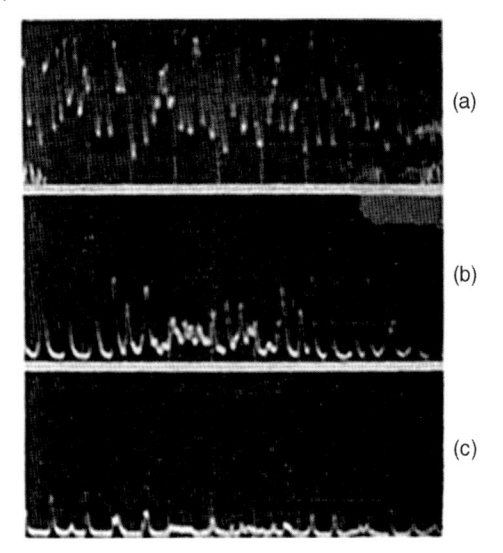

Figure 2.1 Irregular output pulse from the ruby laser on an expanded time scale (10 μs/division). (a), (b), and (c) represent the output approximately 600, 1000, and 1200 μs after the onset of oscillation. The vertical sensitivity and the base line are the same in each case (T. H. Maiman, R. H. Hoskins, I. J. D'Haenens, C. K. Asawa, and V. Evtuhov, (1961). © 1961 APS.)

2.1.1.2 Green Problem in a Solid-State Laser for Second-Harmonic Generation

Another example of instabilities of laser output is the generation of green light (wavelength: 532 nm) from a diode-laser pumped Nd:YAG (neodymium-doped yttrium aluminum garnet) solid-state laser with an intracavity KTP (potassium titanyl phosphate) crystal for second-harmonic generation, as shown in Figure 2.2a (Baer, 1986; Roy et al., 1994). The Nd:YAG laser operates in a stable steady state with an infrared beam (wavelength: 1064 nm) without the intracavity crystal. When the intracavity KTP crystal is used to generate green light for second-harmonic generation, large irregular intensity fluctuations are observed, as shown in Figure 2.2b. This behavior occurs when the laser operates in several longitudinal modes. Sum-frequency generation in the KTP crystal can provide mode–mode coupling that destabilizes the laser output. This is not a desirable situation for practical applications. The unstable behavior of this system has been known as the "green problem."

The chaotic nature of the green laser has been investigated in some detail and connected to the destabilization of relaxation oscillations. Relaxation oscillations are always present in a Class B laser (see Section 2.4.2). They are the result of power exchange between the atoms of the active medium and the photons in the laser cavity and are normally very small in amplitude. It is found that the nonlinear coupling of the modes through sum-frequency generation results in the destabilization of relaxation oscillations in the green laser system.

Several methods have been proposed and implemented to remove the fluctuations. These methods consist of system modifications such as restricting the laser to

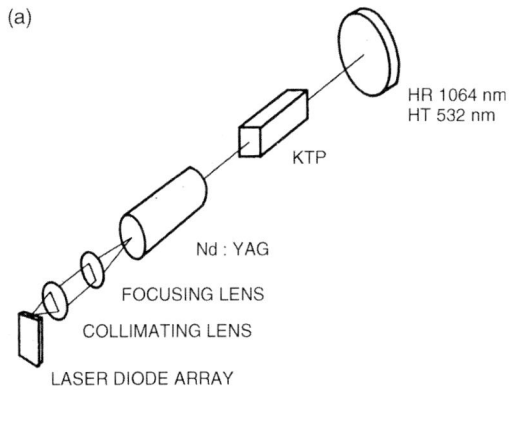

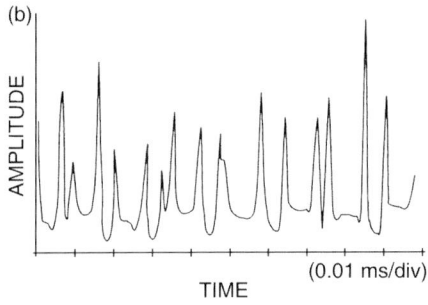

Figure 2.2 (a) Schematic diagram of the laser-diode-array-pumped Nd:YAG laser with an intracavity-doubling crystal. (b) The irregular temporal waveform of the 532-nm green laser output from the intracavity-doubled system. (T. Baer, (1986), © 1986 OSA.)

operate in two orthogonally polarized modes by adding wave plates to the laser cavity or proper orientation of the YAG and KTP crystals. Dynamical control of the chaotic green laser has been developed by using the occasional proportional feedback (OPF) control method, based on the theory of controlling chaos (see Chapter 11) (Roy *et al.*, 1992).

2.1.1.3 Feedback-Induced Instability in a Semiconductor Laser for Optical Disks and Optical Communication Systems

Semiconductor lasers (also referred to as laser diodes) are commonly used for optical data storage systems such as compact disk (CD), digital versatile disk (DVD), and blu-ray disk (BD). One of the important problems is related to the control of intensity fluctuation of semiconductor lasers used inside the optical head for reading and writing the data to the optical disk, as shown in Figure 2.3a. A small fraction of the laser output is invariably fed back into the semiconductor laser because of the reflection occurring at the disk surface (Gray *et al.*, 1993). To minimize the optical feedback into the semiconductor laser, a combination of a polarization beam splitter and a quarter-wave plate is used as an optical isolator. Indeed, the feedback should be absent if the isolator works perfectly for all laser wavelengths and if the reflections

2 Basics of Chaos and Laser

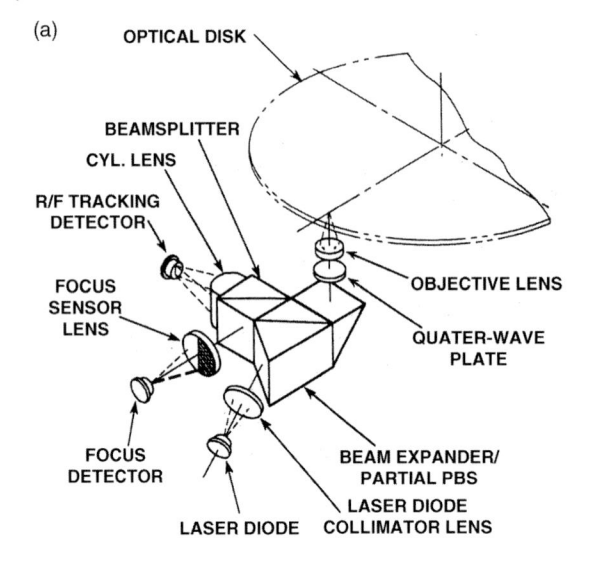

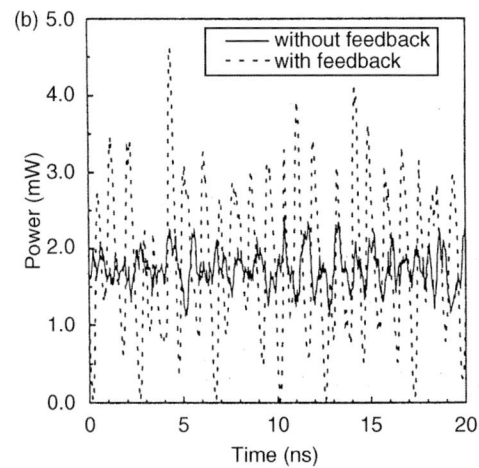

Figure 2.3 (a) Schematic of typical optical recording head. (b) Temporal power variations of a single-mode semiconductor laser with (dotted curve) and without (solid curve) optical feedback. (G. R. Gray, A. T. Ryan, G. P. Agrawal, and E. C. Gage, (1993), © 1993 SPIE.)

from the optical disk are polarization independent. However, the optical substrate is often birefringent, hence, a fraction of the laser output is fed back into the semiconductor laser and the intensity fluctuation of laser output occurs as shown in Figure 2.3b. The increase in intensity fluctuation degrades the signal-to-noise ratio (SNR) and severely impacts the performance of optical recording systems.

Another attempt to suppress the feedback-induced instability has been made by using the high-frequency injection technique. The laser current is modulated

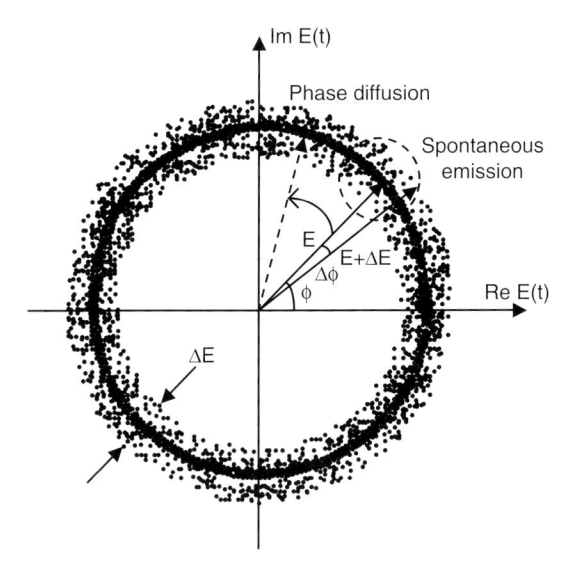

Figure 2.4 Concept of complex electric field. (A. Mooradian, (1985). © 1985 AIP).

sinusoidally at frequencies much higher than the data rate. The intensity fluctuation can be suppressed if the modulation frequency is suitably optimized and if the modulation amplitude is large enough to ensure that the laser is below threshold during a part of the modulation cycle (see Chapter 11).

Similar instability has been observed in optical communication systems. A fraction of the laser beam from the edge of optical fibers is fed back to the laser cavity, and laser intensity becomes unstable. An optical isolator is often used to avoid the reflection of feedback light from the optical fiber edge and to suppress the fluctuation of laser output intensity in the optical communication systems. However, when the amount of feedback light is large, the intensity fluctuation can be observed and the performance of communication systems is degraded. This type of optical feedback-induced instabilities in semiconductor lasers will be discussed in detail in Chapter 4.

2.1.2
Inherent Noise-Induced Instabilities

In the early days of laser physics, the models for laser dynamics developed theoretically had no place for chaotic behavior. The picture of the electric field emitted by a laser was often characterized by the phase-diffusion model. The electric field maintains the average amplitude and fluctuates in phase around the circle in the complex plane as shown in Figure 2.4 (Mooradian, 1985). The laser linewidth is generated primarily by the phase-diffusion process, and not by the amplitude fluctuations. Both amplitude and phase fluctuations are generated by spontaneous emission in the laser models developed; in the absence of spontaneous emission, the laser field would be determined entirely by stimulated emission, and have negligible linewidth. The simplest laser model that explains this type of behavior is a Langevin equation of the type studied for many years in the theory of Brownian motion. The Langevin equation is written as follows,

$$\frac{dE(t)}{dt} = aE(t) - b|E(t)|^2 E(t) + f(t) \qquad (2.1)$$

$$\langle f(t)^* f(t') \rangle = 2D\delta(t-t') \qquad (2.2)$$

where $E(t)$ is the complex electric field. The real laser intensity $I(t)$ (i.e., the number of photons) can be obtained from the square of the complex electric field, that is, $I(t) = |E(t)|^2$. The parameters a (net gain) and b (saturation) govern the dynamical properties of the field, which is driven by the complex random noise term $f(t)$ that represents spontaneous emission fluctuations (Haken, 1985). D is the strength of white noise. The dynamics of the Eq. (2.1) is simple: the laser field grows from random initial conditions (the spontaneous emission background) to a steady state where the amplitude and phase fluctuate in time, with an average amplitude and phase-diffusion rate that are determined by the parameters a and b and the strength of the correlation coefficient D of the noise source.

The dynamics of the atomic or molecular active medium of the laser is contained in the values of these parameters, and the single equation is derived through a process of eliminating the variables that are necessary to describe the active medium evolution in more detail: the dipole moment of the medium (atomic polarization) and the number of inverted atoms/molecules per unit volume (population inversion). It was apparent from the earliest days of laser development that some types of lasers, including the ruby laser, exhibited much more complex dynamics than contained in the simple model above. These observations of "irregular" or "noisy" dynamics are attributed to external noise sources, and extensive efforts have been made by most optical scientists and engineers to eliminate them, or to use just those laser systems that are stable, for engineering applications.

2.1.3
Deterministic Chaos in Lasers and Lorenz-Haken Equations

How was the connection between deterministic chaos and lasers first made? As the theoretical basis for nonlinear dynamical behavior was developed, it became clear that more degrees of freedom than are contained in the Langevin equation of Eq. (2.1) were necessary for chaotic dynamics to be present. It has been proven mathematically by the Poincaré–Bendixson theorem that at least three degrees of freedom (i.e., three independent variables) are necessary to observe deterministic chaos in continuous-time dynamical systems (Strogatz, 1994).

The ideas of both chaos and lasers were proposed in the early 1960s (Maiman, 1960; Lorenz, 1963) and the two research fields were developed individually. The paradigmatic model of chaos in the early 1960s was the Lorenz model (see

Section 2.2.2), developed in the context of weather prediction (Lorenz, 1963; Lorenz, 1993). The model consists on a simple set of coupled ordinary differential equations describing in a reduced way, for numerical integration purposes, the spatiotemporal problem of fluid convection in a cell heated from below and maintained at a lower temperature on top. The model consists of just three equations with a few nonlinear terms but without stochastic noise terms, and has just enough nonlinearity and coupled degrees of freedom to generate chaotic dynamical behavior. For certain ranges of parameters, the Lorenz model shows a very remarkable feature of nonlinear coupled differential equations – strong sensitivity to initial conditions for the variables that result in the trajectories in phase-space diverging exponentially fast away from each other. Lorenz recognized this property as a very important signature of a new kind of dynamical behavior, well known as "deterministic chaos." He indelibly impressed on his audience the significance of chaotic dynamics through the title of his talk to the American Association for the Advancement of Science in 1972 – "Does the flap of a butterfly's wings in Brazil set off a tornado in Texas?" (Lorenz, 1993).

The connection between chaos and lasers was first made by Haken in 1975 (Haken, 1975). He realized that the Maxwell–Bloch equations for the laser model and the Lorenz equations for the chaos model were essentially the same formula by applying variable transformation. The Maxwell–Bloch equations describing the time evolution of the atoms and electric field of a laser system can be put in the same form as the Lorenz equations through a set of transformations of variables and redefinition of parameters. These equations are now known as the Lorenz–Haken equations, consisting of three differential equations. The Lorenz–Haken equations are described as follows (van Tartwijk and Agrawal, 1998),

$$\frac{dx(t)}{dt} = \sigma(y(t) - x(t)) \tag{2.3}$$

$$\frac{dy(t)}{dt} = -x(t)z(t) + rx(t) - (1 - i\delta)y(t) \tag{2.4}$$

$$\frac{dz(t)}{dt} = \text{Re}(x^*(t)y(t)) - bz(t) \tag{2.5}$$

where $x(t)$ is the time evolution of the scaled electric field, $y(t)$ is the induced dipole moment (atomic polarization), and $z(t)$ is the population inversion. The parameters σ and b consist of ratios of the decay rates for the population inversion and photons from the laser cavity, and inversion decay rate to that of the induced dipole moment. The parameters r and δ represent the pump rate and detuning of the atomic resonance frequency from the optical frequency.

Instability of laser output was found in the Lorenz–Haken equations at the condition of high pump power and low Q factor of the laser cavity (bad cavity condition) (Weiss and Vilaseca, 1991). Even though the Lorenz–Haken-type chaos has been observed experimentally, the condition of generating chaos is far from the ordinary operation condition. Therefore, the Lorenz–Haken chaos has proven the stability of lasers at the ordinary operating conditions (i.e., high-Q cavity and intermediate pumping power).

Such a set of three equations, with parameters that correspond to the situation where the decay rates for the field, dipole moment and population inversion are all comparable, describes the dynamics of what is now known as a Class C laser, according to the terminology of (Arecchi *et al.*, 1984) (see Section 2.4.2). For parameters in a certain range, the dynamics displayed by this model is very similar to that of the Lorenz model, and much research followed the publication of Haken's paper (Haken, 1975) to explore the connections between lasers and fluids (Newell and Moloney, 1992).

When there is a large difference in the decay rates, one may progressively eliminate the induced dipole moment equation and then the inversion equation, to finally obtain the single equation for the electric field. This technique is known as "adiabatic elimination." The model with two coupled equations for the dynamics of the electric field and the population inversion is commonly referred to as the Class B laser system, and the model with one variable of the electric field (such as Eq. (2.1) without the stochastic term) corresponds to the model for a Class A laser (Arecchi *et al.*, 1984). This classification based on the dynamical properties of lasers will be described in Section 2.4.2. The dynamics of laser systems has been reviewed comprehensively in the books (Weiss and Vilaseca, 1991; Mandel, 1997; Otsuka, 1999; Kane and Shore, 2005; Ohtsubo, 2008).

It has been proven that at least three independent variables are required to observe deterministic chaos, and it seems that only Class C laser can produce chaotic instabilities. However, in many situations, both Class B and A lasers show chaotic laser output when additional perturbations or modulations are applied to the laser. Solitary Class B or A lasers may also show chaotic intensity fluctuations when they operate in the multimode oscillations due to the nonlinear mode–mode coupling. The generation methods of chaos in lasers with extra degrees of freedom will be classified in Chapter 3.

2.2
Basic Chaos Theory

2.2.1
Logistic Map in Discrete-Time System

2.2.1.1 Recurrence Formula

In this section, the basic concept of deterministic chaos and related terminologies are introduced and described. First, a basic mathematical formula is introduced for the observation of deterministic chaos in discrete time system. A discrete sequence is considered $x_0, x_1, \ldots, x_n, \ldots, x_N$, where $n = 0, 1, \cdots, N$. A recurrence formula (also referred to as a "map") can be defined by a function f that relates x_{n+1} with x_n for all n, that is,

$$x_{n+1} = f(x_n) \tag{2.6}$$

First, x_1 can be obtained from x_0 by using Eq. (2.6) when x_0 is given as an initial condition. Then, x_2 is obtained from x_1, and x_3 is obtained from x_2, and finally all x_n ($n = 1, 2, \cdots, N$) can be calculated from the initial condition of x_0 (i.e., $x_0 \to x_1 \to x_2 \to \cdots \to x_N$). In this sense, the map is completely deterministic by using the rule of Eq. (2.6) and the initial condition x_0.

One of the simplest models to exhibit chaos is the Logistic map. The Logistic map is written as,

$$x_{n+1} = a\, x_n (1 - x_n) \tag{2.7}$$

where $0 \leq x_n \leq 1$ and $0 < a \leq 4$ a is a parameter for the map, and a typical value is $a = 4$ to observe chaos. Source codes of C programming language are listed in Glossary G.2.1 for numerical simulations of chaotic dynamics of the Logistic map.

2.2.1.2 Chaotic Sequence

A sequence of the discrete variable x_n in the Logistic map ($a = 4$) is shown in Figure 2.5a. It is surprising that the time sequence shows an irregular behavior even though Eq. (2.7) is quite simple and completely deterministic. The deterministic rule of Eq. (2.7) can be depicted as a quadratic relationship between x_n and x_{n+1}, as shown in Figure 2.5b. This simple recurrence formula can produce an exotic irregular behavior without external noise term in Eq. (2.7).

It may seem that the irregular sequence shown in Figure 2.5a is predictable because the sequence is derived from the deterministic rule if the rule and the initial condition are known. This is true, however, the initial condition with "infinite" precision is necessary for the prediction. In real chaotic dynamical systems, it is impossible to identify the initial conditions with infinite precision. In addition, there is an important characteristic of deterministic chaos, known as sensitive dependence

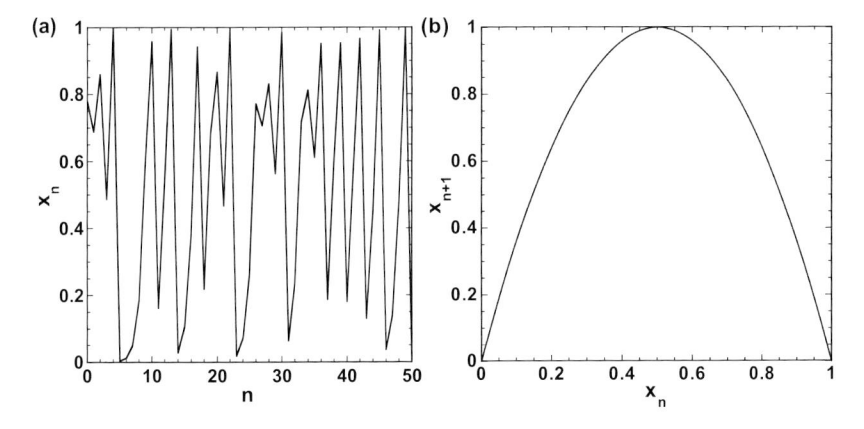

Figure 2.5 (a) Sequence of the discrete variable x_n and (b) quadratic relationship between x_n and x_{n+1} in the Logistic map ($a = 4$). The numerical data in this chapter was provided by Taiki Yamazaki.

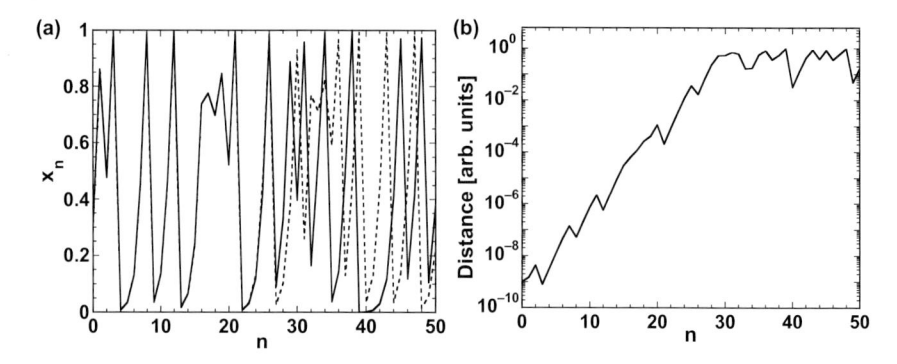

Figure 2.6 (a) Two sequences of x_n starting from slightly different initial conditions in the Logistic map ($a = 4$). (b) The distance of the two sequences in the semilogarithmic scale. This result shows sensitive dependence on initial conditions.

on initial conditions. An example of sensitive dependence on initial conditions is shown in Figure 2.6. Figure 2.6a shows two sequences of x_n of the Logistic map ($a = 4$). The two sequences start from very close but slightly different initial conditions, and the two sequences behave similarly at the beginning, however, they start to diverge in time at $n \approx 25$ and never show the same behavior again, as shown in Figure 2.6a. Figure 2.6b shows the time evolution of the error between the two sequences plotted in a semilogarithmic scale. The exponential divergence can be observed in Figure 2.6b, showing the sensitive dependence on initial conditions. Therefore, a tiny error of the initial conditions makes a chaotic irregular sequence unpredictable. This fact indicates chaos can be unpredictable for long-term duration, even though the deterministic rule exists and only short-term prediction is attainable.

2.2.1.3 Bifurcation Diagram

Another important characteristic of chaos is a transition among different dynamical states (steady state, periodic, and chaotic sequences), referred to as "bifurcation." In the Logistic map, when the parameter value a is changed, different states of sequences can be obtained. Figure 2.7 shows the sequences of different values of a. It is found that different number of values of x_n are obtained at different a: one constant value (Figure 2.7a, $a = 2.5$), two different values (Figure 2.7b, $a = 3.25$), four different values (Figure 2.7c, $a = 3.5$), and irregular values (Figure 2.7d, $a = 4.0$). These dynamical states correspond to period-1, period-2, period-4, and chaotic sequences, respectively.

Figure 2.8 shows the bifurcation diagram of the Logistic map when a is changed continuously. To create the bifurcation diagram, some values of x_n are plotted along the vertical axis at a fixed value of a. Then, a is increased slightly and some values of new x_n are plotted again at a newly fixed a. This procedure of changing a and plotting x_n is repeated. The bifurcation diagram shown in Figure 2.8 clearly indicates different dynamical states of the sequences. For example, x_n is a single value (corresponding to

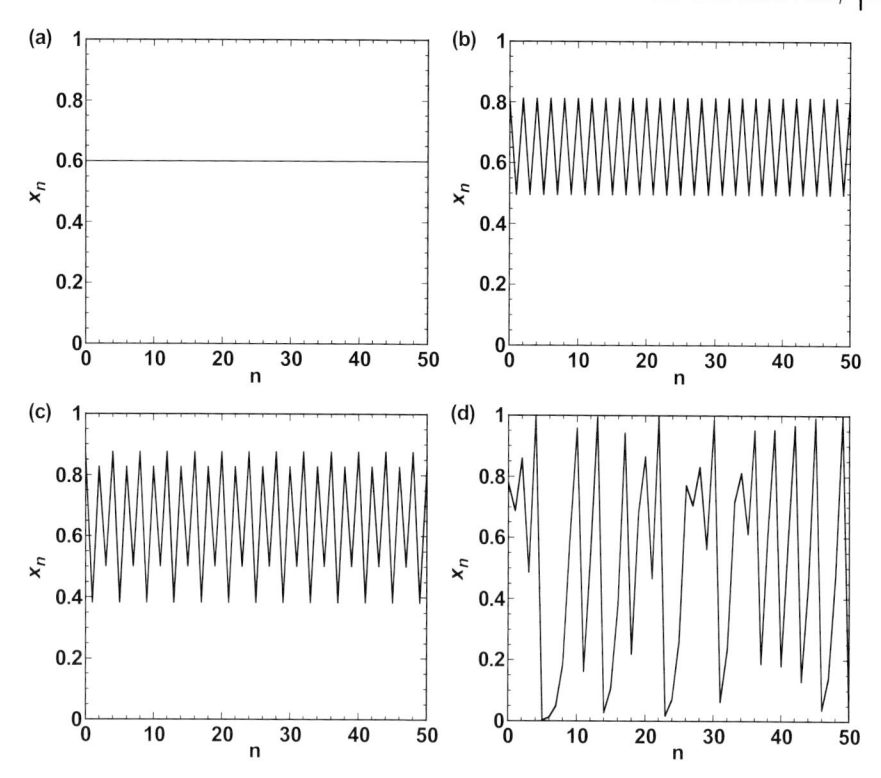

Figure 2.7 Sequences of x_n at different values of a in the Logistic map. (a) Period-1 ($a = 2.5$), (b) period-2 ($a = 3.25$), (c) period-4 ($a = 3.5$), and (d) chaos ($a = 4.0$).

period-1 sequence) at the region of $1 \leq a \leq 3$ in Figure 2.8a. Then, two values of x_n appear (period-2 sequence) in the region of $3 \leq a < 3.4494\ldots$. After the region, four values of x_n is observed (period-4), followed by eight values (period-8), and so on. Finally, irregular values of x_n (chaos) are obtained at the region of $a > 3.5699\ldots$. This

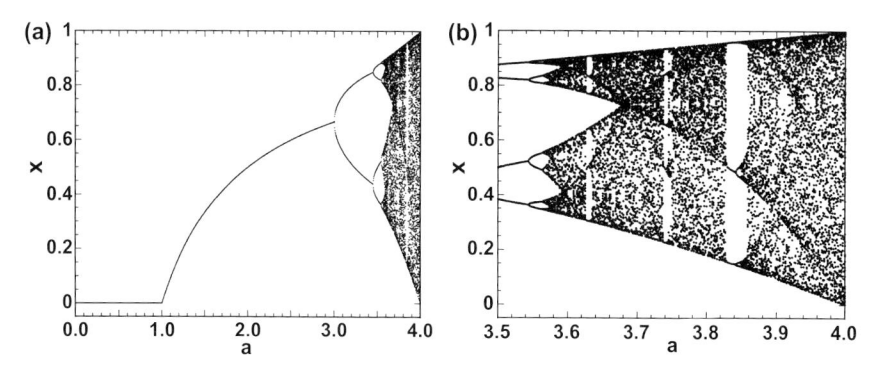

Figure 2.8 (a) Bifurcation diagram of the Logistic map when a is changed continuously from 0.0 to 4.0. (b) Enlargement of (a) from $a = 3.5$ to 4.0.

transition is referred to as "period-doubling" route (scenario) to chaos, since the period of the sequence becomes doubled as the parameter value is increased. After the chaotic state, a period-3 sequence is observed at the region of $a > 3.8284\ldots$ in Figure 2.8b, which is known as the "periodic window." A full chaotic state can be obtained at the region of $a = 4$, in which irregular values of x_n are observed in the whole region of $0 \leq x_n \leq 1$.

The bifurcation (i.e., a transition from a periodic state to chaos through different periodic states) is a typical feature of deterministic chaos, and cannot be observed in a stochastic noise system. Therefore, the observation of bifurcation is often used to identify the existence of deterministic chaos, particularly in laser experiments, if one of the system parameters is accessible and changeable. The period-doubling route to chaos has been observed in many laser systems experimentally and numerically (see Chapter 3).

2.2.1.4 Lyapunov Exponent

The "Lyapunov exponent" is often used as a quantitative measure of the sensitive dependence on initial conditions of chaotic system. The difference in the two chaotic sequences increases exponentially in chaotic dynamical systems, as shown in Figure 2.6. The Lyapunov exponent λ of the recurrence formula of Eq. (2.6) can be defined from the absolute value of the slope (derivative) of the map $|f'(x_i)|$ as,

$$\lambda = \lim_{N \to \infty} \frac{1}{N} \sum_{i=0}^{N-1} \ln|f'(x_i)| \qquad (2.8)$$

where the natural logarithm with the natural base e is used ($\ln = \log_e$). In the case of the Logistic map for $a = 4$, the Lyapunov exponent is analytically calculated as $\lambda = \ln 2$ (Strogatz, 1994). This formula indicates that $\lambda > 0$ is satisified when the average of the absolute value of the slope is satisfied as $|f'(x_i)|_{ave} > 1$ (expansion of the difference in two neighboring sequences), whereas $\lambda < 0$ when $|f'(x_i)|_{ave} < 1$ (contraction of the difference). Therefore, the slope of the map determines the degree of the sensitive dependence on initial conditions. One of the acceptable definitions of deterministic chaos is the existence of a "positive Lyapunov exponent" ($\lambda > 0$).

The Lyapunov exponents for the Logistic map with various values of a are shown in Figure 2.9, which is calculated from Eq. (2.8). For the calculation, the probability density function of x_n (or probability distrubtion, also known as the "invariant measure") needs to be considered when the ensemble average is calculated by Eq. (2.8), because the Logistic map does not show the uniform distribution. It is found that the regions of chaotic sequence in the bifurcation diagram of Figure 2.8 correspond to the regions of positive Lyapunov exponents in Figure 2.9. On the other hand, constant and periodic sequences correspond to zero or negative Lyapunov exponents. It is shown that a positive Lyapunov exponent is thus a good indicator to show the existence of deterministic chaos.

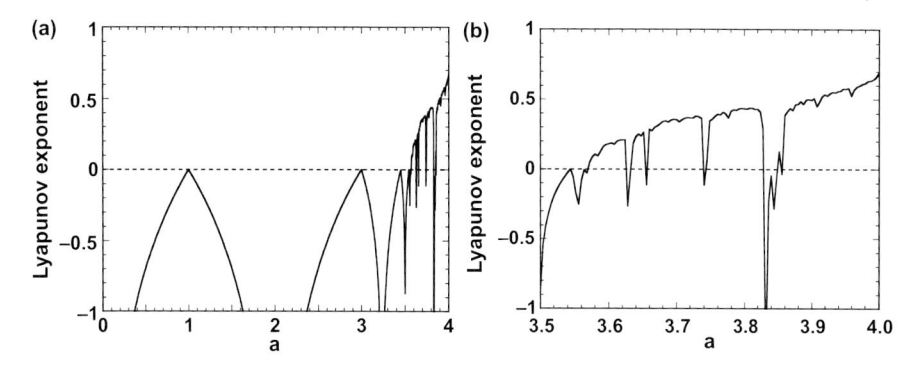

Figure 2.9 Lyapunov exponents of the Logistic map when a is changed continuously. The regions of positive Lyapunov exponents shown in Figure 2.9 correspond to those of chaotic sequences in Figure 2.8.

2.2.2
Lorenz Model in a Continuous-Time System

2.2.2.1 Lorenz Equations

The Logistic map of Eq. (2.7) is a good model for deterministic chaos to understand the essence of chaos, but it may seem like a toy model, since a simple recurrence formula is used to describe the dynamics. In addition, it has been proven that at least three degrees of freedom (i.e., three independent variables) are necessary to observe deterministic chaos in continuous-time dynamical systems (Strogatz, 1994).

A more realistic model can be described by using a simple set of coupled ordinary differential equations in continuous-time dynamical systems. The paradigmatic model of deterministic chaos is the Lorenz model (Lorenz, 1963). The model consists on three coupled ordinary differential equations, describing the fluid convection in a cell heated from below and maintained at a lower temperature on top:

$$\frac{dx(t)}{dt} = \sigma(y(t) - x(t)) \tag{2.9}$$

$$\frac{dy(t)}{dt} = -x(t)z(t) + rx(t) - y(t) \tag{2.10}$$

$$\frac{dz(t)}{dt} = x(t)y(t) - bz(t) \tag{2.11}$$

The variables $x(t)$, $y(t)$, and $z(t)$ represent properties of the convecting fluid flow, and temperature differences between the left- and right-hand sides of the cell of fluid and between its top and bottom, respectively. The parameter σ is the Prandtl number, r is the Rayleigh number, and b is the aspect ratio of the cell. Typical parameter values for the observation of chaos are $\sigma = 10$, $r = 28$, and $b = 8/3$ (Bergé et al., 1984). Source codes of C programming language are listed in Glossary G.2.2 for numerical simulations of chaotic dynamics of the Lorenz equations.

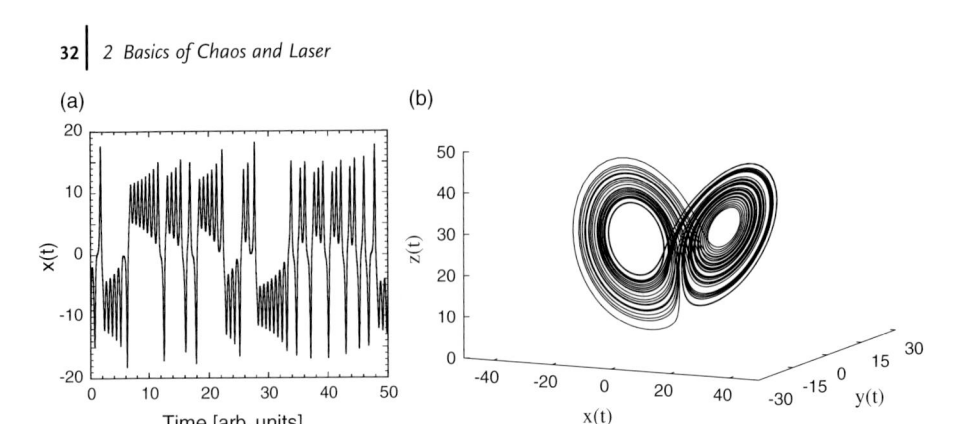

Figure 2.10 (a) Temporal waveform of the variable $x(t)$ and (b) buttefly-shaped chaotic attractor of the Lorenz model.

2.2.2.2 Temporal Waveform and Attractor

Figure 2.10a shows the temporal waveform of one variable $x(t)$, and chaotic irregular fluctuation is clearly observed. The Lorenz model has three degrees of freedom and enough nonlinearity to generate chaotic dynamical behavior, as seen in Figure 2.10a.

Another representation of temporal dynamics is known as an "attractor," which is a temporal history of the variables (called "trajectrory") plotted in a two- or three-dimensional phase space, as seen in Figure 2.10b. The time evolutions of all the three variables ($x(t)$, $y(t)$, and $z(t)$) are plotted in the three-dimensional space consisting of the axes of $x(t)$, $y(t)$, and $z(t)$. The trajectory converges into a certain shape of "attractor" in the phase space at $t \to \infty$. For example, for a constant value of the variables, the attractor becomes a point attractor in phase space, and for a period-1 oscillation the attractor becomes a circle (referred to as a "limit cycle"). By contrast, for a chaotic oscillation, the attractor resembles a strange shape, as seen in Figure 2.10b. This is called a "chaotic attractor" or a "strange attractor." It is known in general that a chaotic attractor consists on "fractal" structure, where multiscale similarity of the density of trajectories exists. As seen in Figure 2.10b, the chaotic attractor obtained from the Lorenz model looks like a butterfly, and it is often called the "butterfly attractor."

2.2.2.3 Sensitive Dependence on Initial Conditions

The Lorenz model shows a remarkable feature of chaos, that is, sensitive dependence on initial conditions that result in the trajectories in phase space diverging exponentially fast away from each other. The sensitive dependence on the initial conditions of the chaotic temporal waveforms is observed, as shown in Figure 2.11a. Two temporal waveforms starting from slightly different initial conditions behave similarly at the beginning, however, they start diverging after $t \approx 25$ in Figure 2.11a.

To investigate the sensitive dependence on initial conditions, the distance $\delta(t)$ between two neighboring trajectories $(x(t), y(t), z(t))$ and $(x'(t), y'(t), z'(t))$ in the phase space is shown for the Lorenz model, as measured by

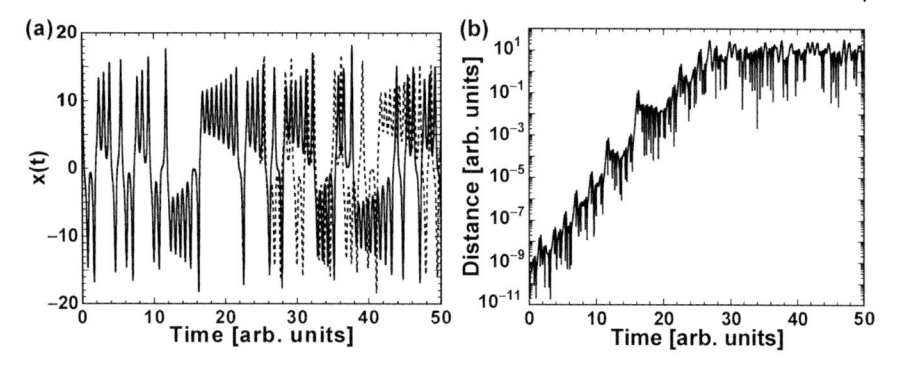

Figure 2.11 (a) Two temporal waveforms of $x(t)$ starting from slightly different initial conditions in the Lorenz model. (b) Time evolution of the distance $\delta(t)$ between the two neighboring trajectories in the phase space plotted in the semilogarithmic scale. Sensitive dependence on initial conditions is evident.

$$\begin{aligned}\delta(t) &= \sqrt{(x(t)-x'(t))^2 + (y(t)-y'(t))^2 + (z(t)-z'(t))^2} \\ &= \delta(0)\exp(\lambda_e t)\end{aligned} \quad (2.12)$$

Figure 2.11b shows the time evolution of the distance $\delta(t)$ plotted on the semilogarithmic scale. It is found that the distance $\delta(t)$ grows exponentially on the average until it saturates. The rate of growth is characterized by the effective Lyapunov exponent for the system, denoted by λ_e.

2.2.2.4 Bifurcation Diagram

Bifurcation diagram can be obtained for the Lorenz model as well as the Logistic map. For the continuous-time system, a bifurcation diagram can be created by sampling the local maxima (or minima) of temporal waveforms. A set of the local maxima is plotted along the vertical axis at a fixed parameter value. Then, the parameter value is slightly increased (or decreased). A new temporal waveform is obtained at a new fixed parameter value and the local maxima of the temporal waveform are sampled and plotted again. This procedure is repeated to create a bifurcation diagram.

The bifurcation diagram for the variable $x(t)$ in the Lorenz model is shown in Figure 2.12a. The parameter value of r is increased continuously. Different periodic oscillations are observed, where a single value of $x(t)$ corresponds to a period-1 oscillation, two values correspond to a period-2 oscillation and so on. The regions of the scattered plots in Figure 2.12a indicate the occurrence of chaotic oscillations. Clear bifurcation from periodic to chaotic oscillations is observed in Figure 2.12a.

2.2.2.5 Lyapunov Exponent

Methods for the calculation of the Lyapunov exponent have been extensively developed for both experimentally measured time series as well as from the integration of model equations for dynamical systems (Abarbanel, 1996; Kantz and Schreiber, 1997). The calculation of the Lyapunov exponent from the "linearized

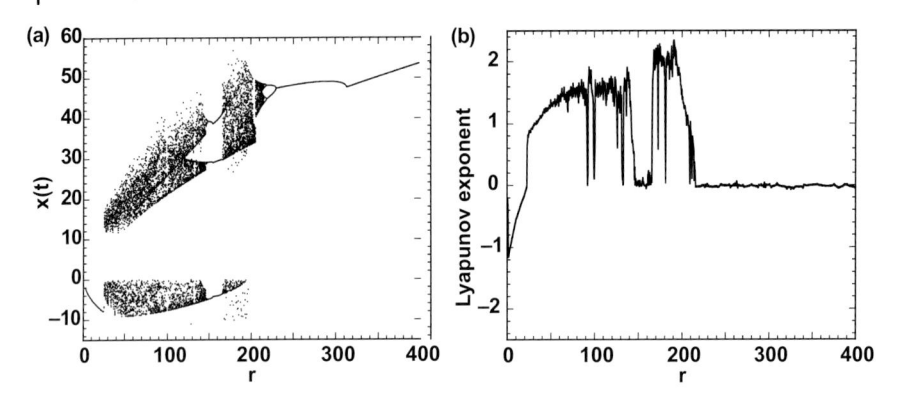

Figure 2.12 (a) Bifurcation diagram and (b) Lyapunov exponent of the Lorenz model when the parameter value of r is changed continuously. The regions for positive Lyapunov exponents shown in Figure 2.12b correspond to those for chaotic temporal waveforms in Figure 2.12a.

equations" is very accurate and recommended, rather than from the original equations or the time series (see Chapter 4 for details). A method for the calculation of the linearized equations of the Lorenz model is described in Appendix 2.A.1. In addition, a method for the calculation of the Lyapnov exponent from the linearized equations is described in Appendix 2.A.2.

The Lyapunov exponents of the Lorenz model for various parameter values of r are shown in Figure 2.12b. It is found that the regions of chaotic oscillations in the bifurcation diagram of Figure 2.12a correspond to the regions of positive Lyapunov exponents in Figure 2.12b. On the other hand, steady states and periodic oscillations correspond to zero or negative Lyapunov exponents, as well as the case for the Logistic map.

More generally, there is the same number of Lyapunov exponents as the number of system variables, that is, three Lyapunov exponents exist for the Lorenz model. The Lyapunov exponent calculated in Figure 2.12b is the maximum value of the three Lyapunov exponents. A comprehensive characterization of the dynamics is done through the computation of all the Lyapunov exponents, which are called the "Lyapunov spectrum". The computation of the Lyapunov spectrum allows to compute the "Kolmogorov–Sinai entropy" and the "Kaplan–Yorke dimension" of a system, which are related to the complexity and dimensionality of a dynamical system (Vicente et al., 2005) (including the correlation dimension (Grassberger and Procaccia, 1983), or the information dimension (Hilborn, 2000)). Detailed methods for the calculation of these quantities will be described in Chapter 4.

Another well-known model for the basic research of deterministic chaos is the "Rössler model" (Bergé et al., 1984). The Rössler model consists of three independent variables and shows the characteristics of deterministic chaos, such as bifurcation to chaos and a positive Lyapunov exponent. Details of the Rössler model are described in Appendix 2.A.3.

2.2.3
Reconstruction of Attractor in Time-Delayed Phase Space

When trying to identify the nonlinear dynamical characteristics of a system, one usually makes measurements in time of the physically accessible variables. In the case of a laser system, the most common variable one can measure is the total intensity of the light output. The induced dipole moment (atomic polarization) and the population inversion are not usually easily measured, and even the phase of the electric field requires more complex instrumentation and measurement techniques. The question then arises as to how to analyze the data and what information can be extracted from the measurements made as best possible. As a case in point, if only the variable $x(t)$ can be measured in the Lorenz system, what can be discovered about the nature of the dynamics of the whole system of coupled variables? This was an important question in the early days of the development of nonlinear dynamics as a scientific discipline, and the answer has been known as the method of phase-space reconstruction, and involved the use of a tool known as "embedding" developed in early 1980s (Takens, 1981; Packard et al., 1980). The basic theorem states that for a multidimensional system, if a single scalar variable (such as the total laser intensity) can be observed at a sequence of times, many important properties of the multivariate phase-space dynamics of the system can be reconstructed by a time-delay embedding process.

As an example of the Lorenz system, Figure 2.13 shows the consequence of plotting the time series measured for $x(t)$ at times t, $t+\tau$, and $t+2\tau$ in the three-dimensional space reconstructed with the axes $x(t)$, $x(t+\tau)$, and $x(t+2\tau)$. The values of τ determine the precise shape of the attractor that has been reconstructed, but it is clear that the similar butterfly attractor is viewed in both Figure 2.13a ($\tau=0.1$) and Figure 2.13b ($\tau=0.2$) in a topological sense. The important questions of what should be the dimensionality of the space that is necessary for reconstruction and the appropriate value for the time delay τ have been discussed in detail (Abarbanel, 1996; Kantz and Schreiber, 1997).

There are many other tools developed by scientists and engineers to extract information about the nature of a dynamical system. These is the notion of the "Poincaré section," which consists of plotting the values of certain variables at

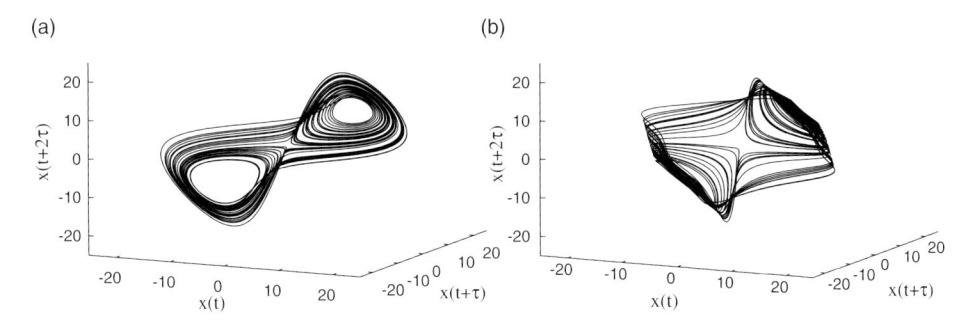

Figure 2.13 Lorenz attractors embedded in the time-delayed phase-space with the axes of $x(t)$, $x(t+\tau)$, and $x(t+2\tau)$. (a) $\tau=0.1$, (b) $\tau=0.2$.

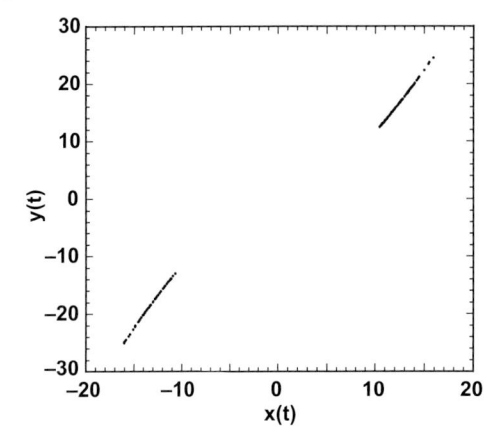

Figure 2.14 Poincaré section of the Lorenz butterfly attractor with a plane parallel to the $x-y$ plane at $z = 27$ when the trajectory goes upwards through it.

selected time intervals and for certain conditions. Figure 2.14 shows the Poincaré section of the Lorenz butterfly attractor with a plane parallel to the $x-y$ plane at $z = 27$ when the trajectory goes upwards through it. If the dynamics is periodic, such a section would show only a fixed number of points. When chaotic dynamics occurs, the section consists of structured clouds of points as shown in Figure 2.14.

The surprising and important aspect of the development of these mathematical tools for the description of nonlinear systems is that one may obtain quantitative characterizations of multidimensional system dynamics through the observation and measurement of a single scalar variable of the system. This fundamental perspective has been carefully tested and verified on many systems and models.

2.2.4
Types of Bifurcation and Route to Chaos

As described above, one of the important features of deterministic chaos is "bifurcation" and "route (scenario) to chaos," which is a transition from steady state to chaos through various periodic states and quasiperiodic states (i.e., two frequencies with incommensurable ratio) when one of the system parameter values is changed. Three types of well-known bifurcations and routes to chaos are described (Bergé et al., 1984). Figure 2.15 shows schematics of the three types of routes to chaos commonly observed in many dynamical systems as one of the system parameter values is changed. These types of routes to chaos have been also observed in many laser systems, and used to identify the existence of deterministic chaos in laser experiments (see Chapter 3).

2.2.4.1 Period-Doubling Route to Chaos
The period-doubling route to chaos starts from a steady state, a period-1 oscillation, a period-2 oscillation, a period-4 oscillation, ..., a period-$2n$ oscillation (n is a positive integer), ..., and reaches a chaotic oscillation, as shown in Figure 2.15a. The distance

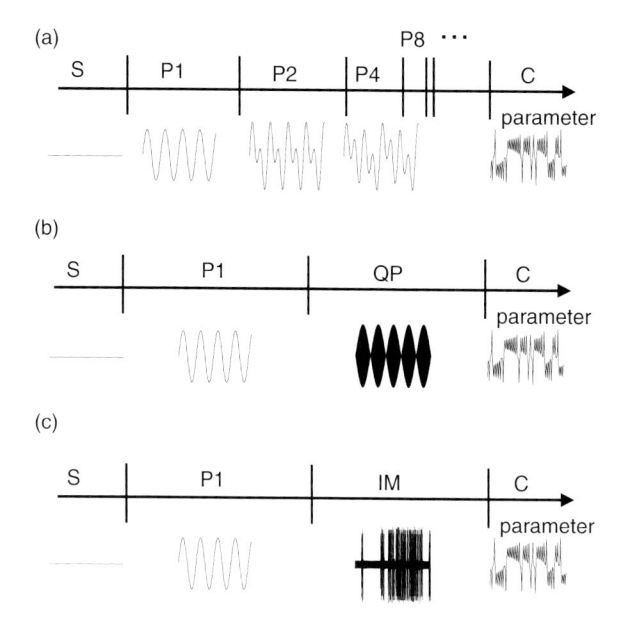

Figure 2.15 Schematics of typical routes to chaos as one of the parameter values is changed. (a) Period-doubling route to chaos, (b) quasiperiodicity route to chaos, and (c) intermittency route to chaos. C, chaos; IM, intermittency; P1, period-1; P2, period-2; P4, period-4; P8, period-8; QP, quasiperiodicity; S, steady state.

between consecutive bifurcation points in terms of the parameter values decreases exponentially, which is known as the Feigenbaum constant ($\delta = 4.6692\ldots$ for the Logistic map). The period-doubling route to chaos is associated with the subharmonic bifurcation (Bergé et al., 1984). The period-doubling route to chaos can be observed in many discrete and continuous-time dynamical systems (see also Figure 2.8), as well as many laser experiments and simulations. This route is also known as the "Feigenbaum scenario to chaos."

2.2.4.2 Quasiperiodicity Route to Chaos

The quasiperiodicity route to chaos starts from a steady state, a period-1 oscillation, a quasiperiodic oscillation, and chaotic oscillation, as shown in Figure 2.15b. Quasiperiodic oscillations can be identified by observing the appearance of two frequencies with incommensurable ratio in the frequency spectrum. When the parameter is changed, the nonlinear interaction between the two frequencies breaks a quasiperiodic attractor (i.e., a "torus" attractor) and produces a chaotic attractor. The quasiperiodicity route to chaos is associated with the Hopf bifurcation (Bergé et al., 1984). This route to chaos is commonly observed in a laser system with external modulation and perturbation, since two characteristic frequencies, which are internal (relaxation oscillation) and external (modulation) frequencies, interact with each other nonlinearly and induce chaotic instabilities. This route is also known as the "Ruelle–Takens–Newhouse scenario to chaos."

2.2.4.3 Intermittency Route to Chaos

The intermittency route to chaos is characterized by the fact that when the control parameter exceeds a critical value, a regular oscillation or a steady state (called the "laminar phase") of the dynamical variable appears to be interrupted at random times by bursts of irregular behavior ("burst phase"), as shown in Figure 2.15c. The duration of the turbulent phases is fairly regular and weakly dependent on the control parameter, but the mean duration of the laminar phases decreases as the control parameter is increased beyond the critical value, and eventually the laminar phases disappear to become a fully chaotic state. The intermittency route to chaos is associated with the saddle-node, Hopf, and subharmonic bifurcations, and can be classified into the type-I, -II, and -III intermittencies, respectively (Bergé et al., 1984). This route is also known as the "Pomeau–Manneville scenario to chaos."

2.3 Basic Laser Theory

2.3.1 Light–Matter Interaction for Laser Radiation

2.3.1.1 Elements of a Laser

The laser is an artificial light source to generate "coherent" light, which indicates the identical phase state of photons (the photon is a fundamental particle consisting of light). The coherent light enhances the brightness of light, the ability to interfere, and the narrow optical spectrum. To make laser oscillations, three key elements are necessary: a "laser medium" for light amplification, a "laser cavity" to confine the generated light, and "pump energy" to start and continue lasing operation. The schematic of the laser configuration is shown in Figure 2.16. The laser medium, which can be semiconductor, fiber, solid-state, or gas, is inserted into the laser cavity that consists of a pair of reflecting mirrors (known as the "Fabry–Perot cavity"). The pump energy is applied to the laser medium by using a flash lump light, laser light, or an electric current. The pump energy excites the atomic state in the laser medium, and a collection of excited atoms in the laser medium amplifies a light signal directed

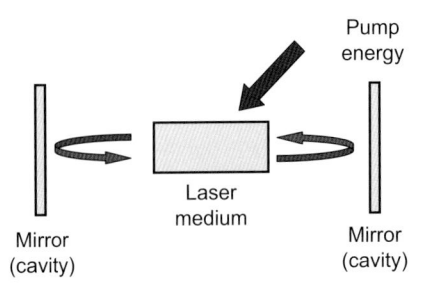

Figure 2.16 Basic configuration of laser system. The laser medium is inserted into the laser cavity that consists of a pair of reflecting mirrors, and pump energy is applied to the laser medium.

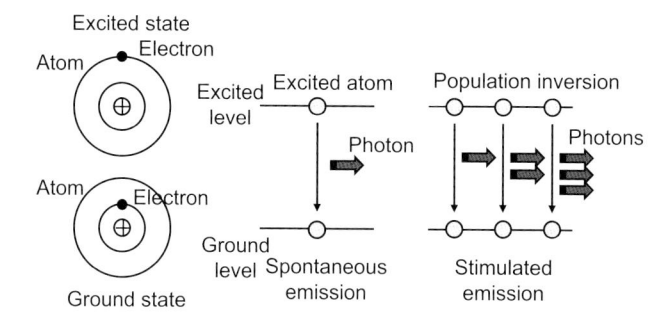

Figure 2.17 Principle of laser oscillation in the two-atomic-level model. By some pumping process, some of these atoms are promoted from a ground state to an excited state. The excited atoms begin radiating spontaneous emission. A spontaneously emitted photon can induce an excited atom to emit another photon of the same frequency and direction as the first. More such photons are produced by stimulated emission.

though it. The amplifying medium is usually enclosed by the cavity that holds the amplified light, in effect redirecting it through the medium for repeated amplifications. The energy of the amplifier that is being converted to light energy needs to be added continuously. The generated light in the form of a beam in the cavity is extracted from one of the mirrors that is partially transparent to the laser light.

2.3.1.2 Two-Atomic-Level Description and Mechanism of Laser Oscillation

Let us consider a process of the generation of coherent light in the laser medium with a two-atomic-level model, as shown in Figure 2.17. Two energy levels of the atoms in the laser medium are considered: ground and excited levels. By some pumping process, such as absorption of light from a pumping source, some of these atoms are promoted from the ground level to the excited level. "Population inversion" is produced when the number of atoms in the excited level is larger than that in the ground level, where population inversion indicates the difference in atomic populations between the excited and ground levels. Note that more than two atomic levels are required to obtain the population inversion in real laser systems (Yariv, 1991). The excited atoms begin radiating "spontaneous emission," for which a photon is produced spontaneously from an excited atom when it decays to the ground level. A spontaneously emitted photon can induce an excited atom to emit another photon of the same frequency, phase, and direction as the first, which is referred to as "stimulated emission." The more such photons are produced by stimulated emission, the faster is the production of still more photons, because the stimulated emission rate is proportional to the flux of photons already in the stimulating electric field (i.e., an avalanche of photons). The mirrors of the laser cavity keep most of the photons from escaping, so that they can be redirected into the active laser medium to stimulate the emission of more photons. By making the mirrors partially transmitting, some of the photons are allowed to escape. They constitute the output laser beam. The intensity of the output laser beam is determined by the rate of production of excited atoms (population-inversion lifetime) and the loss of the photons in the laser cavity and medium (photon lifetime).

When the pumping power is enough to produce the population inversion while photons are generated by stimulated emission, continuous laser output is obtained. However, photons are not generated during some period when the rate of stimulated emission is faster than the generation rate of population inversion. In such a situation, optical pulses are obtained.

The atoms in the laser cavity arise from the strong feedback deliberately imposed by the cavity designer. This feedback means that a small input can be amplified in a straightforward way by the atoms, but not indefinitely. Simple amplification occurs only until the light field in the cavity is strong enough to affect the behavior of the atoms. Then, the strength of the light as it acts on the amplifying atoms must be taken into account in determining the strength of the light itself. The response of the light and the atoms to each other can become so strongly interconnected that they cannot be determined independently but only self-consistently. Strong feedback also means that small perturbation can be rapidly magnified. Thus, it is accurate to anticipate that lasers are potentially highly erratic and unstable devices. In fact, lasers can provide dramatic exhibitions of truly chaotic behavior. The self-consistent interaction of light (photons) and matter (atoms) is important for understanding the dynamics of lasers, as described in the rate equations in Section 2.3.3.

2.3.2
Radiative Recombination of Electron–Hole Pairs in Semiconductor Lasers

For semiconductor lasers, no population inversion exists in the semiconductor device due to the band structure of the energy level. Instead, the electron–hole pairs play a similar role of the population inversion and the coherent light can be obtained from radiative recombination of electron–hole pairs. The basic structure of semiconductor lasers is the p-n junction, formed by bringing a p-type (with many holes) and an n-type (with many electrons) semiconductor into contact with each other (Agrawal and Dutta, 1993). The structure and the energy-band diagram of the p-n junction are shown in Figure 2.18a. When a p-n junction is forward-biased by applying an external voltage, the built-in electric field is reduced, making possible a diffusion of electrons and holes across the junction. In a narrow depletion region both electrons and holes are present simultaneously and can recombine either radiatively or nonradiatively. Photons of energy are emitted during radiative recombination. However, these photons can also be absorbed through a reverse process that generates electron–hole pairs. When the external voltage excess a critical value, a condition known as population inversion is achieved, in which the rate of photon emission exceeds that of absorption. The p-n junction is then able to amplify the electromagnetic radiation. The amount of the electron–hole pairs is also referred to as the "carrier density," which is proportional to the injection current of the semiconductor laser. The carrier density above the lasing threshold can produce coherent laser light.

To obtain high gain of the amplification in semiconductor lasers, p-n heterojunction structure is used. Figure 2.18b shows the structure and the energy-band diagram for a "double-heterostructure" laser wherein the thin p-type active region has a lower bandgap compared to that of the two p-type and n-type cladding layers. Electrons and

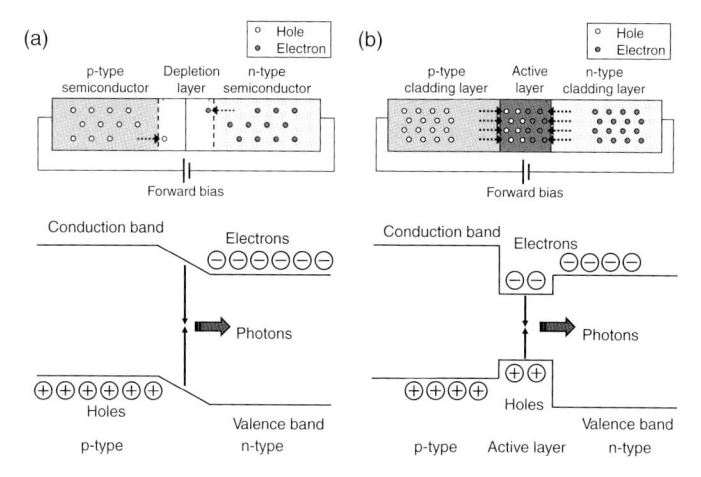

Figure 2.18 Structure and energy-band diagram of (a) a p-n junction in a semiconductor laser and (b) a double-heterostructure semiconductor laser.

holes can move freely to the active region under forward bias. However, once there, they cannot cross over to the other side because of the potential barrier resulting from the bandgap difference. This allows for a substantial build-up of the electron and hole populations inside the active region, where they can recombine to produce optical gain. The width of the gain region is determined by the active-layer thickness, typically 0.1–0.3 μm. The heterostructure scheme results in significantly lower threshold current densities, and leads to the room-temperature operation of semiconductor lasers.

2.3.3 Rate Equations for Laser Dynamics

The temporal dynamics of lasers is described by using a set of coupled ordinary differential equations. The main three physical variables in lasers are the "electric field," the "population inversion," and the "atomic polarization" of the laser medium. The electric-field amplitude E (described by a complex number) can be replaced by the laser intensity I (described by a real number) or the number of photons P (a real number), where the relationship of $I = P = |E|^2$ holds. The complex variable of E is used when the optical phase plays an important role as coherent interaction, such as injection locking or coherent coupling of the electric field, whereas I is used when only the laser intensity is considered for incoherent interaction. For semiconductor lasers the variable of the population inversion is replaced by the carrier density, as described in Section 2.3.2. The rates of these three physical variables are governed by physical laws in the laser cavity and described by a mathematical formula, known as "rate equations." The rate can be increased or decreased, depending on the sign of the terms in the equations. The dynamics of lasers can be described with deterministic equations.

A schematic of light–matter interaction with decay rates is shown in Figure 2.19, where two-level atoms and photons interact with each other. One of the simplest formula for the dynamics of this model with the variables of the laser intensity and the population inversion is described as follows (Yariv, 1991).

2 Basics of Chaos and Laser

Figure 2.19 Schematic of two-atomic-level laser model for describing the rate equations of Eqs. (2.13) and (2.14).

$$\frac{dI(t)}{dt} = gI(t)N(t) - \frac{I(t)}{\tau_p} \tag{2.13}$$

$$\frac{dN(t)}{dt} = R - gI(t)N(t) - \frac{N(t)}{\tau_s} \tag{2.14}$$

where, $I(t)$ is the laser intensity, $N(t)$ is the population inversion, g is the laser gain, R is the pump energy for lasing, τ_p is the photon lifetime, and τ_s is the population lifetime.

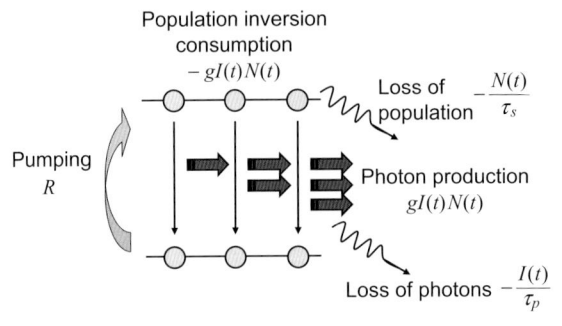

Let us interpret each term of the model in Eqs. (2.13) and (2.14). Equation (2.13) describes the temporal dynamics of laser intensity $I(t)$, which is influenced by the two terms in the right-hand side of the equations. Consider the effect of the first term in the right-hand side and neglect the second term in Eq. (2.13). Equation (2.13) can be changed,

$$\frac{dI(t)}{dt} = gI(t)N(t) \tag{2.15}$$

$$\frac{dI(t)}{I(t)} = gN(t)dt \tag{2.16}$$

Equation (2.16) can be integrated when $N(t)$ is assumed to be nearly constant N_s for simplicity. Then, the solution of Eq. (2.16) can be obtained analytically by integration,

$$\int \frac{dI(t)}{I(t)} = \int gN_s dt \tag{2.17}$$

$$I(t) = A\exp(gN_s t) \quad (A : \text{constant value}) \tag{2.18}$$

Therefore, the first term $gI(t)N(t)$ in Eq. (2.13) indicates the stimulated emissions, where the number of photons increases "exponentially" as described in Eq. (2.18). In fact, the coefficient of the exponential function includes the time-dependent variable $N(t)$, so the dynamics is not a simple exponential growth but is more complicated.

The same argument can be used for the second right-hand-side term of Eq. (2.13). The second term $-I(t)/\tau_p$ indicates the decay of photons in the cavity, where the number of photons decreases exponentially as described by $I(t) = B\exp(-t/\tau_p)$ (B is a constant). This solution is derived from the differential equation of $dI(t)/dt = -I(t)/\tau_p$. The number of photons decays spontaneously at the exponential rate with $-1/\tau_p$, indicating the effect of spontaneous emission of photons. From these arguments, Eq. (2.13) can be interpreted as follows: The dynamics of the laser intensity $I(t)$ is determined by the interaction between the stimulated emission (exponential increase of $I(t)$) and the decay of photons in the laser cavity (exponential decrease of $I(t)$).

Next, let us interpret Eq. (2.14) for the dynamics of the population inversion $N(t)$, which is governed by the three terms in the right-hand side of Eq. (2.14). The first term R indicates the pump energy for lasing, where the population inversion increases "linearly" in time, as described by $N(t) = Rt + C$ (i.e., the solution of $dN(t)/dt = R$, where C is a constant), indicating the effect of pumping for lasing. The second right-hand-side term of Eq. (2.14) $-gI(t)N(t)$ indicates the stimulated emissions, where the population inversion decreases exponentially as described by $N(t) = D\exp(-gI_s t)$ ($I(t) \approx I_s$ is assumed and D is a constant). Note again that the coefficient of the exponential function includes the time-dependent variable $I(t)$, that makes the dynamics complicated. The effect of stimulated emission can be seen in both Eqs. (2.13) and (2.14), that is, $I(t)$ and $N(t)$ are changed at the same time with the same ratio but the opposite sign, indicating the effect of increase or decrease. The third right-hand-side term of Eq. (2.14) $-N(t)/\tau_s$ indicates the spontaneous emission, where the population inversion decreases exponentially as described by $N(t) = E\exp(-t/\tau_s)$ (E is a constant). The dynamics of Eq. (2.14) for the population inversion $N(t)$ is thus determined by the three effects: the pumping for lasing (linear increase of $N(t)$), the stimulated emission (exponential decrease of $N(t)$) and the spontaneous emission (exponential decrease of $N(t)$).

The above interpretation of Eqs. (2.13) and (2.14) is schematically summarized in Figures 2.19 and 2.20. The rate equations can be easily understood by using this schematic relationship of Figure 2.20. The rate equations are very convenient to

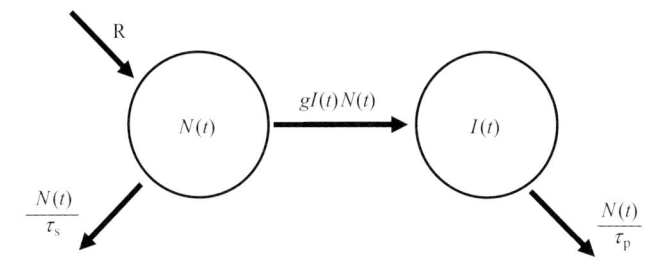

Figure 2.20 Schematic of dynamical coupling between the laser intensity $I(t)$ and the population inversion $N(t)$ described in Eqs. (2.13) and (2.14). The directions of the arrows in the figure correspond to the signs of the right-hand-side terms in Eqs. (2.13) and (2.14).

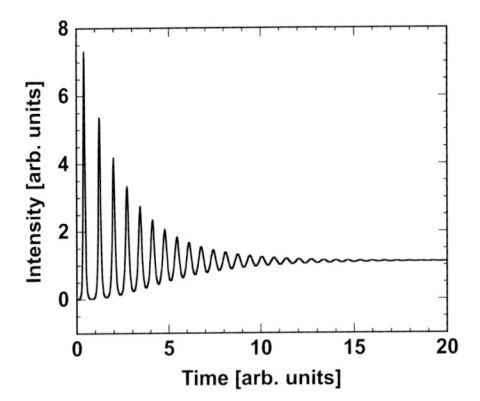

Figure 2.21 Temporal waveform of the relaxation oscillation of the laser intensity when the laser is turned on.

describe the dynamics of physical variables in laser systems. Generally, there are "nonlinear" terms in the rate equations. In Eqs. (2.13) and (2.14), the product of $I(t)$ and $N(t)$ exists as a nonlinear term for the stimulated emission $gI(t)N(t)$. This nonlinearity is the origin of deterministic chaos and complex dynamics in laser dynamical systems. More detailed models for realistic laser systems will be described in Appendices of Chapters 3 and 5.

2.3.4
Relaxation Oscillation Frequency

The relaxation oscillation of laser intensity output is one of the important characteristics to determine the frequency range of the temporal dynamics of lasers. Figure 2.21 shows an example of relaxation oscillation of the laser intensity when the laser is turned on. This oscillation takes place with a period that is considerably longer than the photon lifetime (i.e., class B lasers, see Section 2.4.2). The basic physical mechanism is the interplay between the laser intensity in the cavity and the population inversion. An increase in the laser intensity causes a reduction in the population inversion due to the increased rate of stimulated emission. This causes a reduction in the gain that tends to decrease the laser intensity. Then, the population inversion starts to increase again and the laser intensity follows the growth of the population inversion afterwards. This oscillatory behavior between the laser intensity and the population inversion continues for several or tens of cycles, and relaxes into stable values of the laser intensity and population inversion. Thus, this phenomenon is called relaxation oscillation.

The relaxation oscillation frequency is dependent on the pump current, carrier lifetime, and photon lifetime in the laser cavity. The relaxation oscillation frequency f_r is described as follows,

$$f_r = \frac{1}{2\pi}\sqrt{\frac{p-1}{\tau_p \tau_s}} \qquad (2.19)$$

2.3 Basic Laser Theory

Table 2.1 Relaxation oscillation frequencies for semiconductor, solid-state, and gas lasers.

	Population lifetime (carrier lifetime) [s]	Photon lifetime [s]	Relaxation oscillation frequency [Hz]
Semiconductor lasers	10^{-9}	10^{-12}	$\sim 10^9$
Solid-state lasers	10^{-3}	10^{-9}	$\sim 10^5$
Gas lasers	10^{-8}	10^{-7}	$\sim 10^6$

where p is the pumping power normalized by the lasing threshold value, τ_s is the population inversion lifetime, and τ_p is the photon lifetime in the laser cavity. The detail procedure of the derivation of the relaxation oscillation frequency of Eq. (2.19) from Eqs. (2.13) and (2.14) is described in Appendix 2.A.4.

Typical values of relaxation oscillation frequency are summarized in Table 2.1. It is found that the relaxation oscillation frequency in semiconductor lasers is very fast at around a few GHz due to the short population and photon lifetimes, whereas the relaxation oscillation frequency in solid-state and gas lasers ranges from kHz to MHz. This fast oscillation characteristic of semiconductor lasers allows application of chaotic semiconductor lasers for optical communications and random number generators (see Chapters 7–10).

The relaxation oscillation in the intensity of laser output is usually observed just after the laser is turned on and approaches a steady-state operation. However, perturbations in the pumping power can also cause relaxation oscillations to appear spontaneously. The relaxation time is relatively larger than the oscillation time, which makes the relaxation oscillations readily apparent on an oscilloscope trace of the laser output intensity.

2.3.5
Mechanism of Chaotic Instability in Lasers

The output of a continuously pumped laser sometimes consists of a series of tiny irregular oscillations. This behavior is usually attributed to relaxation oscillations being continuously excited by various inherent noises (i.e., thermal and quantum noises) in a laser medium. This is known as "sustained relaxation oscillation," and is commonly observed in semiconductor lasers and solid-state lasers. The sustained relaxation oscillation also appears by an external perturbation even in the absence of noise. An external feedback or modulation with a frequency that is close to the relaxation oscillation frequency results in a nonlinear frequency mixing in the laser cavity and causes chaotic behavior of laser intensity. Chaotic temporal dynamics can thus be observed in the same frequency range of the relaxation oscillation frequency.

The interaction rate between the photons and population inversion strongly affect the intensity dynamics of lasers. Particularly, when a portion of self-feedback light is injected into the laser cavity, or when an external modulation is applied to the laser medium, the balance of the provision and consumption of the energy between the

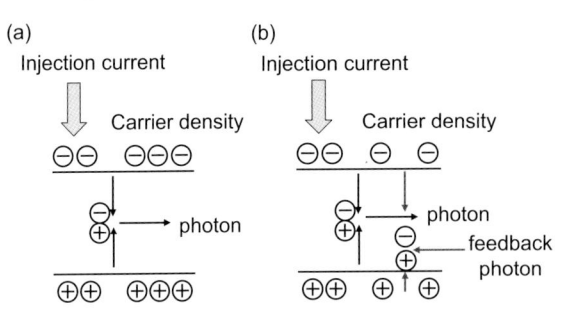

Figure 2.22 Schematic of the carrier–photon interaction in a semiconductor laser (a) without and (b) with optical feedback. The balance of the carrier–photon interaction is destroyed by the feedback photons that produce chaotic instability of laser output.

population inversion (carrier density) and photons can be destroyed, as shown in Figure 2.22, and irregular intensity dynamics is expected. This is a qualitative interpretation for the mechanism of generation of chaotic intensity fluctuations in laser systems with additional perturbation. The atom–photon (carrier–photon) interaction is modulated by the external perturbation, and the nonlinear frequency mixing between the sustained relaxation oscillation and the external modulation results in chaotic intensity fluctuations of the laser output.

2.4
Connection Between Chaos and Lasers

2.4.1
Single-Mode Laser Model Based on Maxwell–Bloch Equations

One of the standard forms of single-mode laser equations is derived from the Maxwell–Bloch equations based on the semiclassical theory, where the electric field is described by the Maxwell's equations, the macroscopic atomic polarization is introduced by using Schrödinger equations, and the phenomenological atomic and photon decays are introduced. The homogeneously broadened single-mode ring laser equations are described as follows (Mandel, 1997),

$$\frac{dE(t)}{dt} = -\kappa[(1+i\Delta)E(t) + AP(t)] \quad (2.20)$$

$$\frac{dP(t)}{dt} = -(1-i\Delta)P(t) - E(t)D(t) \quad (2.21)$$

$$\frac{dD(t)}{dt} = \gamma\left[1 - D(t) + \frac{1}{2}(E^*(t)P(t) + E(t)P^*(t))\right] \quad (2.22)$$

where $E(t)$ is the electric field (complex variable), $P(t)$ is the atomic polarization (complex variable), and $D(t)$ is the population inversion (real variable). $E(t)$ and $P(t)$

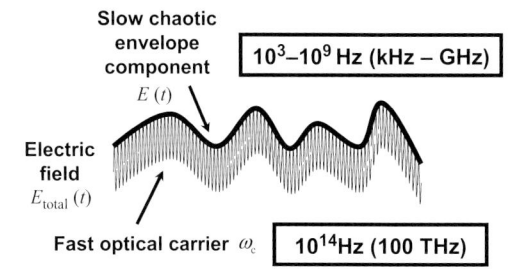

Figure 2.23 Schematics of two different frequency components in the electric field of laser output $E_{\text{total}}(t)$. The dynamics of the slow chaotic envelope component $E(t)$ is treated by using rate equations and the dynamics of the fast optical carrier ω_c is eliminated by slowly varying envelope approximation (SVEA).

are treated as having slowly varying amplitude compared with the optical-carrier frequency ω_c (slowly varying envelope approximation, SVEA), that is,

$$E_{\text{total}}(t) = E(t)\exp(-i\omega_c t) + E^*(t)\exp(i\omega_c t) \tag{2.23}$$

$$P_{\text{total}}(t) = P(t)\exp(-i\omega_c t) + P^*(t)\exp(i\omega_c t) \tag{2.24}$$

It is worth noting that the slow envelope component is only considered as the dynamics of electric field in the rate equations, as shown in Figure 2.23. For coherently coupled lasers, the detuning of optical carrier frequencies between two lasers plays an important role such as injection locking. It is thus important to consider the existence of the detuning frequency of the fast optical carrier oscillation of ω_c.

In Eqs. (2.20)–(2.22), time is normalized by the decay rate of the atomic polarization $\gamma_\perp$. $\gamma_\parallel$ is the decay rate of the population inversion ($\gamma = \gamma_\parallel/\gamma_\perp$), and κ_c is the decay rate of the electric field in the laser cavity ($\kappa = \kappa_c/\gamma_\perp$). Δ is the detuning between the optical-carrier frequency ω_c and the atomic resonant frequency ω_a, that is, $\Delta = (\omega_c - \omega_a)/(\kappa_c + \gamma_\perp)$. A is the gain parameter. The equations are normalized by the pump energy.

For simplicity, it is assumed that the laser is on resonance, $\omega_c = \omega_a$, so that $\Delta = 0$, and the real part of the three variables $E(t)$, $P(t)$, and $D(t)$ are extracted. Equations (2.20)–(2.22) are modified as,

$$\frac{dE(t)}{dt} = -\kappa(E(t) + AP(t)) \tag{2.25}$$

$$\frac{dP(t)}{dt} = -P(t) - E(t)D(t) \tag{2.26}$$

$$\frac{dD(t)}{dt} = \gamma(1 - D(t) + E(t)P(t)) \tag{2.27}$$

These single-mode ring laser equations are used for the classification of lasers in terms of the decay rates of the variables in the next section.

2.4.2
Classification of Laser Models Based on Decay Rates

Lasers are usually classified according to the laser medium that provides the optical amplification: gases, liquids, solids, fibers, and semiconductors. From the laser dynamics point of view, lasers can be classified in another way, indroduced in (Arecchi et al., 1984). Lasers operating in a single emission mode are described by the three differential equations as shown in Eqs. (2.25)–(2.27). The three relevant variables are the electric field, the atomic polarization, and the population inversion. The electric field, the atomic polarization, and the population inversion usually decay on very different time scales, which are given by the decay rates, κ_c (the electric field decay rate), $\gamma_\parallel$ (the population inversion decay rate), and $\gamma_\perp$ (the atomic polarization decay rate), respectively. If one of these rates is larger than the others, the corresponding variable relaxes fast and consequently adiabatically adjusts to the other variables. In fact, because the temporal dynamics of the variable with large relaxation rate is faster than the other variables, this variable is regarded as a dependent variable when compared with the other variables. Therefore, the variable with faster relaxation rate is considered to be dependent on the other variables with slower relaxation rates. The number of equations describing the laser is thus reduced. This is known as "adiabatic elimination" of variables (Mandel, 1997). According to the number of variables, lasers are classified into three types: class A, B, and C lasers.

2.4.2.1 Class C Lasers (Three Variables)
For class C lasers, the relaxation rates of the electric field, the population inversion, and the atomic polarization are of the same order.

$$\kappa_c \approx \gamma_\perp \approx \gamma_\parallel \tag{2.28}$$

The decay rate of the electric field is comparable to those of the atomic polarization and the population inversion. The dynamics of the three variables are described by Eqs. (2.25)–(2.27). This situation indicates that the electric field is not well confined in the laser cavity with low Q factor, which is in the bad cavity (lossy cavity) condition. The lasers with high gain and low Q factor are classified as "class C" lasers. Class C lasers satisfy the necessary condition for generating chaos (i.e., at least three independent variables are necessary for chaos), and chaotic behaviors are observed in the bad cavity condition (Weiss and Vilaseca, 1991). Examples of class C lasers are He-Ne lasers at the 3.39-μm line, He-Xe lasers at the 3.51-μm line, and NH_3 lasers.

2.4.2.2 Class B Lasers (Two Variables)
For class B lasers, the decay rate of the atomic polarization $\gamma_\perp$ is much faster than those of the electric field and the population inversion, and the decay rate of the electric field is faster than that of the population inversion.

$$\gamma_\perp \gg \kappa_c > \gamma_\parallel \tag{2.29}$$

The atomic polarization dynamics is much faster than the two other variables, and the variable of the atomic polarization is regarded as a dependent variable of the two other variables. In this case, the left-hand-side term of Eq. (2.26) is considered as zero ($dP(t)/dt = 0$), and Eq. (2.26) becomes,

$$P(t) = -E(t)D(t) \tag{2.30}$$

Substituting Eq. (2.30) into Eqs. (2.25) and (2.27),

$$\frac{dE(t)}{dt} = \kappa(-1 + AD(t))E(t) \tag{2.31}$$

$$\frac{dD(t)}{dt} = \gamma\left(1 - D(t) - E^2(t)D(t)\right) \tag{2.32}$$

Therefore, the dynamics of lasers are described by the two variables: the electric field $E(t)$ and the population inversion $D(t)$. This is known as a "class B" laser. In Eq. (2.31), the first right-hand-side term indicates the loss of the electric field in the laser cavity, and the second term indicates the stimulated emission. In Eq. (2.32), the first right-hand-side term corresponds to the pump energy, the second term indicates the spontaneous emission, and the third term represents the stimulated emission. In fact, Eqs. (2.31) and (2.32) are essentially the same as Eqs. (2.13) and (2.14), except for the expression of the electric field $E(t)$ instead of the laser intensity $I(t)$ (where $I(t) = |E(t)|^2$).

The interaction between the electric field and population inversion yields the relaxation oscillation, because the population inversion cannot follow the relaxation rate of the electric field ($\kappa > \gamma_\parallel$). In addition, class B lasers have only two variables and they do not satisfy the condition for generation of chaos. Class B lasers are stable in nature. However, they are easily destabilized by the introduction of external perturbations, resulting in the addition of extra degrees of freedom. Chaotic behaviors in class B lasers have been widely observed by adding an external modulation of laser parameters, optical injection, and optical feedback (see Chapter 3).

Many commercial lasers are classified as class B lasers: semiconductor lasers, solid-state lasers, and CO_2 lasers. Examples of chaotic temporal behaviors in class B lasers will be described in Chapter 3.

2.4.2.3 Class A Lasers (One Variable)

For class A lasers, the decay rate of the electric field is much slower than those of the atomic polarization and the population inversion,

$$\gamma_\perp \approx \gamma_\parallel \gg \kappa_c \tag{2.33}$$

The variables of atomic polarization and the population inversion change much faster than that of the electric field, and they are dependent on the electric field. In this

2 Basics of Chaos and Laser

Table 2.2 Classification of laser dynamics based on the decay rates. κ_c is the electric field decay rate, $\gamma_\parallel$ is the population inversion decay rate, and $\gamma_\perp$ is the atomic polarization decay rate.

	Variables to describe dynamics	Conditions	Examples of lasers
Class A	Electric field (photons)	$\gamma_\perp \approx \gamma_\parallel \gg c$ (good-cavity condition)	He-Ne lasers (632.8 nm) Dye lasers
Class B	Electric field (photons) Population inversion (atoms)	$\gamma_\perp \gg c > \gamma_\parallel$	Semiconductor lasers Solid-state lasers CO_2 lasers
Class C	Electric field (photons) Population inversion (atoms) Polarization (matter)	$\kappa_c \approx \gamma_\perp \approx \gamma_\parallel$ (bad-cavity condition)	He-Ne lasers (3.39 μm) He-Xe lasers (3.51 μm) NH_3 lasers

case, the left-hand-side term of Eq. (2.32) is considered as zero ($dD(t)/dt = 0$), and Eq. (2.32) becomes,

$$D(t) = \frac{1}{1 + E^2(t)} \tag{2.34}$$

Substituting Eq. (2.34) into Eq. (2.31),

$$\frac{dE(t)}{dt} = \kappa\left(-1 + \frac{A}{1 + E^2(t)}\right)E(t) \tag{2.35}$$

$$\approx \kappa\left(-1 + A - AE^2(t)\right)E(t) \tag{2.36}$$

Therefore, only one variable of the electric field $E(t)$ is used to describe the dynamics of lasers, which is known as a "class A" laser. This approximation of Eq. (2.36) is satisfied around the lasing threshold, that is, $E(t) \approx 0$. The first right-hand-side term of Eq. (2.36) indicates the loss of electric field in the laser cavity, the second term corresponds to the stimulated emission, and the third term represents the saturation effect. This situation is realized by using a laser cavity with a high Q factor, which is known as the good cavity (well-confined cavity) condition. Class A lasers are the most stable lasers among the three classes, however, they may show chaotic behaviors by external perturbations with two or more extra degrees of freedom, as in the case of class B lasers. Examples of class A lasers are He-Ne lasers at the 632.8-nm line and dye lasers.

The classification of lasers in terms of laser dynamics (i.e., the number of variables that govern the laser dynamics) is summarized in Table 2.2. It is worth noting that this classification is only derived from two-level single-mode lasers. The type of the variables required to describe the dynamics is dependent of the type of lasers, however, the number of each variable may increase for multimode lasers (more than one variable of the electric field) or multilevel lasers (more than one variable of the population inversion). Therefore, chaotic dynamics may appear even without external perturbations for class A and B lasers that have extra degrees of freedom.

It may seem that only class C lasers are good candidates for generating chaotic temporal waveforms. Indeed, most class B and A lasers that have been developed over the years have been operated under conditions and in configurations where they are stable in operation, and emit continuous-wave light suitable for a wide variety of applications. Alternatively, they have been used to generate periodic pulses either through mode-locking or Q-switching techniques. It is, however, possible to generate chaotic waveforms with class B and class A lasers with additional degrees of freedom under a wide variety of conditions easily realized experimentally, as described in Chapter 3.

Appendix

2.A.1
General Formula of the Linearized Equations and the Derivation of the Linearized Equations in the Lorenz Model

The linearized equations can be used for calculating the stability of dynamical systems and Lyapunov exponents. The general formula of the linearized equations is described. Let us consider a set of differential equations

$$\frac{d\mathbf{X}(t)}{dt} = \mathbf{F}(t, \mathbf{X}) \tag{2.A.1}$$

where $\mathbf{X}(t) = (x_1(t) x_2(t) x_3(t) \cdots x_n(t))^T$ is a vector of n variables, $\mathbf{F} = (f_1 f_2 f_3 \cdots f_n)$ is a matrix of n nonlinear functions, and t is time. Let us consider the linear stability analysis and introduce linearized variables as $\mathbf{X}(t) = \mathbf{X}_s + \boldsymbol{\delta}_X(t)$, where $\mathbf{X}_s$ is a vector of the steady-state solutions for $\mathbf{X}(t)$ and $\boldsymbol{\delta}_X(t)$ is a vector of the linearized variables. The linearized equations are obtained as,

$$\frac{d\boldsymbol{\delta}_X(t)}{dt} = \mathbf{D_X F}(t, \mathbf{X})\boldsymbol{\delta}_X(t) \tag{2.A.2}$$

where $\mathbf{D_X F}(t, \mathbf{X})$ is the Jacobian matrix defined as,

$$\mathbf{D_X F}(t, \mathbf{X}) = \begin{pmatrix} \frac{\partial f_1}{\partial x_1} & \frac{\partial f_1}{\partial x_2} & \frac{\partial f_1}{\partial x_3} & \cdots & \frac{\partial f_1}{\partial x_n} \\ \frac{\partial f_2}{\partial x_1} & \frac{\partial f_2}{\partial x_2} & \frac{\partial f_2}{\partial x_3} & \cdots & \frac{\partial f_2}{\partial x_n} \\ \frac{\partial f_3}{\partial x_1} & \frac{\partial f_3}{\partial x_2} & \frac{\partial f_3}{\partial x_3} & \cdots & \frac{\partial f_3}{\partial x_n} \\ \vdots & \vdots & \vdots & \ddots & \vdots \\ \frac{\partial f_n}{\partial x_1} & \frac{\partial f_n}{\partial x_2} & \frac{\partial f_n}{\partial x_3} & \cdots & \frac{\partial f_n}{\partial x_n} \end{pmatrix} \tag{2.A.3}$$

Let us consider three simultaneous ordinary differential equations as a simple example.

$$\frac{dx(t)}{dt} = f(t, x, y, z) \tag{2.A.4}$$

$$\frac{dy(t)}{dt} = g(t, x, y, z) \tag{2.A.5}$$

$$\frac{dz(t)}{dt} = h(t, x, y, z) \tag{2.A.6}$$

To obtain linearized equations, a small linear deviation is considered as,

$$x(t) = x_s + \delta_x(t), \ y(t) = y_s + \delta_y(t), \ z(t) = z_s + \delta_z(t) \tag{2.A.7}$$

The linearized equations are obtained as follows,

$$\frac{d}{dt}\begin{pmatrix} \delta_x(t) \\ \delta_y(t) \\ \delta_z(t) \end{pmatrix} = \begin{pmatrix} \frac{\partial f}{\partial x} & \frac{\partial f}{\partial y} & \frac{\partial f}{\partial z} \\ \frac{\partial g}{\partial x} & \frac{\partial g}{\partial y} & \frac{\partial g}{\partial z} \\ \frac{\partial h}{\partial x} & \frac{\partial h}{\partial y} & \frac{\partial h}{\partial z} \end{pmatrix}\begin{pmatrix} \delta_x(t) \\ \delta_y(t) \\ \delta_z(t) \end{pmatrix} \tag{2.A.8}$$

In other words,

$$\frac{d\delta_x(t)}{dt} = \frac{\partial f}{\partial x}\delta_x(t) + \frac{\partial f}{\partial y}\delta_y(t) + \frac{\partial f}{\partial z}\delta_z(t) \tag{2.A.9}$$

$$\frac{d\delta_y(t)}{dt} = \frac{\partial g}{\partial x}\delta_x(t) + \frac{\partial g}{\partial y}\delta_y(t) + \frac{\partial g}{\partial z}\delta_z(t) \tag{2.A.10}$$

$$\frac{d\delta_z(t)}{dt} = \frac{\partial h}{\partial x}\delta_x(t) + \frac{\partial h}{\partial y}\delta_y(t) + \frac{\partial h}{\partial z}\delta_z(t) \tag{2.A.11}$$

Next, let us consider an example of the Lorenz model. The original Lorenz model is described as (rewriting Eqs. (2.9)–(2.11)),

$$\frac{dx(t)}{dt} = \sigma(y(t) - x(t)) \tag{2.A.12}$$

$$\frac{dy(t)}{dt} = -x(t)z(t) + rx(t) - y(t) \tag{2.A.13}$$

$$\frac{dz(t)}{dt} = x(t)y(t) - bz(t) \tag{2.A.14}$$

2.A.2 General Formula for the Calculation of the Lyapunov Exponent in Nonlinear Dynamical Systems

where $\sigma = 10$, $r = 28$, and $b = 8/3$ are typical parameter values to observe chaos. Using Eq. (2.A.8), the linearized equations of the Lorenz model is written,

$$\frac{d}{dt}\begin{pmatrix} \delta_x(t) \\ \delta_y(t) \\ \delta_z(t) \end{pmatrix} = \begin{pmatrix} -\sigma & \sigma & 0 \\ r-z(t) & -1 & -x(t) \\ y(t) & x(t) & -b \end{pmatrix} \begin{pmatrix} \delta_x(t) \\ \delta_y(t) \\ \delta_z(t) \end{pmatrix} \quad (2.A.15)$$

In other words,

$$\frac{d\delta_x(t)}{dt} = -\sigma\delta_x(t) + \sigma\delta_y(t) \quad (2.A.16)$$

$$\frac{d\delta_y(t)}{dt} = (r-z(t))\delta_x(t) - \delta_y(t) - x(t)\delta_z(t) \quad (2.A.17)$$

$$\frac{d\delta_z(t)}{dt} = y(t)\delta_x(t) + x(t)\delta_y(t) - b\delta_z(t) \quad (2.A.18)$$

The linearized Eqs. (2.A.16)–(2.A.18) need to be solved with the original Eqs. (2.A.12)–(2.A.14) simultaneously for the calculation of the maximum Lyapunov exponent.

2.A.2 General Formula for the Calculation of the Lyapunov Exponent in Nonlinear Dynamical Systems Without Time Delay

To calculate the Lyapunov exponent, the linearized variables $\delta_x(t), \delta_y(t), \delta_z(t)$ obtained in Appendix 2.A.1 are used. Here, it is considered that the "maximum" Lyapunov exponent is derived from the linearized equations. The norm of all the linearized variables $\delta_x(t), \delta_y(t), \delta_z(t)$ are defined as,

$$D(t) = \sqrt{\delta_x^2(t) + \delta_y^2(t) + \delta_z^2(t)} \quad (2.A.19)$$

After one-step integration of the numerical calculation of the linearized equations from time t to $t + h$, the norm is changed as,

$$D(t+h) = \sqrt{\delta_x^2(t+h) + \delta_y^2(t+h) + \delta_z^2(t+h)} \quad (2.A.20)$$

where h is an integration step of numerical simulation. The norm ratio can be calculated as,

$$d_j = \frac{D(t+h)}{D(t)} \quad (2.A.21)$$

The value of d_j is saved for each step of numerical integration. After calculating the norm ratio, the linearized variables need to be normalized as,

$$\delta_{x,\text{new}}(t+h) = \frac{\delta_x(t+h)}{D(t+h)} \tag{2.A.22}$$

$$\delta_{y,\text{new}}(t+h) = \frac{\delta_y(t+h)}{D(t+h)} \tag{2.A.23}$$

$$\delta_{z,\text{new}}(t+h) = \frac{\delta_z(t+h)}{D(t+h)} \tag{2.A.24}$$

The normalization is very important to maintain the linearity of the stability analysis. These new variables of Eqs. (2.A.22)–(2.A.24) are used to calculate the norm ratio between $D(t+h)$ and $D(t+2h)$ in the next step.

The calculation from Eqs. (2.A.19) to (2.A.24) is repeated for a long time. The maximum Lyapunov exponent can be obtained for the N-th repetition of the above-mentioned procedure,

$$\lambda = \frac{1}{Nh} \sum_{j=1}^{N} \ln(d_j) \tag{2.A.25}$$

where ln is natural logarithm with the natural base e.

2.A.3 Rössler Model

The Rössler model is a well-known mathematical model for describing nonlinear dynamics and chaos, as well as the Lorenz model (Bergé et al., 1984). The Rössler model is written as,

$$\frac{dx(t)}{dt} = -(y(t) + z(t)) \tag{2.A.26}$$

$$\frac{dy(t)}{dt} = x(t) + ay(t) \tag{2.A.27}$$

$$\frac{dz(t)}{dt} = b + z(t)(x(t) - \mu) \tag{2.A.28}$$

where $a = 0.2$, $b = 0.2$, and $\mu = 5.7$ are typical parameter values to observe chaos. Figure 2.A.1 shows typical examples of the temporal waveform and the chaotic attractor of the Rössler model in the three-dimensional phase space.

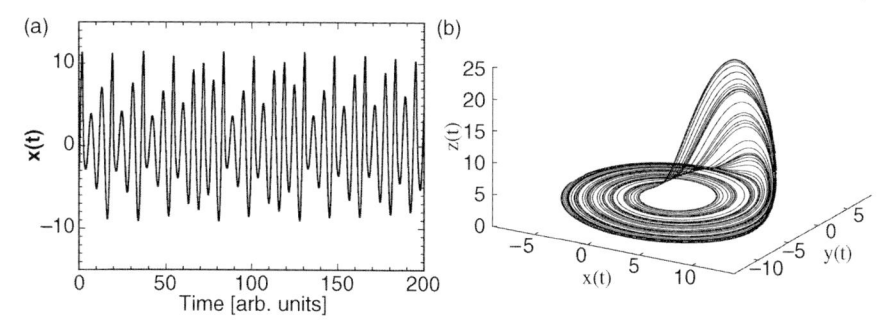

Figure 2.A.1 (a) Temporal waveform and (b) chaotic attractor of the Rössler model.

The linearized equations of the Rössler model can be derived as,

$$\frac{d}{dt}\begin{pmatrix}\delta_x(t)\\ \delta_y(t)\\ \delta_z(t)\end{pmatrix} = \begin{pmatrix} 0 & -1 & -1 \\ 1 & a & 0 \\ z(t) & 0 & x(t)-\mu \end{pmatrix}\begin{pmatrix}\delta_x(t)\\ \delta_y(t)\\ \delta_z(t)\end{pmatrix} \quad (2.A.29)$$

In other words,

$$\frac{d\delta_x(t)}{dt} = -\delta_y(t) - \delta_z(t) \quad (2.A.30)$$

$$\frac{d\delta_y(t)}{dt} = \delta_x(t) + a\delta_y(t) \quad (2.A.31)$$

$$\frac{d\delta_z(t)}{dt} = z(t)\delta_x(t) + (x(t)-\mu)\delta_z(t) \quad (2.A.32)$$

The Lyapnov exponent of the Rössler model can be obtained from the linearized equations of Eqs. (2.A.30)–(2.A.32) by the procedure described in Appendix 2.A.2.

2.A.4 Derivation of the Relaxation Oscillation Frequency

The relaxation oscillation frequency of Eq. (2.19) can be derived from Eqs. (2.13) and (2.14) (see also Section 4.3.2.2 for the detail procedure). Equations (2.13) and (2.14) are rewritten as follows,

$$\frac{dI(t)}{dt} = gI(t)N(t) - \frac{I(t)}{\tau_p} \quad (2.A.33)$$

$$\frac{dN(t)}{dt} = R - gI(t)N(t) - \frac{N(t)}{\tau_s} \quad (2.A.34)$$

First, the steady-state solutions of Eqs. (2.A.33) and (2.A.34) need to be obtained under the condition of $dI/dt = dN/dt = 0$. The steady-state solutions N_s and I_s are obtained when $I_s \neq 0$,

$$N_s = \frac{1}{g\tau_p} \tag{2.A.35}$$

$$I_s = R\tau_p - \frac{1}{g\tau_s} \tag{2.A.36}$$

The linearized variables are introduced,

$$I(t) = I_s + \delta_I(t) \tag{2.A.37}$$

$$N(t) = N_s + \delta_N(t) \tag{2.A.38}$$

The Jacobian matrix $\mathbf{J}$ can be obtained (see Appendix 2.A.1),

$$\begin{pmatrix} \frac{d\delta_I(t)}{dt} \\ \frac{d\delta_N(t)}{dt} \end{pmatrix} = \begin{pmatrix} \frac{\partial f_I}{\partial I} & \frac{\partial f_I}{\partial N} \\ \frac{\partial f_N}{\partial I} & \frac{\partial f_N}{\partial N} \end{pmatrix} \begin{pmatrix} \delta_I(t) \\ \delta_N(t) \end{pmatrix}$$

$$= \begin{pmatrix} 0 & gR\tau_p - \frac{1}{\tau_s} \\ -\frac{1}{\tau_p} & -gR\tau_p \end{pmatrix} \begin{pmatrix} \delta_I(t) \\ \delta_N(t) \end{pmatrix} \tag{2.A.39}$$

The characteristic equation is calculated to obtain the eigenvalues λ of the Jacobian matrix $\mathbf{J}$, where the real and imaginary parts of the eigenvalues of the Jacobian matrix indicate the stability and oscillation frequency of the steady state solutions, respectively.

$$|\mathbf{J} - \lambda \mathbf{I}| = \begin{vmatrix} -\lambda & gR\tau_p - \frac{1}{\tau_s} \\ -\frac{1}{\tau_p} & -gR\tau_p - \lambda \end{vmatrix}$$

$$= \lambda^2 + gR\tau_p \lambda + \left(gR - \frac{1}{\tau_p \tau_s}\right) = 0 \tag{2.A.40}$$

Here, a new pumping factor $p = gR\tau_p\tau_s$ is introduced and substituted into Eq. (2.A.40),

$$\lambda^2 + \frac{p}{\tau_s}\lambda + \frac{1}{\tau_p\tau_s}(p-1) = 0 \tag{2.A.41}$$

2.A.4 Derivation of the Relaxation Oscillation Frequency

The solution is obtained from the quadratic formula,

$$\lambda = -\frac{p}{2\tau_s} \pm i\sqrt{\frac{1}{\tau_p \tau_s}(p-1) - \left(\frac{p}{2\tau_s}\right)^2} \qquad (2.A.42)$$

Here, the following condition is satisfied because of $\tau_p \ll \tau_s$ and $p \approx 1$,

$$\frac{1}{\tau_p \tau_s}(p-1) \gg \left(\frac{p}{2\tau_s}\right)^2 \qquad (2.A.43)$$

Therefore, the solution can be approximated,

$$\lambda = -\frac{p}{2\tau_s} \pm i\sqrt{\frac{p-1}{\tau_p \tau_s}} \qquad (2.A.44)$$

The complex solution is expressed as $\lambda = \lambda_{\text{Re}} \pm i\lambda_{\text{Im}}$, and the relaxation oscillation frequency f_r can be obtained from the imaginary part of λ (i.e., $f_r = \lambda_{\text{Im}}/2\pi$) as follows,

$$f_r = \frac{1}{2\pi}\sqrt{\frac{p-1}{\tau_p \tau_s}} \qquad (2.A.45)$$

This formula is the same as Eq. (2.19).

3
Generation of Chaos in Lasers

3.1
Basics of Generation of Chaos in Lasers

3.1.1
Classification of Generation Techniques of Chaos in Lasers

In this chapter, major techniques for the generation of chaos in various lasers are explained and discussed. For commercially available lasers (e.g., the class B lasers including semiconductor and solid-state lasers), the dynamics of the electric field and the population inversion are treated, and the time scale for exchanging the energy between these two physical variables is characterized by the relaxation oscillation frequency. The relaxation oscillation frequency is determined by a combination of three parameters – the photon lifetime, the population inversion lifetime, and the pumping rate above threshold, as shown in Eq. (2.19). The class B lasers provide stable laser intensity in most situations, since there are only two variables of the electric field and the population inversion that are not satisfied with the condition for generating deterministic chaos. At least one more degree of freedom (i.e., one variable) is required for the generation of chaos (Strogatz, 1994).

An additional degree of freedom can be included in class B lasers intentionally or naturally. Chaotic oscillations can be generated and much effort has been expended to investigate the nonlinear dynamics of these lasers. There are two main motivations for the study on generation of chaos in lasers: basic understanding of nonlinear dynamics in laser systems, and the use for novel engineering applications such as optical chaos communication and random number generation. The techniques for generating chaos with additional degrees of freedom can be classified into the following types:

a) optical feedback (with or without time delay);
b) optical coupling and injection;
c) external modulation (for pump or loss);
d) insertion of nonlinear devices.

In addition, the following lasers and optical systems provide chaotic dynamics spontaneously:

Optical Communication with Chaotic Lasers: Applications of Nonlinear Dynamics and Synchronization, First Edition. Atsushi Uchida.
© 2012 Wiley-VCH Verlag GmbH & Co. KGaA. Published 2012 by Wiley-VCH Verlag GmbH & Co. KGaA.

3 Generation of Chaos in Lasers

Table 3.1 Classification of generation techniques of chaos in various lasers. The number in the table corresponds to the section number describing the generation techniques in Chapter 3. "–" indicates that there is no description for the method in Chapter 3, even though some methods may be available in the literature.

Methods	Feedback	Coupling (Injection)	Modulation	Insertion of nonlinear device	Multimode lasers	Lorenz–Haken type lasers
Semiconductor lasers	3.2.1 3.2.2 3.2.3	3.2.4	3.2.5	—	3.2.6	—
Passive nonlinear devices	3.3.1 ~3.3.4	—	—	—	—	—
Fiber lasers	—	—	3.4.1	—	3.4.2 3.4.3	—
Solid-state lasers	3.5.2	—	3.5.1	3.5.4	3.5.3	—
Gas lasers	3.6.6	3.6.4	3.6.3	3.6.2	3.6.5	3.6.1

e) multimode lasers and high-dimensional laser systems;
f) class C lasers satisfying the condition for Lorenz–Haken chaos;
g) Ikeda-type passive optical systems.

These techniques are summarized in Table 3.1 and Figure 3.1, and discussed below.

3.1.1.1 Optical Feedback

A self-feedback signal to a laser cavity can result in chaotic instability of laser output. The dynamics of semiconductor lasers with time-delayed optical feedback is a typical example of feedback-induced chaos in lasers. An external mirror is placed in front of the laser cavity, and the laser light is reflected back from the external mirror and re-injected into the laser cavity, as shown in Figure 3.1a. The optical self-feedback signal may disturb the balance of the carrier–photon interaction in the laser medium and induce the instability of laser intensity. In this situation, the temporal dynamics is determined by the two dominant frequency components: the relaxation oscillation frequency and the external cavity frequency. The relaxation oscillation frequency is proportional to the square root of the normalized pump power divided by the carrier lifetime and the photon lifetime, as shown in Eq. (2.19), so it is determined by the characteristics of the semiconductor medium. The typical value of the relaxation oscillation frequency of semiconductor lasers is a few GHz. On the other hand, the external cavity frequency f_{ext} depends on the distance between the facet of the laser cavity and the external mirror (i.e., the external cavity length) as,

$$f_{ext} = \frac{c}{2nL_{ext}} \tag{3.1}$$

where L_{ext} is the external cavity length (one-way), n is the refractive index in the external cavity, and c is the speed of light. The external cavity frequency corresponds

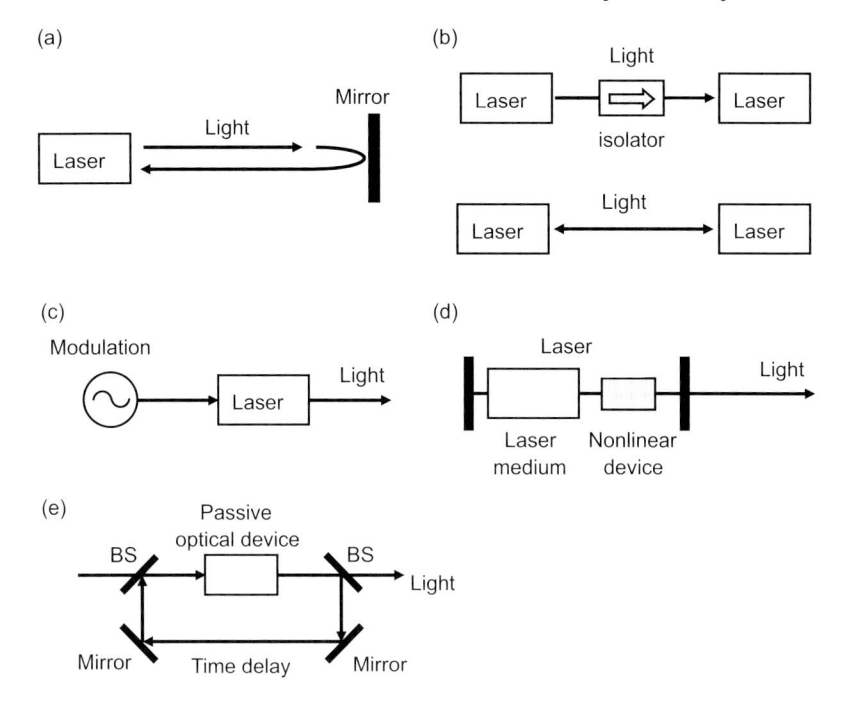

Figure 3.1 Classification of generation techniques of chaos in lasers with an additional degree of freedom. (a) Optical feedback, (b) optical injection and coupling (upper: unidirectional coupling, lower: mutual coupling), (c) external modulation, (d) insertion of optical device, and (e) configuration of Ikeda chaos. (e) Passive optical device and time-delayed feedback are used. BS, beam splitter.

to the inverse of the round-trip time of light propagation in the external cavity. For example, when the external cavity length is $L_{ext} = 0.3$ m (one-way) and $n = 1$, the external cavity frequency is 0.5 GHz. Note that it is useful to memorize that the light takes 1 ns (10^{-9} s) to propagate the distance of 0.3 m in vacuum. The nonlinear interaction between the relaxation oscillation frequency and the external cavity frequency results in the quasiperiodicity route to chaos as the feedback strength increases. In addition, the time delay of the feedback signal enhances the complexity of the laser dynamics, since the time-delayed feedback system is considered as a high-dimensional dynamical system (see Section 4.5.5).

Different types of optical feedback signal can be used for the generation of chaos. Polarization-rotated optical feedback can be useful to generate chaos. In this scheme the polarization of the laser output is rotated at 90° and the two orthogonal polarization modes of the electric fields interact with each other non-linearly. Chaotic dynamics can be generated by using polarization-rotated optical feedback.

An optoelectronic feedback signal can be also used for generation of chaos in semiconductor lasers. The laser output is detected by a photodetector and is converted into an electronic signal. The electronic signal is fed back to the injection current for pumping to induce instability of laser output. The feedback signal of the laser intensity only interacts with the dynamics of the carrier density, but not the electric field of the laser, thus it is considered as incoherent optical feedback. The effect of the bandwidth limitation in the photodetector and electronic components strongly affects the nonlinear dynamics of the laser intensity.

3.1.1.2 Optical Coupling and Injection

The unidirectional or mutual optical coupling from one laser to another laser can generate chaotic instability of laser output, as shown in Figure 3.1b. It is worth noting that the laser intensity has two dominant frequency components (see also Figure 2.23). One is the optical-carrier frequency f_c determined by the optical wavelength λ and the speed of light c as,

$$f_c = \frac{c}{\lambda} \tag{3.2}$$

The optical-carrier frequency ranges in the order of several hundreds of THz (10^{14} Hz) for semiconductor lasers (e.g., $f_c \approx 200$ THz for $\lambda = 1.5\,\mu$m and $c = 2.998 \times 10^8$ m/s.) On the other hand, the other frequency is the relaxation oscillation frequency f_r at the range of kHz–GHz (10^3–10^9 Hz).

When the detuning of the optical carrier frequencies between an injection and injected lasers is set to the order of the relaxation oscillation frequency, the nonlinear interaction between the optical-carrier frequency detuning and the relaxation oscillation frequency can occur and chaotic fluctuation may appear. The control of the detuning of the optical-carrier frequency, as well as the coupling strength, is crucial for chaos generation in this method.

3.1.1.3 External Modulation

When an external modulation is added to the pumping of a laser system as shown in Figure 3.1c (i.e., pump modulation), chaotic instability of laser intensity may appear. The external modulation frequency needs to be set around the relaxation oscillation frequency of the laser, so that the nonlinear interaction between the external modulation frequency and the relaxation oscillation frequency may result in the generation of chaos. The external modulation can also be applied to the loss of the laser cavity, which is referred to as loss modulation. Chaotic dynamics is typically observed through the quasiperiodicity route to chaos as the external modulation strength is increased.

3.1.1.4 Insertion of Nonlinear Element

The insertion of a nonlinear element may cause chaotic dynamics, as shown in Figure 3.1d. For example, chaotic intensity fluctuation is observed in a solid-state laser system with a nonlinear crystal for second-harmonic generation (SHG). The mode–mode interaction between the fundamental wavelength and the SHG wavelength occurs in a nonlinear fashion, and chaotic dynamics may be observed. Also, when a saturable absorber is inserted in a gas-laser system, complex chaotic dynamics may be observed. An additional degree of freedom enhances the nonlinear atom–photon interaction in the laser medium and leads to chaotic intensity fluctuations.

3.1.1.5 Multimode Lasers and High-Dimensional Laser Systems

For some lasers, chaos can be generated without any extra additional components, particularly for multimode lasers. Multimode lasers have already more than two degrees of freedom and satisfy the condition to generate chaos (i.e., at least three degrees of freedom are required) if there is strong nonlinear interaction among the lasing modes. Chaos may be unavoidable in some multimode lasers. Mode–mode interaction can induce chaotic instability of laser intensity or irregular mode-hopping phenomenon. The lasing modes can be longitudinal, transverse, or polarization modes. The control of lasing modes is very important to enhance or stabilize the chaotic dynamics in multimode laser systems.

Spatially distributed lasers such as broad-area semiconductor lasers (BAL) may generate chaotic intensity fluctuations in a solitary operation. The spatial-mode interaction is very strong due to the carrier diffusion, and chaotic oscillations can be easily observed. The control of spatial modes is also important to enhance or stabilize chaotic intensity fluctuations. These types of lasers, which generate chaos in solitary lasers (without extra degrees of freedom), may be considered as 'not well established' lasers from the laser engineering point of view. On the other hand, they can be treated as very rich and complex dynamical optical devices from the nonlinear dynamical point of view, and may be useful as a universal model tool for describing other nonlinear dynamical systems, such as complex fluid dynamics or biological systems.

3.1.1.6 Class C Lasers Satisfying the Condition for Lorenz–Haken Chaos

As described in Sections 2.1.3 and 2.4.2, class C lasers with three physical variables (electric field, population inversion, and atomic polarization) can produce chaotic temporal dynamics. The condition of high pump power and low Q factor of the laser cavity (bad cavity condition) is required for the generation of this type of chaos. Many works have been reported to find the Lorenz–Haken chaos in class C gas lasers experimentally and numerically (see Section 3.6.1).

3.1.1.7 Ikeda-Type Passive Optical Systems

Ikeda-type optical systems consist of a passive nonlinear device and a time-delayed feedback loop, as shown in Figure 3.1e. The time-delayed feedback can induce the nonlinear dynamics of optical systems. This type of chaos is well known as "Ikeda chaos" (or Ikeda instability) and chaos is generated through the bifurcation of the generation of odd-order harmonic oscillations (Ikeda, 1979). A major difference from the above-described schemes for chaos generation is the use of a "passive" nonlinear device. The laser dynamics based on the atom–photon interaction is not used for chaos generation in this scheme. Instead, the laser is considered as a linear light source, and chaos is generated from an additional passive nonlinear device and time-delayed feedback. This dynamics can be thus described by a simple delay differential equation, and can also be simplified as a one-dimensional map at some conditions (see Section 3.3.1). The dynamics of fiber lasers and electro-optic systems are based on the ideas that incorporate the Ikeda instability (VanWiggeren and Roy, 1999; Larger et al., 2004)

3.1.2
Characteristics of Chaos in Lasers

The characteristics of chaos in lasers, compared with other nonlinear dynamical systems (e.g., electronic circuits, chemical dynamics, fluid dynamics, neuronal systems, and biological systems) are summarized as follows. Some of these characteristics of chaotic lasers are advantageous for engineering applications, as well as for tools to investigate universal nonlinear dynamics.

3.1.2.1 High-Frequency Chaos
High-frequency oscillations of chaos can be obtained due to the fast atom–photon interaction in the laser medium. Typical frequency ranges are in GHz (10^9 Hz) for semiconductor lasers and MHz (10^6–10^7 Hz) for solid-state lasers and gas lasers (see Table 2.1). The oscillation frequency of chaotic lasers is mainly dependent on the relaxation oscillation frequency of the lasers.

From the nonlinear dynamics viewpoint, experiments with lasers offer advantages over similar investigation with other nonlinear dynamical systems. Since the basic oscillation of the optical-carrier frequency is of the order of $\sim 10^{14}$ Hz, all characteristic frequencies are very high. Hence, the requirements for parameter stability for measuring the time-dependent behavior for the long periods necessary for quantitative investigations are easier to fulfill than in other nonlinear dynamical systems. Pulsing frequencies are typically much higher than for other systems, with a corresponding reduction in the observation time for lasers. For example, fast laser dynamics may be used for the prediction of spreads of epidemics, whose dynamics change in a very slow time scale and are difficult for prediction by using a small amount of past data (Kim et al., 2005).

3.1.2.2 Existence of Two Oscillation Frequencies in Different Orders of Magnitude
Two fundamental oscillation components exist in chaotic laser intensity dynamics as described above. One is the optical-carrier frequency at several hundred THz (10^{14} Hz), corresponding to the optical carrier oscillation of the electric field, and determined by the optical wavelength (see Eq. (3.2)). This oscillation frequency is too fast to observe directly by a photodetector. Instead, the mean value of the optical wavelength can be measured by using a monochromator or an optical spectrum analyzer.

The other frequency component is the relaxation oscillation frequency in the frequency range from kHz to GHz (10^3–10^9 Hz). The chaotic dynamics can be mainly observed in this frequency range corresponding to the relaxation oscillation frequency, because of the limitation of the detection equipments for temporal dynamics by a photodetector up to tens of GHz ($\sim 10^{10}$ Hz). However, the dynamics of the optical-carrier frequency (10^{14} Hz) sometimes play an important role, particularly for coherently coupled laser systems. For example, synchronization of chaos in lasers requires the match of the optical-carrier frequency between the coupled lasers (known as injection locking), and the degree of chaos synchronization is strongly affected by the characteristics of injection locking.

3.1.2.3 Middle Degrees of Freedom

There are middle numbers of the degrees of freedom in laser chaos in general. The numbers of degrees of freedom, that is equivalent to the dimensionality of the system, are dependent on the number of optical modes, type of lasers, and additional equipments for chaos generation. In the case of time-delayed feedback, the dimensionality can be very high and is considered as infinite dimensional in general.

The number of degrees of freedom of lasers, for example, number of modes, can be easily controlled. This is different from other nonlinear dynamical systems such as fluids, where boundary layers present their own difficult-to-control dynamics. In laser experiments, the relevant parameters are usually well known and controllable, and the experimental conditions are well defined and reproducible.

3.1.2.4 Physical Variables

Many physical variables in lasers can become chaotic simultaneously. For example, laser intensity, optical phase, population inversion, and polarization of light can be chaotic. These chaotic variables can be used for different types of optical chaos communication schemes (see Section 8.6).

One of the observable variables is the laser intensity by using a photodetector. By contrast, unobservable variables also exist in chaotic dynamics of laser systems in experiments. The dynamics of the population inversion and the optical phase cannot be directly measured in experiment. This characteristic is different from the chaos generated in electronic circuits where most of the voltages and currents are accessible.

3.1.2.5 Transmission of Chaos

A chaotic light waveform can be easily propagated through optical fibers for a long distance. Transmission of chaos can be achieved at a distance of hundreds of kilometers (km). This characteristic enables utilization of optical chaos for optical communication applications. Optical amplification and dispersion compensation for chaotic optical signals can be achieved for the applications of optical chaos communications (Argyris *et al.*, 2005).

3.1.2.6 Variety of Laser Models

There have been a variety of numerical models to describe the temporal and spatial dynamics of lasers. The laser dynamics is dominated by the atom–photon interactions in the laser medium and the nonlinear interaction between the optical modes. For example, the Lang–Kobayashi equation is a *de-facto* standard model for describing the dynamics of a single-mode semiconductor laser with optical feedback (see Chapter 4). The Tang–Statz–deMars equations are used for multimode solid-state lasers with spatial hole burning. Multilevel rate equations are used for gas lasers. The model equations for various laser systems are summarized in Appendices 3.A.1-3.A.7.

3.1.3
How to Distinguish between Chaos and Noise from Experimental Data?

In this section, the measurement tools of temporal chaotic dynamics are described from the experimental point of view. Several techniques for distinguishing between deterministic chaos and stochastic noise from experimental data are also discussed.

3.1.3.1 Experimental apparatus for Measurement of Instability in Laser Intensity

To measure temporal waveforms of laser output intensity experimentally, a "photodetector" (also referred to as a photoreceiver or a photodiode) is often used, which converts the laser output intensity to an electric signal. The electric signal can be amplified by an "electric amplifier" for detection. The temporal waveform of the amplified electric signal can be observed by using a "digital oscilloscope", which captures and depicts the temporal waveform of the electric signal as a voltage. A radio-frequency (RF) spectrum of laser intensity is another important measure, which is observed by using a "RF spectrum analyzer". The RF spectrum contains the information on dominant frequency components of the temporal waveform of laser intensity. The measurements of both temporal waveforms and RF spectra are the basis of analysis of chaotic nonlinear dynamics experimentally. The bandwidths of these equipments (photodetectors, electric amplifiers, digital oscilloscopes, and RF spectrum analyzers) need to be sufficiently higher than those of chaotic oscillations to observe the proper temporal dynamics of chaos.

3.1.3.2 Examples of Chaos and Noise from Real Experimental Data

Examples of experimental data for the temporal waveforms of laser intensity output are shown in Figure 3.2a and b, which were measured by using a digital oscilloscope. Is it possible to determine whether they are "chaos" or "noise" from the experimental data? As described in Chapter 2, chaos can be defined as an irregular temporal waveform generated from a deterministic physical law, whereas noise can be described statistically by stochastic process and there are no deterministic rules. Figure 3.2a is an example of "chaos" generated from a semiconductor laser with optical feedback light. By contrast, Figure 3.2b shows an example of "noise" in a semiconductor laser. This type of temporal dynamics of Figure 3.2b is known as the sustained relaxation oscillation, where the tiny fluctuation of laser intensity is observed at around the relaxation oscillation frequency driven by an inherent quantum or thermal noise. As seen in these figures, it is not so simple to distinguish between chaos and noise from direct observation of time series.

The observation of the RF spectrum may be more helpful to distinguish between chaos and noise experimentally, as shown in Figures 3.2c and d, which were measured by using a RF spectrum analyzer. There are some periodic structures of the RF spectrum for chaos in Figure 3.2c, where the peak interval corresponds to the external cavity frequency of the semiconductor laser. On the other hand, there is no characteristic frequency component in the RF spectrum of noise in Figure 3.2d. The nonlinear frequency mixing is one of the important mechanisms to generate

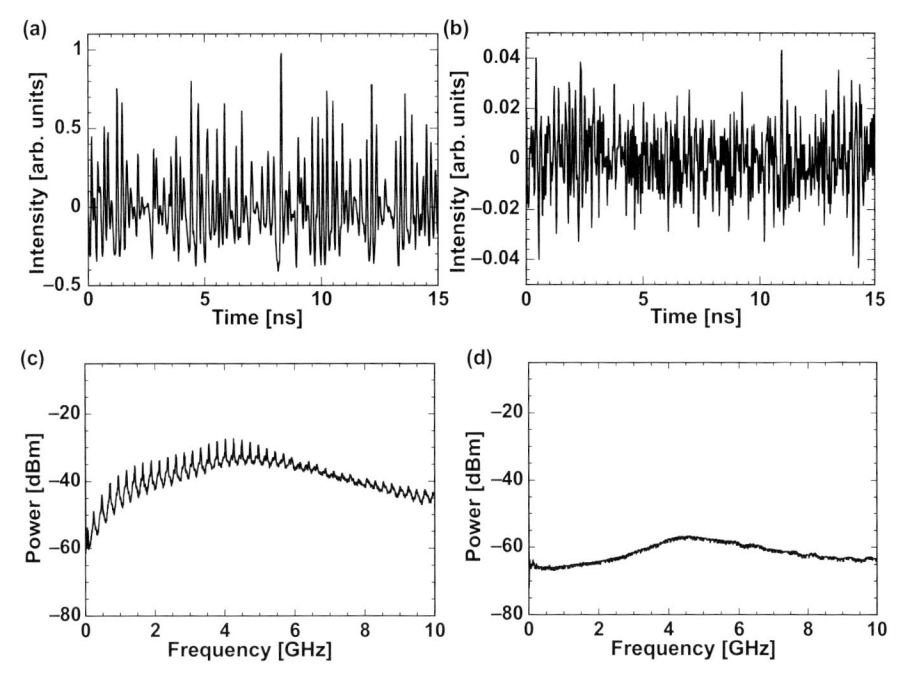

Figure 3.2 (a), (b) Experimental results of temporal waveforms of laser intensity output in the semiconductor laser. (a) Chaos induced by optical feedback and (b) inherent noise (sustained relaxation oscillation). (c), (d) Experimental results of RF spectra of laser intensity output in the semiconductor laser. (c) Chaos induced by optical feedback and (d) inherent noise (sustained relaxation oscillation).

deterministic chaos, as seen in the quasiperiodicity route to chaos. Therefore, the existence of characteristic frequencies related to some physical parameters of lasers can be considered as proof of deterministic chaos.

3.1.3.3 Observation of Bifurcation

How to find the "determinism" of chaos from experimental data? One of the important characteristics of chaos is the transition from one oscillation state to another state, known as "bifurcation", when one of the laser parameter values is changed. Figures 3.3a and b show an example of bifurcation obtained from experiment. When one of the laser parameter values is changed (here, the optical feedback strength is increased in a semiconductor laser), the transition is found from a period-1 oscillation (Figure 3.3a) and a quasiperiodic oscillation (Figure 3.3b) before a chaotic oscillation is observed (Figure 3.2a). This bifurcation is known as the quasiperiodicity route to chaos, and this result strongly indicates the existence of deterministic chaos, since bifurcation can be observed only in deterministic systems, but not in stochastic systems. The observation of bifurcation is thus one of the most convincing pieces of evidence of deterministic chaos if one of the laser parameters is accessible and changeable, particularly for experimental data.

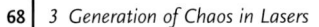

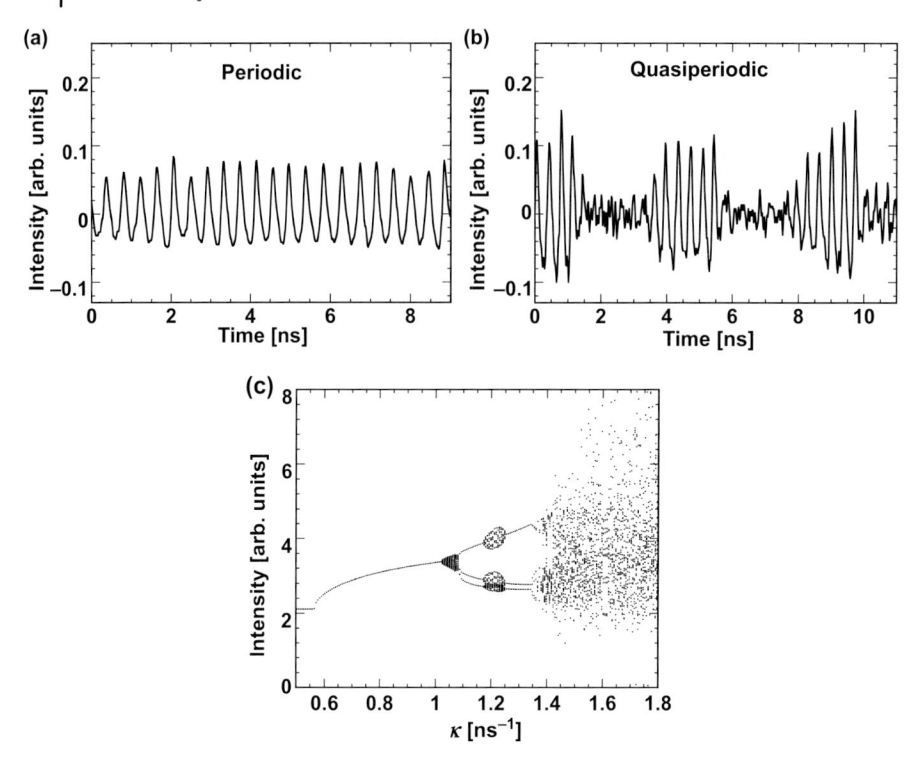

Figure 3.3 (a), (b) Experimental results of temporal waveforms of laser intensity output in the semiconductor laser at various feedback strengths. (a) period-1 oscillation and (b) quasiperiodic oscillation. Bifurcation is observed experimentally. (c) Numerical result of bifurcation diagram as a function of the feedback strength in the semiconductor laser with optical feedback. The peak values of the temporal waveforms are plotted at various feedback strengths.

3.1.3.4 Comparison with Laser Model

To find a stronger indication of chaos, "determinism" itself needs to be found. The existence of a deterministic rule based on nonlinear "differential equations" is a key to distinguish between chaos and noise. Therefore, modeling of laser dynamical systems is very important to identify chaos. The dynamics of lasers are mainly described by using a set of rate equations, in which the time variation of the amount of the laser intensity, the population inversion, and the atomic polarization can be governed (see Section 2.4). Depending on the type of laser medium and laser-mode configuration, a variety of numerical models have been reported as shown in Appendices 3.A.1-3.A.7. The match between experimental data and numerical results obtained from the corresponding laser dynamical model is one of the most important procedures to identify deterministic chaos, and it is a standard approach to analyze chaotic phenomena in nonlinear dynamical systems.

Figure 3.3c shows an example of a bifurcation diagram obtained numerically from the Lang–Kobayashi equations, describing the dynamics of a single-mode semiconductor laser with optical feedback. The peak values of the temporal waveforms are sampled and plotted when the feedback strength is varied. The number of peak values indicates the periodicity of oscillations (e.g., one value corresponds to a period-1 oscillation). For chaotic waveforms, many peak values are scattered in the diagram. In between, scattered plots within certain regions indicate quasiperiodic oscillations. If the bifurcation obtained from an experiment is similar to that calculated from numerical simulation, it is a good indication that deterministic chaos can be observed in the experiment. Qualitative and quantitative coincidences between experimental and numerical results confirm the existence of determinism of chaos. Therefore, one may always need to pay attention to the comparison between experimental and numerical results, and the comparison between experimental and numerical measurements is a standard approach in this interdisciplinary research field of chaos and laser.

3.2
Chaos in Semiconductor Lasers

In the following sections (Sections 3.2–3.6), detailed examples of different methods for generating chaos are described in various laser systems, such as semiconductor lasers, electro-optic systems, fiber lasers, solid-state lasers, and gas lasers. Both experimental and numerical examples from the literature are discussed. In Section 3.2, chaos in semiconductor lasers is treated and described.

Semiconductor lasers (also known as laser diodes) are widely used for optical communications and optical data storages. Instability of laser intensity in semiconductor lasers has been reported in such engineering systems. In Section 3.2, several methods are described to induce chaotic instability in semiconductor lasers: optical feedback, polarization-rotated optical feedback, optoelectronic feedback, optical injection, and injection-current modulation. The dynamics of vertical-cavity surface-emitting lasers is also discussed. The details of nonlinear dynamics in semiconductor lasers with various configurations are described comprehensively in the book (Ohtsubo, 2008). Also, see Chapter 4 for details of analytical, numerical, and experimental analyses on the dynamics of a semiconductor laser with optical feedback.

3.2.1
Semiconductor Laser with Optical Feedback

3.2.1.1 Dynamical Regime and L–I Characteristics
Optical feedback is one of the simplest methods to induce chaotic dynamics in semiconductor lasers. An external mirror is set in front of a semiconductor laser as shown in Figure 3.4 (left). The optical beam is reflected from the external mirror and is fed back to the laser cavity. Depending on the strength of the feedback light, many nonlinear dynamical phenomena are observed, such as bistability, instability, self-pulsation, and coherence collapse. Typical semiconductor lasers have cleaved facets as a laser resonator whose intensity reflectivity is about 30%. This dissipative laser

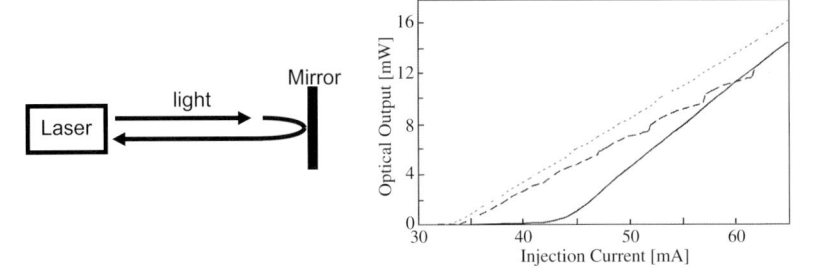

Figure 3.4 (Left) Model for a semiconductor laser with optical feedback. (Right) Experimentally obtained light power *versus* injection current (L–I) characteristics of semiconductor lasers with optical feedback. Solid line: solitary oscillation, dotted line: with optical feedback at the external cavity length of $L = 0.15$ m, broken line: with optical feedback at the external cavity length of $L = 1.5$ m. (J. Ohtsubo, (2008), © 2008 Springer.)

cavity makes semiconductor lasers very sensitive to the external light. An optical isolator is thus normally used for semiconductor lasers in optical communication systems and optical data-storage systems.

The regimes of the dynamics of semiconductor lasers subject to optical feedback are categorized as follows (Tkach and Chraplyvy, 1986; Ohtsubo, 2008),

> **Regime I::** Very small feedback and small effect. The feedback strength of electric-field amplitude is less than 0.01%. The linewidth of the laser oscillation becomes broad or narrow, depending on the feedback fraction.
> **Regime II::** Small, but not negligible effects. The feedback strength is less than 0.1%. Generation of the external modes gives rise to mode hopping among internal and external modes.
> **Regime III::** This is a narrow region when the feedback strength is around ~0.1%. The mode-hopping noise is suppressed and the laser may oscillate with a narrow linewidth.
> **Regime IV::** This regime corresponds to moderate feedback strength around 1%. The relaxation oscillation becomes undamped and the laser linewidth is broadened greatly. The laser shows chaotic behavior and sometimes evolves into unstable oscillations in a coherence-collapse state. The noise level is enhanced greatly under this condition.
> **Regime V::** This is a strong feedback regime where the feedback strength is higher than 10%. The internal and external cavities behave like a single cavity and the laser oscillates in a single mode. The linewidth of the laser is narrowed greatly.

In these regimes, the feedback strength of electric-field amplitude is the actual optical feedback level into the active layer and does not correspond to the reflectivity of the external mirror. There are scattering, absorption, and diffraction losses of light in the laser cavity. However, semiconductor lasers are sensitive enough to destabilize the laser intensity by a small amount of optical feedback less than 1% of the electric-field amplitude. An isolation of 40 dB is typically used for optical communication systems to avoid optical feedback.

The laser output power of semiconductor lasers at solitary oscillations is linearly proportional to the bias injection current. Figure 3.4 (right) shows an experimental example of light power *versus* injection current (L–I) characteristics for a semiconductor laser with and without optical feedback (Ohtsubo, 2008). The solid curve is the L–I characteristics for the solitary oscillation and the dashed and dotted curves are with optical feedback at different external cavity lengths. A typical feature of a semiconductor laser with coherent optical feedback is the threshold reduction, which is related to the gain reduction (Hegarty et al., 1998). The reductions of the threshold current are about 30% in the presence of the optical feedback in Figure 3.4 (right). The angular frequency of the laser oscillation at solitary oscillation is dependent of the bias injection current (Petermann, 1988). Due to the change in the frequency for the increase or decrease of the injection current, the successive external modes are sequentially selected. Then, mode hops occur at certain bias injection currents, which are shown as plateaus in Figure 3.4 (right). These plateaus are referred to as "kinks". Instabilities and chaotic oscillations of the semiconductor laser output are observed in-between the mode jumps in the region of kinks.

3.2.1.2 Coherence Collapse and Chaos in Experiment

Figures 3.5a–c show experimental examples of the temporal waveforms of the output intensity in a semiconductor laser with optical feedback when the feedback strength is changed. Without optical feedback (Figure 3.5a), sustained relaxation oscillation is observed and small fluctuation of the laser output appears. As the feedback strength is increased, a quasiperiodic oscillation is observed as shown in Figure 3.5b. The observed quasiperiodic oscillation consists of two different frequency components: the fast relaxation oscillation frequency with the slow modulation component at the external-cavity frequency. The scheme corresponds to regimes III to IV discussed above. With further increase of optical feedback, this regular oscillation becomes unstable and chaotic oscillation appears, as shown in Figure 3.5c. The oscillation corresponds to regime IV and the coherence of the laser is almost collapsed. The chaotic oscillation becomes more irregular as the feedback strength is increased. The bifurcation from the relaxation oscillation to the chaotic oscillation through the quasiperiodic oscillation indicates the quasiperiodicity route to chaos.

These dynamical transitions can be distinguished more clearly by using RF spectra. Figures 3.5d–f show the RF spectra corresponding to Figures 3.5a–c, respectively. Without optical feedback (Figure 3.5d), there is a small and broad spectral component at around 2.86 GHz that corresponds to the relaxation oscillation frequency of the solitary semiconductor laser. As the feedback strength is increased (Figure 3.5e), this spectral component becomes larger and there are fine structures around the relaxation oscillation frequency, which consists of many spectral peaks with the constant interval of the external cavity frequency of 0.234 GHz. The existence of the two different frequencies, the relaxation oscillation frequency and the external cavity frequency, indicates a quasiperiodic oscillation. As the feedback strength is increased, the spectral components are broadened and show chaotic dynamics (Figure 3.5f). The amplitudes of the spectral

72 | *3 Generation of Chaos in Lasers*

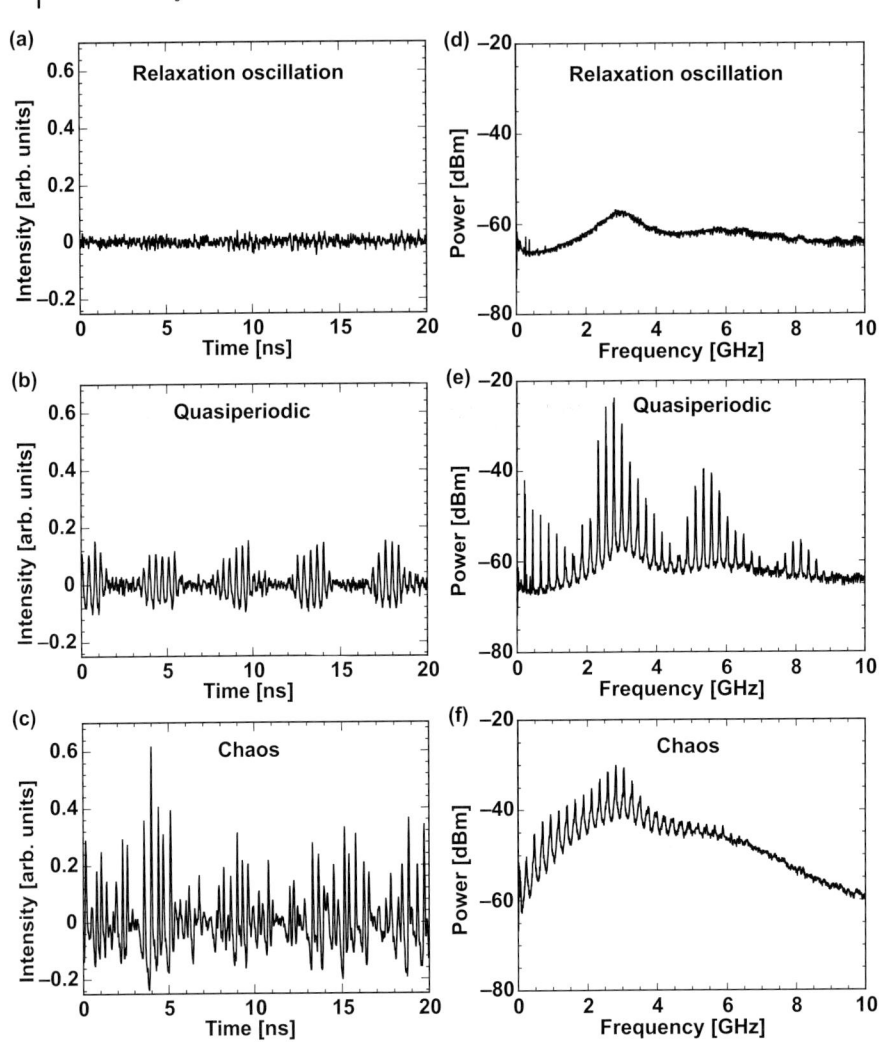

Figure 3.5 Experimentally obtained (a)–(c) temporal waveforms and (d)–(f) corresponding RF spectra in the semiconductor laser with optical feedback when the feedback strength is increased. The injection current is fixed at 13.02 mA ($J = 1.50 J_{th}$) and the external cavity length is fixed at $L = 0.64$ m. (a), (d) Sustained relaxation oscillation, (b), (e) quasiperiodic oscillation, and (c), (f) chaotic oscillation.

peaks decrease as the feedback strength is increased, and a smooth broad spectrum is obtained for the chaotic dynamics. Note that the floor level of the RF spectrum of the chaotic oscillation increases, compared with the quasiperiodic oscillation. Figure 3.5f shows a typical RF spectrum for a chaotic oscillation of laser output intensity in a semiconductor laser with optical feedback. The detail experimental results will be also described in Section 4.1.

3.2.1.3 Numerical Results and Bifurcation Diagram

Numerical simulations have been conducted by using the Lang–Kobayashi equations for a single-mode semiconductor laser with optical feedback (see Section 4.2.1 and Appendix 3.A.1.1). Numerical models for chaotic dynamics in a semiconductor laser are summarized in Appendix 3.A.1. Figure 3.6 shows the numerical result of dynamic behaviors of the laser output power for the variations of the strength of the optical feedback (Ohtsubo, 2008). The left panels in Figure 3.6 are the time series, the middle panels are their attractors, and the right panels are the RF spectra. For a small optical feedback, the laser output power is constant. For the external feedback level of 0.5% (2.5×10^{-5} in intensity), the laser becomes unstable and exhibits a period-1 oscillation as shown in Figure 3.6a. The main frequency of the oscillation is 2.53 GHz and it is very close to the relaxation oscillation frequency of 2.50 GHz at the solitary mode. When the feedback level is raised at 1.0%, a period-2 oscillation appears as shown in Figure 3.6b. Figure 3.6c shows a chaotic oscillation at the feedback level of 2.0%. When the laser output power shows periodic oscillations, clear spectral peaks can be observed, however, at the chaotic state, clear spectral peaks are not observable but the RF spectrum is broadened around the relaxation oscillation frequency, as shown in Figure 3.6c (right).

The attractor is a trajectory in the phase space of the system variables and is frequently used for the analysis of chaotic oscillations (see Section 2.2.2.2). When the laser output power at a stable oscillation is constant, the attractor becomes a fixed point in the phase space of the photon number (laser intensity) and the carrier

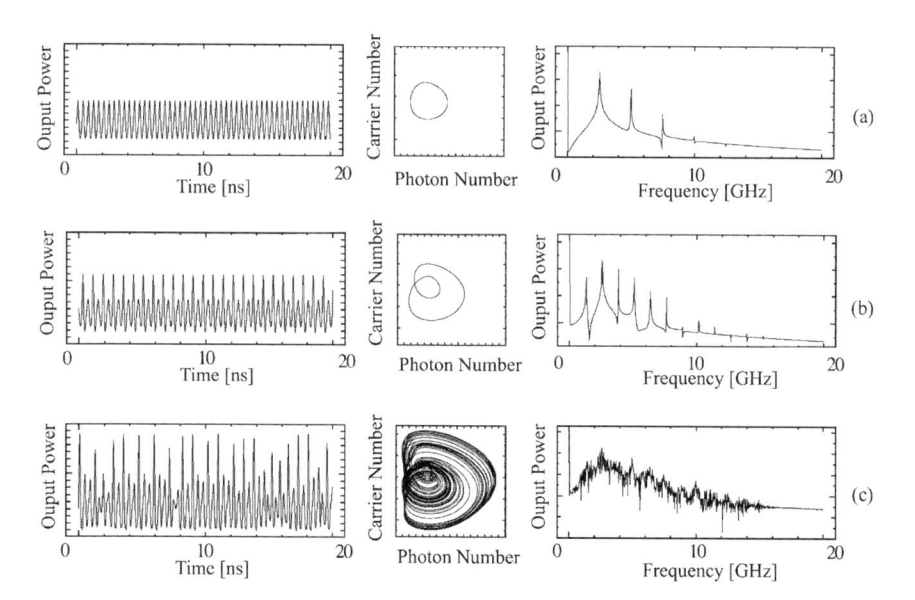

Figure 3.6 Numerically calculated temporal waveforms (left panels), attractors (middle panels), and RF spectra (right panels) for different feedback ratios at the bias injection current of $J = 1.3J_{th}$ and an external cavity length of $L = 0.03$ m. Feedback fractions of the electric-field amplitude are (a) $r = 0.005$, (b) $r = 0.01$, and (c) $r = 0.02$. (J. Ohtsubo, (2008), © 2008 Springer.)

density. For a period-1 signal, the attractor forms a circle (referred to as a "limit-cycle attractor") as shown in Figure 3.6a (middle). The attractor of a period-2 oscillation is a double-loop circle, as shown in Figure 3.6b (middle). However, the chaotic attractor behaves in a rather different way from fixed state or periodic oscillations. In chaotic oscillations, the trajectory goes around thick circles within the closed compact space, as shown in Figure 3.6c (middle). The chaotic attractor of Figure 3.6c (middle) is quite different from other periodic attractors and is referred to as a "strange attractor". In fact, the chaotic trajectory goes around in a multidimensional space and never crosses in such a space (Mørk et al., 1990; Ye et al., 1993).

A bifurcation diagram is frequently used to investigate chaotic evolutions for the change in one of the parameter values (see Section 2.2.4). A bifurcation diagram is obtained from a numerically obtained time series by sampling and plotting local maxima and minima of the temporal waveform for the parameter change. Figure 3.7 shows bifurcation diagrams as a function of the optical feedback strength (Ohtsubo, 2008). The vertical axis is the local maxima and minima of the temporal waveform. The horizontal axis is the parameter of the optical feedback level (the reflectivity of the external mirror r). In Figure 3.7a the laser is stable for an external feedback of less than 0.35%. The state is called a fixed point. Above this value, a relaxation oscillation appears in the laser output and the diagram has two points corresponding to the local maxima and minima of the period-1 oscillation. When the feedback level

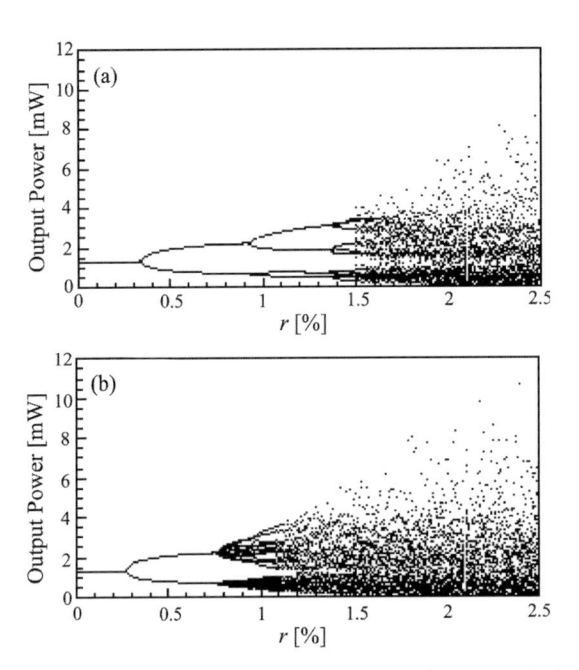

Figure 3.7 Numerically calculated bifurcation diagrams at the bias injection current of $J = 1.3J_{th}$. The external cavity length of (a) $L = 0.09$ m (the period-doubling route to chaos) and (b) $L = 0.15$ m (the quasiperiodicity route to chaos). (J. Ohtsubo, (2008), © 2008 Springer.)

exceeds the value of 0.94%, a period-2 oscillation starts and the output has four states of local maxima and minima. For a further increase of the feedback, the laser evolves into quasiperiodic states and finally chaotic oscillations over the feedback level of 1.36%. The chaotic laser oscillates at the mixed frequencies of the relaxation oscillation frequency and the external cavity frequency. The evolution such as that shown in Figure 3.7a clearly shows the period-doubling route to chaos. A test for period doubling is often used to check a route to chaos and bifurcations of the output in nonlinear dynamical systems for variations of chaotic parameters.

The period-doubling route to chaos is not only a bifurcation but other routes to chaos exist in nonlinear dynamical systems. Figure 3.7b is another example. The external cavity length is different from that in Figure 3.7a, but the other parameters are the same. In Figure 3.7b, the fixed point evolves into period-1 oscillations, however, the laser output becomes a quasiperiodic oscillation, and chaotic states occur immediately after the quasiperiodic oscillation for the increase of the feedback parameter. This bifurcation of Figure 3.7b is known as the quasiperiodicity route to chaos. Another route is an intermittency route to chaos, which can be considered in the state of low-frequency fluctuations in semiconductor lasers (Mørk et al., 1988; Fischer et al., 1996). The type of bifurcation highly depends on system parameter values in nonlinear dynamical laser systems.

For experimental data, it is not easy to obtain a clear bifurcation diagram from temporal waveforms with high-frequency fluctuations such as in a semiconductor laser, since the detected temporal waveforms are often contaminated by detection noise. In such a case, the local maxima and minima of the time series of the laser output have finite widths even in the fixed and periodic states in the bifurcation diagram. Instead of analyzing chaotic temporal waveforms, RF spectra are often used, and the number of spectral peaks and their widths are analyzed to determine the dynamical states in experiments, as already seen in Figure 3.5.

The detail analytical and numerical results of the Lang-Kobayashi equations for a semiconductor laser with optical feedback will be described in Sections 4.2–4.6.

3.2.1.4 Low-Frequency Fluctuations (LFFs)

Low-frequency fluctuation (LFF) is a type of chaotic oscillation known as intermittent chaos observed in semiconductor lasers with optical feedback under the conditions of low injection current and strong optical feedback strength. A typical feature of LFFs is a sudden power dropout with a following gradual power recovery. LFFs occur irregularly in time depending on the system parameters and the frequency of LFFs is of the order of MHz to a hundred MHz. Since the frequency of LFFs is much lower than ordinary chaotic fluctuations related to the relaxation oscillation frequency ($\sim$GHz), the phenomena are called "low-frequency" fluctuations. When temporal waveforms of LFFs are observed by a fast digital oscilloscope, they seem to be continuous signals. However, it is proved that LFFs have very fast time structures within the waveform and they consist of a series of fast pulses on the order of subnanoseconds (Fischer et al., 1996). Although LFFs were first recognized as sudden power dropouts with low frequency in the early days, LFFs have quite different features from the ordinary chaotic behaviors and show a rich variety of

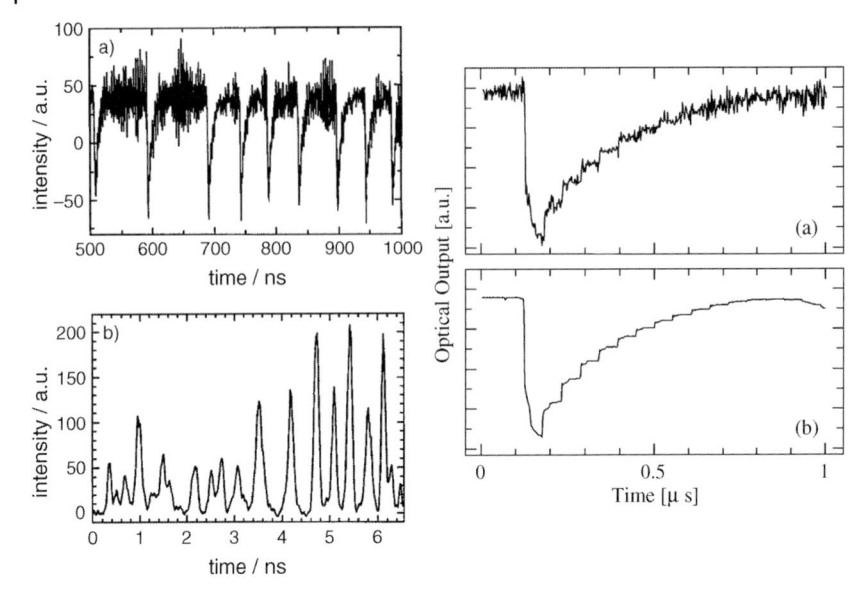

Figure 3.8 (Left) Low-frequency fluctuations in the semiconductor laser with optical feedback at $J = 1.03J_{th}$. (a) Detected by a slow response detector. (b) Observed by a fast streak camera. Delay time is $\tau = 3.6$ ns. (J. Ohtsubo, (2008), © 2008 Springer.) (Right) (a) Single shot of LFFs and (b) one-shot of LFFs averaged over 3000 events. The external cavity length and the bias injection current are $L = 8.10$ m and $J = 1.07J_{th}$. The external reflectivity is $r^2 = 0.12$. (I. Fischer, G. H. M. van Tartwijk, A. M. Levine, W. Elsäßer, E. Göbel, and D. Lenstra, (1996), © 1996 APS.)

dynamics. It is known that LFFs are the deterministic chaos induced by saddle-node instability with time-inverted type-II intermittency in semiconductor lasers with optical feedback (Sacher et al., 1989; Sacher et al., 1992).

Figure 3.8a (left) is an example of typical LFF waveforms obtained from experiment (Fischer et al., 1996). This is a low-pass filtered waveform and a higher component over nanosecond oscillations is not observable. The typical features of LFFs are frequent power dropouts after stationary output and subsequent gradual power recovery, as shown in Figure 3.8a (left). When the laser is operated close to the threshold, the output power breaks down even below the threshold in the presence of optical feedback. The frequency strongly depends on the conditions of the laser parameters and the external mirror for optical feedback. The power dropouts occur irregularly and their average frequency is about 5 MHz in Figure 3.8a (left). Figure 3.8b (left) shows a time-resolved waveform of LFFs observed by a high-speed streak camera. Indeed, the temporal waveform consists of a fast pulse train with an average period of 300 ps.

Figure 3.8a (right) is a one-shot of LFFs (Ohtsubo, 2008). To show the power-recovery process clearly, the one-shot of LFFs is averaged and plotted in Figure 3.8b (right). Stepwise power recovery is clearly visible in Figure 3.8b (right). The time duration of each step in the power recovery is equal to the round-trip time of light in the external cavity. The existence of an external round-trip time in the temporal waveforms is a typical feature of LFFs. In fact, LFFs are composed of three components of different time scales. One is a component of a low-frequency fluctuation with a period of microseconds; the second is a component related to the external cavity length with a period of tens of nanoseconds; the third is a high-frequency component related to the relaxation oscillation with a period of subnanoseconds.

In order to map the dynamics of the system in the parameter space, the feedback strength γ and the injection current I are simultaneously changed, and the dynamical behavior is investigated and classified. Figure 3.9 shows the two-dimensional (2D) map of the dynamical behavior of a semiconductor laser subject to optical feedback in the $\gamma - I$ parameter space (Heil *et al.*, 1998). The most striking aspect of this $\gamma - I$ space diagram is the existence of a large region within the LFF regime, where discrete transitions between LFF and stable emission on a single external-cavity mode occur. A large parameter regime in which the LFF phenomenon coexists with stable emission on a single high-gain external cavity mode is also obtained. It is worth noting that LFFs can be observed at the conditions of strong optical feedback strength and low injection current near the lasing threshold. It has also been reported that LFFs tend to appear in the laser output when the external cavity length is sufficiently long and the external feedback is large enough (Ohtsubo, 2008). The mechanism of LFFs is described with a model in detail in Appendix 3.A.2.

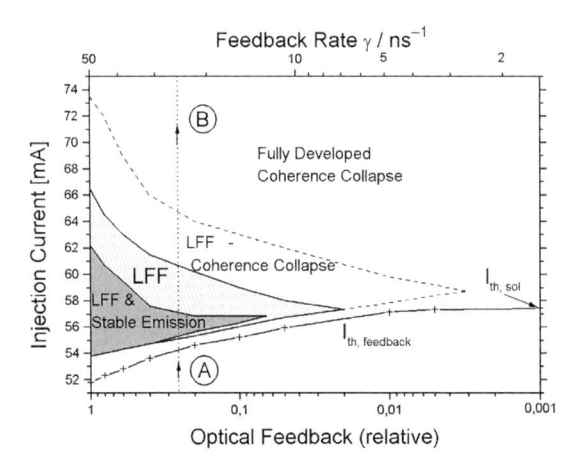

Figure 3.9 Dynamical behavior of the semiconductor laser subject to optical feedback in feedback strength (γ) – injection current (I) space. The LFF regime is depicted in light gray. The dark-gray region embedded in the LFF regime corresponds to the region of coexistence of the stable emission state and the LFF state. The unshaded region encompassed by the dashed line corresponds to the continuous transition between the LFF regime and the fully developed coherence collapse chaotic regime. (T. Heil, I. Fischer, and W. Elsäßer, (1998), © 1998 APS.)

3.2.1.5 Short-Cavity Regime

The dynamics considered so far is the situation of a semiconductor laser with a long external cavity length, where the external cavity frequency f_{ext} (see Eq. (3.1)) is smaller than the relaxation oscillation frequency f_r (see Eq. (2.19)) that is, $f_{ext} < f_r$. When the external cavity length is set in the order of a few cm or shorter, the opposite condition is satisfied, that is, $f_{ext} > f_r$, which is called the short-cavity regime. For the short-cavity regime, the dynamics of "regular pulse package" can be observed, where the fast regular pulse oscillation at the frequency of f_{ext} is obtained, whose amplitude is modulated at the slower frequency of f_r. The short-cavity regime can also be observed in photonic integrated circuits due to its short external cavity length (see Sections 8.5.1 and 10.3.6).

Figure 3.10 shows the results of the dynamic characteristics in a semiconductor laser with optical feedback from a short external cavity (Heil et al., 2001b). Figure 3.10a (left) and b (left) are the experimental results of observed RF spectra. In Figure 3.10a (left), the laser shows a regular pulse package of an LFF frequency of

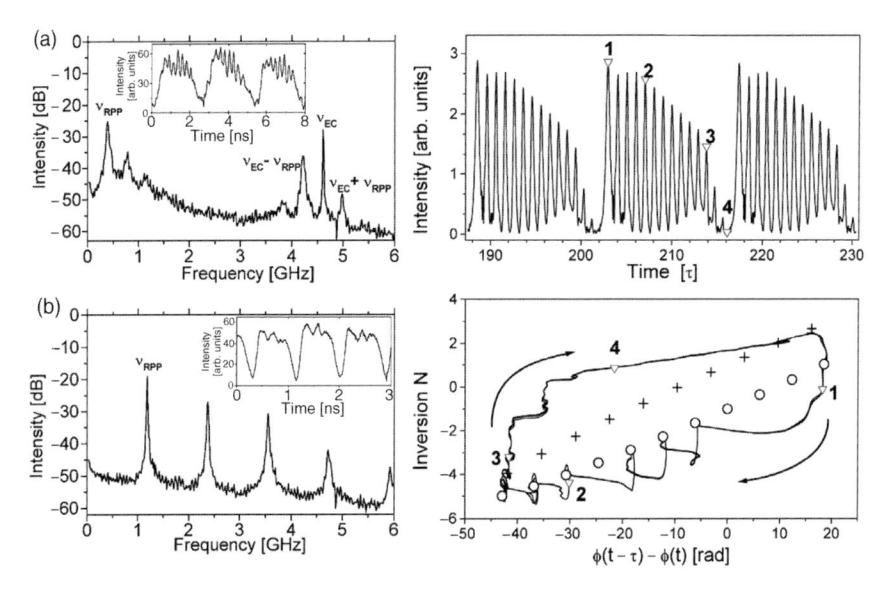

Figure 3.10 (Left) Experimental results of RF spectra and intensity time series (inset) of the semiconductor laser operating in the short-cavity regime. (a) The injection current $I = 1.08I_{th}$ corresponds to the relaxation oscillation frequency of $f_r = 1.1$ GHz, and the external cavity length of $L = 3.3$ cm corresponds to the external cavity frequency of f_{ext} (ν_{EC} in the figure) $= 4.5$ GHz. (b) $I = 1.80I_{th}$ corresponding to $f_r = 3.8$ GHz, and $L = 1.1$ cm corresponds to $f_{ext} = 14.0$ GHz. ν_{RPP} is the frequency of the regular pulse package in the figure. (Right) Numerically computed dynamics of the regular pulse package. Top: Laser intensity plotted against time normalized to the external cavity round-trip time τ. Bottom: Regular pulse package trajectory plotted on the normalized carrier density N and phase difference $\phi(t-\tau)-\phi(t)$ plane. The location of the nodes is indicated by circles, and the saddles by crosses. The temporal evolution occurs clockwise; the numbers provide a one-to-one correspondence of time series and phase-space portrait. (T. Heil, I. Fischer, W. Elsäßer, and A. Gavrielides, (2001b), © 2001 APS.)

390 MHz. The pulse package is very regular and the trajectory variations from a pulse package to the next pulse package are very small. Also, the spectral peak for the external cavity of 4.5 GHz, which corresponds to the short external cavity length of 3.3 cm, is seen. The inset is the direct waveform observed by a digital oscilloscope, but the waveform is smeared out due to the resolution of a digitizer (the bandwidth of 4.5 GHz). The regular pulsating oscillations with fully modulated waveform were confirmed by the observation from a streak camera and each pulse width was observed to be around 10 ps. In Figure 3.10b (left), the frequency of the regular pulse package is 1.195 GHz, which is quite fast, compared with the case of a long external cavity feedback. Only the frequency of the regular pulse package is shown in Figure 3.10b (left), since the external cavity frequency is 14 GHz and the spectral peak is outside the scope of this plot. Under regular pulse package emission, the lasers operate on several external cavity modes, which is similar to ordinary LFFs with long external cavity. In all cases of external cavity length in semiconductor lasers, the dynamics is sensitive to the optical phase compatible to the optical wavelength. However, strong phase sensitivity is observed in the case of short external cavity optical feedback. In this experiment, the regular pulse package occurs only over a certain phase interval for the ranges of $1.3 I_{th} < I < 1.5 I_{th}$. For $I > 1.5 I_{th}$, the regular pulse package is present for all feedback phase.

The top panel of Figure 3.10 (right) is the numerical results for regular pulse package oscillations based on the laser rate equations (Heil *et al.*, 2001b). It is noted that the laser intensity is plotted against time normalized to the photon lifetime τ_p. The bottom panel of Figure 3.10 (right) shows the trajectory of chaotic oscillations. In the phase space, the location of external modes is indicated by circles and the antimodes by crosses. The temporal evolution occurs clockwise and the numbers in the figure provide a one-to-one correspondence of time series and phase-space portrait. One of the remarkable characteristics is that the trajectory always visits the same external cavity modes and the laser shows regular pulses.

3.2.2
Semiconductor Laser with Polarization-Rotated Optical Feedback

Polarization-rotated optical feedback has been proposed as a method to generate chaos, where the polarization of the optical feedback light is rotated at 90 degrees and fed back to the laser cavity. Two orthogonal polarization modes, transverse-electric (TE) mode and transverse-magnetic (TM) mode, are coupled to each other in the semiconductor laser cavity by the polarization rotated feedback. Only the TE mode lases in a solitary semiconductor laser due to the higher gain for the TE mode than that for the TM mode, however, the TM mode can be enhanced by the polarization-rotated optical feedback light. Chaotic instability can occur due to the nonlinear interaction between the TE and TM polarization modes. Chaotic behavior of semiconductor lasers with polarization-rotated optical feedback has been observed experimentally (Houlihan *et al.*, 2001; Saucedo *et al.*, 2002). For the dynamics of the polarization-rotated optical feedback, a model including two polarization modes and the feedback of the complex electric field (amplitude and phase) has been proposed to explain the behavior of chaos synchronization observed in experiments (Heil et al., 2003; Shibasaki et al., 2006). Two polarization modes without explicitly assuming an incoherent feedback effect (i.e., direct feedback to the carrier density) are crucial to reproduce the experimental results of the characteristics of synchronization of chaos (Sukow et al., 2004; Shibasaki et al., 2006; Takeuchi et al., 2008). A numerical model for chaotic dynamics in a semiconductor laser with polarization-rotated optical feedback is described in Appendix 3.A.1.2.

Figure 3.11 (left) shows the experimental setup for a distributed-feedback (DFB) semiconductor laser with polarization-rotated optical feedback (Heil et al., 2003). The optical wavelength is 1537 nm with a single-longitudinal mode. At twice the threshold, the solitary DFB laser exhibits single TE mode emission with TM mode suppression ratios of 1000. (i.e., the power ratio of TE and TM modes is 1000:1 for the solitary laser, and TE and TM modes are orthogonally polarized.) The delayed optical feedback is provided by an external optical loop that polarizes the TE laser beam, rotates this polarization by 90 degrees from the TE to TM polarization direction, and reinjects this polarization-rotated TM beam back into the laser. An optical isolator (ISO) is used to achieve one-way loop propagation with isolation of -60 dB. A half-wave plate ($\lambda/2$) rotates the polarization direction of the laser beam by 90 degree from TE mode to TM mode, and a polarizer (TM-Pol) is used to ensure only the TM mode returns to the laser. The feedback loop is formed by mirrors (M) and a polarization beam splitter (PBS), which feeds the outgoing TE beam into the loop and feeds the returning TM beam back into the laser. A neutral density filter (NDF) controls the strength of optical feedback. The delay time is given by the round-trip time of the light in the loop, and amounts to 7.4 ns, corresponding to a round-trip

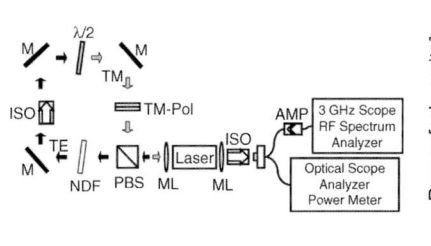

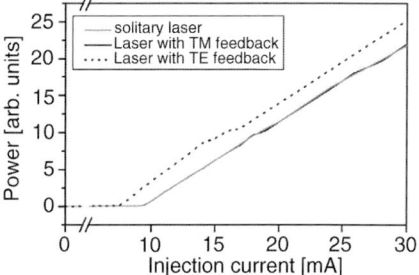

Figure 3.11 (Left) Experimental setup for observation of the dynamics in the DFB semiconductor laser with polarization-rotated optical feedback. The semiconductor laser oscillating mainly in the TE mode is subject to delayed polarization-rotated optical feedback injected into the TM mode of the laser. Amp, amplifier; ISO, optical isolator; M, mirror; ML, microscopic lens; NDF, neutral density filter; PBS, polarization beam splitter; TE, TE-polarization mode; TM, TM-polarization mode; TM-Pol, polarizer along TM direction; $\lambda/2$, half-wave plate. (Right) L–I characteristics for the solitary DFB laser (the gray solid curve), the laser with polarization-rotated TM-mode feedback (the black solid curve), and the laser with coherent TE-mode feedback (the dotted curve), obtained in experiment. (T. Heil, A. Uchida, P. Davis, and T. Aida, (2003), © 2003 APS.)

frequency of 0.135 GHz. The AR-coated facet of the lasers is used to provide the optical feedback, and the light from the uncoated facet is used for detection. The dynamical behavior of the intensity is detected with a photodiode and analyzed with a radio-frequency (RF) spectrum analyzer and a fast digital oscilloscope.

The dynamics of polarization-rotated optical feedback are quite different from those of coherent (normal) optical feedback, as seen in Section 3.2.1. Figure 3.11 (right) shows an experimental observation of the L–I characteristics for the solitary DFB semiconductor laser (the gray solid curve), the laser with polarization-rotated TM-mode feedback (the black solid curve), and the laser with coherent TE-mode feedback (the dotted curve) (Heil et al., 2003). The intensities with optical feedback contain instabilities for the time development, but the plotted L–I characteristics are averaged intensities. All curves are recorded for a similar level of optical feedback. The L–I curve for the coherent feedback is typical for coherent optical feedback (see Section 3.2.1), where there is a typical threshold reduction of 20%. The kink in the L–I curve marks the onset of chaotic fluctuations induced by the coherent optical feedback. However, for the polarization-rotated TM feedback, the L–I curve is similar to that for the solitary laser. Specifically, there is no threshold reduction and almost no change of slope. It is worth noting that the time-averaged intensity of the laser is almost unaffected by the polarization-rotated feedback. However, observing the temporal waveforms of the intensity, TM-mode feedback-induced instabilities can be found.

Figure 3.12 (left) shows the experimental results of the temporal waveforms of the DFB laser with polarization-rotated feedback at various injection currents, and Figure 3.12 (right) displays the corresponding RF spectra (Heil et al., 2003). For low injection currents, the small-amplitude instabilities are observed as shown in Figure 3.12a (left). The corresponding RF spectrum depicted in Figure 3.12a (right) shows a series of equidistant peaks separated by the round-trip frequency of the external cavity (external cavity frequency). The amplitudes of these peaks exhibit a characteristic envelope with a maximum approximately at the relaxation oscillation frequency. As the injection current is increased, the amplitude of the oscillations increases. Figure 3.12b (left) shows a typical example of the temporal waveform, where the instabilities appear to be weakly chaotic oscillations that are close to quasiperiodicity. Figure 3.12b (right) demonstrates that the instabilities are dominated by two basic frequency components: a low-frequency component that is the external cavity frequency and a high-frequency component near the relaxation oscillation frequency. Interestingly, the positions of the peak associated with the external cavity frequency do not shift with increasing injection current. This is in contrast to the dynamics of semiconductor lasers with coherent feedback, where a significant shift and broadening of these peaks occurs for increasing injection current. Finally, as the injection current is increased further, the amplitude of the intensity oscillations decreases again, as shown in Figures 3.12c (left) and (right). The amplitude of the peaks is substantially reduced for large injection currents. Thus, a characteristic behavior of the dependence of the amplitude of the TM-mode feedback-induced instabilities on the injection current is found. The amplitudes are largest for the intermediate injection currents, whereas the instabilities totally

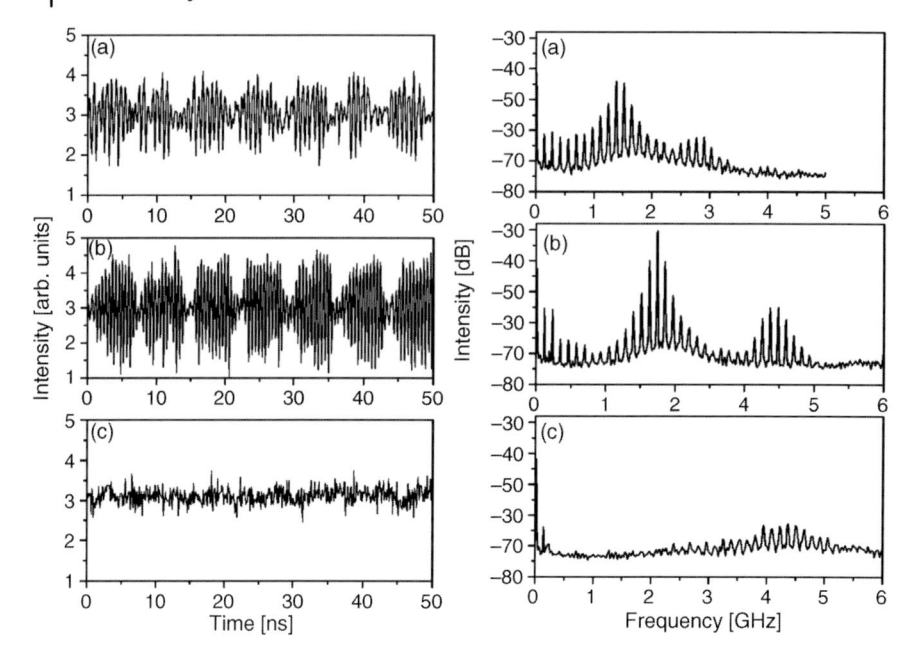

Figure 3.12 (Left) Experimental results of temporal waveforms of the DFB semiconductor laser with polarization-rotated optical feedback at various injection currents. (a) $J = 1.15 J_{th,sol}$, (b) $J = 1.25 J_{th,sol}$, (c) $J = 2.0 J_{th,sol}$. (Right) RF spectra corresponding to Figure 3.12 (left). (T. Heil, A. Uchida, P. Davis, and T. Aida, (2003), © 2003 APS.)

disappear for strong injection currents, where stable steady-state output is observed with flat RF spectra. There are some small kinks on the L–I curve for the TM-mode feedback shown in Figure 3.11 (right). However, unlike the case of coherent feedback, significant changes in dynamical behaviors are not observed at these small kinks.

Figure 3.13 (left) shows the numerical result of the systematic dependence of the dynamics on the injection current using a bifurcation diagram (Heil et al., 2003). The bifurcation diagram is created by sampling the peak values of the temporal waveforms as the injection current parameter is changed. In Figure 3.13 (left), it is clearly seen that the amplitudes of the temporal waveforms increase with injection current to a maximum value around the injection current of 1.55 $J_{th,sol}$, where $J_{th,sol}$ is the threshold value of the injection current. The output stabilizes around the injection current value of 1.8 $J_{th,sol}$. The restabilization of temporal waveforms at high injection current is an interesting phenomenon. The mechanism of restabilization can be interpreted as follows. The coupling term of TE and TM modes needs to be large enough compared with the term for injection current, in order to generate chaotic dynamics. When the injection current is increased, the value of the steady-state solution of the carrier density is increased, and the laser tends to be less sensitive to the feedback light. The bifurcation shown in Figure 3.13 (left) is consistent with the behavior obtained in the experiments shown in Figure 3.12.

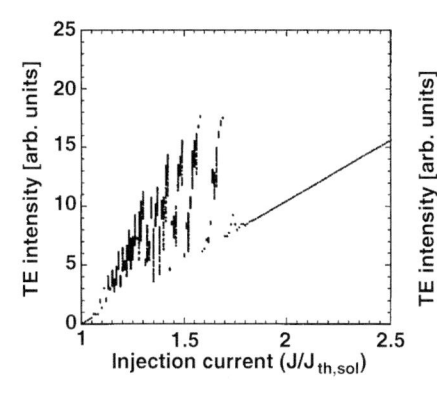

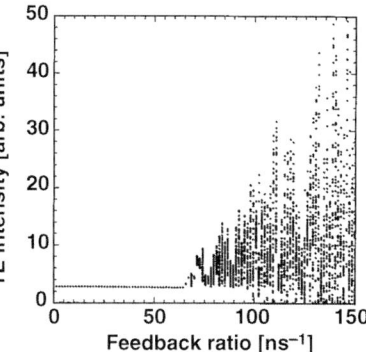

Figure 3.13 (Left) Numerical result of bifurcation diagram of the temporal waveforms of the DFB semiconductor laser with polarization-rotated optical feedback as a function of the normalized injection current, $J/J_{th.sol}$. (Right) Numerical result of bifurcation diagram of the temporal waveforms of the DFB semiconductor laser with polarization-rotated optical feedback as a function of the feedback power ratio. (T. Heil, A. Uchida, P. Davis, and T. Aida, (2003), © 2003 APS.)

To provide an overview of the dependence of the dynamics on the feedback strength, another bifurcation diagram is generated for various feedback strengths, as shown in Figure 3.13 (right). Figure 3.13 (right) demonstrates that a large feedback level is required for the onset of self-oscillations, and that the amplitude of the oscillations increases with the increase of the feedback level. These results are also in good qualitative agreement with the experimental observations.

3.2.3
Semiconductor Laser with Optoelectronic Feedback

The use of semiconductor lasers with "incoherent" optical feedback is a way to generate chaos. Semiconductor lasers subject to incoherent optical feedback indicate that the optical feedback acts only on the carrier density in the laser rather than the electric field. Semiconductor lasers with incoherent optical feedback have been studied theoretically using rate-equation models (Otsuka, 1991; Ishiyama, 1999), where the intensity of the optical feedback directly interacts with the carrier density. These models are similar to the model for semiconductor lasers with optoelectronic feedback, where a signal proportional to the intensity of optical feedback is directly applied to the injection current of the semiconductor laser.

Semiconductor lasers with optoelectronic feedback have been proposed as a way of implementing the incoherent feedback scheme. Figure 3.14 (left) shows a schematic diagram of optoelectronic feedback in a semiconductor laser (Tang and Liu, 2001a). The light emitted from a semiconductor laser is detected by a photodetector and the detected photocurrent is fed back to the injection current through a bias Tee circuit. The feedback may be positive or negative depending on the polarity of the output of the amplifier in the circuit. In optoelectronic feedback, the modulation is not for the

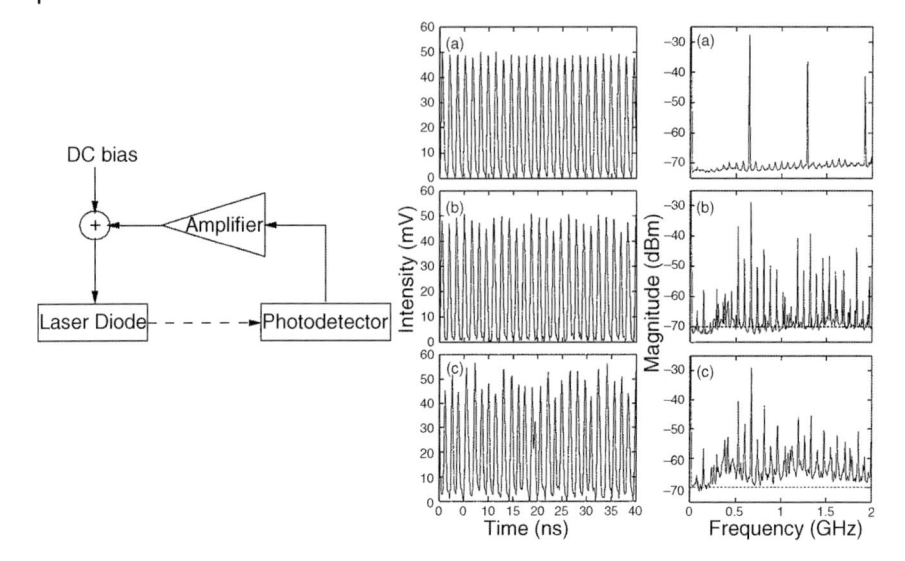

Figure 3.14 (Left) Schematic setup of a semiconductor laser with time-delayed optoelectronic feedback. The dashed and solid lines indicate the optical and electric paths, respectively. (Right) Experimental results of time series (middle panels) and RF spectra (right panels) of different pulsing states at different delay times in the semiconductor laser with optoelectronic feedback. (a) Regular pulsing at $\tau = 7.47$ ns. (b) Two-frequency quasiperiodic pulsing at $\tau = 7.09$ ns. (c) Chaotic pulsing at $\tau = 6.92$ ns. The broadband background in the RF spectrum of (c) is much higher than in that of (b), indicated by the dashed reference lines at -70 dBm. (S. Tang and J. M. Liu, (2001a), © 2001 IEEE.)

electric field but for the carrier density through the disturbance to the injection current. Therefore, the nonlinear dynamics is insensitive to the phase of the electric field and this method is called incoherent feedback. A numerical model for chaotic dynamics in a semiconductor laser with optoelectronic feedback is described in Appendix 3.A.1.3.

A feedback circuit with time response of subnanoseconds is easily available at present and the response is sufficient to follow chaotic variations in semiconductor lasers. However, when the time response of the electronic feedback circuit has the same order as the relaxation oscillation frequency of the laser, the effect of the finite response must be taken into account. The effect of the bandwidth limitation for the feedback signal is strongly influenced by the nonlinear dynamics of semiconductor lasers.

Experimental results of pulsation oscillations have been reported in optoelectronic feedback systems, as shown in Figure 3.14 (right) (Tang and Liu, 2001a). The laser outputs (middle panels) and corresponding RF spectra (right panels) are obtained for certain conditions of the feedback delay (the delay time including the flight of light in the feedback loop of the laser, the detector, and the electric circuit). From Figure 3.14a (right) to c (right), the laser evolves a regular pulsing state of a constant peak intensity into a quasiperiodic pulsing state. Only one fundamental frequency at 650 MHz is excited in the RF spectrum of Figure 3.14a (right), which has the pulsing frequency of the regular pulses seen in the corresponding time series. However, ripples with small amplitude are included in the RF spectrum. The laser is not completely stabilized to a periodic state, although the level of the ripples is below 40 dB. The frequency of 140 MHz corresponding to the delay time of the feedback loop is excited as the second oscillation frequency in Figure 3.14b (right). The two frequencies are incommensurable. Therefore, the laser oscillates at the two fundamental frequencies and the oscillation is in a quasiperiodic pulsing state. In Figure 3.14c (right), the broadband background of the RF spectrum is much higher than that in Figure 3.14b (right), indicated by the dashed reference lines at -70 dBm. Overall, the spectrum is much broadened and the peak heights of the pulsation oscillation vary irregularly, corresponding to chaotic oscillation.

Figure 3.15 (left) shows the numerical results of pulsing states for the variations of the delay time in a positive optoelectronic feedback system (Tang and Liu, 2001a). The

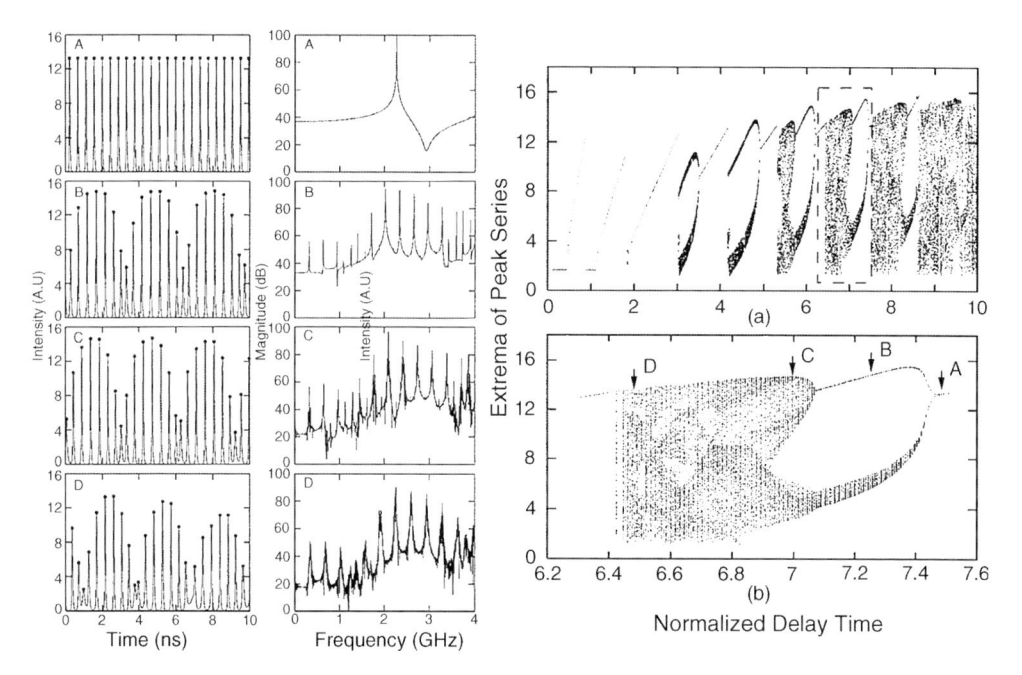

Figure 3.15 (Left) Numerical results of time series (left panels) and RF spectra (middle panels) of different pulsing states. (a) Regular pulsing at $\hat{\tau} = 7.47$. (b) Two-frequency quasiperiodic pulsing at $\hat{\tau} = 7.25$. (c) Three-frequency quasiperiodic pulsing at $\hat{\tau} = 7.00$. (d) Chaotic pulsing at $\hat{\tau} = 6.48$. The pulse peak intensities are marked by the filled circles in the time series. The calculated spectra have relative magnitudes with decibels increment. The numerical results of Figures 3.15(a), (b), and (d) (left) agree well with the experimental results of Figures 3.14(a), (b), and (c) (right), respectively. (Right) (a) Bifurcation diagram of the extrema of the peak series with the normalized delay time $\hat{\tau}$ varying from 0 to 10. (b) Enlargement of the small region indicated by the dashed rectangle in (a) with the identification of various pulsing states. A–D indicate the corresponding pulsing states, as in Figure 3.15 (left). (S. Tang and J. M. Liu, (2001a), © 2001 IEEE.)

delay time is normalized by the inverse of the relaxation oscillation frequency of 2.5 GHz. With the normalized delay time of $\hat{\tau} = 7.47$, the time series shows a sequence of regular pulses with a constant pulsing intensity and interval, as shown in Figure 3.15a (left). The corresponding RF spectrum has only one fundamental pulsing frequency at $f_1 \sim 2.3$ GHz, which is close to the relaxation oscillation frequency of the laser. When the delay time is decreased to $\hat{\tau} = 7.25$ in Figure 3.15b (left), the laser enters a two-frequency quasiperiodic pulsing state with the intensity modulated at a certain frequency f_2. The pulses are clearly modulated and the new modulation frequency is read to be $f_2 \sim 320$ MHz. This f_2 is close to, but slightly less than, the inverse of the delay time of the optoelectronic feedback loop. The appearance of two incommensurable frequencies, f_1 and f_2, is the indication of quasiperiodicity. When the delay time is further decreased to $\hat{\tau} = 7.00$ in Figure 3.15c (left), the laser enters a three-frequency quasiperiodic pulsing state as a third frequency at $f_3 \sim 23$ MHz shows up. The component of this frequency f_3 is very small and the frequency is incommensurable with f_1 and f_2. The appearance of f_3 results from nonlinear interaction between the laser relaxation oscillation and the delayed feedback. Finally, when $\hat{\tau} = 6.48$, the laser enters a chaotic pulsing state, as shown in Figure 3.15d (left). In the chaotic states, not only the pulse height but also the separation becomes chaotic (jitter) and the corresponding RF spectrum is much broadened. The numerical results in Figure 3.15 (left) agree well with the experimental results in Figure 3.14 (right).

To visualize the route to chaos in optoelectronic feedback in a semiconductor laser, the bifurcation diagrams are numerically calculated. Bifurcation diagrams of the extrema of the peak time series *versus* delay time corresponding to Figure 3.15 (left) are plotted in Figure 3.15 (right) (Tang and Liu, 2001a). Figure 3.15a (right) is the bifurcation diagram with the normalized delay time $\hat{\tau}$ varying from 0 to 10. For a small delay time, the effect of the feedback on the dynamics is not distinct and the feedback increases the laser output slightly over that in the free-running condition. With the increase of the delay time, the laser output shows regular pulsing with constant pulse peak intensity and interval. For a larger delay, the laser follows a quasiperiodicity route into chaotic pulsing states. Figure 3.15b (right) is the enlarged bifurcation diagram of a part in Figure 3.15a (right) denoted by the dashed rectangle. A–D in Figure 3.15b (right) can be compared with the pulsing states in Figure 3.15 (left). The different pulsing states are indicated by the corresponding arrows, where A is regular pulsing, B is two-frequency quasiperiodic pulsing, C is three-frequency quasiperiodic pulsing, and D is chaotic pulsing. The quasiperiodicity route to chaos is clearly observed in Figure 3.15b (right).

The feedback delay time and the feedback ratio play crucial roles in the dynamics of optoelectronic feedback systems, as is the case for optical feedback. Two-dimensional maps of the dynamics in the phase space of the delay time and the feedback strength are calculated from the numerical simulation. Figure 3.16 shows the maps of the routes to chaos in the 2D space for the normalized delay time of $\hat{\tau}$ and the feedback strength ξ at a fixed bias injection current (Lin and Liu, 2003). The positive value of ξ is the case for positive feedback (Figure 3.16a) and the negative value of ξ is the case for negative feedback (Figure 3.16b). With the increases of the normalized delay time $\hat{\tau}$ or the feedback strength ξ, the laser output shows very complicated dynamics. The laser evolves into chaotic states through a quasiperiodicity route following regular pulsing (RP), two-frequency quasiperiodic pulsing (Q2), three-frequency quasiperiodic pulsing (Q3), and finally chaotic pulsing states (C). Quasiperiodic states are typically observed for strong feedback strength in the positive feedback system. On the other hand, such states are limited within small feedback strength in the negative feedback system. In the mappings, RP, Q2, and Q3 states spread over large areas in the positive feedback system, while chaotic states have large areas in the negative feedback system. Therefore, chaotic pulsing states can be easily obtained in the negative feedback system at large feedback strengths and long delay times. Another important difference between these two systems is the regions of the frequency-locked pulsing state (FL). In the positive optoelectronic feedback system in Figure 3.16a, the states that separate the chaotic regions are the RP states. However, in the negative optoelectronic feedback system in Figure 3.16b, the states that separate the chaotic regions are the FL states instead. The FL states are clearly observed only in the negative-feedback system. Thus, the dynamics of optoelectronic feedback systems strongly depends on the positive or negative feedback even for the same delay time and the same strength.

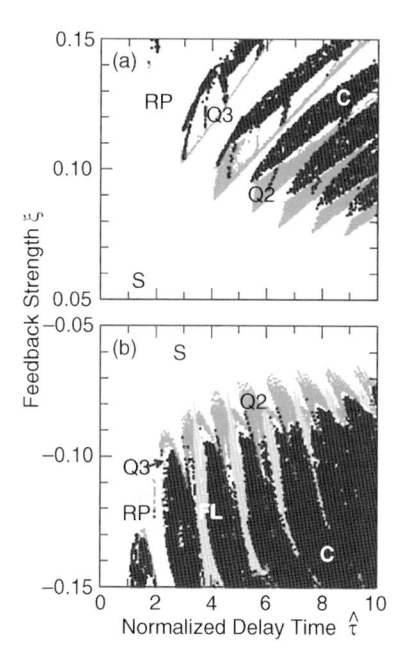

Figure 3.16 Numerical results of two-dimensional (2D) maps of dynamic states of (a) positive optoelectronic feedback and (b) negative optoelectronic feedback systems at $J = 1.33J_{th}$. S: steady states, RP: regular pulsing, Q2: two-frequency quasiperiodic pulsing, Q3: three-frequency quasiperiodic pulsing, PL: frequency-locked pulsing, C: chaotic pulsing. (F. Y. Lin and J. M. Liu, (2003), © 2003 IEEE.)

3.2.4
Semiconductor Laser with Optical Injection and Coupling

3.2.4.1 Temporal Dynamics and Bifurcation Diagram

Various nonlinear dynamics can be observed in a semiconductor laser with optical injection from another laser. Figure 3.17 (top) shows the schematic of two unidirectionally coupled semiconductor lasers with optical injection. This scheme is often used to lock the optical-carrier frequency and stabilize the oscillation of a response laser, known as "injection locking". Injection-locked semiconductor lasers are very

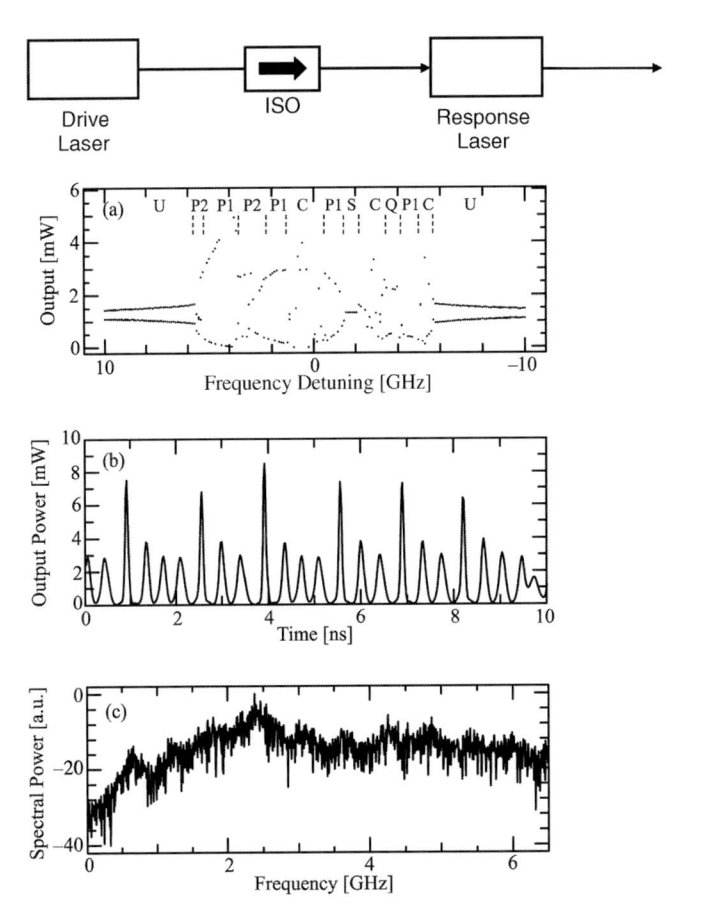

Figure 3.17 (Top) Schematic of a semiconductor laser with optical injection from another semiconductor laser. ISO, optical isolator for unidirectional injection. (Middle and bottom) (a) Bifurcation diagram for the frequency detuning at a bias injection current of $J = 1.3J_{th}$. (b) Example of the time series of chaotic states at the frequency detuning of $\Delta f = 2.56$ GHz. (c) The RF spectrum corresponding to (b). (a) S: stable state, U: unlocking state, P1: period-1 oscillation, P2: period-2 oscillation, Q: quasiperiodic oscillation, C: chaotic oscillation. (J. Ohtsubo, (2008), © 2008 Springer.)

useful for stabilizing the laser, however, they sometimes shows a rich variety of dynamics outside the injection-locking range. Light from a laser (referred to as the drive laser in Figure 3.17 (top)) is injected into the active layer of the other laser (response laser). The optical injection technique was originally developed for the stabilization of the injected laser (Siegman, 1986), so that it may be surprising that the laser is destabilized by the optical injection. Optical injection to a laser is a way to add an extra degree of freedom from the viewpoint of nonlinear dynamics.

Figure 3.17a shows an example of bifurcation diagram of the light-injected response laser for the change in the frequency detuning between the drive and response lasers at a fixed optical injection strength (Ohtsubo, 2008). In Figure 3.17a, the number of sampling points is not large enough to show the states and each bifurcation may not be clear. However, stable and unlocking oscillations, and various unstable oscillation states can be observed for the change in the frequency detuning. Figures 3.17b and c show the time series and the RF spectrum at the frequency detuning of $\Delta f = 2.56$ GHz. A chaotic temporal waveform and a broad RF spectrum can be observed. In fact, periodic and unstable oscillations are observed adjacent to the stable injection-locking range. Unlocking oscillations are distributed for large values of the large frequency detuning.

The property of injection locking is very important to consider the nonlinear dynamics in a semiconductor laser with optical injection. Figure 3.18 shows the areas of optical injection locking (i.e., matching of the optical-carrier frequency due to the frequency pulling effect) in the phase space of the optical carrier-frequency detuning and the injection strength ratio (Ohtsubo, 2008). The solid curves show the boundaries between optical injection locking and unlocking regions. In the unlocking region, various nonlinear dynamics can be expected such as chaotic oscillations and four-wave mixing when the detuning is not so far from zero. Indeed, various

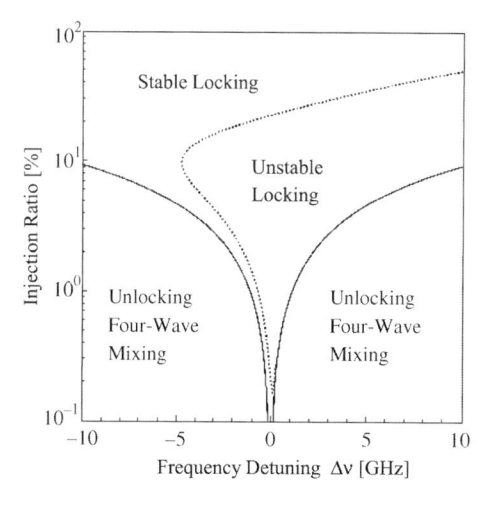

Figure 3.18 Locking and unlocking regions in phase space of optical carrier-frequency detuning and injection strength ratio for an optically injected semiconductor laser. (J. Ohtsubo, (2008), © 2008 Springer.)

dynamics can be observed when the frequency detuning and the injection strength ratio are small in these regions. Within the region of the optical injection locking, there are stable and unstable locking areas. The boundary of the unstable and stable injection locking areas is denoted by a dotted curve. In the unstable injection locking area, chaotic bifurcations can be observed for certain parameter ranges. The asymmetric feature of the stable injection-locking range originates from the fact that the α parameter (the linewidth-enhancement factor) has a nonzero value in semiconductor lasers (Ohtsubo, 2008). Larger values of the α parameter result in a more asymmetric stable locking area in Figure 3.18.

To investigate various nonlinear dynamics, a 2D map has been created experimentally and numerically. Figure 3.19 (left) shows the experimental result for the map in a semiconductor laser with side-mode excitation by optical injection (Simpson et al., 1997; Hwang and Liu, 2000). Note that the vertical and horizontal axes of Figure 3.19 (left) are exchanged from those shown in Figure 3.18. The laser used is a conventional Fabry–Perot-type edge-emitting semiconductor laser with a quantum well structure. For a small injection, the optical injection acts as a perturbation generating weak sidebands at the offset frequency, regenerative

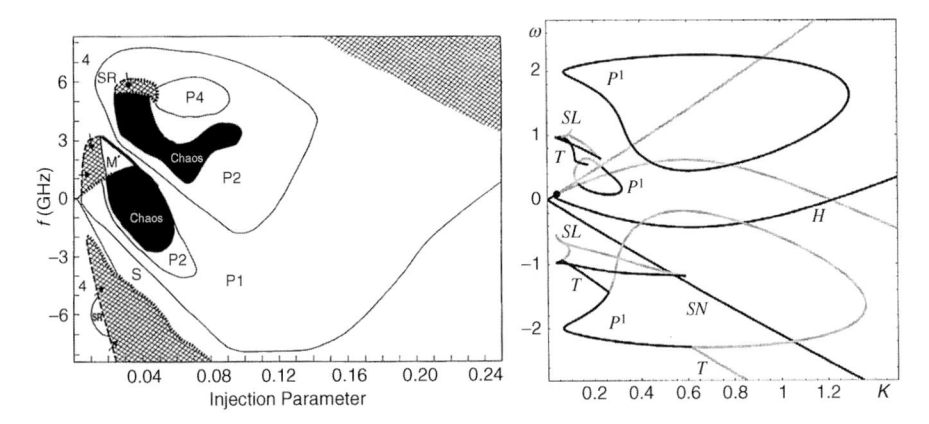

Figure 3.19 (Left) 2D map measured experimentally from optical spectra in a Fabry–Perot semiconductor laser with optical injection. The injection strength ratio and the optical carrier-frequency detuning f are changed. Note that the vertical and horizontal axes are exchanged from those shown in Figure 3.18. The bias injection current is $J = 1.67 J_{th}$. The symbols in the figure are as follows. 4, four-wave mixing sidebands; S, stable injection locking; P1, limit-cycle oscillation; P2, period-doubling; P4, period-quadrupling; chaos, deterministic chaos; M, multiwave mixing; SR, subharmonic resonance; hatched regions, principal output on another longitudinal mode. (S. K. Hwang and J. M. Liu, (2000), © 2000 Elsevier.) (Right) 2D map for the bifurcations showing coexistence states in a phase space of normalized frequency detuning and injection ratio. The vertical axis is the normalized frequency, and the horizontal axis is the normalized injection ratio. P1: period-doubling bifurcations, SL: saddle nodes of limit cycles, T: torus, H: Hopf bifurcation, and SN: saddle-node bifurcation. (S. Wieczorek, B. Krauskopf, and D. Lenstra, (2000), © 2000 Elsevier.)

amplification, and equally and oppositely shifted four-wave mixing. On increasing both the frequency detuning and the injection strength ratio, various instabilities appear in the laser output power. The period-doubling route and chaotic regions are located in the unstable locking region. There is an abrupt mode hop near the locking–unlocking boundary at negative detuning that has a small hysteresis. Analytical studies of the locking–unlocking boundary at negative detuning have shown that there is a region of bistability associated with the locking–unlocking transition. The bistability results from competing attractors representing locked and unlocked solutions for the coupled equations.

Figure 3.19 (right) shows the numerical result of the dynamical map generated from bifurcation analysis in semiconductor lasers subjected to optical injection (Wieczorek et al., 2000). The plot is a similar one in the phase space, as shown in Figure 3.19 (left). The region inside the straight line from zero detuning to negative detuning is the stable injection locking area in Figure 3.19 (right). In the stable region, there exist areas for the saddle-node bifurcation (SN) and the Hopf bifurcation (H). When the black part of SN is crossed, one of the bifurcating stationary points is an attractor. On the other hand, along the gray part of the curve SN, a repellor and a saddle point bifurcate. Along the black part of H, an attracting periodic orbit is born from the attracting stationary point and this corresponds physically to the undamping of the relaxation oscillation. The two saddle nodes of the limit cycle bifurcation curves starting with a cusp at $\omega \approx \pm 1$ represent a resonance between the relaxation oscillation frequency of the laser and the detuning of the injected light from the free-running laser frequency.

3.2.4.2 Bandwidth Enhancement of Chaos by Optical Injection

Optical injection can be used to enhance the bandwidth of chaotic laser output, which can be applied for the applications of optical chaos communication and random number generation (Someya et al., 2009; Hirano et al., 2010) (see Section 10.3.5). The bandwidth of chaos generated in a semiconductor laser is limited by the relaxation oscillation frequency of several GHz. The bandwidth of chaos can be enhanced in the order of tens of GHz by strong optical injection from another semiconductor laser.

Two semiconductor lasers are unidirectionally coupled to each other and one of the two lasers has an external mirror, either the drive or response laser. For both configurations, the chaos induced by the optical feedback can be enhanced due to the optical injection. For bandwidth enhancement of chaos in semiconductor lasers, it is important to set the optical wavelength (i.e., carrier frequency) detuning properly, that is, the wavelength of the drive laser λ_d should be shorter than that of the response laser λ_r ($\lambda_d < \lambda_r$) at the vicinity of (and out of) the injection-locking range so that the two optical wavelengths can be unlocked. Nonlinear frequency mixing between the optical detuning frequency (beat frequency of the two wavelengths) and the relaxation oscillation frequency results in the bandwidth enhancement of chaos in the RF spectrum region.

Figure 3.20 shows an example of the RF spectra of the chaotic laser output for the bandwidth enhancement of chaos (Uchida et al., 2003b). The gray curve indicates the

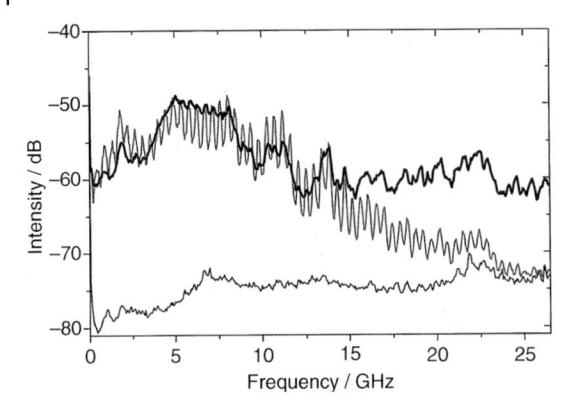

Figure 3.20 Experimentally obtained RF spectra of the chaotic outputs in the drive and response lasers when the chaotic output in the drive laser is injected into the cavity of the response laser. The gray curve indicates the chaotic spectrum of the drive laser, and the black thick curve indicates that of the injected response laser. The black thin curve corresponds to the noise-floor level. (A. Uchida, T. Heil, Y. Liu, P. Davis and T. Aida, (2003), © 2003 IEEE.)

chaotic spectrum of the drive laser, and the black thick curve indicates that of the injected response laser. The thin black curve corresponds to the noise-floor level. The RF spectrum of the drive laser is a typical spectrum for the chaos induced by coherent optical feedback. The spectrum has many peaks whose separation corresponds to the external cavity frequency. The largest peaks of the spectrum appear around 5 GHz, which is the relaxation oscillation frequency of the laser. The peak heights decrease above 5 GHz and are close to the noise-floor level around 20 GHz. On the other hand, the output of the injected response laser has a spectrum that is much broader, extending over 20 GHz with more than 15 dB above the noise-floor level at 20 GHz. Moreover, the spectrum is flat over large sections of the spectral range. The spectrum of the response laser still has some features similar to the drive laser, such as harmonics of the relaxation oscillation frequencies. However, these features are much less prominent compared to the drive laser.

The mechanism of the bandwidth enhancement of chaos can be analytically interpreted by using the coupled rate equations (Ohtsubo, 2008). There have been many reports on the bandwidth enhancement of chaos by using optical injection techniques (Simpson et al., 1995; Takiguchi et al., 2003; Uchida et al., 2003; Wang et al., 2008a; Someya et al., 2009; Hirano et al., 2010).

3.2.5
Semiconductor Laser with Injection Current Modulation

The output from a semiconductor laser follows an injection-current modulation as far as the modulation is small. By contrast, for strong injection-current modulation around the relaxation oscillation frequency, the laser output intensity clearly exhibits a number of nonlinear dynamics; harmonic distortion, pulsation different from the

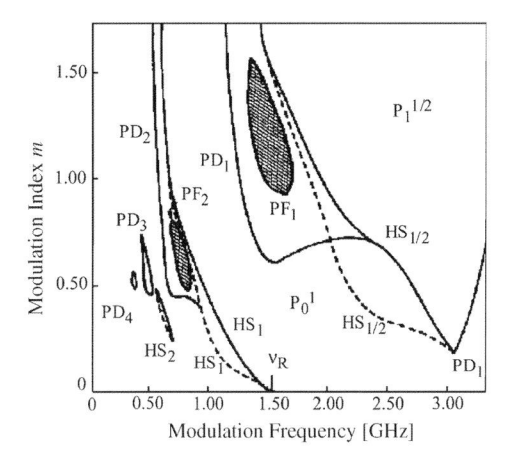

Figure 3.21 2D map of phase diagram for the semiconductor laser with injection current modulation at $J = 1.5J_{th}$ as functions of modulation frequency and index (amplitude). The relaxation oscillation frequency of the solitary laser is $\nu_R = 1.512$ GHz. Curve HS_n is the boundary of the hysteresis jump of the nth spiking state; the section with dashed line denotes the downward jump. Curves PD_n and PF_n are the boundaries of period-2 and period-4 of the nth spiking state. (Y. H. Kao and H. T. Lin, (1993), © 1993 IEEE.)

modulation frequency, bistability, and period-doubling and quasiperiodicity routes to chaos. The bifurcation to chaos in semiconductor lasers with injection-current modulation has also been intensively studied not only for theoretical interest, but also for practical purposes, especially in the area of analog modulation in optical-fiber communications.

The stability and instability for a semiconductor laser with injection-current modulation has been investigated in the space of the modulation frequency ν and index (amplitude) m. The 2D maps for the model of the semiconductor laser with injection current modulation is numerically calculated as functions of modulation frequency and amplitude, as shown in Figure 3.21 (Kao and Lin, 1993). The relaxation oscillation frequency of the solitary laser is $\nu_R = 1.512$ GHz. Curves PD_n and PF_n are the boundaries of period-2 and period-4 of the nth spiking state. The laser oscillation becomes spiky when the modulation current is increased. The laser output shows multiple spikes for $\nu < \nu_R$ and submultiple spikes for $\nu > \nu_R$. The period-doubling route to chaos is observed in this 2D map, and these results are confirmed by experiment (Bennett et al., 1997).

3.2.6
Vertical-Cavity Surface-Emitting Laser (VCSEL)

3.2.6.1 Basic Characteristics
Vertical-cavity surface-emitting lasers (VCSELs) have become essential devices for optical communications and data communication *via* plastic fibers in local-area networks. These semiconductor lasers emit light in the direction perpendicular to the

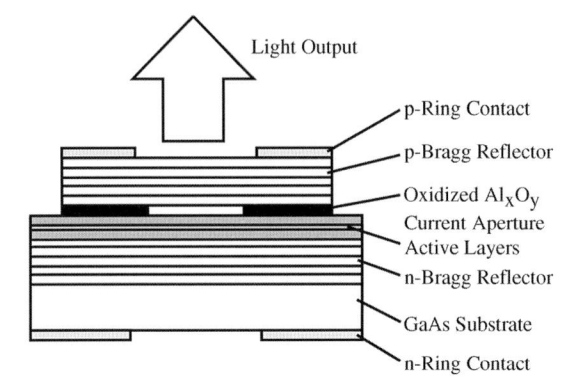

Figure 3.22 Schematic of distributed Bragg reflector (DBR) vertical-cavity surface-emitting laser (VCSEL) structure. (J. Ohtsubo, (2008), © 2008 Springer.)

surface of the active region and therefore, have some advantages compared to edge-emitting (normal) semiconductor lasers: low threshold current, compact, high efficiency, large modulation bandwidth, and wafer-scale integration capability for large array configuration. Because of their short cavity length (a few micrometers), VCSELs emit in a single-longitudinal mode. VCSELs are very sensitive to optical injection or optical feedback due to their short cavity length, in spite of the high reflectivity of their facets, since the photon lifetimes more or less coincide in VCSELs and edge-emitting semiconductor lasers. Typical VCSELs have two polarization modes and many transverse modes, which makes their dynamics very complicated. However, the basic dynamics of VCSELs with optical feedback is very similar to that of edge-emitting semiconductor lasers, including low-frequency fluctuations and coherence collapse with nanosecond-scale chaotic oscillations.

Figure 3.22 shows a schematic of the distributed Bragg reflector (DBR) VCSEL (Ohtsubo, 2008). The thickness of the active layer is approximately equal to the wavelength of light. The top view of the laser looks like a disk and its diameter is several to tens of μm. For special use, a disk diameter over 100 μm has been fabricated. In these devices, the reflectivity of the bottom surface is almost 100% and the top reflectivity of the DBR structure is around 99%. The laser light comes out from the top. Although the internal reflectivity is very high compared with edge-emitting semiconductor lasers (~10%), VCSELs are sensitive to optical feedback and optical injection. The photon number in the active volume is much less than that of edge-emitting lasers, and a few external photons would cause instabilities in the laser oscillations.

The polarization dynamics of VCSELs have been intensively investigated for many years (Sciamanna and Panajotov, 2005, 2006). One of the most common dynamics in VCSELs is spontaneous polarization switching. In a semiconductor medium, there exists the difference in the refractive indices between the components for the principal axis and the orthogonal axis to it because of the distortion and birefringence of the medium. The difference between the indices is very small and it is 10^{-3}–10^{-4}. For ordinary edge-emitting semiconductor lasers, the difference can be ignored due to a large asymmetric configuration for the TE and TM modes in the active layer and the solitary laser operates at only TE mode. However, the difference plays a crucial role for the operations of VCSEL, since it has a circular disk structure of the light-emitting facet. Then, there is an ambiguity for the polarization direction of the laser oscillation. A VCSEL usually oscillates at a polarization mode along the optical axis of the medium (this polarization mode is referred to as the x-polarization mode) when the laser is biased at a low injection current. However, the polarization mode may switch from this mode to the orthogonal one with the increase of the bias injection current. This switching is mainly induced by the distortion or the birefringence of the laser medium. This main oscillation mode after the polarization switching is sometimes referred to as the y-polarization mode, in accordance with the crystal axis of semiconductor laser medium.

Figure 3.23 (left) shows the experimental result of the L–I characteristic of a VCSEL (Ohtsubo, 2008). At a low bias injection current, the fundamental transverse mode (higher frequency mode) starts to lase, then the orthogonal mode (lower-frequency mode) grows after the polarization switching point. Thus, the main oscillation mode switches from x- to y-polarization mode well above the switching point. Usually, the polarization switching has a hysteresis for the increase or the decrease in the bias injection current. In this example, clear switching of the polarization modes from the x- to y-mode is visible at the bias injection current of 9.5 mA. The appearance of clear switching of the polarization modes strongly depends on the characteristics of the laser medium and the device structures. Some VCSELs do not show clear polarization switching for the bias injection current.

The polarization switching can be well reproduced by the numerical simulations, taking into account the birefringence of laser medium (Giudici et al., 1999; Danckaert et al., 2002). At a low bias injection current, the carrier density has a maximum value at the center of the disk in the active area and the carrier density smoothly decreases toward the edge of the disk. However, for a large bias injection current, hole burning of the carriers occurs at the center of the disk. The carrier density takes the maximum

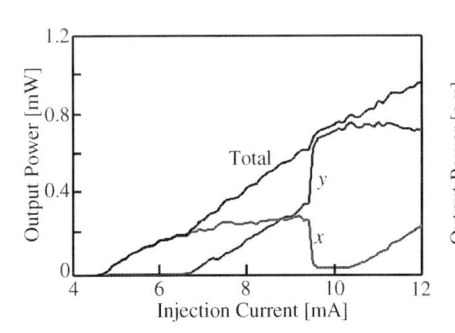

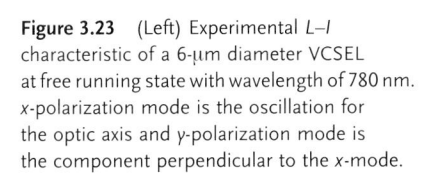

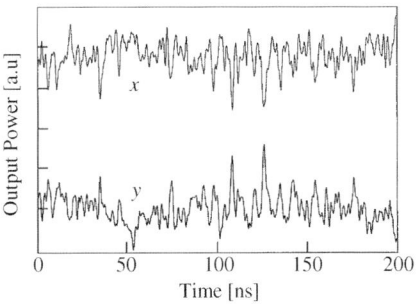

Figure 3.23 (Left) Experimental L–I characteristic of a 6-μm diameter VCSEL at free running state with wavelength of 780 nm. x-polarization mode is the oscillation for the optic axis and y-polarization mode is the component perpendicular to the x-mode. The total power is the addition of the x- and y-modes. (Right) Temporal waveforms of antiphase oscillations of x- and y-modes in VCSEL at solitary oscillation at $J = 1.23 J_{th}$. (J. Ohtsubo, (2008), © 2008 Springer.)

value a little away from the center of the disk. This induces the excitation of the orthogonal mode and the suppression of the original mode, since the hole burning due to the birefringence causes the transfer of the optical energy from the x- to y-mode. The laser oscillation is then switched from the x- to y-mode. A numerical model for chaotic dynamics in a VCSEL is described in Appendix 3.A.3.

Even in solitary oscillations, VCSELs show dynamic characteristics. One such type of dynamics is the antiphase irregular oscillation of the optical power between the two polarization modes. Figure 3.23 (right) shows an experimental example of antiphase oscillations of the x- and y-polarization modes in a VCSEL (Ohtsubo, 2008). Unstable pulsations and bistability are sometimes observed at the switching point of the two polarization modes (Tang et al., 1997). However, at certain bias injection currents different from the switching point, VCSELs show fast unstable oscillations and the two polarization modes oscillate in an antiphase manner in time (*i.e.*, antiphase dynamics). When the output power of the x-mode goes down in time, the output power of the y-mode goes up, and *vice versa*, as shown in Figure 3.23 (right).

3.2.6.2 Optical Feedback

As seen in the edge-emitting semiconductor lasers, optical feedback or optical injection can also induce chaotic temporal dynamics in VCSELs. The switching dynamics from one polarization mode to the other with orthogonal polarization direction has been observed experimentally in the LFF regime for a VCSEL with optical feedback (Tabaka et al., 2006). The existence of the two polarization modes in VCSELs can give rise to an additional polarization-mode competition dynamics in the presence of optical feedback. Figure 3.24 (left) shows the experimental results of the polarization-resolved dynamics for various injection currents. The external cavity length is 65 mm and the relaxation oscillation frequency is comparable to the external cavity frequency, indicating the short cavity regime (see Section 3.2.1.5). In Figure 3.24a (left) the amplitude of the peaks is still small and the shape of the slow envelope oscillations (referred to as pulse package dynamics) is not very regular. However, the envelope of the packages can be clearly identified, which indicates that the pulse packages in the two polarization modes are almost periodic with a characteristic frequency. The pulse package dynamics in the two polarization modes can be much better recognized in Figure 3.24b (left). The polarization-resolved pulse package dynamics is not as regular as for the total intensity. The reason for this is that the polarization-mode competition, underlying the pulse package dynamics, reduces the regularity of the pulse package dynamics in each polarization mode. This mechanism becomes more relevant at a higher injection current, approaching the polarization switching point. A gradual loss of the regularity in the pulse package dynamics as the bias injection current is increased from Figure 3.24b (left) to c (left). In the time series of Figure 3.24a (left) and b (left), the pulse package dynamics temporarily take place in one of the polarization modes only in some cases and the second mode is almost turned off. In other cases of Figure 3.24c (left) and d (left), the pulse package dynamics take place in the two polarization modes simultaneously. Similar interplay of the feedback induced complex dynamics and polarization-mode competition in VCSELs has been found numerically by for LFF dynamics in the long

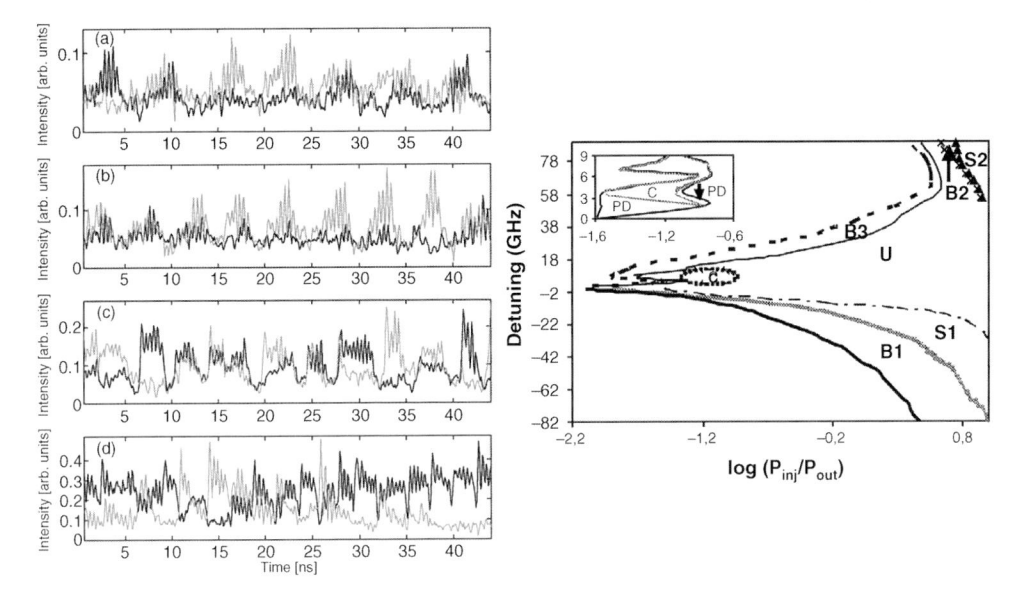

Figure 3.24 (Left) Polarization-resolved temporal dynamics of a VCSEL with optical feedback for various injection currents. (a) $I = 3.2$ mA, (b) $I = 3.4$ mA, (c) $I = 3.8$ mA, and (d) $I = 5.0$ mA. Gray plot corresponds to x-polarization mode and black plot corresponds to y-polarization mode. (A. Tabaka, M. Peil, M. Sciamanna, I. Fischer, W. Elsäßer, H. Thienpont, I. Veretennicoff, and K. Panajotov, (2006), © 2006 APS.) (Right) 2D map of the dynamics of a VCSEL subject to orthogonally-polarized optical injection in the phase space of the optical frequency detuning and the optical injection power. Thin solid line and gray line are polarization-switching boundaries for the increase of the bias injection current. Dashed line and thick solid line are polarization switching boundaries for the decrease of the bias injection current. B1, B2, B3: bistable regions, S1, S2: stable locking regions, U: unlocking region, C: chaotic region. The inset shows period doubling (PD) dynamics around the region of instabilities C. (J. B. Altés, I. Gatare, K. Panajotov, H. Thienpont, and M. Sciamanna, (2006), © 2006 IEEE.)

external cavity regime (Sciamanna *et al.*, 2003) and experimentally confirmed (Naumenko *et al.*, 2003).

3.2.6.3 Optical Injection

Optical injection also induces chaotic temporal dynamics in VCSELs. Nonlinear polarization dynamics in a VCSEL with orthogonally-polarized optical injection has been reported experimentally (Altés *et al.*, 2006; Gatare *et al.*, 2006). The 2D map of the boundaries of different dynamical regimes is drawn in the two-dimensional phase space of the optical frequency detuning and the injection power, as shown in Figure 3.24 (right) (Altés *et al.*, 2006; Gatare *et al.*, 2006). The laser is oscillated at the y-polarization mode above the polarization switching point. The VCSEL is externally injected by the linear polarization light with x-mode, and the y-polarization mode dynamics of the VCSEL is investigated. Two-polarization bistable regions (B1 and B3) are observed in a regime of fundamental mode emission, which correspond to two different ways of polarization switching. For small positive detunings ranging from about 0–10 GHz, complicated dynamics like wave-mixing, subharmonic resonance, sustained limit-cycle oscillation, period-doubling (PD) and chaotic regimes (C) are observed as shown in the inset in Figure 3.24 (right).

3.3
Chaos in Electro-Optic Systems

In Section 3.2, the laser devices are considered in which chaos is induced by the interplay between the lasing electric field and the carrier density (or population inversion) of the laser. Another method of inducing chaotic behavior by feedback exists. Specifically, chaotic behavior can be induced when the light emitted by a laser goes through an optoelectronic feedback loop containing a nonlinear optical device (Larger et al., 2004; Larger and Dudley, 2010). A distinctive feature of these systems is the retardation time of the driving signal in the feedback loop, which is much longer than the time response of the systems. The chaotic dynamics is determined by the nonlinear optical device in the loop, not by the laser itself. The laser is thus treated as a linear component, and no dynamics of the laser device such as the relaxation oscillation is considered in Section 3.3.

The dynamics of this type of setup can be described by a particular class of delay-differential equations where the feedback is modeled by a simple nonlinear function, whose transfer characteristic has at least a maximum or a minimum on the variable range. This class of systems was shown to exhibit oscillating states with higher harmonics appearing successively in the course of transition toward developed chaos (Ikeda et al., 1982; Larger et al., 1998a; Larger et al., 1998b).

3.3.1
Ikeda-Type Nonlinear Delay Dynamics

Nonlinear delay differential dynamics have been investigated since these dynamical systems exhibit complex chaotic behavior with high attractor dimension, although their mathematical description can be as simple as a scalar first-order differential equation for the variable $y(t)$ (Larger et al., 2004),

$$y(t) + \tau \cdot \frac{dy(t)}{dt} = \beta \cdot f[y(t-\tau_R)] \qquad (3.3)$$

The left-hand side is typical of a stable linear first-order dynamics, with a characteristic response time τ; its role is only to limit the fastest oscillations time scale. The right-hand side contains a nonlinear function $f[\cdot]$ applied to the delayed dynamical variable $y(t-\tau_R)$, as shown in Figure 3.25. The nonlinear function is practically bounded for physical reasons. The delay forces the natural dynamic phase space to be infinite dimensional: instead of a single initial condition $y(t_0)$ as usually required for a first-order differential equation to determine a solution uniquely, an infinite number of values is needed to define the necessary functional $y(t)$ over the time interval $[t_0-\tau_R; t_0]$. The importance of the role of the nonlinear transformation

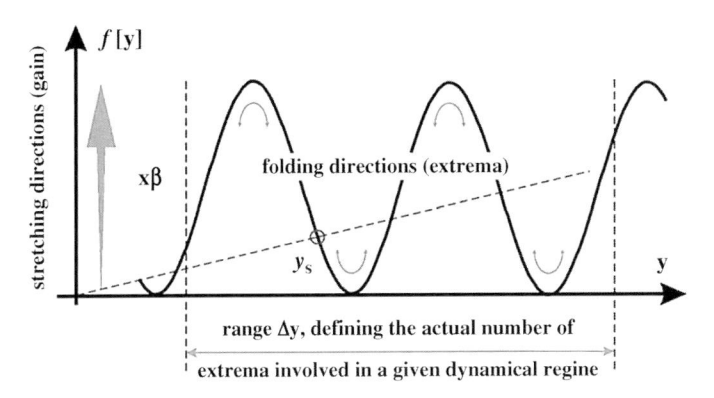

Figure 3.25 Important properties of the nonlinear function acting on the delayed variable in Eq. (3.3). (L. Larger, J.-P. Goedgebuer, and V. Udaltsov, (2004), © 2004 Elsevier.)

in the high complexity chaotic behavior is determined by two main factors: the amplitude of the magnification factor β for the nonlinear delayed feedback terms and the number of extrema of the nonlinear function $f[\cdot]$, as shown in Figure 3.25. In addition to these two factors of the nonlinear transformation (magnified by a factor β, bounded, and at least with one extremum), the delay time τ_R (usually much greater than the response time τ) is also the key element in the generation of a high-dimensional chaotic process. A major advantage of this dynamical system is its easy experimental implementation to electro-optic systems. An example of a numerical model for chaotic dynamics in an electro-optic system is described in Appendix 3.A.4.

The Ikeda model, one of the pioneering optical chaotic systems described by the delay-differential equation of Eq. (3.3), is depicted in Figure 3.26a (top) (Ikeda, 1979; Larger et al., 2004). An input laser beam with constant optical intensity I_0 is an important parameter for the tuning of a given dynamical regime observed at the system output. The coherence of the laser light ensures the existence of interferences between the input light beam, and the one fed back by the cavity after one round-trip. A ring cavity comprises two partial reflecting mirrors, one for the input and one for the output. The length L of the cavity determines a round-trip time of the light beam, which defines the delay time $\tau_R = L/c$ (where c is the speed of light). Intensity or phase modulation observed at the cavity output is fed back to the cavity input with a delay time τ_R. In the ring cavity a two-level atomic cell is inserted as a nonlinear optical medium, in which light–matter interaction occurs. In a simplified model, only the Kerr effect is considered. At the atomic cell input, a two-wave interference occurs between the constant-intensity cavity input beam, and the intracavity feedback beam, whose phase is determined by the time-delayed intensity interference through the Kerr effect in the atomic cell. Under these conditions, the phase of the light beam propagating through the cell is changed proportionally to its intensity $I_{\text{in}}(t)$, that is, this phase change is expressed as $2\pi n_2 I_{\text{in}}(t) l/\lambda$, where l is the medium length, λ is the laser wavelength, and n_2 is the Kerr refractive index coefficient. Note that the dynamics of this light–matter interaction is extremely fast since it is determined

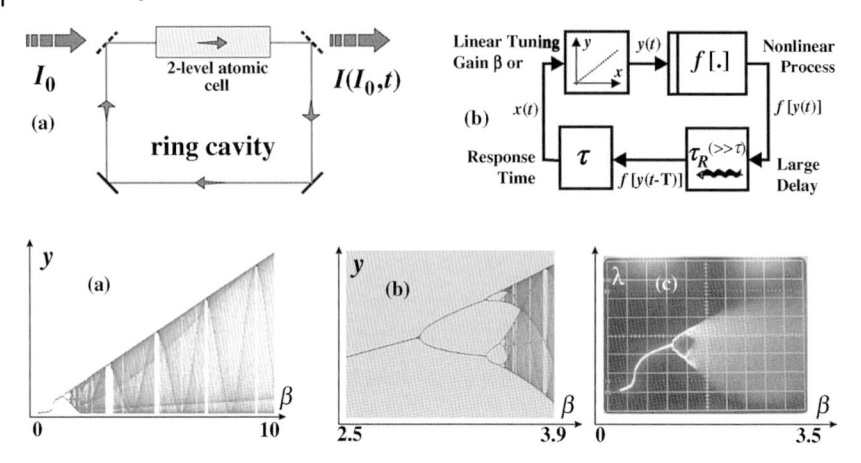

Figure 3.26 (Top) The Ikeda ring cavity: (a) experimental setup; (b) block diagram interpretation. (Bottom) Bifurcation diagrams for (a) the $\sin^2$-map; (b) the logistic map; and (c) experimental $\sin^2$-delay differential dynamics. (L. Larger, J.-P. Goedgebuer, and V. Udaltsov, (2004), © 2004 Elsevier.)

by the level lifetime τ of the atomic cell, thus leading to dynamical fluctuations much faster than the round-trip time τ_R. The dynamics of the cavity output intensity $I(t)$ can be described by the nonlinear delay differential equation in Eq. (3.3), in which the nonlinear function corresponds to the transformation law of the input phase into an output intensity (the intensity of a two-wave interference, typically a sinusoidal function as shown in Figure 3.25).

According to this description, the physical setup appears as an oscillator, with a feedback loop comprising a strong nonlinearity (β and $f(y(t))$), and a delay (τ_R). This delay is large compared to the characteristic response time (τ) of the limiting dynamics. A block diagram can be translated to generalize this oscillation principle, as depicted in Figure 3.26b (top) (Larger et al., 2004). The linear tuning is representative of the optical phase change rate with respect to the optical intensity through the Kerr effect. The nonlinear transformation $f(y(t))$ is physically generated by the interference after the optical feedback at the cavity input. The cavity length determines the delay τ_R, and the dynamics limitation is fixed by the atomic cell level lifetime τ.

The first and simple approach to the oscillator dynamics in the case of large delays ($\tau_R \gg \tau \approx 0$) usually involves the adiabatic approximation. This consists in neglecting the derivative term in Eq. (3.3). The continuous-time dynamics is then expressed as discrete-time dynamics, for which the time evolution is a sequence of discrete values of the dynamical variable y over the time interval τ_R. Labeling each τ_R-time interval with an integer n, the dynamics are reduced to a one-dimensional (1D) discrete mapping,

$$y_{n+1} = \beta \cdot f(y_n) \tag{3.4}$$

where f is the nonlinear function similar to the plot in Figure 3.25, for example, $f(y_n) = \sin^2(y_n - \phi_0)$. The oscillator feedback is then equivalent to an iteration process, returning the vertical axis value y_{n+1} onto the horizontal axis. This operation can be represented graphically with the first bisector straight line in Figure 3.25, which intersects the nonlinear function at the steady-state values (defined as the solutions of $y_s = \beta \cdot f(y_s)$). The stability of these steady states can be determined by a first-order analysis, leading to the following result: the steady state is stable if the absolute value of the slope $|f'(y_s)|$ is lower than 1, otherwise it is unstable (see Section 2.2.1). Increasing the feedback gain β (or the slope of the linear tuning gain in Figure 3.26b (top)) changes the number of the steady states, as well as the slope, at these positions. This is why β is usually considered as a bifurcation parameter of the system. For low values of β, a single steady state exists and is necessarily stable. When increasing β, the steady states loses their stability and periodic regimes are observed. For sufficiently large values of β, high complexity chaotic regimes are observed. These regimes are shown in Figure 3.26a (bottom) (Larger et al., 2004) as the bifurcation diagram with the change in β. Between the low and high β values, a period-doubling route to chaos is observed when increasing β, as shown in Figure 3.26a (bottom). The bifurcation diagram of the 1D Ikeda map of Eq. (3.4) shown in Figure 3.26a (bottom) is similar to that of the well-known logistic map, as shown in Figure 3.26b (bottom) for comparison. When comparing the two bifurcation diagrams for the Ikeda and logistic maps, it can be qualitatively noticed that the multiple extrema nonlinear function for the Ikeda map allows a broad range of chaos for the bifurcation parameter, and high complexity chaotic dynamics are obtained. This result confirms that the Ikeda model with its multiple extrema nonlinear function is a good candidate for complex chaos generation.

The actual dynamics complexity of the Ikeda model is even better when considering a nonzero response time τ. The dynamics is thus no longer a discrete map, it has to fluctuate continuously in time according to Eq. (3.3). An experimental bifurcation diagram of such a continuous-time delay dynamics is represented in Figure 3.26c (bottom). The qualitative profile of the bifurcation diagram is not dramatically changed compared to the discrete-time case (Figure 3.26a (bottom)), however, the dynamics complexity is strongly improved. The phase-space dimension of the dynamical system is indeed increased from 1 to infinity. It is found that there exist the numerous positive Lyapunov exponents, leading to a Kaplan–Yorke dimension as high as 470, thus indicating a high complexity for the chaotic regime (Larger et al., 2004).

3.3.2
Wavelength Chaos in Electro-Optic System

The experimental implementation of the Ikeda-type nonlinear delay dynamics of Eq. (3.3) has been demonstrated for modulating optical wavelength, intensity, or phase in different experimental configurations. The details are described in the following sections.

The generation of chaos in optical wavelength has been demonstrated experimentally (Larger et al., 1998a; Larger et al., 1998b). The advantages of chaos in wavelength are in the high accuracy and high reliability of chaos control. Generating chaos in wavelength relies on the wavelength agility of a laser diode with a feedback loop and a nonlinear element in wavelength.

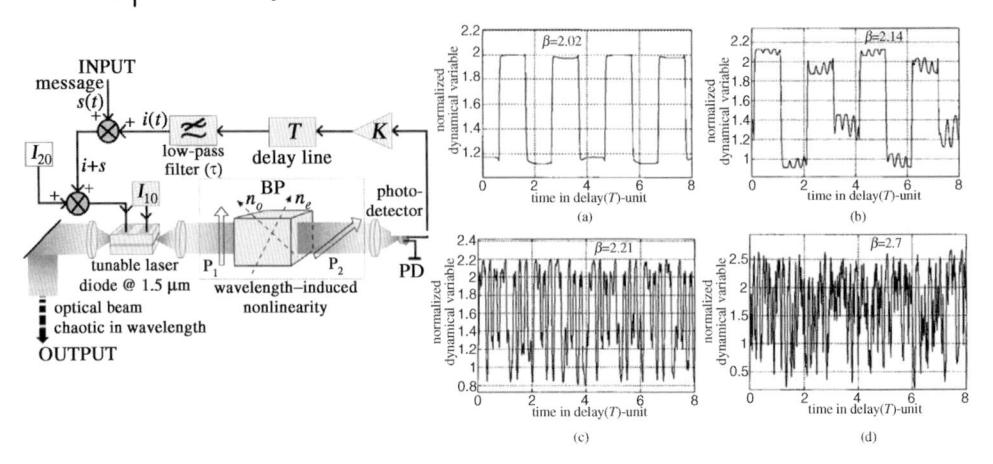

Figure 3.27 (Left) Chaos generator formed by a two-section wavelength-tunable DBR laser diode with a delayed nonlinear feedback loop. (L. Larger, J.-P. Goedgebuer, and F. Delorme, (1998a), © 1998 APS.) (Right) Temporal evolution simulated with the continuous model. (a) T_4 periodic regime. (b) T_4 chaotic regime. (c) Higher-harmonic synchronization. (d) Fully developed chaos. (L. Larger, J.-P. Goedgebuer, and J.-M. Merolla, (1998b), © 1998 IEEE.)

The generator of the chaotic wavelength beam is depicted in Figure 3.27 (left) (Larger et al., 1998a). It consists of an electrically wavelength-tunable DBR multi-electrode laser diode with a feedback loop formed by a delay line and an optical device whose peculiarity is to exhibit nonlinearity in wavelength. Under some conditions, it turns out that the wavelength emitted by the laser diode fluctuates chaotically around its center wavelength.

The double-electrode wavelength-tunable DBR semiconductor laser is used in order to adjust the laser wavelength within a few nanometers (nm) around the center wavelength of 1.55 μm. A 6-cm-long calcite slab placed between two crossed polarizers is used as a birefringent interferometer, whose output interference is scanned according to the laser wavelength. Note that any other spectral filtering (e.g., a more complex multiple-wave interference filter like a Fabry–Perot cavity) can be used to perform the nonlinear transformation, as long as the filter profile exhibits extrema within the wavelength tuning range of the laser. The 1.5-nm continuous range allows one to scan more than 12 extrema of a sinusoidal nonlinear function, as depicted in Figure 3.25. The resulting intensity is detected by a photodiode, from which the electrical signal is delayed by $\tau_R = 512\,\mu$s with an electronic delay line. After amplification and filtering with an electronic first-order low-pass filter of cut-off frequency $f_c = 18$ kHz, the resulting signal serves as the input current for the laser wavelength tuning.

Figure 3.27 (right) shows the time evolution of the wavelength chaos at different nonlinear gain β (Larger et al., 1998b). A square waveform is obtained with a period of $4\tau_R$ is obtained (referred to as T_4), as shown in Figure 3.27a. As β is increased, the periodic waveform with chaotic fluctuations is observed, as shown in Figure 3.27b (referred to as period-4 chaos). For $\beta > 2.166$, so-called higher harmonic synchronization regimes occur (Ikeda, 1982), which are characterized by periodic oscillations with a period of $n\tau_R$ and exhibit fine oscillations with an average period of τ (Figure 3.27c). For further increase of β, full chaos is obtained, as shown in Figure 3.27d. The full chaos is characterized by a colored noise-like spectrum with a cut-off frequency of $1/\tau$. The dimension of the chaos is related to $\beta\tau_R/\tau$. The corresponding bifurcation diagram measured in experiments is already shown in Figure 3.26c (bottom).

3.3.3
Intensity Chaos in Electro-Optic System

Chaos in laser intensity has been reported in an electro-optic system. One of the most straightforward ways to modulate electro-optically an optical interference is to choose a component widely used in ultrafast fiber telecommunication systems, the Mach–Zehnder electro-optic modulator (EOM). Such integrated optics components in lithium niobate (LiNbO$_3$) are commercially available for bit rates of tens of Gb/s. Those devices are usually operating in a weak nonlinear operation, since the applied voltage is typically intended to encode bits 0 and 1 through the switching between destructive and constructive interference conditions. The corresponding voltage-switching amplitude is called V_π; it can be practically as low as a few volts for integrated optics components. However, operating with a larger voltage swing enables scanning practically at least 2 to 3 extrema of the interference transfer function, thus performing a highly nonlinear transformation suitable for high-complexity dynamics in a time-delay system.

An intensity chaos generator can be constructed similarly to the wavelength chaos generator. The setup has been used as an electro-optic demonstrator for the Ikeda ring cavity instabilities (Celka, 1995). Figure 3.28 (top) shows the electro-optic system with EOM (Goedgebuer et al., 2002). The system is formed by an impedance-matched laser diode (LD) driven electrically by a feedback loop. The LD operates above its threshold, in the linear part of its power–current curve. The optical intensity of LD is modulated around a mean intensity by voltage of the feedback signal. The feedback loop consists of a photodetector (D$_1$), an amplifier, an EOM, a delay line (T), and another photodetector (D$_2$). The transmission curve in intensity of the EOM is the nonlinear function of $\cos^2 x$. For generality, it is assumed that the feedback circuitry features a bandwidth with a high and low cut-off frequencies, and behaves as a bandpass filter (BPF), as depicted in Figure 3.28 (top).

Figure 3.28 (bottom) shows the dynamical properties of the intensity chaos setup: (a) RF spectrum of the chaotic optical carrier; (b) optical spectrum for increasing CW laser power, and (c) experimental bifurcation diagram (Larger et al., 2004). Chaotic broadband spectrum over 6 GHz can be observed in Figure 3.28a. The optical spectra also show large bandwidth of the laser output in Figure 3.28b. A large chaotic region is observed in the bifurcation diagram of the intensity chaos as the CW laser power is increased, as shown in Figure 3.28c.

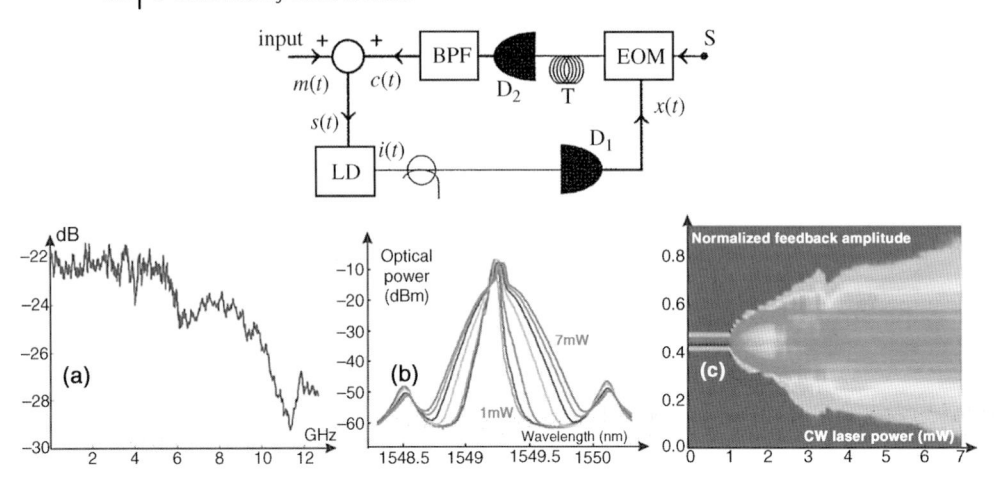

Figure 3.28 (Top) Chaotic generator formed by a laser diode with a time-delayed feedback loop containing an electro-optic modulator (EOM). (J.-P. Goedgebuer, P. Levy, L. Larger, C.-C. Chen, and W. T. Rhodes, (2002), © 2002 IEEE.) (Bottom) Dynamical properties of the intensity chaos setup: (a) RF spectrum of the chaotic optical carrier spread by the nonlinear function and filtered by a 10-GHz photodiode; (b) optical spectrum for increasing CW laser power, from 1 mW to 7 mW with 1 mW step; (c) experimental bifurcation diagram recorded with a 5-GHz oscilloscope. (L. Larger, J.-P. Goedgebuer, and V. Udaltsov, (2004), © 2004 Elsevier.)

The main feature of the intensity chaos generator, compared with the wavelength chaos generator, consists in the large bandpass nature of the dynamical process. Usually, in most of the ultrawide-band communication systems, the low frequencies are filtered out by the electronic feedback, thus yielding a bandpass dynamical behavior. The process involved in the nonlinear feedback is therefore fundamentally different, as well as the dynamical trajectories that can be observed on the bifurcation diagram in Figure 3.28c (to compare with Figure 3.26c (bottom)). The bandpass filtered systems exhibit greater Kaplan–Yorke dimension than the low-pass filtered systems (Larger et al., 2004).

Another implementation of intensity chaos in electro-optic systems has been reported with large time delay (Aida and Davis, 1994). The experimental setup is depicted in Figure 3.29 (left). The nonlinear resonator used for the experiment is an electro-optic (EO) hybrid ring resonator with very large effective delay through an optical fiber, in which a large number of nonlinear oscillation modes are excited. The light intensity of the LD is modulated by the EOM with a sinusoidal nonlinear function. The laser light is transmitted through the long optical fiber with delay time T_r and is detected by a photodiode. The converted electric signal is amplified and low-pass filtered with the cut-off frequency of $1/T_m$, and is fed back to the bias voltage of the EOM. Large effective delay T_r/T_m is needed to give delay-induced bifurcation to chaos with multistable modes as the input optical power μ is increased.

For low input optical power from the LD, the oscillation modes that can be excited in the resonator are the fundamental mode of period $T_1 \approx 2T_r$ and odd nth

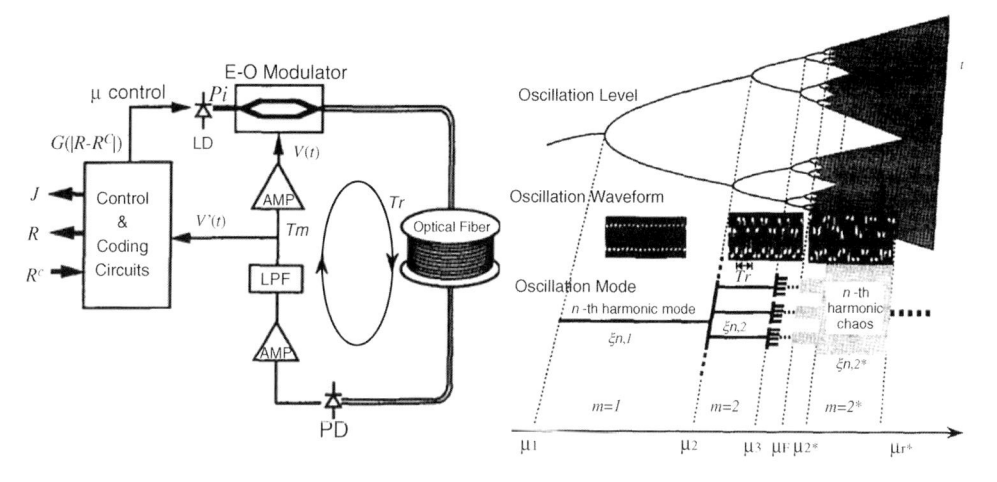

Figure 3.29 (Left) Experimental setup for electro-optic system with delayed feedback. LD, laser diode (wavelength of 1.3 μm); E-O modulator, LiNbO$_3$ waveguide intensity modulator (half-wave voltage of 6 V); Optical fiber, single-mode optical fiber (length of 1 km); PD, pin photodiode; LPF, low-pass filter (response time of 42 ns); AMP, video amplifier (bandwidth of 150 MHz). (Right) Schematic bifurcation diagrams of oscillation level and oscillation mode. Examples of 7th-harmonic oscillation waveforms are shown. (T. Aida and P. Davis, (1994), © 1994 IEEE.)

harmonic modes with period $T_1/n \approx 2T_r/n$. Each of these harmonic modes exhibits bifurcation leading to chaos with increase of input power. The modes of the nonlinear oscillation can be classified with harmonic number n and bifurcation order m as (n, m). Figure 3.29 (right) shows schematic bifurcation diagram of oscillation level and oscillation mode (Aida and Davis, 1994). Successive bifurcations reach the onset of chaos at a certain point μ_F, and after the onset of chaos at μ_F there are inverse bifurcations. In the inverse bifurcation the basins of the multiple chaotic modes merge together, resulting in fluctuations on peak and valley levels of the nth harmonic carrier and intermittent transitions among different (n, m) oscillation waveforms (Ikeda, 1982; Aida and Davis, 1994).

3.3.4
Phase Chaos in Electro-Optic System

Chaotic dynamics of optical phase can be generated in an electro-optic system, which is applicable for fast optical chaos communication system (Genin *et al.*, 2004; Lavrov *et al.*, 2009) (see Section 8.3.3). A scheme of the experimental setup for the electro-optic nonlinear delay phase oscillator is shown in Figure 3.30 (left) (Lavrov *et al.*, 2009). In this setup, a 1.55-μm DFB semiconductor laser is used for injection of continuous-wave (CW) light into the delayed feedback loop of the electro-optic oscillator. A variable attenuator (VA) is used for adjusting the injected intensity. The CW-injected light enters a broadband phase modulator (PM, bandwidth of

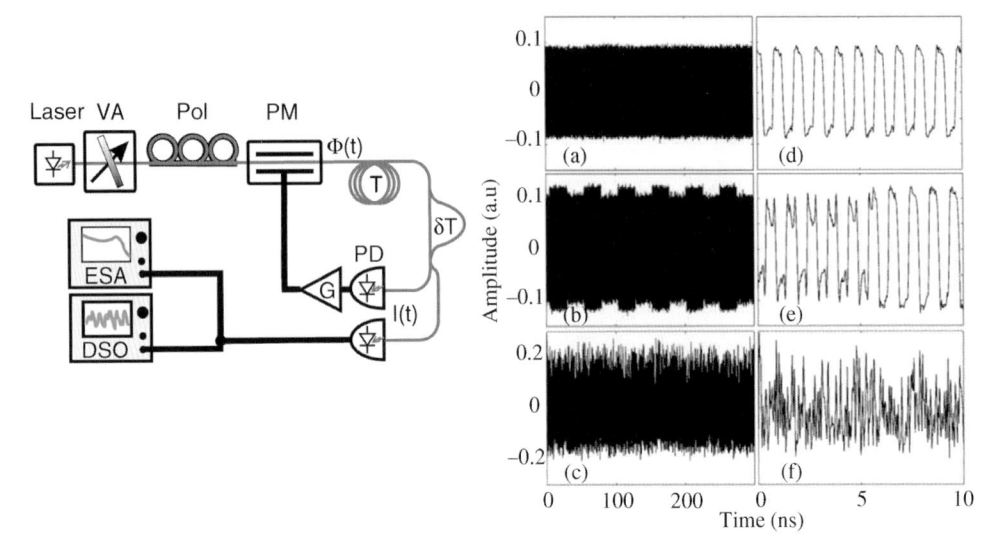

Figure 3.30 (Left) Experimental setup of the electro-optic nonlinear delay phase oscillator. (Right) Experimental time series for different values of the normalized feedback gain β detected at the output of the Mach–Zehnder DPSK-d. (a)–(c) are zoomed in (d)–(f). (a), (d) $\beta = 0.6$, (b), (e) $\beta = 1.3$, and (c), (f) $\beta = 5.1$. (R. Lavrov, M. Peil, M. Jacquot, L. Larger, V. Udaltsov, and J. Dudley, (2009), © 2009 APS.)

20 GHz). The optical phase of the injected CW light is modulated according to the voltage applied to the electrical RF input of the PM, while the intensity of the output light remains constant. This output light is delayed by propagation in a few meters of optical fiber until it enters a fiber-based passive imbalanced interferometer, which is a commercial differential phase-shift keying demodulator (DPSK-d) with a 2.5-GHz free spectral range (FSR, corresponding to a time imbalancing of 400 ps) performed by a fiber-based Mach–Zehnder interferometer (MZI). The interferometer is aimed at performing a nonlinear, dynamical nonlocal in time, phase-to-intensity conversion. The phase-to-intensity conversion is governed by a nonlinear transformation described by the function $\cos^2 x$. The obtained MZI output intensity signal is detected with a broadband amplified photodetector (PD) with a bandwidth of 30 kHz–13 GHz. The PD is the electrical component with the smallest bandwidth in the oscillator loop, therefore it effectively determines the overall bandwidth of the combined bandpass-filtering properties of the system. The converted electrical signal is linearly amplified with a broadband amplifier that drives the RF input of the PM. The optical phase of the light at the PM output is modulated according to its own delayed history. The electro-optic effect closes the delayed feedback loop of the nonlinear oscillator. The total delay time comprising the optical and electrical delays corresponds to 24.35 ns. Since the maximum output voltage of the RF driver is 13.0 V and the half-wave voltage of the PM corresponds to 4.0 V, the phase can be modulated by up to 3.25π rad. Up to three extrema of the nonlinear $\cos^2 x$-interference function can be swept in the DPSK-d. This strong nonlinear operation capability is particularly of interest when complex chaotic

waveforms are desired, typically for chaos communication applications where very low autocorrelation carriers are desired (see Section 8.3.3).

Figure 3.30 (right) depicts examples of measured temporal dynamics, both for long time scales (a few delay times, left figures) and shorter ones (a few DPSK imbalance times, right figures) (Lavrov et al., 2009). These time traces give rise to different dynamical states depending on the feedback gain. When the feedback gain of the nonlinear function β is small, a stable zero steady state is observed as expected. At $\beta = 0.5$, the steady state loses its stability and a limit cycle appears. The usual threshold for this Hopf bifurcation is typically 1 for the Ikeda dynamics. The typical oscillation is represented in Figure 3.30a and d (right) for $\beta = 0.6$; it corresponds to a rapid oscillation between a two-level state, with tilted decreasing plateaus (Figure 3.30d). When the feedback gain is slightly increased, a bifurcation is observed. A typical corresponding time trace is represented in Figure 3.30b and e (right), which is obtained for $\beta = 1.3$. The dynamics is now characterized by both the fast 1 GHz oscillation (Figure 3.30e) as well as a 2T-periodic envelope modulation (Figure 3.30b). The two successively appearing frequencies are related to two independent and very different physical time scales—the DPSK-d imbalancing time (fast oscillations in Figure 3.30e) and the total delay time (slow envelope seen in Figure 3.30b). When higher feedback gains are concerned, the phase chaos generator exhibits significantly complex dynamical behaviors. The time traces observed with the maximum feedback gain $\beta = 5.1$ are represented in Figure 3.30c and f (right). They show large-amplitude chaotic fluctuations corresponding to a broadband spectrum covering the full bandwidth of the electronic feedback up to ~ 13 GHz.

The amplitude probability density function (PDF) is used as a graphical representation of a given temporal dynamics. When finely scanning the bifurcation parameter β, a bifurcation diagram can be plotted to represent a quasicontinuous evolution of the dynamics through its PDF with respect to β. This is represented in Figure 3.31, where the PDF is practically encoded via a logarithmic scaling (white for highly probable amplitudes and dark for low-probability amplitudes) (Lavrov et al., 2009). The same kind of such a bifurcation diagram is drawn for both experimental (top) and numerical (bottom) time traces. It can be noted that both diagrams show a very good resemblance at least qualitatively but also quantitatively in terms of bifurcating values of β. Such diagrams allow a summarized picture to be given of the bifurcation scenario met by the optical phase dynamics, while increasing from the stable steady state to the fully developed chaotic regime. For high gain $\beta \geq 1.9$, the PDFs develop into widespread almost Gaussian-shaped functions. This feature is known to exist for chaotic dynamics of nonlinear delayed feedback systems modeled by standard single delay dynamics.

3.4
Chaos in Fiber Lasers

In fiber lasers the optical gain is provided by rare-earth elements (such as erbium (Er), neodymium (Nd) and ytterbium (Yb)) embedded in silica fiber. Under optical

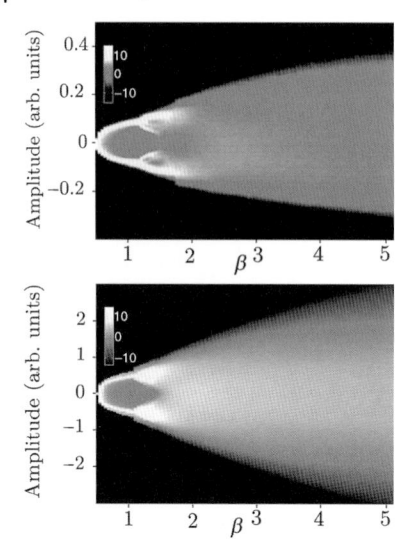

Figure 3.31 Bifurcation diagrams of the amplitude probability density functions (logarithmic scale) of the dynamics *versus* normalized feedback gain β. (Top) from experimental time traces; (Bottom) from numerical time series. (R. Lavrov, M. Peil, M. Jacquot, L. Larger, V. Udaltsov, and J. Dudley, (2009), © 2009 APS.)

pumping, those atoms provide light amplification at a characteristic wavelength; in the particular case of erbium that wavelength is $\sim 1.55\,\mu\text{m}$, which lies within the spectral region of minimal loss of silica fibers. For that reason, erbium-doped fiber amplifiers (EDFAs) have been widely used since the mid-1990s in fiber-optics communication systems (Agrawal, 1997).

In the presence of a feedback mechanism (either by using mirrors, or by closing the fiber on itself forming a fiber ring) laser emission can be obtained. The concurrence of the inherent nonlinear character of both the optical fiber and the light amplification process leads to a rich variety of dynamical instabilities and nonlinear behavior. Erbium-doped fiber lasers, for instance, are very efficient generators of ultrashort pulses and solitons, and exhibit different types of complex nonlinear behavior, including bursting and chaos.

A second defining characteristic of fiber lasers is that, due to the waveguiding properties of optical fibers, their cavities can be very long (of the order of kilometers), several orders of magnitude longer than other lasers. For that reason, the frequency separation between consecutive longitudinal modes is very small (of the order of MHz). Additionally, the amorphous character of the host medium leads to a very broad gain profile (of the order of tens of GHz). As a consequence, a large number of longitudinal cavity modes can experience gain and coexist inside the cavity, coupled through gain sharing. Hence, fiber lasers usually operate in a strongly multimode regime, and consequently their dynamics cannot be described in general by single-mode models, or by models containing a small number of coupled modes. For fiber lengths of tens of meters, the round-trip time taken by the light to travel once along the laser cavity is of the order of hundreds of nanoseconds. This time is much longer than the sampling time of standard oscilloscopes, and therefore intracavity dynamics is straightforward to observe in fiber lasers. As a result, the dependence of the generated radiation field on the propagation direction cannot be neglected, as is usually done in the mean-field approximations that lead to rate-equation descriptions of fiber laser dynamics. A numerical model for chaotic dynamics in a fiber laser is described in Appendix 3.A.5.

3.4.1
External Modulation

In fiber lasers, chaos can appear when one control parameter is modulated. Nonlinear phenomena and more complex dynamical behaviors can be observed by applying a sinusoidal modulation of the pump power near the relaxation oscillation frequencies.

The experimental setup is shown schematically in Figure 3.32 (left) (Lacot et al., 1994). The active medium is an erbium-doped fiber with a length between 1 and 10 m, doped with 40 to 2000 ppm of Er^{3+}. The core diameter is 6.4 μm and the fiber is single mode at the laser operating wavelength (1.538 μm). The pump laser is a

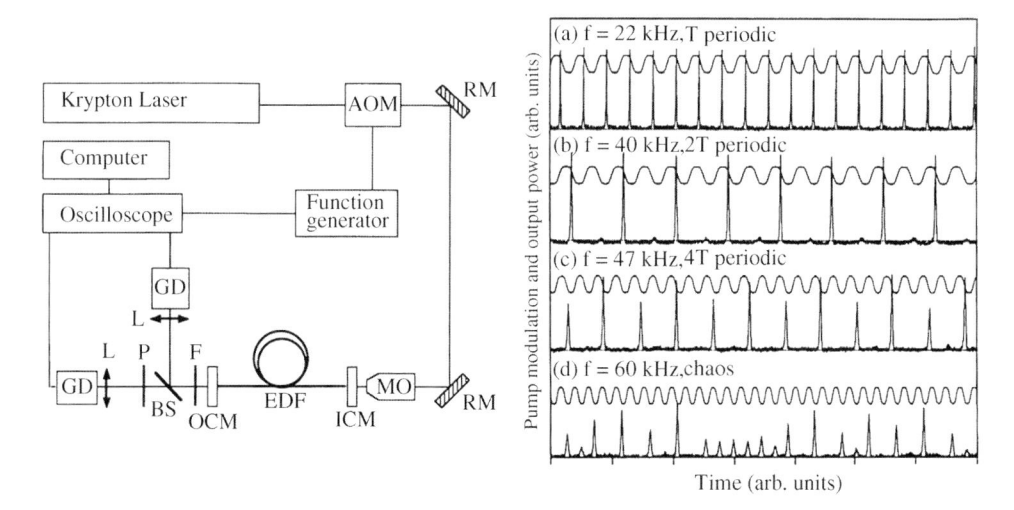

Figure 3.32 (Left) Experimental setup of an erbium-doped fiber laser. AOM: acousto-optic modulator; RM: reflecting mirror; MO: microscope objective; ICM: input coupling mirror ($R_1 > 99\%$ at 1.538 μm); EDF: erbium-doped fiber; OCM: output coupling mirror ($R_2 = 40$–90% at 1.538 μm); F: high-pass filter ($\lambda > 850$ nm); BS: beam splitter; P: polarizer; L: lens; GD: germanium detector. (Right) Experimental time evolution of the total intensity of the fiber laser in response to a sinusoidal modulation of the pump power (upper trace) for different values of the modulation frequency (a) $f = 22$ kHz, T-periodic response; (b) $f = 40$ kHz, 2T-periodic response; (c) $f = 47$ kHz, 4T-periodic response; (d) $f = 60$ kHz, chaotic response. The time scale is different from one plot to another. (E. Lacot, F. Stoeckel, and M. Chenevier, (1994), © 1994 APS.)

krypton ion laser, and the beam passes through an acousto-optic modulator (AOM) controlled by a function generator. The AOM is mainly used to modulate the intensity of the pump beam. The pump wavelength is fixed at 647 nm, and the erbium-doped fiber is pumped longitudinally through a microscope objective and the input dichroic mirror. The output mirror is butt-coupled to the fiber. After a high-pass filter blocking the residual krypton pump beam, a beam splitter and germanium detectors are used to record the total laser intensity. In typical operating conditions, the threshold pump power is of the order of 100 mW at the fiber input. The output power is of the order of a few mW.

Figure 3.32 (right) illustrates the evolution of such a nonlinear dynamic behavior for different values of the modulation frequency in the neighborhood of the highest relaxation oscillation frequency (Lacot *et al.*, 1994). The maximum pump power at the input of the fiber is 450 mW and the relaxation oscillation frequency of the fiber laser is 45 kHz. Far from resonance (Figure 3.32a), the laser resonance is linear, and a sinusoidal modulation of the pump power generates periodic pulses of laser intensity at the same frequency. On the other hand, as soon as the modulation frequency approaches the relaxation oscillation frequency (Figure 3.32b and c) the laser response becomes nonlinear and the laser intensity is modulated with a double period (2T) and afterwards with a quadruple period (4T) of that of the pump power before losing all regularity and varying erratically with time (Figure 3.32d). The occurrence of an erratic regime after a succession of such a period-doubling route to chaos is proof that the fiber laser exhibits deterministic chaos.

Antiphase dynamics of orthogonal polarization modes has been observed in a Nd-doped fiber laser with pump modulation (Bielawski *et al.*, 1992) (see Section 3.5.3 for antiphase dynamics in detail). The temporal dynamics of the two orthogonal polarization modes is investigated when the pump power for the fiber laser is modulated at a frequency close to the relaxation oscillation frequency of the fiber laser. The optical wavelength is centered around 1.08 μm. Figure 3.33 (left) illustrates the temporal evolution of the fiber laser intensities on the two orthogonal polarization directions, that is, I_1 and I_2 when the frequency of the pump modulation is varied (Bielawski *et al.*, 1992). The modulation amplitude is kept constant, and the instantaneous pump power always remains above the lasing threshold. Simultaneous measurements of the two orthogonal polarizations clearly indicate antiphase dynamics, in which large output intensities in one polarization direction correspond to small peaks in the other polarization in the case of a 2T-periodic response as shown in Figure 3.33b (left). The same phenomenon is also observed on the 4T-periodic signals (Figure 3.33c (left)) and in the chaotic regime (Figure 3.33d (left)): the large peaks in one polarization are associated with small peaks in the other polarization. The observation of antiphase phenomenon confirms the existence of a strong coupling effect between the two orthogonally polarization states in the fiber laser through population inversion.

The evolution of the laser dynamics can be summarized in a bifurcation diagram using periodic sampling of the laser output intensity synchronously with the pump modulation, as shown in Figure 3.33 (right) (Bielawski *et al.*, 1992). With this technique, a sampling unit delivers a single-valued output when the response of

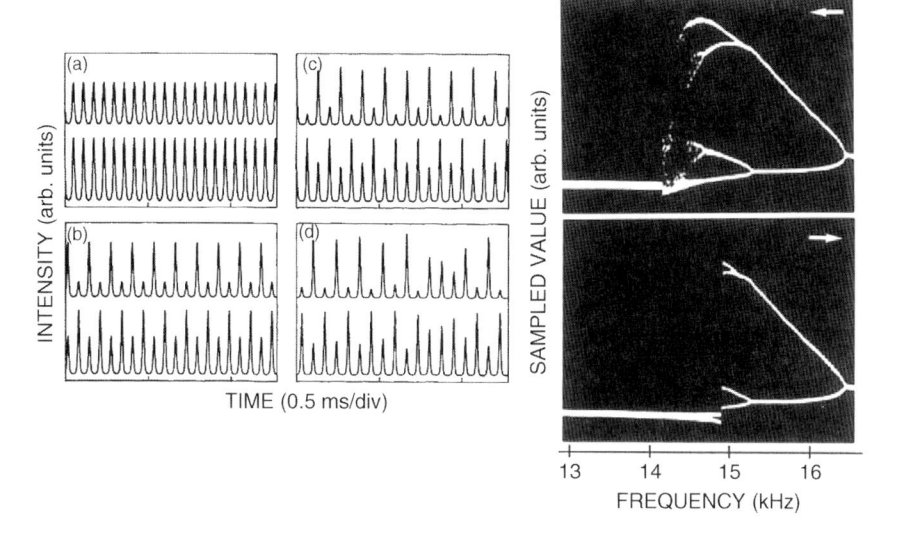

Figure 3.33 (Left) Experimental evidence of antiphase response of the optical fiber laser to pump modulation in different dynamical regimes. The two series of curves are related to the intensity in each polarization eigenstate I_1 (lower traces) and I_2 (upper traces). (a) T response, (b) 2T response, (c) 4T response, (d) chaotic response. (Right) Bifurcation diagram of the optical fiber laser operating at 1.08 μm. The control parameter is the frequency of the pump modulation. The upper diagram corresponds to a decreasing sweep and the lower diagram corresponds to an increase sweep. (S. Bielawski, D. Derozier, and P. Glorieux, (1992), © 1992 APS.)

the system is T-periodic, n different values when its period is nT, and scattered values for a chaotic response. The bifurcation diagrams in the two orthogonal polarization directions do not show any significant difference in the dynamics of I_1 and I_2, supporting the fact that the two polarizations belong to the same dynamics. Because of bistability between attractors, it is necessary to measure the bifurcation diagrams with increasing and decreasing sweeps of the control parameter. Such hysteresis is clearly observed between 14.2 and 14.9 kHz in the two bifurcation diagrams shown in Figure 3.33 (right). The period-doubling route to chaos is observed when the pump modulation frequency is decreased (upper diagram), whereas only periodic oscillations appear with an increase of the modulation frequency in Figure 3.33 (right) (lower diagram).

To investigate the chaotic regimes, the reconstruction of the attractors and Poincaré sections (or projections of these sections) have been used. Polarization-resolved experiments present the advantage of providing measurements of two dynamical variables, the intensities I_1 and I_2 in each polarization eigenstate. A real-time projection of a Poincaré section of the attractor on a two-dimensional plane is readily obtained by using the $X-Y$ mode of an oscilloscope with the values of I_1 and I_2 sampled at a constant phase of pump modulation. The corresponding sections at different bifurcation points of the inverse cascade (C2), in the fully chaotic regime, and

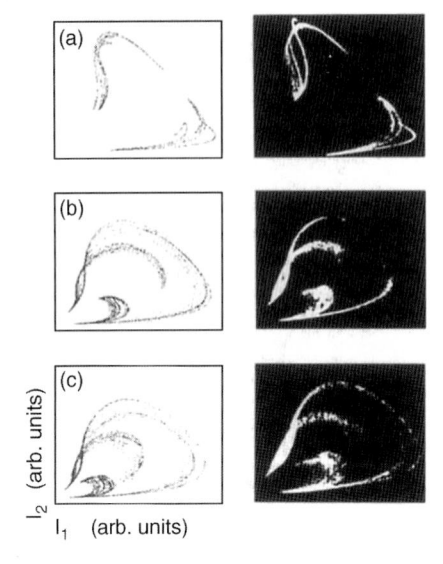

Figure 3.34 Poincaré sections in chaotic regimes corresponding to different points of the inverse cascade. (a) C2; (b) and (c) fully chaotic regimes just after the 2C–C transition and just before the boundary crisis. The right (left) column reports the experimental (numerical) results. (S. Bielawski, D. Derozier, and P. Glorieux, (1992), © 1992 APS.)

just before the boundary crisis are shown in Figure 3.34 (Bielawski et al., 1992). The Poincaré sections for the C2 regimes appear as two clusters of dots periodically visited while they span a wide region of the plane beyond the end of the inverse cascade (Figure 3.34a). These Poincaré sections have special shapes, they are given as a fingerprint of the chaotic attractor with fractal structure. The experimental results of the Poincaré sections agree well with the numerical results, as shown in Figure 3.34.

3.4.2
Multimode (Dual-Wavelength) Dynamics

The quasiperiodicity route to chaos has been observed experimentally in an erbium-doped fiber laser operating at dual wavelengths of 1.550 and 1.536 μm (Sanchez et al., 1995). It is possible to obtain simultaneous oscillation of the laser inside two spectral ranges near 1.550 and 1.536 μm. Each spectral range includes hundreds of longitudinal modes. The fiber has an ion-pair concentration (Er^{3+}, Al^{3+}, and Ge^{4+}) of 7.5%, which has been identified as being responsible for the pulsed behavior of erbium-doped fiber lasers. For lower ion-pair concentrations, the laser system is always stable and CW operation is obtained for the two wavelengths for any pumping rate. On the other hand, for higher ion-pair concentrations, the route to chaos cannot be identified because the system is always chaotic for any pumping rates. Chaotic oscillations result from the coupling between the two wavelengths since no chaotic dynamics is observed under the condition of one lasing wavelength. Antiphase dynamics is always present for periodic and chaotic oscillations.

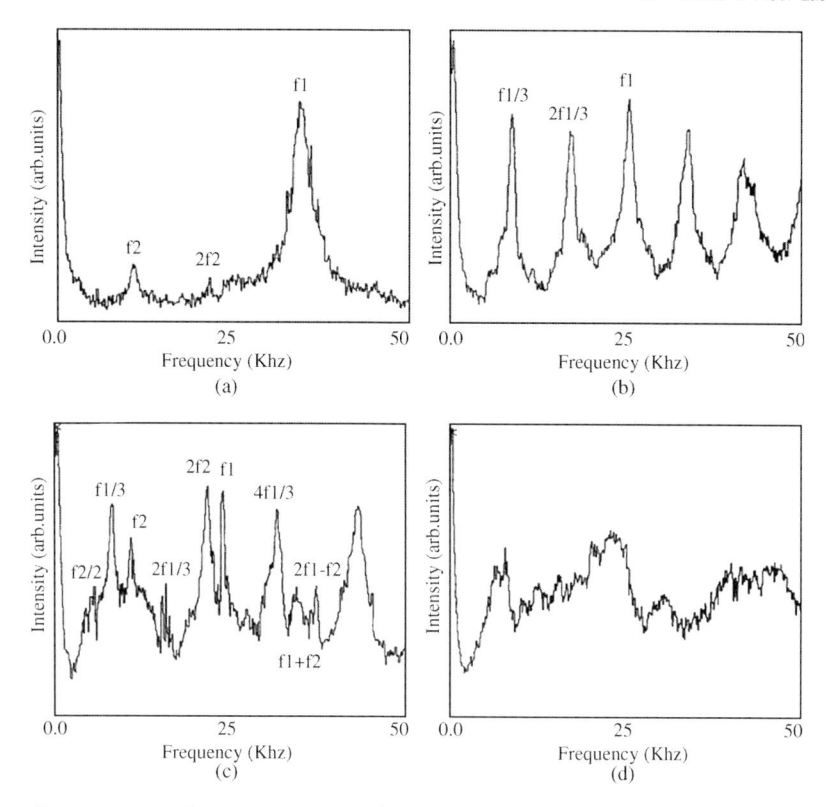

Figure 3.35 Low-frequency RF spectra of the intensity associated to $\lambda = 1.55\,\mu m$: (a) T-periodic, (b) 3T-periodic, (c) quasiperiodic, and (d) chaos. (F. Sanchez, M. LeFlohic, G. M. Stephan, P. LeBoudec, and P.-L. Francois, (1995), © 1995 IEEE.)

The analysis of the low-frequency RF spectra of the laser intensities has been conducted. Typical RF spectra are shown in Figure 3.35, corresponding to different time evolutions by decreasing the pumping ratio (Sanchez et al., 1995). Figure 3.35a gives the spectrum for T-periodic signals. It reveals the existence of two incommensurable frequencies: high frequency f_1 and low frequency f_2. These two frequencies are observed in the quasi-CW regime and in the T-periodic regime. The coupling between the two frequencies is small because the low-frequency spectra do not show any beat frequencies between f_1 and f_2. Figure 3.35b shows the RF spectrum associated with a 3T-periodic oscillation. In this case, a frequency locking of the low frequency occurs on the 1/3 subharmonic of the high frequency. Figure 3.35c shows the RF spectrum associated with the transition region from 2T to 3T. Although this transition is sudden, the frequency locking between f_1 and f_2 does not occur. The system is therefore quasiperiodic. Indeed, the spectral peaks of $f_1 + f_2$ and $2f_1 - f_2$ together with the 1/3 subharmonic of f_1 are observed in Figure 3.35c. Figure 3.35d shows the RF spectrum associated with a chaotic regime. The noise level is greatly enhanced and no sharp peaks emerge in the RF spectrum. The observation of the RF spectra indicates the existence of the quasiperiodicity route to chaos in the dual-wavelength erbium-doped fiber laser.

3.4.3
Fast Polarization Dynamics

The fast temporal dynamics (on the nanosecond time scale) of the Er^{3+}-doped fiber-ring laser (EDFRL) has been reported (Williams and Roy, 1996; Williams et al., 1997; Abarbanel et al., 1999). The fiber laser output beam contains two linearly polarized components. It is within the two groups of orthogonal polarization eigenmodes that the various dynamic states are observed and investigated. Computational results from a model based on coupled delay and differential equations of the Ikeda type provide an explanation of the experimental observations.

A schematic of the experimental configuration is shown in Figure 3.36 (left) (Williams and Roy, 1996). The coherent pump source is the 514.5-nm-wavelength line from an argon-ion (Ar^+) laser. A 6-m length of erbium-doped fiber with an ion concentration of ~240 parts in 10^6 is taken as the gain medium. A Faraday optical isolator is included in the laser cavity to ensure unidirectional operation. An output coupler removes 3% of the intracavity power. The polarization controller functions as a discrete birefringence-inducing element. Overall, the laser cavity is 20 m long, 14 m

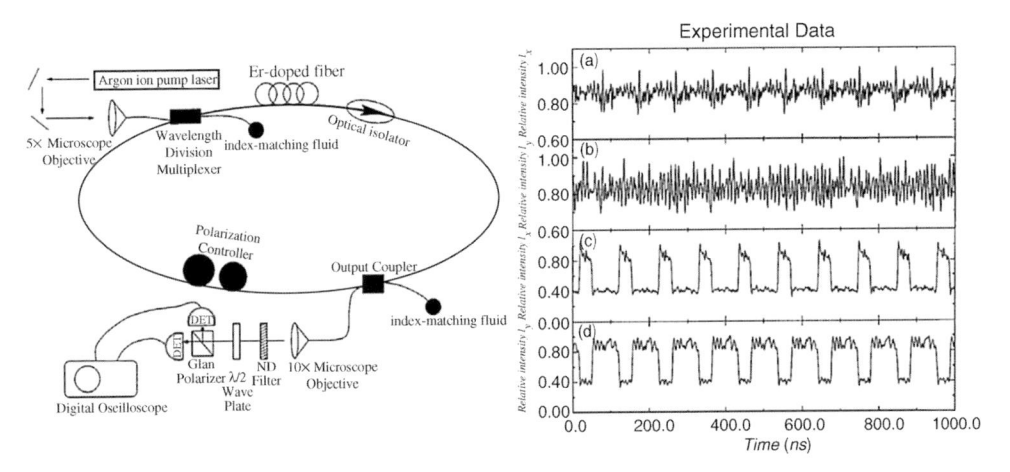

Figure 3.36 (Left) Experimental arrangement: Ar^+-ion laser, $\lambda_p = 514.5$ nm; 514.4–1550 nm wavelength-division multiplexer optical coupler; Faraday optical isolator (not shown); 97/3 coupling ratio output coupler; neutral density (ND) filter with 10% transmission at 1.55 μm; $\lambda/2$-wave plate at 1.55 μm; DET, fast-response InGaAs/p–i–n photodetectors. (Right) Experimentally measured polarization-resolved traces of (a) self-pulsing at the cavity round-trip time in the x-polarization direction from an EDFRL with 10% output coupling, (b) irregular trace in the y-polarization direction. The EDFRL was pumped four times threshold. (c), (d) Antiphase square pulses in the x- and y-polarization directions, respectively, from an EDFRL with 3% output coupling. The EDFRL was pumped at 3.3 times threshold. (Q. L. Williams and R. Roy, (1996), © 1996 OSA.)

being a passive optical fiber. Free ends of the couplers are placed in index-matching fluid to suppress the parasitic Fresnel reflections. The output at the wavelength of 1.561 μm is sent through a half-wave plate and a polarization beam splitter cube, where the orthogonal polarization eigenmodes can be observed simultaneously with high-speed photodetectors. Data are recorded by a fast digital oscilloscope. The round-trip time for the cavity is $\sim$100 ns. The EDFRL lases on a broad 3-dB optical gain bandwidth that is $\sim$10 GHz. The longitudinal mode separation is 9.8 MHz; the number of active oscillating modes is well over 2000. An optical spectrum analyzer reveals that the modes oscillate within orthogonally polarized mode groups that have been modeled as two supermodes.

While the EDFRL is pumped well above threshold (the threshold pump power is $\sim$175 mW), self-pulsing is observed on the nanosecond time scale. Figure 3.36a (right) and b (right) are resolved polarization components of the total output intensity when the 10% output coupler is used (Williams and Roy, 1996). Quasiperiodic behavior is observed in one polarization direction (Figure 3.36a) and nearly random evolution appears in the other (Figure 3.36b). In Figure 3.36a the distinct sharp pulses are separated by the fundamental cavity round-trip time of $\sim$100 ns. Figure 3.36b shows a highly complex time series that is quasiperiodic or nearly perfectly repeating, with a period of $\sim$7 cavity round-trips.

When the output coupling power is decreased from 10 to 3%, (i.e., the light intensity inside the cavity is high enough), the pulsed behavior disappears and square pulses develop in the output intensity of the orthogonal polarization states. This behavior is antiphase in the two polarization states and is periodic at the cavity round-trip time, as shown in Figure 3.36c and d. It is worth noting that the time durations of the plateaus correspond to the lengths of the active and passive part of the fiber. In other words, the 30-ns pulses in Figure 3.36c correspond to the 6-m length of the gain medium, whereas the 70-ns pulses in Figure 3.36d correspond to the 14-m length of the passive fiber within the laser cavity. Another detail to note is the highly structured intensity fluctuations that ride on top of the square pulses and repeat over many round-trips. These experimental observations are well reproduced by numerical simulations (Williams *et al.*, 1997; Abarbanel *et al.*, 1999).

3.5
Chaos in Solid-State Lasers

Solid-state lasers are good tools for testing the fundamental physics of chaos in laser systems. The relaxation oscillation frequency of solid-state lasers ranges from kHz to MHz, much lower than that of semiconductor lasers, which lies around GHz. The low characteristic time scale of solid-state lasers allows experimental detection of chaotic temporal waveforms easily, without using high-speed detection equipments. From the dynamical point of view, the dynamics of single-mode solid-state lasers can be described by a class-B laser model for the electric field and population inversion. The dynamics of solid-state lasers are relatively simple when compared with semiconductor lasers, because there is no coupling between the amplitude of the electric field and its optical phase, which is represented by the linewidth-enhancement factor or α parameter in semiconductor lasers. Because of these characteristics, many works have been devoted to study nonlinear dynamics in solid-state lasers, which will be described in this section. The detailed examples of chaotic dynamics in solid-state lasers are described comprehensively in the books (Otsuka, 1999; Mandel, 1997).

Solid-state lasers that have a short cavity length (typically less than one millimeter) are specifically called microchip lasers. Due to their short cavity length, lasing occurs in a single mode or in a few longitudinal modes. It is thus easy to model the dynamics of microchip lasers precisely. One of the well-known models is called Tang–Statz–deMars equations for describing the dynamics of a multimode solid-state lasers with spatial hole burning of the population inversion (Otsuka, 1999; Mandel, 1997). Some numerical models for chaotic dynamics in a solid-state laser are described in Appendix 3.A.6.

3.5.1
External Modulation

Solid-state lasers with external modulation can generate chaotic intensity fluctuations. Chaotic outputs in microchip lasers are easily obtained by modulating the pump (called pump modulation) or loss of the laser cavity (called loss modulation), which causes a beat oscillation of the order of the relaxation oscillation frequency around a few MHz.

For pump modulation, a laser diode (LD) is typically used as a pumping source for a solid-state laser. The injection current of the LD used for pumping is sinusoidally modulated. The output of the LD is focused onto the solid-state laser crystal, and the population inversion of the solid-state laser is sinusoidally modulated as well. Nonlinear interaction between the relaxation oscillation frequency and the pump modulation frequency causes chaotic intensity fluctuations. Figure 3.37 (left) shows the experimental results of the temporal waveforms of the output intensities of a neodymium-doped yttrium orthovanadate (Nd:YVO$_4$) microchip solid-state lasers with different pump-modulation amplitudes. Different periodic waveforms are observed as the modulation amplitude is changed. Chaotic oscillation is observed at large modulation amplitude (Figure 3.37a (left)), and the oscillations are changed to various periodic and quasiperiodic waveforms as the modulation amplitude is decreased. Finally, a period-1 oscillation is observed for small modulation amplitudes, as shown in Figure 3.37f (left).

Figure 3.37 (right) shows the experimental results of the bifurcation diagram as the modulation amplitude is increased or decreased. As the modulation amplitude is increased, a period-1 oscillation is changed to a period-6 oscillation, a quasiperiodic oscillation with a period-3 oscillation, and a chaotic oscillation. A similar transition is observed when the modulation amplitude is decreased: Chaos, quasiperiodic oscillation with period-3, period-6, period-2, and period-1 oscillations, however, the bifurcation points are different from the case when the mdulation amplitude is increased (*i.e.*, hysteresis). The hysteresis of a bifurcation diagram is commonly observed in many nonlinear dynamical laser systems.

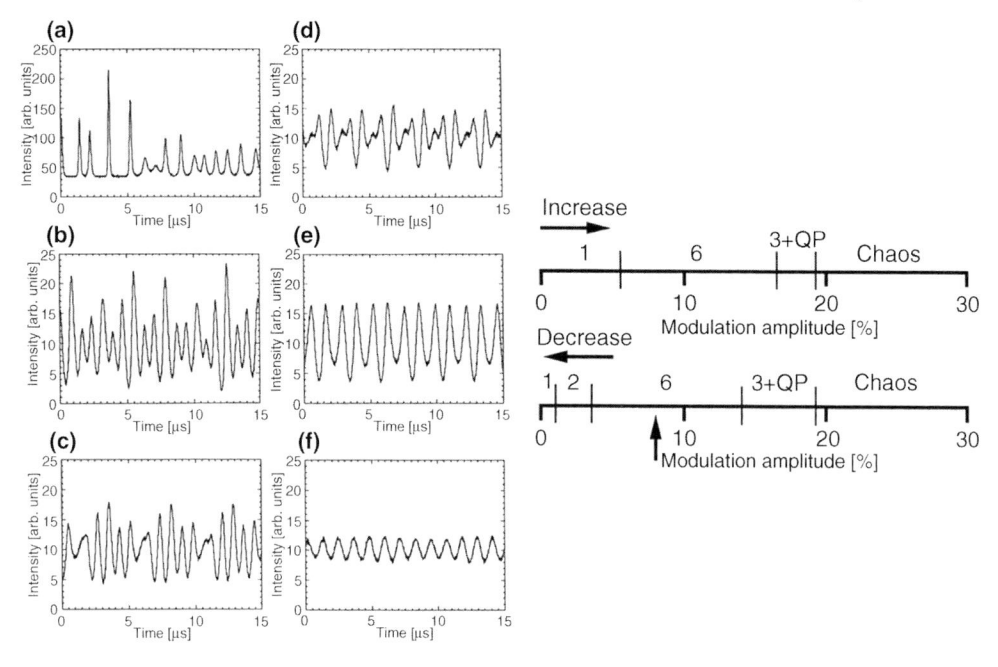

Figure 3.37 (Left) Experimental results of temporal waveforms of Nd:YVO$_4$ solid-state microchip laser output with pump modulation. The modulation amplitude is varied at (a) 21%, (b) 18%, (c) 11%, (d) 5.5%, (e) 1.4%, and (f) 1.0%. The modulation frequency is fixed at 857 kHz. The observed temporal waveforms correspond to (a) chaos, (b) quasiperiod, (c) period-6, (d) period-6, (e) period-2, and (f) period-1 oscillations. (Right) Experimental results of bifurcation diagrams of Nd:YVO$_4$ solid-state microchip laser with pump modulation as a function of modulation amplitude. Hysteresis is seen when the modulation amplitude is increased or decreased.

Chaotic nonlinear dynamics has been also reported in a neodymium pentaphosphate (NdP$_5$O$_{14}$) solid-state laser pumped by a modulated Ar$^+$ laser (Klische et al., 1984). The Ar$^+$ pump laser power is modulated at a frequency approximately corresponding to the relaxation oscillation frequency. Figure 3.38 shows the experimental results of RF spectra of the solid-state laser intensities at different modulation frequencies. A period-doubling route followed by broadband RF spectra demonstrates chaos, including the period-3 and -5 windows in the chaotic range.

3.5.2
Frequency-Shifted Optical Feedback

A frequency-shifted feedback-light injection is used to generate chaotic dynamics in a Nd:YVO$_4$ microchip solid-state laser, as shown in Figure 3.39 (left). The laser light is propagated to a rotating circular paper sheet, and weak scattered light whose center frequency is shifted because of the Doppler effect returns to the laser cavity. The laser

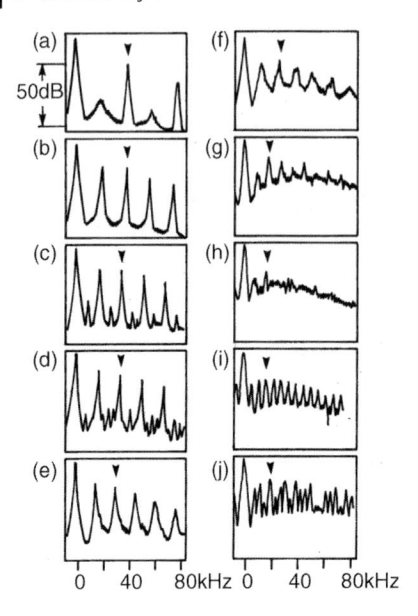

Figure 3.38 Intensity spectra of an NdP$_5$O$_{14}$ solid-state laser modulated near twice the relaxation oscillation resonance frequency of the laser. The solid-state laser is pumped by an Ar$^+$ laser whose intensity is periodically changed by an electro-optic modulator. As the modulation frequency is moved towards the relaxation oscillation frequency, pulsing and period-doubling ending in chaotic dynamics ((h), fully developed chaos) is observed. Within the chaotic region period-3 and period-5 windows are observed ((i) and (j)). The modulation frequency is marked. (W. Klische, H. R. Telle, and C. O. Weiss, (1984), © 1984 OSA.)

intensity is then modulated as a result of self-mixing between the two light components in the cavity.

It is known that the photon statistics of the scattered-light field from a rough surface show a narrow-band Gaussian distribution whose center frequency is shifted owing to the Doppler effect (Otsuka, 1999). Here, the frequency linewidth of the scattered field Δf is much lower than the Doppler-shifted frequency f_D in this experiment ($\Delta f \ll f_D$). As a result, each lasing mode of the laser is found to be modulated effectively at a Doppler-shifted frequency f_D resulting from interference between a lasing field and an extremely weak scattering field, known as the self-mixing laser Doppler velocimetry scheme (Otsuka, 1999).

The dynamics of the laser modulated by the self-mixing laser Doppler velocimetry scheme is investigated. The laser light is modulated at the Doppler-shifted frequency f_D,

$$f_D = 2v \cos \theta_s / \lambda \tag{3.5}$$

where v is the angular velocity of the rotating paper, θ_s is the angle between the laser axis and the velocity vector, and λ is the optical wavelength ($\lambda = 1.064\,\mu$m for the Nd:YVO$_4$ microchip laser). When the intensity of injected scattered light is weak, clear resonance can be observed around $f_D = f_r$, where f_r is the relaxation oscillation frequency of the laser. Chaotic instabilities appear in the laser output when the Doppler-shifted feedback light power is increased or f_D is set close to f_r.

Figure 3.39 (right) shows the RF spectra of the Nd:YVO$_4$ laser output at different Doppler-shifted modulation frequencies. There are two main peaks corresponding to the relaxation oscillation frequency f_r and the Doppler-shifted modulation frequency f_D, as shown in Figure 3.39a (right). With the increase of f_D, the peak of $f_D/2$ appears

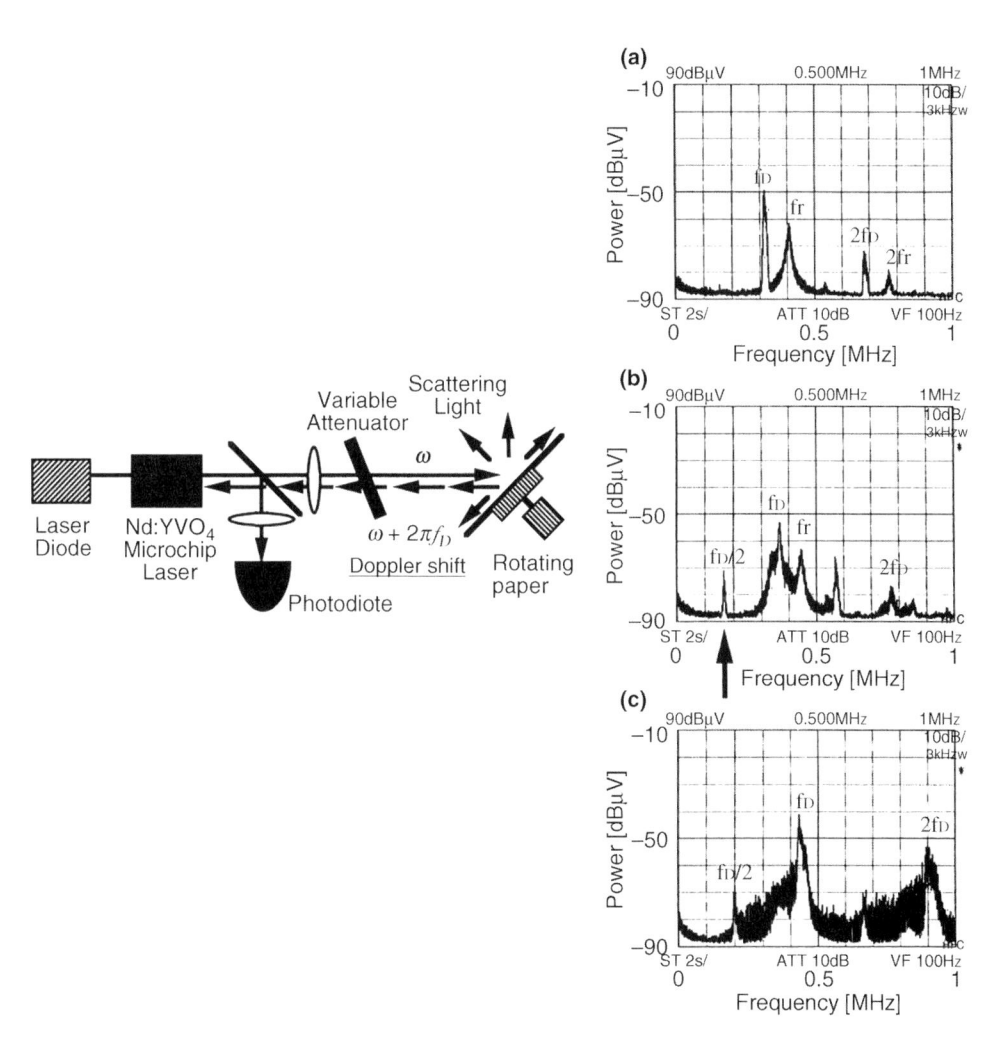

Figure 3.39 (Left) Experimental setup of frequency-shifted feedback light injection using a rotating paper in a Nd:YVO$_4$ microchip solid-state laser. (Right) Experimental results of RF spectra of Nd:YVO$_4$ microchip laser output with frequency-shifted feedback light injection. (a) The Doppler-shifted frequency $f_D = 320$ kHz, (b) $f_D = 380$ kHz, and (c) $f_D = 430$ kHz. The relaxation oscillation frequency is set to $f_r = 420$ kHz. Chaotic broadband spectrum is observed in (c).

(the arrow in Figure 3.39b (right)), showing the period-doubling route to chaos. As f_D approaches f_r further, the mixture of the two frequencies induces continuous components of the RF spectrum (Figure 3.39c (right)), showing chaotic intensity fluctuations.

The rotating paper can be replaced by two acousto-optic modulators (AOMs) to introduce the Doppler-shifted feedback light. A more accurate modulation frequency with narrower frequency linewidth of the scattered field can be achieved by using AOMs. The detuning of the two modulation frequencies for the two AOMs needs to be set around the relaxation oscillation frequency of the laser to observe chaotic instabilities.

3.5.3
Antiphase Dynamics in Multimode Laser

Multimode solid-state lasers can show antiphase dynamics in periodic and chaotic oscillation regimes due to the spatial hole burning effect (Otsuka, 1999; Mandel, 1997), where antiphase dynamics indicate that the temporal waveforms of laser modes oscillate with out-of-phase. Note that the "phase" indicates the phase of the slow temporal waveform of laser intensity (10^3–10^9 Hz), but not the phase of the fast optical carrier of the electric field ($\sim 10^{14}$ Hz).

Multimode experiments are carried out by coupling the lasing beam from an Ar^+ laser-pumped 1-mm flat lithium neodymium tetraphosphate ($LiNdP_4O_{12}$, LNP) microchip solid-state laser (Otsuka, 1999). A schematic illustration of the experimental setup is shown in Figure 3.40 (left), which is similar to Figure 3.39 (left) as the self-mixing laser Doppler velocimetry scheme. To show the multimode dynamics, results for the simplest case of two-mode oscillations are described. However, essentially the same collective behavior featuring antiphase dynamics is observed experimentally for larger number of modes. When the intensity of injected scattered light is weak, clear resonances around $f_D = f_2$, $f_D = f_1$, and $f_D = f_1/2$ are observed, where f_1 and f_2 are the relaxation oscillation frequencies of the first and second lasing modes, respectively ($f_1 > f_2$). Antiphase motions are observed by modulation around the relaxation oscillation frequency of each lasing mode.

The modal and total outputs of temporal waveforms at the different Doppler-shifted frequencies are shown in Figure 3.40 (right) (Otsuka, 1999). In the case of the $f_D = f_2$ resonance (Figure 3.40a (right)), the output intensities of the two modes (s_1 and s_2) show alternate oscillations (indicated by the arrows) at the frequency f_2, showing antiphase dynamics. For $f_D = f_1/2$ (Figure 3.40b (right)), each mode exhibits period-2 sustained relaxation oscillations, in which large peaks in one mode corresponds to small peaks in the other (see the arrows), indicating antiphase dynamics. For $f_D > f_1$ (Figure 3.40c (right)), two successive large peaks and one small peak in one mode are associated with two small peaks and one large peak in the other mode. Antiphase dynamics at f_D are also seen in Figure 3.40c (right).

As the reflected scattering light intensity is increased, period-doubling oscillations leading to chaotic oscillations are observed by the modulation at $f_D \approx f_1$. To find the

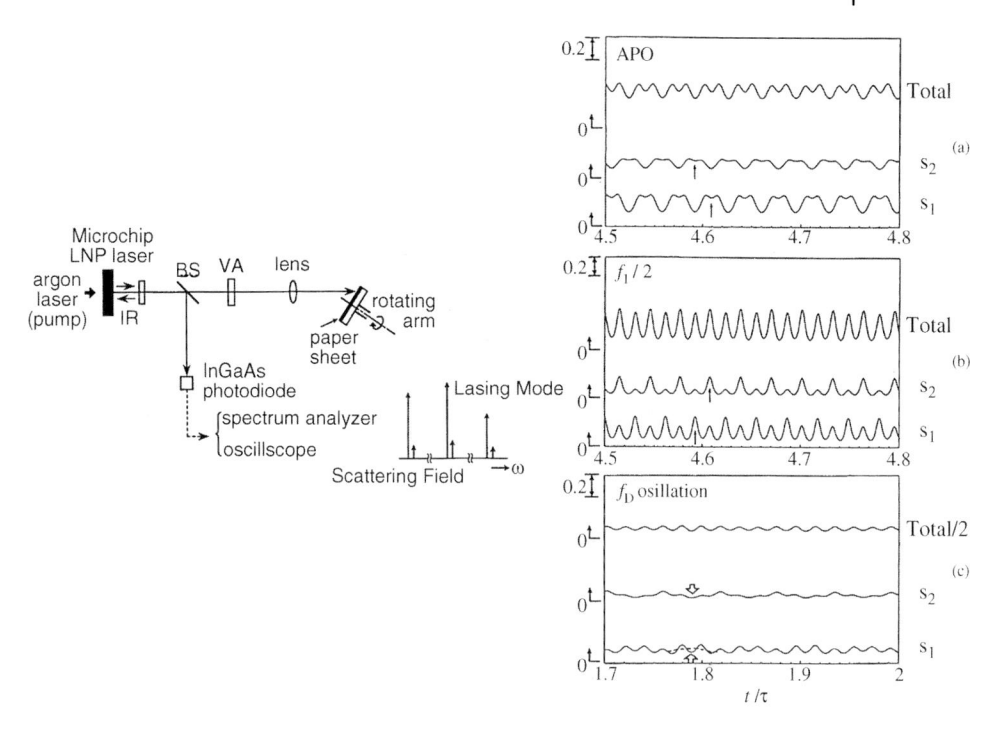

Figure 3.40 (Left) Schematic illustration of modulation experiment utilizing self-mixing laser Doppler velocimetry (SMLDV). (Right) Numerical results of total and modal outputs of temporal waveforms at different modulation frequencies. (a) $f_D = f_2$, (b) $f_D = f_1/2$, and (c) $f_D > f_1$. s_1 and s_2 are the outputs of mode-1 and -2, respectively. Antiphase dynamics are depicted by arrows. (K. Otsuka, (1999), © 1999 KTK.)

interplay between oscillating modes in chaotic regimes, the injected light intensity is increased and RF spectra of chaotic oscillations averaged over a long period of time are measured as well as the temporal waveforms, as shown in Figure 3.41 (Otsuka, 1999). Individual modes (s_1 and s_2) feature strong fluctuation components around antiphase motion frequencies f_2 in addition to f_1, while these low-frequency antiphase motion components f_2 cancel each other out for the total intensity, as seen in the RF spectra (right-hand side) of Figure 3.41. This strongly implies that a self-organized collective behavior based on antiphase dynamics is generic in multimode lasers with cross-saturation of population inversion and exists even in chaotic regimes. In other words, a chaotic attractor of each mode inherits the antiphase dynamics character of the corresponding stationary state from which it is born through bifurcations. Therefore, oscillating modes are not statistically independent in their fluctuations.

For multimode lasers with more than two modes, clustering and breathing motions featuring intermode parametric resonances are demonstrated when the laser is modulated by rational frequencies chosen near-resonant to multiple relaxation oscillation frequencies inherent in multimode lasers. It has been shown that inphase

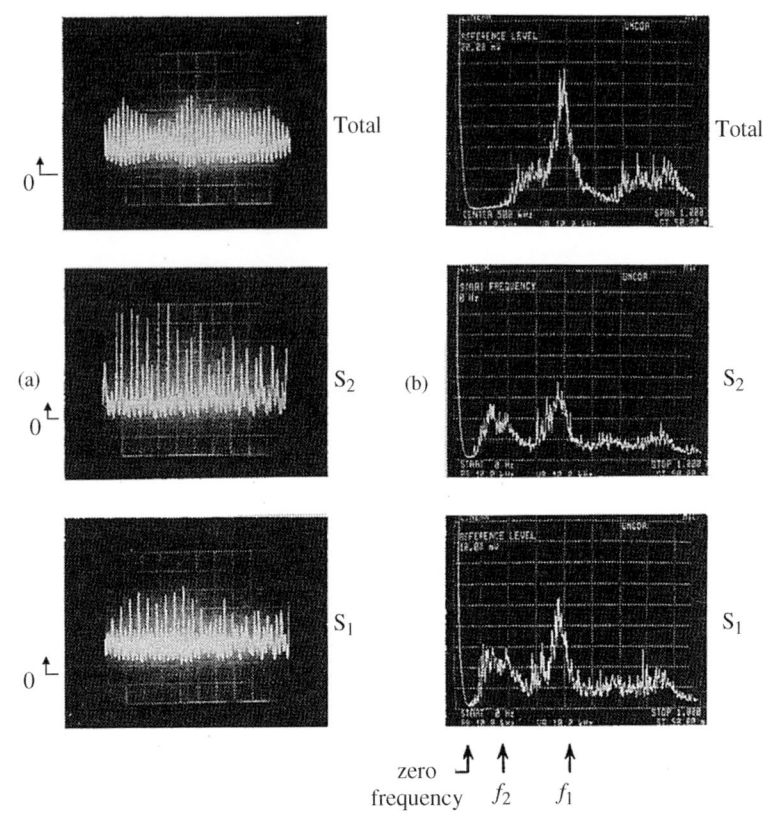

Figure 3.41 Experimental results of the total and modal outputs of (left) temporal waveforms and (right) corresponding RF spectra, indicating antiphase chaotic oscillations in two-mode regimes.
(a) (left) 10 μs/div. (right) 100 kHz/div. The modulation frequency f_D is set near f_1. The RF spectra are obtained by averaging accumulated power densities over 20–30 s. (K. Otsuka, (1999), © 1999 KTK.)

oscillations are shown at the maximum relaxation oscillation frequency while antiphase motions are observed at lower relaxation oscillation frequencies. The total dynamics is similar to a single-mode laser with the maximum relaxation oscillation frequency, while the modal dynamics has lower relaxation oscillation components (Otsuka, 1999).

3.5.4
Insertion of Nonlinear Crystal

An insertion of a nonlinear crystal in a laser cavity is another method to generate chaos in solid-state lasers (Baer, 1986; Bracikowski and Roy, 1991; Roy et al., 1994). A typical example is an insertion of a nonlinear crystal for second-harmonic generation of optical wavelength conversion, known as green problem, as described in Section 2.1.1.2. A schematic of the laser experiment is shown in Figure 3.42 (top)

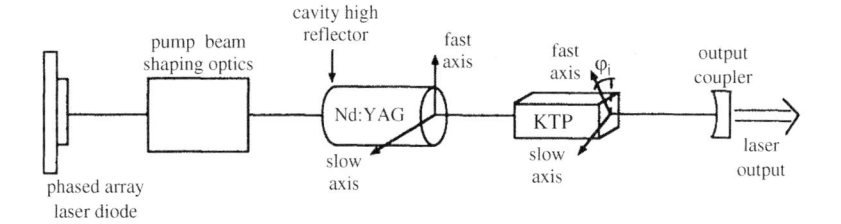

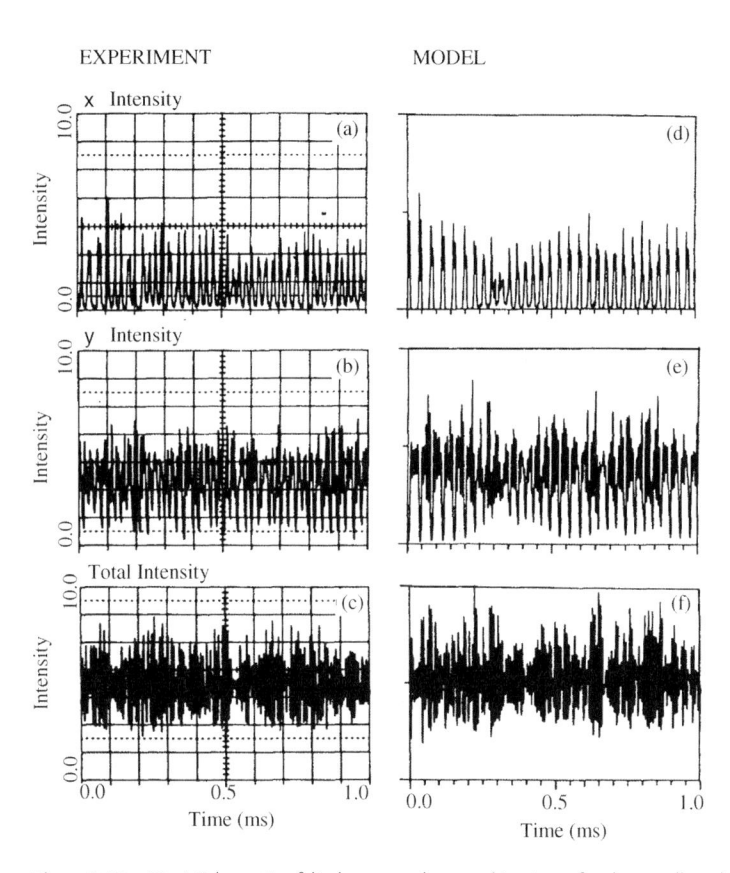

Figure 3.42 (Top) Schematic of diode-pumped Nd:YAG laser with intracavity KTP doubling crystal. The laser cavity is highly reflecting for the fundamental 1064-nm wavelength, but transmits the doubled green light at 532 nm. The relative angular orientation of the Nd:YAG and KTP crystals is adjusted to obtain a given combination of orthogonally polarized modes. (Bottom) (a)–(c) Experimental and (d)–(f) numerical time traces for the (a), (d) x-polarization mode, (b), (e) y-polarization mode, and (c), (f) total intensities. This chaotic instability is known as green problem. (C. Bracikowski and R. Roy, (1991), © 1991 APS.)

(Bracikowski and Roy, 1991). The laser consists of a neodymium-doped yttrium aluminum garnet (Nd:YAG) crystal as a laser medium, a potassium titanyl phosphate (KTP) crystal for the second-harmonic generation (SHG) of the optical wavelength from 1064 nm (infrared light) to 532 nm (green light), and an output coupler to extract the laser output. When the laser is pumped several times above threshold by a phased-array laser diode, many longitudinal modes can be active simultaneously. Chaotic fluctuations in the output intensity are induced by the KTP crystal, which nonlinearly couples the modes through sum-frequency generation. Two orthogonal linear polarization modes (referred to as x- and y-modes) exist in the laser cavity. Some of the longitudinal modes correspond to the x-mode, and other modes correspond to the y-modes. The number of accompanying polarization modes can be varied by changing the pump power and by rotating the alignment of the Nd:YAG and KTP crystals.

In the experiment of Figure 3.42 (top), the laser-diode-pumped Nd:YAG laser is carefully aligned to support a total of five longitudinal modes, composed of one x-polarization mode and four y-polarization modes. The second-harmonic intensity at the wavelength of 532 nm is filtered from the laser output, and only the fundamental wavelength of 1064 nm is incident upon a photodiode. The time trace of the total intensity of the laser output is observed by using the photodiode and stored on a digital oscilloscope. A polarizing prism is also inserted before the photodiode, allowing time traces of the modal intensities of x- and y-polarization modes to be obtained.

Time traces for the single x-mode, y- mode, and total intensities are shown in Figure 3.42 (bottom). Chaotic temporal waveforms are observed for both the modal and total intensities. The experimental results (Figure 3.42a–c) agree well with the numerical results (Figure 3.42d–f). Note that antiphase dynamics is also observed between the x- and y- modes, indicating strong mode–mode interaction through the population inversion. It is found that the nonlinear coupling of the modes through sum-frequency generation results in the destabilization of the relaxation oscillations in the green laser system. The control and stabilization of chaotic second-harmonic green lasers has been reported (Roy et al., 1992; Roy et al., 1994) (see Chapter 11).

3.6
Chaos in Gas Lasers

Gas lasers were frequently used to find deterministic chaos in lasers in the 1980s. Many works have been reported to show the existence of deterministic chaos in gas lasers experimentally and numerically after the connection between chaos and lasers was first made (Haken, 1975). The details of nonlinear dynamics in gas lasers with various configurations are described comprehensively in the book (Weiss and Vilaseca, 1991). A numerical model for chaotic dynamics in a gas laser is described in Appendix 3.A.7.

3.6.1
Lorenz–Haken Chaos

The Lorenz–Haken chaos (see Section 2.1.3) can be observed in gas lasers under the bad-cavity condition, where the electrical field decay rate is larger than the relaxation rates of the population inversion and the atomic polarization (i.e., Class C laser) (Weiss and Vilaseca, 1991). In most popular lasers, this condition is equivalent to a very high resonator loss that requires an extremely high pump parameter (10–20 times the lasing threshold) to reach the lasing threshold (referred to as the "first laser threshold"). Hence, there appeared to be little chance of experimentally observing the dynamics of the Lorenz–Haken chaos.

The pump parameter necessary to reach the threshold for generating chaos (referred to as the "second laser threshold") is proportional to the homogeneous linewidth of the gas medium. In lasers with narrow homogeneous linewidths one may expect to reach the second laser threshold with attainable pump parameters.

The homogeneous linewidth of gas lasers is determined by spontaneous emission and atomic/molecular collisions. The spontaneous emission rate decreases with the optical frequency of the laser transition as the mode density of free space. Long-wavelength laser transitions effectively permit spontaneous emission to be eliminated, and collision rates are proportional to gas pressure. Therefore, long-wavelength infrared lasers operating at low pressure are one possible choice for observing chaotic laser dynamics. Continuously operating lasers in the far-infrared region that are pumped by a laser should permit the second laser threshold to be reached (Weiss et al., 1985). An ammonia (NH_3) laser at 100 μm wavelength meets the bad-cavity condition with a resonator loss of only a few per cent.

The resonator used in the experiment is shown in Figure 3.43 (left) (Weiss and Vilaseca, 1991). It is made up of three reflectors forming a triangular ring cavity. The laser has to be a ring laser since many theoretical models and also the Lorenz–Haken

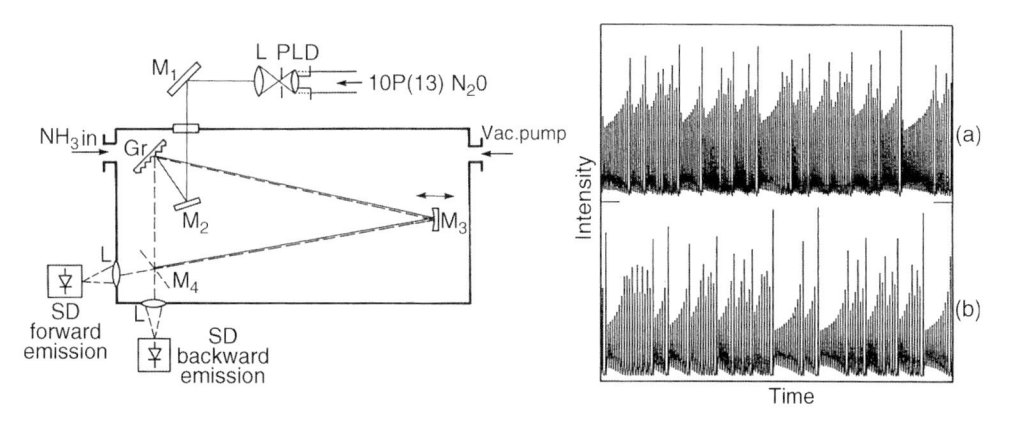

Figure 3.43 (Left) Ammonia (NH_3) ring laser used for the experiments. The ring consists of three reflectors: grating, curved gold mirror and semitransparent (gold-mesh) mirror. The first order of the grating serves to couple the pump laser beam into the resonator, the zero-order reflection of it is utilized for the generated radiation whose wavelength is long compared with the grating period. A combination of spatial filter and diaphragm serves as a pump-intensity attenuator to vary the pump strength. M: mirror, L: lens, P: pinhole, Gr: grating, SD: detector diode, D: diaphragm. (Right) (a) Experimentally measured intensity pulses of the ammonia laser, and (b) intensity pulses calculated from the Lorenz equations with $r = 15$, $b = 0.25$, and $\sigma = 2$. (C. O. Weiss and R. Vilaseca, (1991), © 1991 VCH.)

equations treat traveling waves, but not standing waves with spatially varying intensity. One of the reflectors is a diffraction grating used in zero-th order as a plane far-infrared mirror and simultaneously in first order to couple the pump laser beam into the resonator volume. A gold-mesh reflector of 30 μm grid constant is used as a semitransparent mirror to couple out the generated 81 μm wavelength radiation. It is designed to generate unperturbed TEM$_{00}$ modes, which are the best possible approximation to a plane wave described by the Lorenz–Haken equations. The spatial filter–diaphragm combination permits the pump laser beam power to be changed without changing its geometry, direction or frequency, which is necessary to vary the laser pump parameter in a controlled manner.

Figure 3.43a (right) shows an experimentally measured time series of pulses from the ammonia laser (Weiss and Vilaseca, 1991). For comparison, Figure 3.43b (right) shows a time series of pulses calculated from the Lorenz model (see Section 2.2.2) for parameters realistic for the ammonia laser. Note the similarity between the measured and calculated pulses even of details such as the last large pulse terminating each spiral.

A way of looking at chaotic attractors is by the phase-space representations. These can be obtained by plotting the nth point of the time series against the $(n+\tau)$th point (time-delay embedding in phase space, see Section 2.2.3). Depending on the choice of τ, the plot obtained will look different. Choosing τ to be very small, the nth and $(n+\tau)$th points will be strongly correlated. In this case the points of the plot will lie on or close to a 45-degree straight line. If τ is chosen to be very large, the points will have no correlation at all and no structure of points will be found. For intermediate values of τ (typically of the order of one tenth to one period), the plot shows "structure", reflecting the structure of the n-dimensional attractor underlying the system dynamics. Figure 3.44a (left) shows phase-space plots for two different values of τ calculated from the time series of the Lorenz model (Weiss and Vilaseca, 1991). For comparison, the phase-space plots calculated for the same τ values from the experimentally measured intensity of the NH$_3$ laser are shown in Figure 3.44b (left). Note the similarity between the laser experiment and Lorenz model results, even in detail.

Another indication of deterministic chaos is the bifurcation diagram. Figure 3.44 (right) shows the experimentally measured RF spectra when the pump is set above the second laser threshold and the laser resonator is tuned towards the gain-line center (Weiss and Vilaseca, 1991). When tuned far enough away from the gain-line center, the laser operates continuously. Closer to the line center self-pulsing starts, followed on further tuning towards the line center by the period-doubling route to chaos, recognizable by the appearance of subharmonic frequencies ($f_0, f_0/2, f_0/4$, and so on) in the RF spectra from Figure 3.44a–d (right). The period-doubling route to chaos is followed by the part-chaotic inverse bifurcation until full chaos is reached (Figures 3.44e–h (right)).

Inhomogeneously broadened gas lasers have been extensively studied both theoretically and experimentally (Casperson, 1978; Gioggia and Abraham, 1983). The types of gas lasers broadened inhomogeneously by the Doppler effect and showing enough gain to fulfill the requirement of reaching an instability threshold

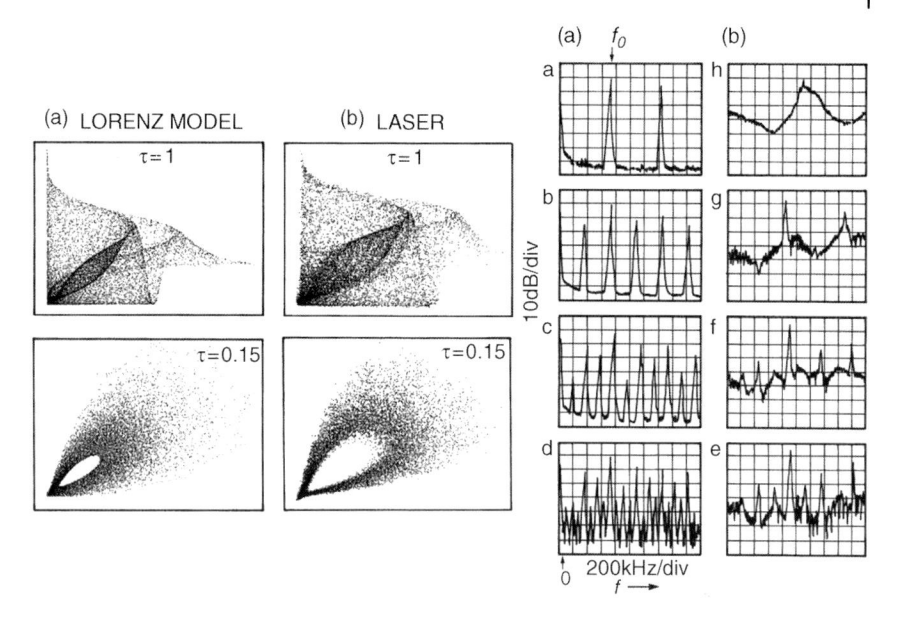

Figure 3.44 (Left) (a) Phase-space projections of Lorenz model intensities for $\tau = 0.15$ and 1 pulsing period, and (b) same for experimentally measured intensity of ammonia laser. (Right) Experimentally measured RF spectra of the intensity of ammonia laser when tuning progressively (a to h) the laser resonator towards the gain-line center. The left column shows the period-doubling bifurcation and the right column shows the inverse cascade bifurcation. (C. O. Weiss and R. Vilaseca, (1991), © 1991 VCH.)

appear to be the He-Xe laser at 3.5 μm, on which consequently most experiments have been done, and the He-Ne laser at 3.39 μm. These two systems show outstanding high gain, larger for the He-Xe than the He-Ne system. Figure 3.45 shows the RF spectra of the He-Xe laser intensity for different laser resonator detunings (Gioggia and Abraham, 1983). The RF spectra show the appearance of the frequency of $f_0, f_0/2$, $f_0/4$ in Figure 3.45a–c, indicating the period-doubling route to chaos. The broad spectrum is obtained, showing the chaotic oscillations (Figure 3.45d). The period-3 oscillations are also observed as a periodic window (Figure 3.45e). The He-Xe laser

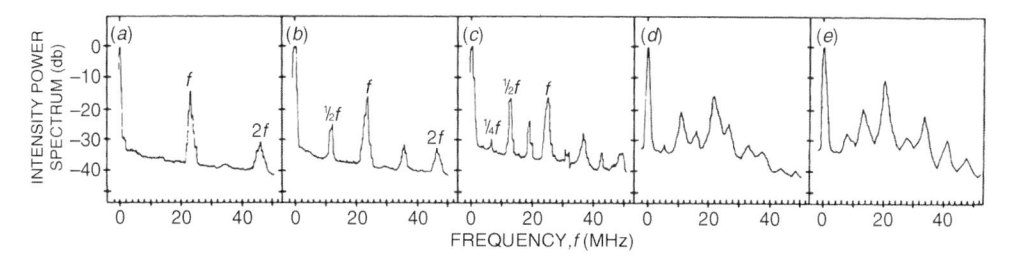

Figure 3.45 RF spectra of He-Xe laser output intensity for different laser resonator detunings. A clear period-doubling bifurcation appears. (a) period-1, (b) period-2, (c) period-4, (d) chaos, and (e) chaos with period-3. (R. S. Gioggia and N. B. Abraham (1983), © 1983 APS.).

also shows two other types of bifurcation such as quasiperiodicity and intermittency routes to chaos (Gioggia and Abraham, 1983).

3.6.2
Insertion of Saturable Absorbers

In a gas laser with a saturable absorber, two gas media are present inside the laser resonator. One of them is pumped so that the atoms have a positive population inversion (active or amplifying medium) and the other is left with a negative population inversion (passive or absorbing medium). The laser field may saturate the absorption, or the field and the absorber may interact in other ways so that time-dependent or bistable laser behavior can occur. Lasers with saturable absorbers are primarily used to obtain pulsed emission with short and intense repetitive pulses (known as "passive Q-switching"). Interest has expanded because a large variety of unstable behaviors have been observed and studied from the more general aspect of nonlinear dynamics in optical systems.

Experiments are conducted with a carbon dioxide (CO_2) laser operating in the 10.6 μm wavelength range. Gaseous iodomethane (methyl iodide, CH_3I) is used as the saturable absorber at low pressure (Dangoisse et al., 1988). For very low pressure (40 mTorr) the laser operates continuously and for high pressure (500 mTorr) the laser is off. In the range between these two pressures, passive Q-switching is obtained which results in a modulated output. The modulation can vary from a weak sinusoidal modulation to the emission of short (< 100 ns) giant pulses with peak power up to 100 times the CW power at 1–100 kHz repetition frequency.

The CO_2 laser is turned first to the gain-line center and the absorber pressure is adjusted to yield a small-amplitude modulation of the laser output, as shown in Figure 3.46a (left) (Dangoisse et al., 1988; Weiss and Vilaseca, 1991). Detuning of the laser results in deeper modulation, which eventually period doubles (Figure 3.46b) and quadruples (Figure 3.46c). After the period-doubling transition, chaos is obtained as shown in Figure 3.46d. After the chaotic range, ranges are found in which a large pulse is followed by one, two, three, or four oscillations. Figure 3.46e shows the case of a large pulse followed by two oscillations (denoted as T_{12}) and Figure 3.46h shows the large pulse followed by one oscillation (T_{11}). Between T_{12} and T_{11} range a chaotic range of irregular pulsing occurs (Figure 3.46f). The chaotic dynamics brought about by competition of the different temporal patterns is similar to the case of the inhomogeneously broadened laser, where chaos occurs through the competition of the symmetric and asymmetric pulsing.

By choosing particular conditions, pulses followed by more than two oscillations ($T_{13}, T_{14}, \ldots$) can be found. They are found to period double to their respective chaotic ranges. The ranges of the different pulse shapes are given in the phase diagram as shown in Figure 3.46 (right). At the limits, the pulse shapes change abruptly (instability boundaries). The period-doubling route to chaos of the individual pulse types is generally observed. For example, Figure 3.46g (left) shows the first subharmonic of T_{12}.

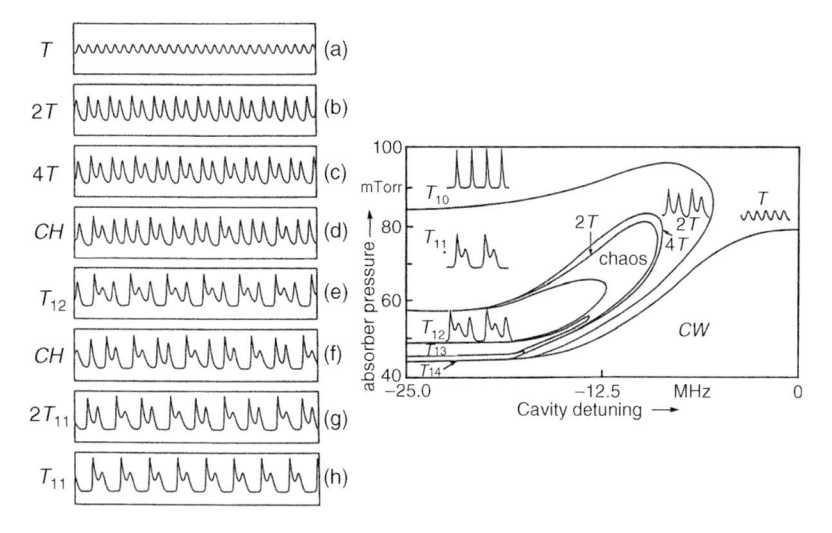

Figure 3.46 (Left) Passive Q-switching pulse shapes as a function of laser resonator detuning. The small-amplitude orbit period doubles to chaos followed by homoclinic cycles, periodic or chaotic. (Right) State diagram of the CO_2 laser with a saturable absorber with the pulse shapes of Figure 3.46 (left). Note the connections between different pulse shapes, for example, $2T \rightarrow T_{11}$. (C. O. Weiss and R. Vilaseca, (1991), © 1991 VCH.)

The experimental findings can be explained by a simple attractor structure in phase space. The dynamics of the system is dominated by two unstable fixed points: the zero-intensity point I_0, which is a saddle point, and the steady-state fixed point I_+, which is a saddle focus, as shown in the phase space representation of Figure 3.47 (Hennequin et al., 1988). Starting at the steady-state value, the system spirals away from the unstable focus point until it touches the zero-intensity point. After sufficient population inversion has build up (since no laser

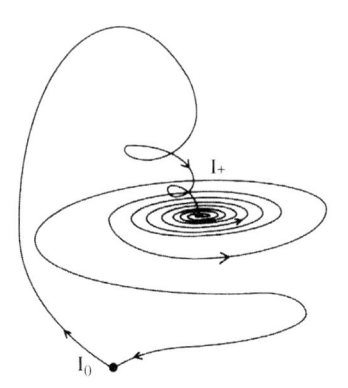

Figure 3.47 Homoclinic cycle representation in phase space connecting two unstable saddle points. This geometry underlies the dynamics of the CO_2 gas laser with a saturable absorber. (D. Hennequin, F. de Tomasi, B. Zambon, and E. Arimondo, (1988), © 1988 APS.)

emission occurs), the system is ejected in a giant pulse from the zero-intensity point and subsequently falls back into the steady-state focus point from which it spirals out again, and so on. The different pulse shapes observed in the experiment differ in the number of orbits around the focus fixed point that are needed to bring the system to the zero-intensity point: In T_{11} it is one, in T_{12} it is two, and so on. The "normal" (periodic) Q-switching pulses correspond to conditions where the steady-state emission point is attractive enough for the system to execute one large loop around the steady-state emission point. The trajectory falls back onto the zero-intensity point, emitting one regular giant Q-switching pulse.

3.6.3
External Modulation

The dynamics of gas lasers with modulated parameters has been studied experimentally and numerically. CO_2 gas lasers have been studied with modulated gain, modulated resonator frequency, and modulated resonator losses. CO_2 lasers with loss modulation shows a period-doubling route to chaos. Figure 3.48 (left) shows RF spectra and phase-space projections of the attractors (Arecchi et al., 1982). The loss of the laser is modulated by an EOM at a frequency near the relaxation oscillation frequency or its subharmonics. The modulation amplitude necessary to destabilize the laser output is lowest near the relaxation oscillation frequency.

Periodic windows within the chaotic range up to period-10 can be observed. Figure 3.48 (right) shows bifurcation diagrams measured by stroboscopically viewing the laser output in a CO_2 laser with loss modulation (Tredicce et al., 1986). The laser intensity is sampled at a fixed point in every period of the modulation. The samples are recorded as a function of the loss modulation amplitude. It is apparent that the bifurcation diagram is different for increasing and decreasing the modulation amplitude. The reason for this hysteresis is multiple coexisting states. In this case, chaotic states coexist with chaotic or periodic states, leading to various kinds of "crises", which are sudden expansions, contractions or disappearances of solutions when two states coalesce, as seen in Figure 3.48 (right).

3.6.4
Optical Injection

The effect of an injected signal into a laser cavity has been studied intensively. The most important effect of an injected optical field on a laser is optical frequency locking, that is, the laser amplification is driven by the injected field and emission occurs at the same optical-carrier frequency of this injected field. Indeed, this method constitutes a widely used technique for controlling the optical-carrier frequency of an unstable powerful laser injected from another laser of high spectral purity or high-frequency stability, known as injection locking (Siegman, 1986). Optical frequency locking occurs for amplitudes of the injected field above a threshold value of injection locking that increases with increasing detuning between the injected and laser fields. Below this threshold, the simultaneous presence of the injected and laser fields

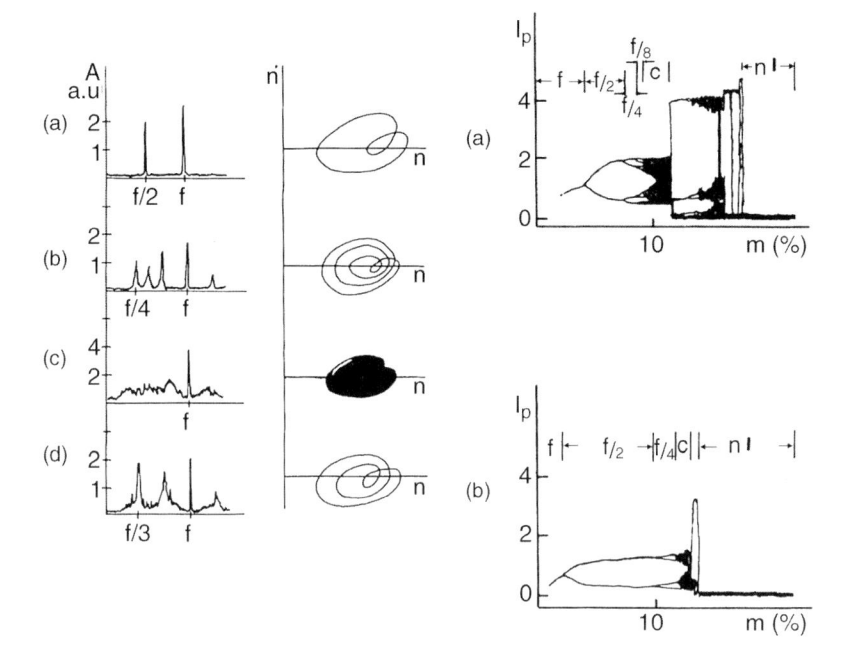

Figure 3.48 (Left) RF spectra of the intensity of a CO_2 laser whose loss is modulated at a frequency near the relaxation oscillation frequency. Period-2 and -4 are visible, followed by a chaotic spectrum and the period-3 window. Also shown is a phase-space projection of the attractors: the dependence of $\dot{n}$ on n (n is the photon number). (J. R. Tredicce, F. T. Arecchi, G. P. Puccioni, A. Poggi, and W. Gadomski, (1986), © 1986 APS.) (Right) Bifurcation diagrams of a loss-modulated CO_2 laser showing the coexistence of attractors by first (a) increasing and then (b) decreasing the modulation amplitude. Coexistence of periodic with chaotic states is seen. Hysteresis of nonlasing state is also evident. (F. T. Arecchi, R. Meucci, G. Puccioni, and J. Tredicce, (1982), © 1982 APS.)

within the laser cavity leads to competition effects that result in complex nonlinear dynamics.

Chaotic oscillations and the period-doubling route to chaos are found in numerical simulations based on a homogeneously broadened single-mode Class B laser with optical field injection. The general findings concerning stable and unstable behaviors are summarized in the phase diagram in Figure 3.49 (top), which gives the areas of different dynamical properties (Tredicce et al., 1985). In the area marked "stable" the laser frequency is locked to the externally injected optical field. The phase difference between the laser and the external signal is zero on the $\theta - \delta = 0$ line and approaches $\pm \pi$ at the stability boundaries.

For example, at the detuning and injection fields in area 6 of Figure 3.49 (top), the period-doubling route to chaos is found. Figure 3.49 (bottom) shows the temporal waveforms of the injected laser output (Tredicce et al., 1985). A period-1 oscillation (with the period of τ_1) is changed to a period-2 oscillation ($2\tau_1$), a period-4 oscillation

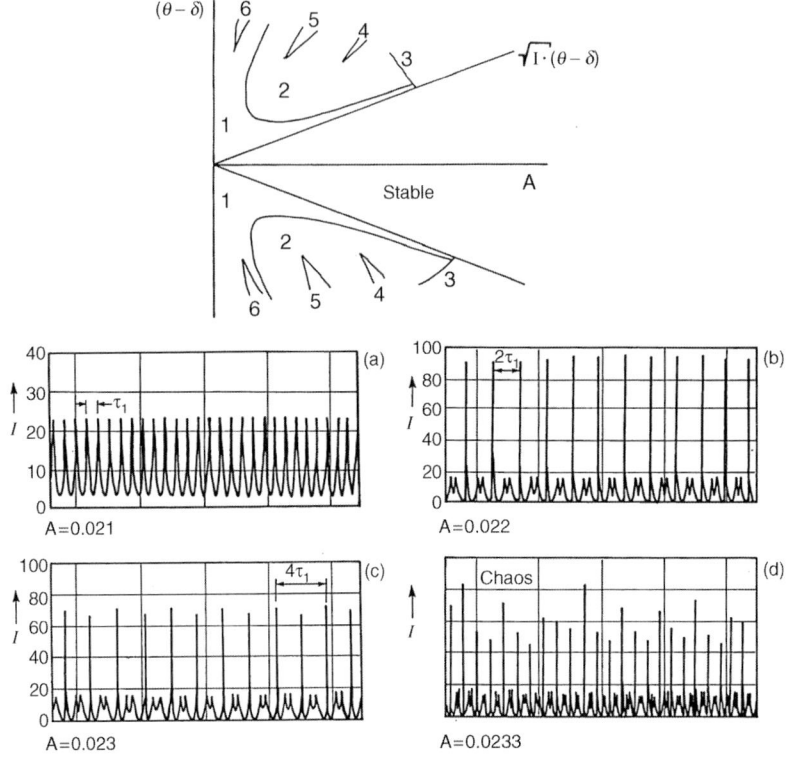

Figure 3.49 (Top) State diagram for a CO_2 laser with an injected field in the detuning – external-field amplitude plane. (Bottom) Temporal waveforms of (a) period-1, (b) period-2, (c) period-4, and (d) chaotic oscillations in area 6 of Figure 3.49 (top). Period-doubling route to chaos is observed. (J. R. Tredicce, F. T. Arecchi, G. L. Lippi, and G. P. Puccioni, (1985), © 1985 OSA.)

($4\tau_1$), and chaotic pulsations. This transition clearly indicates the period-doubling route to chaos.

It is found that this system exhibits multiple coexisting solutions, so that the phase diagram of Figure 3.49 (top) is superimposed with other phase diagrams and the various phases can be reached only by starting from different initial conditions. These numerical results have been confirmed by experimental observations with a CO_2 ring laser with optical injection from another CO_2 laser with active frequency stabilization (Boulnois et al., 1986).

3.6.5
Multimode Laser

Chaotic oscillations can be obtained in multimode gas lasers due to the nonlinear mode–mode interaction. Chaotic dynamics and different types of bifurcations have been observed in a high-gain multimode He-Ne gas lasers at a wavelength of

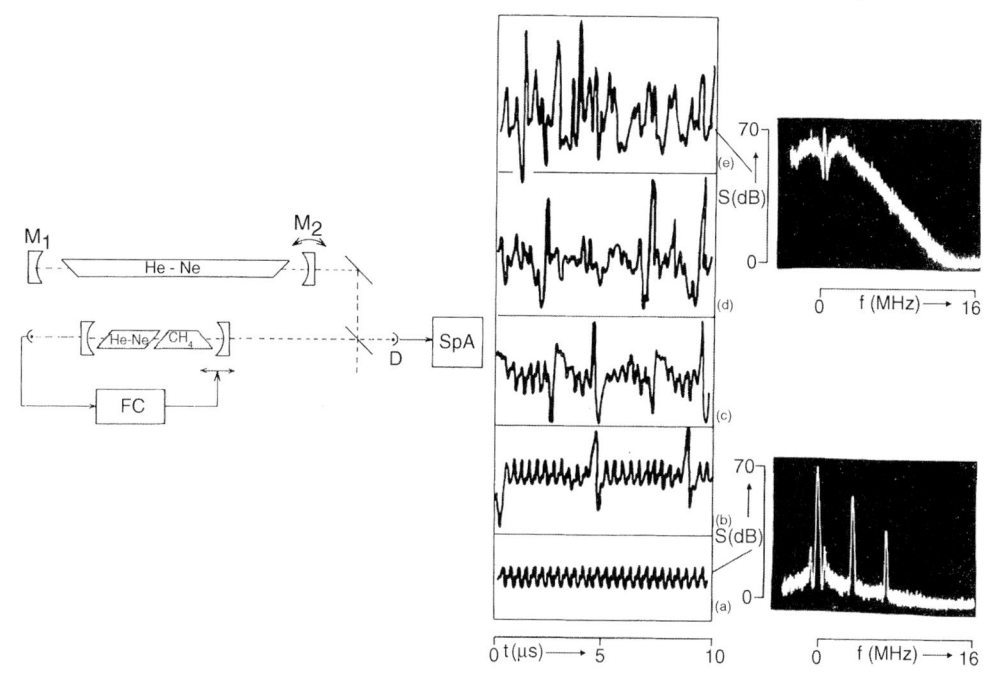

Figure 3.50 (Left) Experimental setup for observation of multimode chaos of a 3.39-μm He-Ne laser, consisting of a 2.5-m long He-Ne laser, heterodyned for precise resonator length control with a methane Lamb-dip-controlled frequency-stable second He-Ne laser. Mirror M_2 is tilted to observe different output pulsing. FC, methane-stabilized laser frequency control; D, photodiode; SpA, radio-frequency spectrum analyzer. (Right) Time dependence of laser output as the control parameter (mirror tilting angle) is varied from (a) stable oscillatory state to (e) chaotic state. RF spectra corresponding to (a) and (e) are also shown. Intermittency route to chaos is found. (C. O. Weiss, A. Godone, and A. Olafsson, (1983), © 1983 APS.)

3.39 μm. The experimental setup is shown in Figure 3.50 (left) (Weiss et al., 1983). The laser frequency is controlled by a reference laser stabilized to a methane Lamb dip. As a control parameter simply the tilting angle of one of the resonator mirrors is used. A tilt increases the relative loss of the various transverse modes. In the experiment only the intensity modulation is detected and analyzed by a RF spectrum analyzer. The observation of an optical spectrum confirms three-mode emission, showing two beats around 60 MHz – the free spectral range of the laser resonator corresponding to three oscillation modes. The difference in the beat frequencies appears as a spectral line near zero frequency, generated by the nonlinear interaction of the three modes.

Figure 3.50 (right) shows the temporal waveforms and RF spectra, representing the intermittency route to chaos (Weiss et al., 1983). A regular, although already nonlinearly distorted, periodic oscillation starts the process. When the mirror tilting angle is varied, sudden bursts of large-amplitude pulses appear in an irregular fashion. With increasing tilt angle these bursts become increasingly frequent until the whole time evolution consists of irregular bursts. As the RF spectrum shows, this

3.6.6
Low-Frequency Dynamics with Optical Feedback

Slow chaotic dynamics less than 1 kHz has been reported in a single-mode class-A He-Ne laser with optical feedback (Kuwashima et al., 1998). An external mirror is placed in front of a He-Ne laser whose wavelength is 632.8 nm (red light). The reflectivity of the mirror is varied by tilting the external mirror. The length of the external cavity between the laser facet and the external mirror is set at 0.23 m. In the presence of optical feedback, the laser remains in a single-mode operation.

Figure 3.51 shows the temporal waveforms of the He-Ne laser intensity when the reflectivity of the external mirror is changed (Kuwashima et al., 1998). The CW steady-state behaviors appear when the tilt of the mirror is −0.7 or 0.9 mrad, as shown in Figures 3.51a and e (the total feedback rate of 0.48%). As the feedback rate increases, the chaotic oscillations occur, as shown in Figure 3.51b–d. Chaotic oscillations with large-amplitude fluctuations are observed at the maximum feedback strength with the tilt of 0.0 mrad (Figure 3.51c). It is worth noting that the center frequency of the chaotic oscillations ranges around ∼100 Hz, which is much slower than the characteristics frequencies of the laser parameters (i.e., the external cavity frequency

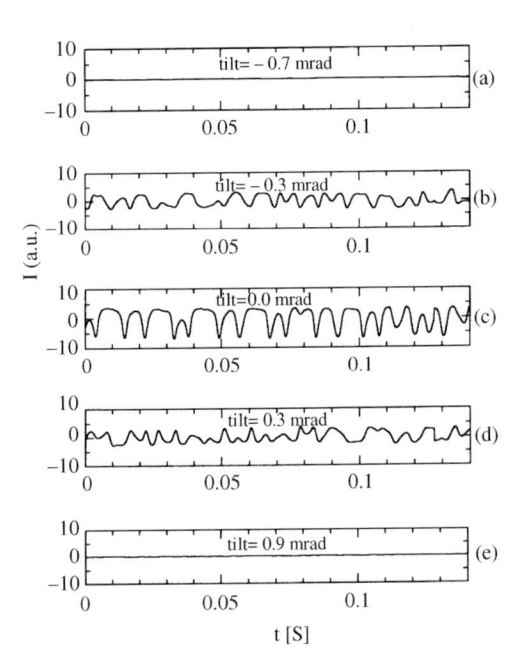

Figure 3.51 Experimental results of laser output intensity from the external mirror. The length of the external resonator is 0.23 m. The tilts of the external mirror are (a) −0.7 mrad, (b) −0.3 mrad, (c) 0.0 mrad, (d) 0.3 mrad, and (e) 0.9 mrad. (F. Kuwashima, I. Kitazima, and H. Iwasawa, (1998), © 1998 JJAP.)

or the relaxation oscillation frequency). This chaotic oscillation frequency is also much slower than those observed in other gas lasers. The He-Ne laser used in the experiment satisfies the good-cavity condition and is classified as a class A laser. It is confirmed experimentally that the class-A He-Ne laser yields chaotic oscillations by means of optical feedback.

Appendix

3.A.1
Numerical Model for Chaotic Dynamics in a Semiconductor Laser

3.A.1.1
Coherent Optical Feedback

The dynamics of intensity fluctuation in a semiconductor laser with time-delayed optical feedback has been modeled, which is called the Lang–Kobayashi equations (Lang and Kobayashi, 1980). The Lang–Kobayashi equations for the real electric amplitude $E(t)$, the real phase $\Phi(t)$, and the carrier density $N(t)$ are written as follows (Ohtsubo, 2008),

$$\frac{dE(t)}{dt} = \frac{1}{2}\left[G_N(N(t)-N_0) - \frac{1}{\tau_p}\right]E(t) + \kappa\, E(t-\tau)\cos\Theta(t) \qquad (3.A.1)$$

$$\frac{d\Phi(t)}{dt} = \frac{\alpha}{2}\left[G_N(N(t)-N_0) - \frac{1}{\tau_p}\right] - \kappa\frac{E(t-\tau)}{E(t)}\sin\Theta(t) \qquad (3.A.2)$$

$$\frac{dN(t)}{dt} = J - \frac{N(t)}{\tau_s} - G_N(N(t)-N_0)E^2(t) \qquad (3.A.3)$$

$$\Theta(t) = \omega\tau + \Phi(t) - \Phi(t-\tau) \qquad (3.A.4)$$

The parameters used in Eqs. (4.3)–(4.6) and their values for numerical simulation are summarized in Table 4.3 in Chapter 4. The laser intensity is calculated from the square of the electric amplitude, that is, $I(t) = E^2(t)$. See the detailed analysis with the Lang–Kobayashi equations in Chapter 4.

3.A.1.2
Polarization-Rotated Optical Feedback

A numerical model for a semiconductor laser with polarization-rotated optical feedback is described. A two-polarization-mode dynamical model is used, allowing for the dynamics of the TM mode as well as the TE mode in the laser. The rate equations for TE and TM modes can be written as follows (Shibasaki et al., 2006),

$$\frac{dE_{te}(t)}{dt} = \frac{1}{2}\{G_{te}(N(t)-N_0) - \gamma_{te}\}E_{te}(t) \qquad (3.A.5)$$

$$\frac{d\Phi_{te}(t)}{dt} = \frac{\alpha}{2}\{G_{te}(N(t)-N_0)-\gamma_{te}\} \qquad (3.A.6)$$

$$\frac{dE_{tm}(t)}{dt} = \frac{1}{2}\{G_{tm}(N(t)-N_0)-\gamma_{tm}\}E_{tm}(t) + \kappa E_{te}(t-\tau)\cos\theta(t) \qquad (3.A.7)$$

$$\frac{d\Phi_{tm}(t)}{dt} = \frac{\alpha}{2}\{G_{tm}(N(t)-N_0)-\gamma_{tm}\} - \kappa\frac{E_{te}(t-\tau)}{E_{tm}(t)}\sin\theta(t) \qquad (3.A.8)$$

$$\frac{dN(t)}{dt} = J - \gamma_s N(t) - (N(t)-N_0)\{G_{te}|E_{te}(t)|^2 + G_{tm}|E_{tm}(t)|^2\} \qquad (3.A.9)$$

$$\theta(t) = -\Delta\omega_{te,tm}t + \omega\tau + \Phi_{tm}(t) - \Phi_{te}(t-\tau) \qquad (3.A.10)$$

The variables E and Φ are the electrical amplitude and the phase, N is the carrier density, and θ is the phase difference between the TM mode of the laser and the feedback or injected light. The subscripts te and tm indicate the TE and TM modes, respectively. G is the gain coefficient, and N_0 is the carrier density at the transparency. $\gamma = 1/\tau_p$ is the inverse of the photon lifetime, $\gamma_s = 1/\tau_s$ is the inverse of the carrier lifetime, α is the linewidth-enhancement factor, J is the injection current density per unit time, and $J_{th,sol} = \gamma_s(N_0 + \gamma_{te}/G_{te})$ is the threshold of the injection current density per unit time. κ is the feedback coefficient for the drive laser, and τ is the propagation time of the external loop in the laser. ω is the angular frequency, $\Delta\omega_{te,tm} = \omega_{te,d} - \omega_{tm,d}$ is the detuning of optical angular frequency between the TE and TM modes of the laser, and $\lambda = 2\pi c/\omega$ is the wavelength. The total intensity of the drive and response lasers can be obtained from $I = |E_{te}|^2 + |E_{tm}|^2$.

Typical parameter values for numerical simulation are as follows:
$G_{te} = 1.374 \times 10^{-12} m^3 s^{-1}$, $G_{tm} = 1.154 \times 10^{-12} m^3 s^{-1}$, $N_0 = 1.400 \times 10^{24} m^{-3}$, $\gamma_{te} = \gamma_{tm} = 8.913 \times 10^{11} s^{-1}$, $\kappa = 1.25 \times 10^{11} s^{-1}$, $\gamma_s = 4.902 \times 10^8 s^{-1}$, $\tau = 6.67 \times 10^{-9} s$, $\alpha = 3.0$, $J = 2.0 J_{th,sol}$, $J_{th,sol} = 1.004 \times 10^{33} m^{-3} s^{-1}$, $\lambda = 1.537 \times 10^{-6} m$, and $\Delta\omega = 0$.

3.A.1.3
Optoelectronic Feedback (Incoherent Feedback)

A way of obtaining an incoherent feedback is via the injection current of a semiconductor laser. The optical power emitted by the single-mode semiconductor laser is detected by a photodiode and converted into an electric current, which is in turn reinjected into the injection current of the laser with time delay. This optoelectronic feedback drives the laser to chaos, provided that the feedback delay is carefully adjusted. This system can be modeled by the following simplified equations (Tang and Liu, 2001a)

$$\frac{dS(t)}{dt} = -\gamma_c S(t) + \Gamma g_0 S(t) + \Gamma g_n (N(t)-N_0) S(t)$$
$$+ \Gamma g_p (S(t)-S_0) S(t) \tag{3.A.11}$$

$$\frac{dN(t)}{dt} = \frac{J}{ed}\left[1 + \frac{\xi S(t-\tau)}{S_0}\right] - \gamma_s N(t) - g_0 S(t)$$
$$-g_n(N(t)-N_0)S(t) - g_p(S(t)-S_0)S(t) \tag{3.A.12}$$

where $S(t)$ is the intracavity photon density and $N(t)$ is the carrier density. S_0 is the free-running intracavity photon density when the laser is not subject to the feedback, N_0 is the free-running carrier density when the laser is not subject to the feedback, g_0 is the optical gain coefficient at free-running condition, g_n is the differential gain parameter, g_p is the nonlinear gain parameter, ξ is the feedback strength, τ is the feedback delay time, J is the bias current density, γ_c is the cavity photon decay rate, γ_s is the spontaneous carrier density rate, Γ is the confinement factor of the laser waveguide, e is the electronic charge constant, and d is the active layer thickness.

The dimensionless equations can be derived from Eqs. (3.A.11) and (3.A.12) as follows (Tang and Liu, 2001a).

$$\frac{d\tilde{s}(t)}{dt} = \frac{\gamma_c \gamma_n}{\tilde{J}\gamma_s}\tilde{n}(\tilde{s}+1) - \gamma_p \tilde{s}(\tilde{s}+1) \tag{3.A.13}$$

$$\frac{d\tilde{n}(t)}{dt} = \gamma_s \xi(1+\tilde{J})[1+\tilde{s}(t-\tau)] - \gamma_s \tilde{n} - \gamma_s \tilde{J}\tilde{s}$$
$$-\gamma_n \tilde{n}(1+\tilde{s}) + \frac{\gamma_s \gamma_p}{\gamma_c}\tilde{J}\tilde{s}(1+\tilde{s}) \tag{3.A.14}$$

Where $\tilde{s}$ is the dimensionless photon density, $\tilde{n}$ is the dimensionless carrier density, and $\tilde{J}$ is the dimensionless injection current. Typical parameter values for numerical simulation are as follows. $\gamma_c = 2.4 \times 10^{11}\,s^{-1}$, $\gamma_s = 1.458 \times 10^9\,s^{-1}$, $\gamma_n = 3\tilde{J} \times 10^9\,s^{-1}$, $\gamma_p = 3.6\tilde{J} \times 10^9\,s^{-1}$, $\tilde{J} = 1/3$, and $\xi = 0.1$.

3.A.2
Mechanism of Low-Frequency Fluctuations (LFFs) in Semiconductor Laser with Optical Feedback

The mechanism of LFFs can be generally explained by employing the model that the laser output power hops around external and antimodes of the laser oscillations due to saddle-node instability generated by the optical feedback (also known as "chaotic itinerancy") (Sano, 1994; Ohtsubo, 2008). The external and antimodes are considered in the phase space of the oscillation frequency and the carrier density. There exist multiple steady-state solutions in semiconductor lasers with optical feedback for the detuning of the optical-carrier frequency $\Delta\omega_{s,0} = \omega_s - \omega_0$ and the carrier density

$\Delta N_{s,th} = N_s - N_{th}$ (see Eqs. (4.69) and (4.70) in Chapter 4). Figure 3.A.1a shows half the distribution of the modes in the phase space of $(\Delta\omega_{s,0}\tau - \Delta N_s)$ (see Eq. (4.74) in Chapter 4) (Ohtsubo, 2008). The laser oscillates at one of the stable modes (lower-half modes on the ellipsoid). The modes of the upper half on the ellipsoid are unstable and they are not generally stable lasing modes. The black dot at the center of the ellipsoid is the mode of the solitary oscillation. The laser without optical feedback oscillates at this single mode. Among the modes in Figure 3.A.1a, the most probable mode of the laser oscillation in the presence of optical feedback is the "maximum gain mode". However, the laser does not always oscillate at this mode. Folding and stretching the variables induced by nonlinear characteristics in the system due to small fluctuations, the laser may be suddenly trapped to a spiky orbit with a large-amplitude oscillation.

The corresponding LFF waveform to Figure 3.A.1a is also shown in Figure 3.A.1b. It is considered that the laser initially oscillates around the maximum gain mode and the state of the laser oscillation fluctuates near this mode with small amplitude. This is not the fixed stable mode. Once the state reaches a point very close to the counterpart antimode, the laser may be trapped in the antimode. The phase remains unchanged, but the carrier density abruptly jumps up to the value of the solitary oscillation (A in Figure 3.A.1). The sudden jump of the carrier density induces the increase of the phase and the laser shows a sudden power dropout. At this state, the laser output power is almost equal to the free running oscillation (B in Figure 3.A.1). This corresponds to a sudden power dropout of LFF. After that, the laser is trapped by one of the external modes close to the solitary oscillation mode (C in Figure 3.A.1). Then, the state goes around the successive external modes toward the maximum gain mode (D in Figure 3.A.1). This corresponds to the power recovery process of LFF. When the laser reaches the maximum gain mode, the above process is repeated. The occurrence of LFFs is not periodic but irregular, since the fluctuations exist in the chaotic itinerary due to the

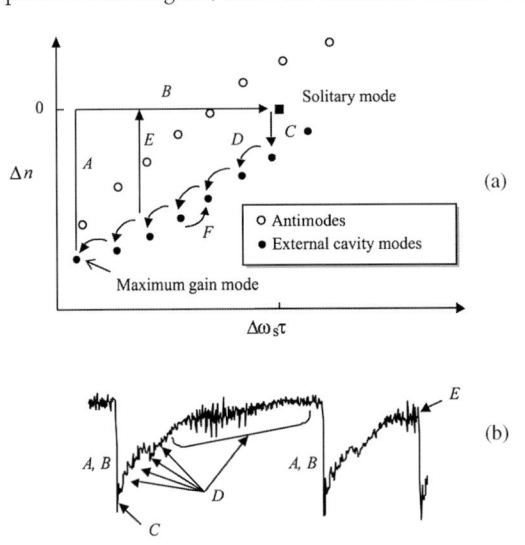

Figure 3.A.1 (a) External- and antimodes in the phase space of frequency and carrier density. Only half of the ellipsoid of the mode distribution is shown in the graph. (b) Corresponding LFF waveform to (a). (J. Ohtsubo, (2008), © 2008 Springer.)

nonlinear effects. For example, when the chaotic oscillation has a large amplitude in the power-recovery process, the laser may be trapped in the associate antimode and a power dropout may occur even before reaching the maximum gain mode (E in Figure 3.A.1). Another case is a reversion of the power recovery and the state goes up against the direction for the maximum gain mode (F in Figure 3.A.1).

3.A.3
Numerical Model for Chaotic Dynamics in a Vertical-Cavity Surface-Emitting Laser (VCSEL)

The rate equations for linearly polarized lights (x- and y-polarized components) of a vertical-cavity surface-emitting laser (VCSEL) can be described as follows (Martin-Regalado, et al., 1997),

$$\frac{dE_x}{dt} = -(\kappa + \gamma_a)E_x - i(\kappa\alpha + \gamma_p)E_x + \kappa(1+i\alpha)(NE_x + inE_y) \quad (3.A.15)$$

$$\frac{dE_y}{dt} = -(\kappa - \gamma_a)E_y - i(\kappa\alpha - \gamma_p)E_y + \kappa(1+i\alpha)(NE_y - inE_x) \quad (3.A.16)$$

$$\frac{dN}{dt} = -\gamma\left[N\left(1 + |E_x|^2 + |E_y|^2\right) - \mu + in\left(E_y E_x^* - E_x E_y^*\right)\right] \quad (3.A.17)$$

$$\frac{dn}{dt} = -\gamma_s n - \gamma\left[n\left(|E_x|^2 + |E_y|^2\right) + iN\left(E_y E_x^* - E_x E_y^*\right)\right] \quad (3.A.18)$$

Where E_x and E_y are the x- and y-polarized linear components of the complex electric fields. N is the total carrier number in excess of its value at transparency normalized to the value of that excess at the lasing threshold, n is the difference in the carrier numbers of the two magnetic sublevels normalized in the same way as N. α is the linewidth enhancement factor, μ is the injection current normalized at the lasing threshold, κ is the decay rate of the electric field in the cavity, γ is the decay rate of the total carrier number, γ_s is the excess in the decay rate over γ, κ is the decay rate of the electric field in the cavity, γ_p is the linear phase anisotropy (birefringence) which represents the effect of different indices of refraction for the orthogonal linearly polarized modes, γ_a is the amplitude anisotropy which represents the effect of different gain-to-loss ratio of the two polarized modes.

3.A.4
Numerical Model for Chaotic Dynamics in an Electro-Optic System

The dynamics of electro-optic systems can be described by a particular class of delay-differential equations where the feedback is modeled by a nonlinear function, whose transfer characteristic has at least a maximum or a minimum on the variable range. An optoelectronic system with a birefringent plate in an open-loop configuration is modeled as (Larger et al., 1998a, 1998b),

$$x(t) + \tau \frac{dx(t)}{dt} = \beta \sin^2[x(t-T) - \Phi_0] \quad (3.A.19)$$

where $x(t)$ is the dynamical variable of the system (i.e., laser intensity, phase, or wavelength), T is the delay time in the optoelectronic feedback loop, τ is the response time of the system, β is the height of the nonlinear function in the optoelectronic feedback loop, and Φ_0 the initial phase of the nonlinear function.

Typical parameter values for numerical simulation are as follows: $\tau = 8.6 \times 10^{-6}$ s, $T = 5.1 \times 10^{-4}$ s, $\Phi_0 = 0.3$, and $\beta = 2.11$.

3.A.5
Numerical Model for Chaotic Dynamics in a Fiber Laser

For a numerical model of a fiber laser, a modeling approach based on delay-differential equations can be used. This approach does not require either single-mode or mean-field approximations, making use instead of the boundary conditions affecting the laser light as it travels around the cavity. Introducing these conditions into the Maxwell–Bloch equations that govern the evolution of the electric field and population inversion, after the material polarization has been adiabatically eliminated, leads to the following delay-differential equation model (Williams et al., 1997),

$$E(t + \tau_R) = E_{\text{inj}} \exp(i\Delta\omega t) + RE(t) \exp\{(\beta + i\alpha)w(t) + i\kappa\} \quad (3.A.20)$$

$$\frac{dw(t)}{dt} = Q - 2\gamma \left(w + 1 + |E(t)|^2 \frac{[\exp\{Gw(t)\} - 1]}{G} \right) \quad (3.A.21)$$

where $E(t)$ is the complex envelope of the electric field, measured at a given reference point inside the cavity, and $w(t)$ is the total population inversion of the nonlinear medium. The propagation round-trip time around the cavity is τ_R, and the fraction of light that remains in the cavity after one round-trip is measured by the return coefficient R. The field acquires a phase κ after each round-trip. Light of constant amplitude E_{inj} is assumed to be injected into the cavity, with $\Delta\omega$ the detuning between the injected and the laser frequencies. Additionally, the gain medium is pumped at a rate Q. γ is the decay time of the atomic transition, α is a detuning parameter, and β and G are gain parameters.

3.A.6
Numerical Model for Chaotic Dynamics in a Solid-State Laser

3.A.6.1
Single-Mode Laser Model

The dynamics of a single-mode solid-state laser can be described by a class-B laser model for the electrical field and population inversion. The rate equations for a single-mode solid-state laser for the slowly varying, complex electric-field amplitude $E_i(t)$ and real gain $G_i(t)$ of laser i are (Fabiny et al., 1993).

$$\frac{dE(t)}{dt} = \frac{1}{\tau_c}[(G(t)-\alpha)E(t)] + i\omega E(t) \tag{3.A.22}$$

$$\frac{dG(t)}{dt} = \frac{1}{\tau_f}\left[p - G(t) - G(t)|E(t)|^2\right] \tag{3.A.23}$$

In these equations, τ_c is the cavity round-trip time, τ_f is the fluorescent time of the upper lasing level, p is the pumping coefficient, α is the cavity loss coefficient, and ω is the detuning of the laser from a common cavity mode i is the imaginary unit.

Chaos can be generated by using either pump modulation or loss modulation, that is,

Pump modulation:

$$p(t) = p_0(1 + b\cos(\Omega t)) \tag{3.A.24}$$

Loss modulation:

$$\alpha(t) = \alpha_0(1 + b\cos(\Omega t)) \tag{3.A.25}$$

where b and Ω are the modulation amplitude and angular frequency.

Typical parameter values for numerical simulation are as follows: $\tau_c = 2.0 \times 10^{-10}$ s, $\tau_f = 2.4 \times 10^{-4}$ s, and $\alpha = 1.0 \times 10^{-2}$.

3.A.6.2
Multimode Laser Model with Spatial Hole Burning (Tang–Statz–deMars Equations)

In solid-state lasers, spatial hole burning usually leads to laser emission in several longitudinal cavity modes (Mandel, 1997). The simple Class B laser model shown in Eqs. (3.A.22)–(3.A.23) is not appropriate for describing the dynamics of multi-longitudinal-mode solid-state lasers. An excellent model describing cross-saturation of population inversion among modes due to the spatial hole-burning effect has been proposed for multimode solid-state lasers (Tang et al., 1963; Mandel, 1997; Otsuka, 1999). A standing wave for each electric-field amplitude is spatially distributed in the laser crystal, which causes a spatial grating of population inversion along the z-axis (propagation direction of laser light). The spatial distribution of the population inversion for an N-mode laser can be decomposed as follows,

$$n_0(t) = \frac{1}{L}\int_0^L n_{\text{total}}(z, t)dz \tag{3.A.26}$$

$$n_i(t) = \frac{2}{L}\int_0^L n_{\text{total}}(z, t)\cos(2k_i z)dz \tag{3.A.27}$$

where $n_0(t)$ and $n_i(t)$ are the space-averaged and the first Fourier components (in space) of population inversion density for the ith mode ($i = 1, 2, \ldots, N$). $n_{\text{total}}(z, t)$ is the total population inversion density inside the cavity. L is the length of the laser cavity. k_i is the wave number of the electric field for the i-mode. The electric field of each laser mode is coupled to other modes through the spatially decomposed population inversions of $n_0(t)$ and $n_i(t)$. The Tang–Statz–deMars equations for a N-mode laser are described with $n_0(t)$ and $n_i(t)$ as follows (Tang et al., 1963; Otsuka, 1999; Ogawa et al., 2002),

$$\frac{dn_0(t)}{dt} = p - n_0(t) - \sum_{k=1}^{N} \gamma_k \left(n_0(t) - \frac{n_k(t)}{2} \right) E_k^2(t) \tag{3.A.28}$$

$$\frac{dn_i(t)}{dt} = \gamma_i n_0(t) E_i^2(t) - n_i(t) \left(1 + \sum_{k=1}^{N} \gamma_k E_k^2(t) \right) \tag{3.A.29}$$

$$\frac{dE_i(t)}{dt} = \frac{K}{2} \left[\gamma_i \left(n_0(t) - \frac{n_i(t)}{2} \right) - 1 \right] E_i(t) \tag{3.A.30}$$

where $E_i(t)$ is the real amplitude of the lasing electric field for the ith mode. p is the pumping parameter scaled to the laser threshold. γ is the gain coefficient. $K = \tau/\tau_p$, where τ is the upper state lifetime and τ_p is the photon lifetime in the laser cavity. Time is scaled by τ. To generate chaos, pump modulation can be used as,

$$p = p_0(1 + A\cos(2\pi\tau f t)) \tag{3.A.31}$$

where A and f are the modulation amplitude and frequency, respectively.

For a two-mode laser ($N = 2$), typical parameter values for numerical simulation are as follows:
$\gamma_1 = 1.0$, $\gamma_2 = 0.875$, $\tau = 8.8 \times 10^{-5}$ s, $\tau_p = 1.15 \times 10^{-9}$ s, $K = \tau/\tau_p = 7.65 \times 10^4$, $p = 5.2$, $A = 0.15$, and $f = 1.08 \times 10^6$ Hz.

3.A.7
Numerical Model for Chaotic Dynamics in a Gas Laser

A CO_2 gas laser with a saturable absorber can be described by a three-level–two-level rate-equation model, as an extension of the class-B laser model, for the normalized photon density $I(t)$, population density in the upper laser level $M_1(t)$, population density in the lower laser level $M_2(t)$, and the difference of the population density in the absorber levels $N(t)$ (Sugawara et al., 1994).

$$\frac{dI(t)}{dt} = [(M_1(t) - M_2(t)) - BN(t) - k] I(t) \tag{3.A.32}$$

$$\frac{dM_1(t)}{dt} = PM(t) - R_{10} M_1(t) - (M_1(t) - M_2(t)) I(t) \tag{3.A.33}$$

$$\frac{dM_2(t)}{dt} = -R_{20} M_2(t) + (M_1(t) - M_2(t)) I(t) \tag{3.A.34}$$

$$\frac{dN(t)}{dt} = -2bN(t)I(t) - r(N(t)-1) \qquad (3.A.35)$$

where B is the normalized rate of absorption in the passive medium, k is the cavity loss rate, P is the rate at which the upper laser level is pumped, R_{10} is the rate of vibrational relaxation from the upper laser level to all other levels except the lower level, R_{20} is the rate of vibrational relaxation from the lower laser level to other levels, b is the normalized cross section of the absorption in the passive medium, and r is the rotational relaxation rate of the absorptive levels.

Typical parameter values for numerical simulation are as follows: $B = 1.5 \times 10^6$ Hz, $k = 2.0 \times 10^6$ Hz, $P = 16.70$ Hz, $R_{10} = 3.0 \times 10^2$ Hz, $R_{20} = 3.8 \times 10^5$ Hz, $b = 900.0$, and $r = 6.0 \times 10^6$ Hz

4
Analysis of Chaotic Laser Dynamics: Example of Semiconductor Laser with Optical Feedback

In this chapter, the methods for nonlinear analysis of chaotic laser dynamics are introduced and discussed in detail. Chaotic semiconductor lasers with time-delayed optical feedback are used as an example for nonlinear dynamical analysis. It is described how to understand and analyze the characteristics of intensity fluctuations in semiconductor lasers in this chapter. Three main approaches will be introduced: experimental, analytical, and numerical analyses. The methods described in this chapter are standard techniques for nonlinear dynamical analysis, and can be applied to other dynamical laser systems.

This chapter, describing typical examples of analysis on the dynamics of a chaotic semiconductor laser with optical feedback, is specially designed for those readers who are interested in starting research works in this interdisciplinary field but do not know where to start. The contents in this chapter (as well as Chapter 6) are very specific and technical, and different from other chapters where many examples of each topic are widely described from the literature. If the readers would like to learn detail techniques of nonlinear analysis in dynamical laser systems by using rate equations and numerical simulations, they may learn many techniques from the sections of analytical and numerical analyses in this chapter. The contents described in this chapter would be good examples for students and researchers to start scientific research of chaotic laser dynamics. If the readers are not familiar with mathematical formula and feel uncomfortable reading through the whole chapter with complicated mathematical equations, they may read only the experimental part of Section 4.1 and may skip the analytical and numerical sections in this chapter.

4.1
Experimental Analysis of Semiconductor Lasers with Optical Feedback

In this section, the dynamics of laser output is observed experimentally. The output of a semiconductor laser is susceptible to self-feedback light from an external mirror or reflector, which are frequently observed in optical storage and optical communication systems. Typical temporal dynamics obtained from experiment are shown in this section, such as quasiperiodic oscillations, chaotic oscillations, and low-frequency fluctuations. RF spectra and autocorrelation functions are shown to identify the dynamical states. A two-dimensional (2D) dynamical map is also created experimentally.

4.1.1
Experimental Setup

The model for generating optical chaos in a semiconductor laser has been already shown in Figures 3.1a and 3.4 (left). To implement this scheme in experiments, two configurations can be considered: space-optic-based and fiber-optic-based laser systems. Figure 4.1 shows the experimental setup for the observation of chaotic dynamics in a semiconductor laser with optical feedback based on space-optic and fiber-optic components. A distributed-feedback (DFB) semiconductor laser is used in this experiment (the optical wavelength of 1547 nm), which was developed for

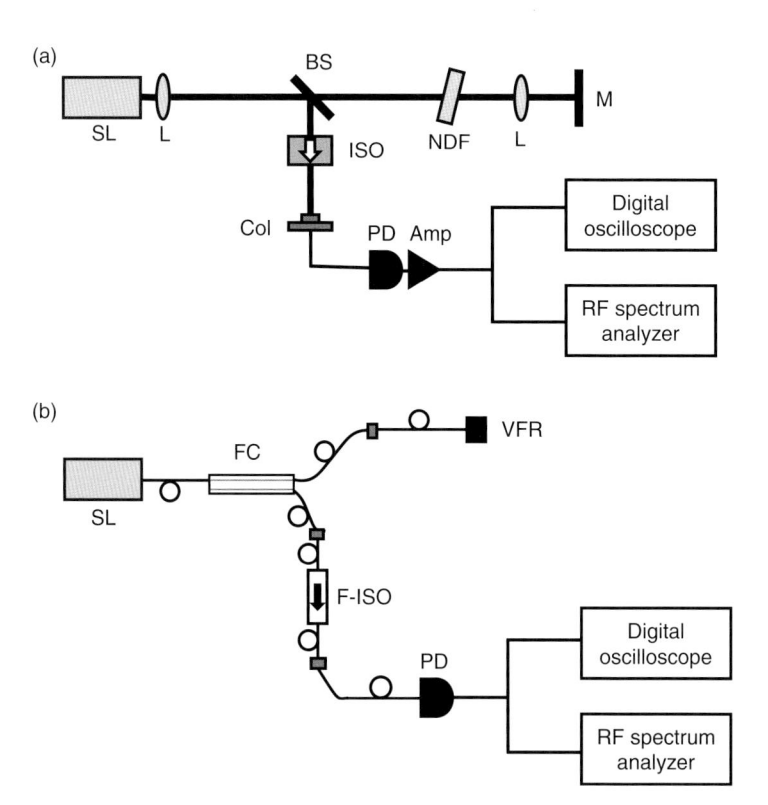

Figure 4.1 Experimental setup for the observation of chaotic dynamics in a semiconductor laser with optical feedback with (a) space-optic and (b) fiber-optic components. Amp, electronic amplifier; BS, beam splitter; Col, fiber collimator; FC, fiber coupler; F-ISO, optical-fiber isolator; ISO, optical isolator; L, lens; M, mirror; NDF, neutral density filter (variable attenuator); PD, photodetector; SL, semiconductor laser; VFR, variable fiber reflector.

conventional optical-fiber communications. For the space-optic configuration in Figure 4.1a, an external mirror (M) is placed in front of the laser, forming an external cavity. A portion of the laser beam from the laser is fed back to the laser cavity to induce chaotic fluctuation of laser output. The feedback power is adjusted by a neutral-density filter (NDF). A portion of the laser output is extracted by a beam splitter (BS), injected into a fiber collimator (Col) through an optical isolator (ISO) and propagated through an optical fiber to be detected by a photodetector (PD). The converted electronic signal is amplified by an electronic amplifier (Amp).

By contrast, for the fiber-optic configuration in Figure 4.1b, the SL is mounted in a butterfly package with an optical-fiber pigtail. The laser is prepared without standard optical isolators to allow optical feedback. The laser is connected to a fiber coupler (FC) and a variable fiber reflector (VFR) that reflects a fraction of the light back into the laser, forming an external cavity. The feedback light induces high-frequency chaotic oscillations of the optical intensity. The amount of optical feedback light is adjusted by the VFR. A portion of the laser output is extracted by the FC and detected by a PD through an optical-fiber isolator (F-ISO). No electronic amplifier is needed due to small optical loss compared with space-optic configuration. Polarization-maintaining (PM) fibers are used for all the optical-fiber components. Polarization controllers are required in the optical feedback loop when single-mode fiber components are used instead of PM fibers. For both configurations, the converted electronic signal is sent to a digital oscilloscope and a radio-frequency (RF) spectrum analyzer to observe temporal waveforms and the corresponding RF spectra, respectively.

There is no essential difference between the space-optic and fiber-optic configurations from the nonlinear dynamics point of view, except for the stability of the system and the external cavity length. The dynamics in fiber-based systems is more robust against mechanical vibration and air turbulence. For the space-optic configuration, it may be required to realign the beam position frequently during experiments. In addition, the external cavity length for fiber-optic systems (1–10 m or more) can be generally larger than that for space-optic systems (0.1–1 m), which have the limitation due to the divergence of the laser beam. The space-optic configurations are traditional laser systems and commonly used in the literature, although some experimental skills are required to implement space-optic experimental setups. By contrast, fiber-optic configurations became more popular because of good stability and ease of handling the optical components, even for beginners.

In Chapter 4.1, the experimental results obtained from the fiber-optic configuration are presented. The specification of the experimental equipment used in this experiment is shown in Table 4.1. The values of the laser parameters used in this experiment are summarized in Table 4.2.

4.1.2
Light Power Versus Injection Current (L–I) Characteristics

One of the basic characteristics of semiconductor lasers is the light power *versus* injection current characteristics (referred to as the *L–I* curve or *L-I* characteristics). Figure 4.2 shows the characteristics of the light power as the injection current is

Table 4.1 Specifications of experimental equipments used in Chapters 4 and 6.

Component	Brand, Product ID	Specifications
Distributed-feedback (DFB) semiconductor laser	NTT Electronics KELD1C5GAAA	Wavelength: 1547 nm Butterfly package with optical-fiber pigtail
Current–temperature controller for semiconductor lasers	Newport 8000-OPT-41-41-41-41	Current resolution: 0.01 mA Temperature resolution: 0.01 K
Photodetector	New Focus 1554-B	12 GHz bandwidth
Digital oscilloscope	Tektronix DPO71604	16 GHz bandwidth, 50 GigaSamples/s
RF spectrum analyzer	Agilent N9010A-526	26.5 GHz bandwidth
Optical spectrum analyzer	Advantest, Q8384	0.01 nm resolution

changed. Without optical feedback, the laser starts lasing at the threshold ($I_{th} = 9.50$ mA) and the light power increases linearly after the threshold. In the presence of optical feedback, however, both the threshold and the slope of the L–I curve are reduced as the feedback light power is increased. For strong feedback, the threshold is reduced to 6.05 mA, which corresponds to 36% threshold reduction. In addition, for the moderate feedback power, there is a discontinuity in the L–I curve around 9.5 mA, which is referred to as a "kink." Complex nonlinear dynamics can be observed at the conditions of the occurrence of the kink.

Table 4.2 Parameter values used in the experiment of Section 4.1 for generation of chaos in the fiber-optic configuration.

Parameter	Parameter Value
Threshold of injection current [mA]	9.50
Injection current [mA]	14.10
Relative injection current to the threshold	1.50
Relaxation oscillation frequency [GHz]	2.80
Laser temperature [K]	293.00
Optical feedback power [μW]	variable
External cavity length [m] (One-way, fiber length)	2.52
Feedback delay time [ns]	24.3
External cavity frequency [MHz]	41.2

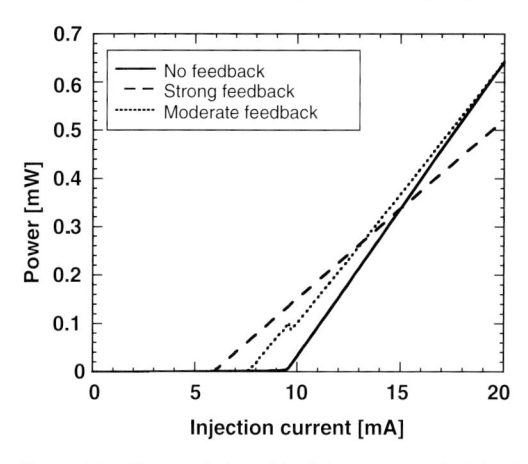

Figure 4.2 Characteristics of the light power as the injection current is changed (L–I curve) for a semiconductor laser. The experimental data in this chapter was provided by Haruka Okumura and Hiroki Aida.

4.1.3
Temporal Waveforms, RF Spectra, Autocorrelations, and Optical Spectra

The observation of temporal waveforms and radio-frequency (RF) spectra of laser output intensity is a first step in nonlinear dynamical analysis. Figure 4.3 shows the temporal waveforms of the semiconductor laser output when the feedback strength is increased. Without optical feedback (Figure 4.3a), sustained relaxation oscillation is observed and slight fluctuation of the laser output appears. As the feedback strength is increased, a quasiperiodic oscillation appears with increase of the feedback strength (Figure 4.3b). This regular oscillation becomes unstable and weak chaotic oscillation appears in Figure 4.5c. The chaotic oscillation becomes more irregular as the feedback strength is increased, as shown in Figure 4.3d (referred to as moderate chaos). For further increase of the feedback strength, the oscillation frequency becomes larger and more irregular chaotic oscillation appears, as shown in Figure 4.3e (referred to as strong chaos). Finally, the laser output becomes stable and dynamical behavior is no longer observed at stronger optical feedback strength, as shown in Figure 4.3f.

When time series are observed in a longer range, quasiperiodic oscillations are clearly observed as shown in Figure 4.4a, which corresponds to the condition for Figure 4.3b. Quasiperiodic oscillation with small sustained relaxation oscillation is observed in Figure 4.4a. The period of the quasiperiodic oscillation is around 25 ns, corresponding to the round-trip time of the optical feedback light in the external cavity (i.e., the inverse of the external cavity frequency). Low-frequency fluctuations are also observed as shown in Figure 4.4b at the condition of the feedback strength between the strong chaos (Figure 4.3e) and stable output (Figure 4.3f). An irregular burst of the laser intensity is observed at a long period of several hundreds of nanoseconds, whose frequency corresponds to a few MHz. This oscillation is very

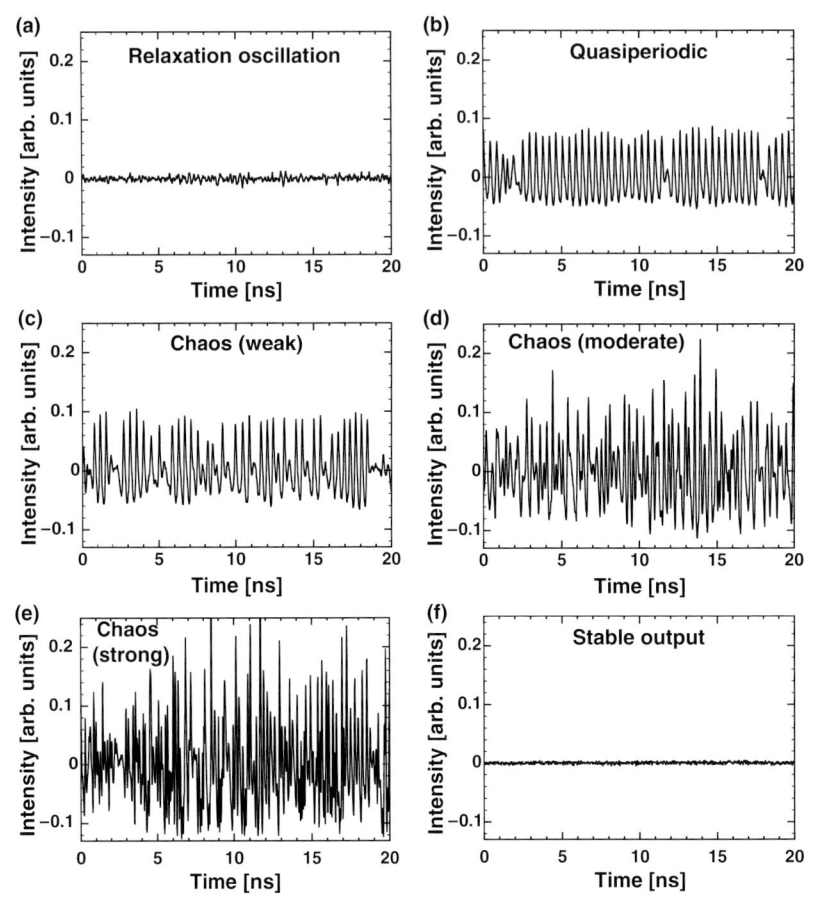

Figure 4.3 Experimental results of temporal waveforms of the semiconductor laser output when the feedback strength is changed. The temporal waveforms are obtained in experiments. (a) Sustained relaxation oscillation (no feedback, $\kappa = 0$, where κ is the ratio of the feedback power to the total laser power), (b) quasiperiodic oscillation (the feedback power of 89 nW, $\kappa = 1.2 \times 10^{-4}$) (c) weak chaos (140 nW, $\kappa = 1.9 \times 10^{-4}$), (d) moderate chaos (679 nW, $\kappa = 9.0 \times 10^{-4}$), (e) strong chaos (8.29 µW, $\kappa = 1.0 \times 10^{-2}$), and (f) stable output (115 µW, $\kappa = 1.5 \times 10^{-1}$).

slow compared with the chaotic oscillations of a few GHz. Low-frequency fluctuations (see also Section 3.2.1.4) are often observed at the condition of small injection current (around the lasing threshold) and strong optical feedback strength.

These dynamical transitions can be distinguished more clearly by using the RF spectra. Figure 4.5 shows the RF spectra corresponding to Figure 4.3 when the feedback strength is increased. Without optical feedback (Figure 4.5a), there is a small and broad RF spectral components at around 2.8 GHz that corresponds to the relaxation oscillation frequency of the solitary semiconductor laser. As the feedback strength is increased (Figure 4.3b), these spectral components become larger and

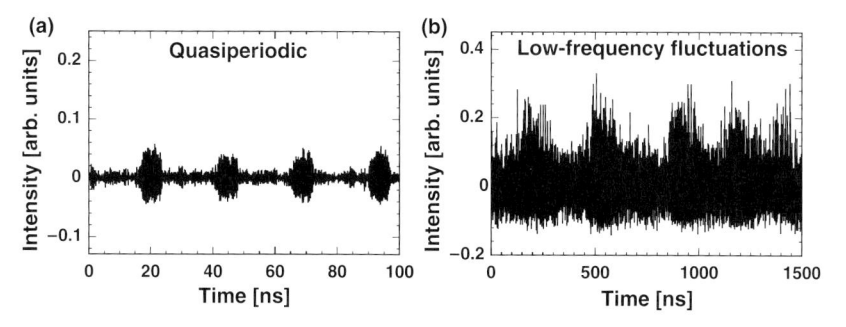

4.1 Experimental Analysis of Semiconductor Lasers with Optical Feedback | 151

Figure 4.4 Experimental results of long temporal waveforms of (a) quasiperiodic oscillation (the feedback power of 89 nW, $\kappa = 1.2 \times 10^{-4}$) and (b) low-frequency fluctuations (89 μW, $\kappa = 1.0 \times 10^{-1}$).

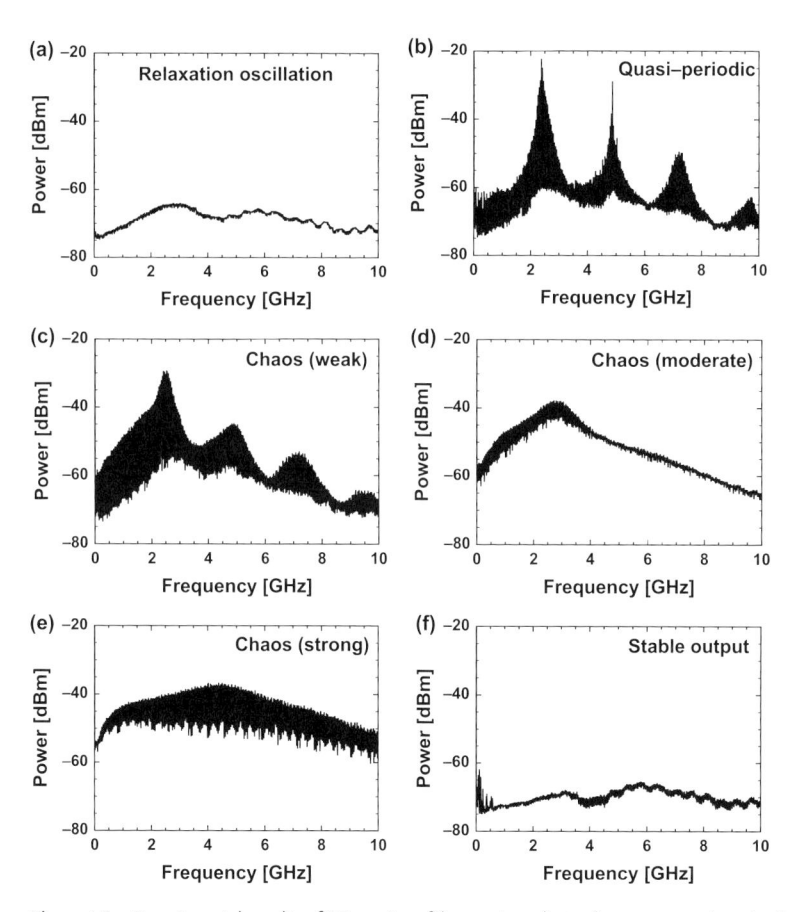

Figure 4.5 Experimental results of RF spectra of the semiconductor laser output when the feedback strength is changed. The RF spectra correspond to the temporal waveforms shown in Figure 4.3.

there are many peaks in the spectral components. These peaks become dominant in the RF spectrum (Figure 4.5b), corresponding to the quasiperiodic oscillation in the time domain. As the feedback strength is increased, the spectral components are broadened and show a weak chaotic dynamical state (Figure 4.5c). Note that the floor level of the chaotic RF spectrum components in Figure 4.5c increases compared with that for the quasiperiodic spectrum in Figure 4.5b, indicating the occurrence of chaos. The difference in the height of the spectral peaks decreases as the feedback strength is increased, and a smooth broader spectrum is obtained for the region of moderate chaos (Figure 4.5d). The peak heights, as well as the peak frequency of the broad RF spectrum, increase as the feedback strength is increased, which is the region of strong chaos (Figure 4.5e). Finally, the spectral components suddenly decrease and no oscillatory behavior is observed at strong optical feedback (Figure 4.5f).

To investigate the RF spectra in more detail, the fine structure in the RF spectra is observed. Figure 4.6 shows the enlargement of the RF spectra of Figure 4.5 at the center frequency of 2.5 GHz in the range of 0.5 GHz. There exist many spectral peaks

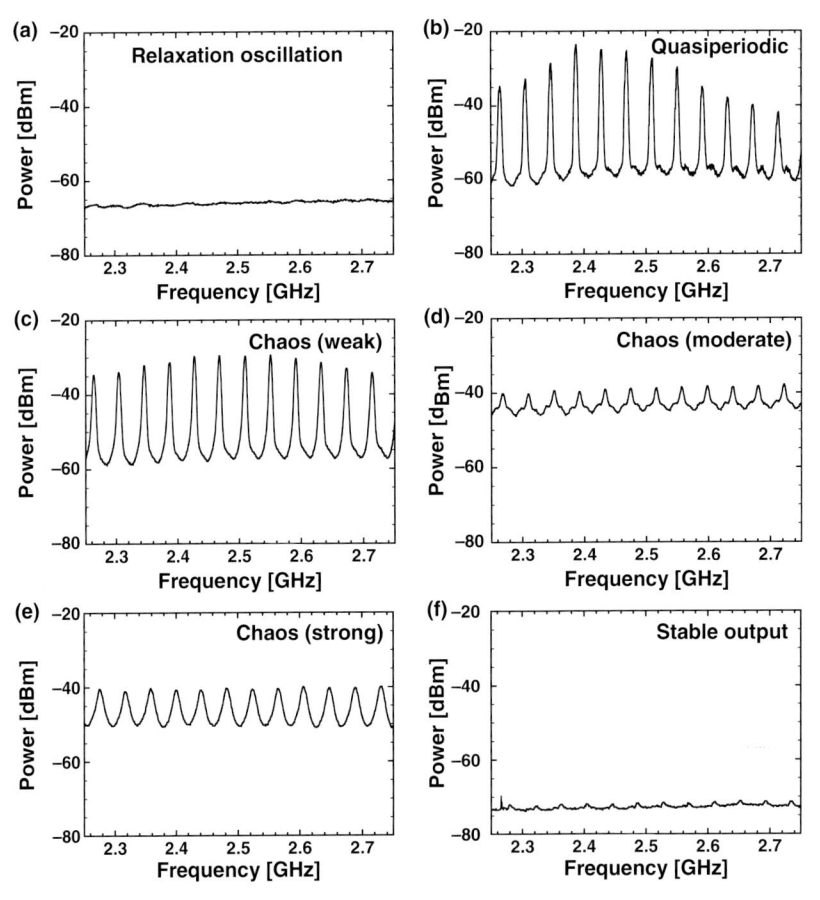

Figure 4.6 Enlargements of the RF spectra shown in Figure 4.5. The center frequency is 2.5 GHz in the range of 0.5 GHz.

in the RF spectra. Note that this peak interval is almost constant as the feedback strength is changed in Figure 4.6b–e, even though the peak heights are changed. The peak interval corresponds to the external cavity frequency, which is the inverse of the round-trip time of the feedback light in the external cavity. The external cavity frequency f_{ext} is calculated from the external cavity length (the same formula as Eq. (3.1)),

$$f_{ext} = \frac{c}{2nL_{ext}} \quad (4.1)$$

where L_{ext} is the external cavity length (one-way), n is the refractive index in the external cavity in the optical fiber, and c is the speed of light ($c = 2.998 \times 10^8$ m/s). For the parameter values used in the experiment ($L_{ext} = 2.52$ m, $n = 1.444$), the external cavity frequency is estimated as 41.2 MHz. This frequency exactly corresponds to the peak interval shown in Figures 4.6b–e. The nonlinear interaction between the relaxation oscillation frequency (2.8 GHz) and the external cavity frequency (41.2 MHz) results in the quasiperiodicity route to chaos as the feedback strength increases. Therefore, the observation of the peak interval corresponding to the external cavity frequency is an important investigation to confirm the existence of deterministic chaos from experimental data.

The autocorrelation functions are also calculated from the temporal waveforms of Figure 4.3 to identify the dynamical regimes, as shown in Figure 4.7. There is one to one correspondence between autocorrelation and RF spectrum. No peaks are found except zero delay time in Figure 4.7a. As the feedback strength is increased, a peak and its harmonics are observed. The first peak of autocorrelation from zero corresponds to the round-trip time τ of the feedback light in the external cavity,

$$\tau = \frac{1}{f_{ext}} = \frac{2nL_{ext}}{c} \quad (4.2)$$

For the parameter values used in the experiment ($L_{ext} = 2.52$ m, $n = 1.444$), the external-cavity round-trip time is estimated as $\tau = 24.3$ ns. The peaks of the autocorrelation functions in Figures 4.7b–e exactly correspond to $m\tau$ (m is an integer). For quasiperiodic oscillations (Figure 4.7b), many peaks of $m\tau$ with the correlation values close to 1 appear. The peaks are repeated with a constant interval of τ. For weak chaotic oscillations (Figure 4.7c), the width of the correlation peaks becomes sharper, even though the correlation peak values are still large. For moderate chaos (Figure 4.7d), the peak values decrease to less than 0.2 and very narrow peaks are obtained, indicating that more irregular chaotic temporal oscillation is obtained. The moderate chaos is in fact useful for synchronization of chaos in optical-communication applications and for random number generation (see Chapters 7–10). For strong chaos (Figure 4.7e), the correlation peak values increase again and less irregularity appears. Finally, stable output is obtained, however, intermittent irregular bursts sometimes occur, and small peaks of the autocorrelation still remain in Figure 4.7f.

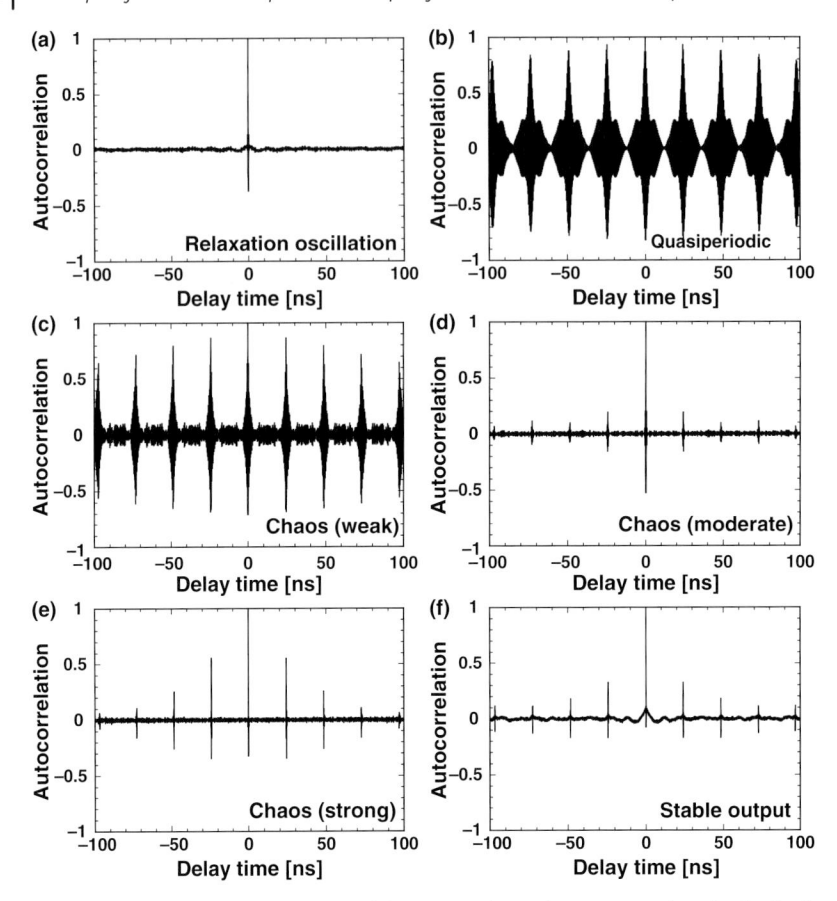

Figure 4.7 Autocorrelation functions of the semiconductor laser output when the feedback strength is changed. The autocorrelation functions are obtained from the temporal waveforms shown in Figure 4.3.

Figure 4.8 shows the optical spectra of the semiconductor laser output without and with optical feedback. Without optical feedback, the optical spectrum with narrow linewidth can be observed, and the corresponding temporal dynamics is sustained relaxation oscillation. By contrast, the optical spectrum is broadened in the presence of optical feedback, where the temporal waveform shows moderate chaos. For both cases, single-longitudinal-mode oscillations are preserved. The change in the linewidth of optical spectra is another indication of the generation of chaos.

The RF spectra and the autocorrelation functions are useful tools to identify the existence of deterministic chaos and bifurcations, even though temporal waveforms of the laser intensity are contaminated by intrinsic noise and the periods of the temporal waveforms are not clearly distinguished. For many experimental studies, the dynamical states of experimental data and the corresponding bifurcation diagram have been identified from the combinations of temporal waveforms, RF spectra, autocorrelation functions, and optical spectra.

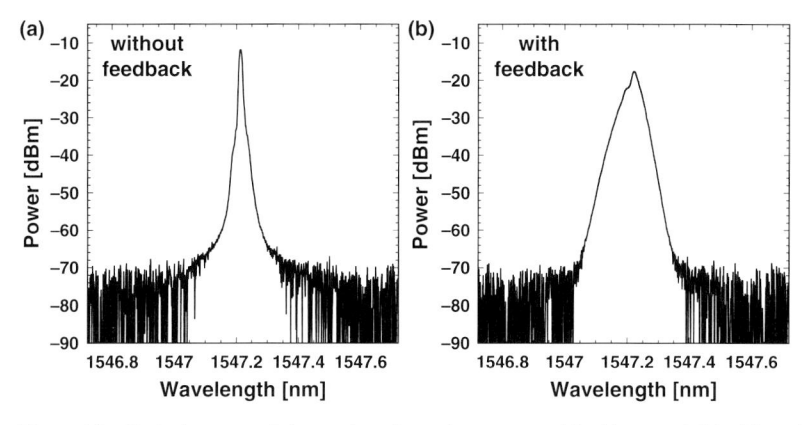

Figure 4.8 Optical spectra of the semiconductor laser output (a) without and (b) with optical feedback. (b) The feedback strength is set to generate moderate chaos.

4.1.4
Two-Dimensional Map

To investigate the nonlinear dynamics of a semiconductor laser with optical feedback systematically, a two-dimensional (2D) map is created. Figure 4.9 shows the 2D

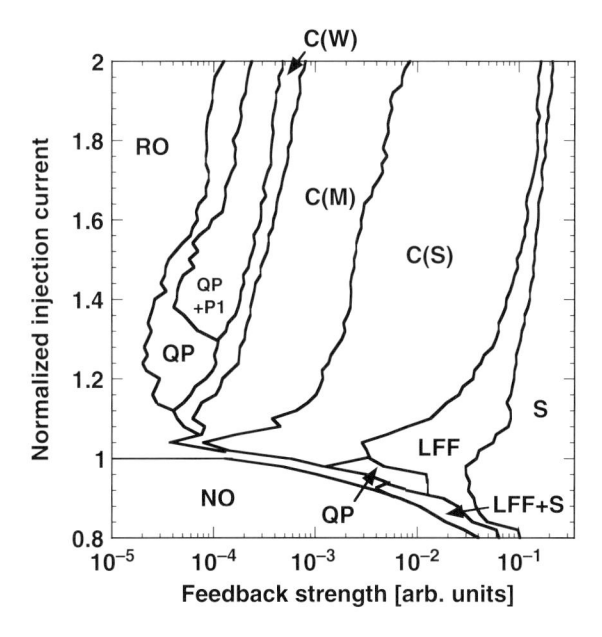

Figure 4.9 Experimental result of two-dimensional dynamical map. Both the feedback strength (measured in intensity) and the injection current (normalized by the threshold value) are changed. C(W), weak chaos; C(M), moderate chaos; C(S), strong chaos; LFF, low-frequency fluctuations; NO, no lasing state; P1, period-1 oscillation; QP, quasiperiodic oscillation; RO, sustained relaxation oscillation; S, stable output; +, mixture of two dynamical states.

dynamical map obtained experimentally. Both the feedback strength (measured in intensity) and the injection current are changed and the transition of the dynamics is investigated. The injection current is normalized by the threshold value for the solitary laser. The feedback strength is normalized by the laser output power in the presence of optical feedback. Many dynamical behaviors are observed in the 2D map. For small feedback strength (10^{-5}–10^{-4}) relaxation oscillation (RO) is widely observed. Quasiperiodic oscillation (QP) is also observed as the feedback strength is increased. The quasiperiodicity route to chaos can be identified in Figure 4.9. The regions of weak, moderate, and strong chaos are widely distributed in moderate feedback strength (10^{-4}–10^{-1}). It is found that a larger feedback strength is required for a larger injection current to obtain chaos. Low-frequency fluctuations (LFF) are also observed at the region of large feedback strengths (10^{-2}–10^{-1}) and small injection currents around (and even below) the lasing threshold (~1). When the feedback strength is larger than 10^{-1}, the laser output becomes stable and instabilities no longer appear. The 2D map is very informative to access different dynamical states in the parameter space in experiments.

4.2
Model for Semiconductor Laser with Optical Feedback

4.2.1
Lang–Kobayashi Equations

The temporal dynamics of intensity fluctuation in a semiconductor laser with optical feedback has been modeled by (Lang and Kobayashi, 1980), which is one of the historical milestone papers for the studies on chaotic dynamics of semiconductor lasers. The model is referred to as the Lang–Kobayashi equations, which include the effect of time-delayed optical feedback. The Lang–Kobayashi equations for the three real variables of the electric-field amplitude $E(t)$, the electric field electric-field phase $\Phi(t)$, and the carrier density $N(t)$ are written as follows (Ohtsubo, 2008),

$$\frac{dE(t)}{dt} = \frac{1}{2}\left[G_N(N(t)-N_0) - \frac{1}{\tau_p}\right]E(t) + \kappa\, E(t-\tau)\cos\Theta(t) \qquad (4.3)$$

$$\frac{d\Phi(t)}{dt} = \frac{\alpha}{2}\left[G_N(N(t)-N_0) - \frac{1}{\tau_p}\right] - \kappa\frac{E(t-\tau)}{E(t)}\sin\Theta(t) \qquad (4.4)$$

$$\frac{dN(t)}{dt} = J - \frac{N(t)}{\tau_s} - G_N(N(t)-N_0)E^2(t) \qquad (4.5)$$

$$\Theta(t) = \omega\tau + \Phi(t) - \Phi(t-\tau) \qquad (4.6)$$

The parameters in Eqs. (4.3)–(4.6) and their values for numerical simulations are summarized in Table 4.3. The laser intensity is calculated from the square of the electric-field amplitude, that is, $I(t) = E^2(t)$.

4.2 Model for Semiconductor Laser with Optical Feedback

Table 4.3 Parameter values used for numerical simulation of the original Lang–Kobayashi equations in Chapter 4.

Symbol	Parameter	Value
G_N	Gain coefficient	$8.40 \times 10^{-13}\,\mathrm{m^3\,s^{-1}}$
N_0	Carrier density at transparency	$1.40 \times 10^{24}\,\mathrm{m^{-3}}$
τ_p	Photon lifetime	$1.927 \times 10^{-12}\,\mathrm{s}$
τ_s	Carrier lifetime	$2.04 \times 10^{-9}\,\mathrm{s}$
τ_{in}	Round-trip time in internal cavity	$8.0 \times 10^{-12}\,\mathrm{s}$
r_2	Reflectivity of laser facet	0.556
r_3	Reflectivity of external mirror	0.01 (variable)
$j = J/J_{th}$	Normalized injection current	1.11
L	External cavity length	0.225 m
α	Linewidth enhancement factor	3.0
λ	Optical wavelength	$1.537 \times 10^{-6}\,\mathrm{m}$
c	Speed of light	$2.998 \times 10^{8}\,\mathrm{m\,s^{-1}}$
$\kappa = \dfrac{(1-r_2^2)r_3}{r_2}\dfrac{1}{\tau_{in}}$	Feedback strength	$1.553 \times 10^{9}\,\mathrm{s^{-1}}$ ($1.553\,\mathrm{ns^{-1}}$) (variable)
$\tau = \dfrac{2L}{c}$	Round-trip time of feedback light in external cavity (feedback delay time)	$1.501 \times 10^{-9}\,\mathrm{s}$
$N_{th} = N_0 + \dfrac{1}{G_N \tau_p}$	Carrier density at threshold	$2.018 \times 10^{24}\,\mathrm{m^{-3}}$
$J_{th} = \dfrac{N_{th}}{\tau_s}$	Injection current at threshold	$9.892 \times 10^{32}\,\mathrm{m^{-3}\,s^{-1}}$
$\omega = \dfrac{2\pi c}{\lambda}$	Optical angular frequency	$1.226 \times 10^{15}\,\mathrm{s^{-1}}$

For the equation of the electric-field amplitude, the first right-hand-side term of Eq. (4.3) including G_N indicates the stimulated emission, the first right-hand-side term of Eq. (4.3) including $-1/\tau_p$ represents the decay of electric-field amplitude due to the cavity loss, and the second right-hand-side term of Eq. (4.3) including κ indicates the time-delayed optical feedback effect. The same argument holds for the phase equation of Eq. (4.4). For the equation of the carrier density, the first right-hand-side term of Eq. (4.5) including J indicates the injection current for pumping, the second right-hand-side term of Eq. (4.5) including $-1/\tau_n$ represents the decay of the carrier density due to the spontaneous emission, and the third right-hand-side term of Eq. (4.5) including G_N indicates the stimulated emission. Equation (4.6) shows the phase difference between the intracavity and feedback electric fields, with the initial phase difference $\omega\tau$.

For numerical simulation of Eq. (4.6), the constant value of $\omega\tau$ is several orders of magnitude larger than the value of the phase difference $\Phi(t)-\Phi(t-\tau)$ (see the values in Table 4.3), and, $\Theta(t)$ can be treated between 0 and 2π. Therefore, the value of an initial phase shift $\Theta_{ini} = \omega\tau \pmod{2\pi}$ (where $0 \leq \Theta_{ini} < 2\pi$) needs to be calculated in advance and $\omega\tau$ is replaced to Θ_{ini} in Eq. (4.6) in numerical calculations to avoid accumulation numerical errors.

The derivation of the rate equations for a semiconductor laser from the laser oscillation conditions is described in Appendix 4.A.1. The derivation of the Lang–Kobayashi equations for a semiconductor laser with optical feedback is explained in Appendix 4.A.2.

4.2.2
Derivation of the Electric-Field Amplitude and Phase of the Lang–Kobayashi Equations from the Complex Electric-Field Equations

The derivation of electric-field amplitude and phase of the laser is described in this section. The complex electric field $\hat{\mathbf{E}}(t)$ including the fast optical carrier in a semiconductor laser with optical feedback is described as follows (Lang and Kobayashi, 1980; Ohtsubo, 2008),

$$\frac{d\hat{\mathbf{E}}(t)}{dt} = \left[\frac{1+i\alpha}{2} \left\{ G_N(N(t)-N_0) - \frac{1}{\tau_p} \right\} + i\omega \right] \hat{\mathbf{E}}(t) + \kappa \, \hat{\mathbf{E}}(t-\tau) \tag{4.7}$$

where i is the imaginary unit. A variable with bold font indicates a complex variable, whereas a variable with italic font represents a real variable. $\hat{\mathbf{E}}$ indicates that the complex variable $\mathbf{E}$ includes the fast optical-carrier frequency component. The dynamics of electric-field amplitude is much slower than the angular frequency ω of the fast optical carrier. The fast oscillation component can be separated from the slow complex electric field $\mathbf{E}(t)$ as follows,

$$\hat{\mathbf{E}}(t) = \mathbf{E}(t)\exp(i\omega t) \tag{4.8}$$

Substituting Eq. (4.8) into Eq. (4.7).

$$\frac{d\mathbf{E}(t)}{dt}\exp(i\omega t) + i\omega \, \mathbf{E}(t)\exp(i\omega t)$$

$$= \left[\frac{1+i\alpha}{2} \left\{ G_N(N(t)-N_0) - \frac{1}{\tau_p} \right\} + i\omega \right] \mathbf{E}(t)\exp(i\omega t) + \kappa \mathbf{E}(t-\tau)\exp(i\omega(t-\tau)) \tag{4.9}$$

All terms are divided by $\exp(i\omega t)$,

$$\frac{d\mathbf{E}(t)}{dt} = \frac{1+i\alpha}{2}\left[G_N(N(t)-N_0) - \frac{1}{\tau_p} \right] \mathbf{E}(t) + \kappa \, \mathbf{E}(t-\tau)\exp(-i\omega\tau) \tag{4.10}$$

Equation (4.10) is the Lang–Kobayashi equation for a slow complex electric field after the elimination of the fast optical carrier component. The deviation of Eq. (4.10) from the laser oscillation conditions is described in detail in Appendices 4.A.1 and

4.A.2 (see Eq. (4.A.45) in Appendix 4.A.2). The slow complex electric field **E**(t) can be written by the real values of the amplitude $E(t)$ and the phase $\Phi(t)$ as follows.

$$\mathbf{E}(t) = E(t)\exp(i\Phi(t)) \tag{4.11}$$

Substituting Eq. (4.11) into Eq. (4.10).

$$\frac{dE(t)}{dt}\exp(i\Phi(t)) + i\frac{d\Phi(t)}{dt}E(t)\exp(i\Phi(t))$$
$$= \frac{1+i\alpha}{2}\left[G_N(N(t)-N_0) - \frac{1}{\tau_p}\right]E(t)\exp(i\Phi(t)) \tag{4.12}$$
$$+ \kappa E(t-\tau)\exp(i\Phi(t-\tau))\exp(-i\omega\tau)$$

All terms are divided by $\exp(i\Phi(t))$.

$$\frac{dE(t)}{dt} + i\frac{d\Phi(t)}{dt}E(t)$$
$$= \frac{1+i\alpha}{2}\left[G_N(N(t)-N_0) - \frac{1}{\tau_p}\right]E(t) + \kappa E(t-\tau)\exp(-i(\omega\tau+\Phi(t)-\Phi(t-\tau)))$$
$$\tag{4.13}$$

Here, the Euler's formula is used,

$$\exp(-ix) = \cos x - i\sin x \tag{4.14}$$

Substituting Eq. (4.14) into Eq. (4.13),

$$\frac{dE(t)}{dt} + i\frac{d\Phi(t)}{dt}E(t)$$
$$= \frac{1+i\alpha}{2}\left[G_N(N(t)-N_0) - \frac{1}{\tau_p}\right]E(t) + \kappa E(t-\tau)\cos(\omega\tau+\Phi(t)-\Phi(t-\tau))$$
$$- i\kappa E(t-\tau)\sin(\omega\tau+\Phi(t)-\Phi(t-\tau)) \tag{4.15}$$

The real and imaginary parts of Eq. (4.15) are separated,

$$\frac{dE(t)}{dt} = \frac{1}{2}\left[G_N(N(t)-N_0) - \frac{1}{\tau_p}\right]E(t) + \kappa E(t-\tau)\cos\Theta(t) \tag{4.16}$$

$$\frac{d\Phi(t)}{dt} = \frac{\alpha}{2}\left[G_N(N(t)-N_0) - \frac{1}{\tau_p}\right] - \kappa\frac{E(t-\tau)}{E(t)}\sin\Theta(t) \tag{4.17}$$

$$\Theta(t) = \omega\tau + \Phi(t) - \Phi(t-\tau) \tag{4.18}$$

These equations are the same formula as Eqs. (4.3), (4.4), and (4.6) with the real values of the electric-field amplitude $E(t)$ and phase $\Phi(t)$.

4.2.3
Derivation of the Real and Imaginary Electric Fields of the Lang–Kobayashi Equations from the Complex Electric-Field Equations

For the Lang–Kobayashi equations with the real amplitude and real phase (Eqs. (4.3) and (4.4)), there exists a problem when a chaotic oscillation becomes strong pulsations. The terms of $E(t-\tau)/E(t)$ in the real-phase equation of Eq. (4.4) (and Eq. (4.17)) becomes infinite when $E(t)$ is close to zero after strong chaotic pulsations. This may cause accumulation errors for numerical integration. Therefore, the formula of the real and imaginary electric fields may be useful, instead of the real amplitude and real phase of the electric field (see also Chapter 6).

The complex electric-field amplitude of Eq. (4.10) is used for the derivation. Instead of using the real amplitude and real phase of Eq. (4.11), the real and imaginary electric fields ($E_R(t)$ and $E_I(t)$) are introduced,

$$\mathbf{E}(t) = E_R(t) + iE_I(t) \tag{4.19}$$

Where $E_R(t)$ $E_I(t)$ are real variables. Substituting Eq. (4.19) into Eq. (4.10).

$$\begin{aligned}\frac{dE_R(t)}{dt} &+ i\frac{dE_I(t)}{dt} \\ &= \frac{1+i\alpha}{2}\left[G_N(N(t)-N_0) - \frac{1}{\tau_p}\right]\{E_R(t)+iE_I(t)\} \\ &+ \kappa\{E_R(t-\tau) + iE_I(t-\tau)\}\exp(-i\omega\tau)\end{aligned} \tag{4.20}$$

Substituting Eq. (4.14) into Eq. (4.20).

$$\begin{aligned}\frac{dE_R(t)}{dt} &+ i\frac{dE_I(t)}{dt} \\ &= \frac{1}{2}\left[G_N(N(t)-N_0) - \frac{1}{\tau_p}\right]\{E_R(t)-\alpha E_I(t)\} \\ &+ i\frac{1}{2}\left[G_N(N(t)-N_0) - \frac{1}{\tau_p}\right]\{\alpha E_R(t)+E_I(t)\} \\ &+ \kappa\{E_R(t-\tau)+iE_I(t-\tau)\}\cos(\omega\tau) \\ &- i\kappa\{E_R(t-\tau)+iE_I(t-\tau)\}\sin(\omega\tau)\end{aligned} \tag{4.21}$$

The real and imaginary parts of Eq. (4.21) are separated,

$$\begin{aligned}\frac{dE_R(t)}{dt} &= \frac{1}{2}\left[G_N(N(t)-N_0)-\frac{1}{\tau_p}\right]\{E_R(t)-\alpha E_I(t)\} \\ &+ \kappa\{E_R(t-\tau)\cos(\omega\tau)+E_I(t-\tau)\sin(\omega\tau)\}\end{aligned} \tag{4.22}$$

$$\begin{aligned}\frac{dE_I(t)}{dt} &= \frac{1}{2}\left[G_N(N(t)-N_0)-\frac{1}{\tau_p}\right]\{\alpha E_R(t)+E_I(t)\} \\ &+ \kappa\{E_I(t-\tau)\cos(\omega\tau)-E_R(t-\tau)\sin(\omega\tau)\}\end{aligned} \tag{4.23}$$

The carrier density can be derived from Eq. (4.5),

$$\frac{dN(t)}{dt} = J - \frac{N(t)}{\tau_s} - G_N(N(t)-N_0)\left(E_R^2(t) + E_I^2(t)\right) \tag{4.24}$$

Equations (4.22)–(4.24) are the real–imaginary expression of the Lang–Kobayashi equations. The laser intensity $I(t)$ and optical phase $\Phi(t)$ can be calculated from $E_R(t)$ and $E_I(t)$ as follows,

$$I(t) = E_R^2(t) + E_I^2(t) \tag{4.25}$$

$$\Phi(t) = \tan^{-1}\left(\frac{E_I}{E_R}\right) \tag{4.26}$$

The formula of the real–imaginary equations is useful for the cases of small electric amplitudes after strong chaotic pulsations (i.e., $E(t) \to 0$) at the conditions of strong optical feedback and weak injection current. In the following sections, however, the real amplitude-phase equations of Eqs. (4.3)–(4.5) are used mainly for convenience. For many situations, the numerical results obtained from Eqs. (4.3)–(4.5) are almost the same as those obtained from Eqs. (4.22)–(4.24). On the contrary, Eqs. (4.22)–(4.24) are more useful for coupled Lang–Kobayashi equations for the calculation of synchronization of chaotic semiconductor lasers, as shown in Chapter 6.

For convenience, the value of an initial phase shift $\Theta_{\text{ini}} = \omega\tau \pmod{2\pi}$ (where $0 \leq \Theta_{\text{ini}} < 2\pi$) needs to be calculated in advance, and $\omega\tau$ is replaced to Θ_{ini} in Eqs. (4.22) and (4.23) in numerical calculations to avoid numerical errors, as described in Section 4.2.1.

4.3
Analytical Approach of Semiconductor Laser with Optical Feedback

4.3.1
Steady-State Solutions without Optical Feedback

In the following sections, the original Lang–Kobayashi equations of Eqs. (4.3)–(4.6) are analytically and numerically investigated. As a first step of the analytical approach of nonlinear dynamical analysis for the Lang–Kobayashi equations, let us consider a simple case where no optical feedback exists (i.e., $\kappa = 0$). The Lang–Kobayashi equations without optical feedback are modified from Eqs. (4.3)–(4.5),

$$\frac{dE(t)}{dt} = \frac{1}{2}\left[G_N(N(t)-N_0) - \frac{1}{\tau_p}\right]E(t) \tag{4.27}$$

$$\frac{d\Phi(t)}{dt} = \frac{\alpha}{2}\left[G_N(N(t)-N_0) - \frac{1}{\tau_p}\right] \tag{4.28}$$

$$\frac{dN(t)}{dt} = J - \frac{N(t)}{\tau_s} - G_N(N(t)-N_0)E^2(t) \tag{4.29}$$

First, steady-state solutions need to be derived from Eqs. (4.27)–(4.29). The steady-state solutions are denoted as,

$$E(t) = E_s, \quad \Phi(t) = (\omega_s - \omega_0)\,t, \quad N(t) = N_s \tag{4.30}$$

where E_s, ω_s, and N_s are the steady-state solutions of the electric-field amplitude, angular optical frequency, and the carrier density, respectively. ω_0 is the angular optical frequency for the solitary laser, and t is time. Here, a lasing steady-state solution is assumed ($E_s \neq 0$). For the steady-state solutions, the following conditions are satisfied,

$$\frac{dE(t)}{dt} = 0, \quad \frac{d\Phi(t)}{dt} = \omega_s - \omega_0, \quad \frac{dN(t)}{dt} = 0 \tag{4.31}$$

Substituting Eqs. (4.30)–(4.31) into Eqs. (4.27)–(4.29),

$$0 = \frac{1}{2}\left[G_N(N_s - N_0) - \frac{1}{\tau_p}\right]E_s \tag{4.32}$$

$$\omega_s - \omega_0 = \frac{\alpha}{2}\left[G_N(N_s - N_0) - \frac{1}{\tau_p}\right] \tag{4.33}$$

$$0 = J - \frac{N_s}{\tau_s} - G_N(N_s - N_0)E_s^2 \tag{4.34}$$

The steady-state solutions are obtained from Eqs. (4.32)–(4.34),

$$E_s = \sqrt{\frac{\tau_p N_{\text{th}}(j-1)}{\tau_s}} \tag{4.35}$$

$$\omega_s = \omega_0 \tag{4.36}$$

$$N_s = N_{\text{th}} \tag{4.37}$$

where $N_{\text{th}} = N_0 + \dfrac{1}{G_N \tau_p}$ and $J = j \cdot J_{\text{th}} = j \cdot \dfrac{N_{\text{th}}}{\tau_s}$ are used. j is the injection current normalized by the threshold value J_{th}.

4.3.2
Linear Stability Analysis for Steady-State Solutions without Optical Feedback

4.3.2.1 Eigenvalues of Jacobian Matrix

It is very important to investigate the stability of the steady-state solutions of Eqs. (4.35)–(4.37). When the steady-state solutions are stable, the laser operates at the steady state and there are no oscillations or instabilities of the laser output at all. By contrast, when the steady-state solutions are unstable, the laser output intensity oscillates periodically or chaotically. "Linear stability analysis" is a powerful tool to investigate the stability of the steady-state solutions. It is worth noting that linear stability analysis for the steady-state solutions only gives an indication of the stability of the steady states, and there is no information on the type of oscillations (i.e., periodic or chaotic oscillations) when the steady states are unstable. A further numerical approach is required to determine the stability of different types of oscillations (see Section 4.4.2).

For linear stability analysis, the "linearized equations" (also referred to as the "variational equations") are derived from the original equations of Eqs. (4.27)–(4.29). Small linear deviations from the steady-state solutions are introduced,

$$E(t) = E_s + \delta_E(t) \tag{4.38}$$

$$\Phi(t) = (\omega_s - \omega_0)\, t + \delta_\Phi(t) \tag{4.39}$$

$$N(t) = N_s + \delta_N(t) \tag{4.40}$$

where $\delta_E(t)$, $\delta_\Phi(t)$, and $\delta_N(t)$ are new variables for the linearized equations (referred to as "linearized variables"), assuming that $E(t) \gg \delta_E(t)$, $\Phi(t) \gg \delta_\Phi(t)$, and $N(t) \gg \delta_N(t)$. Note that $E(t)$ and $\delta_E(t)$ are independent variables, where $E(t)$ is a variable of the electric-field amplitude and $\delta_E(t)$ is a variable of "linear deviation" from the steady-state solution E_s.

Substituting the linearized variables of Eqs. (4.38)–(4.40) into the original equations of Eqs. (4.27)–(4.29), and the linearized equations are obtained,

$$\frac{d\delta_E(t)}{dt} = \frac{1}{2} G_N E_s \delta_N(t) \tag{4.41}$$

$$\frac{d\delta_\Phi(t)}{dt} = \frac{\alpha}{2} G_N \delta_N(t) \tag{4.42}$$

$$\frac{d\delta_N(t)}{dt} = -\frac{2E_s}{\tau_p} \delta_E(t) - \left(G_N E_s^2 + \frac{1}{\tau_s}\right)\delta_N(t) \tag{4.43}$$

where Eq. (4.31) and the steady-state solutions of Eqs. (4.35)–(4.37) are used to simplify the equations. The square terms of the linearized variables can be neglected because $\delta_i(t) \gg \delta_i^2(t)$ and $\delta_i(t) \gg \delta_i(t) \cdot \delta_j(t)$ are assumed (i, j correspond to E, Φ, N).

The linearized equations of Eqs. (4.41)–(4.43) can also be obtained from the formula with the Jacobian matrix (see Appendix 2.A.1 in Chapter 2 for the general formula of the linearized equations),

$$\begin{pmatrix} \dfrac{d\delta_E(t)}{dt} \\ \dfrac{d\delta_\Phi(t)}{dt} \\ \dfrac{d\delta_N(t)}{dt} \end{pmatrix} = \begin{pmatrix} \dfrac{\partial f_E}{\partial E} & \dfrac{\partial f_E}{\partial \Phi} & \dfrac{\partial f_E}{\partial N} \\ \dfrac{\partial f_\Phi}{\partial E} & \dfrac{\partial f_\Phi}{\partial \Phi} & \dfrac{\partial f_\Phi}{\partial N} \\ \dfrac{\partial f_N}{\partial E} & \dfrac{\partial f_N}{\partial \Phi} & \dfrac{\partial f_N}{\partial N} \end{pmatrix} \begin{pmatrix} \delta_E(t) \\ \delta_\Phi(t) \\ \delta_N(t) \end{pmatrix} \tag{4.44}$$

where f_E, f_Φ, and f_N are the right-hand-side terms of Eqs. (4.27)–(4.29), and the 3 × 3 matrix in Eq. (4.44) is called the Jacobian matrix. The sign of the real part of the maximum eigenvalue of the Jacobian matrix determines the stability of the original dynamical system. The linearized equations are calculated from Eqs. (4.27)–(4.29) and (4.44),

$$\begin{pmatrix} \frac{d\delta_E(t)}{dt} \\ \frac{d\delta_\Phi(t)}{dt} \\ \frac{d\delta_N(t)}{dt} \end{pmatrix} = \begin{pmatrix} \frac{1}{2}\left[G_N(N_s-N_0)-\frac{1}{\tau_p}\right] & 0 & \frac{1}{2}G_N E_s \\ 0 & 0 & \frac{\alpha}{2}G_N \\ -2G_N(N_s-N_0)E_s & 0 & -\frac{1}{\tau_s}-G_N E_s^2 \end{pmatrix} \begin{pmatrix} \delta_E(t) \\ \delta_\Phi(t) \\ \delta_N(t) \end{pmatrix}$$

$$= \begin{pmatrix} 0 & 0 & \frac{1}{2}G_N E_s \\ 0 & 0 & \frac{\alpha}{2}G_N \\ -\frac{2E_s}{\tau_p} & 0 & -\frac{1}{\tau_s}-G_N E_s^2 \end{pmatrix} \begin{pmatrix} \delta_E(t) \\ \delta_\Phi(t) \\ \delta_N(t) \end{pmatrix}$$

(4.45)

where $N_s = N_{th} = N_0 + \frac{1}{G_N \tau_p}$ is used. In fact the matrix formula of Eq. (4.45) is the same as Eqs. (4.41)–(4.43).

To analyze the stability of the steady-state solutions, the eigenvalues of the Jacobian matrix of Eq. (4.45) need to be estimated. The sign of the real part of the eigenvalues determines the stability, that is, the system is stable when all the signs of the real parts of the eigenvalues are negative, while the system becomes unstable when the sign of the real part of the maximum eigenvalue is positive. The characteristic equation for the eigenvalues λ of the Jacobian matrix $\mathbf{J}$ is written,

$$|\mathbf{J}-\lambda\mathbf{I}| = \begin{vmatrix} -\lambda & 0 & \frac{1}{2}G_N E_s \\ 0 & -\lambda & \frac{\alpha}{2}G_N \\ -\frac{2E_s}{\tau_p} & 0 & -\frac{1}{\tau_s}-G_N E_s^2-\lambda \end{vmatrix}$$

$$= -\lambda\left[\lambda^2 + \left(\frac{1}{\tau_s}+G_N E_s^2\right)\lambda + \frac{G_N E_s^2}{\tau_p}\right] = 0$$

(4.46)

where **I** is an identity matrix and the determinant of the 3×3 matrix can be obtained (e.g., the rule of Sarrus). The trivial solution $\lambda = 0$ is not considered here, and Eq. (4.46) becomes,

$$\lambda^2 + \left(\frac{1}{\tau_s} + G_N E_s^2\right)\lambda + \frac{G_N E_s^2}{\tau_p} = 0 \quad (4.47)$$

The solutions of λ are obtained from the quadratic formula,

$$\lambda = -\frac{1}{2}\left(\frac{1}{\tau_s} + G_N E_s^2\right) \pm \sqrt{\frac{1}{4}\left(\frac{1}{\tau_s} + G_N E_s^2\right)^2 - \frac{G_N E_s^2}{\tau_p}} \quad (4.48)$$

Considering the parameter values (e.g., see Table 4.3), the following condition is satisfied,

$$\frac{G_N E_s^2}{\tau_p} \gg \frac{1}{4}\left(\frac{1}{\tau_s} + G_N E_s^2\right)^2 \quad (4.49)$$

Therefore, the solution becomes the complex number,

$$\lambda = -\frac{1}{2}\left(\frac{1}{\tau_s} + G_N E_s^2\right) \pm i\sqrt{\frac{G_N E_s^2}{\tau_p} - \frac{1}{4}\left(\frac{1}{\tau_s} + G_N E_s^2\right)^2} \quad (4.50)$$

The complex solution is expressed as $\lambda = \lambda_{\text{Re}} \pm i\lambda_{\text{Im}}$, where

$$\lambda_{\text{Re}} = -\frac{1}{2}\left(\frac{1}{\tau_s} + G_N E_s^2\right) \quad (4.51)$$

$$\lambda_{\text{Im}} = \sqrt{\frac{G_N E_s^2}{\tau_p} - \frac{1}{4}\left(\frac{1}{\tau_s} + G_N E_s^2\right)^2} \quad (4.52)$$

The stability is calculated from Eq. (4.51). The condition of $\lambda_{\text{Re}} < 0$ is always satisfied from Eq. (4.51) because τ_s, G_N, and E_s^2 are positive values. Therefore, it is found that the steady-state solutions of Eqs. (4.35)–(4.37) are always stable.

4.3.2.2 Relaxation Oscillation Frequency
The relaxation oscillation frequency f_r of a solitary semiconductor laser can be calculated from Eq. (4.52), where $2\pi f_r = \lambda_{\text{Im}}$.

$$f_r = \frac{1}{2\pi}\sqrt{\frac{G_N E_s^2}{\tau_p} - \frac{1}{4}\left(\frac{1}{\tau_s} + G_N E_s^2\right)^2} \quad (4.53)$$

The approximation of Eq. (4.49) is used and the second term of the right-hand side in Eq. (4.53) can be ignored,

$$f_r = \frac{1}{2\pi}\sqrt{\frac{G_N E_s^2}{\tau_p}} \tag{4.54}$$

Substituting Eq. (4.35) into Eq. (4.54),

$$f_r = \frac{1}{2\pi}\sqrt{\frac{(j-1)}{\tau_s \tau_p}(1 + G_N N_0 \tau_p)} \tag{4.55}$$

The relaxation oscillation frequency is also denoted with different formula by using the notation of $N_{th} = N_0 + \frac{1}{G_N \tau_p}$ and $J_{th} = \frac{N_{th}}{\tau_s}$,

$$f_r = \frac{1}{2\pi}\sqrt{\frac{G_N N_{th}}{\tau_s}(j-1)} \tag{4.56}$$

$$f_r = \frac{1}{2\pi}\sqrt{G_N J_{th}(j-1)} \tag{4.57}$$

This formula (Eqs. (4.55), (4.56) or (4.57)) shows the relaxation oscillation frequency in a solitary semiconductor laser without optical feedback. It is shown that the relaxation oscillation frequency of a semiconductor laser (Eq. (4.55)) is different from that of a simple class-B laser model (Eq. (2.19)) in Section 2.3.4.).

4.3.3
Steady-State Solutions with Optical Feedback

Next, a semiconductor laser with optical feedback is considered and an analytical approach is conducted. The Lang–Kobayashi equations for a semiconductor laser with optical feedback (Eqs. (4.3)–(4.6)) are rewritten as follows,

$$\frac{dE(t)}{dt} = \frac{1}{2}\left[G_N(N(t)-N_0) - \frac{1}{\tau_p}\right]E(t) + \kappa E(t-\tau)\cos\Theta(t) \tag{4.58}$$

$$\frac{d\Phi(t)}{dt} = \frac{\alpha}{2}\left[G_N(N(t)-N_0) - \frac{1}{\tau_p}\right] - \kappa\frac{E(t-\tau)}{E(t)}\sin\Theta(t) \tag{4.59}$$

$$\frac{dN(t)}{dt} = J - \frac{N(t)}{\tau_s} - G_N(N(t)-N_0)E^2(t) \tag{4.60}$$

$$\Theta(t) = \omega_0 \tau + \Phi(t) - \Phi(t-\tau) \tag{4.61}$$

The notation of the variables and the parameters is the same as shown in Eqs. (4.3)–(4.6).

First, steady-state solutions of Eqs. (4.58)–(4.60) need to be calculated. The steady-state solutions are denoted as,

$$E(t) = E(t-\tau) = E_s, \quad \Phi(t) = (\omega_s - \omega_0)t, \quad \Phi(t-\tau) = (\omega_s - \omega_0)(t-\tau),$$
$$N(t) = N_s, \quad \Theta(t) = \Theta_s \qquad (4.62)$$

Here, a lasing steady state is assumed ($E_s \neq 0$). For steady-state solutions, the following conditions are satisfied.

$$\frac{dE(t)}{dt} = 0, \quad \frac{d\Phi(t)}{dt} = \omega_s - \omega_0, \quad \frac{dN(t)}{dt} = 0 \qquad (4.63)$$

Substituting Eqs. (4.62) and (4.63) into Eqs. (4.58)–(4.61),

$$0 = \frac{1}{2}\left[G_N(N_s - N_0) - \frac{1}{\tau_p}\right]E_s + \kappa E_s \cos\Theta_s \qquad (4.64)$$

$$\omega_s - \omega_0 = \frac{\alpha}{2}\left[G_N(N_s - N_0) - \frac{1}{\tau_p}\right] - \kappa \sin\Theta_s \qquad (4.65)$$

$$0 = J - \frac{N_s}{\tau_s} - G_N(N_s - N_0)E_s^2 \qquad (4.66)$$

$$\Theta_s = \omega_0\tau + (\omega_s - \omega_0)t - (\omega_s - \omega_0)(t-\tau) = \omega_s\tau \qquad (4.67)$$

Equation (4.66) is changed to,

$$E_s^2 = \frac{\left(J - \frac{N_s}{\tau_s}\right)}{G_N(N_s - N_0)} = \frac{jN_{th} - N_s}{\tau_s G_N(N_s - N_0)} \qquad (4.68)$$

At a lasing condition ($E_s \neq 0$), substituting Eqs. (4.64) and (4.67) into Eq. (4.65),

$$\omega_s - \omega_0 = -\kappa(\alpha \cos(\omega_s\tau) + \sin(\omega_s\tau)) \qquad (4.69)$$

Equation (4.64) is changed to,

$$N_s = N_0 + \frac{1}{G_N\tau_p} - \frac{2\kappa \cos(\omega_s\tau)}{G_N} \qquad (4.70)$$

Equations (4.68)–(4.70) are the steady-state solutions (E_s, ω_s, and N_s) for the semiconductor laser with optical feedback. From Eq. (4.69), the steady-state solutions have multiple solutions, depending on the feedback strength κ. Without optical feedback ($\kappa = 0$), $\omega_s = \omega_0$ is satisfied, and E_s and N_s are uniquely determined, as seen in Eqs. (4.35)–(4.37). However, there are several steady-state solutions in the presence of optical feedback ($\kappa \neq 0$) in Eq. (4.69).

Equations (4.69) is changed to,

$$(\omega_s - \omega_0)\tau = -\kappa\tau(\alpha \cos(\omega_s\tau) + \sin(\omega_s\tau))$$
$$= -\kappa\tau\sqrt{1+\alpha^2}\sin(\omega_s\tau + \tan^{-1}\alpha) \quad (4.71)$$

Here, $\Delta\omega_{s,0} = \omega_s - \omega_0$ is used,

$$\Delta\omega_{s,0}\tau + \kappa\tau\sqrt{1+\alpha^2}\sin(\Delta\omega_{s,0}\tau + \omega_0\tau + \tan^{-1}\alpha) = 0 \quad (4.72)$$

Equation (4.72) is simplified as,

$$\Delta\omega_{s,0}\tau + C\sin(\Delta\omega_{s,0}\tau + \varphi_0) = 0 \quad (4.73)$$

where, $C = \kappa\tau\sqrt{1+\alpha^2}$ and $\varphi_0 = \omega_0\tau + \tan^{-1}\alpha$.

The left-hand-side term of Eq. (4.73) (denoted as f) is plotted as a function of $\Delta\omega_{s,0}$ in Figure 4.10. As the feedback strength is increased, the number of cross sections ($f(\Delta\omega_{s,0}) = 0$, the dotted line in Figure 4.10) increases and multiple steady-state solutions appear. These solutions are known as "external cavity modes" since the frequency separation corresponds to the external cavity frequency.

The external cavity modes can be also obtained from Eqs. (4.69) and (4.70) by eliminating the sine and cosine functions as,

$$\left(\Delta\omega_{s,0}\tau - \frac{\alpha\tau}{2}G_N\Delta N_{s,\text{th}}\right)^2 + \left(\frac{\tau}{2}G_N\Delta N_{s,\text{th}}\right)^2 = (\kappa\tau)^2 \quad (4.74)$$

where, $\Delta N_{s,\text{th}} = N_s - N_{\text{th}} = N_s - \left(N_0 + \dfrac{1}{G_N\tau_p}\right)$.

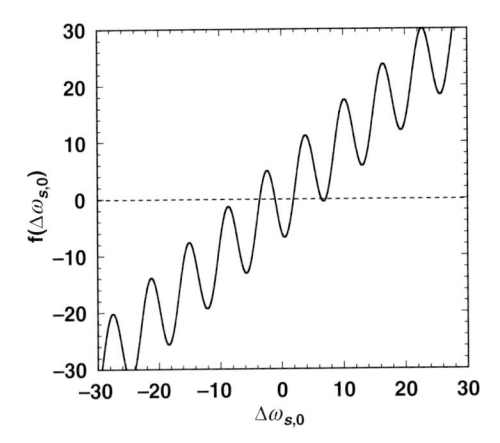

Figure 4.10 The steady-state solutions of Eq. (4.73) are plotted as a function of $\Delta\omega_{s,0}$. The cross sections of the solid curve and the dotted line indicate the steady-state solutions. Multiple steady-state solutions are obtained.

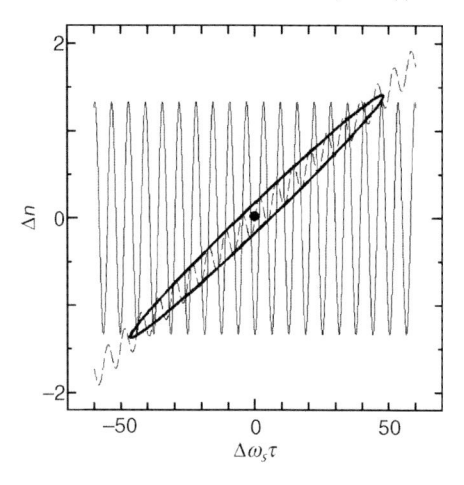

Figure 4.11 Carrier density $\Delta N_{s,th}$ change versus frequency change $\Delta \omega_{s,0}\tau$ for the possible steady states under external feedback, which is derived from Eq. (4.74). Crossing points of the solid and broken sinusoidal waves are the locations of the modes. Modes are on an ellipsoid. The solid dot at the center is the solitary oscillation mode. (J. Ohtsubo, (2008), © 2008 Springer.).

The phase plot of Eq. (4.74) on the ($\Delta \omega_{s,0}\tau$, $\Delta N_{s,th}$) plane is shown in Figure 4.11 (Ohtsubo, 2008). The broken sinusoidal curve in Figure 4.11 denotes the deviation from the steady state of the oscillation angular frequency $\Delta \omega_{s,0}$ ($\Delta \omega_s$ in Figure 4.11) and the other sinusoidal curve represents that of the carrier density $\Delta N_{s,th}$ (Δn in Figure 4.11). The crossing points of these two curves are the locations of possible oscillations and they are on the ellipsoid given by Eq. (4.74) (thick solid curve in Figure 4.11). Those in the lower half are the solutions for stable oscillations (external cavity modes) and those in the upper half are unstable oscillations (antimode). The laser oscillates at one of the external modes and the maximum gain mode (the minimum $\Delta N_{s,th}$ on the ellipsoid) is the most probable mode for laser oscillation. However, when the laser oscillation is unstable due to external feedback, the mode hops around among the external modes and the antimodes, thus the laser exhibits chaotic oscillations. A typical instability of this laser system is the LFF, in which the laser output power shows frequent irregular dropouts having frequencies from a few MHz to hundreds of MHz (Mørk et al., 1988; Fischer et al., 1996). This phase plot of Figure 4.11 is often used to describe the external-mode transition in the LFF regime (see Figure 3.A.1 in Appendix 3.A.2). The solid dot at the center of the ellipsoid in Figure 4.11 is the solution for the laser oscillation in the solitary laser without optical feedback (solitary mode).

4.3.4
Linear Stability Analysis for Steady State Solutions with Optical Feedback

The stability of the steady-state solutions of Eqs. (4.68)–(4.70) is investigated in this section. The linearized equations for a semiconductor laser with optical feedback are derived from the original Lang–Kobayashi equations of Eqs. (4.58)–(4.60). The small linear deviations from the steady-state solutions (Eqs. (4.38)–(4.40)) are used for linear stability analysis. For time-delayed dynamical systems, linearized variables with time delay are also considered,

$$E_\tau(t) = E(t-\tau) = E_s + \delta_E(t-\tau) \tag{4.75}$$

$$\Phi_\tau(t) = \Phi(t-\tau) = (\omega_s - \omega)(t-\tau) + \delta_\Phi(t-\tau) \tag{4.76}$$

To obtain linearized equations, the variables $E(t)$ and $E_\tau(t)$ need to be treated as independent variables for the analysis of time-delayed systems. Substituting the linearized variables of Eqs. (4.38)–(4.40) and (4.75)–(4.76) into the original equations of Eqs. (4.58)–(4.61), and the linearized equations are obtained,

$$\begin{aligned}\frac{d\delta_E(t)}{dt} &= -\kappa \cos(\omega_s\tau)\delta_E(t) - \kappa E_s \sin(\omega_s\tau)\delta_\Phi(t) + \frac{1}{2}G_N E_s \delta_N(t) \\ &+ \kappa \cos(\omega_s\tau)\delta_E(t-\tau) + \kappa E_s \sin(\omega_s\tau)\delta_\Phi(t-\tau)\end{aligned} \tag{4.77}$$

$$\begin{aligned}\frac{d\delta_\Phi(t)}{dt} &= \frac{\kappa}{E_s}\sin(\omega_s\tau)\delta_E(t) - \kappa \cos(\omega_s\tau)\delta_\Phi(t) + \frac{\alpha}{2}G_N\delta_N(t) \\ &- \frac{\kappa}{E_s}\sin(\omega_s\tau)\delta_E(t-\tau) + \kappa \cos(\omega_s\tau)\delta_\Phi(t-\tau)\end{aligned} \tag{4.78}$$

$$\frac{d\delta_N(t)}{dt} = -2G_N(N_s - N_0)E_s\delta_E(t) - \left(\frac{1}{\tau_s} + G_N E_s^2\right)\delta_N(t) \tag{4.79}$$

where the square terms of the linearized variables can be neglected because $\delta_i(t) \gg \delta_i^2(t)$ and $\delta_i(t) \gg \delta_i(t) \cdot \delta_j(t)$ are assumed (i, j correspond to E, Φ, N). The steady-state solutions of Eq. (4.70) are also used for Eq. (4.77).

The linearized equations of Eqs. (4.77)–(4.79) can also be obtained from the formula of the Jacobian matrix. It is worth noting that the formula includes time-delayed linearized variables, unlike Eq. (4.44),

$$\begin{pmatrix}\frac{d\delta_E(t)}{dt} \\ \frac{d\delta_\Phi(t)}{dt} \\ \frac{d\delta_N(t)}{dt}\end{pmatrix} = \begin{pmatrix}\frac{\partial f_E}{\partial E} & \frac{\partial f_E}{\partial \Phi} & \frac{\partial f_E}{\partial N} & \frac{\partial f_E}{\partial E_\tau} & \frac{\partial f_E}{\partial \Phi_\tau} \\ \frac{\partial f_\Phi}{\partial E} & \frac{\partial f_\Phi}{\partial \Phi} & \frac{\partial f_\Phi}{\partial N} & \frac{\partial f_\Phi}{\partial E_\tau} & \frac{\partial f_\Phi}{\partial \Phi_\tau} \\ \frac{\partial f_N}{\partial E} & \frac{\partial f_N}{\partial \Phi} & \frac{\partial f_N}{\partial N} & \frac{\partial f_N}{\partial E_\tau} & \frac{\partial f_N}{\partial \Phi_\tau}\end{pmatrix}\begin{pmatrix}\delta_E(t) \\ \delta_\Phi(t) \\ \delta_N(t) \\ \delta_E(t-\tau) \\ \delta_\Phi(t-\tau)\end{pmatrix} \tag{4.80}$$

where f_E, f_Φ, and f_N are the right-hand-side terms of Eqs. (4.58)–(4.60), respectively. The linearized equations are calculated from Eqs. (4.58)–(4.60) and (4.80) in the matrix formula,

4.3 Analytical Approach of Semiconductor Laser with Optical Feedback

$$\begin{pmatrix} \dfrac{d\delta_E(t)}{dt} \\ \dfrac{d\delta_\Phi(t)}{dt} \\ \dfrac{d\delta_N(t)}{dt} \end{pmatrix}$$

$$= \begin{pmatrix} -\kappa\cos(\omega_s\tau) & -\kappa E_s\sin(\omega_s\tau) & \dfrac{1}{2}G_N E_s & \kappa\cos(\omega_s\tau) & \kappa E_s\sin(\omega_s\tau) \\ \dfrac{\kappa}{E_s}\sin(\omega_s\tau) & -\kappa\cos(\omega_s\tau) & \dfrac{a}{2}G_N & -\dfrac{\kappa}{E_s}\sin(\omega_s\tau) & \kappa\cos(\omega_s\tau) \\ -2G_N(N_s-N_0)E_s & 0 & -\dfrac{1}{\tau_s}-G_N E_s^2 & 0 & 0 \end{pmatrix} \begin{pmatrix} \delta_E(t) \\ \delta_\Phi(t) \\ \delta_N(t) \\ \delta_E(t-\tau) \\ \delta_\Phi(t-\tau) \end{pmatrix}$$

(4.81)

The matrix formula of Eq. (4.81) is in fact the same as a set of equations of Eqs. (4.77)–(4.79).

The eigenvalues λ of the Jacobian matrix in Eq. (4.81) cannot be directly calculated due to the presence of the time-delayed linearized variables. The linearized variables can be expressed by $\delta_E(t) = \bar{\delta}_E\exp(\lambda t)$, $\delta_\Phi(t) = \bar{\delta}_\Phi\exp(\lambda t)$, and $\delta_N(t) = \bar{\delta}_N\exp(\lambda t)$, where $\bar{\delta}_E$, $\bar{\delta}_\Phi$, and $\bar{\delta}_N$ are constant. The time-delayed linearized variables can be written with λ as follows.

$$\delta_E(t-\tau) = \bar{\delta}_E\exp(\lambda(t-\tau)) = \bar{\delta}_E\exp(\lambda t)\exp(-\lambda\tau) \quad (4.82)$$

$$\delta_\Phi(t-\tau) = \bar{\delta}_\Phi\exp(\lambda(t-\tau)) = \bar{\delta}_\Phi\exp(\lambda t)\exp(-\lambda\tau) \quad (4.83)$$

Substituting Eqs. (4.82) and (4.83) into Eqs. (4.81),

$$\begin{pmatrix} \lambda\bar{\delta}_E\exp(\lambda t) \\ \lambda\bar{\delta}_\Phi\exp(\lambda t) \\ \lambda\bar{\delta}_N\exp(\lambda t) \end{pmatrix}$$

$$= \begin{pmatrix} -\kappa\cos(\omega_s\tau)(1-\exp(-\lambda\tau)) & -\kappa E_s\sin(\omega_s\tau)(1-\exp(-\lambda\tau)) & \dfrac{1}{2}G_N E_s \\ \dfrac{\kappa}{E_s}\sin(\omega_s\tau)(1-\exp(-\lambda\tau)) & -\kappa\cos(\omega_s\tau)(1-\exp(-\lambda\tau)) & \dfrac{a}{2}G_N \\ -2G_N(N_s-N_0)E_s & 0 & -\dfrac{1}{\tau_s}-G_N E_s^2 \end{pmatrix} \begin{pmatrix} \bar{\delta}_E\exp(\lambda t) \\ \bar{\delta}_\Phi\exp(\lambda t) \\ \bar{\delta}_N\exp(\lambda t) \end{pmatrix}$$

(4.84)

Equation (4.84) is divided by $\exp(\lambda t)$ and the following equation holds for $\bar{\delta}_E, \bar{\delta}_\Phi, \bar{\delta}_N \neq 0$.

$$\begin{vmatrix} \lambda+\kappa\cos(\omega_s\tau)(1-\exp(-\lambda\tau)) & \kappa E_s\sin(\omega_s\tau)(1-\exp(-\lambda\tau)) & -\dfrac{1}{2}G_N E_s \\ -\dfrac{\kappa}{E_s}\sin(\omega_s\tau)(1-\exp(-\lambda\tau)) & \lambda+\kappa\cos(\omega_s\tau)(1-\exp(-\lambda\tau)) & -\dfrac{a}{2}G_N \\ 2G_N(N_s-N_0)E_s & 0 & \lambda+\dfrac{1}{\tau_s}+G_N E_s^2 \end{vmatrix} = 0$$

(4.85)

The characteristic equation is calculated from Eq. (4.85),

$$\lambda^3 + \left[\left(\frac{1}{\tau_s}+G_N E_s^2\right)+2\kappa\cos(\omega_s\tau)(1-\exp(-\lambda\tau))\right]\lambda^2$$

$$+\left[G_N^2 E_s^2(N_s-N_0)+2\kappa\cos(\omega_s\tau)(1-\exp(-\lambda\tau))\left(\frac{1}{\tau_s}+G_N E_s^2\right)+\kappa^2(1-\exp(-\lambda\tau))^2\right]\lambda$$

$$+\kappa^2(1-\exp(-\lambda\tau))^2\left(\frac{1}{\tau_s}+G_N E_s^2\right)$$

$$+\kappa G_N^2 E_s^2(N_s-N_0)(1-\exp(-\lambda\tau))[\cos(\omega_s\tau)-\alpha\sin(\omega_s\tau)]=0 \quad (4.86)$$

The characteristic equation of Eq. (4.86) cannot be solved analytically. Further analytical treatment can be conducted under the limits of weak and short feedback (i.e., $\kappa\tau \ll 1$), as described in Appendix 4.A.3.

To analyze the stability of the system under strong feedback strength or long feedback time (i.e., $\kappa\tau \geq 1$), it is required to calculate Eq. (4.86) numerically to obtain the eigenvalues λ. The numerical calculations have been conducted and the real and imaginary parts of the eigenvalues are numerically calculated (Liu et al., 1995). Figure 4.12 shows the mode distributions for two typical chaotic output cases. The real and imaginary parts of the eigenvalues λ ($= \lambda_{Re} + i\lambda_{Im}$) are plotted, which are the indicators of the stability and the oscillation frequency, respectively. The vertical axis (Re(γ)) and horizontal axis (Im(γ)) correspond to the real part (λ_{Re}) and imaginary part (λ_{Im}) of the eigenvalues, respectively. The modes that have the lowest frequencies and whose real parts are very close to the stability criterion ($\lambda_R \approx 0$) are plotted. It should be noted that the average separation between adjacent modes on the frequency axis is almost the same as the inverse of the delay time. These calclulation results can be useful for controlling chaos in a semiconductor laser with high-frequency injection (see Section 11.2.2.2).

The calculation of Figure 4.12 by using Eq. (4.86) is not straightforward because multiple steady-state solutions exist. Instead of using direct calculation of the eigenvalues of the Jacobian matrix, the maximum Lyapunov exponent is more useful and will be introduced in Section 4.4.3.

4.4
Numerical Analysis of Semiconductor Laser with Optical Feedback

Numerical analysis is a useful technique to investigate nonlinear dynamics in a laser model. Numerical results of temporal waveforms, frequency spectra, attractors, and bifurcation diagrams are shown in this section. Source codes of C programming language are listed in Glossary G.2.3 for numerical simulations of chaotic dynamics of the Lang–Kobayashi equations for a semiconductor laser with optical feedback.

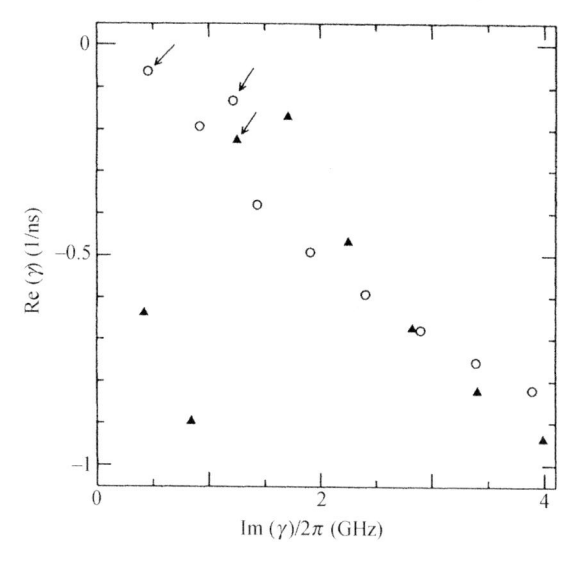

Figure 4.12 Mode distributions for two typical chaotic outputs. The vertical and horizontal axes correspond to the real and imaginary parts of the eigenvalues, respectively. Different parameter values are used for circles and triangles. (Y. Liu, N. Kikuchi, and J. Ohtsubo, (1995), © 1995 APS.).

4.4.1
Numerical Results of Chaotic Dynamics

4.4.1.1 Temporal Waveforms, FFTs, and Attractors

The Lang–Kobayashi equations of Eqs. (4.3)–(4.6) are numerically calculated with the parameter values shown in Table 4.3. The fourth-order Runge–Kutta method is used for the numerical integration (see Appendix 4.A.4 for the Runge–Kutta method). Typical results of temporal waveforms are shown in Figure 4.13 when the feedback strength is changed. Here, the feedback strength κ can be obtained from the reflectivity of the external mirror r_3,

$$\kappa = \frac{(1-r_2^2) r_3}{r_2} \frac{1}{\tau_{in}} \tag{4.87}$$

According to the parameter values shown in Table 4.3, the following relationship between κ and r_3 can be hold,

$$\kappa = 1.553 \times 10^{11} \cdot r_3 = 155.3 \cdot r_3 \, [\text{ns}^{-1}] \tag{4.88}$$

The time evolution of the laser intensity ($I(t) = E^2(t)$) is calculated and plotted in Figure 4.13. The vertical axis of the laser intensity is normalized by 10^{20} for convenience. Without optical feedback $\kappa = 0$ ($r_3 = 0$), the temporal waveform of a steady state is obtained as shown in Figure 4.13a. With the increase of κ, a temporal waveform of a period-1 oscillation is observed at $\kappa = 0.777 \, \text{ns}^{-1}$ ($r_3 = 0.0050$) in

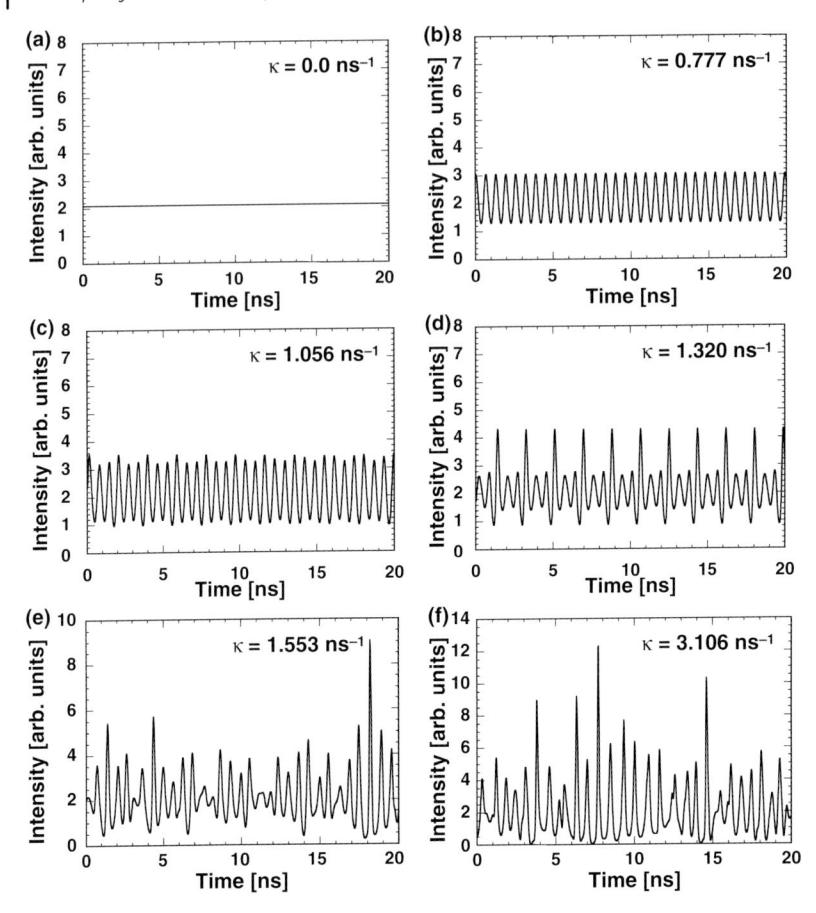

Figure 4.13 Numerical results of temporal waveforms of the laser intensity for various feedback strength κ (or reflectivity of the external mirror r_3). (a) Steady state at $\kappa = 0$ ($r_3 = 0$), (b) the period-1 oscillation at $\kappa = 0.777$ ns^{-1} ($r_3 = 0.0050$), (c) the quasiperiodic oscillation at $\kappa = 1.056$ ns^{-1} ($r_3 = 0.0068$), (d) the period-3 oscillation at $\kappa = 1.320$ ns^{-1} ($r_3 = 0.0085$), (e) the chaotic oscillation at $\kappa = 1.553$ ns^{-1} ($r_3 = 0.0100$), and (f) the chaotic oscillation at $\kappa = 3.106$ ns^{-1} ($r_3 = 0.0200$). The numerical data in this chapter was provided by Kazutaka Kanno.

Figure 4.13b. As κ is increased to $\kappa = 1.056$ ns^{-1} ($r_3 = 0.0068$), another oscillation component appears on the period-1 oscillation, showing a quasiperiodic oscillation in Figure 4.13c. The temporal waveform becomes a period-3 oscillation at $\kappa = 1.320$ ns^{-1} ($r_3 = 0.0085$) in Figure 4.13d. A chaotic oscillation is observed at $\kappa = 1.553$ ns^{-1} ($r_3 = 0.0100$) in Figure 4.13e. As κ is increased further, chaotic oscillations with larger amplitude are observed at $\kappa = 3.106$ ns^{-1} ($r_3 = 0.0200$) in Figure 4.13f. The transition from periodic oscillations to chaotic oscillations through quasiperiodic oscillations is known as the quasiperiodicity route to chaos, indicating the existence of deterministic chaos (Bergé et al., 1984).

The transition from periodic to chaotic oscillations can also be observed in the frequency domain. The frequency components are plotted by calculating fast Fourier transform (FFT) from the temporal waveforms shown in Figure 4.13. Figure 4.14 shows the FFTs corresponding to Figure 4.13. The calculation of FFTs in numerical simulation corresponds to the observation of RF spectra in experiment. For Figure 4.14a at $\kappa = 0$, no peak appears in the FFT because the temporal waveform is stable and there is no oscillatory behavior. As κ is increased to $\kappa = 0.777\,\text{ns}^{-1}$, a large sharp peak at 1.56 GHz, nearly corresponding to the relaxation oscillation frequency, and its harmonics are observed in Figure 4.14b. The relaxation oscillation frequency for a solitary laser without optical feedback is 1.52 GHz, which can be

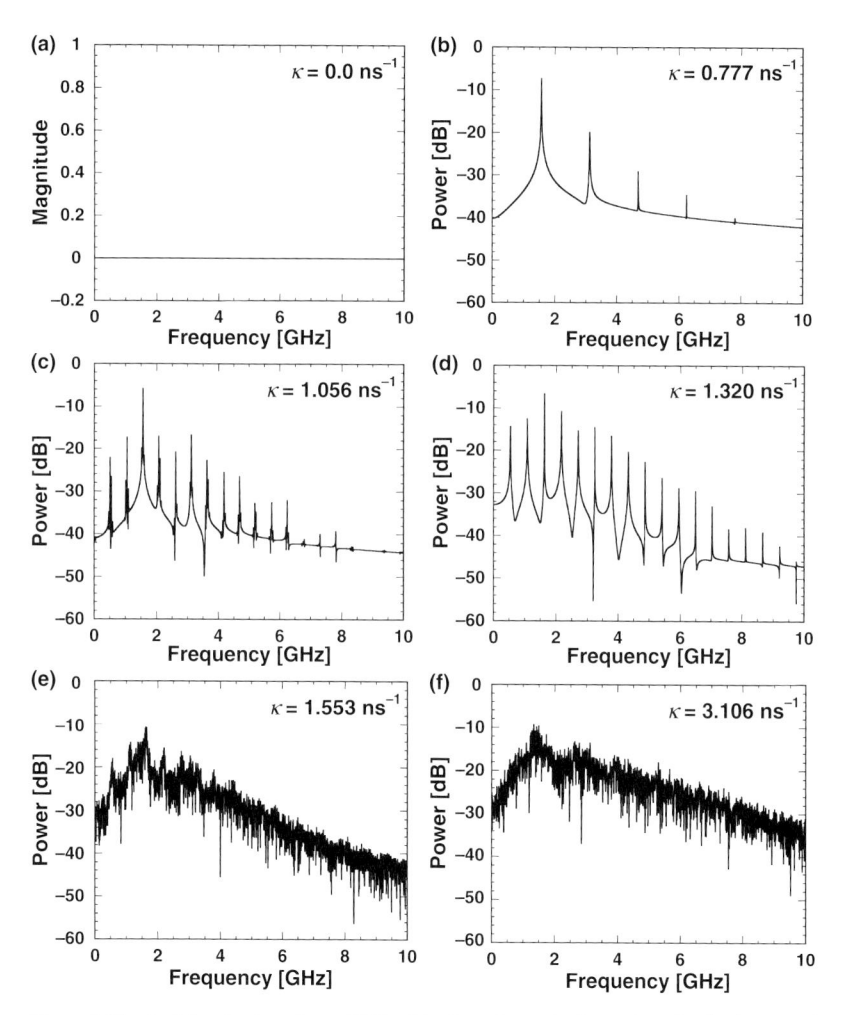

Figure 4.14 Fast Fourier transform (FFT) of the laser intensity for various feedback strength κ. These figures correspond to Figure 4.13.

obtained from Eq. (4.56) with the parameter values shown in Table 4.3. The observed large shape peak at 1.56 GHz is slightly shifted from the original relaxation oscillation frequency due to the optical feedback effect.

With the increase of κ to 1.056 ns^{-1}, many sharp peaks appear in the frequency domain in Figure 4.14c. The two main frequency components (and their harmonics) are observed, which corresponds to the relaxation oscillation frequency and the external cavity frequency. At $\kappa = 1.320$ ns^{-1}, several frequency peaks whose interval corresponds to the 1/3 subharmonics of the largest frequency peak are observed as the period-3 oscillation in Figure 4.14d. As κ is increased to 1.553 ns^{-1}, a broad spectrum is obtained and the chaotic oscillation appears in Figure 4.14e. As κ is increased further, a broader spectrum is observed at $\kappa = 3.106$ ns^{-1} in Figure 4.14f. The nonlinear interaction between the relaxation oscillation frequency and the external cavity frequency results in the quasiperiodicity route to chaos as the feedback strength increases.

Another representation of the nonlinear dynamics is an attractor in phase space. Two variables, the laser intensity $(I(t) = E^2(t))$ and the carrier density $N(t)$, are plotted at the same time on the two-dimensional (2D) phase space as a Lissajous figure. Figure 4.15 shows the attractor plots of the temporal waveforms corresponding to Figure 4.13. The vertical axis of the carrier density is normalized by 10^{24} and the horizontal axis of the laser intensity is normalized by 10^{20} for convenience. The point attractor is observed at $\kappa = 0$ in Figure 4.15a, because the temporal waveform does not show any oscillatory behavior. As κ is increased to 0.777 ns^{-1}, the circular plot is observed in Figure 4.15b, which indicates a period-1 attractor (limit cycle). With increase of κ to 1.056 ns^{-1}, the circular plot has some thickness and looks like an intersection of a tube in Figure 4.15c, which indicates a quasiperiodic attractor (torus attractor). For Figure 4.15d, a triple circle appears at $\kappa = 1.320$ ns^{-1}, indicating a period-3 attractor. A more complex plot is observed at $\kappa = 1.553$ ns^{-1} in Figure 4.15e, indicating a chaotic attractor (strange attractor). The size of the chaotic attractors becomes larger as κ is increased to $\kappa = 3.106$ ns^{-1} in Figure 4.15f. As can be seen in Figure 4.15, the representation of attractors in phase space is useful to distinguish between different dynamical states visually.

4.4.1.2 Bifurcation Diagram and Two-Dimensional Dynamical Map

To investigate the transition from periodic to chaotic oscillations systematically, bifurcation diagram is a very useful and informative tool. To make a bifurcation diagram, one of the laser parameters is selected as a bifurcation parameter (e.g., the feedback strength κ). First, a value of the selected parameter is fixed and a temporal waveform is calculated. The local maxima of the temporal waveform (peaks of the oscillation) are sampled, and the peak values are saved. The peak values are plotted in the vertical direction on the bifurcation diagram, whose horizontal axis corresponds to the selected parameter value. The parameter value is then changed slightly and the peak values of the temporal waveform are plotted again on the bifurcation diagram. Repeating this procedure as the parameter value is increased (or decreased), a bifurcation diagram is obtained as a plot of the peak values of temporal waveforms at various parameter values. The number of peak values indicates the dynamical state:

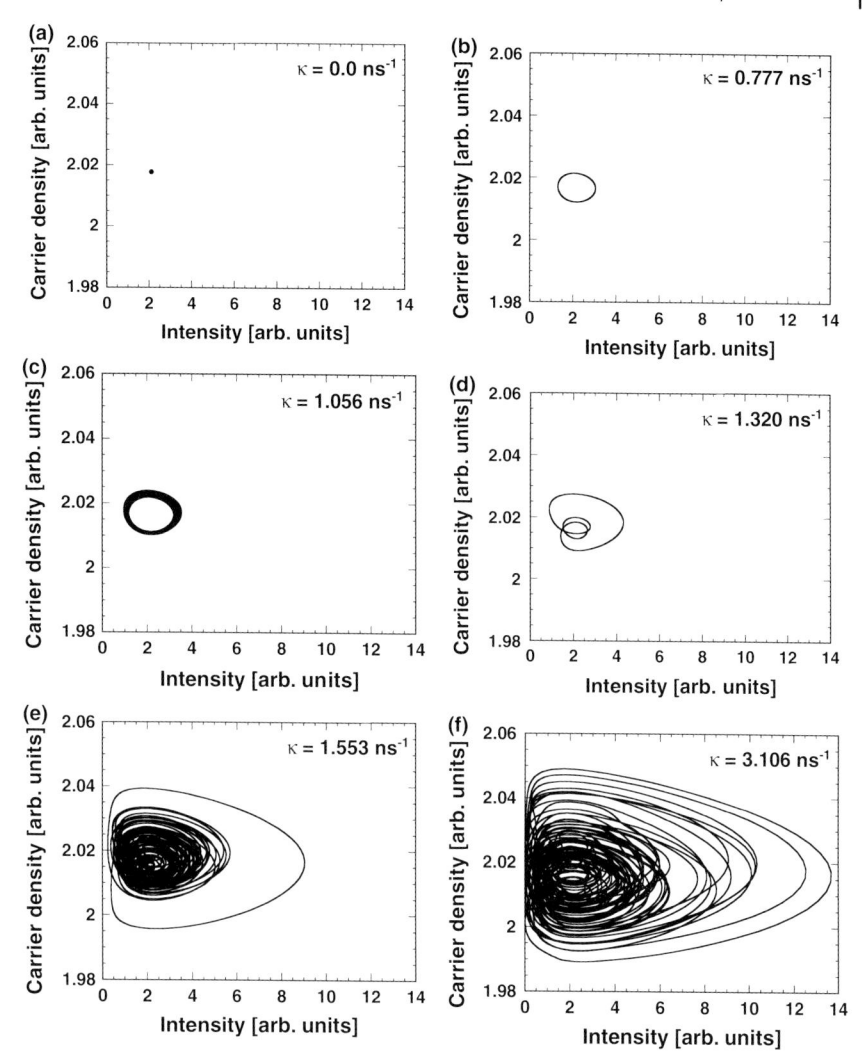

Figure 4.15 Attractors of the laser intensity in the phase space for various feedback strength κ. The two-dimensional phase space consists of the laser intensity $I(t) = E^2(t)$ and the carrier density $N(t)$. These figures correspond to Figure 4.13.

one peak corresponds to steady state or a period-1 oscillation, two peaks corresponds to a period-2 oscillation, many peaks within a limited range indicate a quasiperiodic oscillation, and many peaks with scattered plots over wide range indicate a chaotic oscillation.

Figure 4.16 shows an example of bifurcation diagram as a function of the feedback strength κ. According to the above-mentioned criteria, the bifurcation diagram in Figure 4.16 shows the transition from the steady state ($0 \leq \kappa \leq 0.57$), period-1 oscillation ($0.57 \leq \kappa \leq 1.03$), quasiperiodic oscillation ($1.03 \leq \kappa \leq 1.08$), period-3

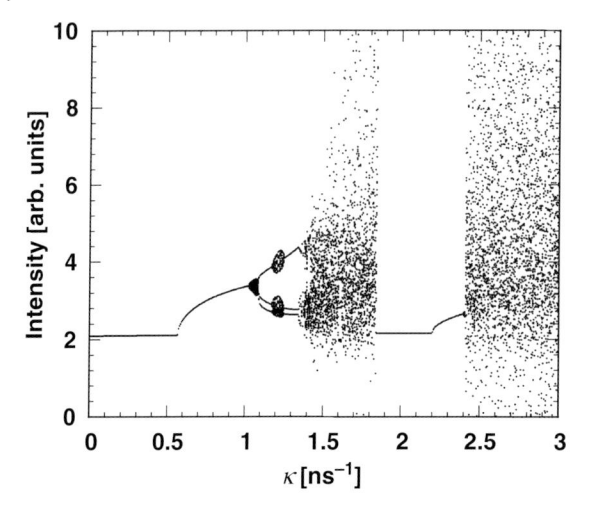

Figure 4.16 Bifurcation diagram as a function of the feedback strength κ. Local maxima of temporal waveforms are sampled and plotted in the bifurcation diagram as κ is increased.

oscillation (1.08 ≤ κ ≤ 1.17), quasiperiodic oscillation with period-3 (1.17 ≤ κ ≤ 1.25), period-3 oscillation (1.25 ≤ κ ≤ 1.35), quasiperiodic oscillation (1.35 ≤ κ ≤ 1.42), and chaotic oscillation (1.42 ≤ κ ≤ 1.84). A periodic window appears (1.60 ≤ κ ≤ 1.62) in the chaotic region. After the chaotic region, the steady state (1.84 ≤ κ ≤ 2.20) and period-1 oscillation (2.20 ≤ κ ≤ 2.41) appear again, and chaotic oscillation suddenly occurs (κ ≥ 2.41). As seen in Figure 4.16, a bifurcation diagram is very convenient to observe the continuous transition of dynamical behaviors in nonlinear dynamical systems. A bifurcation diagram is often used to investigate whether unstable dynamical behaviors result from deterministic chaos, because the existence of bifurcation and the route to chaos is one of the most important characteristics of deterministic chaos.

To investigate the dynamical states more systematically, a 2D bifurcation diagram (frequently called a "dynamical map") is created. Instead of using one parameter, two parameter values are changed simultaneously and the dynamical states are plotted on the 2D space of the dynamical map. First, the feedback strength κ and the normalized injection current j ($= J/J_{\text{th}}$) are changed simultaneously. Figure 4.17 shows the 2D map as functions of κ and j. As κ is increased, the transition from steady state (S), periodic (P), quasiperiodic (QP), to chaotic oscillations (C) is observed clearly for various values of j, indicating the quasiperiodicity route to chaos. The bifurcation points are dependent of j. On the other hand, the dynamics is not dramatically changed with increase of j. Periodic changes in different dynamical states are observed with increase of j at low κ.

These numerical analyses of temporal dynamics (Figure 4.13), FFTs (Figure 4.14), attractors (Figure 4.15), bifurcation diagram (Figure 4.16) and 2D dynamical map (Figure 4.17) are basic tools to investigate the chaotic dynamical behaviors in nonlinear dynamical systems. In the next section, linear stability

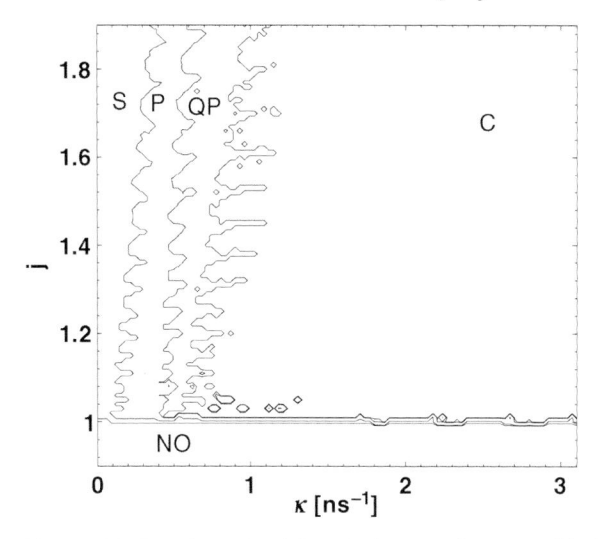

Figure 4.17 Two-dimensional dynamical map as functions of the feedback strength κ and the normalized injection current j. NO, no lasing; S, steady state; P, periodic oscillation; QP, quasiperiodic oscillation; C, chaotic oscillation. The external cavity length is set to $L = 0.6$ m.

analysis around a periodic or chaotic trajectory is introduced and examined to obtain more insight into chaotic nonlinear dynamics in a semiconductor laser with optical feedback.

4.4.2
Linear Stability Analysis for Oscillatory Trajectory

In Sections 4.3.2 and 4.3.4, linear stability analysis for steady-state solutions was performed. In this section, linear stability analysis for oscillatory (periodic or chaotic) trajectory is introduced and described. Lyapunov exponents can be obtained from the linear stability analysis and give the criteria to identify deterministic chaos and to quantify the degree of complexity of chaos.

The linearized equations of the Lang–Kobayashi equations are calculated from Eqs. (4.3)–(4.6). Small linear deviations from the oscillatory trajectory on the attractor are introduced as,

$$E(t) = E_s(t) + \delta_E(t) \tag{4.89}$$

$$\Phi(t) = (\omega_s(t) - \omega)t + \delta_\Phi(t) \tag{4.90}$$

$$N(t) = N_s(t) + \delta_N(t) \tag{4.91}$$

where $\delta_E(t)$, $\delta_\Phi(t)$, and $\delta_N(t)$ are new variables for the linearized equations ($E(t) \gg \delta_E(t)$, $\Phi(t) \gg \delta_\Phi(t)$, $N(t) \gg \delta_N(t)$). $\delta_E(t)$, $\delta_\Phi(t)$, and $\delta_N(t)$ are the variables of small linear deviations from $E_s(t)$, $(\omega_s(t)-\omega)t$ and $N_s(t)$, respectively. It is worth noting that $E_s(t)$, $\omega_s(t)$, and $N_s(t)$ are "time-varying" waveforms of a periodic or chaotic trajectory, unlike Eqs. (4.38)–(4.40). For Eqs. (4.38)–(4.40), E_s, ω_s, and N_s are treated as the steady-state solutions and the stability around the steady-state solutions is considered. By contrast, for Eqs. (4.89)–(4.91) the linear stability around an oscillatory trajectory a periodic or chaotic temporal waveform of $E_s(t)$, $\omega_s(t)$, and $N_s(t)$ is considered in this section. This indicates that the Jacobian matrix becomes time dependent and direct calculation of the eigenvalues of the Jacobian matrix is no longer effective.

For time-delayed dynamical systems, linearized variables with time delay are also introduced as,

$$E_\tau(t) = E(t-\tau) = E_s(t-\tau) + \delta_E(t-\tau) \tag{4.92}$$

$$\Phi_\tau(t) = \Phi(t-\tau) = (\omega_s(t-\tau)-\omega)(t-\tau) + \delta_\Phi(t-\tau) \tag{4.93}$$

To obtain linearized equations, the variables $E(t)$ and $E_\tau(t)$ need to be treated as independent variables for time-delayed systems. It is also important to treat $\delta_E(t)$ and $\delta_E(t-\tau)$ as independent variables when the Jacobian matrix is obtained by derivatives.

The linearized equations of Lang–Kobayashi equations of Eqs. (4.3)–(4.6) are obtained from the following formula,

$$\begin{pmatrix} \frac{d\delta_E(t)}{dt} \\ \frac{d\delta_\Phi(t)}{dt} \\ \frac{d\delta_N(t)}{dt} \end{pmatrix} = \begin{pmatrix} \frac{\partial f_E}{\partial E} & \frac{\partial f_E}{\partial \Phi} & \frac{\partial f_E}{\partial N} & \frac{\partial f_E}{\partial E_\tau} & \frac{\partial f_E}{\partial \Phi_\tau} \\ \frac{\partial f_\Phi}{\partial E} & \frac{\partial f_\Phi}{\partial \Phi} & \frac{\partial f_\Phi}{\partial N} & \frac{\partial f_\Phi}{\partial E_\tau} & \frac{\partial f_\Phi}{\partial \Phi_\tau} \\ \frac{\partial f_N}{\partial E} & \frac{\partial f_N}{\partial \Phi} & \frac{\partial f_N}{\partial N} & \frac{\partial f_N}{\partial E_\tau} & \frac{\partial f_N}{\partial \Phi_\tau} \end{pmatrix} \begin{pmatrix} \delta_E(t) \\ \delta_\Phi(t) \\ \delta_N(t) \\ \delta_E(t-\tau) \\ \delta_\Phi(t-\tau) \end{pmatrix}$$

$$= \begin{pmatrix} \frac{1}{2}\left[G_N(N(t)-N_0)-\frac{1}{\tau_p}\right] & -\kappa E(t-\tau)\sin\Theta(t) & \frac{1}{2}G_N E(t) & \kappa\cos\Theta(t) & \kappa E(t-\tau)\sin\Theta(t) \\ \kappa\frac{E(t-\tau)}{E^2(t)}\sin\Theta(t) & -\kappa\frac{E(t-\tau)}{E(t)}\cos\Theta(t) & \frac{\alpha}{2}G_N & -\kappa\frac{1}{E(t)}\sin\Theta(t) & \kappa\frac{E(t-\tau)}{E(t)}\cos\Theta(t) \\ -2G_N(N(t)-N_0)E(t) & 0 & -\frac{1}{\tau_s}-G_N E^2(t) & 0 & 0 \end{pmatrix} \begin{pmatrix} \delta_E(t) \\ \delta_\Phi(t) \\ \delta_N(t) \\ \delta_E(t-\tau) \\ \delta_\Phi(t-\tau) \end{pmatrix}$$

(4.94)

where f_E, f_Φ, and f_N are the right-hand-side terms of Eqs. (4.3)–(4.5). From Eq. (4.94), the linearized equations of Eqs. (4.3)–(4.5) are described as follows,

$$\frac{d\delta_E(t)}{dt} = \frac{1}{2}\left[G_N(N(t)-N_0) - \frac{1}{\tau_p}\right]\delta_E(t) - \kappa E(t-\tau)\sin\Theta(t)\delta_\Phi(t)$$

$$+ \frac{1}{2}G_N E(t)\delta_N(t)$$

$$+ \kappa\cos\Theta(t)\delta_E(t-\tau) + \kappa E(t-\tau)\sin\Theta(t)\delta_\Phi(t-\tau) \qquad (4.95)$$

$$\frac{d\delta_\Phi(t)}{dt} = \kappa\frac{E(t-\tau)}{E^2(t)}\sin\Theta(t)\delta_E(t) - \kappa\frac{E(t-\tau)}{E(t)}\cos\Theta(t)\delta_\Phi(t) + \frac{\alpha}{2}G_N\delta_N(t)$$

$$-\kappa\frac{1}{E(t)}\sin\Theta(t)\delta_E(t-\tau) + \kappa\frac{E(t-\tau)}{E(t)}\cos\Theta(t)\delta_\Phi(t-\tau) \qquad (4.96)$$

$$\frac{d\delta_N(t)}{dt} = -2G_N(N(t)-N_0)E(t)\delta_E(t) - \left(\frac{1}{\tau_s} + G_N E^2(t)\right)\delta_N(t) \qquad (4.97)$$

where the time-varying waveforms of $E_s(t)$, $N_s(t)$, and $\omega_s(t)$ are replaced by the original variables of $E(t)$, $N(t)$, and $\omega(t)$ for convenience. Note that Eqs. (4.95)–(4.97) include the original variables $E(t)$, $\Phi(t)$, $N(t)$, and $\Theta(t)$. It is thus required to solve both the linearized equations for the variables of $\delta_E(t)$, $\delta_\Phi(t)$, and $\delta_N(t)$ (Eqs. (4.95)–(4.97)) and the original equations (Eqs. (4.3)–(4.6)) simultaneously.

To analyze the stability of the system, the eigenvalues of the Jacobian matrix of Eq. (4.94) need to be estimated. It is, however, not easy to perform analytical calculation of the eigenvalues because time-dependent variables of $E(t)$, $\Phi(t)$, $N(t)$, and $\Theta(t)$ are included in Eq. (4.94). Instead of the calculation of the eigenvalues, a procedure of the calculation of the maximum Lyapunov exponent is introduced in the next section.

4.4.3
Maximum Lyapunov Exponent

The maximum Lyapunov exponent is one of the most important measures for the stability of nonlinear dynamical systems. The maximum Lyapunov exponent indicates the degree of divergence (or convergence) of the distance between two nearby points of temporal waveforms. When the maximum Lyapunov exponent is positive, two nearby points on the attractor start to diverge exponentially in time and the two trajectories are completely different after transient, which indicates the sensitive dependence on initial conditions for chaotic oscillations. By contrast, when the maximum Lyapunov exponent is negative, the two nearby points start to converge and stable (or periodic) temporal dynamics is obtained. The sign of the maximum Lyapunov exponent is thus a measure of chaoticity in nonlinear dynamical systems. The existence of a positive maximum Lyapunov exponent is one of the most widely acceptable evidences of deterministic chaos.

The maximum Lyapunov exponent is calculated from the linearized equations Eqs. (4.95)–(4.97). For dynamical systems without time delay, the norm of all the linearized variables is defined as the Euclidean distance of the vector consisting of all the variables of the linearized equations, that is, $D(t) = \sqrt{\delta_E^2(t) + \delta_\Phi^2(t) + \delta_N^2(t)}$ (see Appendix 2.A.2 in Chapter 2). For dynamical systems with time delay, however, time delay has an extra degree of freedom because the dynamics is governed by all the initial conditions of the variables within the delay time τ (e.g., $E(0) \sim E(\tau)$). It is thus very important to consider all the contributions from the time-delayed variables within the delay loop.

For time-delayed dynamical systems, the norm of the linearized variables from time t to $t + \tau$ can be calculated (τ is the feedback delay time) (Pyragas, 1998). The number of data points normalized by the numerical integration step h in the delay time τ is defined as $M = \tau/h$. It is worth noting that the norm is obtained for each M step for time-delayed dynamical systems, unlike dynamical systems without time delay (see Appendix 2.A.2 in Chapter 2 for comparison). The variables within a time-delayed loop are considered as independent variables.

Figure 4.18 shows the schematic for the calculation of the norm $D(t)$. Suppose the linearized variables of $\delta_E(t)$, $\delta_\Phi(t)$, and $\delta_N(t)$ are already obtained for the time interval between t and $t + 2\tau$ from Eqs. (4.95)–(4.97). The norm of the linearized variables for all the data from time t to $t + \tau$ is defined as,

$$D(t) = \sqrt{\sum_{k=0}^{M-1} \left(\delta_E^2(t+kh) + \delta_\Phi^2(t+kh) + \delta_N^2(t+kh)\right)} \qquad (4.98)$$

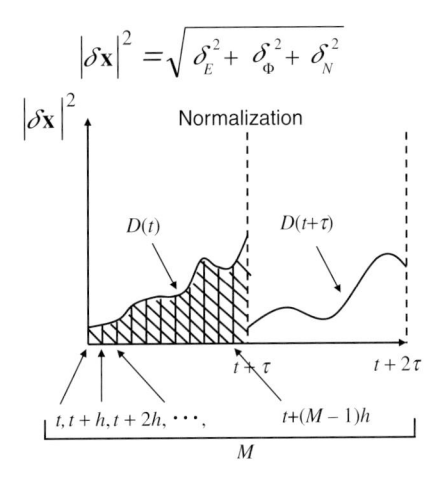

Figure 4.18 Schematic for the calculation of the norm $D(t)$ for time-delayed nonlinear dynamical systems.

The norm from time $t+\tau$ to $t+2\tau$ is also defined as,

$$D(t+\tau) = \sqrt{\sum_{k=0}^{M-1} \left(\delta_E^2(t+\tau+kh) + \delta_\Phi^2(t+\tau+kh) + \delta_N^2(t+\tau+kh)\right)} \quad (4.99)$$

The norm ratio between the time interval from t to $t+\tau$ and the interval from $t+\tau$ to $t+2\tau$ is calculated,

$$d_j = \frac{D(t+\tau)}{D(t)} \quad (4.100)$$

The value of d_j is saved for each M step of numerical integration. After calculating the norm ratio, all the linearized variables from $t+\tau$ to $t+2\tau$ need to be normalized by the norm $D(t+\tau)$ as,

$$\delta_{E,\text{new}}(t+\tau+kh) = \frac{\delta_E(t+\tau+kh)}{D(t+\tau)} \quad (4.101)$$

$$\delta_{\Phi,\text{new}}(t+\tau+kh) = \frac{\delta_\Phi(t+\tau+kh)}{D(t+\tau)} \quad (4.102)$$

$$\delta_{N,\text{new}}(t+\tau+kh) = \frac{\delta_N(t+\tau+kh)}{D(t+\tau)} \quad (4.103)$$

where $k = 0, 1, \ldots, M-1$. The normalization is very important to maintain the linearity of the stability analysis (i.e., small deviation from an oscillatory trajectory). These new variables of Eqs. (4.101)–(4.103) are used as initial conditions to calculate the variables between $t+2\tau$ and $t+3\tau$ and to obtain the next norm ratio between $D(t+\tau)$ and $D(t+2\tau)$ in the next step.

The calculation from Eqs. (4.98) to (4.103) is repeated for a long time period. The maximum Lyapunov exponent is obtained for the Nth repetition of the above-mentioned procedure,

$$\lambda_{\max} = \frac{1}{N\tau} \sum_{j=1}^{N} \ln(d_j) \quad (4.104)$$

where ln is the natural logarithm with the natural base e. The maximum Lyapunov exponent can be obtained as the average ratio of norm change in the logarithmic scale. In this procedure, the Jacobian matrix of the original equations is essentially calculated along the oscillatory trajectory, and the growth of the perturbation from the trajectory in phase space is evaluated by using the linearized equations.

Figure 4.19 shows the numerical result of the maximum Lyapunov exponent of Eq. (4.104) as a function of the feedback strength κ, obtained from the original Lang–Kobayashi equations (Eqs. (4.3)–(4.6)) and the linearized equations (Eqs. (4.95)–(4.97)). The corresponding bifurcation diagram is already shown in Figure 4.16 for comparison. The maximum Lyapunov exponent is almost zero (slightly

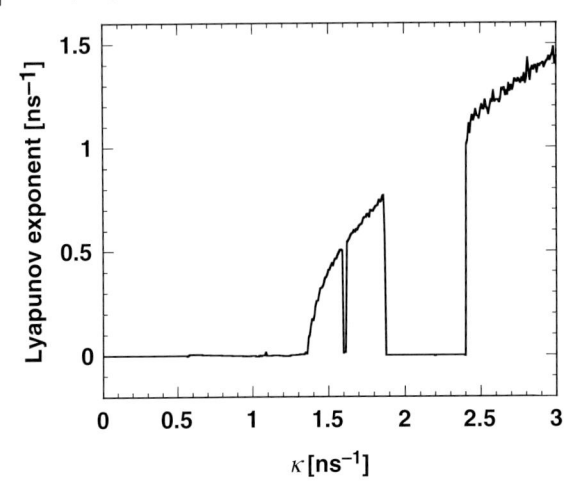

Figure 4.19 Maximum Lyapunov exponent as a function of the feedback strength κ. Positive Lyapunov exponents indicate the regions of chaotic oscillations in the bifurcation diagram shown in Figure 4.16.

negative) for steady-state, periodic oscillation, and quasiperiodic oscillations. By contrast, the maximum Lyapunov exponent becomes positive for chaotic oscillations except for the region of periodic window. The value of the maximum Lyapunov exponent increases monotonically as κ is increased in the chaotic region. The maximum Lyapunov exponent represents a quantitative indicator of complexity of deterministic chaos, and it is found that the complexity increases as the feedback strength is increased (see Section 4.5.5 in detail).

It has been reported that the maximum Lyapunov exponent can be estimated from time-series analysis of original differential equations or experimental data without using linearized equations (Kantz and Schreiber, 1997). It has been, however, known that the value of the maximum Lyapunov exponent estimated from time-series analysis of original temporal waveforms is not accurate, and is strongly dependent of the embedding parameter values such as the delay time for embedding and the dimensionality for the attractor reconstruction in the phase space. It is recommended to use linearized equations to evaluate a reliable value of the maximum Lyapunov exponent if the dynamical laser system can be modeled.

4.5
Dimensionless Equations and Further Nonlinear Analysis

4.5.1
Dimensionless Equations

For the calculation of the maximum Lyapunov exponent, the original Lang–Kobayashi equations of Eqs. (4.3)–(4.6) and the corresponding linearized equations of

Eqs. (4.95)–(4.97) are used in Section 4.4.3. The orders of magnitude of the variables $E(t)$, $\Phi(t)$, and $N(t)$ (also $\delta_E(t)$, $\delta_\Phi(t)$, and $\delta_N(t)$) in the original equations are very different (e.g., $E(t) \sim 10^{10}$, $\Phi(t) \sim 10^0$, and $N(t) \sim 10^{24}$ for the parameter values in Table 4.3). This fact may cause a serious problem for the accuracy of the numerical calculation of the norm value when more than one Lyapunov exponent is calculated. Small components in the norm vector possess tiny numerical errors that result in inaccuracy of the estimation of Lyapunov exponents, particularly for the calculation of a set of Lyapunov exponents (referred to as Lyapunov spectrum). The original Lang–Kobayashi equations thus need to be modified to "dimensionless" equations to match the order of magnitude for all the variables for further treatment of nonlinear dynamical analysis.

In this section, dimensionless Lang–Kobayashi equations are derived and nonlinear analysis is executed further, such as the estimation of the Lyapunov spectrum, Kolmogolov–Sinai entropy, and Kaplan–Yorke dimension. These indicators are useful for the case when the complexity and dimensionality of a laser system need to be measured quantitatively, while the maximum Lyapnov exponent is good enough to justify the occurrence of deterministic chaos.

Dimensionless equations are derived from the normalization. All the variables and the time scale are normalized by predefined constant values indicated by the bars,

$$E(t) = e(t')\bar{E}, \quad \Phi(t) = \phi(t')\bar{\Phi}, \quad N(t) = n(t')\bar{N}, \quad \Theta(t) = \theta(t'),$$
$$t = t'\bar{T}, \quad \tau = \tau'\bar{T}, \quad E(t-\tau) = e(t'-\tau')\bar{E}, \quad \Phi(t-\tau) = \phi(t'-\tau')\bar{\Phi} \tag{4.105}$$

These variables are substituted into the original Lang–Kobayashi equations of Eqs. (4.3)–(4.6), and the dimensionless Lang–Kobayashi equations are obtained as,

$$\frac{de(t')}{dt'} = [g_e(n(t') - n_0) - \gamma_e]e(t') + \kappa_e e(t' - \tau')\cos\theta(t') \tag{4.106}$$

$$\frac{d\phi(t')}{dt'} = g_\phi(n(t') - n_0) - \gamma_\phi - \kappa_\phi \frac{e(t'-\tau')}{e(t')}\sin\theta(t') \tag{4.107}$$

$$\frac{dn(t')}{dt'} = \gamma_n(j\, n_{th} - n(t')) - g_n(n(t') - n_0)e^2(t') \tag{4.108}$$

$$\theta(t') = \omega_\theta \tau' + \phi_\theta(\phi(t') - \phi(t' - \tau')) \tag{4.109}$$

where the parameter values are also normalized as,

$$g_e = \frac{\bar{T}}{2}G_N\bar{N}, \quad g_\phi = \frac{\alpha\bar{T}}{2\bar{\Phi}}G_N\bar{N}, \quad g_n = \bar{T}G_N\bar{E}^2, \quad \gamma_e = \frac{\bar{T}}{2\tau_p}, \quad \gamma_\phi = \frac{\alpha\bar{T}}{2\tau_p\bar{\Phi}},$$
$$\gamma_n = \frac{\bar{T}}{\tau_s}, \quad j = \frac{J}{J_{th}}, \quad \kappa_e = \kappa\bar{T}, \quad \kappa_\phi = \kappa\frac{\bar{T}}{\bar{\Phi}}, \quad n_{th} = \frac{N_{th}}{\bar{N}}, \quad n_0 = \frac{N_0}{\bar{N}},$$
$$\phi_\theta = \bar{\Phi}, \quad \tau' = \frac{\tau}{\bar{T}}, \quad \omega_\theta = \bar{\omega}\bar{T}$$

$$\tag{4.110}$$

4 Analysis of Chaotic Laser Dynamics: Example of Semiconductor Laser with Optical Feedback

Table 4.4 Parameter values used in Sections 4.5 and 4.6. The other parameter values are the same as shown in Table 4.3.

Symbol	Parameter	Value
r_3	Reflectivity of external mirror	0.04
$\kappa = \dfrac{(1-r_2^2)\, r_3}{r_2}\dfrac{1}{\tau_{in}}$	Feedback strength	$6.213 \times 10^9\,\text{s}^{-1}$ ($6.213\,\text{ns}^{-1}$)
$j = \dfrac{J}{J_{th}}$	Normalized injection current	1.44
L	External cavity length	0.3 m
α	Linewidth enhancement factor	5.0

Here, the following relationship is used,

$$N_{th} = N_0 + \frac{1}{G_N \tau_p}, \quad J_{th} = \frac{N_{th}}{\tau_s} \tag{4.111}$$

The constant values for the normalization can be selected arbitrarily. In this section, the normalization constants are set to the following values.

$$\bar{E} = \sqrt{\frac{N_{th}\tau_p}{\tau_s}(j_0-1)},\ j_0 = 1.1,\ \bar{\Phi} = 2\pi,\ \bar{N} = \frac{N_{th}}{100}, \tag{4.112}$$

$$\bar{T} = 1.0 \times 10^{-9}\ (1\,\text{ns})$$

These constant values are selected so that the standard deviations of temporal waveforms for the three normalized variables $e(t)$, $\phi(t)$, and $n(t)$ can be in the same order of magnitude in periodic and chaotic regimes. The time t' of Eqs. (4.106)–(4.109) is replaced by t for convenience in the following section.

The parameter values used for the original equations in this section are shown in Table 4.4. Other parameters are the same as shown in Table 4.3. The values for the normalization constants of Eq. (4.112) are summarized in Table 4.5, which are calculated from the values in Tables 4.3 and 4.4. The parameter values of Eq. (4.110) used for the dimensionless equations of Eqs. (4.106)–(4.109) are also summarized

Table 4.5 Values for normalization constants of the dimensionless Lang–Kobayashi equations in Sections 4.5 and 4.6.

Symbol	Parameter	Value
$\bar{E} = \sqrt{\dfrac{N_{th}\tau_p}{\tau_s}(j_0-1)},\ j_0 = 1.1$	Normalization constant for electric-field amplitude	1.3807×10^{10}
$\bar{\Phi} = 2\pi$	Normalization constant for electric-field phase	6.2832
$\bar{N} = \dfrac{N_{th}}{100}$	Normalization constant for carrier density	2.018×10^{22}
$\bar{T} = 1.0 \times 10^{-9}$	Normalization constant for time scale	1.0×10^{-9}

in Table 4.6. These parameter values are used for the numerical simulations shown in Sections 4.5 and 4.6.

For Eq. (4.109), the constant value of $\omega_\theta \tau'$ is several orders of magnitude larger than the value of $\phi_\theta(\phi(t')-\phi(t'-\tau'))$, and $\theta(t')$ can be treated between 0 and 2π. Therefore, the value of an initial phase shift $\theta_{ini} = \omega_\theta \tau' \pmod{2\pi}$ (where $0 \leq \theta_{ini} < 2\pi$) needs to be calculated in advance and $\omega_\theta \tau'$ is replaced to θ_{ini} in Eq. (4.109) for the numerical calculation, as described in Section 4.2.1.

4.5.2
Numerical Results of Chaotic Dynamics

The dimensionless Lang–Kobayashi equations of Eqs. (4.106)–(4.109) are solved numerically and compared with the results obtained from the original Lang–Kobayashi equations in Section 4.4. Numerical integration is carried out by using the fourth-order Runge–Kutta method (see Appendix 4.A.4).

Table 4.6 Parameter values used for the dimensionless Lang–Kobayashi equations in Sections 4.5 and 4.6.

Symbol	Parameter	Value
$g_e = \dfrac{\bar{T}}{2} G_N \bar{N}$	Gain coefficient for electric-field amplitude	8.4756
$g_\phi = \dfrac{\alpha \bar{T}}{2\bar{\Phi}} G_N \bar{N}$	Gain coefficient for electric-field phase	6.7447
$g_n = \bar{T} G_N \bar{E}^2$	Gain coefficient for carrier density	0.16013
$n_0 = \dfrac{N_0}{\bar{N}}$	Carrier density at transparency	69.376
$n_{th} = \dfrac{N_{th}}{\bar{N}}$	Carrier density at lasing threshold	100.00
$j = \dfrac{J}{J_{th}}$	Normalized injection current	1.44
$\gamma_e = \dfrac{\bar{T}}{2\tau_p}$	Inverse of photon lifetime for electric-field amplitude	259.47
$\gamma_\phi = \dfrac{\alpha \bar{T}}{2\tau_p \bar{\Phi}}$	Inverse of photon lifetime for electric-field phase	206.48
$\gamma_n = \dfrac{\bar{T}}{\tau_s}$	Inverse of carrier lifetime	0.49020
$\kappa_e = \kappa \bar{T}$	Feedback strength for electric-field amplitude	6.2128
$\kappa_\phi = \kappa \dfrac{\bar{T}}{\bar{\Phi}}$	Feedback strength for electric field phase	0.98880
$\phi_\theta = \bar{\Phi}$	Electric-field phase	6.2832
$\tau' = \dfrac{\tau}{\bar{T}}$	Feedback delay time in external cavity (round-trip)	2.0013
$\omega_\theta = \bar{\omega} \bar{T}$	Optical angular frequency	1.226×10^6

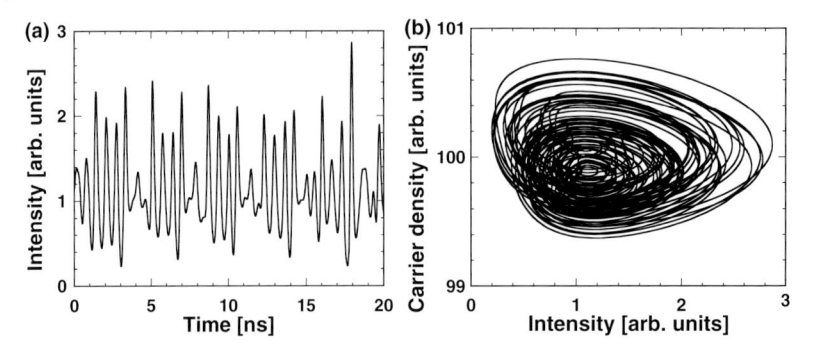

Figure 4.20 Numerical results of (a) chaotic temporal waveform and (b) chaotic attractor obtained from the dimensionless equations Eqs. (4.106)–(4.109). (b) The two-dimensional phase space consists of the normalized laser intensity $e^2(t)$ and the normalized carrier density $n(t)$ for the attractor plot.

Figure 4.20 shows an example of the chaotic temporal waveform and the corresponding chaotic attractor in the phase space. A similar chaotic temporal waveform is observed from the dimensionless equations, as shown in Figure 4.20a. For Figure 4.20b, the oscillation amplitudes (or standard deviations) of the laser intensity ($e^2(t)$) and the carrier density ($n(t)$) are of the same order of magnitude, which is different from Figure 4.15. The dimensionless equations are effective to match the order of magnitude for the amplitudes of all the variables. It is very important to have the same order of magnitude of the fluctuations to avoid numerical errors for the calculation of Lyapunov spectrum, which will be seen in the Section 4.5.4.

4.5.3
Linear Stability Analysis for Oscillatory Trajectory in Dimensionless Equations and Maximum Lyapunov Exponent

The linearized equations of the dimensionless equations are obtained from Eqs. (4.106)–(4.109). Small linear deviations from an oscillatory trajectory are introduced,

$$e(t) = e_s(t) + \delta_e(t) \tag{4.113}$$

$$\phi(t) = (\omega_s(t) - \omega) t + \delta_\phi(t) \tag{4.114}$$

$$n(t) = n_s(t) + \delta_n(t) \tag{4.115}$$

where $\delta_e(t)$, $\delta_\phi(t)$, and $\delta_n(t)$ are new linearized variables for the linearized equations ($e(t) \gg \delta_e(t)$, $\phi(t) \gg \delta_\phi(t)$, $n(t) \gg \delta_n(t)$). The delayed variables are also used as $e(t-\tau) = e_s(t-\tau) + \delta_e(t-\tau)$ and $\phi(t-\tau) = (\omega_s(t-\tau) - \omega)(t-\tau) + \delta_\phi(t-\tau)$.

The linearized equations for the dimensionless Lang–Kobayashi equations can be described as,

$$\frac{d\delta_e(t)}{dt} = [g_e(n(t)-n_0)-\gamma_e]\delta_e(t)-\phi_\theta\kappa_e e(t-\tau)\sin\theta(t)\delta_\phi(t)+g_e e(t)\delta_n(t)$$
$$+\kappa_e\cos\theta(t)\delta_e(t-\tau)+\phi_\theta\kappa_e e(t-\tau)\sin\theta(t)\delta_\phi(t-\tau) \quad (4.116)$$

$$\frac{d\delta_\phi(t)}{dt} = \kappa_\phi\frac{e(t-\tau)}{e^2(t)}\sin\theta(t)\delta_e(t)-\phi_\theta\kappa_\phi\frac{e(t-\tau)}{e(t)}\cos\theta(t)\delta_\phi(t)+g_\phi\delta_n(t)$$
$$-\kappa_\phi\frac{1}{e(t)}\sin\theta(t)\delta_e(t-\tau)+\phi_\theta\kappa_\phi\frac{e(t-\tau)}{e(t)}\cos\theta(t)\delta_\phi(t-\tau) \quad (4.117)$$

$$\frac{d\delta_n(t)}{dt} = -2g_n(n(t)-n_0)e(t)\delta_e(t)-\left(\gamma_n+g_n e^2(t)\right)\delta_n(t) \quad (4.118)$$

These equations are derived similarly to that shown in Section 4.4.2.

The maximum Lyapunov exponent can be calculated from the linearized dimensionless equations of Eqs. (4.116)–(4.118) and the original dimensionless equations of Eqs. (4.106)–(4.109). The algorithm for the calculation of the maximal Lyapunov exponent is the same procedure of Eqs. (4.98)–(4.104) in Section 4.4.3, except for the use of the linearized variables of $\delta_e(t)$, $\delta_\phi(t)$, and $\delta_n(t)$, instead of $\delta_E(t)$, $\delta_\Phi(t)$, and $\delta_N(t)$.

Figure 4.21 shows the bifurcation diagram and the corresponding maximum Lyapunov exponent obtained from the dimensionless equations as a function of the feedback strength κ. The maximum Lyapunov exponent is negative or close to zero for the regions of the steady-state, periodic oscillations, and quasiperiodic oscillations at $\kappa < 1.1$ ns^{-1}. By contrast, the maximum Lyapunov exponent becomes positive and increases monotonically as κ is increased in the chaotic region of $\kappa \geq 1.1$ ns^{-1}.

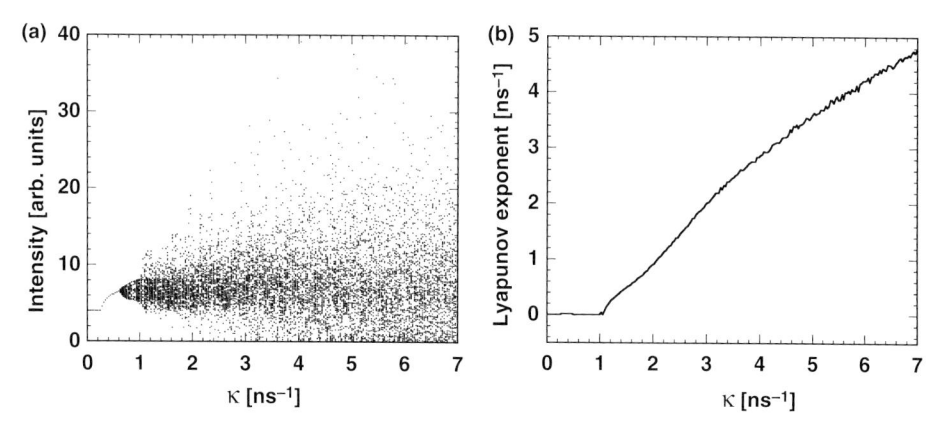

Figure 4.21 (a) Bifurcation diagram and (b) maximum Lyapunov exponent obtained from the dimensionless equations as a function of the feedback strength κ. The parameter values shown in Tables 4.4–4.6 are used.

The region for positive maximum Lyapunov exponent in Figure 4.21b corresponds to chaotic temporal dynamics in Figure 4.21a.

4.5.4
Lyapunov Spectrum

The maximum Lyapunov exponent is very convenient to determine the existence of deterministic chaos. More useful information related to the complexity of the dynamical laser system can be obtained from the calculation of a set of Lyapunov exponents, which is known as the "Lyapunov spectrum". In fact, n Lyapunov exponents exist in the n-dimensional systems (i.e., n variables). It is important to calculate the Lyapunov spectrum to quantify the complexity of chaotic dynamical systems. Some quantities for complexity in dynamical systems, such as Kolmogorov–Sinai (KS) entropy and Kaplan–Yorke (KY) dimension, can be estimated from the Lyapunov spectrum.

For the calculation of the Lyapunov spectrum in n-dimensional systems, n sets of the linearized equations with different initial conditions are required. For dimensionless Lang–Kobayashi equations, the three dimensionless original equations of Eqs. (4.106)–(4.108) and n sets of the three linearized equations of Eqs. (4.116)–(4.118) with different initial conditions are calculated simultaneously. The dimensionality of time-delayed systems is dependent of the delay time (but n is unknown), and large n is required for the estimation of the Lyapunov spectrum for time-delayed systems. It is found that the dimensionality of time-delayed system increases as the delay time is increased (Namajūnas et al., 1995a; Namajūnas et al., 1995b; Vicente et al., 2005).

Suppose that all the equations are integrated from t to $t+\tau$. The norm vector of the ith set of the three linearized equations ($1 \leq i \leq n$), consisting of $3M$ components ($M = \tau/h$ and h is the integration step for numerical calculation), is considered,

$$\mathbf{v_i} = \begin{pmatrix} \delta_e(t) \\ \delta_e(t+h) \\ \vdots \\ \delta_e(t+(M-1)h) \\ \delta_\phi(t) \\ \delta_\phi(t+h) \\ \vdots \\ \delta_\phi(t+(M-1)h) \\ \delta_n(t) \\ \delta_n(t+h) \\ \vdots \\ \delta_n(t+(M-1)h) \end{pmatrix} \quad (4.119)$$

After all the norm vectors $(\mathbf{v}_1, \mathbf{v}_2, \ldots, \mathbf{v}_n)$ are obtained from t to $t+\tau$ for n sets of the three linearized equations, these vectors are orthogonalized to each other by the Gram–Schmidt orthogonalization (see Appendix 4.A.5). The norm is redefined by using the orthogonalized norm vectors $\mathbf{u_i}$. Then, n sets of norm are obtained from $\mathbf{u_i}$ by using the same procedure of Eqs. (4.98)–(4.100), and normalization is achieved for $\mathbf{u_i}$ by using Eqs. (4.101)–(4.103). The procedure of the orthogonalization as well as the normalization is repeated for τ step. n Lyapunov exponents are obtained from the n

sets of norm by using Eq. (4.104), and ordered from the largest to the lowest values, that is, $\lambda_1 \geq \lambda_2 \geq \cdots \geq \lambda_n$. The set of Lyapunov exponents $(\lambda_1, \lambda_2, \ldots, \lambda_n)$ is called the Lyapunov spectrum.

The value of n needs to be equal to or larger than the dimensionality of the dynamical system, however, the dimensionality of time-delayed systems converges to infinity as the delay time is increased. In the calculation of Sections 4.5 and 4.6 n is set to 60 for convenience in numerical calculation. This selection is valid when the obtained dimensionality is less than 60.

The numerical result of Lyapunov spectrum calculated from the dimensionless Lang–Kobayashi equations (Eqs. (4.106)–(4.109) and n sets of Eqs. (4.116)–(4.118)) is shown in Figure 4.22. The Lyapunov exponents are obtained and ordered from the maximum to the minimum value. The Lyapunov exponents λ_i are plotted as a function of the index i up to $i = 20$. Eight positive Lyapunov exponents are obtained, indicating hyperchaos (defined as the system with more than one positive Lyapunov exponent) due to the time-delayed feedback effect. By contrast, negative Lyapunov exponents are obtained for $i \geq 9$.

Figure 4.23a shows the Lyapunov spectrum as a function of the feedback strength κ, and Figure 4.23b shows its enlargement around zero Lyapunov exponent. The curve with the largest value corresponds to the maximum Lyapunov exponent λ_1, (see Fig. 4.21b) and the curve with the second-largest value is λ_2, and so on. The maximum Lyapunov exponent λ_1 changes from negative to positive at $\kappa = 1.1\,\text{ns}^{-1}$ and chaotic dynamics appears. The second-largest Lyapunov exponent λ_2 is also changed from negative to positive with increase of the feedback strength, however, the value is not as large as λ_1. The third-largest Lyapunov exponent λ_3 also behaves similarly, however, the transition point of λ_3 from a negative to positive value (i.e., zero-crossing point) is larger than that of λ_1 and λ_2. The zero-crossing point of λ_i increases as i is increased (Figure

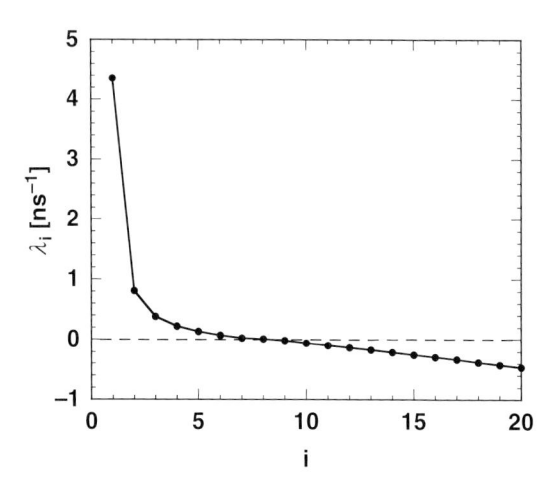

Figure 4.22 Lyapunov spectrum at the feedback strength $\kappa = 6.213\,\text{ns}^{-1}$. Lyapunov exponents λ_i are plotted as a function of the index i up to $i = 20$. Lyapunov exponents are ordered from the maximum to the minimum value.

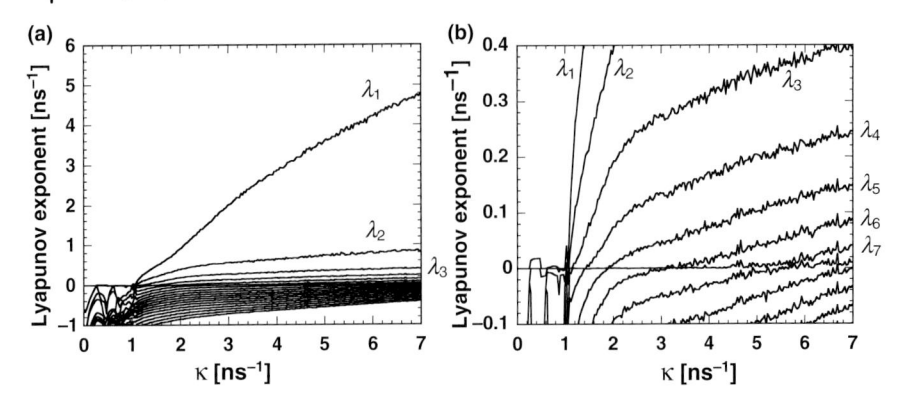

Figure 4.23 (a) Lyapunov spectrum as a function of the feedback strength κ, and (b) its enlargement around zero Lyapunov exponent. Figure 4.23a corresponds to the bifurcation diagram of Figure 4.21a and the maximum Lyapunov exponent of Figure 4.21b.

4.23b). Note that the number of positive Lyapunov exponents increases as the feedback strength is increased, as shown in Figure 4.23b, which is a typical feature for time-delayed dynamical systems (Namajūnas et al., 1995a; Namajūnas et al., 1995b; Vicente et al., 2005).

4.5.5
Kolmogorov–Sinai Entropy and Kaplan–Yorke Dimension

The Lyapunov spectrum can be used to obtain quantitative evaluation of complexity in nonlinear dynamical systems, such as Kolmogorov–Sinai (KS) entropy and Kaplan–Yorke (KY) dimension. KS entropy measures the average loss of information rate, or equivalently is inversely proportional to the time interval over which the future evolution can be predicted (Vicente et al., 2005). Its range of values goes from zero for steady or periodic dynamics, it is positive for chaotic systems, and infinite for perfectly stochastic process. Larger entropy implies more unpredictability of the dynamical system.

The computation of the KS entropy is executed from the sum of all the "positive" Lyapunov exponents,

$$E_{KS} = \sum_{i=1}^{k} \lambda_i, \qquad (4.120)$$

$$\lambda_i > 0 \ (i = 1, 2, \ldots, k), \ \lambda_{k+1} \leq 0,$$

where the Lyapunov exponents are ordered as $\lambda_i \geq \lambda_{i+1}$. In fact, the sum of all the positive Lyapunov exponents indicates an upper bound to the KS entropy, however, Eq. (4.120) holds in general situations and it is a simple way to obtain a good estimation of the KS entropy (Vicente et al., 2005).

Another quantitative measure is the KY dimension, which is a measure of complexity for chaotic attractors (Ott 1993; Vicente et al., 2005). The KY dimension specifies how the amount of information is needed to locate the system in the phase space. The KY dimension is considered to be equal to the information dimension.

The KY dimension is calculated from the Lyapunov spectrum,

$$D_{KY} = j + \frac{\sum_{i=1}^{j} \lambda_i}{|\lambda_{j+1}|} \qquad (4.121)$$

where the integer j, which represents the number of degrees of freedom, satisfies the following condition,

$$\sum_{i=1}^{j} \lambda_i > 0 > \sum_{i=1}^{j+1} \lambda_i \qquad (4.122)$$

where the Lyapunov exponents are ordered as $\lambda_i \geq \lambda_{i+1}$.

The KS entropy and KY dimension are obtained numerically from the Lyapunov spectrum. The numerical results of KS entropy and KY dimension for the dimensionless Lang–Kobayashi equations are shown in Figure 4.24 as a function of the feedback strength κ or the external cavity length L. Figure 4.24a corresponds to the bifurcation diagram of Figure 4.21a. For Figure 4.24a, both the KS entropy and KY dimension increase monotonically in the chaotic region as κ is increased. This result implies that both complexity and dimensionality of the chaotic attractor increase monotonically with increase of the feedback strength κ. For Figure 4.24b, the KY dimension increases almost linearly as the external cavity length is increased, whereas the KS entropy maintains a constant value of ~ 6 ns^{-1} in the chaotic region of $L \geq 0.1$ m. This result indicates that the complexity is almost the same, while the dimensionality is linearly increased with the increase of delay time. This is a typical feature for time-delayed systems (Namajūnas et al., 1995a; Namajūnas et al., 1995b;

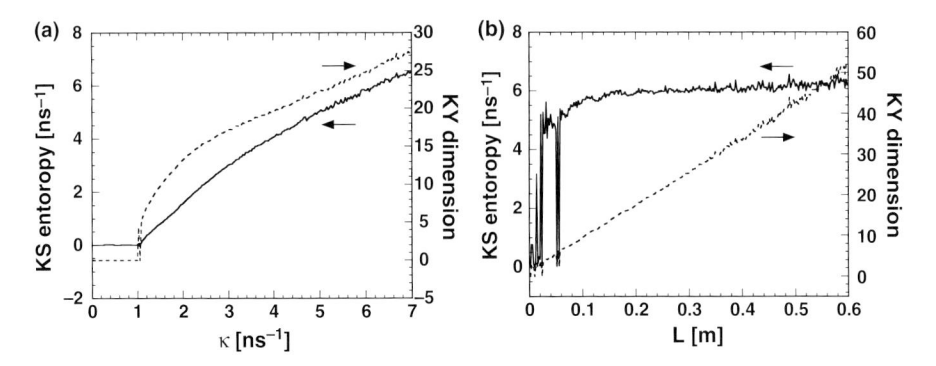

Figure 4.24 Kolmogorov–Sinai (KS) entropy and Kaplan–Yorke (KY) dimension as a function of (a) the feedback strength κ and (b) the external cavity length L. Figure 4.24a corresponds to the bifurcation diagram of Figure 4.21a.

Vicente et al., 2005). The KS entropy and KY dimension are good indicators to quantify the complexity and dimensionality of nonlinear dynamical laser systems.

4.6
Lang–Kobayashi Equations with Gain Saturation

4.6.1
Lang–Kobayashi Equations with Gain Saturation

The original Lang–Kobayashi equations (Eqs. (4.3)–(4.6)) are widely used and many studies have reported that numerical results based on Lang–Kobayashi equations are consistent with experimental results. There is, however, inconsistency between numerical and experimental results, particularly, for the histogram of the laser intensity distribution (Sukow et al., 1999). For example, temporal waveforms obtained from the original Lang–Kobayashi equations behave like strong pulsations at the condition of strong optical feedback, even though these strong pulsations are rarely observed in experiments. Therefore, a modification of the original Lang–Kobayashi equations is required to achieve good agreement between numerical and experimental results. One of the most important phenomenological effects is the gain saturation in the laser medium. The inclusion of the gain-saturation effect in the model produces good agreement of the histogram of laser intensity distribution between the numerical and experimental data, and is required as a proper model for some applications such as random number generators (see Chapter 10).

Strong pulsations of laser intensity are suppressed when the gain-saturation effect is introduced in the model. The gain term is replaced with the phenomenological gain-saturation effect (Ohtsubo, 2008),

$$G_N \rightarrow \frac{G_N}{1+\varepsilon E^2(t)} \tag{4.123}$$

where ε is the gain-saturation coefficient. The equivalent gain decreases as the laser intensity $I(t) = E^2(t)$ is increased due to the gain-saturation effect.

The Lang–Kobayashi equations with gain saturation are described by using Eq. (4.123) as follows,

$$\frac{dE(t)}{dt} = \frac{1}{2}\left[\frac{G_N(N(t)-N_0)}{1+\varepsilon E^2(t)} - \frac{1}{\tau_p}\right]E(t) + \kappa\, E(t-\tau)\cos\Theta(t) \tag{4.124}$$

$$\frac{d\Phi(t)}{dt} = \frac{\alpha}{2}\left[\frac{G_N(N(t)-N_0)}{1+\varepsilon E^2(t)} - \frac{1}{\tau_p}\right] - \kappa\frac{E(t-\tau)}{E(t)}\sin\Theta(t) \tag{4.125}$$

$$\frac{dN(t)}{dt} = J - \frac{N(t)}{\tau_s} - \frac{G_N(N(t)-N_0)}{1+\varepsilon E^2(t)}E^2(t) \tag{4.126}$$

$$\Theta(t) = \omega\tau + \Phi(t) - \Phi(t-\tau) \tag{4.127}$$

Dimensionless equations can also be derived from the normalization of the Lang–Kobayashi equations with gain saturation. The variables are normalized by constant values of Eq. (4.105). The dimensionless equations with gain saturation are described,

$$\frac{de(t')}{dt'} = \left[\frac{g_e(n(t')-n_0)}{1+\varepsilon' e^2(t')} - \gamma_e\right] e(t') + \kappa_e e(t'-\tau')\cos\theta(t') \tag{4.128}$$

$$\frac{d\phi(t')}{dt'} = \frac{g_\phi(n(t')-n_0)}{1+\varepsilon' e^2(t')} - \gamma_\phi - \kappa_\phi \frac{e(t'-\tau')}{e(t')}\sin\theta(t') \tag{4.129}$$

$$\frac{dn(t')}{dt'} = \gamma_n(j\,n_{\text{th}} - n(t')) - \frac{g_n(n(t')-n_0)e^2(t')}{1+\varepsilon' e^2(t')} \tag{4.130}$$

$$\theta(t') = \omega_0\tau' + \phi_0(\phi(t')-\phi(t'-\tau')) \tag{4.131}$$

where the parameter values are normalized as Eq. (4.110) and the following relationship,

$$\varepsilon' = \varepsilon \bar{E}^2 \tag{4.132}$$

The normalization constants are set to the values of Eq. (4.112). The parameter values used for numerical calculations in this section are already shown in Tables 4.4-4.6. The saturation coefficient is set to be $\varepsilon = 2.5 \times 10^{-23}$ ($\varepsilon' = 4.7658 \times 10^{-3}$) in Section 4.6.

4.6.2
Numerical Results and Histogram

The dimensionless Lang–Kobayashi equations with gain saturation are calculated numerically. The temporal waveforms without gain saturation ($\varepsilon = 0$) and with gain saturation ($\varepsilon = 2.5 \times 10^{-23}$) are shown in Figure 4.25. The temporal waveform of Figure 4.25a shows strong spike trains, whereas the temporal waveform of Figure 4.25b looks like continuous oscillations with small amplitude in the presence of gain saturation. The gain-saturation effect suppresses strong pulsations of laser intensity.

Figure 4.26 shows the histogram (probability density function) of the temporal waveform of the laser intensity without and with gain saturation. These two histograms correspond to the temporal waveforms shown in Figure 4.25. The histogram of Figure 4.26a is strongly asymmetric and the peak value is located near zero intensity, which implies strong pulsations. On the other hand, the histogram of Figure 4.26b looks more symmetric and the peak value is located at around 4.5. For

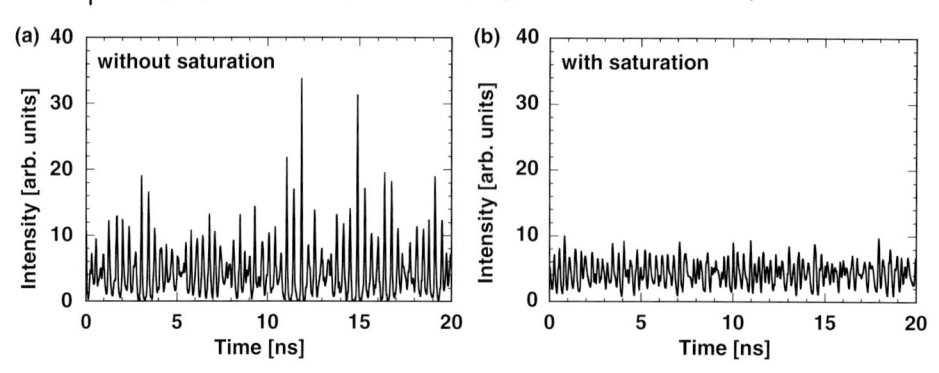

Figure 4.25 Temporal waveforms of laser intensity (a) without gain saturation ($\varepsilon = 0$) and (b) with gain saturation ($\varepsilon = 2.5 \times 10^{-23}$).

comparison with experimental data, the histogram with gain saturation (Figure 4.26b) resembles that obtained from experiment.

The bifurcation diagrams are shown in Figure 4.27 without and with gain saturation as the feedback strength κ is increased. The quasiperiodicity route to chaos is observed for both cases, even though there is a shift of chaotic regions in terms of κ. In addition, the amplitude of the temporal waveforms in the chaotic region is very different, that is, smaller amplitudes are obtained in the presence of gain saturation. Strong spike trains are suppressed due to the gain-saturation effect in Figure 4.27b.

4.6.3
Linear Stability Analysis for Oscillatory Trajectory and Measurement of Complexity

The linearized equations of the dimensionless equations with gain saturation are calculated from Eqs. (4.128)–(4.130). Small linear deviations from the dynamical

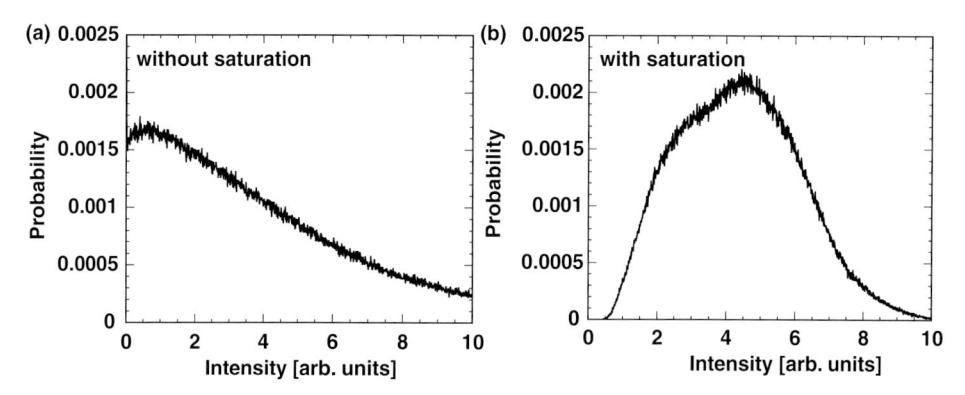

Figure 4.26 Histograms (probability density functions) of the temporal waveform of the laser intensity (a) without and (b) with gain saturation. These two histograms correspond to the temporal waveforms shown in Figure 4.25.

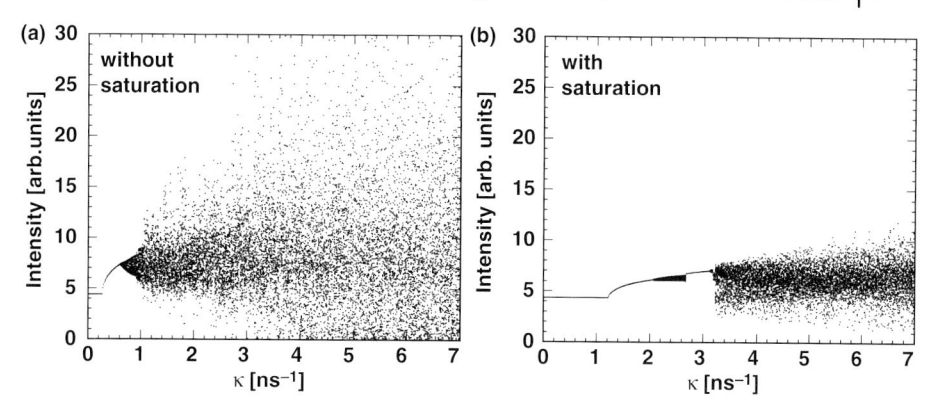

Figure 4.27 Bifurcation diagrams as a function of the feedback strength κ (a) without ($\varepsilon = 0$) and (b) with gain saturation and ($\varepsilon = 2.5 \times 10^{-23}$).

trajectory are introduced as Eqs. (4.113)–(4.115). The linearized equations for Eqs. (4.128)–(4.130) are described as,

$$\frac{d\delta_e(t)}{dt} = \left[\frac{g_e(n(t)-n_0)(1-\varepsilon'e^2(t))}{(1+\varepsilon'e^2(t))^2} - \gamma_e\right]\delta_e(t)$$
$$-\phi_0\kappa_e e(t-\tau)\sin\theta(t)\delta_\phi(t) + \frac{g_e e(t)}{1+\varepsilon'e^2(t)}\delta_n(t) \quad (4.133)$$
$$+\kappa_e\cos\theta(t)\delta_e(t-\tau) + \phi_0\kappa_e e(t-\tau)\sin\theta(t)\delta_\phi(t-\tau)$$

$$\frac{d\delta_\phi(t)}{dt} = \left[-\frac{2\varepsilon' g_\phi(n(t)-n_0)e(t)}{(1+\varepsilon'e^2(t))^2} + \kappa_\phi\frac{e(t-\tau)}{e^2(t)}\sin\theta(t)\right]\delta_e(t)$$
$$-\phi_0\kappa_\phi\frac{e(t-\tau)}{e(t)}\cos\theta(t)\delta_\phi(t) + \frac{g_\phi}{1+\varepsilon'e^2(t)}\delta_n(t) \quad (4.134)$$
$$-\kappa_\phi\frac{1}{e(t)}\sin\theta(t)\delta_e(t-\tau) + \phi_0\kappa_\phi\frac{e(t-\tau)}{e(t)}\cos\theta(t)\delta_\phi(t-\tau)$$

$$\frac{d\delta_n(t)}{dt} = -\frac{2g_n(n(t)-n_0)e(t)}{(1+\varepsilon'e^2(t))^2}\delta_e(t) - \left(\gamma_n + \frac{g_n e^2(t)}{1+\varepsilon'e^2(t)}\right)\delta_n(t) \quad (4.135)$$

The maximum Lyapunov exponent is calculated from the norm of the linearized variables and their time-delayed variables with Eqs. (4.133)–(4.135). The algorithm for the calculation of the maximal Lyapunov exponent is the same procedure of Eqs. (4.98)–(4.104) in Section 4.4.3. Figure 4.28 shows the maximum Lyapunov exponent as a function of the feedback strength κ without (the dotted curve) and with gain saturation (the solid curve). Figure 4.28 corresponds to the bifurcation diagram of Figure 4.27a and b. For both cases, the maximum Lyapunov exponent increases as the feedback strength is increased in the chaotic region. However, the value of the

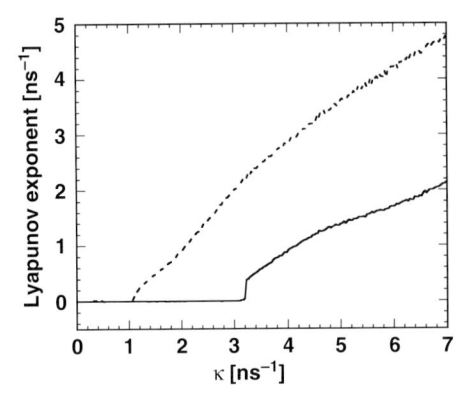

Figure 4.28 Maximum Lyapunov exponent as a function of the feedback strength κ without (dotted curve) and with gain saturation (solid curve). This figure corresponds to the bifurcation diagrams of Figure 4.27.

maximum Lyapunov exponent for the case with gain saturation becomes smaller than the case without gain saturation at the same feedback strength. Therefore, it is expected that the complexity decreases in the presence of gain saturation.

The KS entropy and KY dimension are also estimated numerically from the Lyapunov spectrum obtained from Eqs. (4.133)–(4.135) by using the procedure shown in Sections 4.5.4 and 4.5.5. Figure 4.29 shows the KS entropy and KY dimension as a function of the feedback strength κ without (the dotted curve) and with gain saturation (the solid curve). Both KS entropy and KY dimension increase monotonically as κ is increased for both cases of gain saturation. The values of the KS entropy in the presence of gain saturation are smaller than the case of without gain saturation at the same feedback strength (Figure 4.29a). However, the value of the KY dimension is almost the same for both with and without gain saturation in the chaotic

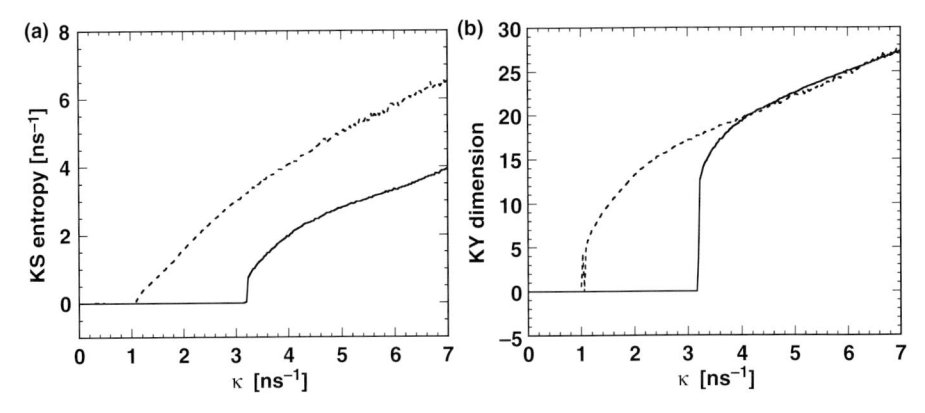

Figure 4.29 (a) KS entropy and (b) KY dimension as a function of the feedback strength κ without (dotted curve) and with gain saturation (solid curve). These figures correspond to the bifurcation diagrams of Figure 4.27.

region for large κ (Figure 4.29b). Therefore, the gain-saturation effect changes the complexity, but not dimensionality, of the dynamical system.

The techniques of the nonlinear dynamical analysis, such as maximum Lyapunov exponent, Lyapunov spectrum, KS entropy, and KY dimension, are very useful for evaluating quantitative complexity and dimensionality, and can be applied for other dynamical laser systems.

Appendix

4.A.1
Derivation of the Rate Equations of Semiconductor Lasers

The rate equations of semiconductor lasers are derived from the laser-oscillation conditions in Appendix 4.A.1 (Petermann, 1988; Rogister, 2001; Ohtsubo, 2008). This derivation was maily organized by Kazutaka Kanno. Figure 4.A.1 shows the model of a semiconductor laser with a Fabry–Perot cavity consisting of a dielectric medium confined between two mirrors. The cavity length is l, and reflectivities of the front and back facets are r_1 and r_2, respectively. The complex electric fields propagating in the forward and backward directions are denoted as $\hat{\mathbf{E}}_f$ and $\hat{\mathbf{E}}_b$, respectively. In the time constant, $\hat{\mathbf{E}}_f$ and $\hat{\mathbf{E}}_b$ at z are written by

$$\hat{\mathbf{E}}_f(z) = \mathbf{E}_{0f} \exp\left\{-ikz + \frac{1}{2}(g-a)z\right\} \tag{4.A.1}$$

$$\hat{\mathbf{E}}_b(z) = \mathbf{E}_{0b} \exp\left\{-ik(l-z) + \frac{1}{2}(g-a)(l-z)\right\} \tag{4.A.2}$$

where g is the gain in the laser medium, a is the total loss due to absorption and scattering in the medium, k is the wave number, and i is the imaginary unit. A variable with bold font indicates a complex variable, whereas a variable with italic font represents a real variable $\hat{\mathbf{E}}$ indicates that the complex variable $\mathbf{E}$ includes the fast optical-carrier frequency component. From the boundary conditions at the facets, $\hat{\mathbf{E}}_f(0) = r_1 \hat{\mathbf{E}}_b(0)$ and $\hat{\mathbf{E}}_b(l) = r_2 \hat{\mathbf{E}}_f(l)$, the steady-state condition for the laser oscillation is given by

$$r_1 r_2 \exp\left\{-2ikl + (g-a)l\right\} = 1 \tag{4.A.3}$$

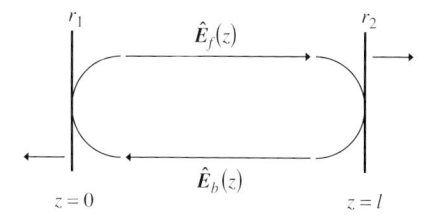

Figure 4.A.1. Model of a semiconductor laser with a Fabry–Perot cavity. l, internal cavity length; r, reflectivity of the mirror.

From the left-hand side above the equation, the round-trip gain is considered as

$$G = r_1 r_2 \exp\{-2ikl + (g-a)l\} \tag{4.A.4}$$

From the real part of Eq. (4.A.3), the amplitude condition of the laser oscillation for the threshold gain g_{th} is given by,

$$g_{th} = a + \frac{1}{l}\ln\left(\frac{1}{r_1 r_2}\right) \tag{4.A.5}$$

The phase condition is also calculated from the imaginary part of Eq. (4.A.3) and reads

$$kl = m\pi \tag{4.A.6}$$

where m is an integer. The refractive index of the laser medium η can be denoted as a function of the carrier density N and the optical frequency ν (or the angular optical frequency $\omega = 2\pi\nu$),

$$\eta = \eta_0 + \frac{\partial\eta}{\partial N}(N - N_{th}) + \frac{\partial\eta}{\partial\omega}(\omega - \omega_{th}) \tag{4.A.7}$$

where ω_{th} is the angular optical frequency at the lasing threshold, N_{th} is the carrier density at the lasing threshold, η_0 is the refractive index at the lasing threshold. The wave number k depends on η and can be expanded by the threshold values of those parameters as,

$$k = \eta\frac{\omega}{c} = \frac{\omega_{th}}{c}\left\{\eta_0 + \frac{\partial\eta}{\partial N}(N - N_{th}) + \frac{\eta_e}{\omega_{th}}(\omega - \omega_{th})\right\} \tag{4.A.8}$$

Substituting Eq. (4.A.8) into Eq. (4.A.4), the gain G is written by the product of the frequency nondependent and dependent terms, G_1 and G_2, as

$$G = G_1 G_2 \tag{4.A.9}$$

$$G_1 = r_1 r_2 \exp\{(g-a)l - i\phi_0\} \tag{4.A.10}$$

$$G_2 = \exp\left[-i\frac{2\omega_{th}l}{c}\left\{\eta_0 + \frac{\eta_e}{\omega_{th}}(\omega - \omega_{th})\right\}\right] \tag{4.A.11}$$

The phase ϕ_0 of the above equation is given by

$$\phi_0 = \frac{2\omega_{th}l}{c}\frac{\partial\eta}{\partial N}(N - N_{th}) \tag{4.A.12}$$

In Eq. (4.A.11), the condition that the phase $2\omega_{th}\eta_0 l/c$ must be equal to integer multiples of 2π is used for the laser oscillation. The round-trip time τ_{in} of the light in cavity is expressed with l, the speed of light c and the effective refractive index η_e as

$$\tau_{in} = \frac{2\eta_e l}{c} \tag{4.A.13}$$

4.A.1 Derivation of the Rate Equations of Semiconductor Lasers

Replacing the quantity $i\omega$ as the equivalent to the operator d/dt and using Eqs. (4.A.13) and (4.A.11) reads

$$G_2 = \exp\{-i\tau_{in}(\omega-\omega_{th})\} = \exp(i\omega_{th}\tau_{in})\exp\left(-\tau_{in}\frac{d}{dt}\right) \qquad (4.A.14)$$

Since frequency and time are a Fourier-transform pair, Eq. (4.A.14) is derived from the equivalence of the equation in the frequency domain with that in the time domain.

To attain the laser oscillations, the complex field after the round-trip within the laser cavity must coincide exactly with the previous field. Assuming the gain in Eqs. (4.A.9)–(4.A.11) as a kind of an operator, the following relation can be obtained,

$$\hat{\mathbf{E}}_f(t) = \mathbf{G}\hat{\mathbf{E}}_f(t) \qquad (4.A.15)$$

Substituting Eqs. (4.A.9) and (4.A.14) into Eq. (4.A.15), the field after the round-trip in the cavity is written as

$$\hat{\mathbf{E}}_f(t) = \mathbf{G}_1\mathbf{G}_2\hat{\mathbf{E}}_f(t) = \mathbf{G}_1\exp(i\omega_{th}\tau_{in})\exp\left(-\tau_{in}\frac{d}{dt}\right)\hat{\mathbf{E}}_f(t) \qquad (4.A.16)$$

With this expression and the fact that the operator $\exp(-\tau_{in}d/dt)$ is equivalent to the time delay effect of τ_{in} and $\hat{\mathbf{E}}_f(t)\exp(-\tau_{in}d/dt)$ is reproduced to $\hat{\mathbf{E}}_f(t-\tau_{in})$, Eq. (4.A.16) yields,

$$\hat{\mathbf{E}}_f(t) = \mathbf{G}_1\exp(i\omega_{th}\tau_{in})\hat{\mathbf{E}}_f(t-\tau_{in}) \qquad (4.A.17)$$

The laser field $\hat{\mathbf{E}}_f(t)$ can be divided into two terms, the term changing with angular frequency ω_{th} and the term $\mathbf{E}_f(t)$ that varies slowly compared with the angular frequency. The field is given by

$$\hat{\mathbf{E}}_f(t) = \mathbf{E}_f(t)\exp(i\omega_{th}t) \qquad (4.A.18)$$

Substituting Eq. (4.A.18) into Eq. (4.A.17),

$$\mathbf{E}_f(t)\exp(i\omega_{th}t) = \mathbf{G}_1\exp(i\omega_{th}\tau_{in})\mathbf{E}_f(t-\tau_{in})\exp\{i\omega_{th}(t-\tau_{in})\} \qquad (4.A.19)$$

Therefore, the field $\mathbf{E}_f(t)$ can be written as

$$\mathbf{E}_f(t) = \mathbf{G}_1\mathbf{E}_f(t-\tau_{in}) \qquad (4.A.20)$$

Equation (4.A.20) means that the field $\mathbf{E}_f(t)$ after the round-trip time of τ_{in} with slowly varying envelope approximation (SVEA) returns as the same form of the field with gain $\mathbf{G}_1$. Here, the field propagation for the positive direction is considered. The same result can be obtained for the field propagation for the negative direction and the same relation as that of Eq. (4.A.19) is derived for the field. Then, the total field $\mathbf{E}(t)$ can be written as the same form as $\mathbf{E}_f(t)$ in Eq. (4.A.20).

When the round-trip time τ_{in} is small enough ($\tau_{in} \ll t$), the total field can be expanded around the delay time τ_{in} as

$$\mathbf{E}(t-\tau_{in}) = \mathbf{E}(t) - \tau_{in}\frac{d\mathbf{E}(t)}{dt} \tag{4.A.21}$$

The differential form for the field can be obtained as follows:

$$\frac{d\mathbf{E}(t)}{dt} = \frac{1}{\tau_{in}}\left(1 - \frac{1}{G_1}\right)\mathbf{E}(t) \tag{4.A.22}$$

Since the gain G_1 is very close to unity for laser oscillation, the gain is approximated from Eq. (4.A.10) as

$$\begin{aligned}\frac{1}{G_1} &= \frac{1}{r_1 r_2}\exp\{-(g-a)l + i\phi_0\} \\ &= \exp\left\{\ln\left(\frac{1}{r_1 r_2}\right) - (g-a)l + i\phi_0\right\} \\ &\approx 1 - (g-a)l + \ln\left(\frac{1}{r_1 r_2}\right) + 2i\frac{\omega_{th} l}{c}\frac{\partial \eta}{\partial N}(N - N_{th})\end{aligned} \tag{4.A.23}$$

where $e^x \approx 1 + x$ is ($x \ll 1$). Substituting Eq. (4.A.23) into Eq. (4.A.22) and using the relations of Eq. (4.A.5) and (4.A.13), one can obtain the following equation.

$$\begin{aligned}\frac{d\mathbf{E}(t)}{dt} &= \frac{1}{\tau_{in}}\left[(g-a)l - \ln\left(\frac{1}{r_1 r_2}\right) - 2i\frac{\omega_{th} l}{c}\frac{\partial \eta}{\partial N}(N - N_{th})\right]\mathbf{E}(t) \\ &= \left[(g-g_{th})\frac{l}{\tau_{in}} - 2i\frac{\omega_{th} l}{c\tau_{in}}\frac{\partial \eta}{\partial N}(N - N_{th})\right]\mathbf{E}(t) \\ &= \left[(g-g_{th})\frac{c}{2\eta_e} - i\frac{\omega_{th}}{\eta_e}\frac{\partial \eta}{\partial N}(N - N_{th})\right]\mathbf{E}(t)\end{aligned} \tag{4.A.24}$$

At this point, the relation between the gain and the carrier density is introduced. The optical gain is observed to increase almost linearly with the total number of electron–hole pairs N for all values of injection current. The gain g can then be approximated as

$$g = \frac{\partial g}{\partial N}(N - N_0) \tag{4.A.25}$$

where N_0 is the carrier density at transparency and $\partial g/\partial N$ is the gain coefficient. g is also often linearized around its threshold value. In this case,

$$g = g_{th} + \frac{\partial g}{\partial N}(N - N_{th}) \tag{4.A.26}$$

4.A.1 Derivation of the Rate Equations of Semiconductor Lasers

Substituting Eq. (4.A.26) into Eq. (4.A.24),

$$\frac{d\mathbf{E}(t)}{dt} = \left[\frac{c}{2\eta_e}\frac{\partial g}{\partial N}(N-N_{\text{th}}) - i\frac{\omega_{\text{th}}}{\eta_e}\frac{\partial \eta}{\partial N}(N-N_{\text{th}})\right]\mathbf{E}(t) \quad (4.A.27)$$

A new parameter, α parameter, is introduced to simplify Eq. (4.A.27). The α parameter is also referred to as the linewidth enhancement factor. The α parameter is defined as,

$$\alpha = \frac{\text{Re}[\chi]}{\text{Im}[\chi]} = -2\frac{\omega\, \partial\eta/\partial N}{c\, \partial g/\partial N} \quad (4.A.28)$$

The real and imaginary parts of the complex electric susceptibility χ are linked by the Kramers–Kronig relations (Petermann, 1988; Ohtsubo, 2008). In semiconductor lasers this parameter is quite large. Its typical value is between 3 and 7 that leads to substantial linewidth broadening with the ratio of $1+\alpha^2$. Substituting Eq. (4.A.28) into Eq. (4.A.27), the following relationship is obtained

$$\begin{aligned}\frac{d\mathbf{E}(t)}{dt} &= \frac{c}{2\eta_e}\frac{\partial g}{\partial N}\left[1-2i\frac{\omega_{\text{th}}}{c}\frac{\partial \eta/\partial N}{\partial g/\partial N}\right](N-N_{\text{th}})\mathbf{E}(t) \\ &= \frac{c}{2\eta_e}\frac{\partial g}{\partial N}(1+i\alpha)(N-N_{\text{th}})\mathbf{E}(t) \\ &= \frac{1+i\alpha}{2}G_N(N-N_{\text{th}})\mathbf{E}(t) \end{aligned} \quad (4.A.29)$$

where we define the linear gain $G_N = v_g \partial g/\partial N$, and $v_g = c/\eta_e$ is the group velocity of light in vacuum. Equation (4.A.29) is the rate equation for the slowly varying envelope of the electric field. Moreover, Eq. (4.A.29) can be rewritten with photon lifetime τ_p that is defined as the inverse of the total loss rate of photon in the cavity.

$$\frac{1}{\tau_p} = v_g g_{\text{th}} \quad (4.A.30)$$

Using Eqs. (4.A.23) and (4.A.24) and above the equation, we find that the photon lifetime and the carrier densities at the threshold have the relation:

$$\begin{aligned}\frac{1}{\tau_p} &= v_g g - v_g \frac{\partial g}{\partial N}(N-N_{\text{th}}) \\ &= v_g \frac{\partial g}{\partial N}(N-N_0) - v_g \frac{\partial g}{\partial N}(N-N_{\text{th}}) \\ &= G_N(N_{\text{th}}-N_0)\end{aligned} \quad (4.A.31)$$

where N_0 is the carrier density at transparency. Transforming Eq. (4.A.29) with Eq. (4.A.31), a rate equation is obtained for the slowly varying envelope of the electric field.

$$\frac{d\mathbf{E}(t)}{dt} = \frac{1+i\alpha}{2}\left[G_N(N(t)-N_0) - \frac{1}{\tau_p}\right]\mathbf{E}(t) \tag{4.A.32}$$

It should be noted that in Appendix 4.A.1 the time-dependent term of the propagation electromagnetic field is defined as $\exp(-ikz)$ in Eq. (4.A.1). The term of $\exp(ikz)$ can be used for the propagating term, instead of $\exp(-ikz)$. In that case, the right-hand-side term of Eq. (4.A.32) related to the α is written by $(1-i\alpha)$, instead of $(1+i\alpha)$.

Equation (4.A.32) alone is not sufficient to describe the behavior of the laser. Indeed, the gain, and therefore the electric field, depends on the carrier density. Assuming charge neutrality, the general form of the rate equation describing the dynamical behavior of the carrier density N (i.e., electron–hole pair number) is

$$\frac{\partial N(t)}{\partial t} = D\nabla^2 N(t) + \frac{J}{e} - \frac{N(t)}{\tau_s} - R_{st}\left(N(t), |\mathbf{E}(t)|^2\right) \tag{4.A.33}$$

The first term on the right-hand side accounts for carrier diffusion, D being the diffusion coefficient. We assume in the following that the carrier distribution does not vary significantly within the active region; the diffusion term in Eq. (4.A.33) is negligible in that case. The second term accounts for the electron–hole pairs that are injected into the active regions by means of the electrical current J. e is the magnitude of the electron charge. The third term accounts for losses of carriers by spontaneous emission and nonradiative transitions. The last term stands for the loss rate of carrier density due to stimulated recombination of electron-hole pairs. The carrier lifetime is approximated by

$$\tau_s(N)^{-1} = A_{nr} + BN + CN^2 \tag{4.A.34}$$

Here, A_{nr}, BN and CN^2 are related to mechanisms such as trap or surface recombination, to radiative recombination and to the Auger recombination process, respectively. Above lasing threshold, τ_s can be considered as a constant to a good approximation because the carrier density is nearly clamped to its threshold value. Finally, R_{st} accounts for the loss of electron–hole pairs due to stimulated recombination. It is given by

$$R_{st}\left(N(t), |\mathbf{E}(t)|^2\right) = G(N(t)-N_0)|\mathbf{E}(t)|^2 \tag{4.A.35}$$

The rate equation for the carrier density can therefore be rewritten as

$$\frac{dN(t)}{dt} = \frac{J}{e} - \frac{N(t)}{\tau_s} - G_N(N(t)-N_0)|\mathbf{E}(t)|^2 \tag{4.A.36}$$

Equations (4.A.32) and (4.A.36) are the basic rate equations for semiconductor lasers.

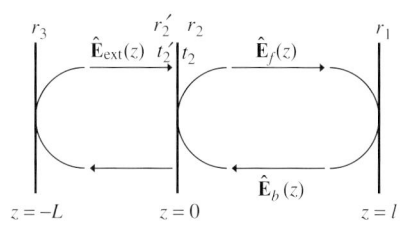

Figure 4.A.2. Model of a semiconductor laser with optical feedback. *l*, internal cavity length; *L*, external cavity length; *r*, reflectivity of the mirror; *t*, transmittance of the mittor.

4.A.2
Derivation of the Rate Equations of Semiconductor Lasers with Optical Feedback

In the following, we derive the rate equation for the electric field of semiconductor laser with delayed optical feedback (Petermann, 1988; Rogister, 2001; Ohtsubo, 2008). Figure 4.A.2 shows the schematic representation of a semiconductor laser subject to coherent optical feedback. In Figure 4.A.2, an external mirror with amplitude transmission coefficient r_3 is positioned at a distance L from the left facet. The round-trip time in the external cavity is $\tau = 2L/c$. $t_2' \hat{\mathbf{E}}_{\text{ext}}$ that enters the laser cavity after reflected by external mirror make a round-trip infinitely in external cavity, so that $t_2' \hat{\mathbf{E}}_{\text{ext}}$ can be written by

$$t_2' \hat{\mathbf{E}}_{\text{ext}}(t) = r_3 t_2 t_2' \hat{\mathbf{E}}_b(t-\tau) + r_2' r_3^2 t_2 t_2' \hat{\mathbf{E}}_b(t-2\tau) + \cdots$$

$$+ \frac{(r_2' r_3)^n}{r_2'} t_2 t_2' \hat{\mathbf{E}}_b(t-n\tau) + \cdots \quad (4.A.37)$$

where r_2' is the amplitude reflection coefficient from outside of the cavity and $\hat{\mathbf{E}}_b(t)$ is the complex electric field propagating in the backward direction that has the relation at $z = 0$

$$\hat{\mathbf{E}}_f(t) = r_2 \hat{\mathbf{E}}_b(t) \quad (4.A.38)$$

The laser field $\hat{\mathbf{E}}_f(t)$ and $\hat{\mathbf{E}}_f(t-n\tau)$ can be divided into two terms, the term changing with angular frequency ω_{th} and the term $\mathbf{E}_f(t)$ that varies slowly compared with the angular frequency. The field is given by

$$\hat{\mathbf{E}}_f(t) = \mathbf{E}_f(t)\exp(i\omega_{\text{th}} t) \quad (4.A.39)$$

$$\hat{\mathbf{E}}_f(t-n\tau) = \mathbf{E}_f(t-n\tau)\exp\{i\omega_{\text{th}}(t-n\tau)\} \quad (4.A.40)$$

Using the relations between the amplitude transmission coefficient and the amplitude reflection coefficient

$$t_2 t_2' = 1 - r_2^2 \quad (4.A.41)$$

and between the amplitude reflection coefficient from inside and outside of the cavity

$$r_2 = -r_2', \tag{4.A.42}$$

The difference equation Eq. (4.A.19) can be rewritten in the presence of optical feedback as

$$\mathbf{E}_f(t)\exp(i\omega_{th}\,t) = \mathbf{G}_1\exp(i\omega_{th}\,\tau_{in})\,\mathbf{E}_f(t-\tau_{in})\exp\{i\omega_{th}\,(t-\tau_{in})\}$$
$$+ \frac{r_2^2-1}{r_2^2}\sum_{n=1}^{\infty}(-r_2r_3)^n\mathbf{E}_f(t-n\tau) \tag{4.A.43}$$

The rate equation for the slowly varying envelope of the electric field $\mathbf{E}(t)$ in semiconductor laser with time-delayed optical feedback can be written as

$$\frac{d\mathbf{E}(t)}{dt} = \frac{1+i\alpha}{2}\left[G_N(N(t)-N_0) - \frac{1}{\tau_p}\right]\mathbf{E}(t)$$
$$+ \frac{1}{\tau_{in}}\frac{r_2^2-1}{r_2^2}\sum_{n=1}^{\infty}(-r_2r_3)^n\mathbf{E}(t-n\tau)\exp(-in\omega_{th}\tau) \tag{4.A.44}$$

When the intensity of the light is weak, the feedback term above the equation can be approximated by taking into account just one round-trip in the external cavity, so that Eq. (4.A.44) can be written as

$$\frac{d\mathbf{E}(t)}{dt} = \frac{1+i\alpha}{2}\left[G_N(N(t)-N_0) - \frac{1}{\tau_p}\right]\mathbf{E}(t) + \kappa\mathbf{E}(t-\tau)\exp(-i\omega_{th}\tau) \tag{4.A.45}$$

where the feedback strength κ is defined as

$$\kappa = \frac{1}{\tau_{in}}\frac{(1-r_2^2)\,r_3}{r_2}. \tag{4.A.46}$$

Equation (4.A.45) is the rate equation for the electric field of a semiconductor laser with delayed optical feedback. Equations (4.A.45) for the electric field and Eq. (4.A.36) for the carrier density are well known as the Lang–Kobayashi equations (see Section 4.2).

4.A.3
Analytical Approach of the Stability of Steady State Solutions for Semiconductor Laser with Optical Feedback Under the Limits of Weak and Short Feedback

In the limit of weak feedback strength and short feedback time, Eq. (4.86) in Section 4.3.4 can be analytically investigated (Ohtsubo, 2008). The following approximation (i.e., weak feedback strength and short feedback time) is used,

$$\kappa\tau \ll 1 \tag{4.A.47}$$

The following approximation can be satisfied,

$$1-\exp(-\lambda\tau) \approx \lambda\tau \tag{4.A.48}$$

Substituting Eq. (4.A.48) into Eq. (4.86) in Section 4.3.4, and the characteristic equation is simplified,

$$(1+\kappa^2\tau^2 + 2\kappa\tau\cos(\omega_s\tau))\lambda^3 + \left(\frac{1}{\tau_s} + G_N E_s^2\right)(1+\kappa^2\tau^2 + 2\kappa\tau\cos(\omega_s\tau))\lambda^2$$
$$+ G_N^2 E_s^2(N_s-N_0)[1+\kappa\tau(\cos(\omega_s\tau)-\alpha\sin(\omega_s\tau))]\lambda = 0 \tag{4.A.49}$$

The complex solution ($\lambda \neq 0$) is expressed as $\lambda = \lambda_{Re} \pm i\lambda_{Im}$, where

$$\lambda_{Re} = -\frac{1}{2}\left(\frac{1}{\tau_s} + G_N E_s^2\right) \tag{4.A.50}$$

$$\lambda_{Im} = \sqrt{\frac{G_N^2 E_s^2(N_s-N_0)[1+\kappa\tau(\cos(\omega_s\tau)-\alpha\sin(\omega_s\tau))]}{1+\kappa^2\tau^2 + 2\kappa\tau\cos(\omega_s\tau)}}$$

$$= \sqrt{\frac{j-1}{\tau_s\tau_p}(1+G_N N_0\tau_p)}\sqrt{\frac{(1-2\kappa\tau_p\cos(\omega_s\tau))[1+\kappa\tau(\cos(\omega_s\tau)-\alpha\sin(\omega_s\tau))]}{1+\kappa^2\tau^2 + 2\kappa\tau\cos(\omega_s\tau)}}$$

$$\tag{4.A.51}$$

The stability is calculated from Eq. (4.A.50). The condition of $\lambda_{Re} < 0$ is always satisfied from Eq. (4.A.50) because τ_s, G_N, and E_s^2 are positive values. Therefore, it is found that the steady-state solutions of Eqs. (4.68)–(4.70) in Section 4.3.3 are always stable. In fact, λ_{Re} obtained with optical feedback (Eq. (4.A.50)) is exactly the same as the case without optical feedback of Eq. (4.51) in Section 4.3.2. This is due to the assumption of weak feedback strength and short feedback time of Eq. (4.A.47). On the other hand, the stability is not identified yet when the conditions of strong optical feedback and long feedback time are satisfied (in fact, chaotic instability occurs).

The relaxation oscillation frequency f_r can be calculated from Eq. (4.A.51), where $2\pi f_r = \lambda_{Im}$

$$f_r = \frac{1}{2\pi}\sqrt{\frac{j-1}{\tau_s\tau_p}(1+G_N N_0\tau_p)}\sqrt{\frac{(1-2\kappa\tau_p\cos(\omega_s\tau))[1+\kappa\tau(\cos(\omega_s\tau)-\alpha\sin(\omega_s\tau))]}{1+\kappa^2\tau^2 + 2\kappa\tau\cos(\omega_s\tau)}}$$

$$\tag{4.A.52}$$

This formula shows the relaxation oscillation frequency in a semiconductor laser with optical feedback under the conditions of weak feedback strength and short feedback time. The relaxation oscillation frequency is changed by the multiplication of the second square-root term of the right-hand-side in Eq. (4.A.52), compared with the case without optical feedback shown in Eq. (4.55) in Section 4.3.2. The relaxation oscillation frequency in Eq. (4.A.52) is dependent on the feedback strength κ, the delay time τ, the photon lifetime τ_p, the alpha parameter α, the and optical feedback phase $\omega_s\tau$.

4.A.4
Runge–Kutta Method for the Integration of Ordinary Differential Equations

A method to solve initial-value problems for a set of ordinary differential equations is described in Appendix 4.A.4 (Press *et al.*, 1992). In initial-value problems all the initial values $x_i(0)$ are given at some starting time t_0, and it is desired to find the $x_i(t_n)$ at some final time t_n. Then N, the number of simultaneous ordinary differential equations are described as,

$$\frac{dx_i(t)}{dt} = f_i(t, x_1, x_2, \ldots, x_N) \quad i = 1, 2, \ldots, N \tag{4.A.53}$$

where the functions f_i on the right-hand side are known. The initial values of the variables $x_i(0)$ are also given.

The underlying idea of any routine for solving the initial-value problem is the following: Rewrite the dx_i and dt in Eq. (4.A.53) as finite steps Δx_i and Δt, and multiply the equations by Δt. This gives algebraic formulas for the change in the function when the independent variable t is stepped by one stepsize Δt. In the limit of making the stepsize very small, a good approximation to the underlying differential equation is achieved. Literal implementation of this procedure results in Euler's method. Euler's method is conceptually important even though it is not recommended for any practical use. Practical methods all come down to this same idea: Add small increments to the functions corresponding to derivatives (right-hand sides of the equations) multiplied by stepsizes.

Runge–Kutta methods are considered as major types of practical numerical methods for solving initial-value problems for ordinary differential equations. Runge–Kutta methods propagate a solution over an interval by combining the information from several Euler-style steps (each involving one evaluation of the right-hand sides of the equations) and then using the information obtained to match a Taylor-series expansion up to some higher order (fourth order is usually recommended).

Let us consider the mathematical formula of Euler and Runge–Kutta methods. Consider the situation where $x_i(t+h)$ needs to be obtained from the given values of $x_i(t)$ and $f_i(t, x_1, \ldots, x_N)$, where h is a finite stepsize of time for numerical integration. The left-hand side of Eq. (4.A.53) is rewritten by the definition of derivatives.

$$\frac{dx_i(t)}{dt} = \lim_{h \to 0} \frac{x_i(t+h) - x_i(t)}{h} \tag{4.A.54}$$

Substitute Eq. (4.A.54) into Eq. (4.A.53), and rewrite $f_i(t, x_1, \ldots, x_N)$ to $f_i(t, x_i)$ for simplicity (x_i includes all the variables $x_1, \ldots, x_N$),

4.A.4 Runge–Kutta Method for the Integration of Ordinary Differential Equations

$$x_i(t+h) = x_i(t) + h f_i(t, x_i) \tag{4.A.55}$$

This is the formula for the Euler method. This formula is unsymmetrical: It advances the solution through an interval h, but uses derivative information only at the beginning of that interval. This means that the step's error is only one power of h smaller than the correction, that is, $O(h^2)$ added to Eq. (4.A.55).

Consider the use of a step like Eq. (4.A.55) to take a trial step to the midpoint of the interval. Then, use the value of both t and x_i at the midpoint to compute the real step across the whole interval. In equations,

$$\begin{aligned} k_{1,i} &= f_i(t, x_i) \\ k_{2,i} &= f_i\left(t + \frac{1}{2}h, \, x_i + \frac{1}{2}hk_{1,i}\right) \\ x_i(t+h) &= x_i(t) + h\, k_{2,i} \end{aligned} \tag{4.A.56}$$

This symmetrization cancels out the first-order error term and the step's error is two powers of h smaller than the correction, that is, $O(h^3)$, making the method second order. (A method is conventionally called nth order if its error term is $O(h^{n+1})$.) In fact, Eq. (4.A.56) is called the second-order Runge–Kutta or midpoint method.

There are many ways to evaluate the right-hand side $f_i(t, x_i)$ that all agree to first order, but that have different coefficients of higher-order error terms. Adding up the right combination of these, the error terms can be eliminated order by order. That is the basic idea of the Runge–Kutta method. The most often used method is the fourth-order Runge–Kutta method which has a certain sleekness of organization about it:

$$\begin{aligned} k_{1,i} &= f_i(t, x_i) \\ k_{2,i} &= f_i(t + \frac{1}{2}h, \, x_i + \frac{1}{2}h\, k_{1,i}) \\ k_{3,i} &= f_i(t + \frac{1}{2}h, \, x_i + \frac{1}{2}h\, k_{2,i}) \\ k_{4,i} &= f_i(t + h, \, x_i + h\, k_{3,i}) \\ x_i(t+h) &= x_i(t) + h\left(\frac{1}{6}k_{1,i} + \frac{1}{3}k_{2,i} + \frac{1}{3}k_{3,i} + \frac{1}{6}k_{4,i}\right) \end{aligned} \tag{4.A.57}$$

The fourth-order Runge–Kutta method requires four evaluations of the right-hand side per step h. This will be superior to the midpoint method of Eq. (4.A.56), and the step's error is four powers of h smaller than the correction, that is, $O(h^5)$.

The Runge–Kutta method treats every step in a sequence of steps in an identical manner. Prior behavior of a solution is not used in its propagation. This is mathematically proper, since any point along the trajectory of an ordinary differential equation can serve as an initial point. The fact that all steps are treated identically also makes it easy to implement the Runge–Kutta method.

4.A.5
Gram–Schmidt Orthogonalization

Gram–Schmidt orthogonalization is used to create vectors that are orthogonal to each other. When a set of nonorthogonal vectors $\mathbf{a}_j$ ($j = 1, 2, \ldots, n$) is given, a set of orthogonal vectors $\mathbf{q}_j$ ($j = 1, 2, \ldots, n$) can be calculated from $\mathbf{a}_j$ as,

$$\mathbf{q}_1 = \mathbf{a}_1$$
$$\mathbf{q}_j = \mathbf{a}_j - \sum_{i=1}^{j-1} \frac{(\mathbf{q}_i \cdot \mathbf{a}_j)}{\|\mathbf{q}_i\|^2} \mathbf{q}_i, \quad (j > 1) \tag{4.A.58}$$

To implement the Gram–Schmidt orthogonalization, two types of the algorithms are shown as follows (Giraud et al., 2005). The modified Gram–Schmidt method is recommended to avoid the accumulation of numerical errors.

(a) Classical Gram–Schmidt method:

$$\begin{aligned}
&\text{do } j = 1, n \\
&\quad \mathbf{q}_j = \mathbf{a}_j \\
&\quad \text{do } i = 1, j-1 \\
&\quad\quad \mathbf{q}_j = \mathbf{q}_j - (\mathbf{q}_i \cdot \mathbf{a}_j)\mathbf{q}_i \\
&\quad \text{end do} \\
&\quad \mathbf{q}_j = \mathbf{q}_j / \|\mathbf{q}_j\| \\
&\text{end do}
\end{aligned}$$

(b) Modified Gram–Schmidt method:

$$\begin{aligned}
&\text{do } j = 1, n \\
&\quad \mathbf{a}_j^{(0)} = \mathbf{a}_j \\
&\quad \text{do } i = 1, j-1 \\
&\quad\quad \mathbf{a}_j^{(i)} = \mathbf{a}_j^{(i-1)} - \left(\mathbf{q}_i \cdot \mathbf{a}_j^{(i-1)}\right)\mathbf{q}_i \\
&\quad \text{end do} \\
&\quad \mathbf{q}_j = \mathbf{a}_j^{(j-1)}; \quad \mathbf{q}_j = \mathbf{q}_j / \|\mathbf{q}_j\| \\
&\text{end do}
\end{aligned}$$

where the vector $\mathbf{a}_j$ is updated in the i-loop, and $\mathbf{a}_j^{(i-1)}$ indicate the updated $\mathbf{a}_j$ at the $i-1$ iteration step.

5
Synchronization of Chaos in Lasers

5.1
Concept of Synchronization of Chaos in Lasers

5.1.1
Introduction

A topic that evolved through the 1990s is the problem of synchronization of chaotic systems. Though the synchronization of clocks (periodic systems) has been studied with great care over centuries (Pikovsky *et al.*, 2001; Strogatz, 2003), it was the surprising discovery of temporal synchronization between two chaotic systems that initiated the field of "chaos communication" (see Chapters 7 and 8). At first, even the notion of synchronization of chaotic systems appears self-contradictory. How can two chaotic systems that will inevitably start from slightly different initial conditions ever be synchronized? One can look at the exponentially fast divergence of phase-space trajectories from ever so slightly different initial conditions, and the issue of sensitivity to initial conditions as one of the hallmarks of chaos. The crucial realization is that when two chaotic systems are coupled to each other in a suitable way, they exert a form of "control" on each other, and it is possible for both systems to synchronize in their dynamics, even when they start from very different initial conditions.

The early studies and demonstrations of synchronization of chaotic systems were done on theoretical models and electronic circuits (Blekhman, 1971; Fujisaka and Yamada, 1983; Afraimovich *et al.*, 1986), and soon led to the question: could the chaotic dynamics of these systems be used for something practical? After the appearance of the pioneering work on the experimental demonstration of synchronization of chaos in electronic circuits (Pecora and Carroll, 1990), the possibility that chaotic lasers could be synchronized was examined in an important numerical study on a linear array of coupled semiconductor lasers (Winful and Rahman, 1990). It was shown that synchronization of chaos could be achieved by mutually coupled nearest-neighbor lasers through injection of their optical fields. This work stimulated experiments a few years later on mutually coupled solid-state lasers (Roy and Thornburg, 1994) and one-way coupled gas lasers (Sugawara *et al.*, 1994). These early studies demonstrated that synchronization of chaotic optical systems could

Optical Communication with Chaotic Lasers: Applications of Nonlinear Dynamics and Synchronization, First Edition.
Atsushi Uchida.
© 2012 Wiley-VCH Verlag GmbH & Co. KGaA. Published 2012 by Wiley-VCH Verlag GmbH & Co. KGaA.

provide a means for communications through free space or optical fibers, and motivated much of the research on synchronization of chaos in laser systems and its communication applications.

Comprehensive reviews on synchronization of chaos in nonlinear dynamical systems have been reported in the literature (Pikovsky et al., 2001; Boccaletti et al., 2002; Strogatz, 2003). Review articles and books on synchronization of chaos in optical and laser systems have also been published (Donati and Mirasso, 2002; Larger and Goedgebuer, 2004; Kane and Shore, 2005; Uchida et al., 2005; Larson et al., 2006; Ohtsubo, 2008).

5.1.2
What is Synchronization?

"Synchronization" can be defined in this chapter as temporal behaviors with a certain relationship among coupled dynamical systems. Identical synchronization of periodic temporal waveforms between two coupled periodic oscillators (called drive and response systems) is a simple example, as shown in Figure 5.1. When the coupling

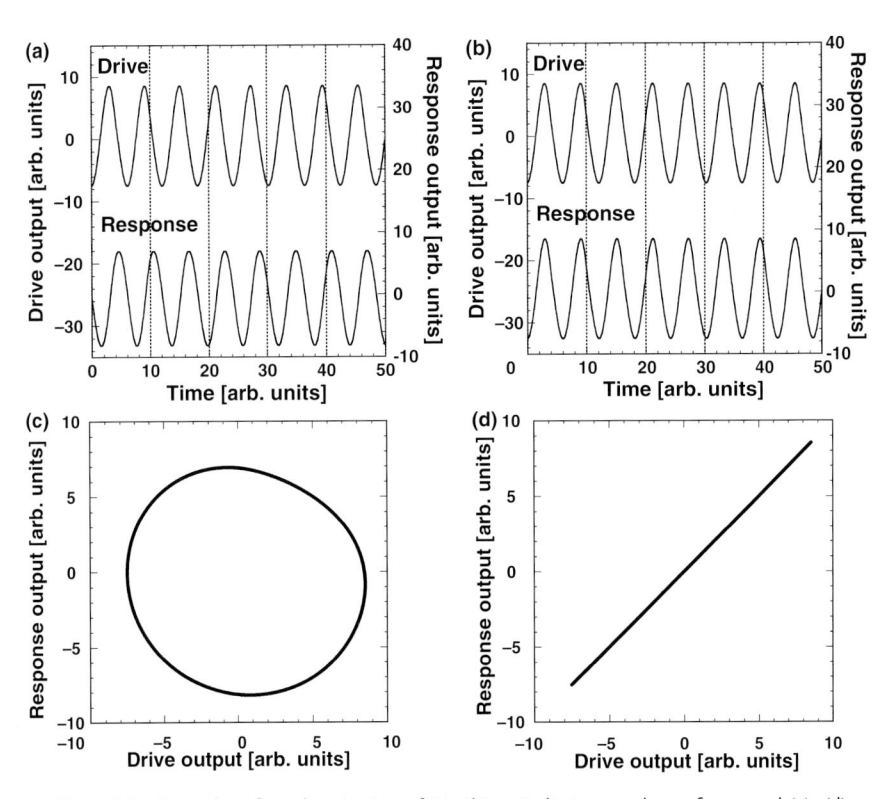

Figure 5.1 Examples of synchronization of (a), (b) periodic temporal waveforms and (c), (d) corresponding correlation plots between drive and response systems. (a), (c) No synchronization. (b), (d) Synchronization of periodic waveforms.

strength is weak, two temporal waveforms oscillate with different frequency and phase (Figure 5.1a). When the coupling strength is increased, the two temporal waveforms become identical with the same frequency and phase (Figure 5.1b). The correlation plot, obtained from the temporal waveform of one oscillator *versus* that of the other coupled oscillator, indicates the transition from a no-synchronization state with a circular correlation plot (Figure 5.1c) to a synchronization state with a 45-degree straight line (Figure 5.1d).

In the seventeenth century, synchronization of periodic oscillations was first observed by Christian Huygens, who discovered that a couple of pendulum clocks hanging from a common support had synchronized, that is, their oscillations coincided perfectly and the pendula moved always in opposite direction (antiphase synchronization) (Pikovsky *et al.*, 2001). Nowadays, many communication systems require synchronization of periodic carrier waveforms to tune an appropriate communication channel and extract an encoded message signal. Synchronization of periodic waveforms has been widely used in many engineering applications. In optics, injection locking (matching of the optical-carrier frequency) between coupled lasers can be regarded as synchronization of periodic optical-carrier frequency.

Instead of synchronization of periodic waveforms, synchronization of chaotic temporal waveforms can also be observed. Figure 5.2 shows an example of synchronization of chaos in coupled chaotic oscillators. When the coupling strength is weak, two independent chaotic oscillations are observed, as shown in Figure 5.2a. As the coupling strength is increased, the two temporal waveforms show identical oscillations and synchronization of chaos can be achieved, as shown in Figure 5.2b. The correlation plots of Figures 5.2c and d show the appearance of synchronization between the two chaotic oscillations. There is no linear correlation between the chaotic temporal waveforms at weak coupling strength (Figure 5.2c). By contrast, a linear correlation is clearly observed in Figure 5.2d, showing identical synchronization of chaos. The cross-correlation values (see Eq. (5.19)) are estimated from Figures 5.2c and d. The cross-correlation value for Figure 5.2c is -0.063, indicating no synchronization. By contrast, the cross-correlation value for Figure 5.2d is 1.0, indicating identical synchronization of chaos.

5.1.3
Why Should Chaos be Synchronized?

Synchronization of chaos is in fact a surprising phenomenon because chaos has a strong dependence on the initial conditions and a small deviation of two temporal waveforms diverges exponentially in time. One of the most remarkable characteristics of chaotic systems is their sensitive dependence on initial conditions. Given this feature, it may seem counterintuitive to expect two coupled chaotic devices (whose initial conditions are in general different) to synchronize, however similar their structures and parameters are. This contrasts with the intuitively clear synchronization routinely observed in periodic systems. In spite of these intuitive expectations, coupled chaotic systems do synchronize their irregular dynamics starting from

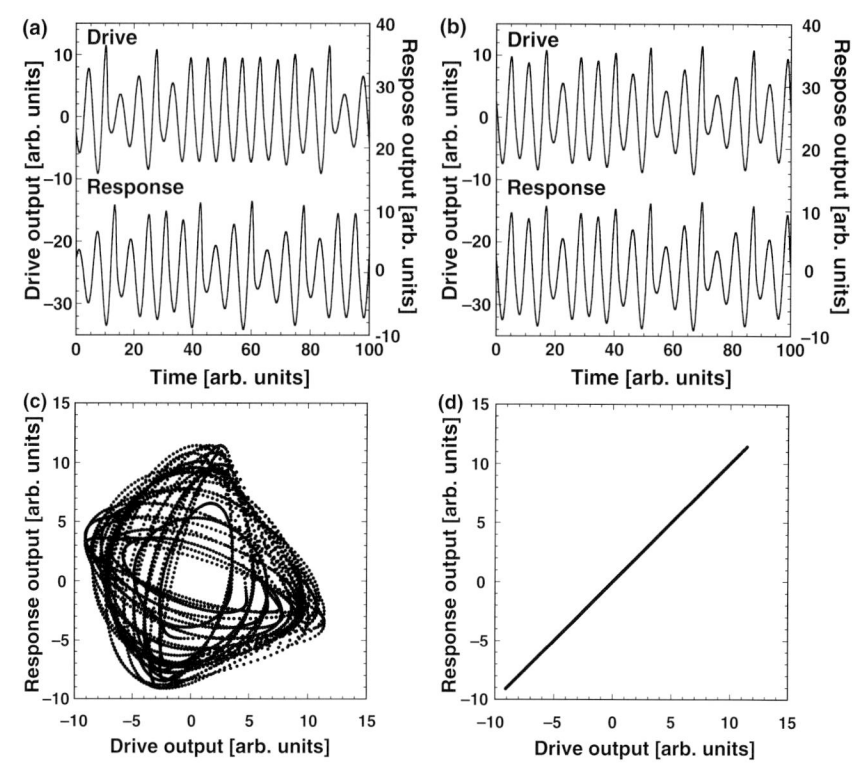

Figure 5.2 Examples of synchronization of (a), (b) chaotic temporal waveforms and (c), (d) corresponding correlation plots between drive and response systems. (a), (c) No synchronization. (b), (d) Synchronization of chaos.

arbitrary initial conditions, provided coupling between them is large enough and the two devices are sufficiently similar.

The conditions for achieving identical synchronization among coupled laser systems are the following:

i) All the laser systems must consist of (almost) identical devices with (almost) identical parameter settings.
ii) All the laser systems must be subject to the (almost) identical drive signal or feedback signal.

The "symmetry" of the coupled laser systems results in an identical synchronous solution of the coupled systems in a mathematical sense, and identical synchronization can be achieved (see Section 5.3.1). In addition to the above two conditions, another important condition is required for achieving synchronization.

iii) The identical synchronous solution must be stable.

The stability of the synchronous solution depends on the particular dynamical system under consideration, but regions of stability in parameter space have

5.1.4
Characteristics of Synchronization of Chaos in Laser Systems

In chaotic laser systems, it is worth noting that there are two oscillation components in the electric field of the laser output, as shown in Figure 5.3. One is the fast optical-carrier frequency in the order of 10^{14} Hz, determined by $f_c = c/\lambda$, where c is the speed of light and λ is the optical wavelength. The other is the slow chaotic envelope component on the electric field due to the photon–carrier interaction in the laser medium in the order of 10^3–10^9 Hz, related to the relaxation oscillation frequency f_r of the laser. The existence of the two frequency components in the different orders of magnitude makes synchronization problems in coupled laser systems more complicated than those in other nonlinear dynamical systems. Chaos can only be observed experimentally for the latter (slower) frequency range because it is required for the observation to convert the temporal fluctuation of the laser intensity into an electrical signal by a photoreceiver that has a limited frequency bandwidth up to tens of GHz ($\sim 10^{10}$ Hz). Synchronization of chaos is thus observed in the slower frequency range of 10^3–10^9 Hz. However, it is very important to synchronize the fast optical-carrier-frequency component ($\sim 10^{14}$ Hz) in coherently coupled lasers in order to achieve synchronization of chaos at lower frequencies (10^3–10^9 Hz). The synchronization phenomenon of the fast optical-carrier frequency among coupled lasers has been well known as "injection locking" in the field of optics, where optical carrier frequencies of two coupled lasers can be exactly matched due to the frequency-pulling effect (Siegman, 1986). Injection locking is necessary to achieve

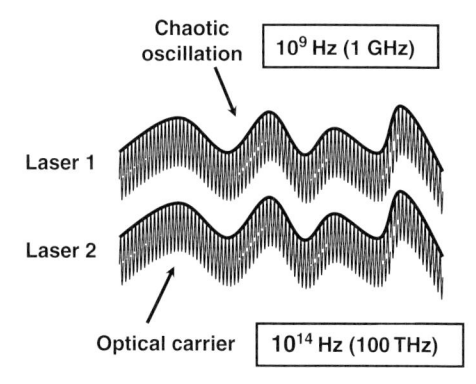

Figure 5.3 Schematic for synchronization of chaotic oscillation of the laser electric field with two different frequency components; slow-envelope component of chaotic oscillation at $\sim 10^9$ Hz and fast optical carrier at $\sim 10^{14}$ Hz. Chaos can only be observed experimentally for the slower frequency range due to the limitation of photoreceivers up to $\sim 10^{10}$ Hz.

synchronization of slow-envelope chaotic oscillations, and synchronization of chaos in coherently coupled lasers is mostly observed in the injection-locking range. It is thus very important to include the detuning of optical-carrier frequencies among coupled lasers to mimic the injection-locking effect in numerical models for synchronization of coupled chaotic lasers. The relationship between injection locking and synchronization of chaos has been intensively investigated and will be discussed in Chapter 6.

5.1.5
Synchronization of Chaos for Communication Applications

The idea of synchronization of chaos leads to a possible application for communication. The application of synchronization of chaos to secret communication systems was suggested in early work (Pecora and Carroll, 1990). It was discovered that a chaotic transmitter could consist of an electronic circuit that generated nonlinear dynamics and chaos. A message to be concealed with small amplitude is added to the chaotic fluctuations of one of the dynamical variables and transmitted to a receiver, while another chaotic variable was separately transmitted (Ditto and Pecora, 1993). The receiver consisted of a subsystem of the circuits in the transmitter that generated the dynamics of the transmitter and was driven by the separately transmitted signal. The receiver synchronized to the chaos of the transmitter for the given operating parameters, and one could recover the message from the chaos through a subtraction at the receiver.

An elegant variation of the method was introduced above that did not require the separate transmission of a driving signal to the receiver (Cuomo and Oppenheim, 1993). It was shown that the receiver could actually synchronize to the chaotic dynamics of the transmitter even when a message was added to the chaotic driving signal from the transmitter. The synchronized output from the receiver was then used to subtract out the information from the transmitted signal. The synchronization was not perfect, and the message, treated as a perturbation of the chaotic signal, had to be small compared to the chaos (see Section 8.1.1).

Following the original implementation of chaotic communications in electronic circuits, a suggestion to use optical systems was made (Colet and Roy, 1994). One does not have access in laser systems to all the system variables in the way possible with electronic circuits, so a somewhat different conceptual approach was developed to synchronize chaotic lasers. One of the major motivations for using lasers was that optical chaotic systems offer the possibility of high-speed data transfer, as shown in early simulations of numerical models that include realistic operational characteristics of the transmitter, receiver and communication channel (Mirasso et al., 1996).

Several schemes for communication with chaotic waveforms were experimentally demonstrated in which the intensity fluctuations generated by erbium-doped fiber lasers (EDFRLs) were used either to mask or carry a message (VanWiggeren and Roy, 1998a). Similar concepts were used to demonstrate optical communication using the chaotic wavelength fluctuations output from an electro-optic system (Goedgebuer et al., 1998). Optical chaotic communication was also demonstrated for semiconductor lasers with time-delayed optical feedback (Sivaprakasam and Shore, 1999b;

Fischer et al., 2000; Kusumoto and Ohtsubo, 2002), semiconductor lasers with optoelectronic feedback (Liu et al., 2002a), and electro-optic systems (Larger et al., 2004b). Bit rates of 2.5–10 Gb/s have been achieved experimentally in commercial optical-fiber networks (Argyris et al., 2005; Argyris et al., 2010c; Lavrov et al., 2010). The details of chaos communication based on the techniques of synchronization of chaos will be described comprehensively in Chapter 8.

5.2
History of Synchronization of Chaos in Lasers

5.2.1
Synchronization of Chaos in Electronic Circuits

5.2.1.1 Pecora–Carroll Method

A historical demonstration of chaos synchronization was carried out in 1990 (Pecora and Carroll, 1990). In this proposal, a nonlinear system (drive) is divided arbitrarily into two subsystems. For example, a subsystem consists of the dynamics of x and z variables and the other subsystem consist of the dynamics of y variable in the Lorenz model. The former subsystem is duplicated into a new subsystem acting as a response, consisting of the dynamics of x' and z' variables as shown in Figure 5.4 (left). The drive and response systems are coupled unidirectionally through a common signal of the y variable. Synchronization of chaos can be achieved when the dynamics of the response subsystem driven by the y variable follows the chaotic

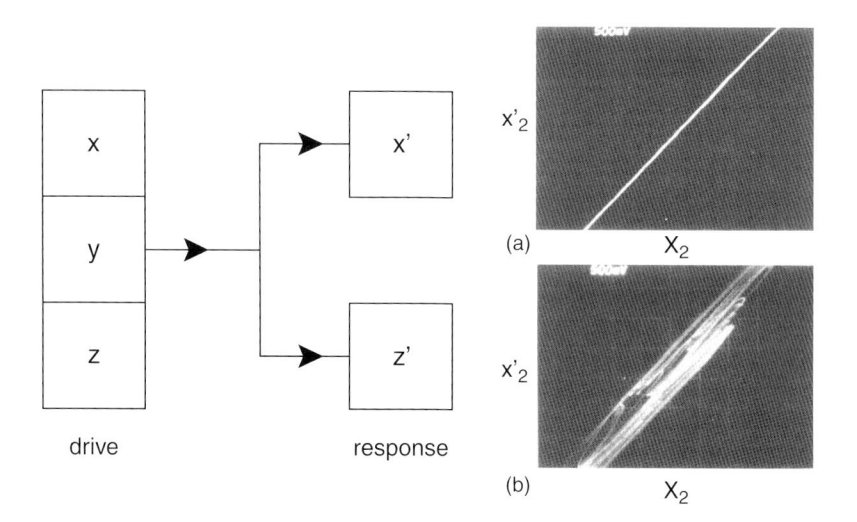

Figure 5.4 (Left) Schematics of the Pecora–Carroll method for synchronizing two Lorenz systems. (Right) Oscilloscope traces of the response voltage vs. its drive counterpart voltage for (a) circuit parameters the same and (b) circuit parameters different by 50%. (L.M. Pecora and T.L. Carroll, (1990), © 1990 APS.)

dynamics of the drive system. The detailed mathematical description of the Pecora–Carroll method for synchronization of chaos is described in Appendix 5.A.1.

Synchronization properties of different coupling schemes were investigated in the Lorenz and the Rössler systems. This idea was also applied to a pair of coupled chaotic electronic circuits, demonstrating synchronization between the chaotic dynamics of the drive and the response after a transient. Figure 5.4 (right) shows oscilloscope traces of one of the variables in the drive system *versus* its response counterpart for two different parameter values. Identical synchronization is observed at identical parameter values (Figure 5.4a (right)), whereas a 50% mismatch in the circuit parameters distorts synchronization substantially (Figure 5.4b (right)). The possibility for applications of chaos synchronization to communications was also pointed out.

The Pecora–Carroll method requires that a chaotic system can be separated into two subsystems, one of which has to be replicated into a response system that must have only negative conditional Lyapunov exponents (see Section 5.3.5) when acted upon by one of the drive's variables. While this ingenious scheme has been implemented in practice in electronic oscillators, it cannot be readily applied to optical systems, since it is basically impossible to separate the elements of a nonlinear optical system in that way.

5.2.1.2 Pyragas Method

Another synchronization scheme that fits better the capabilities of optical systems was proposed, where a method of continuous control of chaos *via* feedback to synchronization of chaos was extended (Pyragas, 1993). The block diagram of the method is presented in Figure 5.5 (left). One of the variables in the drive system is recorded in a memory, and the response system can be forced by using a small feedback perturbation of the difference between two variables for the drive and response (referred to as "diffusive coupling"). The perturbation has to be introduced into the response system as a negative feedback. The important feature of this perturbation is that it vanishes when the transmitter and receiver signals coincide. For identical parameters of the transmitter and the receiver, an identical synchronous

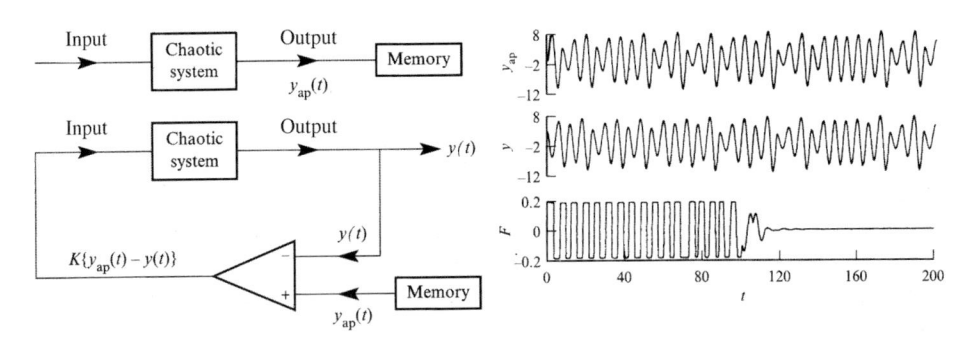

Figure 5.5 (Left) Block diagram of the method of continuous control of chaos *via* feedback to the synchronization of chaos (Pyragas method or diffusive coupling method). (Right) Recorded aperiodic output signal $y_{ap}(t)$, the dynamics of the output signal $y(t)$, and the difference $\Delta y = y_{ap}(t) - y(t)$ for the Rössler system. (K. Pyragas, (1993), © 1993 Elsevier.)

solution exists in a mathematical sense. In some conditions, the synchronized state can be stable due to the coupling term for a sufficiently large coupling strength. Using this approach, synchronization of chaos was demonstrated in the Rössler system after the feedback signal is applied to the receiver, as shown in Figure 5.5 (right).

Most studies of chaos synchronization in electronic circuits have relied on coupling through an operational amplifier as a buffer for unidirectional coupling, or through a resistor for mutual coupling. In both cases, the coupling signal is proportional to the difference between the voltages of the drive and response systems, which is identical to the Pyragas method with diffusive coupling. Synchronization of chaos has been experimentally observed in many different types of chaotic electronic circuits (Endo and Chua, 1991; Anishchenko *et al.*, 1992; Chua *et al.*, 1992).

Source codes of C programming language are listed in Glossary G.2.2.4 for synchronization of chaos in the coupled Lorenz model with diffusive coupling.

5.2.2
Synchronization of Chaos in Lasers

The first numerical observation of synchronization of chaos was reported in an array of coupled semiconductor lasers (Winful and Rahman, 1990). The model system is an array of waveguide lasers coupled by means of their overlapping evanescent fields. In the absence of coupling, each laser operates in a single longitudinal and transverse mode, assumed to be the same for all the lasers. Nearest-neighbor coupling is used to observe chaotic dynamics and synchronization of chaos.

To elucidate the notion of synchronized chaos, consider the case of three identical coupled lasers (referred to as laser 1, 2, and 3). The presence of a spatial grid imposes a spatial symmetry and makes it possible to determine *a priori* which elements in the array are likely to synchronize. For weakly coupled lasers, the stable phase-locked or steady state is one in which the amplitude distribution across the array is nearly uniform. As the coupling strength is increased, the branch of phase-locked solutions loses stability in favor of a self-pulsing solution. The spatial symmetry, however, maintained in the sense that the pulsations in the outer lasers (laser 1 and 3) are synchronous. As the coupling strength is increased further, the period-doubling route to chaos occurs. Throughout this bifurcation sequence, the lasers 1 and 3 remain synchronized with each other.

For strong coupling strength, the temporal evolution of the output of each laser in the array is chaotic. The maximum Lyapunov exponent, indicating a measure of the average exponential divergence of nearby orbits, is positive, and it confirms that the evolution is indeed chaotic. It is remarkable, however, that the temporal evolution of the output intensities in lasers 1 and 3 are identical, as shown in Figure 5.6a (X_1 and X_3). This is the regime of synchronized chaos, as can be seen in the correlation plot of the temporal evolution between the laser 1 and 3 intensities (Figure 5.6b). In the correlation plot between the laser 1 and 2 intensities, however, it is clear that the motion is along a strange attractor and no synchronization is observed (Figure 5.6c). The regime of synchronized chaos thus corresponds to spatial order and temporal chaos. At this condition the maximum "conditional" Lyapunov exponent is negative,

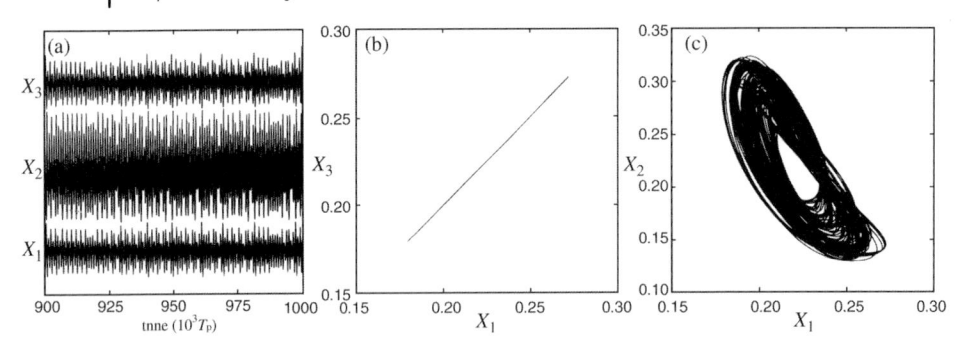

Figure 5.6 Synchronized chaos. (a) Time series for the amplitudes of the three laser outputs, (b) projection of the flow of the amplitudes of the laser 1 and 3 (synchronization of chaos). (c) projection of the amplitudes of the laser 1 and 2 (no synchronization). (H.G. Winful and L. Rahman, (1990), © 1990 APS.)

indicating stable synchronization can be observed between the laser 1 and 3 intensities (see Section 5.3.5). The presence of spatial symmetry does not guarantee synchronization of chaotic orbits. When the coupling strength is increased further, synchronization breaks down and spatiotemporal chaos is observed.

After the numerical demonstration of synchronization of chaos in laser arrays, one of the first experimental observations of synchronization of chaotic lasers was reported by using two neodymium-doped yttrium aluminum garnet (Nd:YAG) solid-state lasers (Roy and Thornburg, 1994). The system consists of two Nd:YAG laser beams of wavelength 1.06 μm generated in the same crystal by two 514.5-nm pump beams of almost equal intensity obtained from an argon (Ar^+) laser, as shown in Figure 5.7 (left). The spatial separation of the parallel pump beams is varied so that the mutual coupling between the two Nd:YAG lasers can be provided by overlap of the intracavity laser fields. One or both lasers are driven chaotic by periodic modulation of their pump beams, and synchronized chaotic intensity fluctuations are observed in both cases when the lasers are sufficiently coupled.

For strong coupling at a small separation of the pump beams, the optical phases of the two lasers are locked. Modulation of the pump beam for laser 1 leads to well-synchronized chaotic fluctuations of the two laser intensities, as shown in the time traces of Figure 5.7a (right). A plot of the intensity of laser 1 *versus* the intensity of laser 2 (Figure 5.7b (right)) is distributed near a 45-degree line and the synchronized nature of the chaotic lasers is evident. Synchronization of the chaotic lasers persists stably, as long as the temperature and other environmental conditions are maintained constant.

The other first experimental observation of synchronization of chaotic lasers was reported in unidirectionally coupled gas lasers (Sugawara et al., 1994). The system consists of two carbon dioxide (CO_2) lasers with gaseous saturable absorbers; one is the drive (master) laser and the other is the response (slave) laser, as shown in Figure 5.8 (top). An intracavity saturable absorber induces self-sustained pulsation in a single-mode oscillation, which is known as passive Q switching. This pulsation can become chaotic following the period-doubling route to chaos. It was demonstrated that the two chaotic pulsations are synchronized when the radiation from

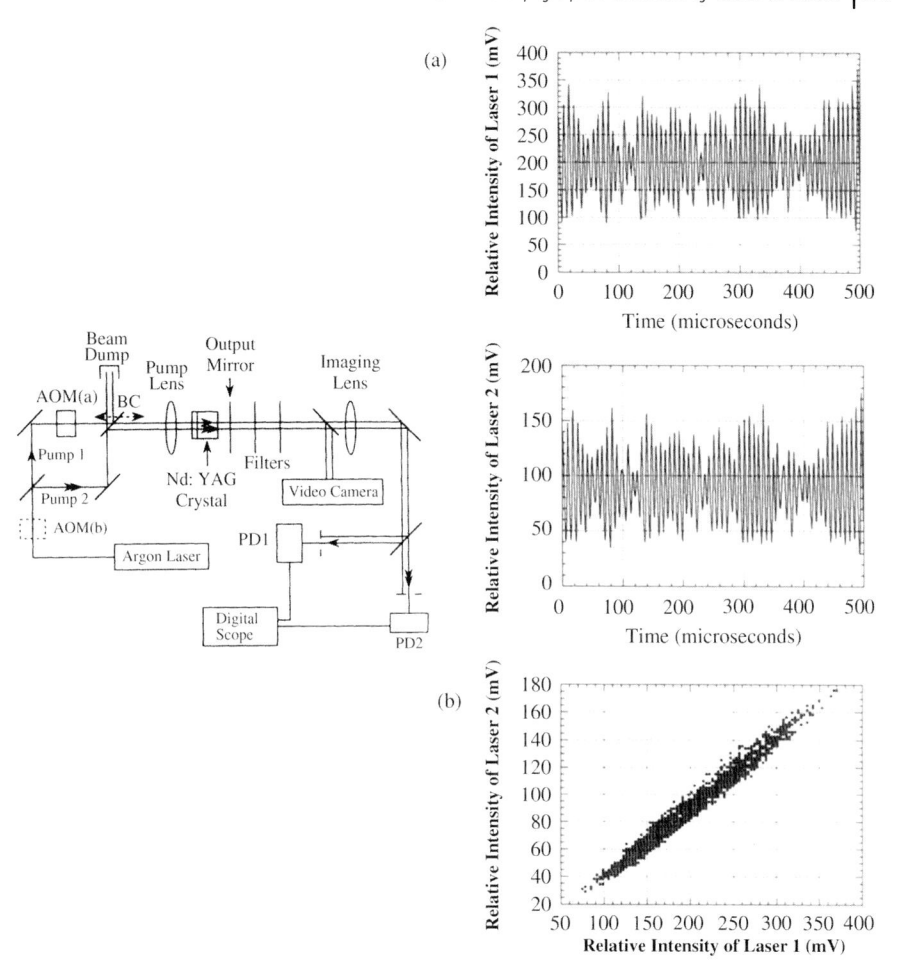

Figure 5.7 (Left) Experimental system for generating two spatially coupled chaotic lasers and monitoring their outputs. An acousto-optic modulator (AOM) can be placed in position (a) to modulate only laser 1, or in position (b) to modulate both lasers simultaneously. Beam splitters divide the argon laser output into two beams, each of which pumps a spatially separate region in the Nd:YAG crystal. The separation of the beams can be varied by monitoring the beam profiles. A lens is used to image the lasers so that the individual beams can be resolved and monitored by the photodiodes PD1 and PD2. The time traces of the two lasers can be displayed and stored by the digital oscilloscope. (Right) Relative intensities of two strongly coupled lasers with the pump beam for laser 1 modulated by the AOM. Note the strong synchronization of the two laser intensities. (b) X–Y plot of the two laser intensities shown in (a). Note the strong linearity of this figure, indicating the synchronized nature of the time traces. (R. Roy and K.S. Thornburg, Jr., (1994). © 1994 APS.)

the drive laser is unidirectionally injected into the saturable absorber of the response laser. Figure 5.8 (bottom) shows time sequences of the chaotic pulsations from the two lasers and correlation plots between the peak heights and between the peak intervals for different values of the injection power of the drive laser beam. When

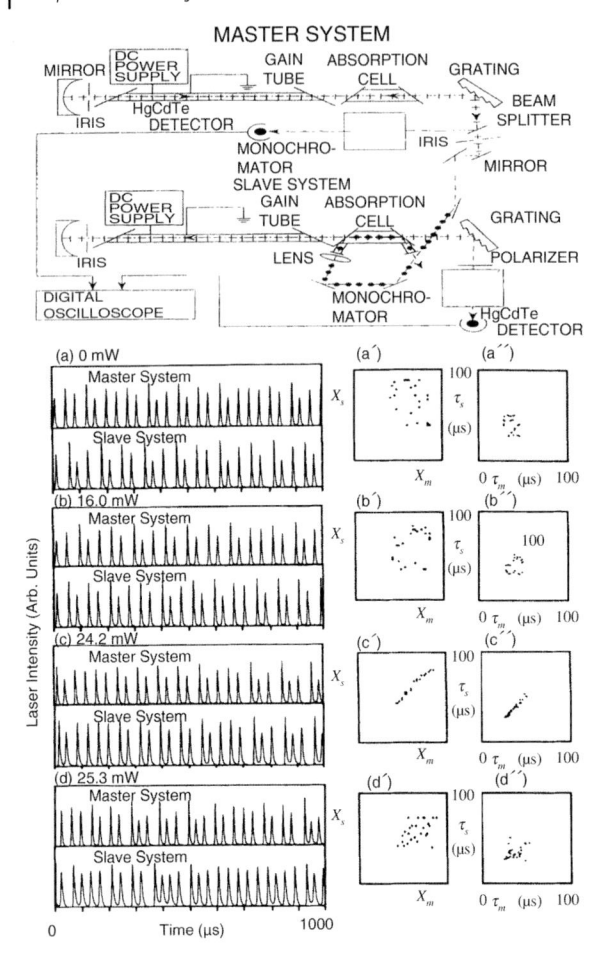

Figure 5.8 (Top) Diagram of the experimental setup for synchronization of chaos in two unidirectionally coupled CO_2 lasers. (Bottom) Observed time sequences of the passive Q-switching pulsation of the master and slave lasers for different powers of the injected beam [(a)–(d)] together with the correlation plots between the peak heights [(a′)–(d′)] and between the peak intervals [(a″)–(d″)]. (T. Sugawara, M. Tachikawa, T. Tsukamoto, and T. Shimizu, (1994). © 1994 APS.)

there is no coupling between the two systems, each laser exhibits chaotic pulsations as shown in Figure 5.8a. In the correlation plots of Figure 5.8a′ and a″, data points are scattered two dimensionally in an erratic manner, indicating that the two lasers are pulsating independently. When the input power is raised, the chaotic pulsation of the response laser is synchronized to the driving pulsation as seen in Figure 5.8c. A linear relation between the corresponding peak heights and intervals is clearly observed in Figures 5.8c′ and c″. When the input power is further increased, the synchronization is destroyed, as shown in Figure 5.8d.

The radiation of the drive laser modulates the absorber's population in the response laser. The two laser intensities interact with each other within the saturable absorber. The coupling is thus incoherent (i.e., independent of the optical phase of the electric field). The dependence of the chaotic pulsations on the strength of the driving force is qualitatively reproduced by a numerical simulation based on a dynamical model of unidirectionally coupled CO_2 lasers (Sugawara *et al.*, 1994).

After these two pioneering experimental demonstrations of synchronization of chaos in laser systems, many research activities on synchronization of chaos have been reported in different laser devices and coupling methods. The details will be described in Sections 5.4 and 5.5.

5.3
Coupling Schemes and Synchronization Types

To synchronize chaotic temporal waveforms, laser systems need to be coupled to each other. Coupling schemes are very important to consider synchronization of chaos. The coupling schemes can be mainly classified into two types: unidirectional (one-way) and mutual (two-way) couplings. The unidirectional coupling is treated mainly in this section for the purpose of the engineering applications of optical communications. The schemes of unidirectional coupling can be modified to mutual coupling, as seen in Section 5.3.4.

One of the simplest coupling schemes for chaos synchronization is a unidirectional injection from one laser (referred to as a drive laser) to another laser (referred to as a response laser). The output of the drive laser is directly injected into the response laser through an optical isolator to achieve unidirectional coherent coupling. Another scheme is the modification of the transmission signal to the response laser with a feedback signal. The subtraction of the outputs between the drive and response lasers can be used as an injection signal to the response laser (i.e., the Pyragas method in Section 5.2.1.2).

These coupling schemes can be extended to a drive laser with a self-feedback signal. The self-feedback signal in the drive laser is used not only to generate chaos but also to maintain the symmetry of the system between the drive and response lasers, as will be seen in Section 5.3.3. The subtraction for the injection signal in the response laser can be replaced by the addition of the feedback signal when introducing the feedback in the drive laser, which is attainable in experiments.

Another classification of coupling schemes can be introduced: coherent and incoherent couplings. For coherent coupling, the electric field of a drive laser output is directly injected into the laser cavity of a response laser. Both the amplitude and phase (frequency) of the electric field interact with each other between the drive and response lasers, and injection locking (matching of the optical-carrier frequency) plays an important role in the achievement of synchronization of chaos. On the other hand, incoherent coupling indicates that only the intensity of the drive laser output interacts with either the laser intensity or the population inversion of the response laser. The phase and frequency of the fast optical carrier of the electric field do not play a crucial role in incoherent coupling. The condition for synchronization of chaos can be determined, depending of the coupling schemes. The details will be described in Sections 5.4 and 5.5.

5.3.1
Identical Synchronization

A synchronization scheme that fits better the capabilities of optical systems has been proposed, based on the Pyragas method (diffusive coupling) as seen in Section 5.2.1.2 (Pyragas, 1993). The block diagram of the method is presented in Figure 5.9a. One of the drive variables is used as an injection signal, and the response system can be forced by using a small feedback perturbation of the difference between the variables for the drive and response systems. This can be represented mathematically by:

$$\frac{dx_d(t)}{dt} = f(x_d(t), \mathbf{y}_d(t)) \tag{5.1}$$

$$\frac{dx_r(t)}{dt} = f(x_r(t), \mathbf{y}_r(t)) + \kappa_{inj}(x_d(t) - x_r(t)) \tag{5.2}$$

where only the equations of the state variables directly involved in the coupling mechanism are shown. κ_{inj} is the injection (coupling) strength of the perturbation, $x_d(t)$ and $x_r(t)$ are the relevant variables of the drive and response systems, with $\mathbf{y}_d(t)$ and $\mathbf{y}_r(t)$ representing the rest of variables of each system, and f is the nonlinear function that governs the dynamics of $x_d(t)$ and $x_r(t)$.

For identical parameters of the drive and response lasers an identical synchronous solution of Eqs. (5.1) and (5.2) exists in a mathematical sense,

$$x_d(t) - x_r(t) \to 0, \ \mathbf{y}_d(t) - \mathbf{y}_r(t) \to 0 \ (t \to \infty) \tag{5.3}$$

This synchronous state may be stabilized by the coupling term for a sufficiently large strength κ_{inj}. The perturbation has to be introduced into the response system as a negative feedback in terms of $x_r(t)$ (i.e., $-\kappa_{inj}x_r(t)$, $\kappa_{inj} > 0$) to stabilize the synchronous solution of Eq. (5.3). Another important feature of this perturbation is that it vanishes when the drive and response signals coincide, that is, $\kappa_{inj}(x_d(t) - x_r(t)) = 0$ when $x_r(t) = x_d(t)$ (i.e., diffusive coupling). Equations (5.1) and (5.2) thus become symmetric under the condition of identical parameter settings and the synchronous state $x_r(t) = x_d(t)$. This type of synchronization is referred to as

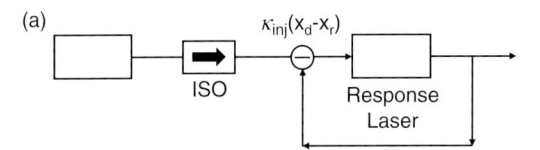

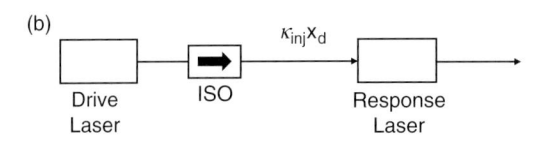

Figure 5.9 Block diagram of the schemes of chaos synchronization. (a) Diffusive coupling and (b) simple injection. κ_{inj} is the injection strength. $x_{d,r}$ are the variables for the drive and response lasers, respectively.

"identical synchronization". The expression of "complete synchronization" is also used for this type of synchronization in the literature.

5.3.2
Generalized Synchronization (with High Correlation)

The diffusive coupling in Eq. (5.2) is very general and applicable to many nonlinear dynamical systems for synchronization. However, when one tries to apply that method to coherently coupled optical systems a problem arises, because it is not easy to make a coherent subtraction between two optical fields (i.e., a subtraction that includes their fast optical phases oscillating at $\sim 10^{14}$ Hz). For that reason, the diffusive coupling is frequently substituted in optical setups by simple injection as shown in Figure 5.9b, which corresponds to replacing Eq. (5.2) by

$$\frac{dx_r(t)}{dt} = f(x_r(t), \mathbf{y}_r(t)) + \kappa_{\text{inj}} x_d(t) \qquad (5.4)$$

In this case the identical synchronous solution $x_r(t) = x_d(t)$ no longer exists since the Eqs. (5.1) and (5.4) are asymmetric under the condition of identical parameter settings and the synchronization state $x_r(t) = x_d(t)$. This problem can be circumvented by introducing an extra (linear) cavity loss in the receiver, which equivalently adds the term $-\kappa_{\text{inj}} x_r(t)$ in Eq. (5.4), so that the injection signal can be compensated by the extra loss (Colet and Roy, 1994; Chen and Liu, 2000). Therefore, a simple injection from the transmitter to the receiver can cause identical synchronization in optical systems under certain conditions, which make it equivalent to the diffusive coupling method.

However, in many experimental implementations (for instance, when using semiconductor lasers as emitting/receiving devices), it is not easy to tune the receiver's cavity loss, hence the simple injection scheme described in Eq. (5.4) has to be used. In addition, it has been known experimentally and numerically that synchronization can still be achieved even in the absence of the identical synchronous solution $x_d(t) = x_r(t)$. This type of synchronization by simple optical injection has been observed in many semiconductor and solid-state lasers (see Sections 5.4 and 5.5), yet the mechanism of synchronization has not been well understood. This type of synchronization is referred to as "generalized synchronization". It has been known that this type of synchronization is much more tolerant to parameter mismatch than identical synchronization (see Section 5.4.1 and Chapter 6).

5.3.3
Synchronization of Chaos in Feedback Systems

5.3.3.1 Open-Loop Configuration
There are other ways of recovering the symmetry between the drive and response equations, so that an identical synchronous solution exists. An example consists in compensating the extra power injected into the response laser by including feedback in the drive laser, as shown in Figure 5.10a. This setup is referred to as an "open-loop configuration" because there is no optical feedback in the response laser, and can be

5.4 [continuation]

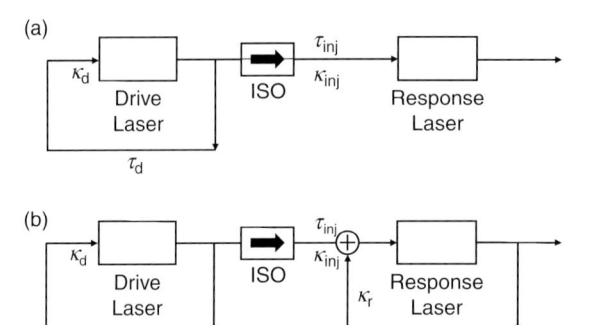

Figure 5.10 Block diagram of the schemes of chaos synchronization with feedback in the drive laser. (a) Open-loop (no feedback) and (b) closed-loop (with feedback) configurations in the response laser. $\kappa_{d,r}$ are the feedback strengths for the drive and response lasers. κ_{inj} is the injection strength. $\tau_{d,r}$ are the feedback delay times for the drive and response lasers. τ_{inj} is the injection delay time.

represented mathematically by:

$$\frac{dx_d(t)}{dt} = f(x_d(t), \mathbf{y}_d(t)) + \kappa_d x_d(t-\tau_d) \tag{5.5}$$

$$\frac{dx_r(t)}{dt} = f(x_r(t), \mathbf{y}_r(t)) + \kappa_{inj} x_d(t-\tau_{inj}) \tag{5.6}$$

where κ_d is the feedback strength of the drive laser, κ_{inj} is the injection (coupling) strength from the drive to response lasers, τ_d is the round-trip time of the feedback loop in the drive laser, and τ_{inj} is the injection (coupling) delay time from the drive to the response laser.

Compared with Eqs. (5.1) and (5.2), a subtraction of the optical field in the response laser is replaced by an addition of the optical field in the drive laser in Eqs. (5.5) and (5.6), which is readily attainable in experiments. In Eqs. (5.5) and (5.6), the delay times associated with the feedback loop of the drive laser and the injection from the drive to response lasers are explicitly introduced, but the results that follow can also be obtained for instantaneous (i.e., zero-delay) feedbacks and coupling.

An identical synchronous solution exists for Eqs. (5.5) and (5.6) since the symmetry of the two equations is obtained. Identical synchronization can be written as follows.

$$x_r(t) = x_d(t) \tag{5.7}$$

where the identical parameter settings for the feedback and coupling terms are required as,

$$\kappa_d = \kappa_{inj}, \quad \tau_d = \tau_{inj} \tag{5.8}$$

The solution Eq. (5.7) has to be stable to observe synchronization of chaos. This type of synchronization corresponds to the identical synchronization described in Section 5.3.1.

For Eqs. (5.7) and (5.8), an identical synchronous solution still exists even if $\tau_d \neq \tau_{inj}$. The identical synchronization with time shift can be written as follows.

$$x_r(t) = x_d(t - \Delta\tau) \tag{5.9}$$

where $\Delta\tau = \tau_{inj} - \tau_d$ is the difference between the injection and feedback delay times. When the feedback delay time of the drive laser is longer than the injection delay time between the two lasers ($\tau_d > \tau_{inj}$ and $\Delta\tau < 0$), the response laser "anticipates" the behavior of the drive laser, which is referred to as anticipative synchronization in the literature (Voss, 2000; Masoller, 2001). In spite of its seemingly counterintuitive character, this type of synchronization is a particular synchronous solution of coupled time-delayed feedback systems at the condition of $\tau_{inj} < \tau_d$. In the opposite case of $\tau_{inj} > \tau_d$, retarded synchronization can be observed with the time delay of $\Delta\tau$ (Tang and Liu, 2003). The anticipative or retarded synchronization can be considered as a case of identical synchronization with time shift (Eq. (5.9)) when $\tau_d \neq \tau_{inj}$ is satisfied in the time-delayed feedback systems.

Similar to the case shown in Figure 5.9b, Eqs. (5.5) and (5.6) can also exhibit the generalized synchronization described in Section 5.3.2. This type of synchronization can be observed under the strong injection strength $\kappa_{inj} > \kappa_d$ and mismatched parameter conditions. Generalized synchronization can be written as follows.

$$x_r(t) \approx a x_d(t - \tau_{inj}) \tag{5.10}$$

This response behavior can be interpreted as a driven waveform, since the time delay between $x_d(t)$ and $x_r(t)$ only depends on the transmission time τ_{inj}, but not τ_d. The internal dynamics of the drive laser just generates the driving waveform and does not enter into the dynamics of the response laser. Therefore, generalized synchronization of Eq. (5.10) can be distinguished from the identical synchronization of Eq. (5.9) by investigating the delay time between $x_d(t)$ and $x_r(t)$ in time-delayed feedback systems: Identical synchronization can be identified by the delay time of $\Delta\tau$ ($= \tau_{inj} - \tau_d$), whereas generalized synchronization can be identified by the delay time of τ_{inj} (Liu et al., 2002b) (see Chapter 6 in detail).

Generalized synchronization is associated with a nonlinear amplification of the drive signal in the response laser *via* injection locking, so the amplitude of the response laser is typically larger than that of the drive laser (i.e., $a \geq 1$). It should be noted that the tolerance to parameter mismatch for generalized synchronization is much larger than that for identical synchronization, therefore, generalized synchronization can be useful for engineering applications of optical chaos communications. It is very important to distinguish between the two types of synchronization shown in Eqs. (5.9) and (5.10) by using the delay time ($\Delta\tau$ or τ_{inj}) for the investigation of synchronization regimes in time-delayed feedback systems (see Section 5.4.1 and Chapter 6).

5.3.3.2 Closed-Loop Configuration
In the open-loop configuration of Figure 5.10a (i.e., Eqs. (5.5) and (5.6)), optical feedback is absent in the response laser. On the other hand, optical feedback can be

introduced in the response laser, which is referred to as the "closed-loop" configuration, as shown in Figure 5.10b. The dynamical equations of Eqs. (5.5) and (5.6) are modified for the closed-loop configuration as,

$$\frac{dx_d(t)}{dt} = f(x_d(t), \mathbf{y}_d(t)) + \kappa_d x_d(t-\tau_d) \tag{5.11}$$

$$\frac{dx_r(t)}{dt} = f(x_r(t), \mathbf{y}_r(t)) + \kappa_r x_r(t-\tau_r) + \kappa_{\text{inj}} x_d(t-\tau_{\text{inj}}) \tag{5.12}$$

where κ_r is the feedback strength of the response laser and τ_r is the round-trip time of the feedback loop in the response laser. An identical synchronous solution exists for the closed-loop configuration as well,

$$x_r(t) = x_d(t) \tag{5.13}$$

where the identical parameter settings for the feedback and coupling terms are required as,

$$\kappa_d = \kappa_r + \kappa_{\text{inj}}, \quad \tau_d = \tau_r = \tau_{\text{inj}} \tag{5.14}$$

Note that an identical synchronous solution with time delay also exists even if $\tau_d \neq \tau_{\text{inj}}$ as,

$$x_r(t) = x_d(t-\Delta\tau) \tag{5.15}$$

where $\Delta\tau = \tau_{\text{inj}} - \tau_d$ and the identical parameter settings are required as,

$$\kappa_d = \kappa_r + \kappa_{\text{inj}}, \quad \tau_d = \tau_r \tag{5.16}$$

Similar to the open-loop configuration, generalized synchronization can be observed in the closed-loop configuration. The open-loop configuration is less sensitive to parameter mismatch than the closed-loop one, since the open-loop does not require feedback in the response laser, and hence no tuning of feedback parameters in the response laser is needed (Vicente et al., 2002). The transient time for synchronization in the closed-loop configuration has been investigated and is much longer than that in the open-loop configuration due to the existence of the feedback in the response laser (Uchida et al., 2004b).

5.3.4
Mutual Coupling

In Sections 5.3.1–5.3.3, the situation is considered in which the coupling signal travels unidirectionally from the drive laser to the response laser. Such a one-way (unidirectional) coupling is well suited for communication applications, which connect two distantly separated chaotic lasers with well-defined drive and response roles, maintaining the original chaotic temporal waveform generated by the transmitter free from the influence of the receiver's dynamics.

There are situations, however, where the coupling signal travels bidirectionally between the two lasers. This type of "mutual coupling" can be implemented *ad hoc* in face-to-face coupled lasers (Heil et al., 2001a), but it also arises naturally in laser arrays, where each laser is coupled with its neighbors *via* the overlap between the evanescent tails of their respective electric fields (Winful and Rahman, 1990). Hence, the coupling is inherently mutual. The case of mutual coupling can be modeled as follows:

$$\frac{dx_d(t)}{dt} = f(x_d(t), \mathbf{y}_d(t)) + \kappa_{inj1} x_r(t) \tag{5.17}$$

$$\frac{dx_r(t)}{dt} = f(x_r(t), \mathbf{y}_r(t)) + \kappa_{inj2} x_d(t) \tag{5.18}$$

where κ_{inj1} is the injection (coupling) strength from the response to drive laser, and κ_{inj2} is the injection strength from the drive to response laser. One-way coupling is the case when $\kappa_{inj1} = 0$ and $\kappa_{inj2} \neq 0$, whereas mutual coupling is the case when $\kappa_{inj1} \neq 0$ and $\kappa_{inj2} \neq 0$. In most optical systems, coupling can be achieved through the electric field (or laser intensity) because it is easy to transmit the electric field between separate laser systems.

The comparison of the characteristics between one-way and mutual couplings has been investigated (Uchida *et al.*, 1998a; Klein *et al.*, 2006b). The coupling strength required for synchronization in the mutual-coupling configuration is smaller than that in the one-way coupling configuration. The effect of mutual coupling sometimes stabilizes chaotic oscillations, which is known as "amplitude death" (Herrero *et al.*, 2000; Kuntsevich and Pisarchik, 2001).

5.3.5
Linear Stability Analysis and Conditional Lyapunov Exponent

In Sections 5.3.1–5.3.3, the existence of an identical synchronous solution is discussed. To observe synchronization, the synchronous solution must be stable. Linear stability analysis of the synchronous solution is required to determine the stability of synchronization of chaos.

The detailed procedures of the linear stability analysis for synchronization are summarized in Appendices 5.A.2 and 5.A.3. Linearized equations can be derived from the coupled original equations (see Appendix 5.A.2 for the general formula of linearized equations). The linearized equation along the synchronous solution is obtained as a new set of equations. A set of linearized equations can be written as a matrix formula with the Jacobian matrix.

To analyze the stability of synchronization, the eigenvalues of the Jacobian matrix for the linearized equations need to be estimated. The sign of the real part of the eigenvalues determines the stability, that is, the synchronous solution is stable when all the signs of the real parts of the eigenvalues are negative, while the synchronization becomes unstable when the sign of the real part of the maximum eigenvalue is positive. It is, however, not easy to perform direct calculation of the eigenvalues because time-dependent variables are included in the Jacobian matrix. Instead of direct calculation of the eigenvalues of the Jacobian matrix, a procedure of numerical calculation of the conditional Lyapunov exponent is introduced (see Appendix 5.A.3).

The "conditional" Lyapunov exponent (also known as "transverse" Lyapunov exponent) is a measure of the stability of an identical synchronous solution in coupled dynamical systems. The conditional Lyapunov exponent indicates the degree of divergence (or convegence) of the distance between two synchronized variables of the coupled dynamical systems on the time series. When the maximum conditional Lyapunov exponent is negative, the variables of the two coupled systems start to converge in time and synchronization is achieved. On the other hand, when the maximum conditional Lyapunov exponent is positive, the variables of the two coupled systems start to diverge in time and synchronization is not obtained (see examples in Sections 6.3.4.3 and 6.6.3.3).

For comparison, *normal* Lyapunov exponent shown in Chapter 4 indicates the degree of divergence (or convergence) of the distance between two nearby points on the time series (i.e., on the attractor). When the maximum Lyapunov exponent is positive, the two nearby points start to diverge in time and chaotic temporal dynamics is observed. When the maximum Lyapunov exponent is negative, the two points start to converge and stable (or periodic) temporal dynamics is obtained.

The conditional Lyapunov exponent λ_c represents "synchronizability" of two coupled dynamical systems, whereas the (normal) Lyapunov exponent λ indicates "chaoticity" of a temporal waveform. Therefore, identical synchronization of chaos can be achieved when satisfying the conditions of $\lambda > 0$ (chaos) and $\lambda_c < 0$ (stable synchronization).

In the presence of noise, the stable synchronous solution may become destroyed, which is known as "bubbling" (Flunkert *et al.*, 2009). The identical solution can be stable, but there may exist one unstable periodic orbit in the synchronization manifold that is transversely unstable. Then, noise can kick the trajectory out of the stable synchronous solution at this point and it ends up with intermittent escapes from the synchronization manifold.

Examples of the derivation of the linearized equations and the calculation of the conditional Lyapunov exponent in coupled semiconductor lasers with optical feedback will be discussed in detail in Sections 6.3 and 6.6.

5.4
Examples of Synchronization of Chaos in Semiconductor Lasers

Several synchronization schemes are described in Section 5.4 for coupled semiconductor lasers that implement some of the different ways to drive lasers into chaos, namely, coherent optical feedback, polarization-rotated optical feedback, optoelectronic feedback, optical injection, and mutual coupling. The detail description of these schemes is presented for synchronization of chaos in coupled semiconductor lasers, because these are the subject of much research. Synchronization of chaos in electro-optic systems and other lasers will be reviewed in Section 5.5.

First, a quantitative measure of the degree of synchronization of chaos is introduced by using the cross-correlation of two temporal waveforms of the drive and response lasers normalized by the product of their standard deviations as,

$$C = \frac{\langle (I_{d,i}-\bar{I}_d)(I_{r,i}-\bar{I}_r) \rangle}{\sigma_d \sigma_r} \tag{5.19}$$

where $I_{d,i}$ and $I_{r,i}$ are the ith sampled point of the temporal waveforms of the drive and response laser intensities. The subscripts d and r indicate the drive and response lasers, respectively. $\bar{I}_d$ and $\bar{I}_r$ are the mean values of $I_{d,i}$ and $I_{r,i}$ defined as,

$$\bar{I}_d = \frac{1}{N}\sum_{i=1}^{N} I_{d,i} \tag{5.20}$$

$$\bar{I}_r = \frac{1}{N}\sum_{i=1}^{N} I_{r,i} \tag{5.21}$$

where N is the total number of points sampled from the temporal waveforms. σ_d and σ_r are the standard deviations of $I_{d,i}$ and $I_{r,i}$,

$$\sigma_d = \sqrt{\frac{1}{N}\sum_{i=1}^{N}(I_{d,i}-\bar{I}_d)^2} \tag{5.22}$$

$$\sigma_r = \sqrt{\frac{1}{N}\sum_{i=1}^{N}(I_{r,i}-\bar{I}_r)^2} \tag{5.23}$$

The angle brackets denote time averaging,

$$\langle (I_{d,i}-\bar{I}_d)(I_{r,i}-\bar{I}_r) \rangle = \frac{1}{N}\sum_{i=1}^{N}\left[(I_{d,i}-\bar{I}_d)(I_{r,i}-\bar{I}_r)\right] \tag{5.24}$$

The value of C in Eq. (5.19) ranges from -1 to 1. The best inphase synchronization of chaos is obtained at the cross-correlation value of $C = 1$. The best antiphase synchronization of chaos is obtained at $C = -1$. No synchronization is observed at $C = 0$. The value of C indicates the quantitative measure of synchronization of chaos. This synchronization measure C is used in the following sections.

5.4.1
Semiconductor Lasers with Coherent Optical Feedback

One of the most well-known synchronization schemes is unidirectional coupling in semiconductor lasers with time-delayed optical feedback. In the system depicted in Figure 5.11, chaos is induced by delayed, coherent optical feedback from a mirror in the drive laser. The response laser, which is unidirectionally coupled to the drive laser through coherent optical injection, may or may not have an external mirror for optical feedback, which corresponds to a closed- or open-loop configuration, respectively. A numerical model for chaos synchronization in unidirectionally coupled semiconductor lasers with optical feedback is described in Appendix 5.A.4.1 and Section 6.2.1.

Both the identical synchronization and the generalized synchronization described in Sections 5.3.1–5.3.3 have been reported in this configuration. One of the first

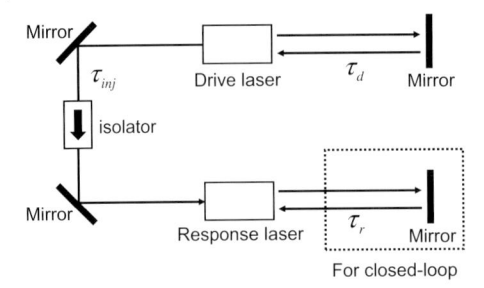

Figure 5.11 Schematics of a unidirectional coupling scheme for chaos synchronization in semiconductor lasers with time-delayed optical feedback. $\tau_{d,r}$ are the feedback delay times for the drive and response lasers. τ_{inj} is the injection delay time.

numerical simulations of chaos synchronization in this scheme corresponds to generalized synchronization (Mirasso et al., 1996). The temporal dynamics of the response output intensity can mimic the dynamics of the drive laser intensity with an amplification ratio after a delay equal to the coupling delay time τ_{inj} (see Eq. (5.10)). By contrast, identical synchronization was theoretically predicted under the parameter-matching conditions (Ahlers et al., 1998). No synchronization error is found for identical synchronization under the parameter-matching condition without noise.

Figure 5.12 shows numerical examples of the temporal waveforms of the drive and response laser intensities and their correlation plots for the comparison between identical and generalized synchronization at the condition of the same delay and injection times ($\tau_d = \tau_{inj}$). For Figures 5.12a and b, identical synchronization with the same amplitude and oscillation phase is observed. The correlation value is 1.0, indicating identical synchronization of chaos. By contrast, the quality of generalized synchronization is not as good as identical synchronization, where the correlation value is 0.951 for optimal synchronization, as shown in Figures 5.12c and d. The amplitude of the response laser intensity is slightly larger than that of the drive-laser intensity, as seen in the correlation plot of Figure 5.12d. In addition, there is a time delay between the drive and response laser intensities by the injection delay time τ_{inj}, as seen in Figure 5.12c. The investigation of the delay time of the temporal waveforms between the drive- and response-laser intensities can distinguish between identical (no time delay, $\Delta\tau = \tau_{inj} - \tau_d = 0$) and generalized synchronization (the time delay of τ_{inj}), as seen in Eqs. (5.9) and (5.10).

To investigate the parameter regions for the two types of synchronization, the two-dimensional (2D) map of the synchronization error is created as functions of the optical frequency detuning and the injection strength between the two lasers, as shown in Figure 5.13 (Murakami and Ohtsubo, 2002). Generalized synchronization can be obtained (shown as the amplification area in Figure 5.13) in wide ranges of large injection strengths and negative frequency detunings $\Delta f = f_d - f_r < 0$, and for rather large mismatches between the drive and response parameters. Generalized synchronization is typically observed in parameter ranges where injection locking is achieved, as shown in Figure 5.13. Generalized synchronization is associated with a nonlinear amplification of the drive signal in the response laser *via* injection locking.

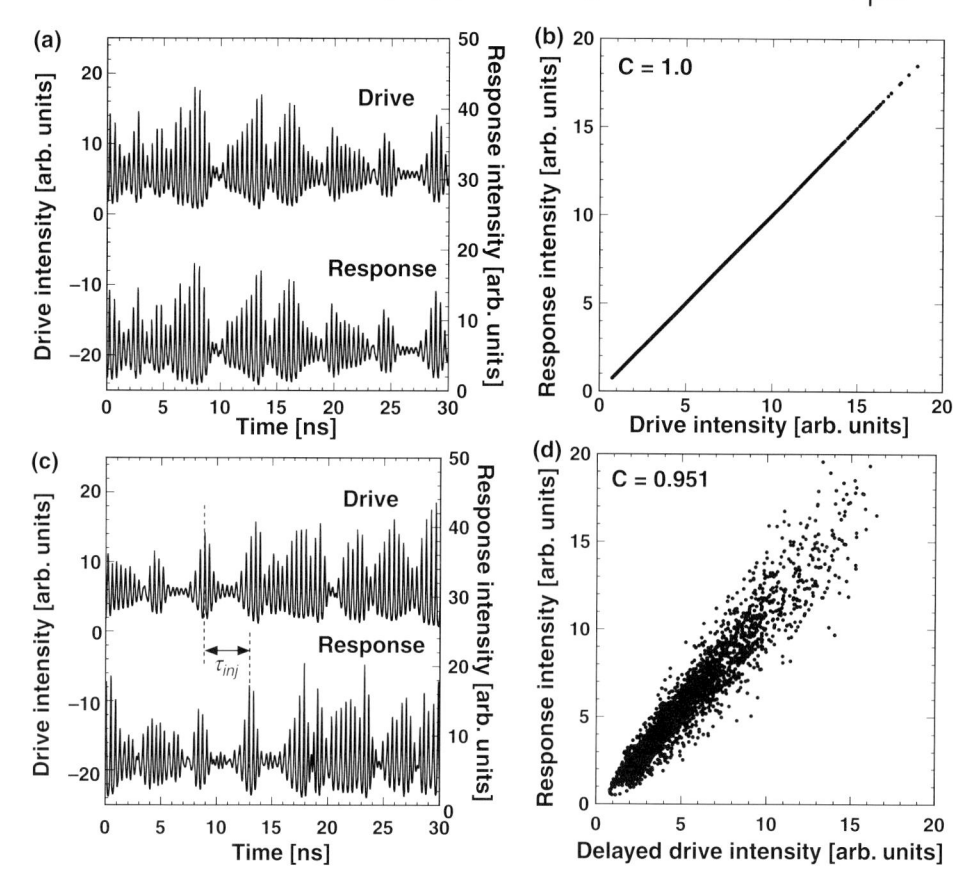

Figure 5.12 Numerical examples of (a), (c) temporal waveforms of the laser intensities and (b), (d) corresponding correlation plots in unidirectionally coupled semiconductor lasers with time-delayed optical feedback for the two types of synchronization. (a), (b) Identical synchronization with the correlation value of 1.0. (c), (d) Restricted type of generalized synchronization with the correlation value of 0.951. The chaotic oscillation of the response laser intensity is delayed to that of the drive laser intensity by the injection delay time of $\tau_{inj} = 4.0$ ns in (c). The drive waveform is shifted in time to obtain optimal synchronization in (d).

By contrast, identical synchronization is obtained (shown as complete synchronization area in Figure 5.13) in the area of the parameter-matching condition, where $\Delta f = 0$ GHz and $R_{inj} = 1.5 \times 10^{-2}$%. The optical frequency detuning needs to be zero (or close to zero) for identical synchronization.

In experiments on chaos synchronization in semiconductor lasers with optical feedback, generalized synchronization has been reported on millisecond time scales in a closed-loop configuration (Sivaprakasam and Shore, 1999a) and in the low-frequency fluctuation regime (Takiguchi et al., 1999). The synchronization of chaos on nanosecond time scales has been reported in the open-loop (Fischer et al., 2000) and closed-loop (Fujino and Ohtsubo, 2000) configurations. Figure 5.14 show synchronized output intensity time series of both drive laser (transmitter) and

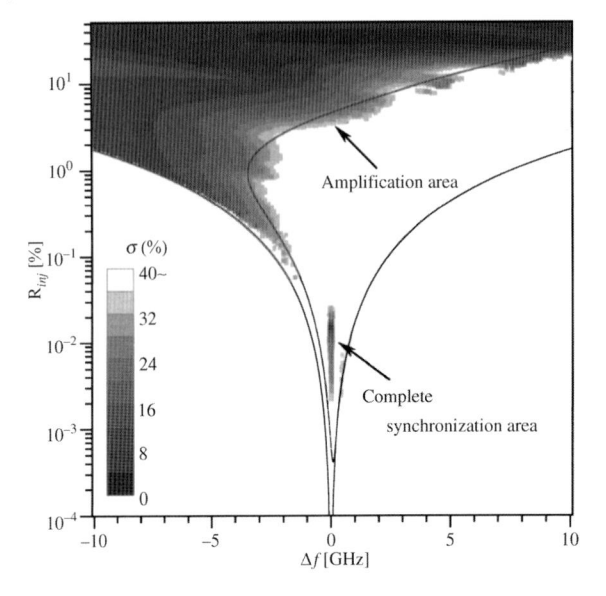

Figure 5.13 Calculated synchronization error for variation in the frequency detuning $\Delta f (= f_d - f_r)$ and the injection rate R_{inj}. Quality of synchronization is represented by using a grayscale with synchronization error (i.e., black corresponds to good synchronization). The boundary denoted by the solid line represents an injection-locking range for constant-intensity injection into the receiver laser. Coincidence between the area for generalized synchronization and the injection-locking range is found. (A. Murakami and J. Ohtsubo, (2002), © 2002 APS.)

response laser (receiver), and the corresponding correlation plot obtained from the experiment in the open-loop configuration (Fischer et al., 2000). Good synchronization is achieved between the two unidirectionally coupled semiconductor lasers, as seen in the temporal waveforms of Figure 5.14 (left) and the correlation plot of

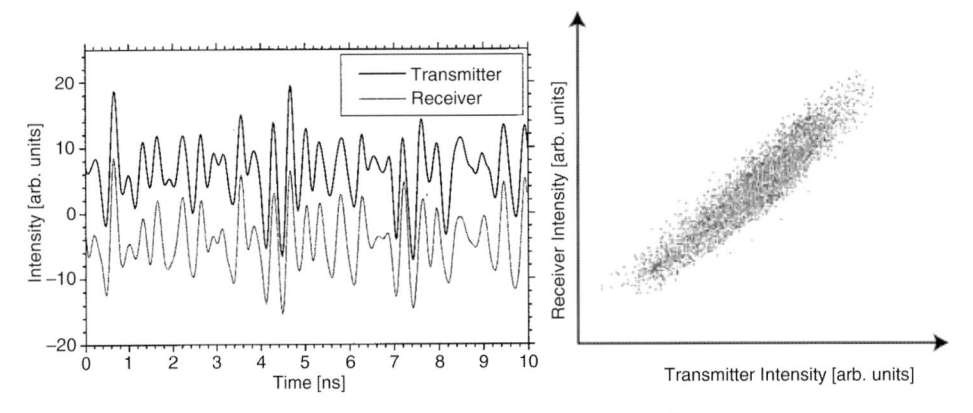

Figure 5.14 (Left) Synchronized output intensity time series of transmitter (upper trace) and receiver lasers (lower trace). (Right) Correlation plot of transmitter and receiver intensity. (I. Fischer, Y. Liu, and P. Davis, (2000), © 2000 APS.)

Figure 5.14 (right). Synchronization of chaos in multimode semiconductor lasers with optical feedback was also reported experimentally (Uchida et al., 2001a) and numerically (Buldú et al., 2004).

For the closed-loop configuration, the relative optical feedback phase strongly determines the quality of synchronization (Peil et al., 2002). The relative optical feedback phase $\Phi_{rel} - \Phi_d - \Phi_r$ is gradually varied by changing the phase of the feedback light between the drive and response lasers from 0 to 2π, where Φ_{rel} are the phase difference of the feedback light between the drive and response lasers in the closed-loop configuration. Figure 5.15 (left) shows the cross-correlation coefficient of the intensity time series of the two unidirectionally coupled semiconductor lasers in the closed-loop configuration in dependence on Φ_{rel}. High correlation coefficients are obtained for the adjusted phase of $\Phi_{rel} \approx 0$, where the regime of chaos synchronization is located. On gradually increasing Φ_{rel}, the correlation coefficients are slowly decreasing. The minimal correlation of the intensity time series is obtained around $\pi \sim 1.2\pi$, and a steep increase for the correlation coefficients on further increasing Φ_{rel} to 2π regaining chaos synchronization. The experimental data (black squares) agrees well with the numerically calculated data (triangles) in Figure 5.15 (left).

Identical synchronization has been observed experimentally in semiconductor lasers with optical feedback (Liu et al., 2002b). It was found that the time lag between the drive and response waveforms depends on the feedback delay time of the drive laser to distinguish identical synchronization from generalized synchronization (see Eqs. (5.9) and (5.10)). Figure 5.15 (right) shows the experimental result of the variation of the time lag $\Delta\tau$ between the drive and response laser intensities measured by changing the feedback delay time of the drive laser τ_d (Liu et al., 2002b). The relationship $\Delta\tau = \tau_{inj} - \tau_d$ can be found from Figure 5.15 (right).

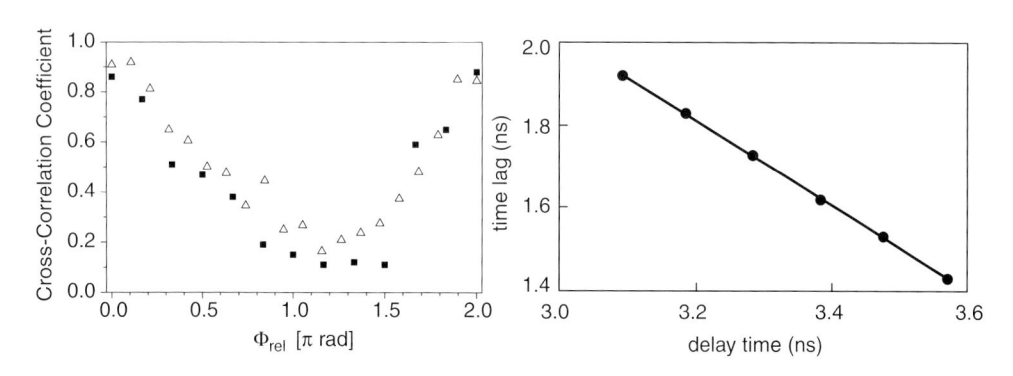

Figure 5.15 (Left) Cross-correlation coefficients of the intensity time series of the drive and response lasers versus the relative optical feedback phase $\Phi_{rel} = \Phi_d - \Phi_r$ in the closed-loop configuration. The squares depict the values obtained from experimental data, the triangles show the results for numerically obtained time series with corresponding parameters. (M. Peil, T. Heil, I. Fischer and W. Elsäßer, (2002), © 2002 APS.) (Right) Experimentally measured time lag as a function of time delay. Identical synchronization of chaos is confirmed in experiment. (Y. Liu, Y. Takiguchi, P. Davis, S. Saito, and J.M. Liu, (2002), © 2002 AIP.)

indicating an experimental evidence of identical synchronization, where $\tau_{inj} = 5.0$ ns is the injection delay time (see Eq. (5.9)). Identical synchronization requires precise matching of all the laser parameters between the drive and response lasers, including the fast optical-carrier frequency. Identical synchronization is strongly sensitive to noise and frequency detuning between the lasers.

The details of the characteristics of the identical and generalized synchronization will be described experimentally and numerically in Chapter 6. The comparison between the open- and closed-loop configurations will also be discussed in Chapter 6.

5.4.2
Semiconductor Lasers with Polarization-Rotated Optical Feedback

A scheme for synchronization of chaos in a semiconductor laser with polarization-rotated optical feedback has been proposed, as shown in Figure 5.16 (left) (Rogister et al., 2001). In that setup, the linearly polarized output field of the transmitter (drive) laser first undergoes a $\pi/2$ polarization rotation through an external cavity formed by a Faraday rotator (FR) and a mirror. It is then split by a nonpolarizing beam splitter (BS) into two parts: one is fed back into the transmitter laser and the other is injected into the receiver (response) laser. Polarization directions of feedback and injection fields are orthogonal to those of the transmitter and receiver output fields, respectively. Therefore, the transmitter laser is subject to polarization-rotated optical feedback while the receiver laser is subjected to polarization-rotated optical injection. A linear polarizer (LP) is placed between the Faraday rotator and the mirror to prevent coherent feedback induced by a second round-trip in the external cavity after reflection on the transmitter laser front facet. The receiver is necessarily open loop,

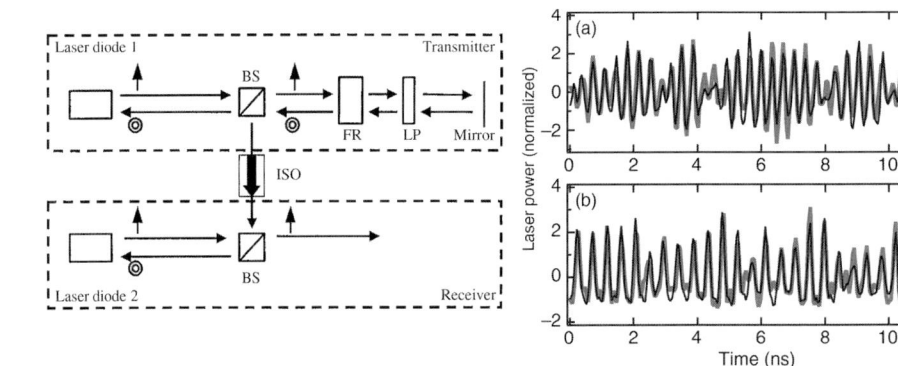

Figure 5.16 (Left) Synchronization setup with laser diodes subject to polarization-rotated optical feedback and injection. (F. Register, A. Locquet, D. Pieroux, M. Sciamanna, O. Deparis, P. Mégret and M. Blondel, (2001), © 2001 OSA.) (Right) Synchronized experimental laser intensity time series of transmitter (thick gray curve) and receiver (thin black curve). Time axes are shifted by 4.95 ns to illustrate optimal synchronization. (a) Weak and (b) strong feedback strengths of the transmitter. (D.W. Sukow, K.L. Blackburn, A.R. Spain, K.J. Babcock, J.V. Bennett, and A. Gavrielides, (2004), © 2004 OSA.)

because a feedback at the receiver would induce a beating with the coupling field. Numerical simulations for semiconductor lasers subject to polarization-rotated optical feedback predict identical synchronization between the two lasers (Rogister et al., 2001). A dynamical model that takes into account both modes of polarization is required to describe the synchronization characteristics including generalized synchronization in coupled semiconductor lasers with polarization-rotated optical feedback (Shibasaki et al., 2006; Takeuchi et al., 2008). A numerical model for chaos synchronization in unidirectionally coupled semiconductor lasers with polarization-rotated optical feedback is described in Appendix 5.A.4.2.

In an experiment with a similar setup it was shown that generalized synchronization occurs with a time shift close to the coupling delay time between the transmitter and receiver outputs (Sukow et al., 2004). Experimental time-series data are shown in Figure 5.16 (right). Each graph shows two traces of laser intensity *versus* time; thick gray curves and thin black curves represent the transmitter and the receiver intensities, respectively. For ease of visual comparison the vertical and horizontal axes are scaled and shifted for each pair of time-series data. In particular, the time axes between the transmitter and receiver lasers are shifted such that synchronized behavior between the lasers appears simultaneously on the graphs, accounting for the coupling delay time in the system. With these scalings, synchronization is apparent. Figure 5.16a (right) shows the case in which weak feedback and injection strengths are used. For Figure 5.16b (right) the feedback strength is increased but the injection strength is held at the same level. Both feedback levels place the transmitter well into the chaotic regime and high-quality synchronization of chaos is obtained.

5.4.3
Semiconductor Lasers with Optoelectronic Feedback

Another situation for synchronization is considered in which the feedback and injected fields act on the carrier density in the diode active layers, but do not interact with the intracavity lasing fields. As a consequence, the phases of the feedback and injection fields do not intervene on the laser dynamics. For that reason, these incoherent feedback schemes require no fine tuning of the optical frequencies.

A way of obtaining an incoherent feedback is the introduction of an optoelectronic feedback *via* the injection current of the laser. An example is shown in the experimental setup displayed in Figure 5.17 (top) (Tang and Liu, 2001b). The optical power emitted by the single-mode distributed-feedback (DFB) transmitter (drive) laser is detected by a photodiode and converted into an electric current, which is in turn reinjected into the laser through the injection current. This optoelectronic feedback drives the laser to chaos, provided that the feedback delay is carefully adjusted. The transmitter laser has an optoelectronic feedback loop that drives the laser into chaotic pulsing at certain delay times. At the receiver (response) DFB laser, the coupled signal from the transmitter, together with the optoelectronic feedback signal from the receiver, is used to drive and synchronize the receiver laser. Two identical chaotic systems can be synchronized when they are driven by the same

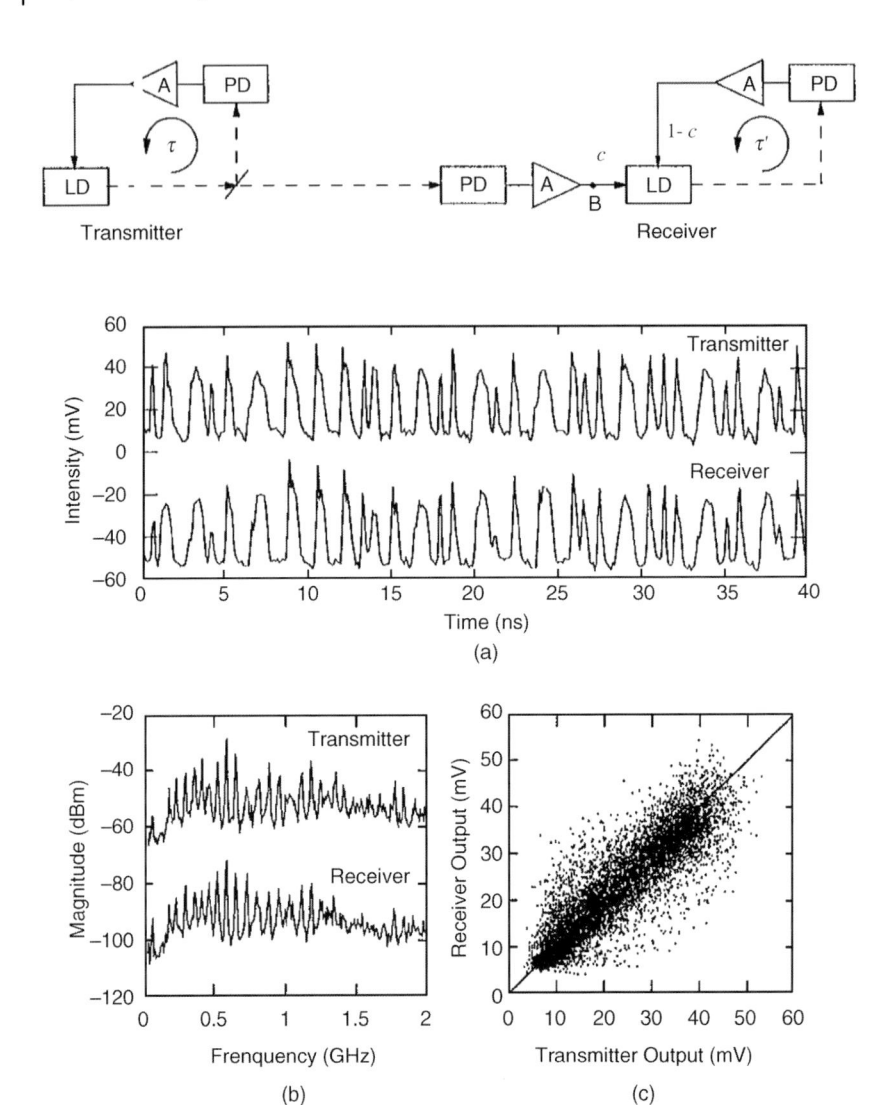

Figure 5.17 (Top) Schematic of the experimental setup for the synchronization of chaotic pulsing in two semiconductor lasers with delayed optoelectronic feedback. LDs, laser diodes; PDs, photodetectors; As, amplifiers. (Bottom) Experimental results of synchronization of chaos in two semiconductor lasers with delayed optoelectronic feedback when the transmitter signal constitutes 80% of the driving signal. (a) Time series of transmitter and receiver. (b) Power spectra of transmitter and receiver. (c) Correlation diagram of the transmitter output *versus* the receiver output. (S. Tang and J.M. Liu, (2001b), © 2001 OSA.)

chaotic drive signal. Therefore, the total effect of both driving and feedback signals to the receiver laser is required to be equivalent to the feedback signal in the transmitter laser in the closed-loop configuration. A numerical model for chaos synchronization in unidirectionally coupled semiconductor lasers with optoelectronic feedback is described in Appendix 5.A.4.3.

Identical synchronization can be achieved when the strength of the driving signal to the receiver (composed of transmitted signal and self-feedback) is close to the feedback strength of the transmitter (see Eq. (5.14)). Figure 5.17 (bottom) shows the experimental results of the synchronization of high-frequency chaotic pulsing between the transmitter and receiver when the transmitter signal constitutes 80% of the driving signal (Tang and Liu, 2001b). The relaxation oscillation frequencies of the two DFB lasers are both ~ 2.5 GHz. The time series are compared in Figure 5.17a, and the radio-frequency (RF) spectra are compared in Figure 5.17b. The time series of the transmitter and the receiver are almost identical, and so are their RF spectra. The correlation diagram of the transmitter and the receiver is shown in Figure 5.17c, which is obtained by plotting the output intensity waveform of the transmitter *versus* that of the receiver. The correlation diagram shows a distribution along the 45-degree diagonal, indicating that the two chaotic lasers are synchronized.

The synchronization quality is investigated to increase with the fraction of the receiver driving signal coming from the transmitter, and to be optimal when the receiver operates in the open-loop configuration. Anticipative and retarded synchronization have also been observed by changing the delay time of the feedback loop in the transmitter (see Eq. (5.9) and Section 5.3.3.1) (Tang and Liu, 2003).

5.4.4
Semiconductor Lasers with Optical Injection

Optical injection from one semiconductor laser to another semiconductor laser induces chaotic fluctuations of laser intensity. Synchronization can be achieved in coupled semiconductor lasers with optical injection. Figure 5.18 (top) shows the synchronization scheme for optically injected semiconductor lasers (Annovazzi-Lodi *et al.*, 1996). The synchronization scheme is based on the subtraction of the optical field as in Section 5.2.1.2 (Pyragas, 1993). The transmitter (drive) and receiver (response) semiconductor lasers, which are ideally identical, are optically injected and driven to chaos by external isolated lasers. The output fields of the transmitter and receiver, namely E_1 and E_2 in Figure 5.18 (top) are combined with a relative π phase shift and injected into the receiver semiconductor laser. This input acts as a control signal: it vanishes when the transmitter and receiver synchronize. By contrast, it contributes to resynchronize the lasers when a small deviation between their respective dynamics occurs. It should be noted that the coupling delay time in the receiver feedback loop must be small with respect of the inverse of the chaotic waveform bandwidth and that the relative phase difference between the input fields E_1 and E_2 must not differ sensibly from π to allow synchronization (Annovazzi-Lodi *et al.*, 1996).

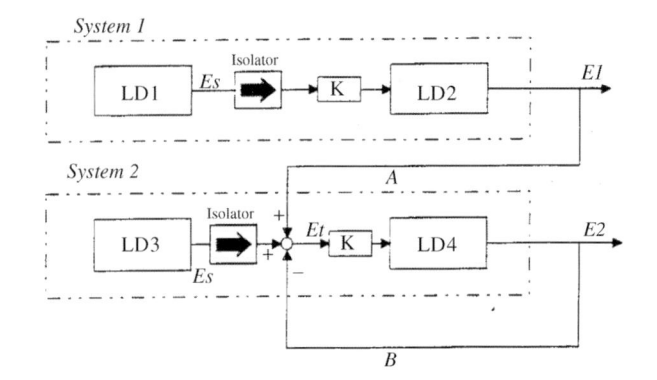

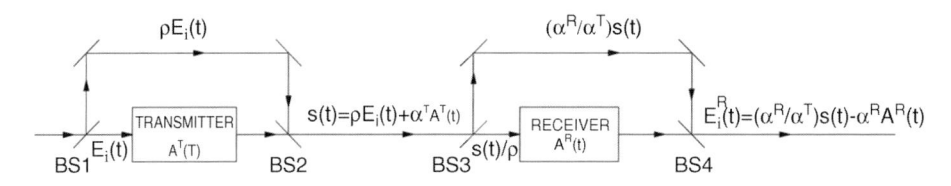

Figure 5.18 (Top) Synchronization scheme for two optically injected semiconductor lasers. (V. Annovazzi-Lodi, S. Donati, and A. Scirè, (1996), © 1996 IEEE.) (Bottom) Open-loop synchronization scheme for two optically injected semiconductor lasers. (H.F. Chen and J.M. Liu, (2000), © 2000 IEEE.)

It is not easy to implement the coherent subtraction of the fast optical fields described above. Another scheme based on optical injection is proposed and depicted in Figure 5.18 (bottom) (Chen and Liu, 2000). In this scheme, an optical field is split into two parts. The first part drives the transmitter laser to chaos. The second part is amplified or attenuated and added to the output of the transmitter laser. The sum is injected into the receiver laser that synchronizes to the transmitter, provided that both laser parameters are identical. Indeed, the photon decay rate of the receiver laser needs to be adjusted in order to compensate a photon excess since more photons are injected in the receiver laser than in the transmitter. This adjustment can be done by changing the coating on the laser facets. This scheme has been experimentally investigated and different regimes of synchronization have been characterized (Chen and Liu, 2004).

5.4.5
Semiconductor Lasers with Mutual Coupling

5.4.5.1 Symmetry Breaking and Leader–Laggard Relationship

Chaotic intensity fluctuations and synchronization can be observed when semiconductor lasers are mutually coupled. Synchronization in mutually coupled semiconductor lasers with time delay has been investigated intensively. One of the first

5.4 Examples of Synchronization of Chaos in Semiconductor Lasers

numerical observations of chaos synchronization was reported in mutually coupled laser arrays, as shown in Section 5.2.2 (Winful and Rahman, 1990). For experiments, generalized synchronization with chaotic oscillations was observed in mutually coupled semiconductor lasers in a low-frequency fluctuation regime (Heil et al., 2001a). In this experiment, mutual coupling induces chaotic oscillations on a nanosecond time scale. A symmetry-breaking and leader–laggard relationship are observed by the time lag, corresponding to the coupling delay time between the two lasers.

A scheme of the experimental setup for mutually coupled semiconductor lasers is depicted in Figure 5.19 (left) (Heil et al., 2001a). The system consists of two uncoated Fabry–Perot semiconductor lasers. Guaranteed by the polarizer (Pol.), the lasers are coherently coupled *via* the dominant TE component of the optical field. The neutral density filter (NDF) controls the mutual coupling strength. The coupling delay time τ is determined by the propagation of the light between the lasers and has been varied between 3.8 and 5 ns. Thus, τ is significantly larger than the internal time scales of the semiconductor lasers, that is, the relaxation oscillations, or the laser cavity round-trip resonance. On pumping both lasers at their solitary thresholds, coupling-induced instabilities are observed.

Figure 5.19 (right) shows a snapshot of the intensity time series of both of the mutually coupled semiconductor lasers; the lower trace is plotted inverted to ease the comparison between the two signals. The coupling-induced instabilities are characterized by fast intensity fluctuations on a subnanosecond time scale combined with pronounced intensity dropouts occurring on much slower time scales. This dynamical behavior closely resembles the low-frequency fluctuations (LFF) phenomenon of a single semiconductor laser subject to delayed optical feedback (see Section 3.2.1.4). In addition, Figure 5.19 (right) exhibits another remarkable dynamical phenomenon. The low-frequency intensity dropouts always occur strongly correlated in both lasers, however, with a constant time lag between the two signals. This delay between the leading laser (leader) and the lagging laser (laggard) that exactly

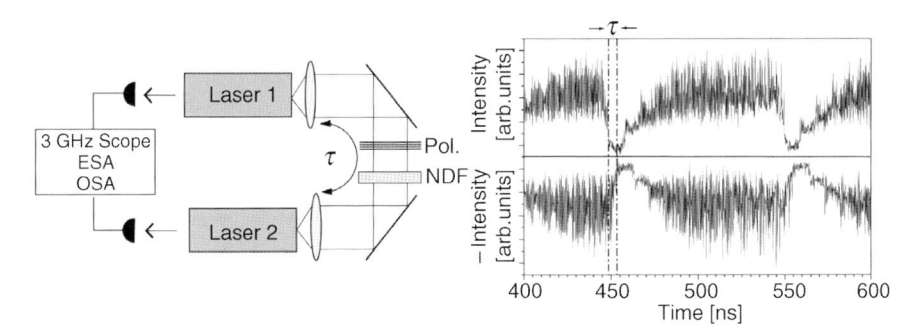

Figure 5.19 (Left) Scheme of the experimental setup for two mutually coupled semiconductor lasers. (Right) Intensity time series of the two mutually coupled semiconductor lasers. The lower trace shows the inverted time series. The coupling delay time is $\tau = 4.75$ ns. (T. Heil, I. Fischer, W. Elsäßer, J. Mulet, and C.R. Mirasso, (2001a), © 2001 APS.)

corresponds to the coupling delay time τ is the manifestation of a spontaneous symmetry-breaking in the system (i.e., generalized synchronization with delay time τ). By correlation analysis of a large number of pairs of time series, it is confirmed that the synchronization and the delay are robust against reasonable variations of the injection current, the coupling strength, and the coupling time; the delay between leader and laggard remains equal to the coupling time τ. A leader–laggard-type conjunction of spontaneous symmetry-breaking with chaos synchronization is observed, which is linked to the delay present in the coupling. From numerical simulations, it is concluded that the zero-lag synchronous solution is unstable, and spontaneous emission prevents the system from operating in such a state (Heil et al., 2001a; D'Huys et al., 2010).

5.4.5.2 Zero-Lag Synchronization

As seen in the above-described example, it is difficult to observe zero-lag synchronization in mutually coupled semiconductor lasers with time delay. However, stable zero-lag (identical) synchronization can be observed in which a third laser mediates the other two lasers in three coupled laser systems (Fischer et al., 2006). The experimental system consists of three similar lasers coupled bidirectionally along a line, in such a way that the central laser acts as a relay of the dynamics between the outer lasers. A central laser diode (LD2) is bidirectionally coupled to two outer lasers (LD1) and (LD3) by mutual injection. The central laser, which does not need to be carefully matched to the other two, mediates their dynamics *via* their lasing TE-polarized fields.

All the mutually coupled lasers exhibit chaotic outputs. Remarkably, both outer lasers synchronize with zero lag, while the central laser either leads or lags the outer lasers. Figure 5.20 shows the time series of the output intensities (left column), in pairs, and the corresponding cross-correlation functions $C_{ij}(\Delta t)$ (right column) for two of the three lasers. Zero-lag synchronization between the intensities of the outer lasers (LD1 and LD3) can be clearly seen in Figure 5.20a, and also manifests itself in the cross-correlation function shown in Figure 5.20d, which presents an absolute maximum of 0.86 at $\Delta t_{max} = 0$ (i.e., at zero delay). The correlation between the central laser and the outer ones (Figure 5.20b and c) is not as high, and exhibits a nonzero time lag, as can be seen from the cross-correlation functions shown in Figures 5.20e and f, which yield maxima of 0.56 and 0.59, respectively, placed at $\Delta t_{max} = -3.65$ ns. This lag coincides with the coupling delay time τ between the lasers. The fact that Δt_{max} is negative means that the central laser dynamically lags the two outer lasers. Therefore, it can be excluded that the outer lasers are simply driven by the central one. This zero-lag synchronization solution is quite robust against spectral detuning of the lasers. Note that the synchronization also remains robust for positive detuning, however, then the central laser leads the dynamics.

Zero-lag synchronization has also been observed in mutually coupled semiconductor lasers with the optical feedback in the closed-loop configuration (Klein et al., 2006a), by delayed-shared feedback coupling (Peil et al., 2007), and with two mutual coupling delay times with certain allowed integer ratios (Englert et al., 2010).

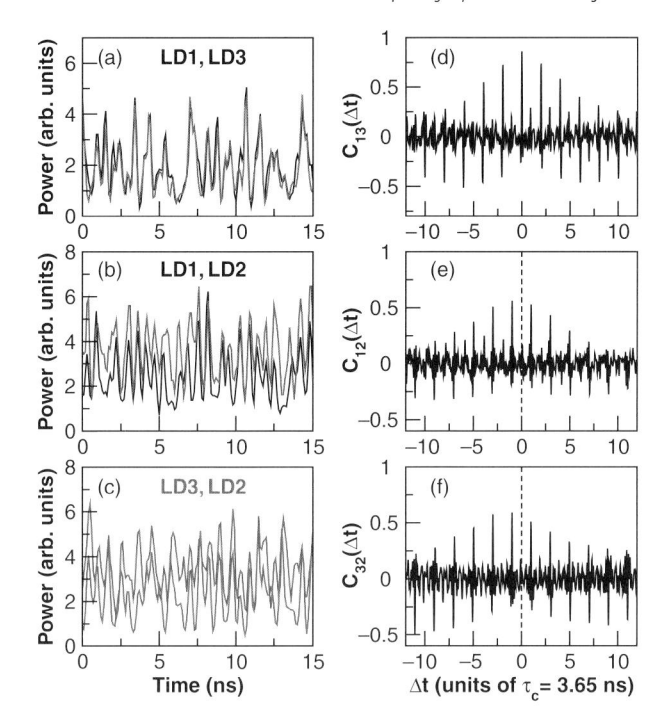

Figure 5.20 (a), (b), (c) Time series (in pairs) of the output intensity of the lasers, for the case of a central laser with negative detuning $\Delta f = f_2 - f_{1,3} = -4.1$ GHz. (d), (e), (f) Cross-correlation functions of the corresponding time series. The time series of the central laser have been shifted by the coupling delay time $\tau_c = 3.65$ ns to allow an easier comparison. (I. Fischer, R. Vicente, J.M. Buldú, M. Peil, C.R. Mirasso, M.C. Torrent, and J. García-Ojalvo, (2006), © 2006 APS.)

5.4.6
Vertical-Cavity Surface-Emitting Lasers (VCSELs)

Synchronization of chaos in unidirectionally coupled vertical-cavity surface-emitting lasers (VCSELs) has been reported numerically (Spencer *et al.*, 1998). A traveling-wave model was used in a strong optical-feedback regime, which induces chaotic oscillations in VCSELs. Synchronization of chaos was observed over a large range of parameters.

Experimental observation of chaos synchronization in VCSELs has been reported (Fujiwara *et al.*, 2003). Two standalone VCSELs are mutually coupled and chaotic dynamics are obtained in a low-frequency fluctuation regime. Two polarization modes are adjusted between the two VCSELs. Synchronization of chaos for *x*-polarization components is obtained as shown in Figure 5.21a (left) when the wavelengths of the *x*-polarization modes are locked to each other by injection locking (Fujiwara *et al.*, 2003). However, synchronization is also observed between the *y*-polarization modes of the two VCSELs without locking the optical frequency, as shown in

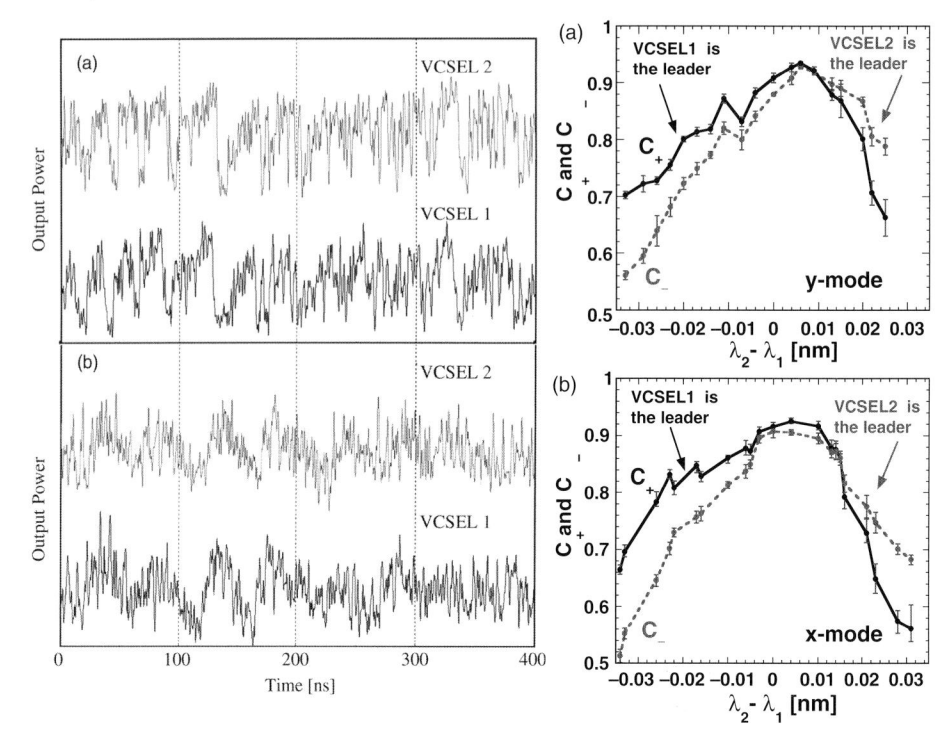

Figure 5.21 (Left) Time series of VCSEL outputs at synchronization: (a) x-polarization mode, (b) y-polarization mode. (N. Fujiwara, Y. Takiguchi, and J. Ohtsubo, (2003), © 2003 OSA.) (Right) Experimental result of the leader–laggard relationship between the synchronized chaotic waveforms of the two VCSELs for (a) y-mode and (b) x-mode intensities when the optical wavelength detuning $\Delta\lambda$ between the solitary VCSEL 1 and 2 is changed, where $\Delta\lambda = \lambda_2 - \lambda_1$ and λ_i is the optical wavelength of the solitary VCSEL i. When the solid curve is higher than the dotted curve VCSEL 1 is the leader and VCSEL 2 is the laggard, and *vice versa*. (M. Ozaki, H. Someya, T. Mihara, A. Uchida, S. Yoshimori, K. Panajotov, and M. Sciamanna, (2009), © 2009 APS.)

Figure 5.21b (left). Since the two polarization modes are strongly anticorrelated (i.e., antiphase dynamics, see Section 3.2.6), synchronization of both modes can be achieved with locking either of the two polarization modes between two VCSELs.

In a further development, synchronization of chaos in unidirectionally coupled VCSELs in a fully developed fast chaotic state (a few GHz) has been reported (Hong *et al.*, 2004). One of the two polarization modes (x-mode) is selected in the transmitter and the polarization direction is orthogonally rotated to couple into the y-polarization mode of the receiver. Synchronization is observed between the injected beam and the y-polarization of the receiver. Generalized synchronization is achieved through injection locking, because the output of the receiver lags only by the coupling delay time between the two VCSELs. The transmission signal and the x-polarization component of the receiver show synchronization, although the gradient of the correlation plots is negative, which implies antiphase synchronization.

The occurrence of antiphase synchronization can be explained by the antiphase dynamics between the two polarization modes in the receiver.

The leader–laggard relationship has been reported in mutually coupled VCSELs with time delay (Ozaki et al., 2009). The two solitary VCSELs (referred to as VCSEL 1 and 2) are mutually coupled to each other. The distance between the two VCSELs is set to 0.96 m, corresponding to the one-way coupling delay time of $\tau = 3.2$ ns. Each of the two polarization intensities of VCSEL 1 is coherently coupled to that of VCSEL 2. The polarization-resolved temporal dynamics of the two VCSELs are observed.

To investigate the leader–laggard relationship, the two delay times of $\pm\tau$ are selected and shift the temporal waveform of VCSEL 2 ($I_2(t)$) with respect to VCSEL 1 ($I_1(t)$) by the coupling delay time of τ to both the forward and backward directions in time (i.e., $I_2(t+\tau)$ and $I_2(t-\tau)$). The cross-correlation value between $I_1(t)$ and $I_2(t+\tau)$ (referred to as C_+), and that between $I_1(t)$ and $I_2(t-\tau)$ (referred to as C_-) are calculated. Then, the values of C_+ and C_- are compared to each other. When C_+ is higher than C_- it can be determined that VCSEL 1 is the leader and VCSEL 2 is the laggard, and *vice versa*. Figure 5.21 (right) shows the leader–laggard relationship of the y-mode and x-mode temporal waveforms between VCSEL 1 and 2 when the initial optical wavelength detuning ($\Delta\lambda = \lambda_2 - \lambda_1$) between the solitary VCSEL 1 and 2 is changed (Ozaki et al., 2009). Note that wavelength matching between the two coupled VCSELs occurs under the coupling due to injection locking in the detuning region shown in Figure 5.21 (right). When the solid curve for C_+ is larger than the dotted curve for C_-, VCSEL 1 is the leader and VCSEL 2 is the laggard, and *vice versa*. It is found that VCSEL 1 becomes the leader at negative $\Delta\lambda$ ($\lambda_2 < \lambda_1$) and VCSEL 2 is the leader at positive $\Delta\lambda$ ($\lambda_2 > \lambda_1$). Therefore, the VCSEL with longer wavelength becomes the leader. This result coincides with the injection locking characteristics, where injection locking is achieved mostly at the negative detuning $\Delta\lambda < 0$ due to the α parameter of VCSELs (see Figure 5.13).

5.5
Examples of Synchronization of Chaos in Electro-Optic Systems and Other Lasers

In this section, examples of synchronization of chaos in electro-optic systems (with passive optical devices) and other lasers (fiber lasers, solid-state lasers, and gas lasers) are described in detail from the literature.

5.5.1
Electro-Optic Systems

For electro-optic systems, a laser is treated as a linear component and an insertion of a passive nonlinear device with an electro-optic feedback loop can induce chaotic fluctuations, as described in Section 3.3. Synchronization of chaos in unidirectionally coupled electro-optic systems has been reported by implementing integrated Mach–Zehnder (MZ) modulators (Goedgebuer et al., 2002). In this system, shown in Figure 5.22 (left), the transmitter contains a laser diode operating in the linear part of

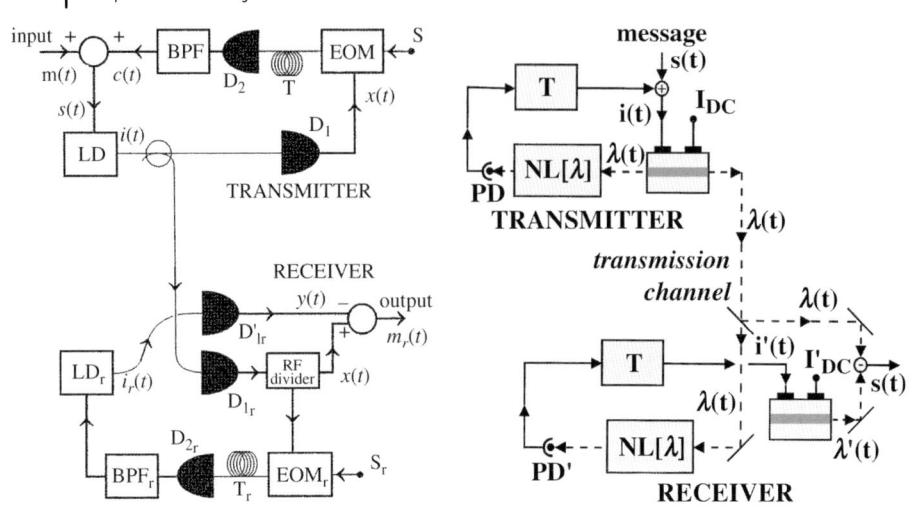

Figure 5.22 (Left) Optoelectronic feedback system with Mach–Zehnder modulators. The chaotic transmitter is formed by a laser diode (LD) with a time-delayed feedback loop containing an electro-optic modulator (EOM) (powered by an auxiliary source S) operating nonlinearly. The receiver is formed by the same elements but operates with its feedback loop open. (J.-P. Goedgebuer, P. Levy, L. Larger, C.-C. Chen, and W.T. Rhodes, (2002), © 2002 IEEE.) (Right) Schematic diagram of the two electro-optic wavelength oscillators for synchronization of chaos. NL stands for the nonlinear function. (J.-P. Goedgebuer, L. Larger, and H. Porte, (1998), © 1998 APS.)

its L–I curve and driven by a feedback signal. The electric current that results from the detection of the laser output is amplified and injected in a MZ modulator (an electro-optic modulator (EOM) in the figure) that in turn modulates nonlinearly a continuous-wave (CW) optical source. The modulator output then feeds the laser. The receiver is similar to the transmitter, except that the feedback loop is open. The effect of parameter mismatch on synchronization quality has also been reported in this system. A numerical model for synchronization of chaos in unidirectionally coupled electro-optic systems is described in Appendix 5.A.5.

An optical cryptosystem based on synchronization between two electro-optic wavelength oscillators has been demonstrated, as shown in Figure 5.22 (right) (Goedgebuer et al., 1998). In this system, a transmitter and receiver are composed of a wavelength-tunable distributed-Bragg-reflector (DBR) semiconductor laser and a feedback loop containing an optical component that exhibits a nonlinearity in wavelength. The latter is implemented by a birefringent plate placed between two crossed linear polarizers. The variation range of the wavelength is small enough to ensure that the DBR lasers always oscillate in a single longitudinal mode. At the transmitter, the laser output goes through the nonlinear optical component. Its variable output power is converted into an electric current by a photodiode that in turn is injected into the tunable-wavelength laser diode. As a consequence, the delayed feedback induces chaotic fluctuations in the laser wavelength for adequately chosen control parameters. The receiver is a replica of the transmitter, except that its feedback loop is open and is directly fed by the transmitter laser light. Synchronization is achieved between the transmitter and the receiver. This scheme was used for an early demonstration of optical chaos communication (see Section 8.1.2).

Coupled electro-optic systems with wavelength-tunable DBR lasers and filters have been proposed with a tuning range over many longitudinal modes (Liu and Davis, 2000a). Optical bandpass filters covering several laser modes are used as nonlinear wavelength functions. The wavelength of each laser is a piecewise-continuous function of the current: a large change in current can induce mode hopping, even though the optical power remains nearly constant. Figure 5.23 (left) shows the corresponding synchronization scheme (Liu and Davis, 2000a). At the transmitter, the light emitted by the laser goes through an optical tunable filter after retardation in a fiber line. The output power of the filter, which depends nonlinearly on the laser wavelength, is detected by a photodiode whose electric current is amplified and fed back into the DBR section of the laser diode. This feedback induces on-off intensity modulation of the individual modes with a chaotic pattern, the lasing intensity being concentrated in just one mode at any one time. Coupling between the transmitter and receiver is achieved by injecting the transmitter filter output in the receiver feedback loop, which is closed. The resulting driving signal (i.e., the transmitted signal plus the receiver's own feedback) induces chaotic mode hopping in the receiver laser. Synchronization of chaotic mode hopping is achieved when the coupling is strong enough.

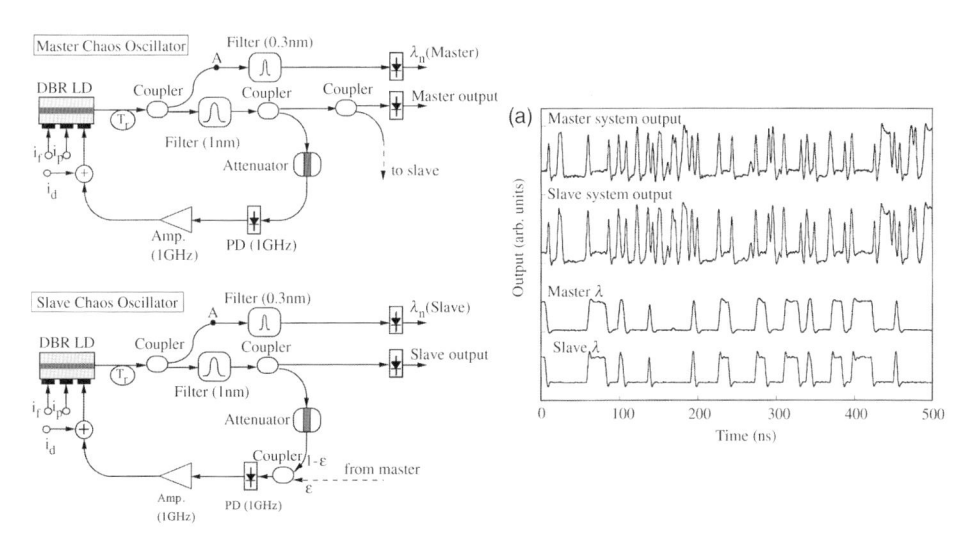

Figure 5.23 (Left) Schematic diagram of experimental setup for synchronization of chaotic wavelength hopping: i_d, offset DBR current; i_p, phase-control current; i_f, laser-diode (LD) pump injection current; ε, coupling coefficient. The total laser output is measured at A before the wavelength filter. T_r, time delay; PD, photodiodes. (Right) Typical waveforms of the synchronized transmitter (master) and receiver (slave) systems. All traces are on the same relative scale. Waveforms of the system output and mode 1 ($\lambda = 1548.25$ nm) are shown. (Y. Liu and P. Davis, (2000a), © 2000 OSA.)

Figure 5.23 (right) shows examples of synchronized chaotic mode hopping. The top two waveforms correspond to the light output of the transmitter and the receiver, and the two bottom waveforms show the signals of one particular wavelength of the transmitter and the receiver. High-quality synchronization of chaos is achieved for both the total and modal intensities at optimal parameter-matching conditions.

5.5.2
Fiber Lasers

The first experimental observation of synchronization between chaotic fiber lasers was made by using the experimental setup shown in Figure 5.24 (left) (VanWiggeren and Roy, 1999). The transmitter is an erbium-doped fiber-ring laser, whose dynamical regime can be manipulated by means of an intracavity polarization controller. The laser is set to operate in a chaotic regime, from which 10% of the intracavity radiation is extracted *via* a 90/10 output coupler and transmitted to the receiver. 50% of that transmitted signal is injected into an erbium-doped fiber amplifier (EDFA) whose physical characteristics (e.g., fiber length, dopant concentration, and pump diode

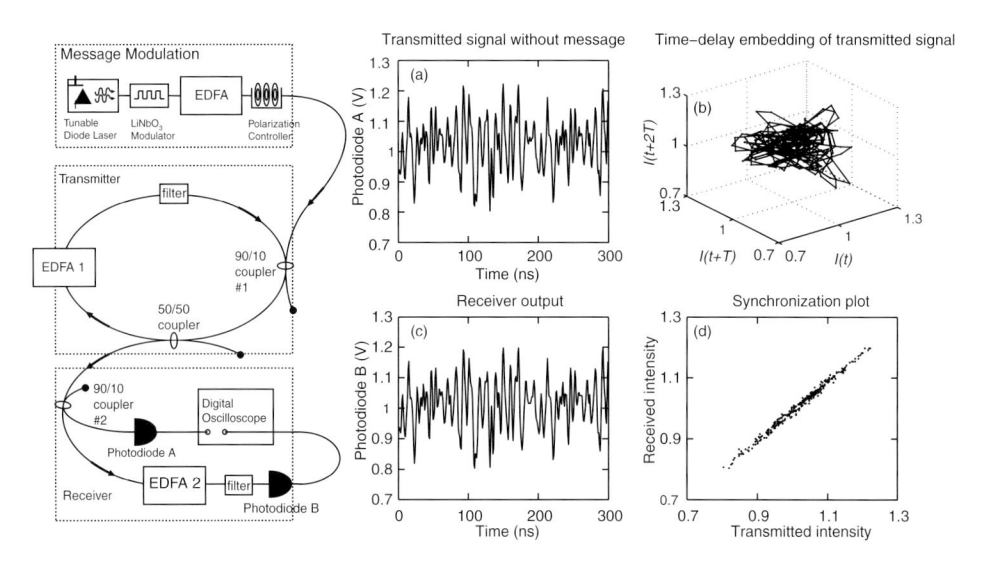

Figure 5.24 (Left) Experimental system for optical chaos synchronization and communication, consisting of three parts. In the message modulation unit, CW laser light is intensity modulated to produce a message signal consisting of a series of digital bits. The message signal is injected into the transmitter where it is mixed with the chaotic lightwaves produced by the erbium-doped fiber-ring laser. The message, now masked by the chaotic light, propagates through the communication channel to the receiver where the message is recovered from the synchronized chaos. (Right) (a) Transmitted signal measured by photodiode. (b) Time-delay-embedding plot of the data in (a); the lack of structure indicates that the data is not low dimensional. (c) Signal simultaneously detected by photodiode B. (d) Correlation plot between (a) and (c). The signals recorded by the photodiodes are clearly synchronized, as shown in (d). (G.D. VanWiggeren and R. Roy, (1999), © 1999 World Scientific.)

laser) are matched as closely as possible to those of the transmitter EDFA. It should be noted that the receiver output is not reinjected back into itself, that is, the system operates in open loop. A second 90/10 coupler in the transmitter cavity allows the introduction of an external signal in the context of chaotic communications (see Section 8.1.2 for chaos communication with this setup).

Figure 5.24a (right) shows a sample time trace of the transmitted laser intensity in the absence of an external signal. The phase-space reconstruction shown in Figure 5.24b exhibits no low-dimensional structure of the dynamics displayed in Figure 5.24a. The signal detected by photodiode B (see experimental setup in Figure 5.24 (left)) after passing through the receiver's EDFA is shown in Figure 5.24c. This signal has been shifted in time ($\tau = 51$ ns), equal to the time mismatch corresponding to the fiber length difference between the paths leading to photodiodes A and B. Visual inspection indicates that the time series of the transmitter and receiver are very similar. The existence of synchronization is confirmed by plotting the output of the receiver *versus* that of the transmitter, as shown in Figure 5.24d; the resulting straight line with slope unity in this synchronization plot is a clear indicator of the occurrence of synchronization.

The experimental setup described above is analogous to the open-loop version of the configuration (Abarbanel and Kennel, 1998). The experimentally observed synchronization described above can be interpreted as identical synchronization, since after compensating for fiber-length differences between the transmitter and the receiver, the receiver output can be expected to reproduce exactly the state of the transmitter without time delay. Indeed, Figure 5.24d reveals a very good quality of synchronization. A numerical model for chaos synchronization in unidirectionally coupled fiber lasers is described in Appendix 5.A.6.

Synchronization of chaos has also been observed in erbium-doped fiber lasers with external modulation (Luo *et al.*, 2000b). The experimental setup is schematically shown in Figure 5.25 (left). Two erbium-doped fiber lasers of a ring resonator are connected through a 10-m fiber and a polarization controller. Each fiber resonator consists of two fiber couplers, an erbium-doped fiber and an electro-optical modulator. The two modulators have the same property and they provide optical amplitude modulation, which induces chaotic intensity fluctuations. Both the erbium-doped fibers came from the same fiber. The fiber has a polarization-maintaining property and the lasing light is highly polarized. In the ring resonator of the transmitter, the lasing light goes through an EOM, which favors transmission in the direction. The light is split by a coupler in the transmitter and a part of the lasing light intensity in the drive (master) laser goes into the response (slave) laser system for synchronization. There are two fiber ends in the drive and response lasers, respectively, to introduce the pump laser lights at 980 nm. All other spare ends are put into index oil to remove the reflections.

When the lasers are modulated by the EOMs, the lasing is chaotic even at a modulation index as low as 1%. This is a typical property of an external modulated chaotic system that chaos exists when it is modulated around its inherent frequencies and their harmonic frequencies (see Section 3.4.1). The polarization controller is adjusted to match the polarization direction of the drive laser to that of the response

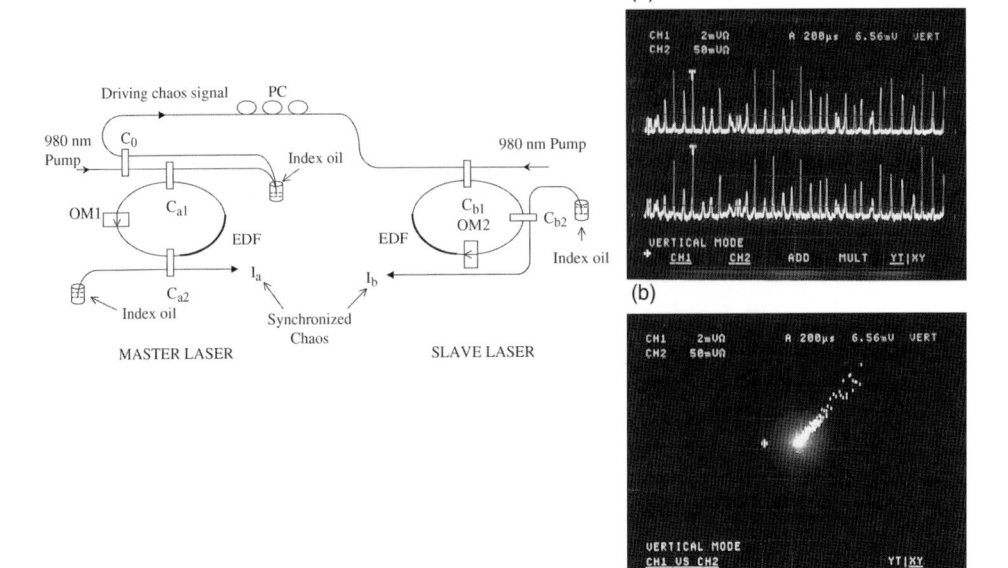

Figure 5.25 (Left) Experimental setup for chaos synchronization with two erbium-doped fiber-ring lasers. EDF, erbium-doped fiber; OM1 and OM2, electro-optical modulators; PC, polarization controller. (Right) Chaos synchronization of two fiber lasers. (a) In the time domain: lower trace, the drive (master) laser light; upper trace, the response (slave) laser light. Y-axis: arbitrary unit; X-axis: 200 μs/div. (b) The drive laser light *vs.* the response laser light. (L. Luo, P.L. Chu, T. Whitbread, and R.F. Peng, (2000b), © 2000 Elsevier.)

laser; the chaotic signals in the drive and response lasers are then synchronized. This is shown in Figure 5.25 (right). It can clearly be seen that chaotic signals in the drive and response lasers are in synchronization (Figure 5.25a (right)) and the correlation plot of drive *versus* response laser lights shows a strong linearity (Figure 5.25b (right)), which indicates good synchronization. Proper adjustment of the respective pump power in the drive and response lasers can improve the synchronization quality. It is worth noting that a single-frequency response laser with a wavelength very close to that of the drive laser would require a smaller injection intensity for injection locking to achieve synchronization of chaos.

5.5.3
Solid-State Lasers

One of the first experimental observations of synchronization of chaotic lasers was reported in 1994 by using two Nd:YAG solid-state lasers, as described in Section 5.2.2 (Roy and Thornburg, 1994). Synchronization of chaos in one-way coupled microchip solid-state lasers has also been reported (Uchida *et al.*, 2000). In that case, two neodymium-doped yttrium orthovanadate ($Nd:YVO_4$) microchip lasers are used as

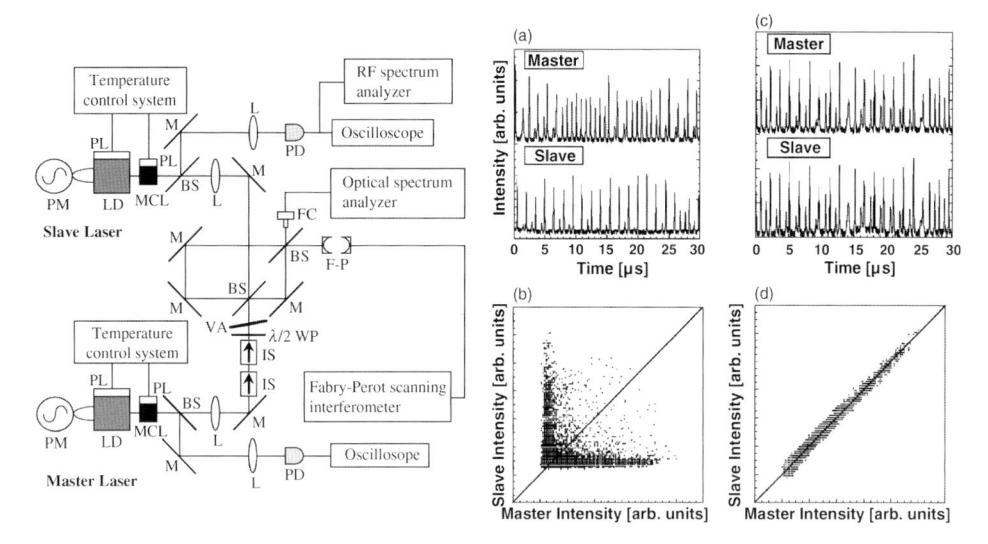

Figure 5.26 (Left) Experimental setup for chaos synchronization in two Nd:YVO$_4$ microchip lasers with pump modulation in a unidirectional coupling configuration: BS, beam splitter; L, lens; M, mirror; VA, variable attenuator (neutral-density filter); LD, laser diode; MCL, Nd:YVO$_4$ microchip laser; PL, Peltier device; IS, optical isolator; PD, photodiode; FC, fiber coupler; PM, pump modulation; $\lambda/2$ WP, $\lambda/2$ wave plate; F-P, Fabry–Perot etalon. (Right) Experimentally obtained chaotic temporal waveforms and correlation plots for the two laser outputs: (a), (b) without synchronization and (c), (d) with synchronization. (A. Uchida, T. Ogawa, M. Shinozuka, and F. Kannari, (2000). © 2000 APS.)

laser sources (wavelength of 1064 nm), as shown in Figure 5.26 (left). Chaotic outputs are obtained by modulating the pumping at frequencies of the order of the sustained relaxation oscillation frequency at a few MHz (see Section 3.5.1). A fraction of the drive (master) laser output is unidirectionally and coherently coupled to the response (slave) laser cavity for chaos synchronization. The optical frequencies of the two microchip lasers are precisely controlled with thermoelectric coolers in order to achieve injection locking, where the optical frequencies of two individual lasers are perfectly matched when the frequency difference is set within an injection-locking range.

Figure 5.26 (right) shows the chaotic temporal waveforms and the correlation plots between the two laser outputs. In the absence of coupling, there is no correlation at all between the chaotic pulsations of the two lasers, as shown in Figures 5.26a and b. When a fraction of the drive laser output is injected into the response laser cavity, injection locking is achieved between the two lasers and the chaotic oscillations are synchronized, as shown in Figure 5.26c. The linear correlation between the two laser outputs, shown in Figure 5.26d, confirms the existence of synchronization. The synchronization can be maintained as long as injection locking of the two laser frequencies is preserved. This experiment shows that the condition to achieve synchronization of chaos is almost equivalent to that for injection locking. In other words, synchronization of chaos in lasers can be interpreted as nonlinear regeneration of the chaotic laser output through the mechanism of injection locking (i.e., matching of fast optical-carrier frequency) (Uchida et al., 2000). A numerical model for synchronization of chaos in unidirectionally coupled solid-state lasers is described in Appendix 5.A.7.

Synchronization of mutually coupled chaotic solid-state lasers has been demonstrated in a microchip $LiNdP_4O_{12}$ (LNP) laser array with self-mixing feedback modulation (Otsuka et al., 2000). In that case, a pumping beam from an argon-ion (Ar^+) laser is divided into two beams and is focused on the input surface of the LNP crystal by a common focusing lens, as shown in Figure 5.27 (left). The coherent coupling between the two LNP lasers occurs through the overlap of their lasing fields, which have a spot size of 200 μm each. The two parallel beams from the LNP lasers are incident to and scattered from the turntable. The interference between the lasing and frequency-shifted feedback fields modulates the losses of the microcavity lasers, which is known as self-mixing laser-Doppler-velocimetry feedback (see Sections 3.5.2 and 3.5.3).

Figure 5.27a (right) shows a plot of the correlation in amplitude between the two laser signals when the distance between the two beams is 1.5 mm. To examine the phase correlation of the chaotic pulsations, the time interval between the nth peak and the subsequent peak for laser 2 is plotted against that for laser 1 in Figure 5.27b.

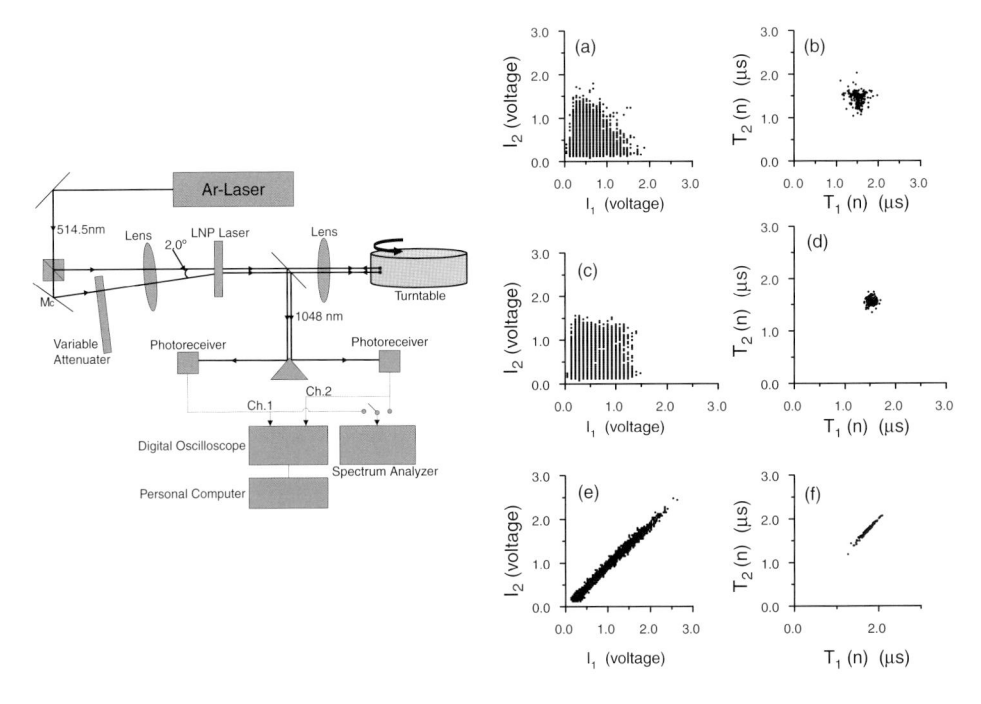

Figure 5.27 (Left) Experimental setup of a LNP laser array subjected to Doppler-shifted light injections. (Right) Signal correlations of the LNP laser array: (a),(b) at weak coupling (distance of $d=1.5$ mm); (c), (d) at $d=0.85$ mm; (e), (f) at stronger coupling ($d=0.8$ mm). (a), (c), (e) are for amplitude correlation, and (b), (d), (f) are for phase correlation. (K. Otsuka, R. Kawai, S.-L. Hwong, J.-Y. Ko, and J.-L. Chern, (2000), © 2000 APS.)

In this case, asynchronous chaotic fluctuations in both amplitude and phase are apparent, and the two lasers are found to behave independently. When the separation between the two lasers is decreased, mutual interaction appears and the phase fluctuations of the two lasers are squeezed while their amplitudes remain uncorrelated, just before the onset of chaos synchronization, as shown in Figures 5.27c and d. When the separation is decreased further, synchronized chaotic states, in which the two lasers exhibit chaotic pulsations with both strong amplitude and phase correlation, are obtained as shown in Figures 5.27e and f.

5.5.4
Gas Lasers

Gas lasers provided one of the first experimental observations of synchronization of chaotic lasers, as shown in Section 5.2.2 (Sugawara et al., 1994). Synchronization of chaotic Q-switching dynamics has been experimentally observed in two mutually coupled CO_2 lasers with an intracavity saturable absorber (Liu et al., 1994). Figure 5.28 (left) shows the temporal waveforms of the output pulses of the two CO_2 lasers and

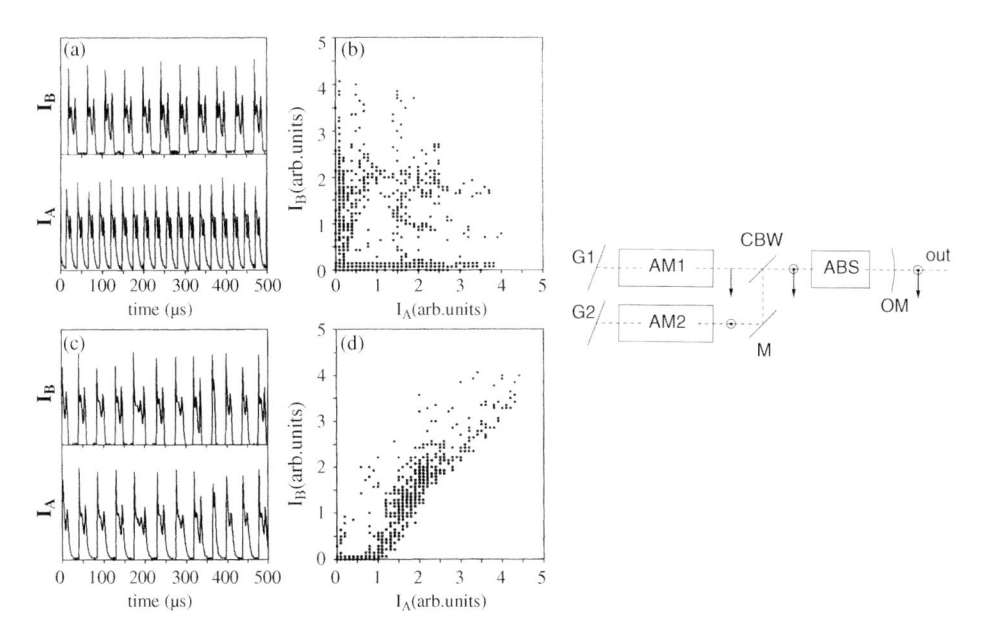

Figure 5.28 (Left) Passive Q-switching of the CO_2 lasers. (a) Before coupling, the two lasers are periodically pulsing but unsynchronized. (b) Intensities' correlation of uncoupled lasers. (c) After coupling the chaotic lasers become synchronized and their intensities show (d) a positive degree of correlation. The optical frequency detuning is 2 MHz. (Y. Liu, P.C. de Oliveira, M.B. Danailov, and J.R. Rios Leite, (1994), © 1994 APS.) (Right) Schematic representation of the experimental set-up. AM1, AM2: active media, ABS: absorber, G1, G2: diffraction gratings, CBW: coated Brewster window, M: mirror, OM: partly transparent output mirror. (A. Barsella, C. Lepers, D. Dangoisse, P. Glorieux, and T. Erneux, (1999), © 1999 Elsevier.)

the corresponding correlation plots. Without coupling, the temporal waveforms are in the periodic regime and no synchronization is observed (Figures 5.28a and b (left)). When the coupling is introduced, synchronized chaotic oscillations are observed, as shown in Figures 5.28c and d (left). The achievement of synchronization is strongly dependent of the optical frequency detuning between the two lasers. A numerical model for synchronization of chaos in unidirectionally coupled gas lasers is described in Appendix 5.A.8.

Another way of synchronizing two mutually coupled gas lasers is through a common intracavity saturable absorber as shown in Figure 5.28 (right) (Barsella *et al.*, 1999). The system is composed of two laser cavities with two separate active media (AM1 and AM2). The two lasers operate on orthogonal polarization states and allow the incoherent interaction to happen only in the saturable absorber cell (ABS), where the two beams are superimposed. This mutual coupling system allows synchronization between the two chaotic temporal waveforms with time delay. The experimental results are well reproduced by the numerical simulation based on the rate-equation model (Barsella *et al.*, 1999).

5.6
Specific Types of Synchronization

5.6.1
Phase Synchronization

So far in Chapter 5 synchronization has been considered as a regime in which temporal waveforms of two laser intensities have a well-defined relation at all times. But different forms of synchronizations exist. Consider, for instance, that the chaotic oscillations can be decomposed in terms of an amplitude and a phase. Under certain conditions, it can happen that the amplitudes of the chaotic oscillations of the two lasers are desynchronized, while a clear synchronization exists between the phases. This is called phase synchronization (Rosenblum *et al.*, 1996). Note that "phase" in this context indicates the phase of chaotic oscillations of laser intensity corresponding to the slow-envelope component, but not the optical phase of the fast-carrier electric field. Therefore, the frequency of the phase oscillation is the same order as that of the amplitude oscillation at the relaxation oscillation frequency of 10^3–10^9 Hz.

Phase synchronization can be defined as,

$$\lim_{t \to \infty} \left(\phi_y(t) - \phi_x(t) \right) \to \varepsilon \tag{5.25}$$

where $\phi_{x,y}(t)$ are the phases of two laser intensities. ε is a constant and remains within certain bounds.

The phase can be defined from the time series of the laser intensity by interpolating linearly between two adjacent crossing points of a predetermined threshold value, as shown in Figure 5.29a, so that the phase gains 2π with each crossing of the threshold value as (Pikovsky *et al.*, 2001),

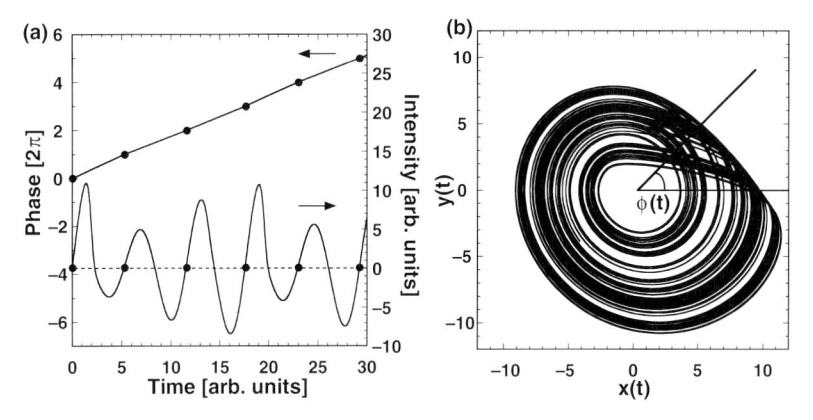

Figure 5.29 Definition of phase in chaotic oscillations. (a) Linear interpolation of the time series between the crossing of the threshold value (Eq. (5.26)), and (b) phase of the polar coordinates in the two-dimensional phase space (Eq. (5.27)).

$$\phi(t) = 2\pi \frac{t-t_n}{t_{n+1}-t_n} + 2\pi n, \quad t_n \leq t \leq t_{n+1}, \tag{5.26}$$

where t_n is the time of the nth crossing of the threshold. This definition is ambiguous because it crucially depends on the choice of the threshold value. Nevertheless, it is convenient to extract the phase from an experimentally measured temporal waveform of the laser intensity by using Eq. (5.26).

Another definition of the phase is the use of the chaotic attractor in the 2D phase space, as shown in Figure 5.29b. The phase can be defined as the angle of the trajectory in the 2D phase space *via* transformation to the polar coodinates with the origin coinciding with the center of rotation (Pikovsky *et al.*, 2001),

$$\phi(t) = \tan^{-1}\left(\frac{y(t)}{x(t)}\right) \tag{5.27}$$

where $x(t)$ and $y(t)$ are the variables of the x- and y-axes in the 2D phase space. This definition is suitable for chaotic attractors with circular shape, such as the Rössler model. A more general definition of the phase for chaotic laser intensity has been also used, based on the Hilbert transform, as shown in Appendix 5.A.9.

Phase synchronization has been observed experimentally in mutually coupled two Nd:YAG solid-state lasers (Volodchenko *et al.*, 2001). The experimental setup is shown in Figure 5.30. Each laser output is injected into another laser cavity for mutual coupling, and the intensity of the injection beam is controlled with a Glan–Thompson polarizer that is set on the beam path. For the polarized laser outputs, Brewster windows are inserted into each laser cavity so that the coupling strength between the two lasers can be controlled. The optical signals coming from the back mirrors are detected with fast pin photodiodes and stored in a digital oscilloscope.

When the pumping currents of the two lasers are set near the threshold, the CW Nd:YAG lasers produce chaotic pulsations. As the pumping current increases, the pulsations appear more frequently, and finally a CW laser output with chaotic

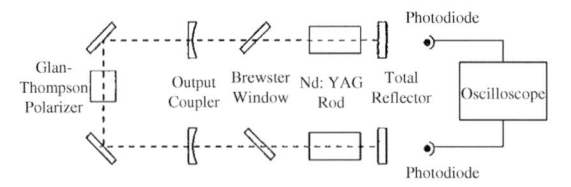

Figure 5.30 Schematic diagram of the experiment for phase synchronization in two coupled Nd: YAG lasers. (K.V. Volodchenko, V.N. Ivanov, S.-H. Gong, M. Choi, Y.-J. Park, and C.-M. Kim, (2001), © 2001 OSA.)

pulsations is generated. Figure 5.31 (left) shows the temporal behaviors of the two laser intensities for different coupling strengths (Volodchenko et al., 2001). When the lasers are uncoupled, it is evident that the temporal behavior of one laser is different from that of the other laser (Figure 5.31a (left)). However, when the two lasers are coupled, the temporal behaviors of the two lasers become more similar as the coupling strength increases (Figures 5.31b and c (left)). When the lasers are fully coupled, the phases of the two chaotic oscillations are locked with each other, as shown in Figure 5.31d (left).

To observe the transition to phase synchronization, the phase of the chaotic pulsations is measured. Figure 5.31 (right) shows the phase difference in the two laser intensities when the coupling strength is increased. As shown in Figure 5.31a (right), the phase difference in the two lasers increases and decreases irregularly without coupling. As the coupling strength increases, the phase difference is locked at $\pm 2\pi n$ (n is an integer) for a rather long time and jumps intermittently, as shown in Figure 5.31b (right). For strong coupling, the phase difference is locked within $\pm 2\pi$ without phase jumps (Figure 5.31c (right)), which implies the occurrence of phase synchronization. It is found that when the coupling strength is above the critical

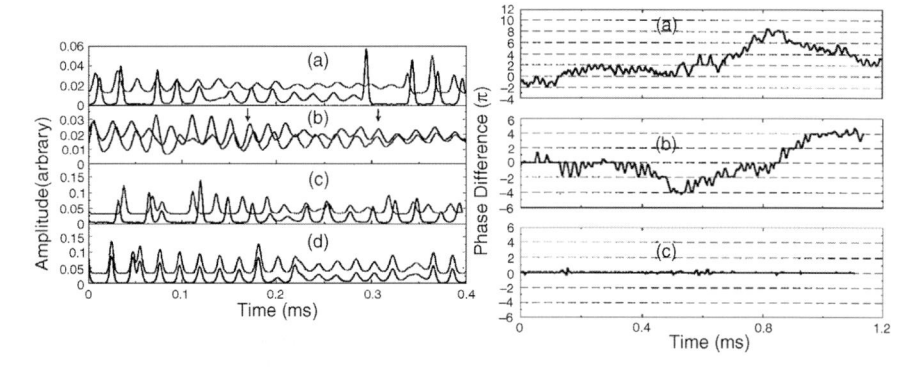

Figure 5.31 (Left) Temporal behaviors of the two laser outputs depending on the coupling strength: the angles of the Glan–Thompson polarizer are (a) 90 degree (no couping strength), (b) 45 degree, (c) 15 degree, and (d) 0 degree (maximum coupling strength). (Right) Temporal behaviors of phase difference: (a) nonsynchronous state when the angle of the Glan–Thompson polarizer is 90 degree (no coupling strength), (b) phase jump state when the angle is 45 degree, and (c) phase synchronization when the angle is 15 degree. (K.V. Volodchenko, V.N. Ivanov, S.-H. Gong, M. Choi, Y.-J. Park, and C.-M. Kim, (2001), © 2001 OSA.)

value, the coupled lasers exhibit phase synchronization, whereas when they are below the critical value they exhibit $\pm 2\pi$ phase jumps irregularly.

Phase synchronization between mutually coupled chaotic lasers has been examined experimentally for the case of a linear laser array (DeShazer et al., 2001). The chaotic system consists of three parallel, laterally coupled single-mode Nd:YAG laser arrays (referred to as lasers 1, 2, and 3). Coupling through the electric fields of the individual beams exists only for adjacent pairs. Experimental intensity measurements of the three laser intensities are displayed in Figure 5.32a–c (top). The two

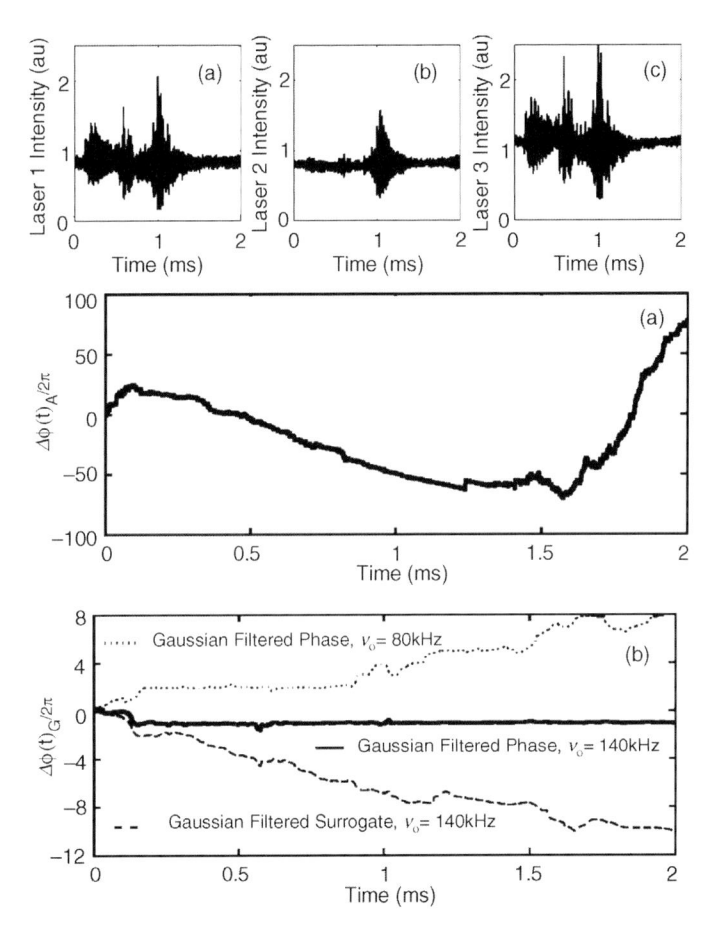

Figure 5.32 (Top) Experimental intensity time series, showing chaotic bursts and similarity of the dynamics of lasers 1 and 3. (a) Laser 1, (b) laser 2 (middle of the laser arrays), and (c) laser 3. (Middle and Bottom) Time series for the relative phases. (a) Analytic phase and (b) Gaussian filtered phase for lasers 1 and 2. Synchronization is not discernible in (a), while (b) shows phase synchronization when the Gaussian filter is centered at $\nu_0 = 140\,\text{kHz}$ (solid curve). Intermittent periods of phase synchronization are observed with $\nu_0 = 80\,\text{kHz}$ (dotted curve). The dashed curve is the relative phase computed with a surrogate (randomized) time series; as expected, it does not show synchronization. In all cases in (b) the bandwidth is 15 kHz for the Gaussian bandpass filter. (D.J. DeShazer, R. Breban, E. Ott, and R. Roy, (2001), © 2001 APS.)

outer lasers in the array (lasers 1 and 3) have nearly identical intensity fluctuations. However, no synchronization relationship is obvious between the center laser (laser 2) and the outer lasers (lasers 1 and 3), even though the center laser mediates identical synchronization of the outer lasers.

To test for interdependence between the time series of the outer and center lasers, an analytic phase and a Gaussian filtered phase are introduced by using the Hilbert transform (see Appendix 5.A.9). The Gaussian filtered phase is extracted from the time series of the laser intensities filtered by a bandpass filter with a Gaussian profile in order to measure the phase of a particular frequency component in the chaotic time series (DeShazer et al., 2001). Figure 5.32a (middle) shows the difference in the analytic signal phases for these lasers, which has a large range of variation ($\sim$130 rotations). Phase synchronization is not discernible. Next, in Figure 5.32b (bottom), the difference in the Gaussian filtered phase for these two laser outputs is plotted at different frequencies of the Gaussian bandpass filter. Synchronization of the side and central lasers in the frequency regime of 140 kHz is immediately apparent, since the flat portion of this plot extends across essentially the entire time of observation (solid curve). Periods of phase synchronization and phase slipping are found in the less-correlated frequency regime of 80 kHz (dotted curve). No indication of synchronization is found when one of the component phases is replaced with a surrogate (randomized) phase extracted from another experimental data set taken from this array under identical conditions (dashed curve). These results illustrate that the detection of phase synchronization may require careful consideration of the frequency component of the time series measured. The time series considered in this experiment are of a distinctly nonstationary nature, and it is clearly advantageous to introduce a Gaussian filtered phase variable. It is possible to quantitatively assess phase synchronization for different frequency components of the dynamics.

5.6.2
Generalized Synchronization (with Low Correlation)

Generalized synchronization of chaos can be considered as a more general concept than the two types of synchronization described in Sections 5.3.1 and 5.3.2 (Rulkov et al., 1995). The generalized synchronization described in this section is based on a more "general" relationship between the drive and response lasers even with low correlation, whereas the generalized synchronization described in Section 5.3.2 is limited to the synchronization with high correlation. The generalized synchronization treated in this section may have low correlation between the two coupled laser intensities, however, it may still have a certain relationship between the two laser intensities. In fact, the generalized synchronization shown in Section 5.3.2 is a special case of the generalized synchronization described in this section.

Generalized synchronization can be defined as the existence of a functional relationship $\mathbf{F}$ between the drive and response variables $\mathbf{x}(t)$ and $\mathbf{y}(t)$,

$$\mathbf{y}(t) = \mathbf{F}(\mathbf{x}(t)) \tag{5.28}$$

Identical synchronization is a special case of generalized synchronization when **F** is the identity function, and the generalized synchronization described in Section 5.3.2 is also a special case when **F** is a nearly identity function with a time-delayed relationship. However, **F** can be a more complex function in general. In fact, it is difficult to identify **F** directly from the time series of the system variables $\mathbf{x}(t)$ and $\mathbf{y}(t)$, since the function **F** itself can be very complicated. Generalized synchronization may be much more prevalent in nature than realized so far. It may have important applications in methods for noninvasive testing (Moniz et al., 2003) and encoded communications systems (Terry and VanWiggeren, 2001).

Generalized synchronization has been shown to exist through the predictability (Rulkov et al., 1995) or the existence of a functional relationship (Brown, 1998) between the drive and response systems. These approaches are often difficult to implement in experimental measurements, due to the presence of noise and the lack of precision in measurements. When replicas or duplicates of the response system are available, "auxiliary system approach" can be used for detecting generalized synchronization (Abarbanel et al., 1996). In this method, two or more response systems are coupled with the drive system. If the response systems, starting from different initial conditions, display identical synchronization between them after transients have disappeared, one can conclude that the response signal is synchronized with the drive in a generalized way.

The auxiliary system approach is described in detail (Abarbanel et al., 1996). The concept of the auxiliary system approach is shown in Figure 5.33. A copy of the response system (response-2) with identical parameter values (but different initial conditions) is prepared and coupled with the drive system as for the original response system. A new response variable $\mathbf{y}'(t)$ is described,

$$\mathbf{y}'(t) = \mathbf{F}(\mathbf{x}(t)) \tag{5.29}$$

Therefore, the dynamics of the variables of the response system and its replica can become identical from Eqs. (5.28) and (5.29),

$$\mathbf{y}'(t) = \mathbf{y}(t) \tag{5.30}$$

To detect generalized synchronization in the auxiliary system approach, identical synchronization between the two identical response systems that are driven by a

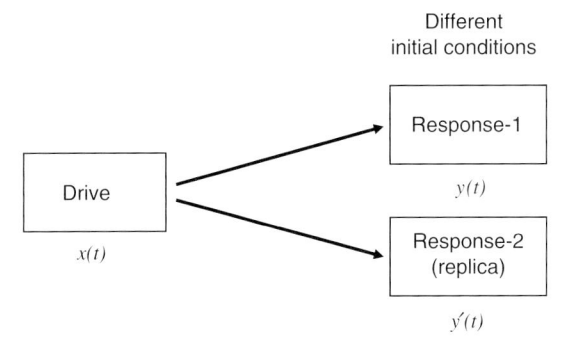

Figure 5.33 Concept of the auxiliary system approach for detecting generalized synchronization.

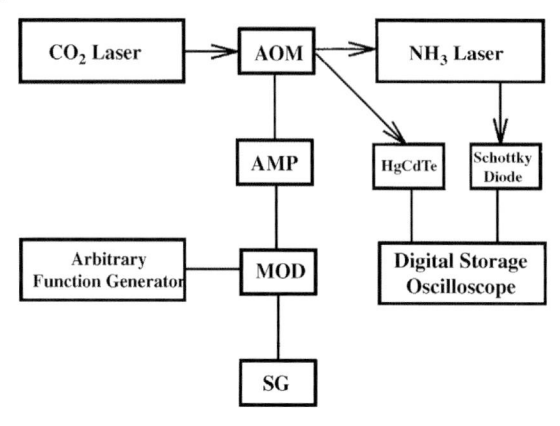

Figure 5.34 Schematic block diagram of the experiment setup for generalized synchronization in a NH_3 laser. SG, signal generator; MOD, modulator; AMP, amplifier; AOM, acousto-optic modulator. (D.Y. Tang, R. Dykstra, M.W. Hamilton, and N.R. Heckenberg, (1998), © 1998 APS.)

common drive signal needs to be observed. This configuration is also known as common-signal-induced synchronization (Yamamoto et al., 2007; Oowada et al., 2009), since the two response lasers that are subject to a common drive signal can be synchronized to each other, as shown in Figure 5.33.

The first experimental observation of generalized synchronization of chaos in laser systems was reported in an optically pumped single-mode NH_3 laser in a ring-cavity configuration, as shown in Figure 5.34 (Tang et al., 1998). The laser is operated in the parameter regime where it behaves similarly to the dynamics of the Lorenz–Haken laser equations (see Sections 2.1.3 and 3.6.1). An acousto-optic modulator (AOM) is used to modulate the pump intensity of the laser. A chaotic intensity waveform emitted by the laser is recorded as a drive signal. This chaotic waveform is then stored in the memory of an arbitrary function generator, which produces an analog signal with exactly the same waveform as the stored one. This analog signal is then used to modulate the amplitude of the RF driving signal of the AOM, which in turn transferred this modulation signal to the intensity of the pump laser beam. The undiffracted beam from the AOM is used as the pump of the NH_3 laser. The auxiliary system approach is used to test generalized synchronization. Instead of using two response systems and comparing their dynamics, the chaotic laser is repeatedly driven by the same chaotic signal under the same experimental conditions, and the detected chaotic temporal waveforms of the laser are compared to each other for different repetitions.

Figure 5.35 (left) shows temporal waveforms exhibiting generalized synchronization (Tang et al., 1998). The chaotic dynamics of the laser in response to a driving Lorenz-like chaotic signal, after a transient, is shown in Figure 5.35a (left). These chaotic intensity evolutions are different from the driving signal, shown in Figure 5.35b (left). Figure 5.35 (right) shows the comparison of the laser dynamics under repeated driving signals. Figure 5.35a (right) is the correlation plot of the repeatedly injected driving signals between different repetitions and Figure 5.35b (right) represent the corresponding outputs of the response laser driven by the same drive signal. The chaotic dynamics of the laser repeats itself completely in successive

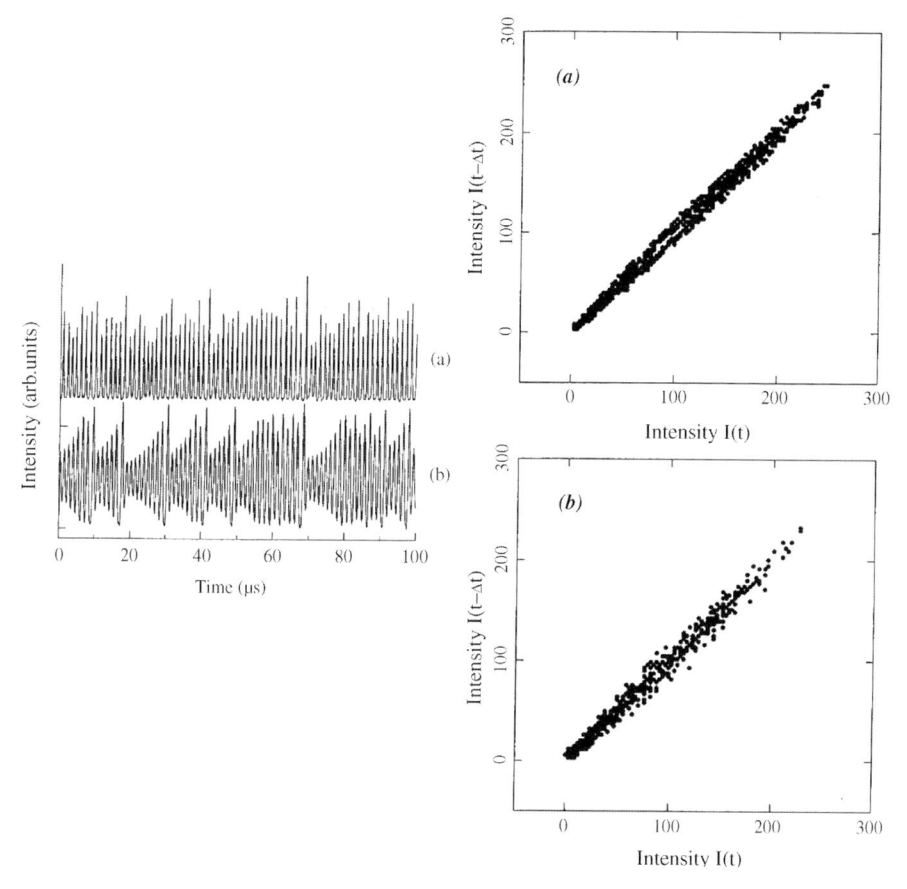

Figure 5.35 (Left) Chaotic dynamics of the laser under the generalized synchronization of chaos to the driving signal. (a) Chaotic laser intensity evolution (response). (b) Evolution of the chaotic driving signal measured (drive). (Right) Relationship plot of the laser dynamics shown in Figure 5.35 (left) under repeated drivings. Δt is the repetition period of driving signal, which is 150 μs in the experiment. (a) Relationship plot of the driving signal (drive). (b) Relationship plot of the laser dynamics (response). (D.Y. Tang, R. Dykstra, M.W. Hamilton, and N.R. Heckenberg, (1998), © 1998 APS.)

events, showing that its dynamics is now completely controlled by the driving signal and is insensitive to the noise and initial conditions. The dynamics shown in Figure 5.35b (right) is therefore an example of generalized synchronization of chaos in the sense of the auxiliary system approach.

It is observed experimentally that the functional relation in a state of generalized synchronization is not fixed, but varies with the driving signal strength. When the driving strength is increased, the dynamical relationship between the drive and response becomes simpler, and eventually becoming a linear relation, which implies that the state of identical synchronization of chaos appears.

The examples of generalized synchronization discussed in the literature consist of situations where the drive and response systems are different from each other, or

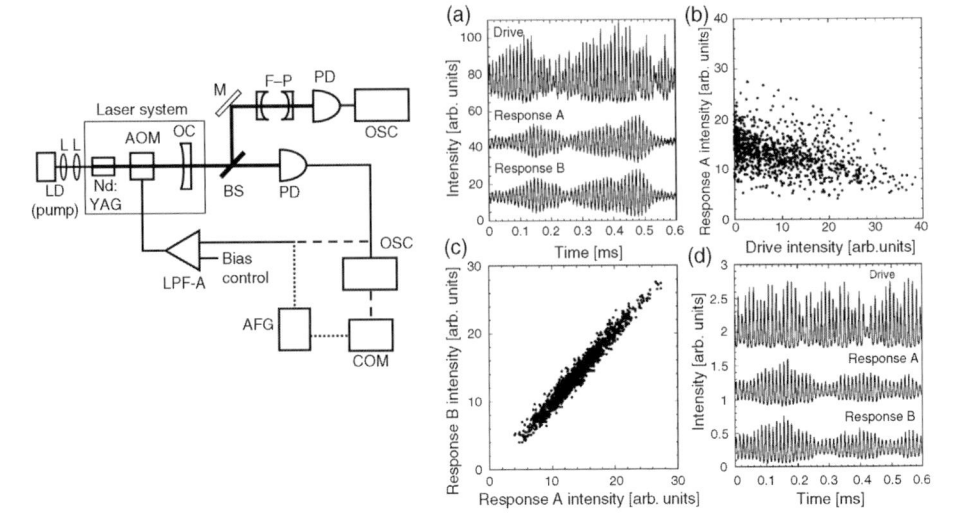

Figure 5.36 (Left) Experimental setup of a diode-pumped Nd:YAG microchip laser with optoelectronic feedback for generalized synchronization. The dashed line corresponds to the closed-loop drive system, and the dotted line corresponds to the open-loop response system. AFG, arbitrary function generator; AOM, acousto-optic modulator; BS, beam splitter; COM, computer; F-P, Fabry–Perot interferometer; L, lens; LD, laser diode for pumping; LPF-A, low-pass filter and amplifier; M, mirror; Nd:YAG, Nd:YAG laser crystal; OC, output coupler; OSC, digital oscilloscope; PD, photodetector. (Right) (a) Temporal waveforms of experimentally measured total intensity of the drive and two response systems. (b) Correlation plots between the drive and response outputs. (c) Correlation plots between the two response outputs. (b) and (c) are obtained from (a). (d) Temporal waveforms of total intensity obtained from numerical simulation. (A. Uchida, R. McAllister, R. Meucci, and R. Roy, (2003e), © 2003 APS.)

of the same system operated at different parameter values (Lewis et al., 2000). One may expect that strongly coupled identical systems with similar parameter values will display identical synchronization, if they synchronize at all. However, generalized synchronization of chaos can occur with identical drive and response systems with similar parameter values when the laser system has hidden internal degrees of freedom. This was shown by using a chaotic two-longitudinal-mode Nd:YAG microchip laser (Uchida et al., 2003e).

The experimental setup is shown in Figure 5.36 (left). The total intensity of the laser output is detected by a photodiode and the voltage signal is fed back into an intracavity AOM in the laser cavity, through an electronic low-pass filter with an amplifier. The loss of the laser cavity is modulated by the self-feedback signal through the AOM, which induces chaotic oscillations. Temporal waveforms of the laser output are measured by a digital oscilloscope and stored in a computer, for later use as a drive signal. In order to test for generalized synchronization, the same laser is used as a response system, which ensured identical parameter settings between drive and response. The recorded drive signal is sent to the AOM in the same laser cavity, using an arbitrary function generator connected with the computer. The original feedback loop is disconnected (dashed line in Figure 5.36 (left)), that is, the open loop configuration (dotted line in Figure 5.36 (left)) is used for the response system. The total intensity of the laser output is detected with the digital oscilloscope.

Typical temporal waveforms of the drive and the response are shown in Figure 5.36a (right) (Uchida *et al.*, 2003e). There is no obvious correlation between the drive and response outputs. Indeed, the correlation plot between the drive and response waveforms shows no evidence of identical synchronization (Figure 5.36b (right)). Then, the drive signal is sent to the response laser repeatedly, in order to apply the auxiliary system approach. The middle and bottom plots of Figure 5.36a (right) show two response outputs driven by the same drive signal at different times. The correlation plot between the two response outputs shows a linear correlation, as shown in Figure 5.36c (right). This result implies that the response laser driven by the same drive signal always generates identical outputs, independent of initial conditions. Since the dynamics of the response laser are repeatable and reproducible, generalized synchronization can be stably achieved in this system. Numerical results agree well with the experimental observations, as shown in Figure 5.36d (right). It turns out that two-mode dynamics is responsible for the occurrence of generalized synchronization in the identical systems (Uchida *et al.*, 2003e).

Generalized synchronization of optical 2D patterns has also been observed experimentally in a spatiotemporal chaotic system with a liquid-crystal spatial-light modulator with optoelectronic feedback (Rogers *et al.*, 2004).

5.7
Consistency

The study of generalized synchronization leads to a more general concept of "consistency", which is the ability of a nonlinear dynamical system to produce identical response outputs after some transient period, when the system is driven by a repeated complex drive signal (Uchida *et al.*, 2004a).

5.7.1
What is Consistency?

Many nonlinear dynamical systems have an ability to generate consistent outputs when driven by a repeated external signal. "Consistency" is defined as the reproducibility of response waveforms in a nonlinear dynamical system driven repeatedly by a signal, starting from different initial conditions of the system. The concept of consistency is illustrated in Figure 5.37 (Uchida *et al.*, 2004a). Any complex waveform such as deterministic chaos or stochastic noise can be used as a drive signal. This drive signal is sent repeatedly to a nonlinear dynamical system (called the response system) starting from arbitrary initial conditions. Complex temporal waveforms of the response system are obtained at each repetition of the drive signal. Consistency is the ability of a system to produce identical response outputs after some transient period, when the system is driven by a repeated complex drive signal.

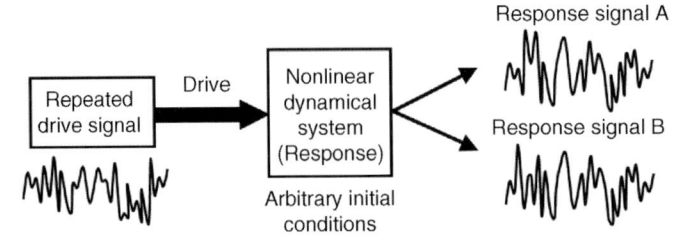

Figure 5.37 Concept of consistency. (A. Uchida, R. McAllister, and R. Roy, (2004a), © 2004 APS.)

Several phenomena related to consistency have been studied in various nonlinear dynamical systems. Generalized synchronization has been observed in which there is a functional relation between the dynamics of a drive and response system, but the dynamics may differ greatly in character (see Section 5.6.2). Noise-induced synchronization is a phenomenon where two identical nonlinear systems driven by a common noise signal can be identically synchronized to each other (Zhou et al., 2003). Reliability of spike timing in neurons has been investigated, where neurons that are repeatedly driven by a random drive signal can fire a consistent spike train with high temporal precision (Mainen and Sejnowski, 1995). The characteristics of these three examples of drive–response systems can be commonly interpreted from the viewpoint of consistency: Reproducibility of the response outputs with respect to a repeated drive signal is essential for all the three phenomena.

The viewpoint of synchronization is based on the "relationship" among coupled dynamical systems. The viewpoint of consistency is based on the "reproducibility" or "stability" of a response system with respect to a complex drive signal, including chaos and noise. Depending on the type of drive signal, synchronization can be classified into different categories: identical synchronization with an identical chaotic drive signal, generalized synchronization with the chaotic drive signal generated from different chaotic system or parameter-mismatched chaotic system, and noise-induced synchronization with a noise drive signal. These three phenomena can be interpreted from the same viewpoint: the stability of the response system with respect to "any" type of complex drive signals. For the viewpoint of consistency, there is no restriction on the types of the drive signal and the reproducibility of the response system with respect to the drive signal can be treated, regardless of the type of drive signal.

Some general questions for consistency are the following: is it possible to obtain consistency of the response system for "any" kind of drive signals? How to design the drive signal to obtain "optimal" consistency with less energy of the drive signal? The concept of consistency is free from restrictions concerning the type of drive signals and can be used to consider the universal concept of susceptibility of driven response systems with respect to arbitrary external drive signals.

In stochastic resonance and coherence resonance, it is the periodicity or coherence of the system response that is of interest, and this is modified by an external noise signal (Lindner et al., 2004). Here, the focus is on the consistency of the system response to a repeated, complex waveform drive signal. The response signal may or may not have a functional relationship to the drive signal; its consistency is a measure of the ability of the external drive to interact with and excite the system degrees of freedom in a reproducible fashion.

The viewpoint of synchronized responses strongly emphasizes the similarity of behavior in time – a certain waveform feature in the response system corresponds to a certain waveform feature in the drive system, or another response system, with a fixed time relation. This viewpoint is particularly useful for viewing the drive–response relations in oscillatory systems. However, for many drive–response systems in social, behavioral and medical sciences the timing relation does not seem to be so strict, and so other aspects of the response may be more relevant and useful. The consistency viewpoint emphasizes the reproducibility of responses in driven systems.

The conditional Lyapunov exponent can be used to analyze the occurrence of consistency, as in the case of synchronization. The stability of the response system can be described by linearized equations for the response system, thus the mathematical formula of the linearized equations for the response system looks independent of the drive signals. However, the linearized equations include the original variables of the response system that are also affected by the drive signal and the resultant change in the trajectory of the response system. The stability of the response system thus strongly depends on the type of drive signals.

5.7.2
Examples of Consistency in Laser Systems

In this section a quantitative measure of consistency is observed experimentally in the dynamics of a laser system driven repeatedly by a complex waveform such as chaos and noise (Uchida et al., 2004a). A nonlinear dynamical system used is a laser-diode-pumped Nd:YAG microchip laser, similar to Figure 5.36 (left). Two longitudinal modes are observed in the output of the laser in a wide range of pump power. An AOM is inserted in the laser cavity to modulate the loss of the laser cavity. The digitized drive signal stored in a computer is sent to the AOM in the laser system through an amplifier and a low-pass filter to smooth the signal by using an arbitrary function generator connected with the computer. The response laser system is driven repeatedly by the same drive signal, and the temporal waveform of the response laser output is detected by using a digital oscilloscope and photodiode. The detected signals are compared for different repetitions to observe the consistency of the response waveforms.

One of the advantages of the consistency viewpoint is that different types of drive signals, including deterministic chaos and stochastic noise, can be used to test the consistency of the response system. In this experiment, chaotic and colored-noise temporal waveforms are used as drive signals. First, a chaotic signal generated by the same laser system with closed feedback loop is used as a drive signal. The chaotic signal is sent to the response laser repeatedly. Temporal waveforms of the chaotic drive signal and two response laser outputs obtained from the experiment are shown in Figure 5.38a. Two consistent response outputs are clearly observed after a transient of ∼1 ms in Figure 5.38a, even though the drive and response signals are totally different. Next, a colored noise signal generated by a numerical algorithm is used as a

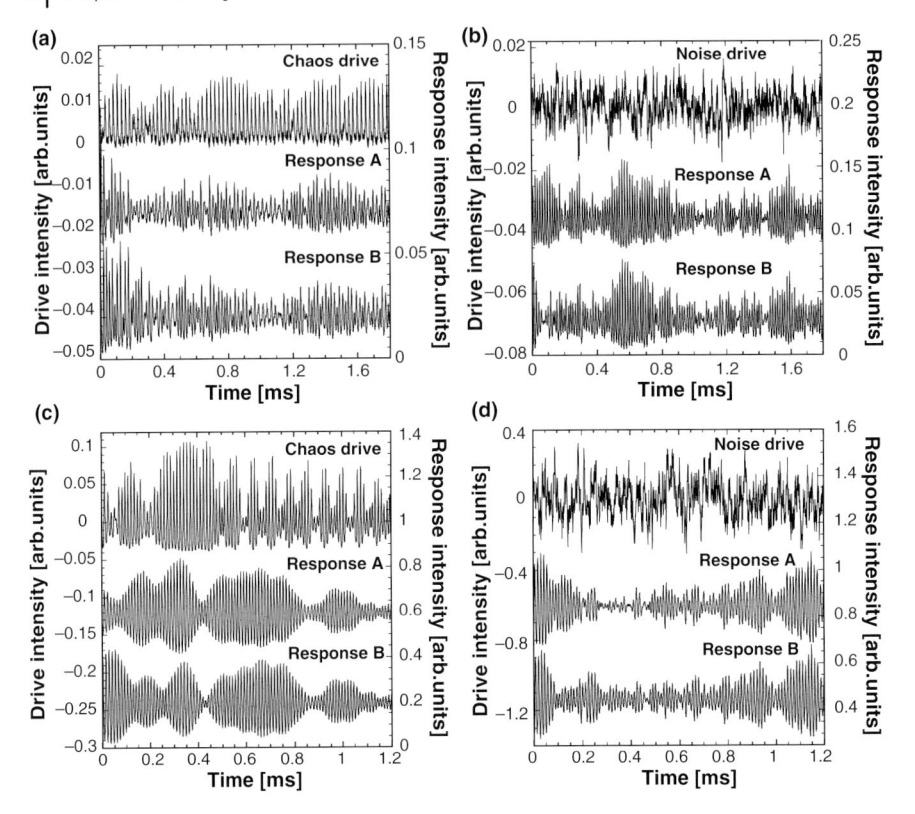

Figure 5.38 Experimental results of temporal waveforms of (a) chaotic drive signal and (b) colored noise drive signal, and two corresponding response laser outputs. Numerical results of temporal waveforms of (c) chaotic drive signal and (d) colored noise drive signal, and two corresponding response laser outputs. Consistent outputs are observed for all the figures after transients. (A. Uchida, R. McAllister, and R. Roy, (2004a), © 2004 APS.)

drive signal and the noise signal is sent to the laser system repeatedly. The inverse of the correlation time for the colored noise is set to be 40 kHz. Temporal waveforms obtained from the experiment are shown in Figure 5.38b. Consistent outputs of the response laser driven by the same colored noise signal are obtained after a transient as well. The corresponding numerical results for chaotic and colored noise drive signals are shown in Figures 5.38c and d, respectively. Two response outputs starting from different initial conditions converge to the consistent outputs after transients in both cases. These numerical results agree well with the experimental observation shown in Figures 5.38a and b.

To investigate the characteristics of consistency, the consistency parameter C is quantitatively defined as the cross-correlation of two response temporal waveforms normalized by the product of their standard deviations, as the same formula for synchronization shown in Eq. (5.19).

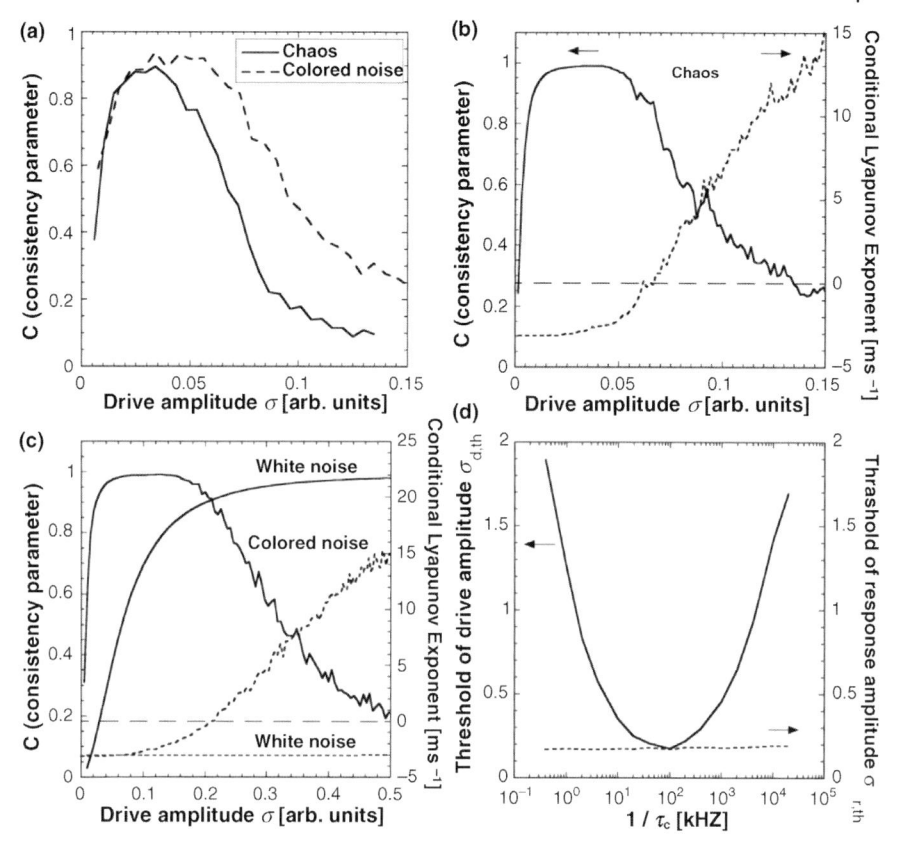

Figure 5.39 (a) Experimental result of the consistency parameter C as a function of the amplitude (measured as the standard deviation) of the chaos (solid curve) and colored noise (dashed curve) drive signals. (b) Numerical result of C (solid curve) and the conditional Lyapunov exponent (dotted curve) as a function of the amplitude of the chaos drive signal. (c) Numerical result of C (solid curve) and the conditional Lyapunov exponent (dotted curve) as a function of the amplitude of colored noise and white noise drive signals. (d) Thresholds for inconsistency with drive amplitude (solid curve) and with response amplitude (dotted line) as a function of the inverse of the correlation time $1/\tau_c$ for the colored noise drive signal. Note that the minimum of the solid curve coincides with the laser relaxation oscillation frequency. (A. Uchida, R. McAllister, and R. Roy, (2004a), © 2004 APS.)

The amplitude of the drive waveform is varied to measure the characteristics of consistency. The average of the consistency parameter C is computed after the transient period. The experimental results for the consistency as a function of the amplitude of the drive waveform are shown in Figure 5.39a for chaos and colored noise drive signals (Uchida et al., 2004a). There is a maximum of the consistency curve for both drive signals as the amplitude is increased. In the very small amplitude region ($\sigma < 0.02$, where σ is the standard deviation of the drive signal as a measure of the amplitude), consistent outputs are not observed because the internal noise-driven relaxation oscillations are dominant in the laser output. As the amplitude is increased, the drive signal overcomes the internal noise and optimal, consistent outputs are observed. When the amplitude of the drive signal is increased further, the consistency decreases and inconsistent outputs are observed.

Corresponding numerical results indicate a similar curve with the maximum peak of consistency as shown in Figure 5.39b (solid curve) for the chaos drive signal. In order to explain the development of inconsistency in the case of strong driving, the conditional Lyapunov exponent (see Sections 5.3.5 and 6.3.4) along the trajectory of the response laser output is also estimated without the internal noise terms (dotted curve of Figure 5.39b). It is found that inconsistency appears at large drive amplitudes when the sign of the conditional Lyapunov exponent changes from negative to positive. This result suggests that the dynamical response of the nonlinear system generates inconsistency for large-amplitude drive waveforms. On the other hand, inconsistency appears at small drive amplitudes even though the conditional Lyapunov exponent is negative. In this region, internal noise dominates and prevents consistency. Consistency is thus optimized between these two regimes for some intermediate amplitude of drive waveform.

The threshold for inconsistency is defined as the drive amplitude at which the sign of the conditional Lyapunov exponent changes from negative to positive values. The threshold is dependent on the type of drive waveform. Numerical results show that the threshold for the chaos drive signal ($\sigma_{\text{th,chaos}} = 0.061$) is smaller than that for the colored noise drive signal ($\sigma_{\text{th,noise}} = 0.200$), as shown in the dotted curves of Figures 5.39b and c. Since the chaotic signal is generated from the same laser system, it excites the dynamics of the laser system more effectively than the colored noise signal. This chaotic excitation thus leads more quickly to instability and reduces the threshold for inconsistency with the chaotic drive signal than with the colored noise drive.

In the limit of the correlation time for the colored noise $\tau_c \to 0$, which represents white noise, a threshold for inconsistency is not found, as shown in Figure 5.39c. This point is further clarified when the correlation time τ_c is changed for the colored noise systematically. The threshold for inconsistency for the colored noise is plotted as a function of $1/\tau_c$ (i.e., the characteristic frequency of the colored noise) as shown in the solid curve of Figure 5.39d. The threshold value changes as $1/\tau_c$ is changed. The solid curve of the threshold for inconsistency displays a clear minimum around a frequency of 80 kHz that corresponds to the relaxation oscillation frequency of the laser. As $1/\tau_c$ differs further from the relaxation oscillation frequency, the inconsistency threshold value becomes larger. In the high-bandwidth (white noise) limit, the inconsistency threshold appears to become infinite and consistency is always maintained even with strong driving.

The amplitude of the response laser outputs is also measured at the threshold for inconsistency in the dotted line of Figure 5.39d. It is worth noting that the amplitude of the response outputs at the threshold for inconsistency is almost constant when $1/\tau_c$ of the colored noise is changed, as shown in Figure 5.39d. Therefore, the threshold for inconsistency is dominated by the amplitude of the response laser output, $\sigma_{\text{th,res}} = 0.18$, that is, by the nature of its deterministic dynamics in response to the drive signal.

5.7.3
Application of Consistency

The concept of consistency could be applied for engineering applications. The analysis of nonlinear dynamical systems in terms of drive–response characteristics is widely applicable to physics, chemistry, biology, and engineering. For example, consistency of dynamics is essential for information transmission in biological and physiological systems and for reproduction of spatiotemporal patterns in nature. Consistency tests could be applied in noninvasive diagnostic procedures to detect changes in system parameters due to aging, catastrophic events, or other system changes. Lasers systems can provide good experimental platforms for studying such consistent complex dynamics.

The response of nonlinear dynamical systems to complex drive signals can be observed in numerical or laboratory experiments, and the usefulness of these will depend critically on the design of the drive waveform and the measures used to evaluate or characterize the response. In the following, some possible applications based on consistency are discussed (Uchida *et al.*, 2008b).

5.7.3.1 Noninvasive Testing
The response of structures ranging from biomolecules to large aircrafts to complex drive waveforms may provide a means for noninvasive testing (Moniz *et al.*, 2003) and evaluation of the integrity of these structures and changes due to aging or damage (Daido and Nakanishi, 2004). This approach can be extended to systems with internal dynamics, such as electronic circuits, laser systems and even the brain.

5.7.3.2 Analysis of Brain Dynamics and Learning Process in the Brain
In past work on neuronal systems, the concept of reliability has been used to describe the consistency of the evoked spike-sequences of rat cortical neurons in response to repeated presentations of the same noise waveform (Mainen and Sejnowski, 1995). However, characterizing the responses of nonlinear dynamical systems with complex time evolution presents many challenges. It may appear at first sight that the wide variety of possible responses for a system with complex dynamics would make this task hopeless. On the other hand, it is well known from everyday experience that a system as complex as the brain can have responses with various degrees of consistency when subjected to repeated inputs of information and stimulation. Consistency could be a useful tool to analyze the brain dynamics and learning process in the brain (Pérez and Uchida, 2011).

5.7.3.3 Physical One-Way Function
Another example is the generation of secure tokens using physical one-way functions. One approach has been demonstrated using the speckle patterns generated by light scattering from a rough surface (Pappu *et al.*, 2002). For this function it is essential that the drive–response relation is complex yet reproducible. It is an open question whether the complexity and robustness characteristics can be improved by using dynamical devices with consistency (Kanno and Uchida, 2011).

5.7.3.4 Teaching-Learning Methodologies

The general concept of consistency and the measures introduced here should be very widely applicable to systems in the social, behavioral and medical sciences as well. The implementation and evaluation of teaching–learning methodologies, as developed in the context of modern cognitive science, provide many examples of drive–response relationships that are complex and involve many variables, as well as the use of feedback and memory functions. Rapidly developing understanding of cognition, aided by the development of sophisticated neuronal modeling and measurements, is another area for the application of the basic ideas of consistency.

5.7.3.5 Design of Drug Delivery

Psychological tests for the emergence of Alzheimers disease rely on the accuracy and consistency of the answers received in response to a large set of questions designed to test different aspects of brain function. The design of appropriate drug delivery and simultaneous time-dependent monitoring of parameters of the human body to combat disease is another example where consistency in all its implications is of prime importance to determine effective therapeutic strategies.

The above-described research fields are examples of the applications of the consistency viewpoint. Consistency analysis would be applied for nonlinear dynamical systems with drive–response relationship in many interdisciplinary research fields.

Appendix

5.A.1
Pecora–Carroll Method for Synchronization of Chaos

The Pecora–Carroll method for synchronization of chaos can be represented mathematically by considering an autonomous n-dimensional dynamical system,

$$\frac{d\mathbf{u}}{dt} = \mathbf{f}(\mathbf{u}) \tag{5.A.1}$$

where $\mathbf{u} = (u_1, u_2, \ldots, u_n)$ is an n-dimensional vector variable describing the state of the system, which can be divided arbitrarily into two subsystems:

$$\frac{d\mathbf{v}}{dt} = \mathbf{g}(\mathbf{v}, \mathbf{w}) \tag{5.A.2}$$

$$\frac{d\mathbf{w}}{dt} = \mathbf{h}(\mathbf{v}, \mathbf{w}) \tag{5.A.3}$$

with $\mathbf{u} = (u_1, u_2, \ldots, u_m)$ and $\mathbf{w} = (u_{m+1}, u_{m+2}, \ldots, u_n)$ representing the reduced state vectors of the two subsystems, and $\mathbf{g} = (f_1(\mathbf{u}), f_2(\mathbf{u}), \ldots, f_m(\mathbf{u}))$ and $\mathbf{h} = (f_{m+1}(\mathbf{u}), f_{m+2}(\mathbf{u}), \ldots, f_n(\mathbf{u}))$ being the respective forces. One now creates a response system as a copy $\mathbf{w}'$ of the $\mathbf{w}$ subsystem with the following dynamics:

$$\frac{d\mathbf{w}'}{dt} = \mathbf{h}(\mathbf{v}, \mathbf{w}') \tag{5.A.4}$$

Hence, the dynamics of the response subsystem $\mathbf{w}'$ depends on the driving variable $\mathbf{v}$. This corresponds to a unidirectional coupling between the drive and response subsystems. The stability of the synchronized solution $\mathbf{w} = \mathbf{w}'$ can be determined by analyzing the linear equation

$$\frac{d\xi}{dt} = \mathbf{D_w h}(\mathbf{v}, \mathbf{w})\xi \qquad (5.A.5)$$

where $\xi = \mathbf{w} - \mathbf{w}'$ and $\mathbf{D_w h}$ is the Jacobian of the drive vector field with respect to $\mathbf{w}$. The eigenvalues of this Jacobian are called the conditional Lyapunov exponents of the coupled subsystem, and their signs determine the stability of the synchronized solution: the subsystems $\mathbf{w}$ and $\mathbf{w}'$ synchronize only if all the conditional Lyapunov exponents are negative.

As a simple example, let us consider the Lorenz system:

$$\frac{dx(t)}{dt} = \sigma(y(t) - x(t)) \qquad (5.A.6)$$

$$\frac{dy(t)}{dt} = rx(t) - y(t) - x(t)z(t) \qquad (5.A.7)$$

$$\frac{dz(t)}{dt} = -bz(t) + x(t)y(t) \qquad (5.A.8)$$

We can divide this set of equations into two subsystems given by the variables $(y(t))$ and $(x(t), z(t))$, and consider a response subsystem $(x'(t), z'(t))$ with dynamics:

$$\frac{dx'(t)}{dt} = \sigma(y(t) - x'(t)) \qquad (5.A.9)$$

$$\frac{dz'(t)}{dt} = -bz'(t) + x'(t)y(t) \qquad (5.A.10)$$

so that $y(t)$ is the driving signal in this case. This coupling scheme is depicted schematically in Figure 5.4 (left). The Jacobian $\mathbf{D_w h}$ is in this case:

$$\mathbf{D_w h} = \begin{pmatrix} -\sigma & 0 \\ y & -b \end{pmatrix} \qquad (5.A.11)$$

Hence, the two conditional Lyapunov exponents in this case $(-\sigma, -b)$ are negative and the synchronized solution is stable.

5.A.2
General Formula of Linearized Equations for Coupled Differential Equations

Identical synchronization is considered in Section 5.3.3.1, and original dynamical equations with the identical feedback and injection signals are generalized as,

$$\frac{d\mathbf{X}(t)}{dt} = \mathbf{F}(t, \mathbf{X}) + \mathbf{G}(t, \mathbf{X}) \qquad (5.A.12)$$

$$\frac{d\mathbf{Y}(t)}{dt} = \mathbf{F}(t,\mathbf{Y}) + \mathbf{G}(t,\mathbf{X}) \tag{5.A.13}$$

where $\mathbf{X}(t) = (x_1(t)\ x_2(t)\ x_3(t)\ \ldots\ x_n(t))^T$ is a vector of n variables for the drive system, $\mathbf{Y}(t) = (y_1(t)\ y_2(t)\ y_3(t)\ \ldots\ y_n(t))^T$ is a vector of n variables for the response system, $\mathbf{F} = (f_1\ f_2\ f_3\ \ldots\ f_n)$ is a matrix of n nonlinear functions without feedback and coupling terms, $\mathbf{G} = (g_1\ g_2\ g_3\ \ldots\ g_n)$ is a matrix that describes the feedback terms for the drive system and the coupling terms for the response system, and t is time.

The identical synchronous solution is obtained as,

$$\mathbf{X}(t) = \mathbf{Y}(t) \tag{5.A.14}$$

Let us consider linear stability analysis of the identical synchronous solution, and introduce linearized variables as,

$$\Delta(t) = \mathbf{Y}(t) - \mathbf{X}(t) \tag{5.A.15}$$

where $\Delta(t) = (y_1(t) - x_1(t),\ y_2(t) - x_2(t),\ \ldots,\ y_n(t) - x_n(t))^T$ is a vector of the linearized variables for the synchronous solution. Subtracting Eq. (5.A.12) from Eq. (5.A.13),

$$\begin{aligned}\frac{d(\mathbf{Y}(t) - \mathbf{X}(t))}{dt} &= \mathbf{F}(t,\mathbf{Y}) + \mathbf{G}(t,\mathbf{X}) - \mathbf{F}(t,\mathbf{X}) - \mathbf{G}(t,\mathbf{X}) \\ &= \mathbf{F}(t,\mathbf{Y}) - \mathbf{F}(t,\mathbf{X})\end{aligned} \tag{5.A.16}$$

Note that the coupling terms $\mathbf{G}(t,\mathbf{X})$ are eliminated in Eq. (5.A.16) due to the symmetry of the equations. $\Delta(t)$ of Eq. (5.A.15) can be substituted into Eq. (5.A.16) and the linear approximation is applied (i.e., $|\Delta(t)| \ll |\mathbf{X}(t)|, |\mathbf{Y}(t)|$, $|\Delta(t)|^2 \ll |\Delta(t)|$). The linearized equations are obtained as,

$$\frac{d\Delta(t)}{dt} = \mathbf{D}_{\mathbf{X}}\mathbf{F}(t,\mathbf{X})\Delta(t) \tag{5.A.17}$$

where $\mathbf{D}_{\mathbf{X}}\mathbf{F}(t,\mathbf{X})$ is the Jacobian matrix defined as,

$$\mathbf{D}_{\mathbf{X}}\mathbf{F}(t,\mathbf{X}) = \begin{pmatrix} \frac{\partial f_1}{\partial x_1} & \frac{\partial f_1}{\partial x_2} & \frac{\partial f_1}{\partial x_3} & \cdots & \frac{\partial f_1}{\partial x_n} \\ \frac{\partial f_2}{\partial x_1} & \frac{\partial f_2}{\partial x_2} & \frac{\partial f_2}{\partial x_3} & \cdots & \frac{\partial f_1}{\partial x_n} \\ \frac{\partial f_3}{\partial x_1} & \frac{\partial f_3}{\partial x_2} & \frac{\partial f_3}{\partial x_3} & \cdots & \frac{\partial f_1}{\partial x_n} \\ \vdots & \vdots & \vdots & \ddots & \vdots \\ \frac{\partial f_n}{\partial x_1} & \frac{\partial f_n}{\partial x_2} & \frac{\partial f_n}{\partial x_3} & \cdots & \frac{\partial f_n}{\partial x_n} \end{pmatrix} \tag{5.A.18}$$

Note that the Jacobian matrix is obtained only from the nonlinear function $\mathbf{F}$ without coupling terms in this case. This indicates that a synchronous solution is obtained from the nonlinear equations without coupling terms when the coupling terms are identical between the drive and response lasers.

5.A.3
Procedure for the Calculation of the Conditional Lyapunov Exponent

The conditional Lyapunov exponent is calculated from the linearized equations Eq. (5.A.17) for the identical synchronous solutions of Eq. (5.A.14). Only the maximum value of the conditional Lyapunov exponents is considered to investigate synchronizability. The norm is simply defined as the Euclidean distance of the vector consisting of all the variables of the linearized equations, that is

$$D_c(t) = |\Delta(t)| = \sqrt{\Delta_1^2(t) + \Delta_2^2(t) + \cdots + \Delta_n^2(t)} \quad (5.A.19)$$

$D_c(t)$ can be obtained at each step of numerical integration. After a short evolution of the trajectory from t to $t + \tau_{ev}$ (τ_{ev} is the evolution time for norm calculation, and it is recommended to use a small value for τ_{ev} such as the integration step of numerical calculation to maintain the validity of the linearized equations), the variation of the norm is calculated as,

$$d_j = \frac{D_c(t + \tau_{ev})}{D_c(t)} \quad (5.A.20)$$

where d_j is the ratio of the norms after the short evolution at the jth procedure of the above calculation of the norm ratio. Next, all the linearized variables $\Delta_i(t + \tau_{ev})$ ($i = 1, 2, \ldots, n$) are rescaled by $D_c(t + \tau_{ev})$ after the calculation of the norm ratio as,

$$\Delta_{i,new}(t + \tau_{ev}) = \frac{\Delta_i(t + \tau_{ev})}{D_c(t + \tau_{ev})} \quad (5.A.21)$$

This rescaling procedure is very important to maintain a small deviation of the linearized variables from synchronous trajectory. The integration of the trajectory resumes for the next evolution of τ_{ev}. After this procedure is repeated N times, where $N = T/\tau_{ev}$ and T is the total calculation time, the conditional Lyapunov exponent is calculated as,

$$\lambda_c = \frac{1}{N\tau_{ev}} \sum_{j=1}^{N} \ln(d_j) \quad (5.A.22)$$

where ln is the natural logarithm with the base e. The conditional Lyapunov exponent can be obtained as the average ratio of the norm change in the natural logarithmic scale. In this procedure, the Jacobian matrix of the original equations (but without coupling terms) is essentially calculated along the trajectory of the attractor, and the growth of perturbation between two synchronous trajectories is evaluated.

5.A.4
Numerical Model for Synchronization of Chaos in Unidirectionally Coupled Semiconductor Lasers

5.A.4.1
Coherent Optical Feedback

A numerical model for chaotic dynamics in a semiconductor laser with optical feedback is described in Appendix 3.A.1.1 and Chapter 4, known as the Lang–Kobayashi equations (Lang and Kobayashi, 1980). The coupled Lang–Kobayashi equations of unidirectionally coupled semiconductor lasers with optical feedback in the open-loop configuration are described as follows (see in Chapter 6 for details),

$$\frac{dE_d(t)}{dt} = \frac{1}{2}\left[G_{N,d}(N_d(t)-N_{0,d}) - \frac{1}{\tau_{p,d}}\right]E_d(t) + \kappa_d\, E_d(t-\tau_d)\cos\Theta_d(t) \quad (5.A.23)$$

$$\frac{d\Phi_d(t)}{dt} = \frac{\alpha_d}{2}\left[G_{N,d}(N_d(t)-N_{0,d}) - \frac{1}{\tau_{p,d}}\right] - \kappa_d\frac{E_d(t-\tau_d)}{E_d(t)}\sin\Theta_d(t) \quad (5.A.24)$$

$$\frac{dN_d(t)}{dt} = J_d - \frac{N_d(t)}{\tau_{s,d}} - G_{N,d}(N_d(t)-N_{0,d})E_d^2(t) \quad (5.A.25)$$

$$\Theta_d(t) = \omega_d\tau_d + \Phi_d(t) - \Phi_d(t-\tau_d) \quad (5.A.26)$$

$$\frac{dE_r(t)}{dt} = \frac{1}{2}\left[G_{N,r}(N_r(t)-N_{0,r}) - \frac{1}{\tau_{p,r}}\right]E_r(t) + \kappa_{\text{inj}}\, E_d(t-\tau_{\text{inj}})\cos\Theta_{\text{inj}}(t) \quad (5.A.27)$$

$$\frac{d\Phi_r(t)}{dt} = \frac{\alpha_r}{2}\left[G_{N,r}(N_r(t)-N_{0,r}) - \frac{1}{\tau_{p,r}}\right] - \kappa_{\text{inj}}\frac{E_d(t-\tau_{\text{inj}})}{E_r(t)}\sin\Theta_{\text{inj}}(t) \quad (5.A.28)$$

$$\frac{dN_r(t)}{dt} = J_r - \frac{N_r(t)}{\tau_{s,r}} - G_{N,r}(N_r(t)-N_{0,r})E_r^2(t) \quad (5.A.29)$$

$$\Theta_{\text{inj}}(t) = -\Delta\omega\, t + \omega_d\tau_{\text{inj}} + \Phi_r(t) - \Phi_d(t-\tau_{\text{inj}}) \quad (5.A.30)$$

where $E(t)$ is the real electric amplitude, $\Phi(t)$ is the real optical phase, and $N(t)$ is the carrier density. The symbols and parameters are summarized in Tables 6.2 and 6.3 in Chapter 6. The subscripts d and r indicate the drive and response lasers, respectively.

5.A.4.2
Polarizaiton-Rotated Optical Feedback

A two-polarization-mode model for describing the dynamics of the TE and TM polarization modes in a semiconductor laser with polarization rotated optical feedback is described in Appendix 3.A.1.2. The rate equations for unidirectionally coupled semiconductor lasers with polarization-rotated optical feedback and injection can be written as follows (Shibasaki et al., 2006)

$$\frac{dE_{te,d}(t)}{dt} = \frac{1}{2}\left\{G_{te,d}\left(N_d(t)-N_{0,d}\right)-\gamma_{te,d}\right\}E_{te,d}(t) \tag{5.A.31}$$

$$\frac{d\Phi_{te,d}(t)}{dt} = \frac{\alpha_d}{2}\left\{G_{te,d}\left(N_d(t)-N_{0,d}\right)-\gamma_{te,d}\right\} \tag{5.A.32}$$

$$\frac{dE_{tm,d}(t)}{dt} = \frac{1}{2}\left\{G_{tm,d}\left(N_d(t)-N_{0,d}\right)-\gamma_{tm,d}\right\}E_{tm,d}(t) + \kappa_d E_{te,d}(t-\tau_d)\cos\theta_d(t) \tag{5.A.33}$$

$$\frac{d\Phi_{tm,d}(t)}{dt} = \frac{\alpha_d}{2}\left\{G_{tm,d}\left(N_d(t)-N_{0,d}\right)-\gamma_{tm,d}\right\}-\kappa_d\frac{E_{te,d}(t-\tau_d)}{E_{tm,d}(t)}\sin\theta_d(t) \tag{5.A.34}$$

$$\frac{dN_d(t)}{dt} = J_d - \gamma_{s,d}N_d(t) - (N_d(t)-N_{0,d})\left\{G_{te,d}\left|E_{te,d}(t)\right|^2 + G_{tm,d}\left|E_{tm,d}(t)\right|^2\right\} \tag{5.A.35}$$

$$\theta_d(t) = -\Delta\omega_{te,tm}t + \omega_d\tau_d + \Phi_{tm,d}(t) - \Phi_{te,d}(t-\tau_d) \tag{5.A.36}$$

$$\frac{dE_{te,r}(t)}{dt} = \frac{1}{2}\left\{G_{te,r}\left(N_r(t)-N_{0,r}\right)-\gamma_{te,r}\right\}E_{te,r}(t) \tag{5.A.37}$$

$$\frac{d\Phi_{te,r}(t)}{dt} = \frac{\alpha_r}{2}\left\{G_{te,r}\left(N_r(t)-N_{0,r}\right)-\gamma_{te,r}\right\} \tag{5.A.38}$$

$$\frac{dE_{tm,r}(t)}{dt} = \frac{1}{2}\left\{G_{tm,r}\left(N_r(t)-N_{0,r}\right)-\gamma_{tm,r}\right\}E_{tm,r}(t) + \kappa_r E_{te,d}(t-\tau_r)\cos\theta_r(t) \tag{5.A.39}$$

$$\frac{d\Phi_{tm,r}(t)}{dt} = \frac{\alpha_r}{2}\left\{G_{tm,r}\left(N_r(t)-N_{0,r}\right)-\gamma_{tm,r}\right\}-\kappa_r\frac{E_{te,d}(t-\tau_r)}{E_{tm,r}(t)}\sin\theta_r(t) \tag{5.A.40}$$

$$\frac{dN_r(t)}{dt} = J_r - \gamma_{s,r}N_r(t) - (N_r(t)-N_{0,r})\left\{G_{te,r}\left|E_{te,r}(t)\right|^2 + G_{tm,r}\left|E_{tm,r}(t)\right|^2\right\} \tag{5.A.41}$$

$$\theta_r(t) = -\Delta\omega_{d,r}t + \omega_d\tau_r + \Phi_{tm,r}(t) - \Phi_{te,d}(t-\tau_r) \tag{5.A.42}$$

The variables E and Φ are the electrical amplitude and the phase, N is the carrier density, and θ is the phase difference between the TM mode of the laser and the feedback or injected light. The subscripts te and tm indicate the TE- and TM-modes, respectively. The subscripts d and r indicate the drive and response lasers, respectively. G is the gain coefficient, and N_0 is the carrier density at the transparency. γ is the inverse of the photon lifetime, γ_s is the inverse of the carrier lifetime, α is the linewidth-enhancement factor, J is the injection current density per unit time, and $J_{th,sol} = \gamma_s(N_0 + \gamma_{te}/G_{te})$ is the threshold of the injection current density per unit time. κ_d is the feedback coefficient for the drive laser, κ_r is the injection coefficient from the drive to the response laser, τ_d is the propagation time of the external loop in the drive laser, and τ_r is the propagation time from the drive to the response laser. ω is the angular frequency, $\Delta\omega_{d,r} = \omega_d - \omega_r = 2\pi\Delta f$ is the detuning of the optical angular frequency between the drive and response lasers, Δf is the detuning of the optical frequency between the drive and response lasers, and $\lambda = 2\pi c/\omega$ is the wavelength. $\Delta\omega_{te,tm} = \omega_{te,d} - \omega_{tm,d}$ is the detuning of optical angular frequency between the TE and TM modes of the drive laser. The total intensity of the drive and response lasers can be obtained from $I_d = |E_{te,d}|^2 + |E_{tm,d}|^2$ and $I_r = |E_{te,r}|^2 + |E_{tm,r}|^2$, respectively.

Typical parameter values for numerical simulation are as follows: $G_{te} = 1.374 \times 10^{-12} m^3 s^{-1}$, $G_{tm} = 1.154 \times 10^{-12} m^3 s^{-1}$, $N_0 = 1.400 \times 10^{24} m^{-3}$, $\gamma_{te} = \gamma_{tm} = 8.913 \times 10^{11} s^{-1}$, $\kappa_d = \kappa_r = 1.25 \times 10^{11} s^{-1}$, $\gamma_s = 4.902 \times 10^8 s^{-1}$, $\tau_d = \tau_r = 6.67 \times 10^{-9} s$, $\alpha = 3.0$, $J = 2.0 J_{th,sol}$, $J_{th,sol} = 1.004 \times 10^{33} m^{-3} s^{-1}$, $\lambda = 1.537 \times 10^{-6} m$ and $\Delta f = 0\ GHz$.

5.A.4.3
Optoelectronic Feedback (Incoherent Feedback)

A numerical model for chaotic dynamics in a semiconductor laser with optoelectronic feedback is described in Appendix 3.A.1.3. Two unidirectionally coupled semiconductor lasers with optoelectronic feedback in the closed-loop configuration can be modeled as (Tang and Liu, 2001a; Liu et al., 2002a),

$$\frac{dS_d(t)}{dt} = -\gamma_{c,d} S_d(t) + \Gamma_d g_{0,d} S_d(t) + \Gamma_d g_{n,d}(N_d(t) - N_{0,d}) S_d(t) \quad (5.A.43)$$
$$+ \Gamma_d g_{p,d}(S_d(t) - S_{0,d}) S_d(t)$$

$$\frac{dN_d(t)}{dt} = \frac{J_d}{ed}\left[1 + \frac{\xi_d S_d(t-\tau_d)}{S_{0,d}}\right] - \gamma_{s,d} N_d(t) - g_{0,d} S_d(t) \quad (5.A.44)$$
$$- g_{n,d}(N_d(t) - N_{0,d}) S_d(t) - g_{p,d}(S_d(t) - S_{0,d}) S_d(t)$$

$$\frac{dS_r(t)}{dt} = -\gamma_{c,r} S_r(t) + \Gamma_r g_{0,r} S_r(t) + \Gamma_r g_{n,r}(N_r(t) - N_{0,r}) S_r(t) \quad (5.A.45)$$
$$+ \Gamma_r g_{p,r}(S_r(t) - S_{0,r}) S_r(t)$$

$$\frac{dN_r(t)}{dt} = \frac{J_r}{ed}\left[1 + \frac{\xi_c S_d(t-\tau_c) + \xi_r S_r(t-\tau_r)}{S_{0,r}}\right] - \gamma_{s,r} N_r(t) - g_{0,r} S_r(t) \quad (5.A.46)$$
$$- g_{n,r}(N_r(t) - N_{0,r}) S_r(t) - g_{p,r}(S_r(t) - S_{0,r}) S_r(t)$$

where the subscripts d, r indicate the drive and the response lasers, respectively. $S(t)$ is the intracavity photon density and $N(t)$ is the carrier density. S_0 is the free-running intracavity photon density when the laser is not subject to the feedback, N_0 is the free-running carrier density when the laser is not subject to the feedback, g_0 is the optical gain coefficient at free-running condition, g_n is the differential gain parameter, g_p is the nonlinear gain parameter, ξ_d and ξ_r are the feedback strengths of the drive and response lasers, ξ_c is the coupling strenght, τ_d and τ_r are the feedback delay times of the drive and response lasers, τ_c is the coupling delay time, J is the bias current density, γ_c is the cavity photon decay rate, γ_s is the spontaneous carrier density rate, Γ is the confinement factor of the laser waveguide, e is the electronic charge constant, and d is the active layer thickness. The dimensionless equations and the corresopnding parameter values are shown in Appendix 3.A.1.3.

5.A.5
Numerical Model for Synchronization of Chaos in Unidirectionally Coupled Electro-Optic Systems

A numerical model for chaotic dynamics of an electro-optic system is introduced in Appendix 3.A.4. Two unidirectionally coupled optoelectronic systems with birefringent plates in an open-loop configuration is modeled as (Larger et al., 1998a, 1998b),

$$x_d(t) + \tau_d \frac{dx_d(t)}{dt} = \beta_d \sin^2[\kappa_d x_d(t - T_d) - \Phi_{0,d}] \tag{5.A.47}$$

$$x_r(t) + \tau_r \frac{dx_r(t)}{dt} = \beta_r \sin^2[\kappa_c x_d(t - T_c) + \kappa_r x_r(t - T_r) - \Phi_{0,r}] \tag{5.A.48}$$

where the subscripts d, r indicate the drive and the response systems, respectively. $x(t)$ is the dynamical variable of the system (i.e., laser intensity, phase, or wavelength), T_d and T_r are the delay times in the optoelectronic feedback loops of the drive and response systems, T_c is the coupling delay time between the two systems, τ is the response time of the system, β is the height of the nonlinear function in the optoelectronic feedback loop, Φ_0 the initial phase of the nonlinear function, κ_d and κ_r are the feedback strengths of the drive and response systems, and κ_c is the couling strength from the drive to the response systems.

There is a mathematical synchronous solution of $x_d(t) = x_r(t)$ under the parameter matching condition because of the symmetry of these two equations. Identical synchronization can be achieved with an adequate coupling strength between the transmitter and the receiver.

Typical parameter values for numerical simulation are as follows: $\tau_d = \tau_r = 8.6 \times 10^{-6}$ s, $T_d = T_r = T_c = 5.1 \times 10^{-4}$ s, $\Phi_{0,d} = \Phi_{0,r} = 0.3$, $\kappa_d = \kappa_c = 1$, $\kappa_r = 0$, and $\beta_d = \beta_r = 2.11$.

5.A.6
Numerical Model for Synchronization of Chaos in Unidirectionally Coupled Fiber Lasers

A numerical model for chaotic dynamics of a fiber laser is introduced in Appendix 3.A.5. The numerical model of delay-differential equations is rewritten as (Williams et al., 1997),

$$E(t+\tau_R) = E_{inj}\exp(i\Delta\omega t) + RE(t)\exp\{(\beta+i\alpha)w(t)+i\kappa\} \tag{5.A.49}$$

$$\frac{dw(t)}{dt} = Q - 2\gamma\left(w+1+|E(t)|^2\frac{[\exp\{Gw(t)\}-1]}{G}\right) \tag{5.A.50}$$

where $E(t)$ is the complex envelope of the electric field, measured at a given reference point inside the cavity, and $w(t)$ is the total population inversion of the nonlinear medium. The propagation round-trip time around the cavity is τ_R, and the fraction of light that remains in the cavity after one round-trip is measured by the return coefficient R. The field acquires a phase κ after each round-trip. Light of constant amplitude E_{inj} is assumed to be injected into the cavity, with $\Delta\omega$ the detuning between the injected and the laser frequencies. Additionally, the gain medium is pumped at a rate Q. γ is the decay time of the atomic transition, α is a detuning parameter, and β and G are gain parameters.

The behavior of two coupled erbium-doped fiber-ring lasers operating in a high-dimensional chaotic regime can be modeled, showing that the chaotic dynamics of the two lasers becomes synchronized when a sufficient amount of light from the first laser is injected into the second (Abarbanel and Kennel, 1998). The process is represented schematically by

$$E_t(t+\tau_R) = f(E_t(t), w_t(t)) \tag{5.A.51}$$

$$\frac{dw_t(t)}{dt} = g\left(|E_t(t)|^2, w_t(t)\right) \tag{5.A.52}$$

$$E_r(t+\tau_R) = f(cE_t(t) + (1-c)E_r(t), w_r(t)) \tag{5.A.53}$$

$$\frac{dw_r(t)}{dt} = g\left(|cE_t(t) + (1-c)E_r(t)|^2, w_r(t)\right) \tag{5.A.54}$$

where f and g represent the right-hand-side terms of Eqs. (5.A.49) and (5.A.50), respectively, and the subindices t and r denote the corresponding variables of the transmitter and receiver lasers, respectively. c defines the fraction of power from the transmitter injected into the receiver; the same fraction is subtracted from the self-injected power in order to keep the total injected power constant. Under this condition, the equations for transmitter and receiver are the same when $E_t(t) = E_r(t)$, indicating that identical synchronization is possible in this system. In the particular case $c = 1$ (open-loop configuration, no power of the receiver is reinjected into itself), it can be shown in a straightforward way that the synchronized state is stable (Abarbanel and Kennel, 1998).

Let us assume that the transmitted field is perturbed an amount $\zeta(t)$. From Eqs. (5.A.52) and (5.A.54), it is easily found that the difference between the population inversions of transmitter and receiver obeys the equation

$$\frac{d[w_t(t)-w_r(t)]}{dt} = -2\gamma\left(w_t(t)-w_r(t) + |E_t(t)+\zeta(t)|^2 e^{Gw_r(t)}\frac{[e^{G(w_t(t)-w_r(t))}-1]}{G}\right) \tag{5.A.55}$$

Since $e^A - 1 \geq A$ for any real A,

$$\frac{d[w_t(t)-w_t(t)]}{dt} \leq -2\gamma(w_t(t)-w_r(t))\left(1 + |E_t(t)+\zeta(t)|^2 e^{Gw_r(t)}\right) \tag{5.A.56}$$

so that $|w_t(t)-w_r(t)|$ tends to 0 faster than $e^{-2\gamma t}$, and hence $E_t(t)$ tends to $E_r(t)$ as well.

5.A.7
Numerical Model for Synchronization of Chaos in Unidirectionally Coupled Solid-State Lasers

5.A.7.1
Single-Mode Laser Model

A numerical model for a chaotic dynamics of a solid-state laser is introduced in Appendix 3.A.6. The rate equations for two unidirectionally coupled single-mode solid-state lasers for the slowly varying, complex electric-field amplitude $E(t)$ and real gain $G(t)$ of lasers are (Fabiny et al., 1993).

$$\frac{dE_d(t)}{dt} = \frac{1}{\tau_c}[(G_d(t)-\alpha_d)E_d(t)] + i\omega_d E_d(t) \tag{5.A.57}$$

$$\frac{dG_d(t)}{dt} = \frac{1}{\tau_f}\left[p_d - G_d(t) - G_d(t)|E_d(t)|^2\right] \tag{5.A.58}$$

$$\frac{dE_r(t)}{dt} = \frac{1}{\tau_c}[(G_r(t)-\alpha_r)E_r(t) + \kappa E_d(t)] + i\omega_r E_r(t) \tag{5.A.59}$$

$$\frac{dG_r(t)}{dt} = \frac{1}{\tau_f}\left[p_r - G_r(t) - G_r(t)|E_r(t)|^2\right] \tag{5.A.60}$$

In these equations, the subscripts d and r correspond to the drive and response lasers, respectively. τ_c is the cavity round-trip time, τ_f is the fluorescent time of the upper lasing level, p is the pumping coefficient, α is the cavity loss coefficient, ω is the detuning of the laser from a common cavity mode, and κ is the coupling coefficient. i is the imaginary unit. Chaos can be generated by using either pump modulation or loss modulation, that is, $p(t) = p_0(1+b\cos(\Omega t))$ or $\alpha(t) = \alpha_0(1+b\cos(\Omega t))$, respectively, where b and Ω are the modulation amplitude and angular frequency.

The term of $\kappa E_d(t)$ in Eq. (5.A.59) represents the unidirectional coupling from the drive to response lasers through the electric fields, which is a coherent (optical-phase-dependent) coupling. Synchronization of chaos in coupled solid-state lasers has been observed by using this model.

Typical parameter values for numerical simulation are as follows: $\tau_c = 2.0 \times 10^{-10}$ s, $\tau_f = 2.4 \times 10^{-4}$ s, $a_d = a_r = 1.0 \times 10^{-2}$, $\omega_r - \omega_d = 5.0 \times 10^3$, rad/s, and $2\kappa/\tau_c = -5 \times 10^4$ s^{-1}.

5.A.7.2
Multimode Laser Model with Spatial Hole Burning (Tang–Statz–deMars Equations)

For multimode solid-state lasers, the Tang–Statz–deMars model is introduced in Appendix 3.A.6.2, describing cross-saturation of population inversion among modes due to the spatial hole-burning effect for multimode solid-state lasers (Tang et al., 1963; Mandel, 1997; Otsuka, 1999). The spatial distribution of the population inversion for an N-mode laser can be decomposed as $n_0(t)$ and $n_i(t)$, which correspond to the space-averaged and the first Fourier components (in space) of population inversion density for the ith mode ($i = 1, 2, \ldots, N$), respectively (see Eqs. (3.A.28) and (3.A.29) in Appendix 3.A.6.2 of Chapter 3). The electric field of each laser mode $E_i(t)$ is coupled to other modes through the spatially decomposed population inversions of $n_0(t)$ and $n_i(t)$. The Tang–Statz–deMars equations for one-way coupled two N-mode lasers are described with $n_0(t)$ and $n_i(t)$ as follows (Ogawa et al., 2002),

$$\frac{dn_{0,d}(t)}{dt} = p_d - n_{0,d}(t) - \sum_{k=1}^{N} \gamma_{k,d}\left(n_{0,d}(t) - \frac{n_{k,d}(t)}{2}\right) E_{k,d}^2(t) \tag{5.A.61}$$

$$\frac{dn_{i,d}(t)}{dt} = \gamma_{i,d} n_{0,d}(t) E_{i,d}^2(t) - n_{i,d}(t)\left(1 + \sum_{k=1}^{N} \gamma_{k,d} E_{k,d}^2(t)\right) \tag{5.A.62}$$

$$\frac{dE_{i,d}(t)}{dt} = \frac{K_d}{2}\left[\gamma_{i,d}\left(n_{0,d}(t) - \frac{n_{i,d}(t)}{2}\right) - 1\right] E_{i,d}(t) \tag{5.A.63}$$

$$\frac{dn_{0,r}(t)}{dt} = p_r - n_{0,r}(t) - \sum_{k=1}^{N} \gamma_{k,r}\left(n_{0,r}(t) - \frac{n_{k,r}(t)}{2}\right) E_{k,r}^2(t) \tag{5.A.64}$$

$$\frac{dn_{i,r}(t)}{dt} = \gamma_{i,r} n_{0,r}(t) E_{i,r}^2(t) - n_{i,r}(t)\left(1 + \sum_{k=1}^{N} \gamma_{k,r} E_{k,r}^2(t)\right) \tag{5.A.65}$$

$$\frac{dE_{i,r}(t)}{dt} = \frac{K_r}{2}\left[\gamma_{i,r}\left(n_{0,r}(t) - \frac{n_{i,r}(t)}{2}\right) - 1\right] E_{i,r}(t) + \frac{K_r}{2}\kappa_i E_{i,d}(t)\cos(2\pi\tau \Delta\mu_i(t)) \tag{5.A.66}$$

$$\frac{d\Delta\mu_i(t)}{dt} = 2\pi\tau \Delta\nu_i - \frac{K_r}{2}\kappa_i \frac{E_{i,d}(t)}{E_{i,r}(t)} \sin(2\pi\tau \Delta\mu_i(t)) \tag{5.A.67}$$

where $E_i(t)$ is the real amplitude of the lasing electric field for the ith mode. $\Delta u_i(t) = u_{i,d}(t) - u_{i,r}(t)$ is the optical-phase difference between the drive and response electrical fields for the ith mode. The subscripts d and r indicate the drive and response lasers, respectively. p is the pumping parameter scaled to the laser threshold. γ is the gain coefficient. $K = \tau/\tau_p$, where τ is the upper-state lifetime and τ_p is the photon lifetime in the laser cavity. κ_i is the coupling strength from the drive to the response laser. $\Delta \nu_i = \nu_{i,d} - \nu_{i,r}$ is the detuning of the lasing frequency between the two lasers for the ith mode. Time is scaled by τ. Pump modulation can be used to generate chaos in the drive laser, that is, $pd = p_0(1 + A\cos(2\pi \tau ft))$, where A and f are the modulation amplitude and frequency, respectively.

For two-mode lasers ($N = 2$), typical parameter values for numerical simulation are as follows: $\gamma_{1,d} = \gamma_{1,r} = 1.0$, $\gamma_{2,d} = \gamma_{2,r} = 0.875$, $\tau = 8.8 \times 10^{-5}$ s, $\tau_p = 1.15 \times 10^{-9}$ s, $K_d = K_r = \tau/\tau_p = 7.65 \times 10^4$, $p_d = p_r = 5.2$, $\kappa_1 = \kappa_2 = 0.08$, $\nu_1 = \nu_2 = 3.0 \times 10^6$ Hz, $\Delta\nu_1 = \Delta\nu_2 = 0$ Hz, $A = 0.15$, and $f = 1.08 \times 10^6$ Hz.

5.A.8
Numerical Model for Synchronization of Chaos in Unidirectionally Coupled Gas Lasers

A numerical model for chaotic dynamics in a CO_2 gas laser is introduced in Appendix 3.A.7. Unidirectionally Coupled CO_2 lasers with saturable absorber can be described by three-level–two-level rate equations as, (Sugawara et al., 1994).

$$\frac{dI_d(t)}{dt} = \left[(M_{1,d}(t) - M_{2,d}(t)) - B_d N_d(t) - k_d\right] I_d(t) \tag{5.A.68}$$

$$\frac{dM_{1,d}(t)}{dt} = P_d M_d(t) - R_{10,d} M_{1,d}(t) - (M_{1,d}(t) - M_{2,d}(t)) I_d(t) \tag{5.A.69}$$

$$\frac{dM_{2,d}(t)}{dt} = -R_{20,d} M_{2,d}(t) + (M_{1,d}(t) - M_{2,d}(t)) I_d(t) \tag{5.A.70}$$

$$\frac{dN_d(t)}{dt} = -2b_d N_d(t) I_d(t) - r_d(N_d(t) - 1) \tag{5.A.71}$$

$$\frac{dI_r(t)}{dt} = \left[(M_{1,r}(t) - M_{2,r}(t)) - B_r N_r(t) - k_r\right] I_r(t) \tag{5.A.72}$$

$$\frac{dM_{1,r}(t)}{dt} = P_r M_r(t) - R_{10,r} M_{1,r}(t) - (M_{1,r}(t) - M_{2,r}(t)) I_r(t) \tag{5.A.73}$$

$$\frac{dM_{2,r}(t)}{dt} = -R_{20,r} M_{2,r}(t) + (M_{1,r}(t) - M_{2,r}(t)) I_r(t) \tag{5.A.74}$$

$$\frac{dN_r(t)}{dt} = -2b_r N_r(t)(I_r(t) + CI_d(t)) - r_r(N_r(t) - 1) \tag{5.A.75}$$

Here, the subscripts d and r refer to the variables and parameters of the drive and response laser systems, respectively. $I(t)$ is the normalized photon density, $M_1(t)$ is the population density in the upper laser level, $M_2(t)$ is the population density in the lower laser level, and $N(t)$ is the difference of the population density in the absorber levels. B is the normalized rate of absorption in the passive medium, k is the cavity loss rate, P is the rate at which the upper laser level is pumped, R_{10} is the rate of vibrational relaxation from the upper laser level to all other levels except the lower level, R_{20} is the rate of vibrational relaxation from the lower laser level to other levels, b is the normalized cross section of the absorption in the passive medium, and r is the rotational relaxation rate of the absorptive levels.

The radiation of the drive laser modulates the absorber's population in the response laser through the second term on the right side of Eq. (5.A.75), where C is the coupling coefficient. The two laser intensities interact with each other within the saturable absorber. The coupling is thus incoherent (i.e., independent of the optical phase). The dependence of the chaotic pulsations on the strength of the driving force is qualitatively reproduced by a computer simulation based on this model (Sugawara et al., 1994).

Typical parameter values for numerical simulation are as follows: $B_d = B_r = 1.5 \times 10^6$ Hz, $k_d = k_r = 2.0 \times 10^6$ Hz, $P_d = 16.70$ Hz, $P_r = 16.45$ Hz, $R_{10,d} = R_{10,r} = 3.0 \times 10^2$ Hz, $R_{20,d} = R_{20,r} = 3.8 \times 10^5$ Hz, $b_d = b_r = 900.0$, $r_d = r_r = 6.0 \times 10^6$ Hz, and $C = 0.0533$ ($0 \le C \le 1$).

5.A.9
Definition of Phase in Chaotic Temporal Waveform by the Hilbert Transform

A general definition of the phase based on the Hilbert transform has been proposed to observe phase synchronization of chaos, as described in Section 5.6.1. A fairly general means is considered by which a phase may be associated with a real scalar signal such as the laser intensity $I(t)$ (DeShazer et al., 2001). $I(t)$ can be represented in terms of its Fourier transform $\hat{I}(\nu)$:

$$I(t) = \frac{1}{2\pi} \int_{-\infty}^{\infty} \exp(i\nu t) \hat{I}(\nu) d\nu \tag{5.A.76}$$

The variation of each Fourier component, $\hat{I}(\nu)\exp(i\nu t)$, is a complex number whose phase continually increases (decreases) with time for $\nu > 0$ ($\nu < 0$). Thus, one way to introduce a phase is to suppress the negative ν components by replacing $\hat{I}(\nu)$ by $2\theta(\nu)\hat{I}(\nu)$ (where $\theta(\nu)$ is the unit step function, $\theta(\nu) = 1$ for $\nu > 0$ and $\theta(\nu) = 0$ for $\nu < 0$). In this case, a superposition of rotating complex numbers is obtained, all of which have increasing phase,

$$V_A(t) = \frac{1}{\pi} \int_0^{\infty} \exp(i\nu t) \hat{I}(\nu) d\nu \tag{5.A.77}$$

Thus, $V_A(t)$ may be reasonably expected to execute a rotation in the complex plane with continually increasing phase. The function $V_A(t)$ is Gabor's analytic signal, which has been recently introduced for the purpose of the study of phase synchronization of chaos (Rosenblum et al., 1996). Noting that the inverse transform of $2\theta(\nu)$ is $\delta(t) + \frac{i}{\pi}P\frac{1}{t}$, the analytic signal can be expressed as,

$$V_A(t) = I(t) + iI_H(t) = I(t) + \frac{i}{\pi}I(t)^* P\frac{1}{t} \qquad (5.A.78)$$

where $*$ denotes convolution, $P\frac{1}{t}$ is the principal part of $1/t$, $I_H(t)$ and $I(t)$ are related by the Hilbert transform, defined as

$$I_H(t) = \frac{1}{\pi}P\int_{-\infty}^{\infty}\frac{I(t')}{(t-t')}dt' \qquad (5.A.79)$$

Writing,

$$V_A(t) - \langle V_A \rangle = R_A(t)\exp(i\Phi_A(t)) \qquad (5.A.80)$$

where $R_A(t)$ and $\Phi_A(t)$ are real, and $\langle V_A \rangle$ is the time average of $V_A(t)$, $\Phi_A(t)$ is called the analytic phase (DeShazer et al., 2001). Phase synchronization can be tested by extracting the analytic phase from the chaotic time series of laser intensity.

It is useful to consider more general choices for the derivation of the Gaussian filtered phase. In particular, $u(\nu)$ in the original Fourier transform can be replaced by $f(\nu)u(\nu)$, where $f(\nu)$ is suitably chosen (e.g., Eq. (5.A.78) corresponds to $f(\nu) = 2\theta(\nu)$). Specifically, a Gaussian for $f(\nu)$ can be chosen as,

$$f(\nu) = \exp\left[-\frac{(\nu-\nu_0)^2}{2\sigma^2}\right] \qquad (5.A.81)$$

This gives

$$V_G(t) = \frac{1}{2\pi}\int_{-\infty}^{\infty}d\nu\, u(\nu)\exp(i\nu t)\exp\left[-(\nu-\nu_0)^2/(2\sigma^2)\right] \qquad (5.A.82)$$
$$= I(t)*F(t)$$

where

$$F(t) = \frac{\sigma}{\sqrt{2\pi}}\exp\left[-i\nu_0 t + \frac{\sigma^2 t^2}{2}\right] \qquad (5.A.83)$$

The Gaussian filtered phase $\Phi_G(t)$ can be defined by

$$V_G(t) - \langle V_G \rangle = R_G(t)\exp[i\,\Phi_G(t)] \qquad (5.A.84)$$

The frequency ν_0 and the Gaussian's width σ are parameters in the definition of $\Phi_G(t)$. Note that, similar to the choice $f(\nu) = 2\theta(\nu)$ resulting in the analytic signal, the choice of a Gaussian again emphasizes positive frequencies ($\nu_0 > 0$). In fact, the application of a frequency bandpass filter produces similar results for the Gaussian filtered phase.

6
Analysis of Synchronization of Chaos: Example of Unidirectionally Coupled Semiconductor Lasers with Optical Feedback

In this chapter, both experimental and numerical examples of synchronization of chaos are described in detail. Synchronization of chaos is observed experimentally and the numerically in two unidirectionally coupled semiconductor lasers. Linear stability analysis and conditional Lyapunov exponent are used as a tool for analyzing identical synchronization. The comparison between the "open-loop" (without optical feedback in a response laser) and the "closed-loop" (with optical feedback in a response laser) configurations are performed. Generalized synchronization of chaos is also observed experimentally and numerically in three semiconductor lasers; one is used for a drive laser and the other two lasers are used for response lasers that are injected by a common chaotic drive signal.

This chapter, describing typical examples of analysis on synchronization of chaotic semiconductor lasers with optical feedback, is specially designed for the readers who are interested in starting research works on synchronization of chaotic lasers but do not know where to start. This style (as well as Chapter 4) is different from other chapters where many examples of each topic are widely introduced and explained from the literature. The contents described in Chapter 6 would be good starting points of scientific research in synchronization of chaotic lasers for students and researchers if they would like to work in this research field. If the readers feel uncomfortable reading through the whole chapter with complicated mathematical equations, they may read the experimental part of Sections 6.1 and 6.4 and skip the analytical and numerical sections.

6.1
Experimental Analysis on Synchronization of Chaos in Two Semiconductor Lasers with Optical Feedback

6.1.1
Experimental Setup for Synchronization of Chaos

In this section, synchronization of chaos in a semiconductor laser with optical feedback is observed in experiments. Figure 6.1 shows the experimental setup with

Optical Communication with Chaotic Lasers: Applications of Nonlinear Dynamics and Synchronization, First Edition. Atsushi Uchida.
© 2012 Wiley-VCH Verlag GmbH & Co. KGaA. Published 2012 by Wiley-VCH Verlag GmbH & Co. KGaA.

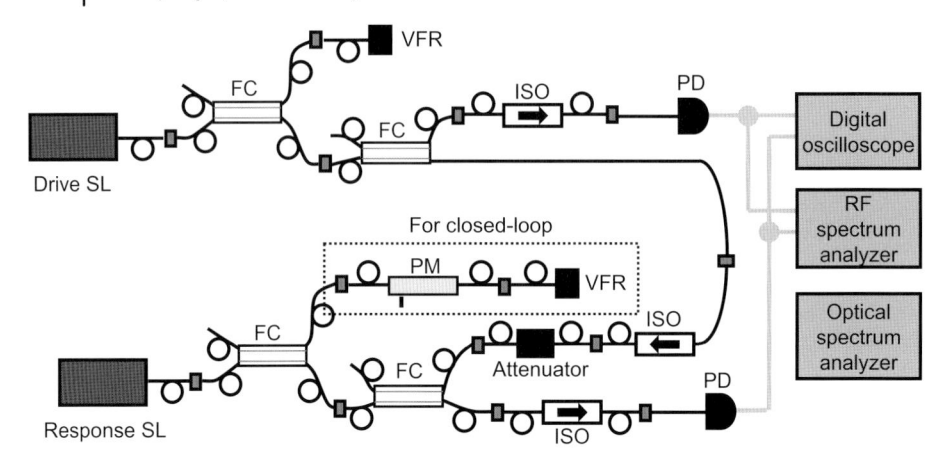

Figure 6.1 Experimental setup for synchronization of chaos in unidirectionally coupled semiconductor lasers. The dotted frame indicates the closed-loop configuration in the response laser. FC, fiber coupler; ISO, optical-fiber isolator; PD, photodetector; PM, phase modulator; SL, semiconductor laser; VFR, variable fiber reflector.

fiber-optic components for synchronization of chaos in unidirectionally coupled semiconductor lasers. The specification of the experimental equipments used for the experiments in Chapter 6 is the same as shown in Table 4.1 in Chapter 4. Two distributed-feedback (DFB) semiconductor lasers mounted in butterfly packages with optical-fiber pigtails are used. The two semiconductor lasers were fabricated from the same wafer, so they have similar laser parameter values. One laser is used as a drive laser and the other laser is used as a response laser. The injection current and the temperature of the semiconductor lasers are adjusted by a current–temperature controller. The optical wavelengths of the two lasers are precisely controlled by the temperature of the laser. Both the drive and response lasers are prepared without standard optical isolators, to allow optical feedback and injection.

The drive laser is connected to a fiber coupler (FC) and a variable fiber reflector (VFR) that reflects a fraction of the light back into the laser, inducing high-frequency chaotic oscillations of the optical intensity. The amount of optical feedback light is adjusted by using VFR. A portion of the chaotic beam of the drive laser is injected into the response laser. An optical-fiber isolator (ISO) is used to achieve one-way coupling from the drive to the response laser. The response laser is connected to a VFR and a phase modulator (PM) to introduce optical feedback, which is referred to as the closed-loop configuration (see the dotted frame in Figure 6.1). The amount of optical feedback light is adjusted by the VFR. Chaotic intensity fluctuation is observed in the response laser in the presence of the optical feedback without optical coupling. By contrast, no optical feedback is introduced into the response laser by disconnecting the VFR and PM for the case of the open-loop configuration. In this case, stable laser output is obtained in the response laser without optical coupling. The wavelengths of the drive and response lasers are precisely adjusted in order to obtain generalized synchronization. Polarization-maintaining (PM) fibers are used for all the optical-fiber components. If single-mode fiber components are used, polarization controllers are required in the feedback loop and the injection path.

A portion of each laser output is extracted by a fiber coupler, and the chaotic waveform of laser intensity is detected by an AC-coupled photodetector. The converted electronic signal is sent to a digital oscilloscope and a radio-frequency (RF) spectrum analyzer to observe temporal waveforms and the corresponding RF spectra, respectively. The optical wavelengths of the two lasers are measured by an optical spectrum analyzer.

6.1.2
Experimental Results of Synchronization of Chaos

The values of the laser parameters used in this experiment are summarized in Table 6.1. The relaxation oscillation frequencies are set at 3.00 and 1.82 GHz for the drive and response lasers, respectively, by adjusting the injection current of the lasers. Figure 6.2 shows the optical spectra of the drive and response lasers without and with optical coupling in the open-loop configuration. The optical wavelength of the response laser is detuned towards the shorter direction with respect to that of the drive laser by controlling the temperature of the two lasers, that is, the optical wavelengths are set to 1547.263 nm for the drive laser with optical feedback and 1547.176 nm for the solitary response laser, respectively, as shown in Figure 6.2a. The width of the optical spectrum for the drive laser is larger than that for the response laser, due to the chaotic intensity output induced by optical feedback. The initial

Table 6.1 Parameter values used in the experiment for synchronization of chaos in Section 6.1.

Parameter	Drive Laser	Response Laser
Threshold of injection current [mA]	10.48	9.38
Injection current [mA]	15.20	12.00
(Relative to the threshold)	(1.45)	(1.28)
Relaxation oscillation frequency [GHz]	3.00	1.82
Wavelength without coupling [nm]	1547.263	1547.176
(detuning from drive laser [nm])		(−0.087)
Wavelength with coupling [nm]	1547.263	1547.263
(detuning from drive laser [nm])		(0.0)
Temperature [K]	290.00	293.90
Optical feedback power [µW]	0.76	0.315
(Maximum feedback power [µW])		(11.4) for closed-loop
Optical injection power from drive to response lasers [µW]	N/A	25.3
(Maximum injection power [µW])		(100)
External cavity length [m]	2.52	3.67 for closed-loop
(One-way, fiber length)		
Feedback delay time [ns]	24.3	35.3
(External cavity frequency [MHz])	(41.2)	(28.3) for closed-loop
Injection delay time from drive laser [ns]	N/A	40.0

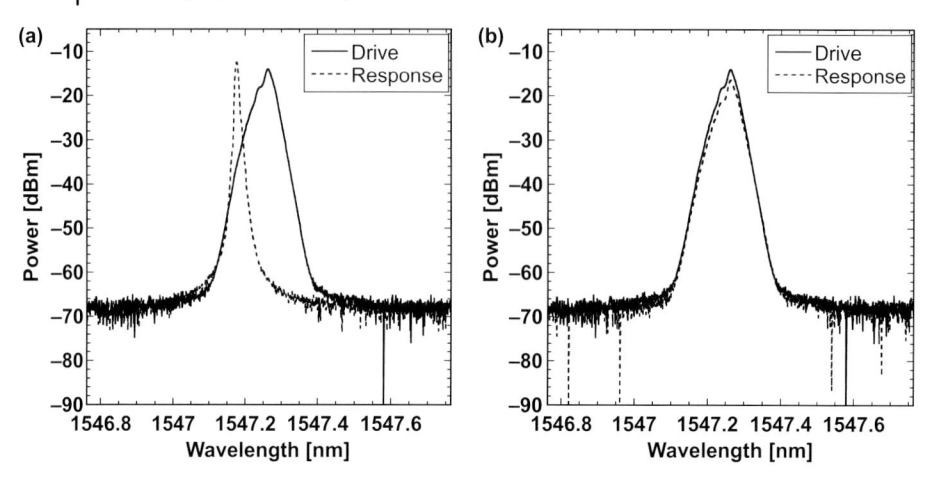

Figure 6.2 Optical spectra of the drive and response lasers (a) without and (b) with optical coupling in the open-loop configuration. Injection locking (matching of the optical wavelengths) is achieved in (b). Solid curve, the drive laser; dotted curve, the response laser. The experimental data in this chapter was provided by Haruka Okumura and Hiroki Aida.

optical wavelength detuning for the solitary (uncoupled) lasers is defined as $\Delta\lambda_{sol} = \lambda_{r,sol} - \lambda_d$, where $\lambda_{r,sol}$ and λ_d indicate the optical wavelengths of the solitary response and drive lasers without optical coupling, respectively. $\Delta\lambda_{sol}$ is set to -0.087 nm, which corresponds to the negative detuning condition ($\lambda_{r,sol} < \lambda_d$).

When the drive laser output is injected into the response laser at this detuning condition, the optical wavelength for the response laser is shifted to 1547.263 nm and exactly matched to that of the drive laser, as shown in Figure 6.2b. This result indicates that injection locking is achieved between the drive and response lasers, where injection locking is defined as the matching of optical wavelength (i.e., optical-carrier frequency) between two coupled lasers due to coherent unidirectional coupling (Siegman, 1986). When the optical-wavelength detuning is small enough, the optical wavelength of the response laser is shifted and exactly matched with that of the drive laser due to injection locking.

Figure 6.3 shows the temporal waveforms of the drive and response lasers without and with optical coupling, and their corresponding correlation plots in the open-loop configuration. The temporal waveform of the response laser is shifted with respect to the drive laser by the injection delay time ($\tau = 40.0$ ns) to obtain the optimal synchronization (i.e., generalized synchronization with high correlation shown in Section 5.3.2). Without optical coupling, the drive and response lasers are not synchronized at all and only the small sustained relaxation oscillation is observed in the response-laser output, as shown in Figure 6.3a. By contrast, with optical coupling, the two temporal waveforms of the drive and response lasers are well synchronized to each other at the condition of injection locking, as shown in Figure 6.3b. Figure 6.3c and d show the correlation plots of the drive and response lasers without and with optical coupling, respectively. The linear correlation on the 45-degree line is found in the presence of the optical coupling as shown in

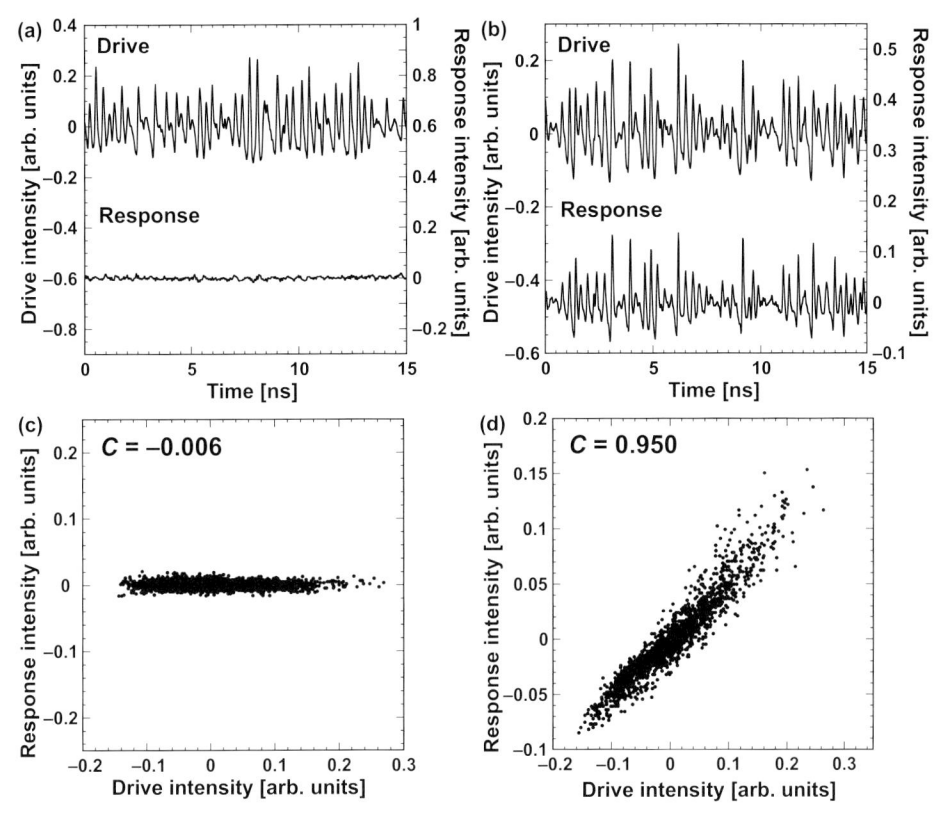

Figure 6.3 Temporal waveforms of the drive and response lasers (a) without and (b) with optical coupling, and the corresponding correlation plots (c) without and (d) with optical coupling in the open-loop configuration. The cross-correlation values are (c) $C = -0.006$ and (d) $C = 0.950$.

Figure 6.3d, whereas there is no correlation between the two laser outputs without optical coupling in Figure 6.3c. Good synchronization of chaos is also achieved in the closed-loop configuration.

The degree of synchronization of chaos is quantitatively measured by Eq. (5.19), that is, the cross-correlation C of two temporal waveforms of the drive and response lasers normalized by the product of their standard deviations. The best (inphase) synchronization of chaos is obtained at the cross-correlation coefficient of $C = 1$, whereas no synchronization appears at $C = 0$. The cross-correlation values are estimated from Figures 6.3c and d. The cross-correlation value for Figure 6.3c is -0.006, indicating no synchronization. By contrast, the cross-correlation value for Figure 6.3d is 0.950, indicating high-quality synchronization of chaos.

Figure 6.4 shows the RF spectra of the drive and response lasers without and with optical coupling. Without optical coupling, the RF spectrum of the response laser shows a peak value at the relaxation oscillation frequency of 1.82 GHz, as shown in

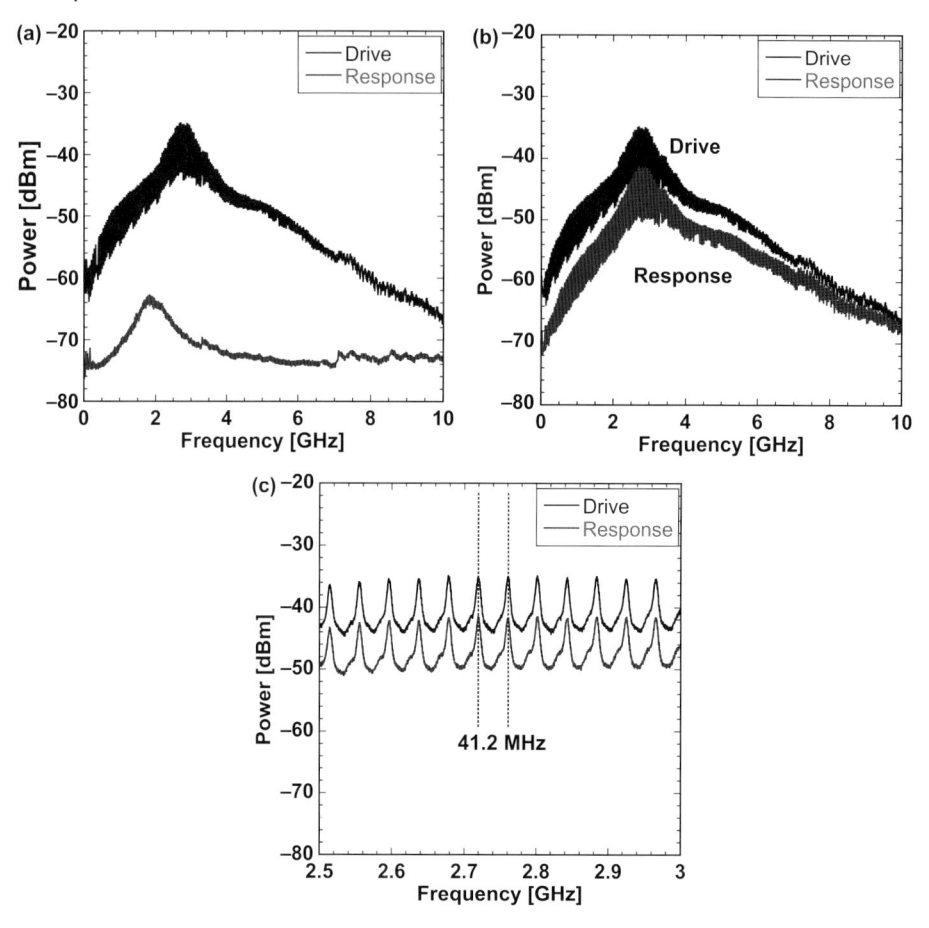

Figure 6.4 RF spectra of the drive and response lasers (a) without and (b) with optical coupling, and (c) the enlargement of (b) in the open-loop configuration. Black curve, the drive laser; dark gray curve, the response laser.

Figure 6.4a. With optical injection, the RF spectrum of the response laser resembles that of the drive laser as shown in Figure 6.4b. For the enlargement of the RF spectra in Figure 6.4c, the peak interval of the external cavity frequency (41.2 MHz) is matched between the drive and response lasers.

The results of Figures 6.2–6.4 confirm that high-quality synchronization of chaos between the drive and response lasers can be achieved experimentally.

6.1.3
Parameter Dependence of Synchronization of Chaos

The parameter dependence of synchronization of chaos is investigated systematically when one of the parameter values for the drive or response laser is changed. The accuracy of synchronization is quantitatively defined by the cross-correlation C of Eq. (5.19) in Chapter 5.

First, the optical injection power from the drive to response laser is changed in the open-loop configuration. The initial optical wavelength detuning without optical coupling ($\Delta\lambda_{sol} = \lambda_{r,sol} - \lambda_d$) is fixed to -0.086 nm. Here, the optical wavelength detuning between the drive and response lasers "with optical coupling" is defined as $\Delta\lambda_{inj} = \lambda_{r,inj} - \lambda_d$, where $\lambda_{r,inj}$ and λ_d indicate the optical wavelengths of the response and drive lasers under the condition of optical coupling, respectively. $\lambda_{r,inj}$ can be shifted from $\lambda_{r,sol}$ due to the injection-locking effect, therefore $\Delta\lambda_{inj}$ can be changed under the condition of optical coupling. When $\Delta\lambda_{inj} \approx 0$ is satisfied, the two optical wavelengths are matched to each other and it is determined that injection locking is achieved. The value of $\Delta\lambda_{inj}$ can be measured experimentally from the difference in the mean values (weighted by the optical spectrum components) of the optical spectra between the two coupled lasers, as seen in Figure 6.2.

The cross-correlation value C and the optical wavelength detuning under the optical coupling $\Delta\lambda_{inj}$ are measured as the optical injection power from the drive to response laser κ_{inj} (normalized by the maximum injection power obtained in the experiment) is increased, as shown in Figure 6.5a. The value of C increases and reaches over 0.80 as the optical injection power is increased at $\kappa_{inj} \geq 0.16$. The value of C slightly decreases as the injection power is increased further. In addition, $|\Delta\lambda_{inj}|$ is less than 0.01 nm (corresponding to the optical frequency detuning of 1.25 GHz) for large optical injection power of $\kappa_{inj} \geq 0.16$. Therefore, high-quality synchronization of chaos is achieved under the condition of the optical wavelength matching of the two coupled lasers by injection locking.

Next, the initial optical wavelength detuning without optical coupling $\Delta\lambda_{sol}$ is changed. The values of C and $\Delta\lambda_{inj}$ are measured as $\Delta\lambda_{sol}$ is increased, as shown in Figure 6.5b. The value of C increases suddenly and reaches over 0.80 as $\Delta\lambda_{sol}$ is increased. The region of $C \geq 0.80$ is observed at the region of -0.095 nm $\leq$

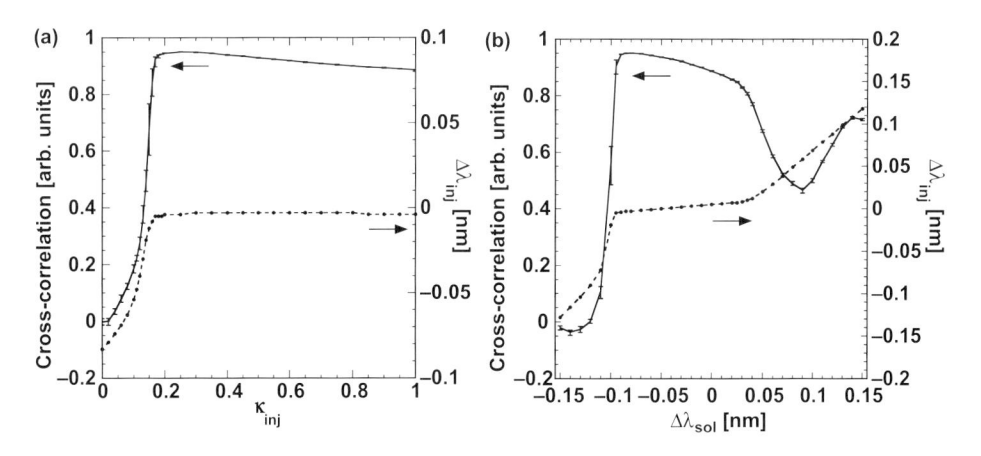

Figure 6.5 Cross-correlation value C (solid curve with error bars) and the optical wavelength detuning between the drive and response lasers under the optical coupling $\Delta\lambda_{inj}$ (dotted curve with solid circles) as a function of (a) the optical injection power κ_{inj} (normalized by the maximum injection power obtained in the experiment) and (b) the initial optical wavelength detuning without optical coupling $\Delta\lambda_{sol}$ in the open-loop configuration.

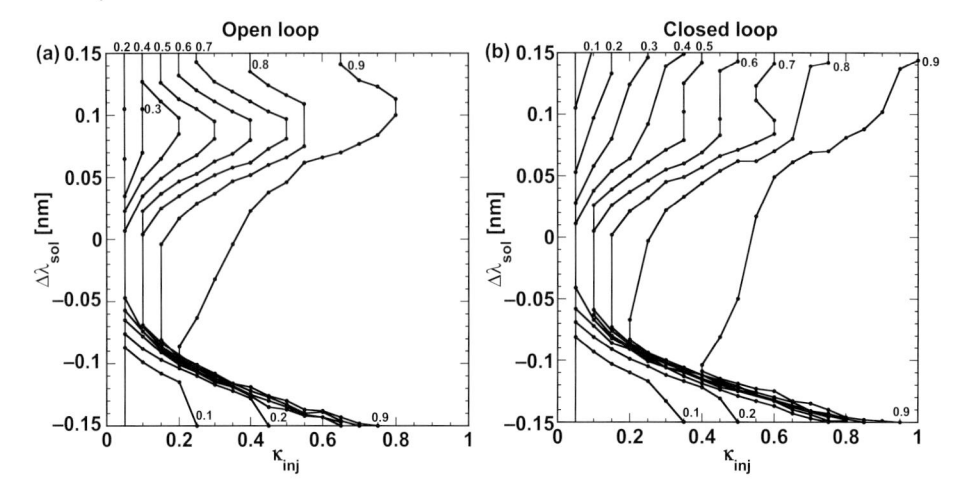

Figure 6.6 Two-dimensional (2D) maps of the cross-correlation values C as functions of the optical injection power κ_{inj} and the initial optical wavelength detuning $\Delta\lambda_{sol}$ in the (a) open-loop and (b) closed-loop configurations.

$\Delta\lambda_{sol} \leq 0.035$ nm. In this region the optical wavelength detuning after optical coupling is $|\Delta\lambda_{inj}| \leq 0.01$ nm, indicating the injection-locking range. The asymmetry of the curve for C in terms of $\Delta\lambda_{sol}$ results from the amplitude-phase coupling effect, indicated as the "α parameter" of semiconductor lasers (see the model in Sections 6.2 and 6.3). It is found that injection locking is crucial for achieving synchronization of chaos in coherently coupled semiconductor lasers.

To investigate the synchronization characteristics more systematically, both κ_{inj} and $\Delta\lambda_{sol}$ are changed simultaneously. Figure 6.6 indicates the two-dimensional (2D) map of the cross-correlation values C as functions of κ_{inj} and $\Delta\lambda_{sol}$ for the open- and closed-loop configurations obtained from the experiment. The curves indicate the contours of the cross-correlation values. It is found that high C is obtained at large κ_{inj}. Also the region for high C is asymmetrically located at the negative $\Delta\lambda_{sol}$ side due to the α parameter. The overall tendency of the 2D map for the closed-loop configuration (Figure 6.6b) is very similar to that for the open-loop configuration (Figure 6.6a), except for the fact that the cross-correlation values for the closed-loop configuration are slightly decreased in the entire parameter region.

For the comparison between the open- and closed-loop configurations, the optical feedback power of the response laser is changed. Figure 6.7 shows C and $\Delta\lambda_{inj}$ as a function of the optical feedback power of the response laser κ_r (normalized by the maximum feedback power obtained in the experiment) in the closed-loop configuration. Zero feedback power ($\kappa_r = 0$) corresponds to the open-loop configuration. As seen in Figure 6.7, the best value of C is obtained at zero κ_r. This experimental result indicates that the quality of synchronization of chaos for the open-loop configuration is better than that for the closed-loop configuration.

In summary, synchronization of chaos is experimentally observed in unidirectionally coupled two semiconductor lasers with optical feedback. Injection locking is a

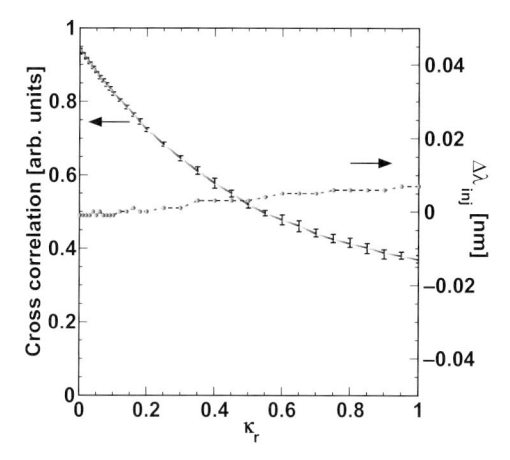

Figure 6.7 Cross-correlation value C (solid curve with error bars) and the optical wavelength detuning under the optical coupling $\Delta\lambda_{inj}$ (dotted curve with solid circles) as a function of the optical feedback power of the response laser κ_r (normalized by the maximum feedback power obtained in the experiment) in the closed-loop configuration. Zero feedback strength ($\kappa_r = 0$) corresponds to the open-loop configuration.

necessary condition to achieve synchronization of chaotic temporal waveforms between coherently coupled lasers. Synchronization is achieved at the regions of large injection power and negative wavelength detuning, which corresponds to the injection-locking range. The performance of synchronization in the open-loop configuration is similar (but slightly better) than that in the closed-loop configuration.

6.2 Model for Synchronization of Chaos in Two Coupled Semiconductor Lasers with Optical Feedback

6.2.1 Lang–Kobayashi Equations for Synchronization of Chaos in Unidirectionally Coupled Semiconductor Lasers with Optical Feedback

6.2.1.1 Coupled Lang–Kobayashi Equations

In this section, a set of rate equations based on a model for two unidirectionally coupled semiconductor lasers is introduced. The model is shown in Figure 6.8. One laser is referred to as a drive laser, and the other laser is referred to as a response laser. An optical feedback light is reflected back from an external mirror into the drive laser to produce chaotic intensity fluctuations. The two lasers are coherently and unidirectionally coupled to each other by injecting the chaotic output from the drive laser to the response laser through an optical isolator. There is no external mirror and no optical feedback for the response laser in the open-loop configuration, whereas the optical feedback is introduced by another external mirror in the response laser in the closed-loop configuration (see the dotted frame in Figure 6.8).

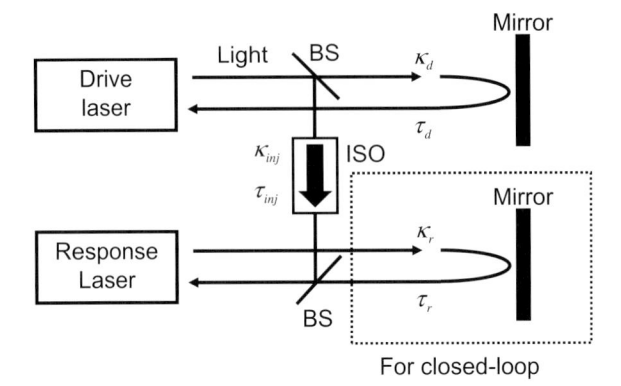

Figure 6.8 Model for synchronization of chaos in unidirectionally coupled semiconductor lasers. The dotted frame indicates the closed-loop configuration in the response laser. BS, beam splitter; ISO, optical isolator for unidirectional coupling. $\kappa_{d,r}$ are the feedback strengths for the drive and response lasers. κ_{inj} is the injection strength. $\tau_{d,r}$ are the feedback delay times for the drive and response lasers. τ_{inj} is the injection delay time.

The Lang–Kobayashi equations (Lang and Kobayashi, 1980; Uchida et.al., 2004b; Ohtsubo, 2008) of unidirectionally coupled semiconductor lasers in the open-loop configuration are described as follows,

Drive laser:

$$\frac{dE_d(t)}{dt} = \frac{1}{2}\left[G_{N,d}(N_d(t)-N_{0,d}) - \frac{1}{\tau_{p,d}}\right]E_d(t) + \kappa_d\, E_d(t-\tau_d)\cos\Theta_d(t) \quad (6.1)$$

$$\frac{d\Phi_d(t)}{dt} = \frac{\alpha_d}{2}\left[G_{N,d}(N_d(t)-N_{0,d}) - \frac{1}{\tau_{p,d}}\right] - \kappa_d\frac{E_d(t-\tau_d)}{E_d(t)}\sin\Theta_d(t) \quad (6.2)$$

$$\frac{dN_d(t)}{dt} = J_d - \frac{N_d(t)}{\tau_{s,d}} - G_{N,d}(N_d(t)-N_{0,d})E_d^2(t) \quad (6.3)$$

$$\Theta_d(t) = \omega_d\tau_d + \Phi_d(t) - \Phi_d(t-\tau_d) \quad (6.4)$$

Response laser:

$$\frac{dE_r(t)}{dt} = \frac{1}{2}\left[G_{N,r}(N_r(t)-N_{0,r}) - \frac{1}{\tau_{p,r}}\right]E_r(t) + \kappa_{inj}\, E_d(t-\tau_{inj})\cos\Theta_{inj}(t) \quad (6.5)$$

$$\frac{d\Phi_r(t)}{dt} = \frac{\alpha_r}{2}\left[G_{N,r}(N_r(t)-N_{0,r}) - \frac{1}{\tau_{p,r}}\right] - \kappa_{inj}\frac{E_d(t-\tau_{inj})}{E_r(t)}\sin\Theta_{inj}(t) \quad (6.6)$$

$$\frac{dN_r(t)}{dt} = J_r - \frac{N_r(t)}{\tau_{s,r}} - G_{N,r}(N_r(t)-N_{0,r})E_r^2(t) \quad (6.7)$$

$$\Theta_{inj}(t) = -\Delta\omega\, t + \omega_d\tau_{inj} + \Phi_r(t) - \Phi_d(t-\tau_{inj}) \quad (6.8)$$

Table 6.2 Parameter values that are common between the drive and response lasers used for the numerical simulation in Sections 6.2 and 6.3.

Symbol	Parameter	Value
$G_{N,d}$, $G_{N,r}$	Gain coefficient	8.40×10^{-13} m^3 s^{-1}
$N_{0,d}$, $N_{0,r}$	Carrier density at transparency	1.40×10^{24} m^{-3}
$\tau_{p,d}$, $\tau_{p,r}$	Photon lifetime	1.927×10^{-12} s
$\tau_{s,d}$, $\tau_{s,r}$	Carrier lifetime	2.04×10^{-9} s
$\tau_{in,d}$, $\tau_{in,r}$	Round-trip time in internal cavity	8.0×10^{-12} s
$r_{2,d}$, $r_{2,r}$	Reflectivity of laser facet	0.556
$J_d/J_{th,d}$, $J_r/J_{th,r}$	Normalized injection current	1.3
α_d, α_r	Linewidth-enhancement factor	3.0
λ_d	Optical wavelength of drive laser	1.537×10^{-6} m
c	Speed of light	2.998×10^{8} m s^{-1}
$N_{th,d}$, $N_{th,r}$	Carrier density at threshold	2.018×10^{24} m^{-3}
$J_{th,d}$, $J_{th,r}$	Injection current at threshold	9.892×10^{32} m^{-3} s^{-1}
ω_d	Optical angular frequency of drive laser	1.226×10^{15} s^{-1}

The subscripts d and r indicate the drive and response lasers, respectively.

where $E(t)$ is the electric-field amplitude, $\Phi(t)$ is the electric-field phase, and $N(t)$ is the carrier density. $E(t)$, $\Phi(t)$, $N(t)$ are the real variables. The symbols and parameters are summarized in Tables 6.2 and 6.3. The subscripts d and r indicate the drive and response lasers, respectively. $J_{th} = N_{th}/\tau_s$ is the injection current at the threshold, $N_{th} = N_0 + 1/(G_N \tau_p)$ is the carrier density at the threshold, and $\omega = 2\pi c/\lambda$ is the optical angular frequency. $\kappa_d = (1 - r_{2,d}^2) r_{3,d}/(r_{2,d} \tau_{in,d})$ is the optical feedback strength of the drive laser, and κ_{inj} is the optical injection strength from the drive to response lasers. $\tau = 2L/c$ is the round-trip time in the external cavity, L is the external cavity length, c is the speed of light. $\Delta\omega = \omega_d - \omega_{r,sol} = 2\pi(f_d - f_{r,sol}) = 2\pi \Delta f_{sol}$ is the detuning of the angular optical-carrier frequencies between the drive and response lasers, where f_d and $f_{r,sol}$ are the optical-carrier frequencies of the drive and response lasers ($f = c/\lambda$, λ is the optical wavelength). The inclusion of the term $\Delta\omega t$ in Eq. (6.8) is very important to model the effect of injection locking.

As described in Section 4.2.1, for numerical simulation of Eqs. (6.4) and (6.8), the constant values of $\omega_d \tau_d$ and $\omega_d \tau_{inj}$ are several orders of magnitude larger than the value of the phase difference $\Phi(t) - \Phi(t-\tau)$ (see Tables 6.2 and 6.3), and $\Theta(t)$ can be treated between 0 and 2π. Therefore, the value of an initial phase shift $\Theta_{ini} = \omega \tau \pmod{2\pi}$ (where $0 \leq \Theta_{ini} < 2\pi$) needs to be calculated in advance and $\omega \tau$ is replaced to Θ_{ini} for numerical calculation.

6.2.1.2 Identical Synchronous Solution

The identical internal parameter values between the drive and response lasers are assumed, as shown in Table 6.2 (i.e., $G_{N,d} = G_{N,r}$, $N_{0,d} = N_{0,r}$, and so on). The rate equations for the drive and response lasers in the open-loop configuration

Table 6.3 Parameter values used in numerical simulation for identical synchronization in the open-loop configuration in Sections 6.2 and 6.3.

Symbol	Parameter	Value
$r_{3,d}$	Reflectivity of external mirror of drive laser	0.008
$\kappa_d = \dfrac{(1-r_{2,d}^2)\, r_{3,d}}{r_{2,d}}\dfrac{1}{\tau_{in,d}}$	Feedback strength of drive laser	$1.243 \times 10^9\,\mathrm{s}^{-1}$ $(1.243\,\mathrm{ns}^{-1})$
κ_{inj}	Injection strength from drive to response lasers	$1.243 \times 10^9\,\mathrm{s}^{-1}$ $(1.243\,\mathrm{ns}^{-1})$
L_d	External cavity length of drive laser	0.6 m (one-way) 1.2 m (round-trip)
L_{inj}	Distance from drive to response lasers	1.2 m
τ_d	Round-trip time in external cavity of drive laser (feedback delay time)	$4.003 \times 10^{-9}\,\mathrm{s}$
τ_{inj}	Injection delay time from drive to response lasers	$4.003 \times 10^{-9}\,\mathrm{s}$
$\Delta f_{sol} = f_d - f_{r,sol}$	Initial detuning of optical-carrier frequency between drive and response lasers	0.0 GHz
$\Delta\omega = \omega_d - \omega_{r,sol}$ $= 2\pi(f_d - f_{r,sol})$ $= 2\pi\,\Delta f_{sol}$	Initial detuning of optical angular frequency between drive and response lasers	$0.0\,\mathrm{s}^{-1}$

The subscripts *d* and *inj* indicate the drive laser and injection (coupling) signal, respectively. Other parameter values are used in Table 6.2.

can be identical under the following conditions for the optical injection and feedback lights,

$$\kappa_d = \kappa_{inj},\ \tau_d = \tau_{inj},\ \Delta\omega = 0 \qquad (6.9)$$

The parameter values shown in Table 6.3 satisfy the conditions of Eq. (6.9). Identical synchronous solutions can be obtained under these conditions as follows,

$$E_d(t) = E_r(t),\quad \Phi_d(t) = \Phi_r(t),\quad N_d(t) = N_r(t) \qquad (6.10)$$

6.2.2
Derivation of the Electric-Field Amplitude and Phase of the Coupled Lang–Kobayashi Equations from the Complex Electric-Field Equations

The derivation of electric-field amplitude and phase of the laser output is described in this section. For the drive laser, the derivation process is exactly the same as shown in Section 4.2.2 (Add the subscript *d* for all the variables and parameters, for example, $E(t)$ is changed to $E_d(t)$). On the other hand, the derivation of the electric-field amplitude and phase for the response laser is a little complicated due to the term of

6.2 Model for Synchronization of Chaos in Two Coupled Semiconductor Lasers with Optical Feedback

the optical injection from the drive laser and the presence of the optical-frequency detuning. The derivation of the complex electric-field amplitude and phase of the response laser is treated and Eqs. (6.5)–(6.8) are derived in this section.

The complex electric field for the response laser ($\hat{\mathbf{E}}_r(t)$) including the fast optical carrier in the open-loop configuration is described as follows (Lang and Kobayashi, 1980; Ohtsubo, 2008),

$$\frac{d\hat{\mathbf{E}}_r(t)}{dt} = \left[\frac{1+i\alpha_r}{2}\left\{G_{N,r}\left(N_r(t)-N_{0,r}\right)-\frac{1}{\tau_{p,r}}\right\} + i\omega_r\right]\hat{\mathbf{E}}_r(t) + \kappa_{\text{inj}}\,\hat{\mathbf{E}}_d(t-\tau_{\text{inj}}) \tag{6.11}$$

A variable with bold font indicates a complex variable, whereas a variable with italic font represents a real variable. $\hat{\mathbf{E}}$ indicates that the complex variable $\mathbf{E}$ includes the fast optical-carrier frequency component. For Eq. (6.11), the feedback term $\kappa\,\hat{\mathbf{E}}(t-\tau)$ of Eq. (4.7) is replaced by the injection term $\kappa_{\text{inj}}\,\hat{\mathbf{E}}_d(t-\tau_{\text{inj}})$. It is assumed that the dynamics of the electric-field amplitude is much slower than the angular frequencies ω_r of the fast optical carrier. The fast oscillation component can be separated from the slow complex electric fields $\mathbf{E}_d(t)$ and $\mathbf{E}_r(t)$ as follows,

$$\hat{\mathbf{E}}_d(t) = \mathbf{E}_d(t)\exp(i\omega_d t) \tag{6.12}$$

$$\hat{\mathbf{E}}_r(t) = \mathbf{E}_r(t)\exp(i\omega_r t) \tag{6.13}$$

Substituting Eqs. (6.12) and (6.13) into Eq. (6.11).

$$\begin{aligned}&\frac{d\mathbf{E}_r(t)}{dt}\exp(i\omega_r t) + i\omega_r\mathbf{E}_r(t)\exp(i\omega_r t) \\ &= \left[\frac{1+i\alpha_r}{2}\left\{G_{N,r}\left(N_r(t)-N_{0,r}\right)-\frac{1}{\tau_{p,r}}\right\} + i\omega_r\right]\mathbf{E}_r(t)\exp(i\omega_r t) \\ &\quad + \kappa_{\text{inj}}\,\mathbf{E}_d(t-\tau_{\text{inj}})\exp(i\omega_d(t-\tau_{\text{inj}}))\end{aligned} \tag{6.14}$$

All terms of Eq. (6.14) are divided by $\exp(i\omega_r t)$,

$$\begin{aligned}\frac{d\mathbf{E}_r(t)}{dt} &= \frac{1+i\alpha_r}{2}\left[G_{N,r}\left(N_r(t)-N_{0,r}\right)-\frac{1}{\tau_{p,r}}\right]\mathbf{E}_r(t) \\ &\quad + \kappa_{\text{inj}}\,\mathbf{E}_d(t-\tau_{\text{inj}})\exp\{i\left((\omega_d-\omega_r)\,t-\omega_d\tau_{\text{inj}}\right)\}\end{aligned} \tag{6.15}$$

Equation (6.15) is the Lang–Kobayashi equation for the slow complex electric field of the response laser after the approximation to eliminate the fast optical-carrier component. The slow complex electric fields $\mathbf{E}_d(t)$ and $\mathbf{E}_r(t)$ can be written by the real values of the amplitudes $E_d(t)$ and $E_r(t)$ and the real values of the phases $\Phi_d(t)$ and $\Phi_r(t)$, respectively, as follows.

$$\mathbf{E}_d(t) = E_d(t)\exp(i\Phi_d(t)) \tag{6.16}$$

$$\mathbf{E}_r(t) = E_r(t)\exp(i\Phi_r(t)) \tag{6.17}$$

Substituting Eqs. (6.16) and (6.17) into Eq. (6.15).

$$\frac{dE_r(t)}{dt}\exp(i\Phi_r(t)) + i\frac{d\Phi_r(t)}{dt}E_r(t)\exp(i\Phi_r(t))$$
$$= \frac{1+i\alpha_r}{2}\left[G_{N,r}(N_r(t)-N_{0,r}) - \frac{1}{\tau_{p,r}}\right]E_r(t)\exp(i\Phi_r(t)) \quad (6.18)$$
$$+ \kappa_{inj}\, E_d(t-\tau_{inj})\exp(i\Phi_d(t-\tau_{inj}))\exp\{-i\left(-(\omega_d-\omega_r)t+\omega_d\tau_{inj}\right)\}$$

All terms are divided by $\exp(i\Phi_r(t))$.

$$\frac{dE_r(t)}{dt} + i\frac{d\Phi_r(t)}{dt}E_r(t)$$
$$= \frac{1+i\alpha_r}{2}\left[G_{N,r}(N_r(t)-N_{0,r}) - \frac{1}{\tau_{p,r}}\right]E_r(t)$$
$$+ \kappa_{inj}\, E_d(t-\tau_{inj})\exp\{-i\left(-(\omega_d-\omega_r)t+\omega_d\tau_{inj}+\Phi_r(t)-\Phi_d(t-\tau_{inj})\right)\}$$
(6.19)

Here, the Euler's formula is used,

$$\exp(-ix) = \cos x - i\sin x \quad (6.20)$$

Substituting Eq. (6.20) into Eq. (6.19), and the optical angular frequency detuning between the drive and response lasers is introduced as $\Delta\omega = \omega_d - \omega_{r,sol}$,

$$\frac{dE_r(t)}{dt} + i\frac{d\Phi_r(t)}{dt}E_r(t)$$
$$= \frac{1+i\alpha_r}{2}\left[G_{N,r}(N_r(t)-N_{0,r}) - \frac{1}{\tau_{p,r}}\right]E_r(t) \quad (6.21)$$
$$+ \kappa_{inj}\, E_d(t-\tau_{inj})\cos\left(-\Delta\omega t + \omega_d\tau_{inj} + \Phi_r(t) - \Phi_d(t-\tau_{inj})\right)$$
$$- i\kappa_{inj}\, E_d(t-\tau_{inj})\sin\left(-\Delta\omega t + \omega_d\tau_{inj} + \Phi_r(t) - \Phi_d(t-\tau_{inj})\right)$$

The real and imaginary parts of Eq. (6.21) are separated,

$$\frac{dE_r(t)}{dt} = \frac{1}{2}\left[G_{N,r}(N_r(t)-N_{0,r}) - \frac{1}{\tau_{p,r}}\right]E_r(t) + \kappa_{inj}\, E_d(t-\tau_{inj})\cos\Theta_{inj}(t) \quad (6.22)$$

$$\frac{d\Phi_r(t)}{dt} = \frac{\alpha_r}{2}\left[G_{N,r}(N_r(t)-N_{0,r}) - \frac{1}{\tau_{p,r}}\right] - \kappa_{inj}\frac{E_d(t-\tau_{inj})}{E_r(t)}\sin\Theta_{inj}(t) \quad (6.23)$$

$$\Theta_{inj}(t) = -\Delta\omega\, t + \omega_d\tau_{inj} + \Phi_r(t) - \Phi_d(t-\tau_{inj}) \quad (6.24)$$

Equations (6.22), (6.23), and (6.24) are the same formula as Eqs. (6.5), (6.6), and (6.8), respectively, with the real values of the electric-field amplitude $E_r(t)$ and phase $\Phi_r(t)$ of the response laser.

It is important to incorporate the optical angular frequency detuning $\Delta\omega = 2\pi\,\Delta f$ in the coupled rate equations, so that the effect of injection locking can be modeled

6.2 Model for Synchronization of Chaos in Two Coupled Semiconductor Lasers with Optical Feedback

properly. For the consideration of $\Delta\omega$, the rate equations need to be derived from the complex electric field $\hat{\mathbf{E}}(t)$ including the fast optical carrier.

6.2.3 Derivation of the Real and Imaginary Electric Fields of the Coupled Lang–Kobayashi Equations from the Complex Electric-Field Equations

For the coupled Lang–Kobayashi equations with the real amplitude and real phase, there exists the same problem as described in Section 4.2.3 when a chaotic oscillation becomes strong pulsations. The terms of $E(t-\tau)/E(t)$ in the real phase equation of Eqs. (6.2) and (6.6) becomes infinite when $E(t) \to 0$ after strong chaotic pulsations. This effect causes accumulation errors in numerical integration. The formula of the real and imaginary electric fields can be more useful than that of the real amplitude and real phase of the electric field for the coupled Lang–Kobayashi equations, because a large detuning of the optical-carrier frequency results in strong pulsations of laser intensity.

The derivation of the real and imaginary electric fields for the drive laser is exactly the same as shown in Section 4.2.3. Only the derivation for the response laser is treated in this section. The complex electric-field amplitude of Eq. (6.15) is used for the derivation. Instead of using the real amplitude and real phase of Eqs. (6.16) and (6.17), the real and imaginary electric fields ($E_R(t)$ and $E_I(t)$) are introduced,

$$\mathbf{E}_d(t) = E_{R,d}(t) + iE_{I,d}(t) \tag{6.25}$$

$$\mathbf{E}_r(t) = E_{R,r}(t) + iE_{I,r}(t) \tag{6.26}$$

Where $E_{R,d}(t)$, $E_{I,d}(t)$, $E_{R,r}(t)$ and $E_{I,r}(t)$ are real variables. Substituting Eqs. (6.25) and (6.26) into Eq. (6.15).

$$\begin{aligned}\frac{dE_{R,r}(t)}{dt} + i\frac{dE_{I,r}(t)}{dt} \\ = \frac{1+i\alpha_r}{2}\left[G_{N,r}\left(N_r(t)-N_{0,r}\right)-\frac{1}{\tau_{p,r}}\right]\left\{E_{R,r}(t)+iE_{I,r}(t)\right\} \\ + \kappa_{\text{inj}}\left\{E_{R,d}(t-\tau_{\text{inj}})+iE_{I,d}(t-\tau_{\text{inj}})\right\}\exp\left\{-i\left(-\Delta\omega\, t+\omega_d\tau_{\text{inj}}\right)\right\}\end{aligned} \tag{6.27}$$

where $\Delta\omega = \omega_d - \omega_r$. Substituting Eq. (6.20) into Eq. (6.27).

$$\begin{aligned}\frac{dE_{R,r}(t)}{dt} + i\frac{dE_{I,r}(t)}{dt} \\ = \frac{1}{2}\left[G_{N,r}\left(N_r(t)-N_{0,r}\right)-\frac{1}{\tau_{p,r}}\right]\left\{E_{R,r}(t)-\alpha_r E_{I,r}(t)\right\} \\ + i\frac{1}{2}\left[G_{N,r}\left(N_r(t)-N_{0,r}\right)-\frac{1}{\tau_{p,r}}\right]\left\{\alpha_r E_{R,r}(t)+E_{I,r}(t)\right\} \\ + \kappa_{\text{inj}}\left\{E_{R,d}(t-\tau_{\text{inj}})+iE_{I,d}(t-\tau_{\text{inj}})\right\}\cos(-\Delta\omega t+\omega_d\tau_{\text{inj}}) \\ - i\kappa_{\text{inj}}\left\{E_{R,d}(t-\tau_{\text{inj}})+iE_{I,d}(t-\tau_{\text{inj}})\right\}\sin(-\Delta\omega t+\omega_d\tau_{\text{inj}})\end{aligned} \tag{6.28}$$

The real and imaginary parts of Eq. (6.28) are separated as,

$$\frac{dE_{R,r}(t)}{dt} = \frac{1}{2}\left[G_{N,r}(N_r(t)-N_{0,r}) - \frac{1}{\tau_{p,r}}\right]\{E_{R,r}(t) - \alpha_r E_{I,r}(t)\} \qquad (6.29)$$
$$+ \kappa_{inj}\{E_{R,d}(t-\tau_{inj})\cos\theta(t) + E_{I,d}(t-\tau_{inj})\sin\theta(t)\}$$

$$\frac{dE_{I,r}(t)}{dt} = \frac{1}{2}\left[G_{N,r}(N_r(t)-N_{0,r}) - \frac{1}{\tau_{p,r}}\right]\{\alpha_r E_{R,r}(t) + E_{I,r}(t)\} \qquad (6.30)$$
$$+ \kappa_{inj}\{E_{I,d}(t-\tau_{inj})\cos\theta(t) - E_{R,d}(t-\tau_{inj})\sin\theta(t)\}$$

$$\theta(t) = -\Delta\omega t + \omega_d \tau_{inj} \qquad (6.31)$$

The carrier density can be derived from Eq. (6.7),

$$\frac{dN_r(t)}{dt} = J_r - \frac{N_r(t)}{\tau_{s,r}} - G_{N,r}(N_r(t)-N_{0,r})\left(E_{R,r}^2(t) + E_{I,r}^2(t)\right) \qquad (6.32)$$

Equations (6.29)–(6.32) are the real–imaginary expression of the coupled Lang–Kobayashi equations for the response laser. The laser intensity $I_r(t)$ and optical phase $\Phi_r(t)$ of the response laser can be calculated as follows,

$$I_r(t) = E_{R,r}^2(t) + E_{I,r}^2(t) \qquad (6.33)$$

$$\Phi_r(t) = \tan^{-1}\left(\frac{E_{I,r}}{E_{R,r}}\right) \qquad (6.34)$$

It is recommended to use the formula of the real-imaginary equations (Eqs. (6.29)–(6.32)) for the calculation of coupled semiconductor lasers when the temporal waveforms of the laser intensity show strong pulsating behaviors, because numerical errors may occur at small electric amplitudes after strong chaotic pulsations (i.e., $E(t) \to 0$) for the amplitude-phase equations of Eqs. (6.5)–(6.8).

6.3
Numerical Analysis on Synchronization of Chaos in Unidirectionally Coupled Semiconductor Lasers with Optical Feedback

6.3.1
Measures for Synchronization of Chaos

6.3.1.1 Two Types of Synchronization and Cross-Correlation Values
In this section, the two types of synchronization of chaos, identical synchronization and generalized synchronization with high correlation, are numerically observed and compared (see Sections 5.3.1 and 5.3.2). Identical synchronization of chaos is observed at the identical parameter settings of the drive and response lasers. All

the parameter values are set to be identical between the drive and response lasers, including zero optical frequency detuning, as shown in Tables 6.2 and 6.3. These parameter settings satisfy the condition of Eq. (6.9) in order to observe the identical synchronous solutions of Eq. (6.10) in numerical simulation.

The temporal waveforms of the laser intensities for the drive and response lasers ($I_d(t)$ and $I_r(t)$) can be obtained as the square of the electric-field amplitude of Eqs. (6.1) and (6.5), that is,

$$I_d(t) = |E_d(t)|^2, I_r(t) = |E_r(t)|^2 \tag{6.35}$$

The identical synchronization of chaos between the drive and response laser intensities is defined as follows.

$$I_r(t) = I_d(t) \tag{6.36}$$

By contrast, the other type of synchronization, that is, generalized synchronization with high correlation, can also be observed in numerical simulation. For generalized synchronization, the temporal waveforms of the laser intensities of the drive and the response lasers are calculated and shifted by the injection time delay τ_{inj} as follows.

$$I_r(t) \approx A I_d(t - \tau_{inj}) \tag{6.37}$$

where A is the amplification ratio. Note that the temporal waveform of the drive laser intensity is time delayed by the injection delay time τ_{inj} from the drive to response lasers τ_{inj} for generalized synchronization (see Section 5.3.2). The two types of synchronization can be distinguished by measuring the delay time of the temporal waveforms between the drive and response lasers to obtain high correlation values. The parameter values used for the observation of the generalized synchronization are summarized in Table 6.4 as well as Table 6.2. Note that κ_{inj} is much larger than κ_d, and Δf_{sol} is a non-zero value in Table 6.4.

For systematic investigation, the cross-correlation values for the two types of synchronization are introduced as follows.

Cross-correlation for identical synchronization:

$$C_{is} = \frac{\langle (I_d(t) - \bar{I}_d)(I_r(t) - \bar{I}_r) \rangle}{\sigma_d \sigma_r} \tag{6.38}$$

Cross-correlation for generalized synchronization:

$$C_{gs} = \frac{\langle (I_d(t - \tau_{inj}) - \bar{I}_d)(I_r(t) - \bar{I}_r) \rangle}{\sigma_d \sigma_r} \tag{6.39}$$

The Eqs. (6.38) and (6.39) are the same as shown in Eqs. (5.19)–(5.24). Note that the cross-correlation for identical synchronization is calculated from $I_d(t)$ and $I_r(t)$, whereas that for generalized synchronization is obtained from $I_d(t - \tau_{inj})$ and $I_r(t)$.

Table 6.4 Parameter values used in numerical simulation for generalized synchronization with high correlation in the open-loop configuration in Section 6.3.

Symbol	Parameter	Value
$r_{3,d}$	Reflectivity of external mirror of drive laser	0.008
$\kappa_d = \dfrac{(1-r_{2,d}^2)\, r_{3,d}}{r_{2,d}}\dfrac{1}{\tau_{in,d}}$	Feedback strength of drive laser	$1.243 \times 10^9\,\mathrm{s}^{-1}$ $(1.243\,\mathrm{ns}^{-1})$
κ_{inj}	Injection strength from drive to response lasers	$1.553 \times 10^{10}\,\mathrm{s}^{-1}$ $(15.53\,\mathrm{ns}^{-1})$
L_d	External cavity length of drive laser	0.6 m (one-way) 1.2 m (round-trip)
L_{inj}	Distance from drive to response lasers	1.2 m
τ_d	Round-trip time in external cavity of drive laser (feedback delay time)	$4.003 \times 10^{-9}\,\mathrm{s}$
τ_{inj}	Injection delay time from drive to response lasers	$4.003 \times 10^{-9}\,\mathrm{s}$
$\Delta f_{sol} = f_d - f_{r,sol}$	Detuning of optical-carrier frequency between drive and response lasers	$-4.0\,\mathrm{GHz}$
$\Delta\omega = \omega_d - \omega_{r,sol}$ $= 2\pi(f_d - f_{r,sol})$ $= 2\pi\,\Delta f_{sol}$	Detuning of optical angular frequency between drive and response lasers	$-2.513 \times 10^{10}\,\mathrm{s}^{-1}$

The subscripts d and inj indicate the drive laser and injection (coupling) signal, respectively. Other parameter values are used in Table 6.2.

6.3.1.2 Optical Frequency Detuning

The detuning of the optical-carrier frequency between the drive and response lasers is calculated for the investigation of injection locking. The initial optical frequency detuning Δf_{sol} between the drive laser with optical feedback and the solitary response laser "without" optical coupling is simply defined as,

$$\Delta f_{sol} = f_d - f_{r,sol} \qquad (6.40)$$

where f_d is the optical frequency of the drive laser and $f_{r,sol}$ is the optical frequency of the solitary response laser without optical coupling. The initial optical detuning Δf_{sol} is given as an initial parameter for numerical simulation.

When the drive laser output is injected into the response laser under the optical coupling, the optical frequency of the response laser is shifted from $f_{r,sol}$ to $f_{r,inj}$ due to the injection-locking effect. The optical frequency detuning under the optical coupling is obtained from,

$$\Delta f_{inj} = f_d - f_{r,inj} \qquad (6.41)$$

For numerical calculation, Δf_{inj} can be obtained from the instantaneous phases of the drive and response lasers ($\Phi_d(t)$ and $\Phi_r(t)$ in Eqs. (6.2) and (6.6)) and the initial frequency detuning Δf_{sol},

$$\Delta f_{\text{inj}} = \Delta f_{\text{sol}} + \frac{1}{2\pi}\left[\frac{d\Phi_d(t)}{dt} - \frac{d\Phi_r(t)}{dt}\right] \tag{6.42}$$

The optical frequency detuning under the coupling Δf_{inj} can be estimated from Eq. (6.42) in numerical simulation. The injection-locking effect can be identified by using Δf_{inj} and it is considered that injection locking is achieved when $\Delta f_{\text{inj}} \approx 0$ is obtained.

The detuning of the optical-carrier frequency Δf_{inj} estimated in numerical simulation is directly related to the optical wavelength detuning $\Delta \lambda_{\text{inj}} = \lambda_{r,\text{inj}} - \lambda_d$ measured in the experiment of Section 6.1. The relationship between Δf_{inj} and $\Delta \lambda_{\text{inj}}$ can be written as,

$$\begin{aligned}\Delta f_{\text{inj}} &= f_d - f_{r,\text{inj}} = \frac{c}{\lambda_d} - \frac{c}{\lambda_{r,\text{inj}}}\\ &= \frac{(\lambda_{r,\text{inj}} - \lambda_d)c}{\lambda_d \lambda_{r,\text{inj}}} = \frac{c}{\lambda_d \lambda_{r,\text{inj}}}\Delta\lambda_{\text{inj}}\end{aligned} \tag{6.43}$$

where c is the speed of light. Note that $\Delta f_{\text{inj}}(=f_d - f_{r,\text{inj}})$ and $\Delta\lambda_{\text{inj}}(=\lambda_{r,\text{inj}} - \lambda_d)$ are defined as the opposite direction in order to match the signs of Δf_{inj} and $\Delta\lambda_{\text{inj}}$, since $f = c/\lambda$ holds. For the parameters used in experiment (see Table 6.1), the following relationship is obtained.

$$\Delta f_{\text{inj}}\,[\text{GHz}] \approx 125\,\Delta\lambda_{\text{inj}}\,[\text{nm}] \tag{6.44}$$

For example, $\Delta\lambda_{\text{inj}} = 0.01$ nm in experiment corresponds to $\Delta f_{\text{inj}} = 1.25$ GHz in numerical simulation. This relationship is convenient for the comparison between the experimental and numerical results in Sections 6.1 and 6.3.

6.3.2
Numerical Results of Temporal Waveforms and Correlation Plots

The coupled Lang–Kobayashi equations are calculated for the investigation of numerical simulation in this section. Source codes of C programming language are listed in Glossary G.2.4 for numerical simulations of synchronization of chaos in the coupled Lang–Kobayashi equations.

Figure 6.9 shows the temporal waveforms and the correlation plots of the drive and response lasers for the two types of synchronization in the open-loop configuration. For the identical parameter settings ($\kappa_{\text{inj}} = \kappa_d$), the temporal waveforms of the two lasers are synchronized to each other, as shown in Figure 6.9a. The cross-correlation plot with the straight line at 45 degrees in Figure 6.9b shows the evidence of identical synchronization. The cross-correlation value for Figure 6.9b is 1.0. This result indicates the identical synchronous solutions of Eq. (6.10) are observed numerically.

Figure 6.9c and d show the temporal waveforms and the correlation plots of the time-delayed drive and response laser intensities ($I_d(t-\tau_{\text{inj}})$ and $I_r(t)$) for the

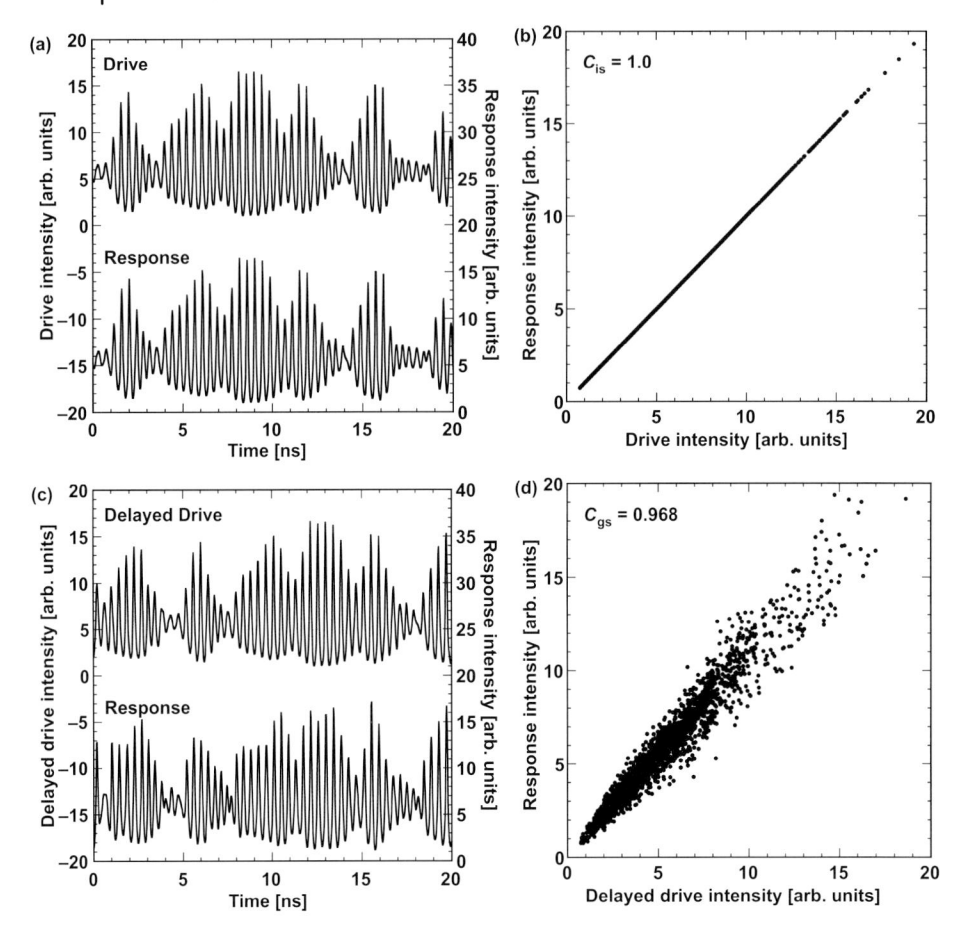

Figure 6.9 (a), (c) Temporal waveforms and (b), (d) corresponding correlation plots of the drive and response laser intensities in the open-loop configuration. (a), (b) identical synchronization and (c), (d) generalized synchronization with high correlation. (c), (d) The temporal waveform of the drive laser is shifted to show optimal synchronization. The cross-correlation values are (b) $C = 1.0$ and (d) $C = 0.968$. The numerical data in this chapter was provided by Kazutaka Kanno.

generalized synchronization with high correlation in the open-loop configuration. When the injection power from the drive to response lasers κ_{inj} is increased from the matching condition of 1.243 ns^{-1} to 15.53 ns^{-1} (see Table 6.4), $I_r(t)$ is synchronized with the time-delayed drive laser intensity $I_d(t-\tau_{inj})$, as shown in Figure 6.9c. The cross-correlation plot of Figure 6.9d shows a linear correlation with some deviations from the 45-degree line. The cross-correlation value for Figure 6.9d is 0.968, which is lower than the case of identical synchronization in Figure 6.9b. In addition, the amplitude of the response laser intensity is slightly larger than that of the time-delayed drive-laser intensity (Figure 6.9d). This indicates that the chaotic temporal waveform from the drive laser is amplified dynamically in the response laser cavity as soon as the injection-laser signal arrives at the response laser, which causes the time delay of τ_{inj} between the drive and response temporal waveforms. Since this process differs from the synchronous solution on the rate equations, the cross-correlation value is slightly degraded from 1.0 for the generalized synchronization.

6.3.3
Parameter Dependence of the Two Types of Synchronization

To investigate the dependence of the two types of synchronization on the laser parameter values, the two cross-correlation values C_{is} and C_{gs} are measured as one of the laser parameter values is detuned between the two lasers. Figure 6.10a shows the

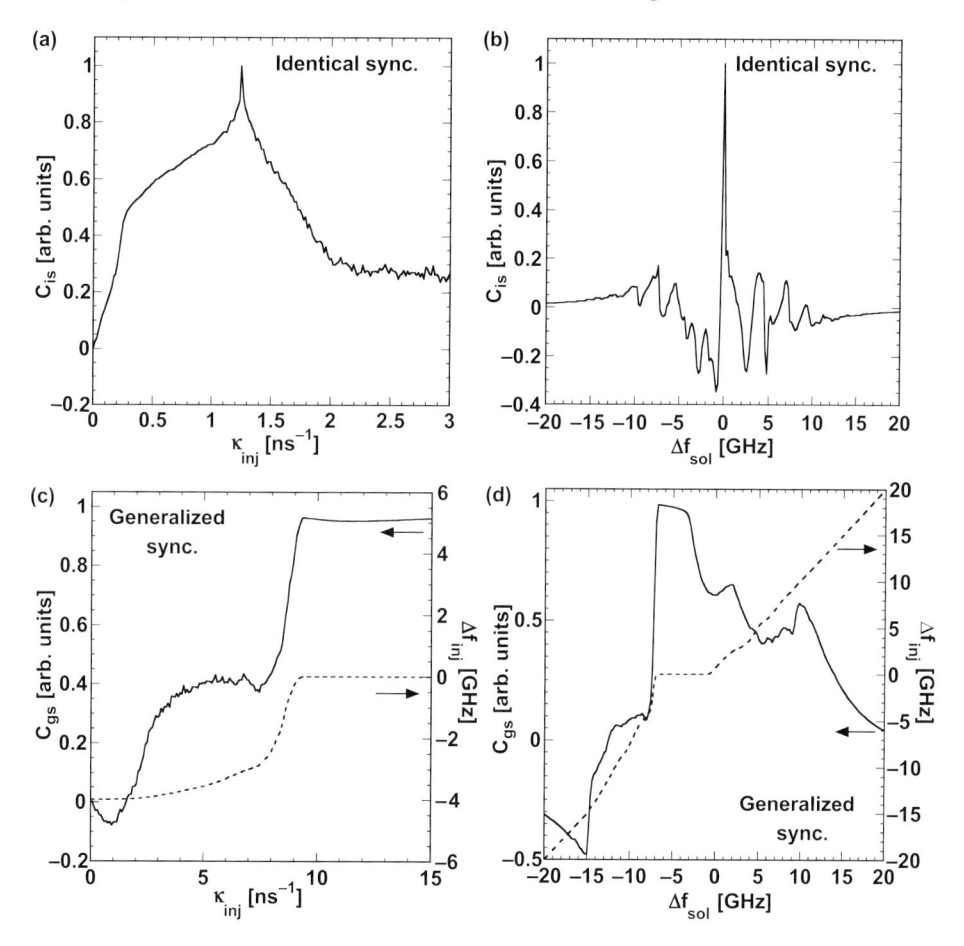

Figure 6.10 Cross-correlation values of (a), (b) identical synchronization C_{is} and (c), (d) generalized synchronization C_{gs} (solid curve) as a function of (a), (c) the optical injection strength κ_{inj} and (b), (d) the initial optical frequency detuning Δf_{sol} in the open-loop configuration. The optical frequency detuning under the optical coupling Δf_{inj} (dotted curve) is also plotted in (c) and (d).

cross-correlation values for identical synchronization C_{is} as the optical injection strength κ_{inj} is changed. The feedback strength of the drive laser is fixed at $\kappa_d = 1.243\,\text{ns}^{-1}$. Identical synchronization with the correlation value of 1.0 is obtained at $\kappa_{inj} = \kappa_d = 1.243\,\text{ns}^{-1}$, as shown in Figure 6.10a. As κ_{inj} is detuned from $1.243\,\text{ns}^{-1}$, the cross-correlation decreases suddenly and identical synchronization is degraded. This result shows the sensitive dependence of the quality of identical synchronization on parameter mismatch.

Figure 6.10b shows C_{is} as the initial optical-carrier frequency detuning Δf_{sol} is changed. Identical synchronization with the correlation of 1.0 is achieved at $\Delta f_{sol} = 0$. However, the correlation decreases rapidly as Δf_{sol} is slightly detuned. It is found that identical synchronization is strongly dependent of Δf_{sol} and can be observed only at $\Delta f_{sol} \approx 0$.

The dependence of generalized synchronization on parameter mismatch is also investigated. Figure 6.10c (solid curve) shows the cross-correlation values for generalized synchronization C_{gs} as the optical injection strength κ_{inj} is changed. At the parameter-matching condition of $\kappa_{inj} = \kappa_d = 1.243\,\text{ns}^{-1}$, no synchronization is observed between $I_d(t-\tau_{inj})$ and $I_r(t)$. As κ_{inj} is increased, the cross-correlation increases and reaches ~ 0.9 at $\kappa_{inj} \approx 9\,\text{ns}^{-1}$. At large injection power of $\kappa_{inj} \geq 9\,\text{ns}^{-1}$, the cross-correlation is almost constant at ~ 0.9 and good synchronization is achieved.

The detuning of the optical-carrier frequency under the optical coupling Δf_{inj} is plotted as a function of κ_{inj} for generalized synchronization, as shown in the dotted curve of Figure 6.10c. Δf_{inj} is almost zero at $\kappa_{inj} \geq 9\,\text{ns}^{-1}$, indicating the optical-carrier frequency matching between the two lasers due to injection locking. Therefore, it is found that high quality of generalized synchronization is achieved in the injection-locking range.

The relationship between the quality of synchronization and the injection-locking range is clearly seen in Figure 6.10d, in which the cross-correlation C_{gs} and Δf_{inj} are plotted as the initial detuning of the optical-carrier frequency without optical coupling Δf_{sol} is changed. The value of C_{gs} more than 0.80 is obtained at the region of $-6.8\,\text{GHz} \leq \Delta f_{sol} \leq -2.8\,\text{GHz}$. In this region injection locking is achieved, where $|\Delta f_{inj}| \leq 0.01\,\text{GHz}$. Therefore, high quality synchronization is obtained in the injection-locking range. Slightly higher cross-correlation is obtained at larger negative value within the injection-locking range.

A systematic investigation is performed by changing κ_{inj} and Δf_{sol} simultaneously. Figures 6.11a and b shows the 2D map of the absolute value of the cross-correlation C_{is} and C_{gs}, respectively, as functions of κ_{inj} and Δf_{sol}. The black region indicates high-quality synchronization and the white region indicates no synchronization. Identical synchronization is only observed at the small region of $\kappa_{inj} \approx 1.243\,\text{ns}^{-1}$ and $\Delta f \approx 0$, as shown in Figure 6.11a. Identical synchronization can be observed at around the parameter-matching condition between the drive and response lasers. By contrast, a large region for high-quality synchronization is observed for generalized synchronization at the conditions of large injection strengths and negative frequency detunings, as shown in Figure 6.11b. Figure 6.11c shows the 2D map of the absolute value of Δf_{inj} as functions of κ_{inj} and Δf_{sol}. The black region in Figure 6.11c

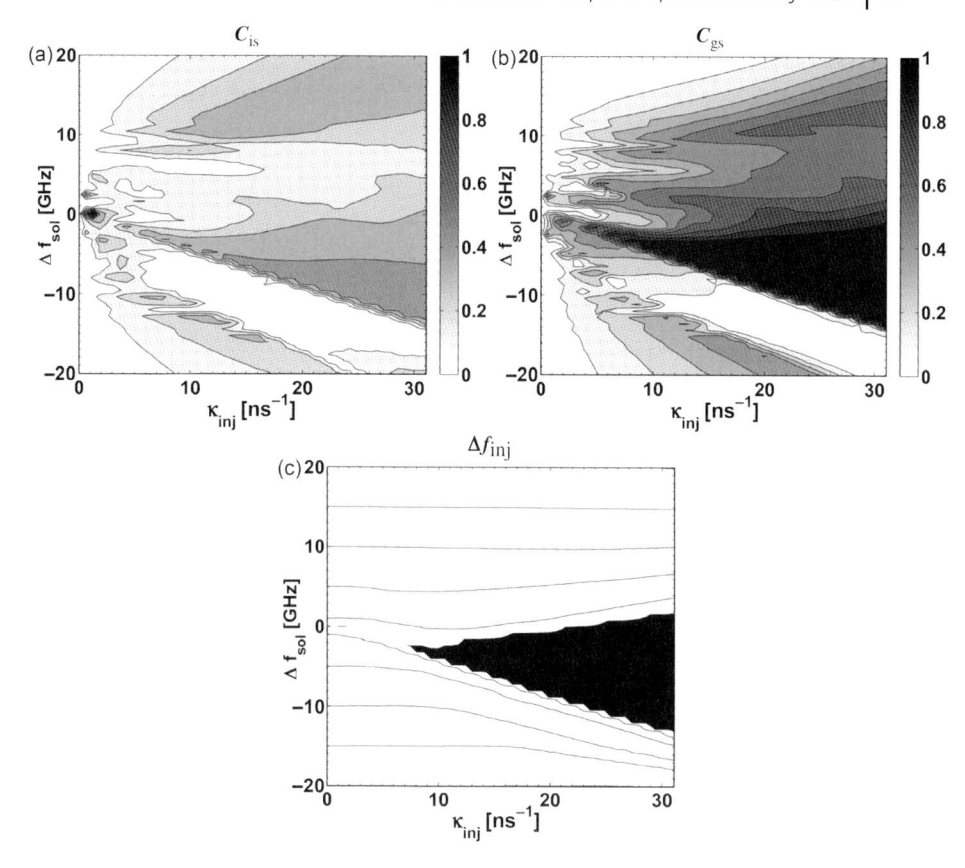

Figure 6.11 Two-dimensional maps of the cross-correlation values of (a) identical synchronization C_{is} and (b) generalized synchronization C_{gs} as functions of the optical injection strength κ_{inj} and the initial optical frequency detuning without optical coupling Δf_{sol} in the open-loop configuration. The black region indicates high-quality synchronization of chaos. (c) Optical frequency detuning under optical coupling Δf_{inj} as functions of κ_{inj} and Δf_{sol} in the open-loop configuration. The black region indicates the injection-locking range.

corresponds to the region of $|\Delta f_{inj}| \leq 0.01$ GHz, indicating the injection-locking range. The region of good synchronization with high correlation values in Figure 6.11b overlaps the injection-locking range in Figure 6.11c. Thus, the matching of the optical-carrier frequency due to injection locking is a necessary condition for achieving the generalized synchronization with high correlation.

In summary, identical synchronization can be observed at the parameter-matching conditions between the two semiconductor lasers. The requirement of parameter matching is strict and identical synchronization is degraded as the parameter mismatch is increased. By contrast, generalized synchronization with high correlation can be observed in large parameter regions of large injection power and small negative frequency detuning, even though parameter mismatch exists between the two lasers. The achievement of generalized synchronization is associated with injection locking, and is very robust against parameter mismatch, as long as the injection locking condition is satisfied. Generalized synchronization can be more easily observed in experiment than identical synchronization, as seen in Section 6.1.

6.3.4
Linear Stability Analysis of Synchronous Oscillatory Solutions for Identical Synchronization

6.3.4.1 Linearized Equations

The stability of the identical synchronous solutions of Eq. (6.10) is important to observe identical synchronization of chaos. Synchronization is achieved when the synchronous solutions are stable, whereas no synchronization is observed when the synchronous solutions are unstable. The general formula of the linearized equations for synchronous solution of identical synchronization is described in Appendix 5.A.2 of Chapter 5.

Linearized variables are introduced to test the stability of the synchronous oscillatory solutions as,

$$\Delta_E(t) = E_d(t) - E_r(t) \tag{6.45}$$

$$\Delta_\Phi(t) = \Phi_d(t) - \Phi_r(t) \tag{6.46}$$

$$\Delta_N(t) = N_d(t) - N_r(t) \tag{6.47}$$

Note that the linearized variables of Eqs. (6.45)–(6.47) are defined as the difference in the variables between the drive and response lasers, whereas the linearized variables shown in Chapter 4 indicate the distance of the two neighboring trajectories in a single laser. The linearized equations (also known as variational equations) for synchronous oscillatory solutions can be obtained by calculating the difference in the variables in the corresponding equations between the drive (Eqs. (6.1)–(6.3)) and response (Eqs. (6.5)–(6.7)) lasers in the open-loop configuration under the identical parameter conditions of Eq. (6.9) as follows,

$$\frac{d\Delta_E(t)}{dt} = \frac{1}{2}\left[G_{N,r}(N_r(t) - N_{0,r}) - \frac{1}{\tau_{p,r}}\right]\Delta_E(t) - \kappa_{inj} E_d(t - \tau_{inj})\sin\Theta_{inj}(t)\Delta_\Phi(t)$$
$$+ \frac{1}{2}G_{N,r} E_r(t)\Delta_N(t)$$

$$\tag{6.48}$$

$$\frac{d\Delta_\Phi(t)}{dt} = \kappa_{inj}\frac{E_d(t-\tau_{inj})}{E_r^2(t)}\sin\Theta_{inj}(t)\Delta_E(t) - \kappa_{inj}\frac{E_d(t-\tau_{inj})}{E_r(t)}\cos\Theta_{inj}(t)\Delta_\Phi(t)$$
$$+ \frac{\alpha_r}{2}G_{N,r}\Delta_N(t)$$

$$\tag{6.49}$$

$$\frac{d\Delta_N(t)}{dt} = -2G_{N,r}E_r(t)(N_r(t) - N_{0,r})\Delta_E(t) - \left(\frac{1}{\tau_{s,r}} + G_{N,r}E_r^2(t)\right)\Delta_N(t) \tag{6.50}$$

6.3 Numerical Analysis on Synchronization of Chaos

To derive the liearized equations of Eqs. (6.48)–(6.50), the Taylor expansion up to the order of Δ_x ($x = E$, Φ, N) is used and the terms whose order is higher than Δ_x are eliminated for the approximation as,

$$f(x+\Delta_x) = f(x) + f'(x)\Delta_x + O(\Delta_x^2) \quad (x \gg \Delta_x) \tag{6.51}$$

Therefore, the following approximations are used to derive the linearized equations of Eqs. (6.48)–(6.50),

$$\cos(x+\Delta_x) \approx \cos x - \Delta_x \sin x \quad (x \gg \Delta_x) \tag{6.52}$$

$$\sin(x+\Delta_x) \approx \sin x + \Delta_x \cos x \quad (x \gg \Delta_x) \tag{6.53}$$

$$\frac{1}{x+\Delta_x} \approx \frac{1}{x} - \frac{\Delta_x}{x^2} \quad (x \gg \Delta_x) \tag{6.54}$$

Note that the linearized equations (Eqs. (6.48)–(6.50)) for synchronous oscillatory solutions between two coupled lasers are different from those for the steady-state solution in a single laser (Eqs. (4.95)–(4.97)) in Section 4.4.2. In Eqs. (6.48)–(6.50), there are no time-delayed linearized variables (e.g., $\Delta_E(t-\tau_d)$ and $\Delta_\Phi(t-\tau_d)$), since the feedback and coupling terms are symmetric between the drive and response lasers under the identical synchronous solution and can be eliminated by subtracting the drive equations from the response equations. No linearized variables for the feedback and coupling terms remain in Eqs. (6.48)–(6.50). Therefore, the linearized equations for the synchronous oscillatory solutions in coupled lasers are different from those for the oscillatory trajectory in a single laser (in Section 4.4.2) for the treatment of the feedback and coupling terms. Note that this is the case for the open-loop configuration. The time-delayed terms remain in the linearized equations for the case of the closed-loop configuration (see Sections 6.5 and 6.6).

6.3.4.2 Conditional Lyapunov Exponent

The conditional Lyapunov exponent is numerically calculated from the linearized equations of Eqs. (6.48)–(6.50) for the identical synchronous solutions of Eq. (6.10) in the following procedure (see also Appendix 5.A.3 of Chapter 5). In this chapter, only the maximum value of the conditional Lyapunov exponents is considered to investigate synchronizability. Most procedures for the calculation of the conditional Lyapunov exponent are the same as the normal maximum Lyapunov exponent described in Section 4.4.3, except for the treatment of the norm. Unlike the linearized equations for a single laser of Eqs. (4.95)–(4.97) in Section 4.4.3, Eqs. (6.48)–(6.50) do not include time-delayed linearized variables ($\Delta_E(t-\tau_d)$ and $\Delta_\Phi(t-\tau_d)$). Therefore, Eqs. (6.48)–(6.50) is a three-dimensional system and the norm is simply defined as the Euclidean distance of the vector consisting of all the three variables of the linearized equations, that is

$$D_c(t) = \sqrt{\Delta_E^2(t) + \Delta_\Phi^2(t) + \Delta_N^2(t)} \tag{6.55}$$

$D_c(t)$ can be obtained at each step of numerical integration. Short evolution of the trajectory from t to $t+\tau_{ev}$ is obtained, where τ_{ev} is the evolution time for norm calculation, and it is recommended to use a small value for τ_{ev} such as the integration step of numerical calculation to maintain the validity of the linearized equations. The variation of the norm is calculated as,

$$d_j = \frac{D_c(t+\tau_{ev})}{D_c(t)} \quad (6.56)$$

where d_j is the norm ratio after the short evolution from t to $t+\tau_{ev}$ at the jth procedure of the calculation of the norm ratio. Next, all the linearized variables $\Delta_E(t+\tau_{ev})$, $\Delta_\Phi(t+\tau_{ev})$, and $\Delta_N(t+\tau_{ev})$ are rescaled by $D_c(t+\tau_{ev})$ after the calculation of the norm ratio as,

$$\Delta_{E,\,new}(t+\tau_{ev}) = \frac{\Delta_E(t+\tau_{ev})}{D_c(t+\tau_{ev})} \quad (6.57)$$

$$\Delta_{\Phi,\,new}(t+\tau_{ev}) = \frac{\Delta_\Phi(t+\tau_{ev})}{D_c(t+\tau_{ev})} \quad (6.58)$$

$$\Delta_{N,\,new}(t+\tau_{ev}) = \frac{\Delta_N(t+\tau_{ev})}{D_c(t+\tau_{ev})} \quad (6.59)$$

This rescaling procedure is very important to maintain a small deviation of the linearized variables from the identical synchronous trajectory. The integration of the trajectory resumes for the next evolution of τ_{ev}, and $d_{j+1} = D_c(t+2\tau_{ev})/D_c(t+\tau_{ev})$ is calculated, and $\Delta_E(t+2\tau_{ev})$, $\Delta_\Phi(t+2\tau_{ev})$, $\Delta_N(t+2\tau_{ev})$ are normalized by $D_c(t+2\tau_{ev})$, and so on. After this procedure is repeated N times, where $N = T/\tau_{ev}$ and T is the total calculation time, the conditional Lyapunov exponent is calculated as,

$$\lambda_c = \frac{1}{N\tau_{ev}} \sum_{j=1}^{N} \ln(d_j) \quad (6.60)$$

where ln is the natural logarithm with the natural base e. The conditional Lyapunov exponent can be obtained as the average ratio of the norm change in the natural logarithmic scale.

6.3.4.3 Numerical Results of Conditional Lyapunov Exponent

For identical synchronization of chaos, the conditional Lyapunov exponent can be calculated from the procedure of Eqs. (6.55)–(6.60) by using the linearized equations of Eqs. (6.48)–(6.50) along with the coupled Lang–Kobayashi equations of Eqs. (6.1)–(6.8). It is important to satisfy the parameter-matching conditions to maintain the mathematical synchronous solutions, because the conditional Lyapunov exponent represents linear stability of the identical synchronous solutions. Therefore, the estimation of the conditional Lyapunov exponent cannot be applied

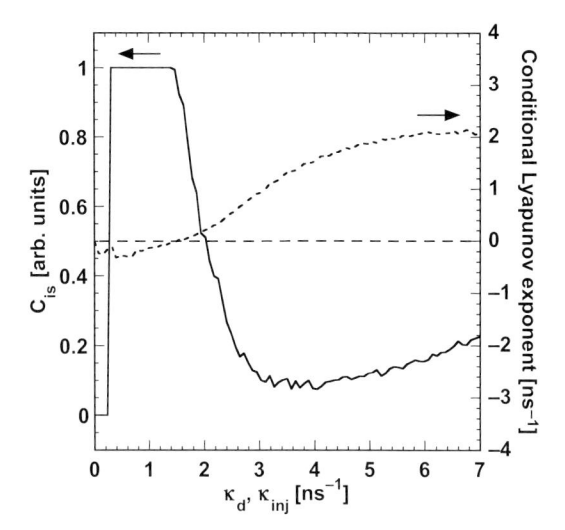

Figure 6.12 Cross-correlation value of identical synchronization C_{is} (solid curve) and conditional Lyapunov exponent λ_c (dotted curve) as a function of the optical feedback strength of the drive laser κ_d and optical injection strength κ_{inj}. κ_d and κ_{inj} are changed simultaneously under the condition of $\kappa_d = \kappa_{inj}$. The dashed line indicates $\lambda_c = 0$.

for the generalized synchronization with high correlation due to the lack of mathematical synchronous solutions. The conditional Lyapunov exponent for identical synchronization is calculated when the optical feedback strength of the drive laser κ_d and the optical injection strength from the drive to response lasers κ_{inj} are changed simultaneously by maintaining the relationship of $\kappa_d = \kappa_{inj}$.

Figure 6.12 shows the cross-correlation values for identical synchronization C_{is} and the conditional Lyapunov exponents λ_c as both κ_d and κ_{inj} are changed simultaneously. At the small κ_d and κ_{inj} near zero, the cross-correlation is zero because no oscillation of temporal waveforms is found for tiny optical feedback strength of the drive laser. In the region of $0.31 \text{ ns}^{-1} \leq \kappa_d, \kappa_{inj} \leq 1.48 \text{ ns}^{-1}$, the cross-correlation is 1.0 and identical synchronization is achieved. In this region, the conditional Lyapunov exponent has a negative value, indicating stable synchronous solutions of Eq. (6.10). By contrast, as κ_d and κ_{inj} are increased at $\kappa_d, \kappa_{inj} > 1.48 \text{ ns}^{-1}$, the cross-correlation starts to decrease to zero. In this region, the conditional Lyapunov exponent becomes positive, indicating unstable synchronous solutions. Identical synchronization is not achieved at large κ_d and κ_{inj} because the conditional Lyapunov exponent increases monotonically even though identical synchronous solution exists at the parameter-matching condition. This fact indicates that parameter matching does not guarantee the achievement of identical synchronization, and the stability of identical synchronization is important to judge the achievement of identical synchronization. The conditional Lyapunov exponent is a good indicator for the criteria of the stability of identical synchronization.

Table 6.5 Parameter values used in numerical simulation for generalized synchronization with high correlation in the closed-loop configuration in Section 6.3.

Symbol	Parameter	Value
$r_{3,r}$	Reflectivity of external mirror of response laser	0.008
$\kappa_r = \dfrac{(1-r_{2,r}^2)\, r_{3,r}}{r_{2,r}} \dfrac{1}{\tau_{in,r}}$	Feedback strength of response laser	$1.243 \times 10^9\,\text{s}^{-1}$ ($1.243\,\text{ns}^{-1}$)
L_r	External cavity length of response laser	0.6 m (one-way) 1.2 m (round-trip)
τ_r	Round-trip time in external cavity of response laser (feedback delay time)	$4.003 \times 10^{-9}\,\text{s}$

The subscript r indicates the response laser. Other parameter values are used in Tables 6.2 and 6.4.

6.3.5
Open- Versus Closed-Loop Configurations

In this section, generalized synchronization in the closed-loop configuration is investigated, where an external mirror is placed in front of the response laser and chaotic dynamics is generated in the response laser (see Figure 6.8). The sensitivity of synchronization quality on parameter mismatch is investigated and compared with the case for the open-loop configuration. The parameter values for the closed-loop configuration are summarized in Table 6.5. Other parameter values are used in Tables 6.2 and 6.4. The rate equations for the closed-loop configuration is shown in Appendix 6.A.1.

Figure 6.13a shows the 2D map of the absolute value of the cross-correlation C_{gs} as functions of κ_{inj} and Δf_{sol} for the closed-loop confguration. The black region

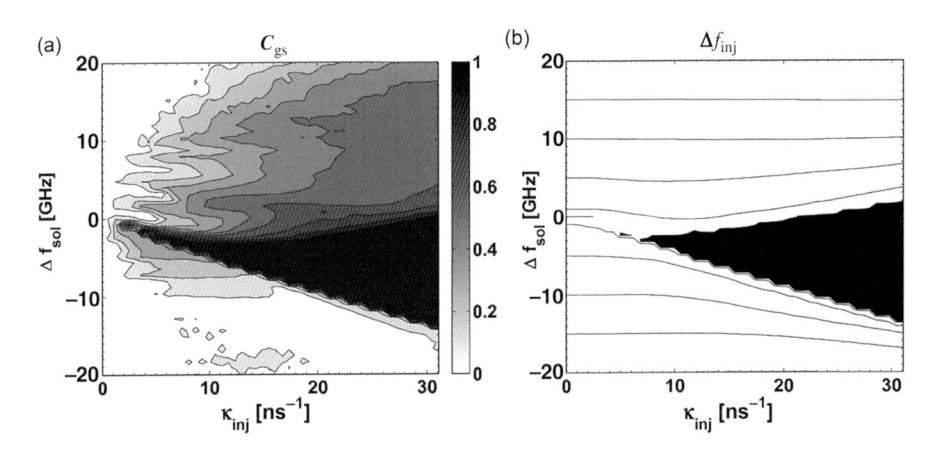

Figure 6.13 Two-dimensional maps of (a) the cross-correlation values of generalized synchronization C_{gs} and (b) optical frequency detuning under optical coupling Δf_{inj} as functions of the optical injection strength κ_{inj} and the initial optical frequency detuning Δf_{sol} in the closed-loop configuration. (a) The black region indicates high-quality synchronization of chaos. (b) The black region indicates the injection-locking range.

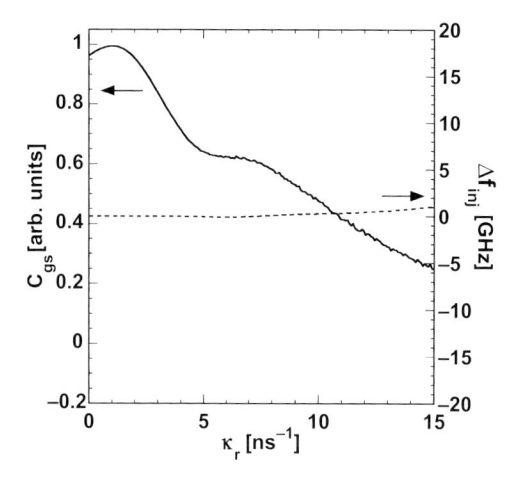

Figure 6.14 Cross-correlation value of generalized synchronization C_{gs} (solid curve) and optical frequency detuning under optical coupling Δf_{inj} (dotted curve) as a function of the optical feedback strength of the response laser κ_r in the closed-loop configurations. No feedback strength ($\kappa_r = 0$) corresponds to the open-loop configuration.

in Figure 6.13a indicates high-quality synchronization. Figure 6.13b shows the 2D map of the absolute value of Δf_{inj} as functions of κ_{inj} and Δf_{sol}. The black region in Figure 6.13b indicates the region of $|\Delta f_{inj}| \leq 0.01$ GHz, corresponding to the injection-locking range. The region of high-quality synchronization with high C_{gs} in Figure 6.13a overlaps the injection-locking range in Figure 6.13b. The result for the closed-loop configuration (Figure 6.13a) is very similar to that for the open-loop configuration (Figure 6.11b) for the generalized synchronization.

Figure 6.14 shows the cross-correlation values for generalized synchronization C_{gs} and the optical frequency detuning under the coupling Δf_{inj} as the optical feedback strength of the response laser κ_r is changed. The optical feedback strength of the drive laser is fixed at $\kappa_d = 1.243$ ns^{-1}. As κ_r is increased, C_{gs} gradually decreases to ~ 0.2, whereas injection locking ($\Delta f_{inj} \approx 0$) is achieved in the whole region of κ_r in Figure 6.14. The best correlation value of C_{gs} is obtained at the parameter-matching condition of $\kappa_r = \kappa_d = 1.243$ ns^{-1}. For overall dynamics, the quality of the generalized synchronization is almost the same between the open- and closed-loop configurations for small κ_r.

6.4
Experimental Analysis on Generalized Synchronization with Low Correlation in Three Semiconductor Lasers in the Auxiliary System Approach

In this section, experimental observation of generalized synchronization with low correlation is investigated by using the auxiliary system approach (see Section 5.6.2).

The model for the auxiliary system approach with coupled lasers is already shown in Figure 5.33. In addition to a drive-response laser system, a replica of the response laser (referred to as the response-2 laser) with identical parameter settings (but different initial conditions) is prepared. A chaotic signal from the drive laser is divided into two signals and each of them is unidirectionally injected into each of the two response lasers. The original response laser (the response-1 laser) is coupled with the drive laser and the response-2 laser is coupled to the drive laser, respectively. There is, however, no coupling between the response-1 and -2 lasers. Synchronization between the response-1 and -2 lasers can be observed and high-quality synchronization of chaos between the response-1 and -2 lasers indicates the existence of generalized synchronization between the drive and response lasers, even though the quality of synchronization (e.g., cross-correlation value) between the drive and response lasers is relatively low.

6.4.1
Experimental Setup for Generalized Synchronization with Low Correlation in the Auxiliary System Approach

Figure 6.15 shows the experimental setup for generalized synchronization with low correlation in unidirectionally coupled semiconductor lasers with optical feedback in the auxiliary system approach. The experimental setup is similar to that used in Figure 6.1, except the use of an auxiliary response laser. Three DFB semiconductor lasers are used, which are mounted in butterfly packages with optical-fiber pigtails. One laser is used for a drive laser and the other two lasers are used for response lasers (referred to as response-1 and response-2 lasers, respectively). The response-1 and -2 lasers are set to have similar setup and parameter values (but different initial conditions) so that the response-2 laser is considered as a replica of the response-1 laser for the auxiliary system approach for detecting generalized synchronization (see the dashed frame in Figure 6.15). On the other hand, some parameter values of the drive laser are detuned from those of the two response lasers.

The drive laser is connected to a fiber coupler (FC) and a variable fiber reflector (VFR) which reflects a fraction of the light back into the laser, inducing high-frequency chaotic oscillations of the optical intensity. A portion of the chaotic intensity beam of the drive laser is divided into two beams by a FC. Each of the beams is injected into each response laser for generalized synchronization. Optical isolators (ISOs) are used to achieve one-way coupling from the drive to the response lasers. The power of the injected light is adjusted by using fiber attenuators so that the same optical power is injected into the two response lasers. Note that the drive and each of the two response lasers are coupled unidirectionally from the drive to the response lasers, whereas there is no optical coupling between the two response lasers.

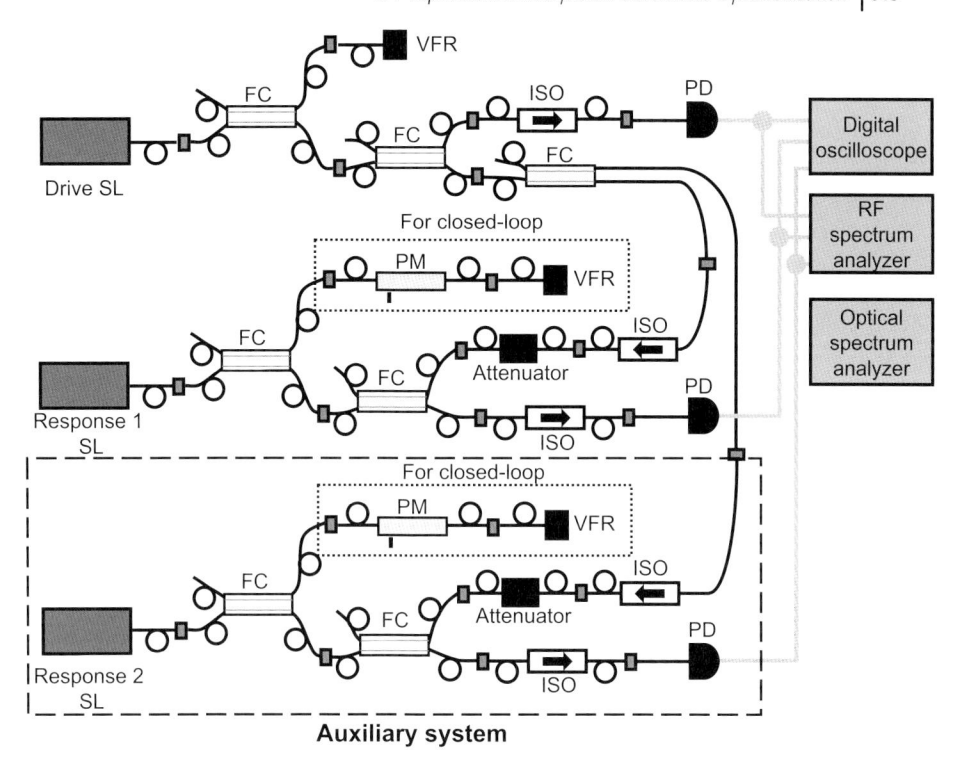

Figure 6.15 Experimental setup for generalized synchronization with low correlation in the auxiliary system approach. Three semiconductor lasers (drive, response-1, and response-2 lasers) are used. The dotted frames indicate the closed-loop configuration in the response lasers. The dashed frame indicates the auxiliary system (the response-2 laser system). FC, fiber coupler; ISO, optical-fiber isolator; PD, photodetector; PM, phase modulator; SL, semiconductor laser; VFR, variable fiber reflector.

Each response laser does or does not have optical feedback, which corresponds to the closed- and open-loop configurations, respectively. For the closed-loop configuration, the response laser is connected to a VFR and a phase modulator (PM) to introduce optical feedback (see the dotted frame in Figure 6.15), and chaotic intensity fluctuation is observed in the response laser in the presence of the optical feedback even without optical coupling. By contrast, the VFR and PM are disconnected for the open-loop configuration, and stable laser output with small fluctuation due to the relaxation oscillation is obtained in the response laser without optical coupling.

The optical wavelengths of the drive and two response lasers are precisely adjusted in order to obtain generalized synchronization. The temporal waveform and the RF spectrum of the laser intensity are measured by a digital oscilloscope and a RF spectrum analyzer through a AC-coupled photodetector, respectively. The optical wavelength of the laser is measured by an optical spectrum analyzer.

Table 6.6 Parameter values used in the experiment for generalized synchronization with low correlation in the auxiliary system approach in Section 6.4.

Parameter	Drive Laser	Response-1 Laser	Response-2 Laser
Threshold of injection current [mA]	10.48	9.38	9.49
Injection current [mA] (relative to the threshold)	11.15 (1.06)	12.30 (1.31)	12.68 (1.34)
Relaxation oscillation frequency [GHz]	1.52	2.00	2.00
Wavelength without coupling [nm] (detuning from drive laser [nm])	1547.224	1547.188 (−0.036)	1547.188 (−0.036)
Wavelength with coupling [nm] (detuning from drive laser [nm])	1547.224	1547.224 (0.0)	1547.224 (0.0)
Laser temperature [K]	290.00	294.19	294.70
Feedback power [μW]	0.084	1.12 for closed-loop	1.03 for closed-loop
Injection power from drive to response lasers [μW] (Maximum injection power [μW])	N/A	10.69 (10.69)	9.57 (9.57)
External cavity length [m] (One-way, fiber length)	2.53	3.67 for closed-loop	3.67 for closed-loop
Feedback delay time [ns] (external cavity frequency [MHz])	24.3 (41.2)	35.3 (28.3) for closed-loop	35.3 (28.3) for closed-loop
Injection delay time from drive laser [ns]	N/A	45.4	45.4

6.4.2
Experimental Results of Generalized Synchronization

The parameter values used for the experiment of generalized synchronization of chaos are summarized in Table 6.6. All the parameter values for the response-1 and -2 lasers are set to be matched as similarly as possible. The relaxation oscillation frequencies for the two response lasers are set to 2.00 GHz by adjusting the injection current of the lasers. By contrast, the relaxation oscillation frequency for the drive laser is set to 1.52 GHz to detune a parameter value between the drive and response lasers.

Figure 6.16 shows the optical spectra of the three lasers without and with optical coupling in the open-loop configuration. The optical wavelengths of the two response lasers are set at an almost identical value, which is a negative detuning with respect to that of the drive laser, that is, the optical wavelengths are set to 1547.224 nm for the drive laser and 1547.188 nm for the response-1 and -2 lasers, respectively, as shown in

6.4 Experimental Analysis on Generalized Synchronization | 317

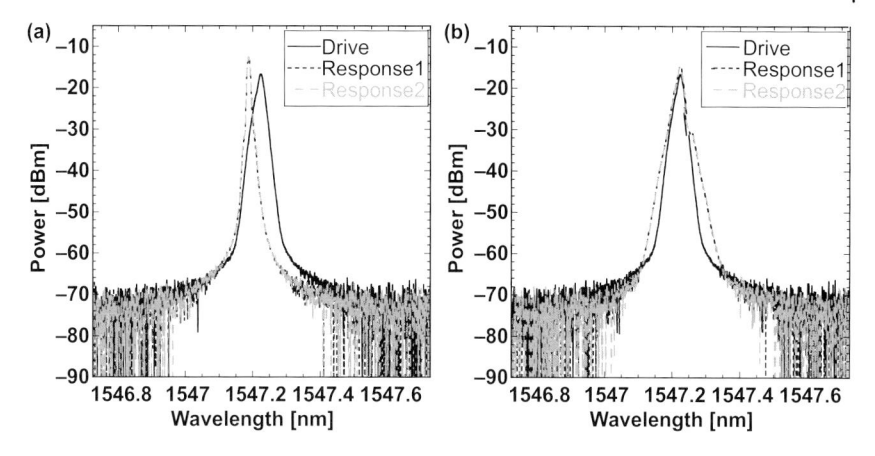

Figure 6.16 Optical spectra of the drive, response-1, and response-2 lasers (a) without and (b) with optical coupling in the open-loop configuration. Injection locking (matching of the optical wavelengths) is achieved in (b). Solid curve, the drive laser; dotted curve, the response-1 laser; dashed gray curve, the response-2 laser.

Figure 6.16a. The initial optical wavelength detuning is defined as $\Delta\lambda_{sol} = \lambda_{r,sol} - \lambda_d$ (same as shown in Section 6.1.2). $\Delta\lambda_{sol}$ is set to -0.036 nm (-4.5 GHz in frequency), which corresponds to the negative detuning condition ($\lambda_{r,sol} < \lambda_d$).

When the drive laser output is injected into the two response lasers at this detuning condition, the optical wavelengths for the two response lasers are shifted to 1547.224 nm and exactly matched to that of the drive laser, as shown in Figure 6.16b. It is found that injection locking is achieved between the drive and each of the two response lasers. When the optical wavelength detuning is small enough, the optical wavelengths of the two response lasers are matched with that of the drive laser due to injection locking, and the wavelengths for all the three lasers become identical.

Without optical coupling, the drive and each of the response lasers are not synchronized at all and only the small sustained relaxation oscillations are observed in the response laser outputs. Synchronization of the temporal waveforms between the two response lasers is also not observed. By contrast, Figure 6.17 shows the temporal waveforms of the drive and two response lasers with optical coupling, and their corresponding correlation plots between two of the three lasers in the open-loop configuration. The temporal waveform of the drive laser is delayed with respect to the response lasers by the injection delay time (45.4 ns) to obtain the optimal synchronization between the drive and each of the response lasers. In the presence of optical coupling, the two temporal waveforms of the drive and each of two response lasers are partially synchronized, as shown in the correlation plots of Figure 6.17b and c. The cross-correlation values C (Eq. (5.19)) between the drive and each of the two response lasers are 0.660 and 0.655 for Figures 6.17b and c, respectively. More interestingly, the temporal waveforms of the two response lasers are well synchronized to each other, as shown in Figures 6.17a and d. The cross-correlation value between the two

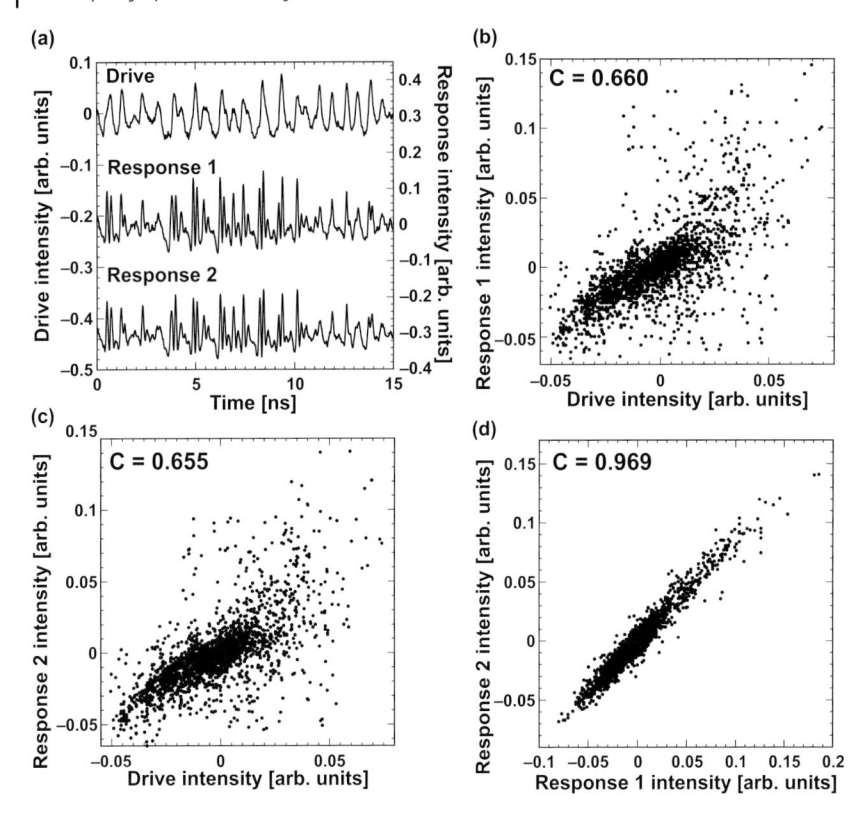

Figure 6.17 (a) Temporal waveforms of the drive, response-1, and response-2 laser intensities and the corresponding correlation plots (b) between the drive and response-1 lasers, (c) between the drive and response-2 lasers, and (d) between the response-1 and response-2 lasers under the optical coupling in the open-loop configuration. The cross-correlation values are (b) $C = 0.660$, (c) $C = 0.655$, and (d) $C = 0.969$.

response lasers is 0.969, close to 1, in Figure 6.17d. Therefore, high-quality synchronization of chaos is achieved between the response-1 and -2 lasers, even though partial synchronization is observed between the drive and each of the two response lasers. This result suggests the existence of generalized synchronization between the drive and response lasers, according to the auxiliary system approach (Abarbanel et al., 1996). Generalized synchronization of chaos can be also achieved in the closed-loop configuration in the experiment.

Figure 6.18 shows the RF spectra of the drive and two response lasers without and with optical coupling in the open-loop configuration. Without optical coupling, the RF spectra of the two response lasers show a peak value at the relaxation oscillation frequency of 2.0 GHz as shown in Figure 6.18a. With optical coupling, two broad peaks appear at around 1.5 and 4.0 GHz in the RF spectra of the two response lasers (Figure 6.18b). A part of the RF spectra is overlapping between the drive and response lasers at ∼1.5 GHz, however, the shape of the RF spectra is different between the drive and two response lasers at ∼4.0 GHz. In the enlargement of the RF spectra at

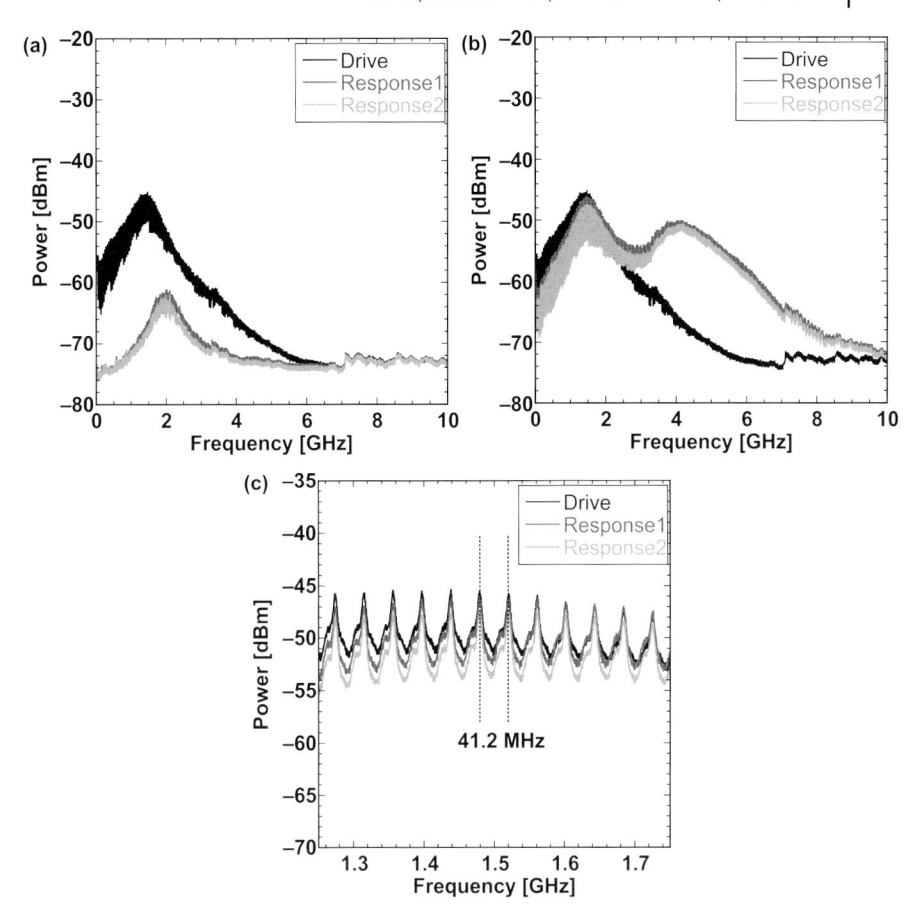

Figure 6.18 RF spectra of the drive, response-1, and response-2 lasers (a) without and (b) with optical coupling, and (c) the enlargement of (b) in the open-loop configuration. Black curve, the drive laser; gray curve, the response-1 laser; light gray curve, the response-2 laser.

~1.5 GHz in Figure 6.18c, the periodic spectral peaks are observed, whose interval corresponds to the external cavity frequency of the drive laser (41.2 MHz). The structures of the RF spectra of the drive and response lasers are thus very similar in the frequency region of ~1.5 GHz, even though the spectra in the region of ~4.0 GHz are quite different. It is speculated that the spectrum component at ~4.0 GHz appears due to the optical frequency detuning between the drive and response lasers, therefore, it only exists in the two response lasers. By contrast, the shape of the RF spectra for the two response lasers resemble each other, as shown in Figure 6.18b. These results indicate the observation of generalized synchronization, for which high-quality synchronization is achieved between the response-1 and -2 lasers, while partial synchronization is observed between the drive and each of the two response lasers.

6.4.3
Parameter Dependence of Generalized Synchronization

The parameter dependence of generalized synchronization of chaos is investigated when one of the parameters for the drive or response lasers is changed. When a response parameter value is changed, both of the parameter values for the response-1 and -2 lasers are maintained the same and changed simultaneously. The accuracy of synchronization is quantitatively defined as the cross-correlation C of Eq. (5.19).

First, the optical injection power from the drive to the response lasers is changed. The initial optical wavelength detuning without optical coupling $\Delta\lambda_{sol}$ is fixed to -0.034 nm. Here, the optical wavelength detuning between the drive and response lasers under the optical coupling is defined as $\Delta\lambda_{inj} = \lambda_{r,inj} - \lambda_d$ (the same as shown in Section 6.1.3). The optical wavelengths of the response-1 and -2 lasers are maintained to be identical by controlling the temperature of the response lasers. When $\Delta\lambda_{inj} \approx 0$ is obtained, the optical wavelengths between the drive and each of the response lasers are completely matched to each other due to injection locking.

For the measurement of cross-correlation, the temporal waveform of the drive laser is delayed by the injection delay time (45.4 ns) to obtain the optimal synchronization between the drive and each of the response lasers. Two cross-correlation values are measured between the time-delayed drive and response-1 lasers (denoted as $C_{d\tau,r1}$, see Eq. (6.80)) and between the response-1 and -2 lasers (denoted as $C_{r1,r2}$, see Eq. (6.79)), as the optical injection power from the drive to each of the response lasers κ_{inj} is increased, as shown in Figure 6.19a. κ_{inj} is normalized by the maximum

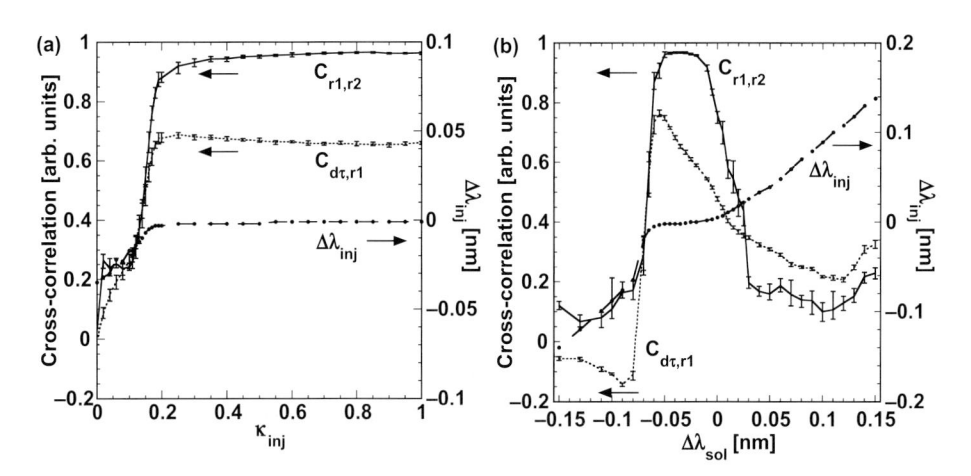

Figure 6.19 Cross-correlation value between the response-1 and -2 lasers $C_{r1,r2}$ (solid curve with error bars), cross-correlation value between the time-delayed drive and response-1 lasers $C_{d\tau,r1}$ (dotted curve with error bars), and the optical wavelength detuning between the drive and response lasers under the optical coupling $\Delta\lambda_{inj}$ (dashed curve with solid circles) in the open-loop configuration. (a) The optical injection power from the drive to each of the response lasers κ_{inj} is changed. κ_{inj} is normalized by the maximum injection power obtained in the experiment. and (b) The initial optical wavelength detuning between the drive and each of the response lasers without optical coupling $\Delta\lambda_{sol}$ is changed.

injection power obtained in the experiment. The optical injection powers injected to the response-1 and -2 lasers are maintained to be identical. The optical wavelength detuning with optical coupling $\Delta\lambda_{\text{inj}}$ is also plotted in Figure 6.19a. The value of $C_{r1,r2}$ increases and reaches over 0.9 as the optical injection power is increased. At the same time $\Delta\lambda_{\text{inj}}$ becomes almost zero at the region of large $C_{r1,r2}$. Therefore, high-quality synchronization of chaos is achieved between the response-1 and -2 lasers under the condition of the optical wavelength matching between the drive and each of the response lasers by injection locking. By contrast, the value of $C_{d\tau,r1}$ between the drive and response-1 lasers reaches only 0.6–0.7 at large optical injection power. Therefore, good synchronization is achieved between the two response lasers, whereas partial synchronization is observed between the drive and each of the response lasers under injection locking.

Next, the initial optical wavelength detuning $\Delta\lambda_{\text{sol}}$ is changed. The cross-correlation values $C_{d\tau,r1}$ and $C_{r1,r2}$ are measured as $\Delta\lambda_{\text{sol}}$ is increased, as shown in Figure 6.19b. The wavelengths of the two response lasers are changed simultaneously to maintain the identical value. $\Delta\lambda_{\text{inj}}$ is also plotted in Figure 6.19b. The region of $C_{r1,r2} > 0.8$ is observed at the region of $-0.060 \text{ nm} \leq \Delta\lambda_{\text{sol}} \leq 0.005 \text{ nm}$. In this region the optical wavelength detuning with optical coupling is $|\Delta\lambda_{\text{inj}}| \leq 0.005 \text{ nm}$, indicating the injection-locking range. Therefore, injection locking between the drive and each of the response lasers is crucial to achieve high-quality synchronization between the two response lasers. By contrast, the value of $C_{d\tau,r1}$ between the drive and response-1 lasers increases up to $\sim$0.75 and decreases even within the injection-locking range. Therefore, high-quality synchronization with large $C_{r1,r2}$ is achieved even though $C_{d\tau,r1}$ is relatively low ($C_{d\tau,r1} = 0.5-0.7$) within the injection-locking range.

The two-dimensional (2D) maps of the cross-correlation values $C_{r1,r2}$ and $C_{d\tau,r1}$ are created experimentally as functions of κ_{inj} and $\Delta\lambda_{\text{sol}}$, as shown in Figures 6.20a and b, respectively, in the open-loop configuration. The curves in the maps indicate the contours of the cross-correlation values. It is found that high $C_{r1,r2}$ is obtained at large κ_{inj} in Figure 6.20a. The region for high $C_{r1,r2}$ is also located at the region with small negative $\Delta\lambda_{\text{sol}}$ due to the amplitude-phase coupling effect (the α parameter). The region for high correlation is similar between $C_{r1,r2}$ and $C_{d\tau,r1}$, as shown in Figures 6.20a and b, even though the maximum value of $C_{d,r1}$ is less than 0.7. The overall region for generalized synchronization with high $C_{r1,r2}$ is located at large κ_{inj} and slightly negative $\Delta\lambda_{\text{sol}}$.

For comparison, the 2D maps of $C_{r1,r2}$ and $C_{d\tau,r1}$ are also generated for the closed-loop configuration, as shown in Figure 6.21. Note that the optical phase of the feedback light between the response-1 and -2 lasers is matched by controlling the voltage of the PMs in the feedback loop. High $C_{r1,r2}$ is obtained at the region with large κ_{inj} and small negative $\Delta\lambda_{\text{sol}}$ in Figure 6.21a, which is similar to that in Figure 6.20a. However, the region for high $C_{r1,r2}$ in Figure 6.21a becomes narrower than that in Figure 6.20a. In addition, the region for $C_{r1,r2} < 0.1$ is broadened in Figure 6.21a. This effect can be interpreted as the fact that the optical feedback in the response lasers disturbs the synchronization between the two response lasers at small κ_{inj}. The region for high $C_{r1,r2}$ is similar to that for high $C_{d,r1}$, as shown in Figure 6.21b, even though the maximum value of $C_{d,r1}$ is less than 0.6. For both cases of the open- and closed-loop configurations, the region for generalized synchronization is located at large κ_{inj} and

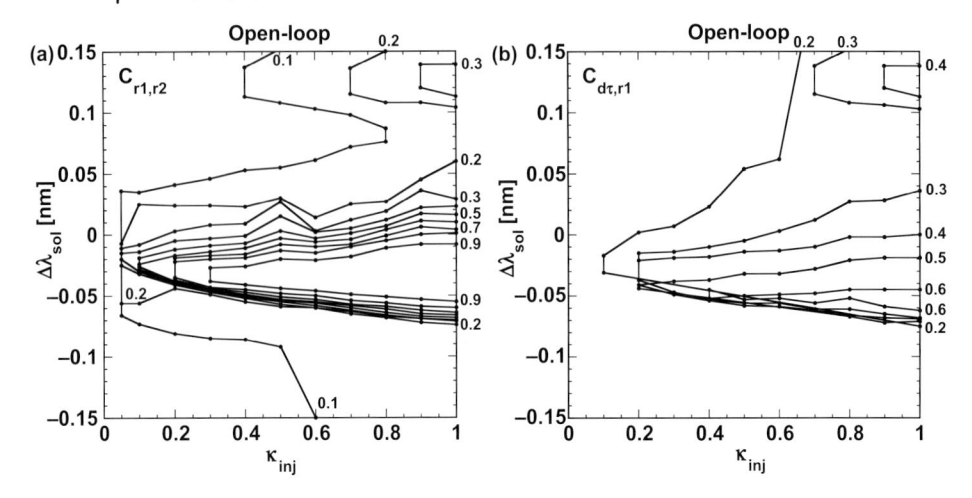

Figure 6.20 Two-dimensional (2D) maps of (a) cross-correlation value between the response-1 and -2 lasers $C_{r1,r2}$, and (b) cross-correlation value between the time-delayed drive and response-1 lasers $C_{d\tau,r1}$ as functions of κ_{inj} and $\Delta\lambda_{sol}$ in the open-loop configuration.

small negative $\Delta\lambda_{sol}$, corresponding to the injection-locking range between the drive and response lasers.

6.4.4
Dependence of Generalized Synchronization on Optical Phase of Feedback Light

The result of Figure 6.21 is obtained under the condition where the optical phase of the feedback light between the two response lasers is matched. When the optical

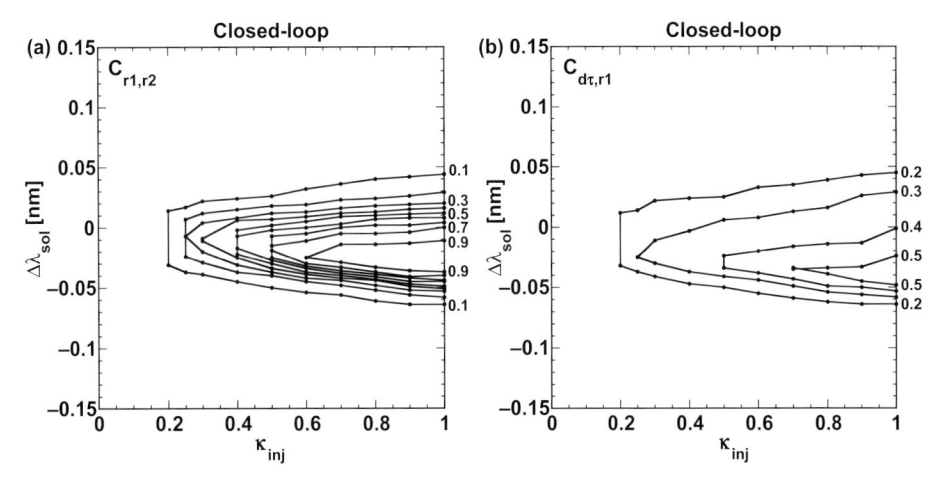

Figure 6.21 2D maps of (a) $C_{r1,r2}$ and (b) $C_{d\tau,r1}$ as functions of κ_{inj} and $\Delta\lambda_{sol}$ in the closed-loop configuration.

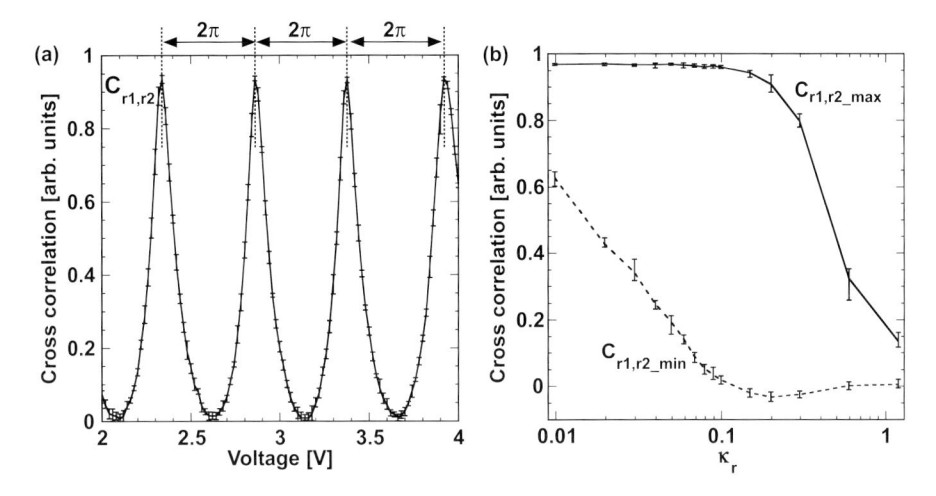

Figure 6.22 (a) Cross-correlation value between the response-1 and -2 lasers $C_{r1,r2}$ as a function of the difference in the optical feedback phase between the response-1 and -2 lasers (with error bars). (b) Maximum ($C_{r1,r2\_max}$, solid curve with error bars) and minimum ($C_{r1,r2\_min}$, dotted curve with error bars) values of the cross-correlation between the response-1 and -2 lasers as a function of the optical feedback power of the two response lasers κ_r in the closed-loop configuration. κ_r is normalized by the optical injection power from the drive to response lasers. $C_{r1,r2\_max}$ and $C_{r1,r2\_min}$ are obtained by scanning the difference in optical feedback phase of the two response lasers from 0 to 2π.

phase is mismatched, the synchronization quality is strongly degraded. To investigate the dependence of generalized synchronization on the optical phase, the difference in the optical phase of the feedback light between the two response lasers is continuously changed by controlling the DC voltage of the PM in the feedback loop of the response-1 laser (see Figure 6.15). The DC voltage of the PM for the response-2 laser is fixed. The cross-correlation value $C_{r1,r2}$ is measured as a function of the difference in the optical feedback phase between the two response lasers, as shown in Figure 6.22a. The value of $C_{r1,r2}$ changes periodically as the difference in the optical phase is changed (indicated by the voltage of the PM in Figure 6.22a). The period of the curve in Figure 6.22a corresponds to the phase shift of 2π. It is found that the maximum correlation is obtained when the phase difference is zero (i.e., the phase-matching condition), whereas the minimum correlation is observed when the phase difference is π. Therefore, it is found that the synchronization quality between the two response lasers is strongly dependent on the matching of the optical phase of the feedback light between the two response lasers.

The maximum and minimum cross-correlation values between the reponse-1 and -2 lasers (referred to as $C_{r1,r2\_max}$ and $C_{r1,r2\_min}$) can be obtained when the difference in the optical phase of the feedback light is changed continuously from 0 to 2π. The values of $C_{r1,r2\_max}$ and $C_{r1,r2\_min}$ are systematically measured as the optical feedback ratio κ_r is changed, as shown in Figure 6.22b. κ_r is defined as the optical feedback power of the two response lasers normalized by the optical injection power from the drive to response lasers. The optical feedback powers for the response-1 and -2 lasers are changed simultaneously to maintain the identical feedback power. For Figure 6.22b, as κ_r is increased, $C_{r1,r2\_min}$ decreases monotonically, while $C_{r1,r2\_max}$ is almost constant at ~0.97. As κ_r is increased further, $C_{r1,r2\_min}$ reaches ~0, and $C_{r1,r2\_max}$ also starts to decrease. For larger κ_r, $C_{r1,r2\_max}$ decreases gradually and synchronization is degraded. This region corresponds to outside the injection-locking range.

6.5
Model for Generalized Synchronization with Low Correlation in Three Semiconductor Lasers in the Auxiliary System Approach

As described in Section 5.6.2, the auxiliary system approach is useful to detect generalized synchronization with low correlation, where a replica of the response system is introduced and identical synchronization between the response and the replica systems indicates the existence of generalized synchronization. In Section 6.5, a model for generalized synchronization by using the auxiliary system approach is introduced. A model for the auxiliary system approach with three semiconductor lasers is shown in Figure 6.23. One laser is referred to as a drive laser, and the other two lasers are referred to as response-1 and response-2 lasers. The response-2 laser is

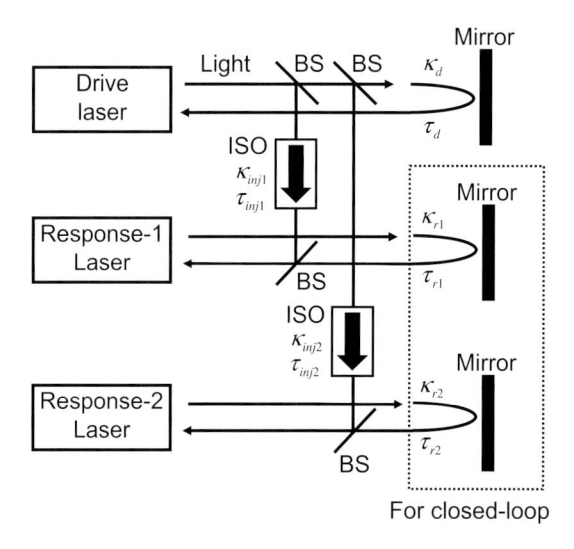

Figure 6.23 Model for generalized synchronization with low correlation in the auxiliary system approach. Three semiconductor lasers (drive, response-1, and response-2 lasers) are used. The chaotic drive signal induced by optical feedback is injected into each of the two response lasers unidirectionally, whereas there is no coupling between the response-1 and -2 lasers. The dotted frames indicate the closed-loop configuration in the response lasers. BS, beam splitter; ISO, optical isolator for unidirectional coupling. κ_d, κ_{r1}, and κ_{r2} are the feedback strengths for the drive, response-1 and response-2 lasers. κ_{inj1} and κ_{inj2} are the injection strengths from the drive to the response-1 and -2 lasers. τ_d, τ_{r1}, and τ_{r2} are the feedback delay times for the drive, response-1, and response-2 lasers. τ_{inj1} and τ_{inj2} are the injection delay times from the drive to the response-1 and -2 lasers.

considered as a replica of the response-1 laser, so they have identical parameter values (but different initial conditions). By contrast, the parameter values between the drive and each of the response lasers do not need to be matched.

Optical feedback light is reflected back from an external mirror into the drive laser to produce chaotic intensity fluctuation. The chaotic signal from the drive laser is divided into two beams by beam splitters and each of them is unidirectionally injected into each of the two response lasers through an optical isolator. Thus, the response-1 laser is coupled with the drive laser and the response-2 laser is coupled with the drive laser, however, there is no optical coupling between the response-1 and -2 lasers. An external mirror is replaced in front of each of the response lasers for the closed-loop configuration, whereas the external mirrors are eliminated and there is no optical feedback for either of the response lasers for the open-loop configuration. High-quality synchronization of chaos between the response-1 and -2 lasers indicates the existence of generalized synchronization between the drive and each of the response lasers, even though the quality of synchronization between the drive and response lasers is relatively low.

6.5.1
Coupled Lang–Kobayashi Equations for Generalized Synchronization in the Auxiliary System Approach

To describe the auxiliary system approach for generalized synchronization, three sets of coupled Lang–Kobayashi equations for the drive, response-1 and response-2 lasers are described in the closed-loop configuration as follows,

Drive laser:

$$\frac{dE_d(t)}{dt} = \frac{1}{2}\left[G_{N,d}(N_d(t)-N_{0,d}) - \frac{1}{\tau_{p,d}}\right]E_d(t) + \kappa_d\, E_d(t-\tau_d)\cos\Theta_d(t) \qquad (6.61)$$

$$\frac{d\Phi_d(t)}{dt} = \frac{\alpha_d}{2}\left[G_{N,d}(N_d(t)-N_{0,d}) - \frac{1}{\tau_{p,d}}\right] - \kappa_d\,\frac{E_d(t-\tau_d)}{E_d(t)}\sin\Theta_d(t) \qquad (6.62)$$

$$\frac{dN_d(t)}{dt} = J_d - \frac{N_d(t)}{\tau_{s,d}} - G_{N,d}(N_d(t)-N_{0,d})E_d^2(t) \qquad (6.63)$$

$$\Theta_d(t) = \omega_d\tau_d + \Phi_d(t) - \Phi_d(t-\tau_d) \qquad (6.64)$$

Response-1 laser:

$$\frac{dE_{r1}(t)}{dt} = \frac{1}{2}\left[G_{N,r1}(N_{r1}(t)-N_{0,r1}) - \frac{1}{\tau_{p,r1}}\right]E_{r1}(t) + \kappa_{inj1}\, E_d(t-\tau_{inj1})\cos\Theta_{inj1}(t) \\ + \kappa_{r1}\, E_{r1}(t-\tau_{r1})\cos\Theta_{r1}(t) \qquad (6.65)$$

$$\frac{d\Phi_{r1}(t)}{dt} = \frac{\alpha_{r1}}{2}\left[G_{N,r1}\left(N_{r1}(t)-N_{0,r1}\right)-\frac{1}{\tau_{p,r1}}\right] - \kappa_{inj1}\frac{E_d(t-\tau_{inj1})}{E_{r1}(t)}\sin\Theta_{inj1}(t)$$
$$-\kappa_{r1}\frac{E_{r1}(t-\tau_{r1})}{E_{r1}(t)}\sin\Theta_{r1}(t)$$

(6.66)

$$\frac{dN_{r1}(t)}{dt} = J_{r1} - \frac{N_{r1}(t)}{\tau_{s,r1}} - G_{N,r1}\left(N_{r1}(t)-N_{0,r1}\right)E_{r1}^2(t) \tag{6.67}$$

$$\Theta_{inj1}(t) = -\Delta\omega_1 t + \omega_d\tau_{inj1} + \Phi_{r1}(t) - \Phi_d(t-\tau_{inj1}) \tag{6.68}$$

$$\Theta_{r1}(t) = \omega_{r1}\tau_{r1} + \Phi_{r1}(t) - \Phi_{r1}(t-\tau_{r1}) \tag{6.69}$$

Response-2 laser:

$$\frac{dE_{r2}(t)}{dt} = \frac{1}{2}\left[G_{N,r2}\left(N_{r2}(t)-N_{0,r2}\right)-\frac{1}{\tau_{p,r2}}\right]E_{r2}(t) + \kappa_{inj2} E_d(t-\tau_{inj2})\cos\Theta_{inj2}(t)$$
$$+\kappa_{r2}E_{r2}(t-\tau_{r2})\cos\Theta_{r2}(t)$$

(6.70)

$$\frac{d\Phi_{r2}(t)}{dt} = \frac{\alpha_{r2}}{2}\left[G_{N,r2}\left(N_{r2}(t)-N_{0,r2}\right)-\frac{1}{\tau_{p,r2}}\right] - \kappa_{inj2}\frac{E_d(t-\tau_{inj2})}{E_{r2}(t)}\sin\Theta_{inj2}(t)$$
$$-\kappa_{r2}\frac{E_{r2}(t-\tau_{r2})}{E_{r2}(t)}\sin\Theta_{r2}(t)$$

(6.71)

$$\frac{dN_{r2}(t)}{dt} = J_{r2} - \frac{N_{r2}(t)}{\tau_{s,r2}} - G_{N,r2}\left(N_{r2}(t)-N_{0,r2}\right)E_{r2}^2(t) \tag{6.72}$$

$$\Theta_{inj2}(t) = -\Delta\omega_2 t + \omega_d\tau_{inj2} + \Phi_{r2}(t) - \Phi_d(t-\tau_{inj2}) \tag{6.73}$$

$$\Theta_{r2}(t) = \omega_{r2}\tau_{r2} + \Phi_{r2}(t) - \Phi_{r2}(t-\tau_{r2}) \tag{6.74}$$

The notation of the variables and parameters are the same as shown in Section 6.2.1. The subscripts d, r1, and r2 indicate the drive, response-1, and reponse-2 lasers, respectively. The symbols and parameter values are summarized in Tables 6.7 and 6.8. $\kappa_{r1,r2}$ is the feedback strength, and $\kappa_{r1,r2} \neq 0$ corresponds to the closed-loop configuration, whereas the open-loop configuration can be realized at the condition of $\kappa_{r1,r2} = 0$. For the response-1 and -2 lasers, $\tau_{r1,r2}$ is the feedback delay time, $\omega_{r1,r2} = \omega_d - \Delta\omega_{1,2}$ is the optical angular frequency, and $\Theta_{r1,r2}(t)$ is the phase difference of the feedback light.

Table 6.7 Parameter values that are common among the drive, response-1, and response-2 lasers used for the numerical simulation of generalized synchronization with low correlation in the auxiliary system approach in Sections 6.5 and 6.6.

Symbol	Parameter	Value
$G_{N,d}, G_{N,r1}, G_{N,r2}$	Gain coefficient	$8.40 \times 10^{-13} \, \text{m}^3 \, \text{s}^{-1}$
$N_{0,d}, N_{0,r1}, N_{0,r2}$	Carrier density at transparency	$1.40 \times 10^{24} \, \text{m}^{-3}$
$\tau_{p,d}, \tau_{p,r1}, \tau_{p,r2}$	Photon lifetime	$1.927 \times 10^{-12} \, \text{s}$
$\tau_{s,d}, \tau_{s,r1}, \tau_{s,r2}$	Carrier lifetime	$2.04 \times 10^{-9} \, \text{s}$
$\tau_{in,d}, \tau_{in,r1}, \tau_{in,r2}$	Round-trip time in internal cavity	$8.0 \times 10^{-12} \, \text{s}$
$r_{2,d}, r_{2,r1}, r_{2,r2}$	Reflectivity of laser facet	0.556
$\alpha_d, \alpha_{r1}, \alpha_{r2}$	Linewidth-enhancement factor	3.0
λ_d	Optical wavelength of the drive laser	$1.537 \times 10^{-6} \, \text{m}$
c	Speed of light	$2.998 \times 10^8 \, \text{m s}^{-1}$
$N_{th,d}, N_{th,r1}, N_{th,r2}$	Carrier density at threshold	$2.018 \times 10^{24} \, \text{m}^{-3}$
$J_{th,d}, J_{th,r1}, J_{th,r2}$	Injection current at threshold	$9.892 \times 10^{32} \, \text{m}^{-3} \, \text{s}^{-1}$
ω_d	Optical angular frequency of the drive laser	$1.226 \times 10^{15} \, \text{s}^{-1}$

The subscripts d, r1, and r2 indicate the drive, response-1, and response-2 lasers, respectively.

6.5.2
Synchronous Solutions for Generalized Synchronization in the Auxiliary System Approach

To observe generalized synchronization, all the parameter values between the response-1 and -2 lasers need to be identical, whereas the parameter values between the drive and response lasers do not need to be matched. The rate equations for the response-1 and -2 lasers can be identical in the closed-loop configuration under the following conditions for the optical injection and feedback lights,

$$\kappa_{inj1} = \kappa_{inj2}, \quad \kappa_{r1} = \kappa_{r2}, \quad \tau_{inj1} = \tau_{inj2}, \quad \tau_{r1} = \tau_{r2}, \quad \Delta\omega_1 = \Delta\omega_2 \quad (6.75)$$

Identical synchronous solutions between the response-1 and -2 lasers can be obtained under the identical parameter values as,

$$E_{r1}(t) = E_{r2}(t), \quad \Phi_{r1}(t) = \Phi_{r2}(t), \quad N_{r1}(t) = N_{r2}(t) \quad (6.76)$$

Note that Eq. (6.76) is the identical synchronous solutions between the response-1 and -2 lasers, unlike Eq. (6.10) showing the identical synchronous solutions between the drive and response lasers.

6.6
Numerical Analysis on Generalized Synchronization of Chaos in Three Semiconductor Lasers in the Auxiliary System Approach

In this section, generalized synchronization of chaos is numerically observed in the auxiliary system approach in the closed-loop configuration. The parameter values

Table 6.8 Parameter values used in numerical simulation for generalized synchronization with low correlation in the auxiliary system approach in the closed-loop configuration in Sections 6.5 and 6.6.

Symbol	Parameter	Value
$r_{3,d}$	Reflectivity of external mirror of drive laser	0.008
$r_{3,r1}$, $r_{3,r2}$	Reflectivity of external mirror of response-1 or response-2	0.020
κ_d	Feedback strength of drive	$1.243 \times 10^9\,\text{s}^{-1}$ ($1.243\,\text{ns}^{-1}$)
κ_{r1}, κ_{r2}	Feedback strength of response-1 or response-2 laser	$3.106 \times 10^9\,\text{s}^{-1}$ ($3.106\,\text{ns}^{-1}$)
$\kappa_{\text{inj}1}$, $\kappa_{\text{inj}2}$	Injection strength from drive to response-1 or response-2 laser	$3.106 \times 10^{10}\,\text{s}^{-1}$ ($31.06\,\text{ns}^{-1}$) $25\kappa_d$
$j_d = J_d/J_{\text{th},d}$	Normalized injection current of drive laser	1.3
$j_{r1} = J_{r1}/J_{\text{th},r1}$ $j_{r2} = J_{r2}/J_{\text{th},r2}$	Normalized injection current of response-1 or response-2 laser	1.6
L_d, L_{r1}, L_{r2}	External cavity length of drive or response laser	0.6 m (one-way) 1.2 m (round-trip)
$L_{\text{inj}1}$, $L_{\text{inj}2}$	Distance from drive to response-1 or response-2 laser	1.2 m
τ_d, τ_{r1}, τ_{r2}	Round-trip time in external cavity of drive or response laser (feedback delay time)	$4.003 \times 10^{-9}\,\text{s}$
$\tau_{\text{inj}1}$, $\tau_{\text{inj}2}$	Injection delay time from drive to response-1 or response-2	$4.003 \times 10^{-9}\,\text{s}$
$\Delta f_{1,\text{sol}} = f_d - f_{r1,\text{sol}}$ $\Delta f_{2,\text{sol}} = f_d - f_{r2,\text{sol}}$	Detuning of optical-carrier frequency between drive and response-1 (or -2) lasers	$-4.0\,\text{GHz}$
$\Delta \omega_1 = 2\pi \Delta f_{1,\text{sol}}$ $\Delta \omega_2 = 2\pi \Delta f_{2,\text{sol}}$	Detuning of optical angular frequency between drive and response-1 (or -2) lasers	$-2.513 \times 10^{10}\,\text{s}^{-1}$

The subscripts d, $r1$, $r2$, inj1, inj2 indicate the drive, response-1, response-2, injection to response-1, and injection to response-2 lasers, respectively. Other parameter values are used in Table 6.7.

used in the simulation for generalized synchronization in the closed-loop configuration are summarized in Tables 6.7 and 6.8. All the parameter values are set to be identical between the response-1 and -2 lasers, whereas some parameters are set to be detuned between the drive and each of the two response lasers. In particular, the normalized injection currents are set to 1.3, 1.6, and 1.6 for the drive, response-1, and response-2 lasers, respectively, resulting in different relaxation oscillation frequencies between the drive and response lasers. The optical feedback is introduced to each of the two response lasers in the closed-loop configuration and the feedback strengths of the two response lasers are set to be identical $\kappa_{r1} = \kappa_{r2} = 3.106\,\text{ns}^{-1}$ as shown in Table 6.8.

6.6.1
Temporal Waveforms

The temporal waveforms of the laser intensities for the drive, response-1 and response-2 lasers ($I_d(t)$, $I_{r1}(t)$, and $I_{r2}(t)$, respectively) can be obtained as the square of the electric-field amplitudes from Eqs. (6.61), (6.65) and (6.70).

$$I_d(t) = |E_d(t)|^2, I_{r1}(t) = |E_{r1}(t)|^2, I_{r2}(t) = |E_{r2}(t)|^2 \tag{6.77}$$

The identical synchronization of chaos between the response-1 and -2 laser intensities is satisfied as follows,

$$I_{r1}(t) = I_{r2}(t) \tag{6.78}$$

Generalized synchronization of chaos between the time-delayed drive laser intensity and the response-1 laser intensity ($I_d(t-\tau_{inj1})$ and $I_{r1}(t)$) is also observed. The variables of $I_d(t-\tau_{inj1})$ and $I_{r1}(t)$ may not be highly correlated, however, it is found that the correlation becomes the maximum value when the temporal waveform of the drive laser intensity is time delayed with respect to the response-1 laser intensity by the injection delay time τ_{inj1}.

The temporal waveforms of the three laser intensities are calculated numerically from Eqs. (6.61)–(6.74) in the closed-loop configuration. Figure 6.24 shows the temporal waveforms of $I_d(t-\tau_{inj1})$, $I_{r1}(t)$, and $I_{r2}(t)$, and the correlation plots between two of the three laser intensities with optical coupling in the closed-loop configuration. The chaotic output of the drive laser is injected into each of the two response lasers with the injection power of $\kappa_{inj} = 31.06$ ns^{-1} as shown in Table 6.8. All the three lasers behave chaotic, as shown in Figure 6.24a. The temporal waveforms between the drive and each of the two response lasers are not well synchronized, as seen in the correlation plots of Figures 6.24b and c. Both of the correlation values between the drive and each of the response lasers are 0.182. By contrast, the temporal waveforms between the response-1 and -2 are perfectly synchronized to each other, as shown in Figure 6.24d, where the correlation value between the response-1 and -2 lasers is 1.0. The achievement of identical synchronization between the reponse-1 and -2 lasers and partial synchronization between the drive and each of the two response lasers indicate the existence of generalized synchronization in the auxiliary system approach.

6.6.2
Parameter Dependence of Generalized Synchronization

For systematic investigation, two cross-correlation values between the response-1 and -2 lasers and between the time-delayed drive and response-1 lasers are introduced. The cross-correlation between the response-1 and -2 is written as follows.

Cross-correlation between the response-1 and -2 lasers:

$$C_{r1.r2} = \frac{\langle (I_{r1}(t) - \bar{I}_{r1})(I_{r2}(t) - \bar{I}_{r2}) \rangle}{\sigma_{r1}\sigma_{r2}} \tag{6.79}$$

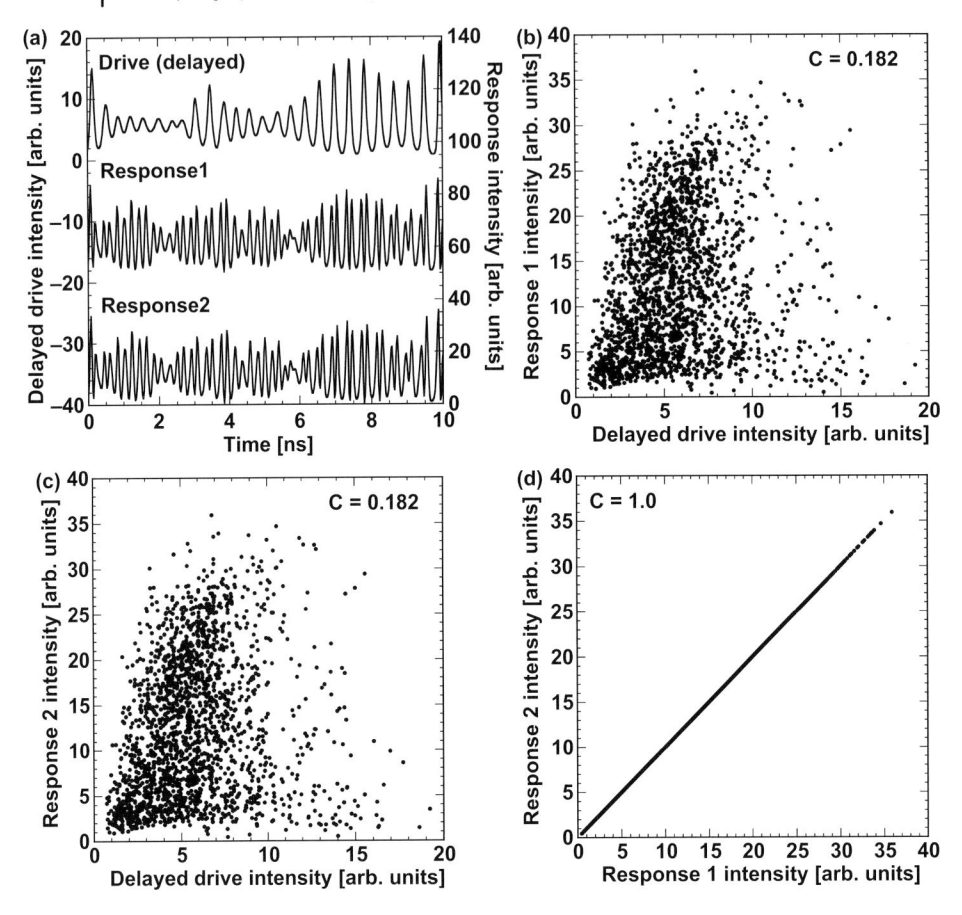

Figure 6.24 Numerical results of (a) temporal waveforms of the time-delayed drive, response-1, and response-2 laser intensities and the corresponding correlation plots (b) between the time-delayed drive and response-1 lasers, (c) between the time-delayed drive and response-2 lasers, and (d) between the response-1 and response-2 lasers under the optical coupling in the closed-loop configuration. The cross-correlation values are (b) $C = 0.182$, (c) $C = 0.182$, and (d) $C = 1.0$.

$$C_{d\tau,r1} = \frac{\left\langle \left(I_d(t-\tau_{1inj1}) - \bar{I}_d\right)\left(I_r(t) - \bar{I}_{r1}\right) \right\rangle}{\sigma_d \sigma_{r1}} \tag{6.80}$$

The subscripts d, $r1$, and $r2$ indicate the drive, response-1, and response-2 lasers, respectively. The notation is the same as Eqs. (5.19)–(5.24) in Section 5.4. Note that the cross-correlation $C_{r1,r2}$ is calculated from $I_{r1}(t)$ and $I_{r2}(t)$, whereas $C_{d\tau,r1}$ is obtained from $I_d(t-\tau_{inj1})$ and $I_{r1}(t)$. Generalized synchronization can be confirmed by the auxiliary system approach when $C_{r1,r2} \approx 1$ is satisfied, even though $C_{d\tau,r1}$ is relatively low.

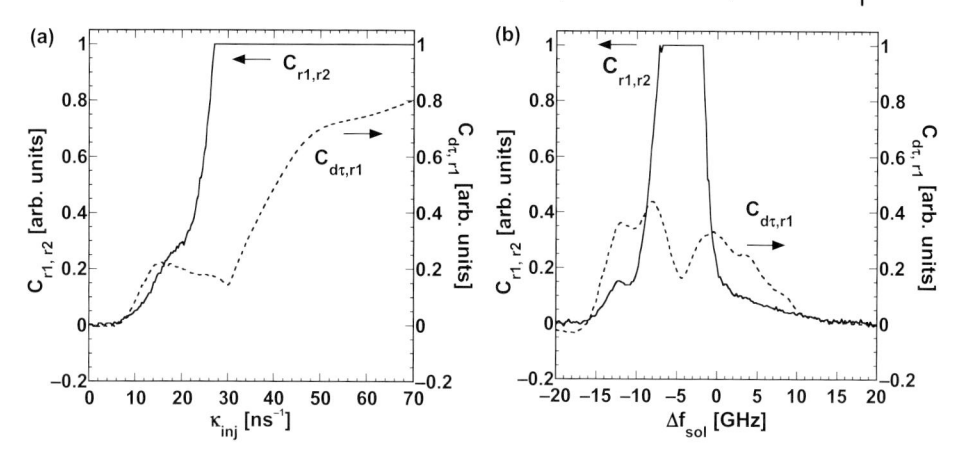

Figure 6.25 Cross-correlation value between the response-1 and -2 lasers $C_{r1,r2}$ (solid curve) and cross-correlation value between the time-delayed drive and response-1 lasers $C_{d\tau,r1}$ (dotted curve) as a function of (a) the optical injection strength from the drive to each of the response lasers κ_{inj} and (b) the initial optical frequency detuning between the drive and each of the response lasers without optical coupling Δf_{sol} in the closed-loop configuration. (a) Δf_{sol} is fixed at -4.0 GHz, and (b) κ_{inj} is fixed at $31.06\,\text{ns}^{-1}$.

The dependence of generalized synchronization on parameter mismatch is investigated as one of the laser parameters is changed. Figure 6.25a shows the two cross-correlation values ($C_{r1,r2}$ and $C_{d\tau,r1}$) of Eqs. (6.79) and (6.80) as the optical injection strength κ_{inj} is changed in the closed-loop configuration. For small κ_{inj}, both $C_{r1,r2}$ and $C_{d\tau,r1}$ are close to zero because all the three laser intensities are chaotic independently. As κ_{inj} is increased, the value of $C_{r1,r2}$ increases and reaches 1.0 at $\kappa_{inj} \geq 27.0\,\text{ns}^{-1}$, even though the value of $C_{d\tau,r1}$ is only ~ 0.2. The value of $C_{r1,r2}$ is maintained at 1.0 at large κ_{inj}. The value of $C_{d\tau,r1}$ is always lower than $C_{r1,r2}$, however, it increases up to ~ 0.8 as κ_{inj} is increased. This result indicates that generalized synchronization is achieved at strong injection strength.

Next, the initial detuning of the optical-carrier frequency between the drive and response-1 lasers without optical coupling ($\Delta f_{sol} = f_d - f_{r1,sol}$) is changed and the correlation characteristics are investigated. The optical carrier frequencies of the two response lasers are simultaneously changed to maintain the identical parameter conditions ($\Delta f_{sol} = f_d - f_{r1,sol} = f_d - f_{r2,sol}$). Figure 6.26a shows the values of $C_{r1,r2}$ and $C_{d\tau,r1}$ as Δf_{sol} is changed. The value of $C_{r1,r2}$ is 1.0 in the region of $-6.8\,\text{GHz} \leq \Delta f_{sol} \leq -2.0\,\text{GHz}$, while the value of $C_{d\tau,r1}$ is less than 0.45 in this range.

6.6.3
Linear Stability Analysis of Synchronous Oscillatory Solutions for Generalized Synchronization

6.6.3.1 Linearized Equations

The stability of the synchronous solutions of Eq. (6.76) between the two response lasers can be estimated from the linearized equations and the conditional Lyapunov

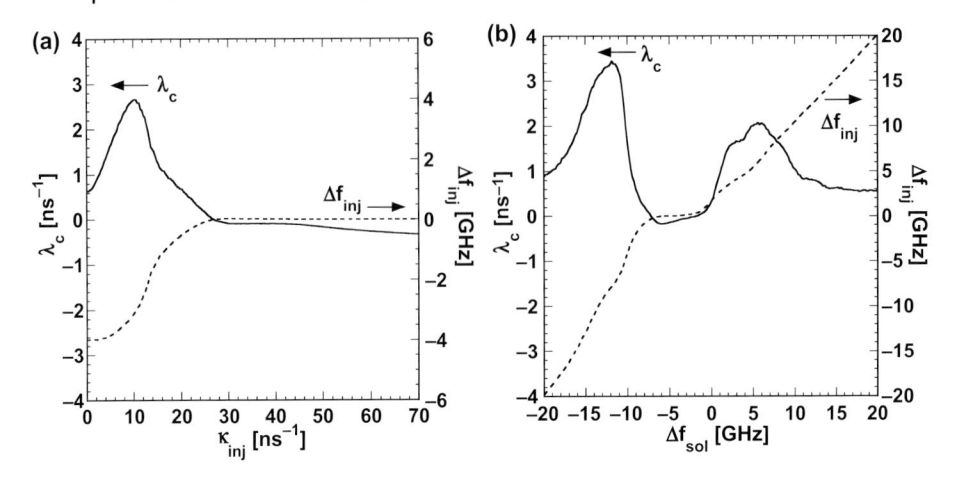

Figure 6.26 Conditional Lyapunov exponent λ_c (solid curve) and optical frequency detuning under the optical coupling Δf_{inj} (dotted curve) as a function (a) κ_{inj} and (b) Δf_{sol} in the closed-loop configuration. (a) Δf_{sol} is fixed at -4.0 GHz, and (b) κ_{inj} is fixed at 31.06 ns^{-1}.

exponent, as shown in Section 6.3.4. The linearized variables are introduced as,

$$\Delta_{E,r}(t) = E_{r2}(t) - E_{r1}(t) \tag{6.81}$$

$$\Delta_{\Phi,r}(t) = \Phi_{r2}(t) - \Phi_{r1}(t) \tag{6.82}$$

$$\Delta_{N,r}(t) = N_{r2}(t) - N_{r1}(t) \tag{6.83}$$

The time-delayed linearized variables are written as,

$$\Delta_{E,r}(t-\tau_r) = E_{r2}(t-\tau_r) - E_{r1}(t-\tau_r) \tag{6.84}$$

$$\Delta_{\Phi,r}(t-\tau_r) = \Phi_{r2}(t-\tau_r) - \Phi_{r1}(t-\tau_r) \tag{6.85}$$

The linearized equations for the synchronous solution of generalized synchronization in the closed-loop configuration can be obtained by calculating the difference in the corresponding variables for the two response lasers with identical parameter values between Eqs. (6.65)–(6.67) and Eqs. (6.70)–(6.72).

$$\begin{aligned}\frac{d\Delta_{E,r}(t)}{dt} &= \frac{1}{2}\left[G_{N,r}\left(N_{r1}(t)-N_{0,r}\right) - \frac{1}{\tau_{p,r}}\right]\Delta_{E,r}(t) \\ &\quad - \left[\kappa_{inj}E_d(t-\tau_{inj})\sin\Theta_{inj1}(t) + \kappa_r E_{r1}(t-\tau_r)\sin\Theta_{r1}(t)\right]\Delta_{\Phi,r}(t) \\ &\quad + \frac{1}{2}G_{N,r}E_{r1}(t)\Delta_{N,r}(t) \\ &\quad + \kappa_r \cos\Theta_{r1}(t)\Delta_{E,r}(t-\tau_r) \\ &\quad + \kappa_r E_{r1}(t-\tau_r)\sin\Theta_{r1}(t)\Delta_{\Phi,r}(t-\tau_r)\end{aligned} \tag{6.86}$$

$$\frac{d\Delta_{\Phi,r}(t)}{dt} = \left[\kappa_{inj}\frac{E_d(t-\tau_{inj})}{E_{r1}^2(t)}\sin\Theta_{inj1}(t) + \kappa_r\frac{E_{r1}(t-\tau_r)}{E_{r1}^2(t)}\sin\Theta_{r1}(t)\right]\Delta_{E,r}(t)$$

$$-\left[\kappa_{inj}\frac{E_d(t-\tau_{inj})}{E_{r1}(t)}\cos\Theta_{inj1}(t) + \kappa_r\frac{E_{r1}(t-\tau_r)}{E_{r1}(t)}\cos\Theta_{r1}(t)\right]\Delta_{\Phi,r}(t)$$

$$+ \frac{\alpha_r}{2}G_{N,r}\Delta_{N,r}(t)$$

$$- \kappa_r \frac{1}{E_{r1}(t)}\sin\Theta_{r1}(t)\Delta_{E,r}(t-\tau_r)$$

$$+ \kappa_r \frac{E_{r1}(t-\tau_r)}{E_{r1}(t)}\cos\Theta_{r1}(t)\Delta_{\Phi,r}(t-\tau_r)$$

(6.87)

$$\frac{d\Delta_{N,r}(t)}{dt} = -2G_{N,r}E_{r1}(t)\left(N_{r1}(t) - N_{0,r}\right)\Delta_{E,r}(t) - \left(\frac{1}{\tau_{s,r}} + G_{N,r}E_{r1}^2(t)\right)\Delta_{N,r}(t)$$

(6.88)

where all the parameter values are set to be identical between the response-1 and -2 lasers, such as $G_{N,r} = G_{N,r1} = G_{N,r2}$, $N_{0,r} = N_{0,r1} = N_{0,r2}$, and so on. The subscript $r1$ is replaced with r for the parameters. The following identical parameter conditions are also satisfied (see Table 6.8).

$$\kappa_{inj} = \kappa_{inj1} = \kappa_{inj2}, \tau_{inj} = \tau_{inj1} = \tau_{inj2}, \kappa_r = \kappa_{r1} = \kappa_{r2}, \tau_r = \tau_{r1} = \tau_{r2}, \Delta\omega = \Delta\omega_1 = \Delta\omega_2$$

(6.89)

Note that the linearized equations for synchronous solution of generalized synchronization in the closed-loop configuration (Eqs. (6.86)–(6.88)) are different from those in the open-loop configuration, since the feedback term of the response laser $E_{r1}(t-\tau_r)$ are used to deriver the linearized variables of $\Delta_{E,r}(t-\tau_r)$ and $\Delta_{\Phi,r}(t-\tau_r)$, which makes the calculation of the conditional Lyapunov exponent more complicated.

6.6.3.2 Conditional Lyapunov Exponent

The conditional Lyapunov exponent is calculated from the linearized equations (Eqs. (6.86)–(6.88)) for the synchronous solutions of generalized synchronization of Eq. (6.76) in the closed-loop configuration. Unlike the linearized equations in the open-loop configuration, Eqs. (6.86)–(6.88) include time-delayed linearized variables ($\Delta_{E,r}(t-\tau_r)$ and $\Delta_{\Phi,r}(t-\tau_r)$) and careful consideration is required to estimate the conditional Lyapunov exponent. For time-delayed linearized equations, the norm of the linearized variables from time t to $t+\tau_r$ needs to be calculated, as shown in Section 4.4.3 (Pyragas, 1998). The linearized equations of Eqs. (6.86)–(6.88) with delay time τ_r are integrated with the original equations of Eqs. (6.61)–(6.74). The number of data points normalized by the numerical integration step h in the delay time τ_r is defined as $M = \tau_r/h$. It is worth noting that the norm is obtained for each

M step for time-delayed dynamical systems, unlike the case of the open-loop configuration without time delay, because the variables within a time-delayed loop are considered to be independent.

The method for the calculation of the conditional Lyapunov exponent is the same as described in Section 4.4.3. The linearized variables $\Delta_E(t)$, $\Delta_\Phi(t)$, $\Delta_N(t)$, and the delay time τ_r need to be used, instead of $\delta_E(t)$, $\delta_\Phi(t)$, $\delta_N(t)$, and τ in Eq. (4.98) to calculate the norm of the linearized variables. The procedure from Eqs. (4.98) to (4.104) is executed repeatedly and the conditional Lyapunov exponent λ_c can be obtained, instead of $\lambda_{\max}$ in Eq. (4.104).

6.6.3.3 Numerical Results of Conditional Lyapunov Exponent

The conditional Lyapunov exponent λ_c is calculated and plotted as κ_{inj} is increased in Figure 6.26a (solid curve). λ_c becomes negative at $\kappa_{\text{inj}} \geq 27.0$ in Figure 6.26a, where $C_{r1,r2}$ is 1.0 in Figure 6.25a. The detuning of the optical-carrier frequency between the drive and response-1 lasers with optical coupling ($\Delta f_{\text{inj}} = f_d - f_{r1,\text{inj}}$) is also plotted in Figure 6.26a (dotted curve). The detuning becomes almost zero, that is, $|\Delta f_{\text{inj}}| \leq 0.01$ GHz at $\kappa_{\text{inj}} \geq 27.0$, therefore this region can be interpreted as the injection-locking range. It is concluded that high $C_{r1,r2}$ is obtained at the negative λ_c within the injection-locking range.

Figure 6.26b shows λ_c and Δf_{inj} as the initial optical-frequency detuning Δf_{sol} is changed. The value of λ_c has a negative value in the region of $-6.8\,\text{GHz} \leq \Delta f_{\text{sol}} \leq -2.0\,\text{GHz}$ in Figure 6.26b, corresponding to the same region of $C_{r1,r2} = 1.0$ in Figure 6.25b. In addition, the injection-locking range ($|\Delta f_{\text{inj}}| \leq 0.01$ GHz) is found at the same region in Figure 6.26b. Therefore, injection locking ($|\Delta f_{\text{inj}}| \approx 0$) and negative conditional Lyapunov exponent ($\lambda_c < 0$) are necessary for generalized synchronization.

6.6.4
Two-Dimensional Map

The two-dimensional (2D) maps of the values of $|C_{r1,r2}|$, $|C_{d\tau,r1}|$, λ_c, and Δf_{inj} are generated as functions of κ_{inj} and Δf_{sol} in the closed-loop configuration, as shown in Figures 6.27a–d, respectively. The black region in Figure 6.27a indicates identical synchronization between the two response lasers. The triangle-shaped region of $C_{r1,r2} = 1.0$ is obtained with large κ_{inj} and negative Δf_{sol}. There is a good correspondence between the region for $C_{r1,r2} = 1.0$ (Figure 6.27a) and negative λ_c (Figure 6.27c). There is also a good correspondence between the region for $C_{r1,r2} = 1.0$ in Figure 6.27a and the injection-locking range ($|\Delta f_{\text{inj}}| \leq 0.01$ GHz) in the black region of Figure 6.27d. The region of relatively low values of $C_{d\tau,r1}$ (0.3–0.8) in Figure 6.27b also corresponds to the region for $C_{r1,r2} = 1.0$ in Figure 6.27a. These 2D maps indicate the necessary condition of $|\Delta f_{\text{inj}}| \approx 0$ and $\lambda_c < 0$ for achieving generalized synchronization ($C_{r1,r2} = 1.0$, $C_{r1,r2} > C_{d\tau,r1}$) in the κ_{inj}–Δf_{sol} parameter space.

Next, the feedback strengths of the two response lasers are changed and synchronization quality is investigated for the comparison between open- and closed-loop

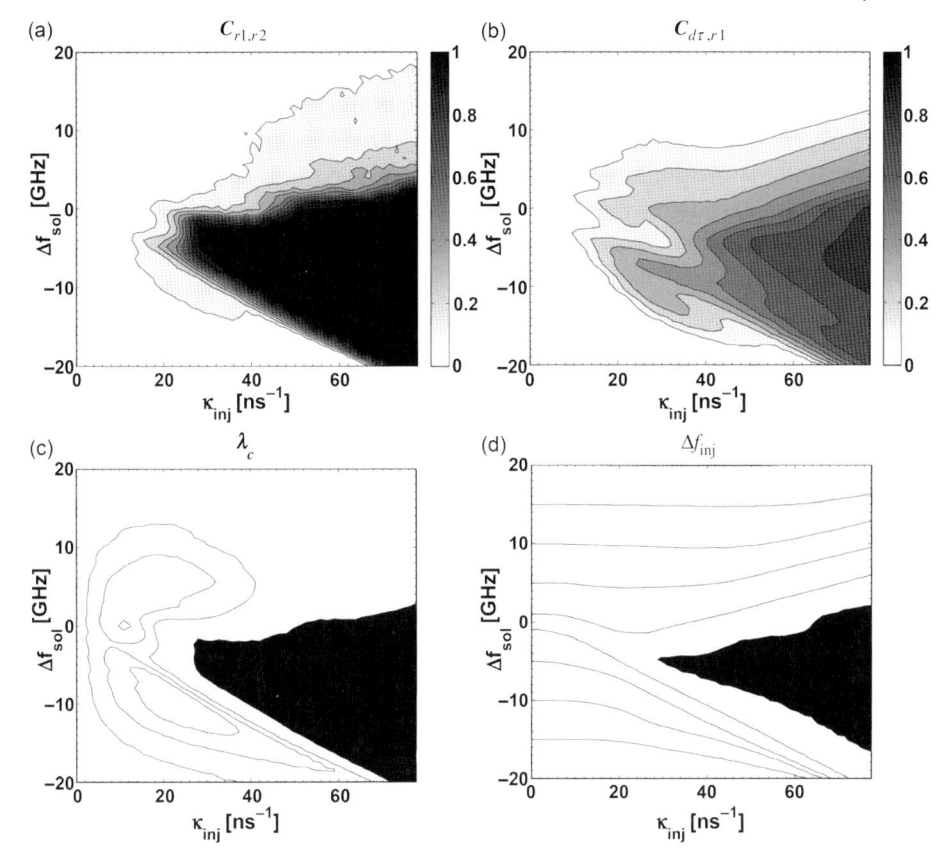

Figure 6.27 Two-dimensional maps of (a) absolute value of cross-correlation between the response-1 and -2 lasers $C_{r1,r2}$, (b) absolute value of cross-correlation between the time-delayed drive and response-1 lasers $C_{d\tau,r1}$, (c) conditional Lyapunov exponent λ_c, and (d) optical frequency detuning between the drive and each of the response lasers under the optical coupling Δf_{inj} as functions of the optical injection strength κ_{inj} and the initial optical frequency detuning Δf_{sol} in the closed-loop configuration. (c) Black region corresponds to $\lambda_c < 0$. (d) Black region corresponds to $|\Delta f_{inj}| < 0.01$ GHz.

configurations. Figure 6.28a shows $C_{r1,r2}$ and $C_{d\tau,r1}$ as the feedback strengths of the response-1 and -2 lasers (κ_{r1} and κ_{r2}) are changed simultaneously ($\kappa_r = \kappa_{r1} = \kappa_{r2}$). The condition for $\kappa_r = 0$ corresponds to the open-loop configuration. The values of λ_c and Δf_{inj} are also plotted in Figure 6.28b. $C_{r1,r2} = 1.0$ is obtained in the region with small κ_r ($\kappa_r < 6.2$ ns^{-1}). λ_c is negative and $|\Delta f_{inj}| \approx 0$ is obtained in this region. It is worth noting that $C_{r1,r2}$ is almost the same for the open-loop (i.e., $\kappa_r = 0$) and closed-loop ($\kappa_r \neq 0$) configurations at small κ_r, whereas slightly better $C_{d\tau,r1}$ for the open-loop configuration is obtained at $\kappa_r = 0$ than that for the closed-loop configuration in Figure 6.28a.

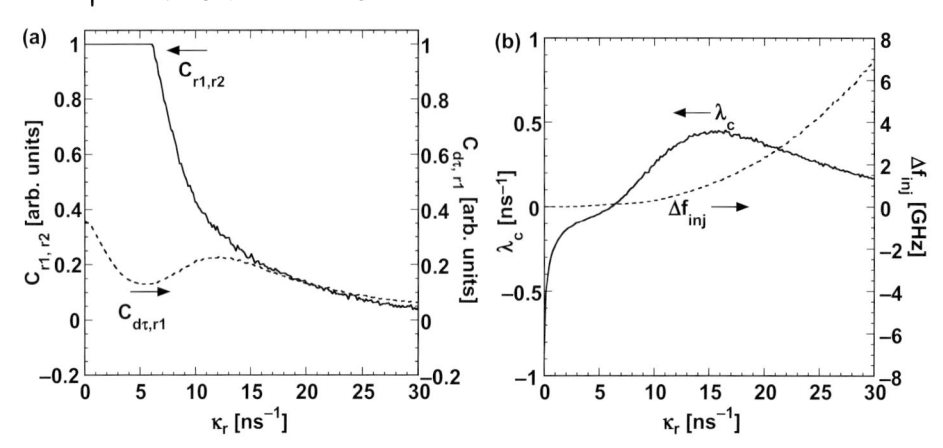

Figure 6.28 (a) Cross-correlation value between the response-1 and -2 lasers $C_{r1,r2}$ (solid curve) and cross-correlation value between the time-delayed drive and response-1 lasers $C_{d\tau,r1}$ (dotted curve) as a function of the optical feedback strength of the two response lasers κ_r ($=\kappa_{r1}=\kappa_{r2}$) in the closed-loop configuration. (b) Conditional Lyapunov exponent λ_c (solid curve) and optical frequency detuning under the optical coupling Δf_{inj} (dotted curve) as a function of κ_r. No feedback strength ($\kappa_r = 0$) corresponds to the open-loop configuration. The feedback strength of the drive laser is fixed at $\kappa_d = 4.659$ ns^{-1}. The injection strengths from the drive to the response lasers are also fixed at $\kappa_{inj} = \kappa_{inj1} = \kappa_{inj2} = 31.06$ ns^{-1}.

6.6.5
Dependence of Synchronization Quality on Optical Phase of Feedback Light in the Closed-Loop Configuration

In the previous numerical simulations, the optical phase of the feedback light in the two response lasers is set to be identical. Strong sensitivity of synchronization quality on the optical phase is found in the experimental results shown in Figure 6.22. The influence of the optical phase of the feedback light in the response lasers on the synchronization quality is investigated by numerical simulation.

The difference in the optical phase of the feedback light between the response-1 and -2 lasers (defined as $\Delta\phi = \phi_{r1} - \phi_{r2}$, where $\phi_{r1,r2}$ is the optical phase of the feedback light in the response-1 and -2 lasers, respectively) is continuously changed by assuming a voltage change in the phase modulator of the feedback loop of the response-1 laser. The value of $C_{r1,r2}$ is measured as $\Delta\phi$ is changed, as shown in Figure 6.29a. The value of $C_{r1,r2}$ changes periodically as $\Delta\phi$ is changed. The period of the curve in Figure 6.29a corresponds to 2π. The maximum correlation is obtained when $\Delta\phi = 2n\pi$ (n is an integer) (the phase-matching condition), whereas the minimum correlation is observed when $\Delta\phi = (2n+1)\pi$. Therefore, it is found that the synchronization quality between the two response lasers is sensitively dependent on the matching of the optical phase of the feedback light between the two response lasers, as observed in the experiment in Figure 6.22.

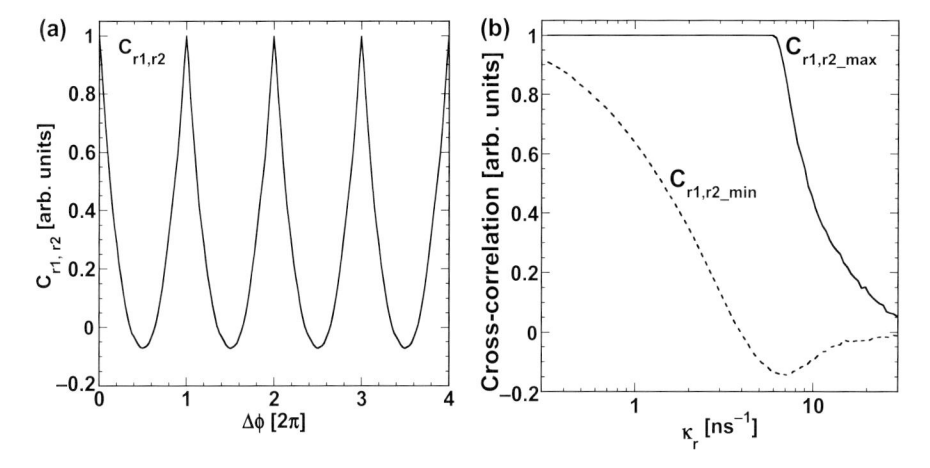

Figure 6.29 (a) Cross-correlation value between the response-1 and -2 lasers $C_{r1,r2}$ (solid curve) as a function of the difference in the optical feedback phase $\Delta\phi$ between the response-1 and -2 lasers in the closed-loop configuration. The feedback strengths of the response lasers are set to $\kappa_r = 4.659$ ns^{-1}. (b) Maximum ($C_{r1,r2\_max}$, solid curve) and minimum ($C_{r1,r2\_min}$, dashed curve) values of the cross-correlation between the response-1 and -2 lasers, obtained by scanning the difference in the optical feedback phases of the two response lasers from 0 to 2π, as a function of the optical feedback strength of the response lasers $\kappa_r (= \kappa_{r1} = \kappa_{r2})$.

The maximum and minimum cross-correlations between the reponse-1 and -2 lasers ($C_{r1,r2\_max}$ and $C_{r1,r2\_min}$) are measured by scanning $\Delta\phi$ from 0 to 2π at various feedback strengths of the two response lasers κ_r. Figure 6.29b shows $C_{r1,r2\_max}$ and $C_{r1,r2\_min}$ as κ_r is changed, where the identical parameter condition of $\kappa_r = \kappa_{r1} = \kappa_{r2}$ is maintained. As κ_r is increased, $C_{r1,r2\_min}$ starts to decrease while $C_{r1,r2\_max}$ remains constant at 1.0. As κ_r is increased further, $C_{r1,r2\_min}$ reaches ~ 0, and $C_{r1,r2\_max}$ also starts to decrease at $\kappa_r \geq 6.2$ ns^{-1}. At large κ_r, $C_{r1,r2\_max}$ decreases gradually and identical synchronization is no longer achieved. In this region, no injection locking is achieved between the drive and response lasers. It is found that the synchronization quality between the response-1 and -2 lasers is strongly sensitive to the difference in the optical phase of the feedback light between the response-1 and -2 lasers. These numerical results are consistent with the experimental observation shown in Figure 6.22.

Appendix

6.A.1
Rate Equations for Identical Synchronization in Unidirectionally Coupled Semiconductor Lasers with Optical Feedback in the Closed-Loop Configuration

Model equations for unidirectionally coupled two semiconductor lasers in the closed-loop configuration are introduced and explained. As seen in Figure 6.8, an optical feedback light is reflected back from an external mirror into the drive laser to produce chaotic intensity fluctuation. The two lasers are coherently and unidirectionally coupled to each other by injecting the output of the chaotic drive signal to the response laser through an optical isolator. An external mirror is placed in front of the response laser and an optical feedback light is reflected back from the external mirror to induce chaotic intensity fluctuation in the response laser. The system in which both of the lasers have optical feedback is referred to as the closed-loop configuration.

The Lang–Kobayashi equations of unidirectionally coupled semiconductor lasers in the closed-loop configuration are described as follows,

Drive laser:

$$\frac{dE_d(t)}{dt} = \frac{1}{2}\left[G_{N,d}\left(N_d(t)-N_{0,d}\right)-\frac{1}{\tau_{p,d}}\right]E_d(t) + \kappa_d\,E_d(t-\tau_d)\cos\Theta_d(t) \qquad (6.A.1)$$

$$\frac{d\Phi_d(t)}{dt} = \frac{\alpha_d}{2}\left[G_{N,d}\left(N_d(t)-N_{0,d}\right)-\frac{1}{\tau_{p,d}}\right]-\kappa_d\frac{E_d(t-\tau_d)}{E_d(t)}\sin\Theta_d(t) \qquad (6.A.2)$$

$$\frac{dN_d(t)}{dt} = J_d - \frac{N_d(t)}{\tau_{s,d}} - G_{N,d}\left(N_d(t)-N_{0,d}\right)E_d^2(t) \qquad (6.A.3)$$

$$\Theta_d(t) = \omega_d\tau_d + \Phi_d(t)-\Phi_d(t-\tau_d) \qquad (6.A.4)$$

Response laser:

$$\frac{dE_r(t)}{dt} = \frac{1}{2}\left[G_{N,r}\left(N_r(t)-N_{0,r}\right)-\frac{1}{\tau_{p,r}}\right]E_r(t) + \kappa_{\text{inj}}\,E_d(t-\tau_{\text{inj}})\cos\Theta_{\text{inj}}(t)$$
$$+ \kappa_r E_r(t-\tau_r)\cos\Theta_r(t) \qquad (6.A.5)$$

$$\frac{d\Phi_r(t)}{dt} = \frac{\alpha_r}{2}\left[G_{N,r}\left(N_r(t)-N_{0,r}\right)-\frac{1}{\tau_{p,r}}\right]-\kappa_{\text{inj}}\frac{E_d(t-\tau_{\text{inj}})}{E_r(t)}\sin\Theta_{\text{inj}}(t)$$
$$-\kappa_r\frac{E_r(t-\tau_r)}{E_r(t)}\sin\Theta_r(t) \qquad (6.A.6)$$

$$\frac{dN_r(t)}{dt} = J_r - \frac{N_r(t)}{\tau_{s,r}} - G_{N,r}\left(N_r(t)-N_{0,r}\right)E_r^2(t) \qquad (6.A.7)$$

$$\Theta_{\text{inj}}(t) = -\Delta\omega t + \omega_d\tau_{\text{inj}} + \Phi_r(t)-\Phi_d(t-\tau_{\text{inj}}) \qquad (6.A.8)$$

$$\Theta_r(t) = \omega_r\tau_r + \Phi_r(t)-\Phi_r(t-\tau_r) \qquad (6.A.9)$$

All the parameter values and symbols are shown in Tables 6.2 and 6.3. The subscripts d and r indicate the drive and response lasers, respectively. κ_r is the feedback strength of the response laser, and τ_r is the round-trip time in the external cavity of the response laser (feedback delay time). $\omega_r = \omega_d - \Delta\omega$ is the optical angular frequency of the response laser.

6.A.1 Rate Equations for Identical Synchronization

The rate equations for the drive and response lasers in the closed-loop configuration can be identical under the following conditions,

$$\kappa_d = \kappa_{\text{inj}} + \kappa_r, \ \tau_d = \tau_{\text{inj}} = \tau_r, \ \Delta\omega = 0 \tag{6.A.10}$$

Then identical synchronous solutions can be obtained as,

$$E_d(t) = E_r(t), \ \Phi_d(t) = \Phi_r(t), \ N_d(t) = N_r(t) \tag{6.A.11}$$

The linearized equations and the conditional Lyapunov exponent for the identical synchronous solutions in the closed-loop configuration can be obtained from the same procedure of the open-loop configuration, as described in Section 6.3.4. However, it should be noted that the time-delayed terms of the linearized variables need to be included in the linearized equations due to the existence of the optical feedback in the response laser. See details in Section 6.6.3 for the derivation of the linearized equations in the closed-loop configuration.

Part Two
Application of Chaotic Lasers to Optical Communication and Information Technology

7
Basic Concept of Optical Communication with Chaotic Lasers

Many communication systems in today's technology world are based on the synchronization between periodic oscillators. In that scenario, tuning a receiver into synchronization with an emitter allows the recovery of information that was encoded in a periodic carrier. Instead of the periodic carrier, a chaotic carrier can be used for the same purpose if synchronization is achievable. The purpose of Chapter 7 is to present basic concepts of secret communication and main ideas of optical communication with chaotic lasers. The detailed techniques for the implementation of optical communication with chaotic lasers will be reviewed comprehensively in Chapter 8.

7.1
History of Secret Communication

7.1.1
Cryptography

The history of secret communication started thousands of years ago in order to protect a message in communication so that only the intended recipient can read it. One of the most sophisticated techniques developed in the long history is known as "cryptography", derived from the Greek word "kryptos" meaning "hidden", and "graphein" meaning "to write" (van der Lubbe, 1998; Singh, 2000). To render a message unintelligible, it is scrambled according to a particular protocol (procedure) with a key that is agreed beforehand between the sender and the intended recipient, as shown in Figure 7.1. Thus, the recipient can reverse the scrambling protocol and make the message comprehensible. The advantage of cryptography is that if the enemy intercepts an encrypted message, then the message is unreadable. Without knowing the scrambling protocol and the key, the enemy should find it difficult to recreate the original message from the encrypted text.

Cryptography can be classified into two main branches, known as "transposition" and "substitution". Transposition corresponds to rearranging letters, such as anagram. For example, "CHAOS" can be encrypted as "HCAOS", "SAOCH", or "ASHCO". On the other hand, substitution indicates replacing words or letters.

Optical Communication with Chaotic Lasers: Applications of Nonlinear Dynamics and Synchronization, First Edition. Atsushi Uchida.
© 2012 Wiley-VCH Verlag GmbH & Co. KGaA. Published 2012 by Wiley-VCH Verlag GmbH & Co. KGaA.

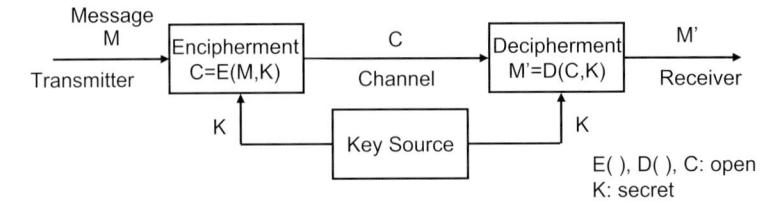

Figure 7.1 Block diagram of cryptography. C, channel signal; D( ), function for decoding; E( ), function for encoding; K, secret key; M, original message; M', decoded message.

For example, in the Caesar shift cipher, which was used in Julius Ceasar's Gallic Wars, each letter in a message is replaced with the letter that is three places further down the alphabet, that is, "CHAOS" can be encrypted as "FKDRV" ("C" is shifted to "F" (C → D → E → F), and "H" becomes "K", and so on).

Substitution can be also categorized into two techniques: "cipher" and "code". Cipher is a technique to replace letters into other letters, as seen in the example of the Caesar shift cipher. By contrast, code is defined as substitution at the level of words or phrases. For example, "CHAOS" can be replaced into "DOG", and "COMMUNICATION" can be changed to "FOOD".

Codes seem to offer more security than ciphers, because words are much less vulnerable to frequency analysis than letters. To decipher a monoalphabetic cipher an attacker needs only identify the true value of each of the 26 characters, whereas to decipher a code an attacker needs to identify the true value of thousands of codewords. However, if one examines codes in more detail, one can see that they suffer from two major practical failings when compared with ciphers. First, once the sender and receiver have agreed upon the 26 letters in the cipher alphabet (which is the key), they can encipher any message, but to achieve the same level of flexibility using a code they would need to go through the painstaking task of defining a codeword for every one of the thousands of possible plaintext words. Compiling a codebook is a major task, and carrying it around is a major inconvenience. Secondly, the consequences of having a codebook captured by the attacker are devastating. Immediately, all the encoded communications would become transparent to the attacker. The senders and receivers would have to go through the painstaking process of having to compile an entirely new codebook, and then this codebook would have to be distributed to everyone in the communication network. In comparison, if the attacker succeeds in capturing a cipher key, then it is relatively easy to compile a new cipher alphabet of 26 letters, which can be memorized and easily redistributed.

7.1.2
Steganography

Another important technique in the history of secret communication is known as "steganography", derived from the Greek words "steganos" meaning "covered", and "graphein" meaning "to write" (Singh, 2000; Wayner, 2009). The aim of steganography is to hide the "existence" of message, not to hide its "meaning" as cryptog-

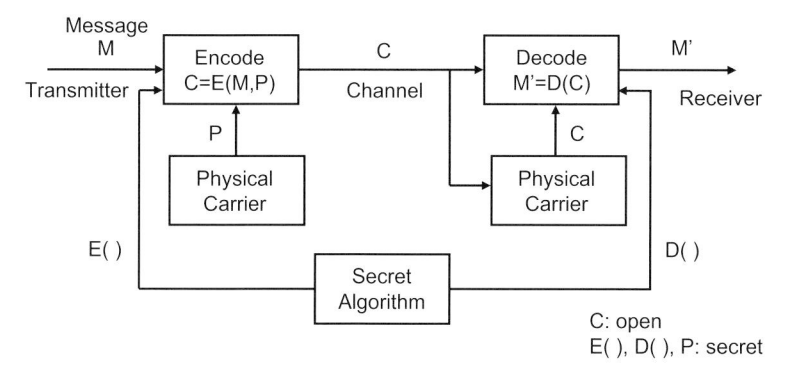

Figure 7.2 Block diagram of steganography. C, channel signal; D( ), function for decoding; E( ), function for encoding; M, original message; M', decoded message; P, physical carrier.

raphy. A scheme of steganography is shown in Figure 7.2. Steganography using physical materials (called physical steganography) has been widely used in ancient historical times and even the present day. A message is concealed in a physical material and the material is sent to the receiver. The message is extracted from the physical material by the secret procedure. The algorithm of encoding and decoding needs to be secret for physical steganography.

The first recorded uses of steganography can be traced back to 440 BC when Herodotus mentions two examples of steganography in "The Histories of Herodotus" (Singh, 2000). Demaratus sent a warning about a forthcoming attack to Greece by writing it directly on the wooden backing of a wax tablet before applying its beeswax surface. Wax tablets were in common use as reusable writing surfaces, sometimes used for shorthand. Another ancient example is that of Histiaeus, who shaved the head of his most trusted slave and tattooed a message on it. After his hair had grown the message was hidden. The purpose was to instigate a revolt against the Persians.

Other examples of physical steganography are shown as follows (Singh, 2000): Messages are hidden on paper written in secret inks, under other messages or on the blank parts of other messages. Messages are written in Morse code on knitting yarn and then knitted into a piece of clothing worn by a courier. Messages are hidden on the back of postage stamps. Microdots can be used to send information, which are typically minute, about or less than the size of the period produced by a typewriter, and needed to be embedded in the paper.

As can be seen the above examples, the aim of steganography is to conceal the "existence" of the message by using physical material, whereas the aim of cryptography is to hide the "meaning" of the message by scrambling the message itself. The advantage of steganography, over cryptography alone, is that messages do not attract attention to themselves. Plainly visible encrypted messages (no matter how unbreakable) will arouse suspicion, and may in themselves be incriminating in countries where encryption is illegal. Therefore, whereas cryptography protects the contents of a message, steganography can be said to protect both messages and communicating parties. By contrast, steganography suffers from a fundamental weakness. If the messenger is searched and the message is discovered, then the contents of the secret

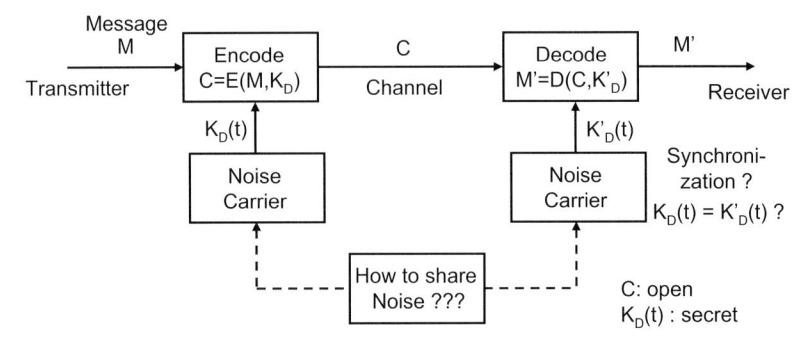

Figure 7.3 Block diagram of noise communication. C, channel signal; D( ), function for decoding; E( ), function for encoding; K_D, randomly time-varying carrier; M, original message; M', decoded message.

communication are revealed at once. Interception of the message immediately compromises all security. Although cryptography and steganography are independent, it is possible to both scramble and hide a message to maximize security.

7.1.3
Noise Communication

Cryptography has been more advanced and frequently used than steganography in modern communication systems. However, an impressive idea of utilizing a technique for steganography with a time-varying noisy signal was introduced in secure telephone speech in a Bell Telephone report (Singh, 2000). It was proposed that the sender should mask his/her speech by adding noise to the line, as shown in Figure 7.3. The recipient could subtract the noise afterward if he/she had succeeded to share it. The noise signal is a randomly time-varying carrier and it is very difficult to reproduce the temporal evolution of the noise signal by attackers since noise is totally random. This technique is called "noise communication," and probably considered as one of the first attempts of using randomly time-varying carrier in the communication channel. One of the practical disadvantages of this system was the difficulty of sharing the same noise signal between the distant users, and prevented it being actually used. However, the use of noise may enhance the security of secret communication systems as a one-time pad if noise can be synchronized.

7.2
Concept of Chaos Communication

7.2.1
Basic Idea of Chaos Communication

The idea of "chaos communication" has been provided to solve the synchronization problem in noise communication. Instead of using noise as a carrier of communication, chaos may be more useful since chaos can synchronize to each other due to

7.2 Concept of Chaos Communication

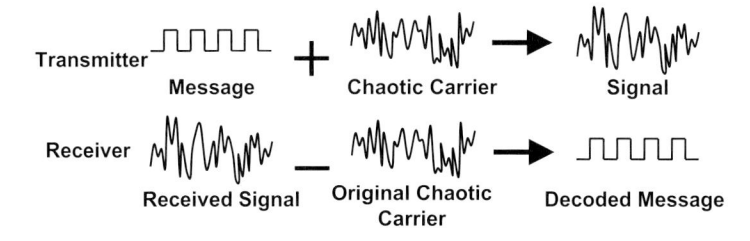

Figure 7.4 Schematics of chaos communication.

its deterministic nature, as described in Chapter 5. The basic concept of chaos communication is shown in Figure 7.4. A message is additively (or multiply) encoded in a chaos carrier in a transmitter, and the mixed signal consisting of the chaos and message is sent to a receiver laser. The technique of synchronization of chaos is used to reproduce the original chaos carrier in the receiver. The message can be recovered by subtracting (or dividing) the synchronized chaos carrier from the transmission signal. The quality of chaos synchronization strongly affects the degree of the recovered message signal.

The block diagram of chaos communication is shown in Figure 7.5. The purpose of chaos communication is to hide the existence of a message by a chaotic carrier waveform, therefore, it is considered as a type of steganography (Figure 7.2) rather than cryptography (Figure 7.1). In addition, the message is concealed in a time-varying chaotic carrier in chaos communication, which is similar to the scheme of noise communication (Figure 7.3).

One of the most important techniques in chaos communication is to share the same chaotic carrier between the distant users, which is called synchronization of chaos (see Chapter 5). To achieve synchronization of chaos, similar hardware systems are required in a transmitter and a receiver for producing the same chaos. The chaotic system in the transmitter produces a chaotic carrier to mask the message signal. The encoded signal is sent to the receiver and it is used for both decoding the message and achieving synchronization of chaos. In the receiver, the similar chaotic system can

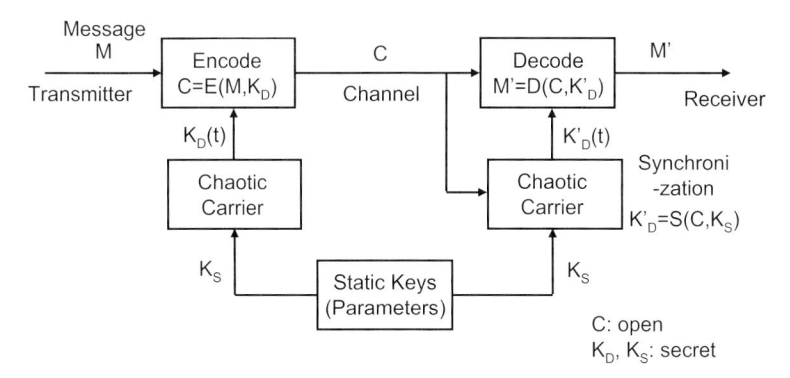

Figure 7.5 Block diagram of chaos communication. C, channel signal; D( ), function for decoding; E( ), function for encoding; K_D, chaotic carrier; K_S, static key (parameter values); M, original message; M', decoded message.

reproduce a nearly identical chaotic carrier by adjusting with a set of the static parameter values that are considered as "static keys" K_S to be shared beforehand. If the receiver succeeds in synchronizing chaos by using the similar hardware and the same static keys, the receiver can reproduce an almost identical chaotic carrier and succeed in decoding the original message by chaos cancelation.

7.2.2
Features of Chaos Communication

Some of the main features of chaos communication, compared with cryptography, steganography, and noise communication described in Section 7.1, are pointed out in the following:

a) use of a chaotic carrier to conceal a message;
b) use of synchronization of chaos to recover the message;
c) use of nearly identical hardware (laser) to achieve synchronization of chaos.

The time-varying chaotic carrier may be very effective to mask the message signal and it may be very difficult to extract the message from the chaotic carrier by eavesdroppers. To extract the message for the intended recipient, the technique of synchronization of chaos is required in the receiver. Also, nearly identical hardware, as well as its parameter values, needs to be shared to achieve synchronization of chaos between the transmitter and the receiver. These characteristics may provide a new way of secret communication, compared with the conventional cryptographic techniques. The chaotic carrier needs to be shared between distant users, and the requirement of similar hardware with similar parameter settings is considered as "hardware keys", since it has to be shared beforehand secretly.

7.2.3
Hardware Keys

Let us discuss some examples of hardware keys in history. Hardware keys have been used not only for steganography but also authentication in the history of secret communication. In the fifteenth and sixteenth centuries, the Japanese ruler, Yoshimitsu Ashikaga, sent trading ships to China (called Ming at that time). In order to legitimize this trade system, the Chinese created official licenses (called Kangou-fu in Japanese, meaning checking and verifying licenses) for approved trading ships. Official Japanese trading ships were given these licenses to distinguish from pirate boats. In order to verify genuine licenses, the Chinese used a special authentication system of writing in a registration book with the edge of the license covering half of the writing surface, as shown in Figure 7.6 (Kangou-fu trade, 2011). When the license was removed, only half of each written character remained in the book. The Chinese officials could check the authenticity of any license by placing it next to its corresponding half in the registration book. Only a real license would match exactly. Using this verification method, the Chinese were able to determine which of the trade ships were officially approved ships. These genuine trading licenses were typical examples of hardware keys for secure authentification systems in the middle age of Asian countries.

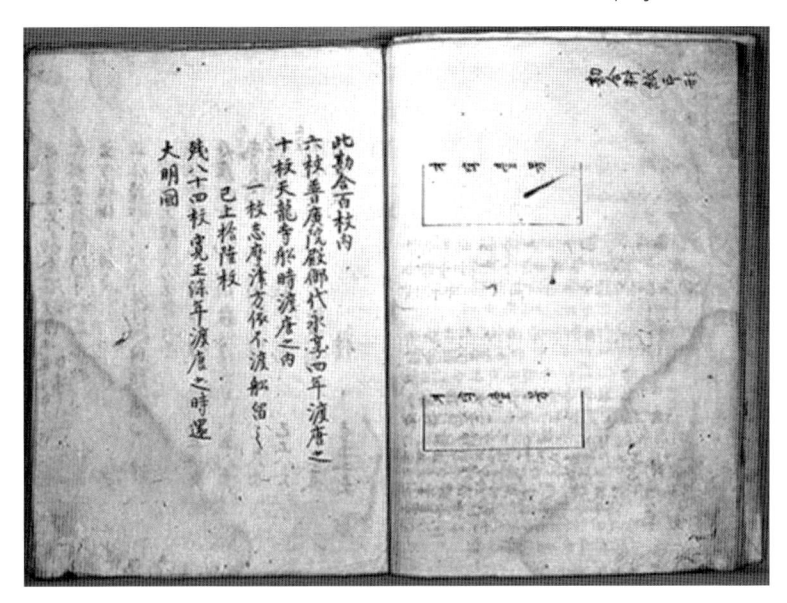

Figure 7.6 Registration book for the special trading authentication system between ancient Japan and China, as an example of hardware key (called Kangou-fu in Japanese). The edge of the license covering half of the writing surface is shown (right-hand side). Kangou-fu trade, (2011), (© Kyoto National Museum, Japan.)

Another example of a hardware key is "bivalve shells" (such as clam), as shown in Figure 7.7 (Bivalve shells, 2011). A half-shell only matches the corresponding pair of the other half-shell. In Japanese ancient time, bivalve shells were used as concentration game (also known as memory game or pairs game), instead of playing cards. Also, in the seventeenth and eighteenth centuries of the Edo era in Japan, bivalve shells were considered as a symbol of love and fidelity in marital relationship, since only one pair of matched bivalve shells existed all over the world. Bivalve shells are typical examples of a unique hardware key.

As seen in the above examples, hardware keys are "static," that is, the system characteristics are unchanged after sharing the hardware keys between the authorized users. This steady structure guarantees the uniqueness of shared one-pair systems, however, it may have an inherent weakness for attack as secret communication systems since attackers may be able to estimate all the structure and mimic the system in a limited time. Therefore, randomly "time-varing" features may be required that are still capable to be shared between legitimate users.

7.2.4
Synchronization for Chaos Communication

Chaos communication is considered as a hardware key system with a randomly changing carrier, where an irregular temporal signal is used to hide a message signal, and the irregular carrier can be shared between the users by using the synchronization technique between similar hardware systems. The characteristics of

Figure 7.7 Bivalve shells as an example of hardware key (called Awase-gai in Japanese). They are considered as a symbol of love and fidelity in marital relationship, since only one pair of matched bivalve shells existed all over the world. They were also used as concentration game (pairs game). (Bivalve shells, (2011), © Hoshino Koubou, Japan.).

"synchronizability" of deterministic chaos allows the use of chaos for secret optical communication. By contrast, statistical noise, such as thermal noise and quantum noise, cannot be applied for this type of communication systems, since synchronization of noise can be hardly achieved. The determinism of chaos is an important feature to guarantee synchronizability of a randomly fluctuating carrier between distant users.

Some security systems have a similar concept to chaos communication, sharing a randomly fluctuating temporal carrier. As an example of stream cipher, pseudorandom number generators and a single seed signal (i.e., a set of initial conditions) are distributed for the users *a priori*, and common random number sequences are generated. By contrast, the initial conditions of the hardware systems do not need to be shared in advance in chaos communication, but it is only necessary to share the similar hardware and parameter values. Thanks to the synchronizability of deterministic chaos, common random signals can be distributed without the information of initial conditions. The two concepts of "time-varying" and "hardware keys" coexist in chaos communication schemes, which may result in promising secure communication systems.

One of the important problems in chaos communication is how to synchronize chaos between the transmitter and receiver by using the transmission signal that consists of "chaos + message". The message component is already included in the transmission signal and synchronization needs to be achieved by using the mixed transmission signal (i.e., chaos + message), but not by using the original chaotic carrier. The synchronization techniques for chaos communication are more complex than the synchronization schemes shown in Chapter 5, where only a chaotic signal is simply sent to the receiver to obtain synchronization, that is, the transmission signal from the transmitter and synchronized output in the receiver is identical. Several encoding and decoding techniques have been proposed and demonstrated by using different mechanisms of chaos extraction from the mixed transmission signal. The encoding and decoding techniques will be discussed in detail in Section 7.4.

7.3 Characteristics of Chaos Communication

Based on the description in Sections 7.1 and 7.2, the characteristics of chaos communication are summarized in this section as follows.

7.3.1 Hardware-Based Communication

Chaos communication is characterized as a hardware-based communication system: encryption is directly applied to the physical layer of the communication protocol. The concept of hardware-based communication is also similar to quantum cryptography, which is a secure-key exchange protocol (see Chapter 9), while chaos communication is a direct encoding and decoding scheme of a message signal in a communication system.

One of the ultimate goals for chaos communication is to have a "one-pair device" with some hidden internal parameters that cannot be copied by any other devices. The one-pair device can allow synchronization of chaos between the specialized pair, but not with any other devices. The hidden parameters may be artificially embedded in the device during the fabrication process. It would be extremely secure if chaos synchronization can be achieved only between the one-pair device. This type of devices could be required for extremely secure communication channels, such as the hotline between presidents of countries, CEOs of companies, and so on.

There have been a huge number of attempts of hardware-based secure communication schemes before the era of today's digital world. The approach with hardware keys may be promising, and a well-designed one-pair device can produce time-varying chaotic sequences that are capable of being synchronized to each other.

7.3.2 Chaos-Synchronization-Based Communication

Synchronization of chaotic temporal dynamics is a major impact of chaos communication. In order to recover the message, synchronization of chaos is indispensable between the transmitter and receiver. The receiver system consists of not only a passive photoreceiver, but also an active laser device that can be synchronized.

Synchronization needs to be obtained by using the transmission signal (chaos + message), not by the chaos signal, as described in Section 7.2. Therefore, more sophisticated techniques for synchronization of chaos are required in chaos communication (see Section 7.4). In addition, parameter matching between the transmitter and receiver lasers is crucial to achieve synchronization of chaos.

7.3.3
Privacy

One of the main motivations for optical chaos communication is to obtain privacy in communication. Complex chaotic carriers may offer a certain degree of intrinsic privacy in the hardware layer of communication protocol. There have been many works on quantification of security in communication systems with chaotic lasers (see Section 8.4 for details).

Privacy in chaos communication systems results from the fact that an eavesdropper must have the proper hardware and parameter settings in order to decode and recover the message. The message encoded on the chaotic carrier can be recovered with a receiver by using synchronization of chaos, and the configuration and parameter settings need to be matched between the transmitter and the receiver for synchronization. The robustness of synchronization of chaos against parameter mismatch is thus one of the measures for privacy, since the robustness of synchronization determines the difficulty of the reproduction of chaos by attackers. There is an essential trade-off between the robustness of synchronization of chaos and the privacy: larger robustness of synchronization of chaos implies lower privacy, since attackers can mimic the set of static laser parameters by searching all the combination of available parameter settings (see Section 8.4.3 for details).

It is interesting to note that using a chaotic carrier to dynamically encode information does not preclude the use of more traditional digital encryption schemes as well. Dynamical encoding with a chaotic waveform can thus be considered as an additional layer of encryption.

7.3.4
Compatibility

One of the advantages for chaos communication is the compatibility with currently available optical-fiber components used in conventional optical communications: Optical-fiber components and transmitter/receiver devices for conventional optical communications can be directly applied to chaos communication. A transmission signal can be amplified by erbium-doped fiber amplifiers (EDFA), and the chromatic dispersion in fibers can be compensated by using dispersion-shifted fibers (DSF) for dispersion management used in the conventional optical-fiber communication. Unlike quantum cryptography, the development of single-photon sources and single-photon detectors is not necessary for chaos communication systems. Several experimental demonstrations of chaos communications in commercial optical-fiber networks have been reported as field experiments, as described in Sections 8.2 and 8.3.

7.3.5
Analog Communication

Chaos communication is an analog communication scheme. A transmission signal consists of the mixture of an analog chaotic carrier and a digital message signal. Therefore, the analog feature of chaos may result in the degradation of the transmission signal during the propagation in the optical fibers, that is, the degeneration of bit error rate (BER) and signal-to-noise ratio (SNR), as well as the addition of channel noise on the chaotic carrier. In addition, synchronization of analog chaotic signal needs to be robust against the existence of channel noise.

7.3.6
Subcarrier Communication

A chaotic signal is considered as a "subcarrier" to hide a message signal in optical communication. The main optical carrier consists of fast oscillation of an electric field whose frequency is of the order of hundreds of THz (10^{14} Hz), as already shown in Figures 2.23 and 5.3. By contrast, the oscillation of a chaos signal corresponds to an envelope component of the electric field at much lower frequency of several GHz (10^9 Hz). This chaotic subcarrier is used to hide a message signal in chaos communication.

7.3.7
Coherent Communication

Chaos communication is a type of "coherent" communication. It requires an active laser device in the receiver to reproduce the chaotic carrier. The optical wavelength needs to be well controlled between the transmitter and receiver lasers, and injection locking (matching of optical wavelength) needs to be achieved for synchronization of chaos in all-optical chaos communication schemes. By contrast, conventional optical communication systems are based on "incoherent" communication, which requires optical-intensity detection by a passive photoreceiver in the receiver. Recent optical-communication schemes utilize the modulation of optical phase to increase the capacity of optical communications as coherent communications, and many techniques have been developed for high-speed optical coherent communication schemes (Ho, 2005).

7.3.8
Multiplexing and Noise Tolerance

The purpose of chaos communications is not only limited to obtain privacy in communication, but also aimed at multiplexing and noise tolerance. A broadband chaotic carrier may be capable to increase communication capacity, such as code-division multiple access (CDMA) with chaos (Kennedy *et al.*, 2000; Rontani *et al.*, 2011). Another possibility of chaos communication may aim at noise tolerance. A broadband chaotic carrier may enhance the robustness against interferometric noise among multiplexing communication channels. The characteristics of noise tolerance are required for noisy wireless communication environment (Larson et al., 2006).

Historically, chaos communication with optical systems is mainly motivated for the purpose of private communication, whereas chaos communication with electronic circuit systems is motivated for the purpose of multiplexing communication, such as spread-spectrum communication (Kennedy et al., 2000).

7.4
Encoding and Decoding Techniques

As described in Section 7.2, synchronization of chaos is required for chaos communication. However, the synchronization technique required for chaos communication is not so simple, since the transmission signal consists of chaos and message signals, and the original chaos signal needs to be reproduced from the transmission signal by using synchronization. Here, a question arises: What is a mechanism to extract the "chaos" component from the "chaos + message" signal by using synchronization?

To answer this question, several schemes of encoding and decoding techniques have been introduced and demonstrated. The techniques for extracting chaos from chaos + message by using synchronization are a very important innovation that enables utilization of chaos for communication applications. These schemes are mainly categorized into three techniques: "chaos masking", "chaos modulation", and "chaos shift keying". The classification of these chaos communication schemes is introduced in this book for convenience, even though different terminology has been sometimes used for these chaos communication schemes in the literature. The comparison of the performance of chaos communication among the three techniques will be reviewed in Section 8.3.4.

7.4.1
Chaos Masking

One of the chaos communication schemes is called "chaos masking". Figure 7.8 shows the chaos masking scheme. In this scheme, an information-bearing message signal is simply added (or multiplied) to a chaotic carrier "outside" the transmitter. Assuming that the receiver's dynamics synchronizes with the original chaotic signal (i.e., the transmitter signal without a message), information recovery follows from a simple subtraction (or division). This procedure works well when synchronization is not very sensitive to perturbations in the original chaotic signal. In particular, the amplitude of the message must be small enough with respect to the chaotic carrier.

It is important to generate the original chaotic waveform from the transmission signal including the message in the receiver. A chaotic signal $C(t)$ and a message $m(t)$ is added in the transmittter as a transmission signal $C(t) + m(t)$. The transmission signal $C(t) + m(t)$ is sent to a receiver laser to obtain a synchronized chaotic signal $C'(t)$. Message $m(t)$ can be recovered by subtracting $C'(t)$ from $C(t) + m(t)$ if $C'(t)$ nearly equals to $C(t)$. However, it has not been mathematically proven whether the

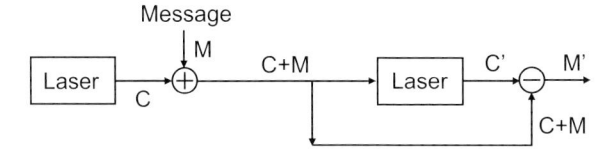

Figure 7.8 Block diagram of chaos masking. C, chaos; C', synchronized chaos; M, original message; M', decoded message.

synchronized signal $C'(t)$ can be identically reproduced from the mixed signal $C(t) + m(t)$.

The reproduction process of $C'(t)$ from the mixed signal of $C(t) + m(t)$ is called the "chaos pass filtering" effect (Fischer et al., 2000), where the receiver is capable of separating the chaotic fluctuations from any other superimposed signals in the same frequency range as the chaotic carrier. Experimental and numerical observations of the chaos pass filtering effect have been reported. One of the experimental examples is shown in Figure 7.9a, which depicts the RF spectrum of the transmitter laser with a current modulation at a frequency of 581.5 MHz as a message signal. This spectrum

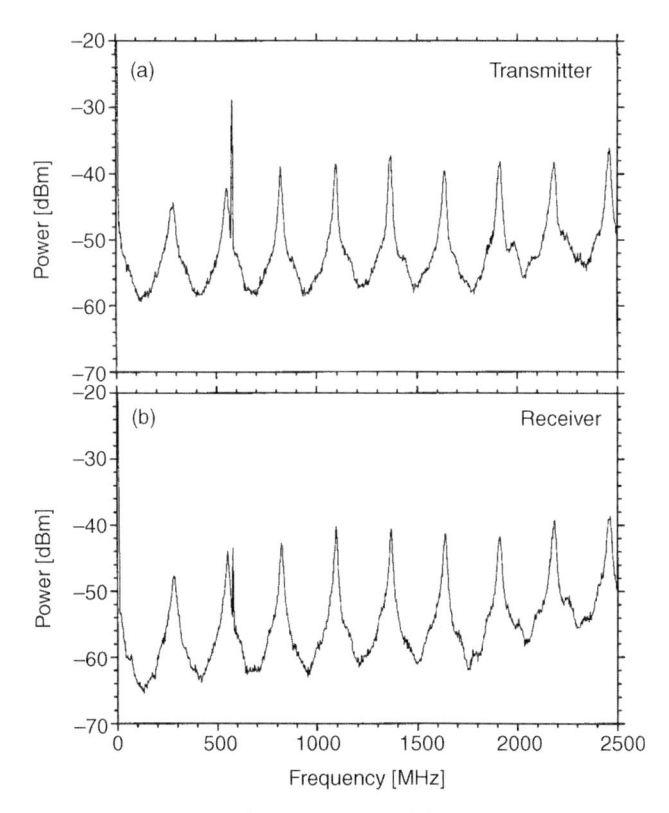

Figure 7.9 RF spectra of (a) transmitter and (b) receiver laser in the synchronization regime. The injection current of the transmitter laser has been modulated at a frequency of $f_{mod} = 581.5$ MHz. (I. Fischer, Y. Liu, and P. Davis, (2000), © 2000 APS.)

is typical for optical-feedback-induced chaos in semiconductor lasers (see Sections 3.2.1.2 and 4.1.3). The broad maxima of the spectrum correspond to the round-trip frequency of the light in the external cavity at 275 MHz and the harmonics of it. The sharp peak at 581.5 MHz corresponds to the frequency of the external current modulation for the message signal. Figure 7.9b depicts the corresponding RF spectrum of the receiver laser. The main features of the chaotic spectrum are very well reproduced in the receiver and correspond to those of the transmitter output. However, the external modulation peak is found to be considerably weaker in the receiver output compared to the transmitter output. At some frequencies, the amplitude of the modulation of the transmitter's emission, caused by the external modulation, can become even larger than the amplitude of the chaotic fluctuations, but still the receiver laser mainly recovers the fluctuations related to the chaotic dynamics and only to a much smaller extent of the external modulation features. The modulation signal (i.e., the message) can be easily recovered by normalizing the transmitter and receiver signals and subtracting them from one another.

The mechanism of the chaos pass filtering effect has been explained by numerical analysis (Uchida et al., 2003d). The transmission function for small perturbation is defined as the small-signal response of a semiconductor laser subject to optical injection around the stable injection-locking state. The transmission function is calculated from the analytical formula, as shown in Figure 7.10. This function has a peak near the relaxation oscillation frequency of 1.68 GHz, which is slightly shifted from the relaxation oscillation frequency of the solitary laser (1.45 GHz) due to optical injection. The transmission function monotonically increases from the low frequencies and reaches the maximum at the shifted relaxation oscillation frequency, where the transmittance reaches the peak value. The transmittance decreases again for frequencies higher than the shifted relaxation oscillation frequency. The characteristics of the transmission function resemble all the cases where an external mod-

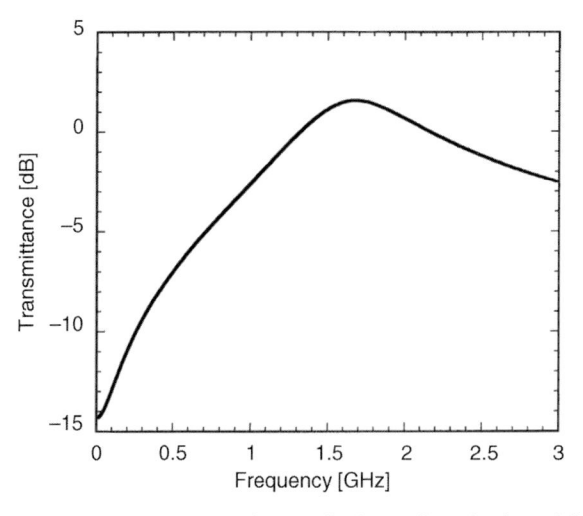

Figure 7.10 Transmission function for the small amplitude modulation calculated by small-signal analysis. (A. Uchida, Y. Liu, and P. Davis, (2003d), © 2003 IEEE.)

ulation by an electro-optic modulator or by an injection current is added to the chaotic carrier or the stable carrier (Uchida et al., 2003d). The transmission function thus indicates that the message recovery can be achieved more easily in the lower-frequency regions than the relaxation oscillation frequency. The transmission characteristic for the imposed small signal has the greatest suppression in the low-frequency regions, whereas the transmission function for the chaotic carrier is almost flat due to synchronization. Hence, the difference in the transmittance for the modulation signal and the chaotic carrier is larger in the lower frequency regions, making signal recovery easier. This result indicates that the maximum transmission speed of the message is limited by the relaxation oscillation frequency for the chaos masking scheme (Argyris et al., 2005).

7.4.2
Chaos Modulation

As opposed to external encoding *via* chaos masking, in internal encoding (*via* both modulation and injection in a feedback loop) the message is mixed nonlinearly with the chaotic carrier. This encoding scheme is called "chaos modulation", as shown in Figure 7.11. For the chaos modulation scheme, a message does not need to be much weaker than a chaotic carrier. In particular, when this method is applied to feedback systems in such a way that the message is encoded inside the feedback loop, perfect message recovery can be achieved under the parameter-matching conditions, regardless of the amplitude of the message.

There exists a synchronous solution between the transmitter and the receiver even in the presence of the message when two identical systems are used as the transmitter and the receiver. Synchronization between the transmitter and the receiver can be achieved when the synchronous solution is stable. The message can be recovered by subtracting (or dividing) the synchronized signal from the transmission signal without any errors in principle (perfect decoding). For time-delayed feedback systems, when the delay time is much longer than the time scale of the chaotic oscillations, the time delay must be compensated in the receiver between the synchronized signal and the transmission signal for achieving perfect decoding.

The encoding and decoding procedure for chaos modulation can be mathematically explained as follows (see also Table 7.1 and Figure 7.12). A message signal $m(t)$

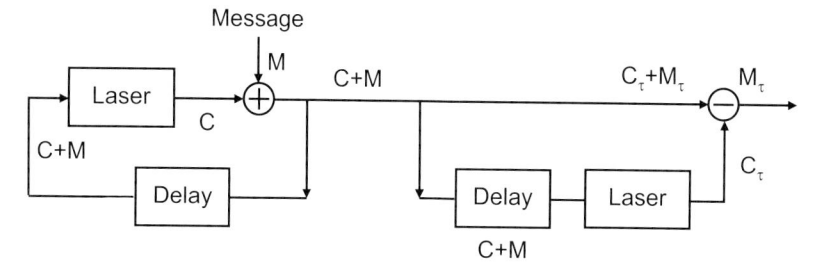

Figure 7.11 Block diagram of chaos modulation. C, chaos; C_τ, synchronized chaos with time delay; M, original message; M_τ, decoded message with time delay.

Table 7.1 Encoding and decoding procedure for the chaos modulation scheme.

Time	Drive Output (1)	Drive Input (2)	Message (3)	Transmission signal (4)	Response Input (before delay) (5)	Response Input (after delay) (6)	Response Output (7)	Transmission signal (8)	Recovered Message (9)
0	$C(0)$	—	$m(0)$	$C(0)+m(0)$	$C(0)+m(0)$	—	—	$C(0)+m(0)$	—
τ	$C(\tau)$	$C(0)+m(0)$	$m(\tau)$	$C(\tau)+m(\tau)$	$C(\tau)+m(\tau)$	$C(0)+m(0)$	$C(\tau)$	$C(\tau)+m(\tau)$	$m(\tau)$
2τ	$C(2\tau)$	$C(\tau)+m(\tau)$	$m(2\tau)$	$C(2\tau)+m(2\tau)$	$C(2\tau)+m(2\tau)$	$C(\tau)+m(\tau)$	$C(2\tau)$	$C(2\tau)+m(2\tau)$	$m(2\tau)$
3τ	$C(3\tau)$	$C(2\tau)+m(2\tau)$	$m(3\tau)$	$C(3\tau)+m(3\tau)$	$C(3\tau)+m(3\tau)$	$C(2\tau)+m(2\tau)$	$C(3\tau)$	$C(3\tau)+m(3\tau)$	$m(3\tau)$
4τ	$C(4\tau)$	$C(3\tau)+m(3\tau)$	$m(4\tau)$	$C(4\tau)+m(4\tau)$	$C(4\tau)+m(4\tau)$	$C(3\tau)+m(3\tau)$	$C(4\tau)$	$C(4\tau)+m(4\tau)$	$m(4\tau)$

The numbers in the first row of the table correspond to the positions shown in Figure 7.12. The transmission delay time T is set to be zero for simplicity and the compensation of the feedback delay time τ is introduced in the receiver. Time is discretized by τ for easy understanding in the table. To obtain the drive (and response) output from the input, the relationship of $C(t+\tau) = F(C(t)+m(t))$ is used.

7.4 Encoding and Decoding Techniques

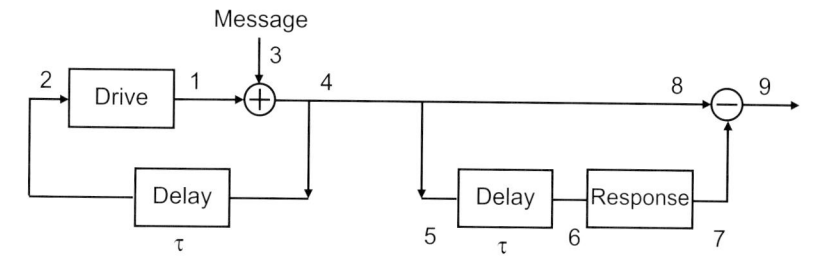

Figure 7.12 Block diagram for the prodecure of chaos modulation. The numbers corresponds to the positions described in Table 7.1.

is added (or multipled) to a chaotic carrier $C(t)$ inside the feedback loop of the transmitter, and the mixed signal $C(t) + m(t)$ is fed back to the nonlinear element of the transmitter to generate a new chaotic signal $C(t + \tau)$ after the feedback delay time τ, that is,

$$C(t + \tau) = F(C(t) + m(t)) \tag{7.1}$$

where F is the nonlinear function of the transmitter. τ can be zero for instantaneous feedback. The mixed signal $C(t) + m(t)$ is also sent to the receiver as a transmission signal, and the receiver generate a new chaotic signal $C'(t + T)$ after the transmission delay time T,

$$C'(t + T) = F'(C(t) + m(t)) \tag{7.2}$$

where F' is the nonlinear function of the receiver. When all the parameters are set to be identical between the transmitter and the receiver (i.e., $F = F'$), the chaotic signals of the transmitter and the receiver are perfectly synchronized with the time delay of $T-\tau$,

$$C'(t + T) = C(t + \tau) \tag{7.3}$$

Therefore, the time-delayed message signal $m(t+\tau)$ can be obtained in the receiver by subtraction of the chaotic carrier $C(t+\tau)$ from the time-delayed transmission signal $C(t+\tau) + m(t+\tau)$ by compensating the delay time $T-\tau$. Table 7.1 shows the summary of the procedure of chaos modulation for the time step of the feedback delay time τ when the transmission delay time T is set to zero for simplicity. The delay-time compensation of τ is required in the receiver, as shown in Figure 7.12.

The essential point of this encoding/decoding method is to generate a new chaotic dynamics by using the mixed signal of the chaos and the message $C(t) + m(t)$, that is, the message signal directly "modulates" the chaotic dynamics. Note that the message modulates the chaotic "dynamics", but not the chaotic "waveform". Chaos is generated from the feedback signal of $C(t) + m(t)$ in both the transmitter and the receiver. This symmetric architecture between the transmitter and the receiver provides perfect synchronization and high-quality decoding performance, because chaotic dynamics is modified by the message signal for both the transmitter and the receiver. Error-free decoding can be achieved mathematically by using the chaos-

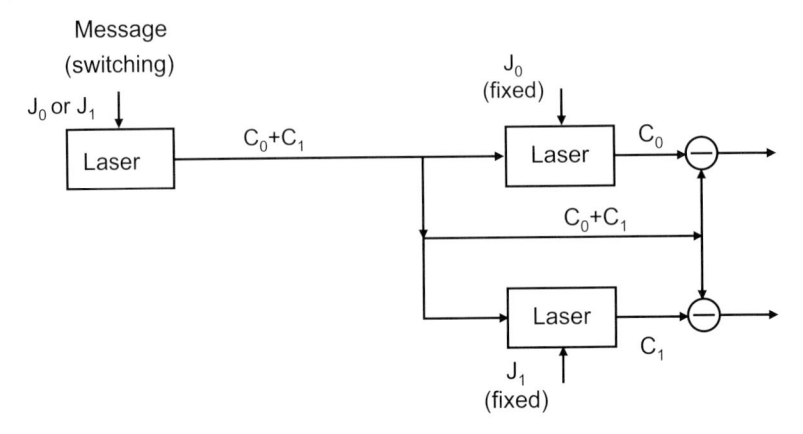

Figure 7.13 Block diagram of chaos shift keying. C_0, chaos generated at J_0; C_1, chaos generated at J_1; J_0, injection current corresponding to bit 0; J_1, injection current corresponding to bit 1.

modulation scheme. The performance of chaos communication using chaos modulation is thus the best among the three encoding/decoding schemes described in this section (Liu et al., 2002a) (see also Section 8.3.4).

For chaos modulation, the quality of message decoding does not depend on the amplitude of the message. The message, however, must not be too large to prevent much change in the dynamics of the original chaotic system. In addition, for large message amplitudes, it may be easy for unauthorized users to detect the message signal by directly observing the transmission signal and by analyzing the change in the dynamics from the original chaotic dynamics (see Sections 8.3 and 8.4).

7.4.3
Chaos Shift Keying

Another scheme for chaos communication has been proposed, known as "chaos shift keying". Figure 7.13 shows the chaos shift keying scheme. The basic idea of this scheme is the modulation of a system parameter of the transmitter laser to generate an information-bearing waveform. One of the laser parameters of the transmitter is modulated (or shifted) according to a digital message $m(t)$, taking two different values for 0 and 1 bits (e.g., the two values of the injection current J_0 and J_1). The chaotic waveform generated from the transmitter with one of the two parameter values is sent to the receiver, consisting of two lasers whose selected parameter values are fixed to the transmitter's 0-bit and 1-bit values (J_0 and J_1), respectively. At the receiver, the modulation of the parameter values produces synchronization between the transmitter and one of the receiver lasers whose selected parameter value is matched to the transmitter's value. The modulation of the message bits can be detected by investigating which receiver laser is synchronized.

A simpler version of the chaos shift keying scheme has been proposed, known as "chaos on-off keying", for which one of the parameter values is switched on or off according to the binary message to be transmitted (Uchida et al., 2001b). Figure 7.14 shows the chaos on-off keying scheme. The receiver consists of a single chaotic

7.4 Encoding and Decoding Techniques | 361

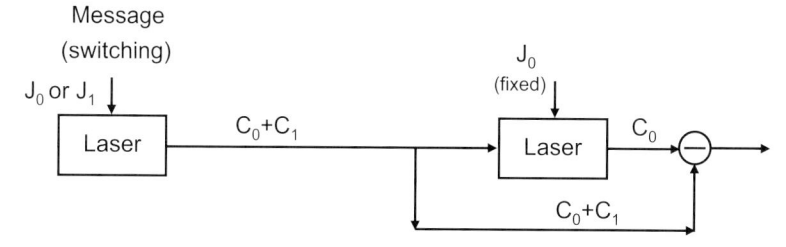

Figure 7.14 Block diagram of chaos on-off keying. C_0, chaos generated at J_0; C_1, chaos generated at J_1; J_0, injection current corresponding to bit 0; J_1, injection current corresponding to bit 1.

generator whose selected parameter is matched to one of the two selected parameters in the transmitter (the one corresponding to the bits 0, J_0 for instance). At the receiver, the modulation of the parameter values (J_0 or J_1) produces a synchronization error between the transmission signal and the receiver's regenerated signal. Message recovery can be accomplished by analyzing the synchronization error, which will be close to zero when the bit corresponding to the matching value of the coding parameter (J_0) and large for the other bit (J_1).

One of the implementation of chaos on-off keying has been demonstrated as "on-off phase shift keying" (Heil et al., 2002), where optical feedback phase is shifted in semiconductor lasers with optical feedback in the closed-loop configuration (see Sections 5.4.1, 6.4.4, and 6.6.5). The fundamental idea consists in using the receiver system as a sensitive detector of the phase dynamics of the transmitter. The privacy of the on-off phase shift keying systems is enhanced from the phase nature of the keying: while the amplitude dynamics of the transmitter is almost unaffected. Figure 7.15a depicts the synchronization error when a train of pseudorandom bits at 64 Mb/s is

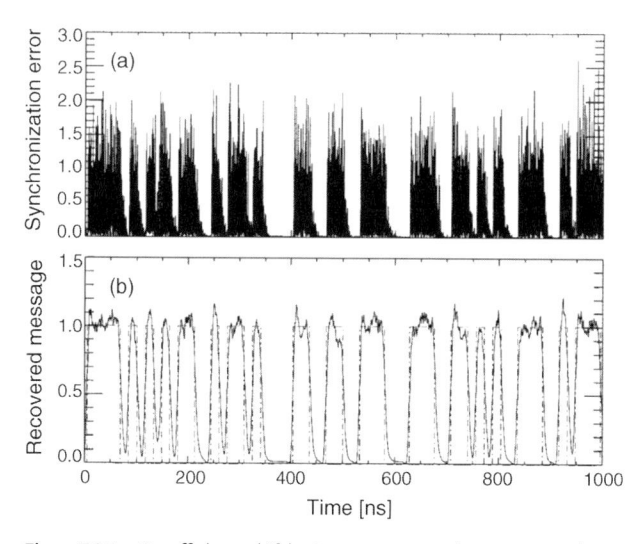

Figure 7.15 On-off phase shift keying encryption scheme. (a) Synchronization error. (b) Recovered digital message at 64 Mb/s after a filtering process. (T. Heil, J. Mulet, I. Fischer, C.R. Mirasso, M. Peil, P. Colet, and W. Elsäßer, (2002), © 2002 IEEE.)

applied. Figure 7.15a demonstrates that the synchronization error is almost zero when the two lasers are phase matched. On the other hand, the synchronization very rapidly degrades when the transmitter feedback phase is switched. Figure 7.15b demonstrates that the binary message can be recovered by applying a low-pass filter to the synchronization error signal, removing the high-frequency chaotic components.

The synchronization transient sets an upper limit for the data transmission rate for chaos shift keying and chaos on-off keying. The synchronization transients represent a limitation of the maximum achievable bit rate. Therefore, a comprehensive understanding of the synchronization process is necessary in order to reduce the transient times, and speed up the transmission (Uchida et al., 2004b).

7.5
Tools for Quantitative Evaluation of Performance of Chaos Communication

7.5.1
Bit Error Rate, Q Factor, and Signal-to-Noise Ratio

In this section, several tools for the quantitative evaluation of the performance of optical communication systems are introduced. These tools are commonly used for conventional digital communication systems. The performance criterion for digital receivers in communication systems is governed by the bit error rate (BER), defined as the probability of incorrect identification of a bit by the decision circuit of the receiver (Agrawal, 1997). For example, a BER of 1×10^{-6} corresponds to on average 1 error per million bits. A commonly used criterion for digital receivers requires BER less than 1×10^{-9}. Error-free operation is defined as the condition where BER is less than 1×10^{-12} in optical communication systems.

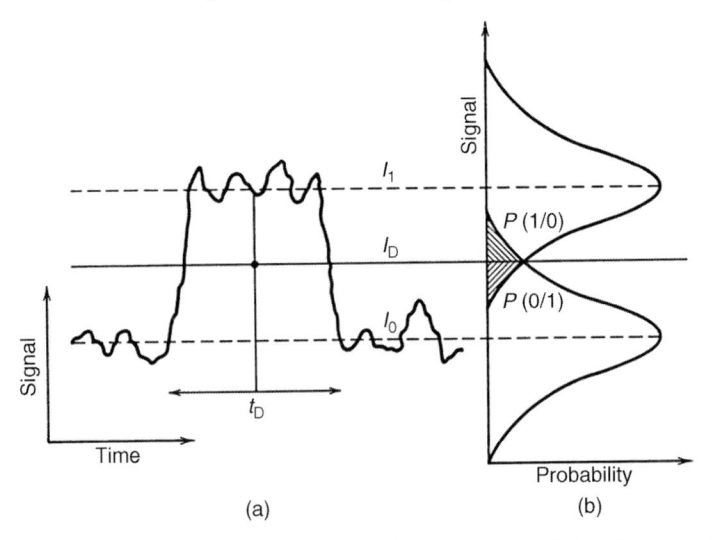

Figure 7.16 (a) Fluctuating signal generated at the receiver. (b) Gaussian probability densities of 1 and 0 bits. The dashed region shows the probability of incorrect identification. (G.P. Agrawal, (1997). © 1997 Wiley.)

7.5 Tools for Quantitative Evaluation of Performance of Chaos Communication

Figure 7.16a shows schematically the fluctuating signal received by the decision circuit, which samples it at the decision instant t_D determined through clock recovery (Agrawal, 1997). The sampled value I fluctuates from bit to bit around an average value I_1 or I_0, depending on where the bit corresponds to 1 or 0 in the bit stream. The decision circuit compares the sampled value with a threshold value I_D and calls it bit 1 if $I > I_D$ or bit 0 if $I < I_D$. An error occurs if $I < I_D$ for bit 1 because of receiver noise. An error also occurs if $I > I_D$ for bit 0. Both sources of errors can be included by defining the error probability as,

$$\text{BER} = p(1)P(0|1) + p(0)P(1|0) \tag{7.4}$$

Where $p(1)$ and $p(0)$ are the probabilities of receiving bits 1 and 0, respectively, $P(0|1)$ is the probability of deciding 0 when 1 is received, and $P(1|0)$ is the probability of deciding 1 when 0 is received. Since 1 and 0 bits are equally likely to occur, $p(1) = p(0) = 1/2$, and the BER becomes

$$\text{BER} = \frac{1}{2}[P(0|1) + P(1|0)] \tag{7.5}$$

Figure 7.16b shows how $P(0|1)$ and $P(1|0)$ depend on the probability density function $p(I)$ of the sampled value I. The functional form of $p(I)$ depends on the statistics of noise sources responsible for current fluctuations. The noise in the receivers, including thermal noise and shot noise, is well described by Gaussian statistics with zero mean and variance σ^2. A common approximation treats the noise as a Gaussian random variable for digital receivers. The conditional probabilities for Gaussian noise are given by,

$$P(0|1) = \frac{1}{\sigma_1\sqrt{2\pi}}\int_{-\infty}^{I_D}\exp\left(-\frac{(I-I_1)^2}{2\sigma_1^2}\right)dI = \frac{1}{2}\text{erfc}\left(\frac{I_1-I_D}{\sigma_1\sqrt{2}}\right) \tag{7.6}$$

$$P(1|0) = \frac{1}{\sigma_0\sqrt{2\pi}}\int_{I_D}^{\infty}\exp\left(-\frac{(I-I_0)^2}{2\sigma_0^2}\right)dI = \frac{1}{2}\text{erfc}\left(\frac{I_D-I_0}{\sigma_0\sqrt{2}}\right) \tag{7.7}$$

where I_1 and I_0 are the means of the sampled values for 1 and 0 bits, σ_1^2 are σ_0^2 are the corresponding variances for 1 and 0 bits, I_D is the threshold value for decision circuit. erfc() stands for the complementary error function defined as,

$$\text{erfc}(x) = \frac{2}{\sqrt{\pi}}\int_{x}^{\infty}\exp(-y^2)dy \tag{7.8}$$

Substituting Eqs. (7.6) and (7.7) into Eq. (7.5), and the BER is given by

$$\text{BER} = \frac{1}{4}\left[\text{erfc}\left(\frac{I_1-I_D}{\sigma_1\sqrt{2}}\right) + \text{erfc}\left(\frac{I_D-I_0}{\sigma_0\sqrt{2}}\right)\right] \tag{7.9}$$

Equation (7.9) shows that the BER depends on the decision threshold I_D. In practice, I_D is optimized to minimize the BER. The minimum occurs when I_D is chosen such that

$$\frac{I_1-I_D}{\sigma_1} = \frac{I_D-I_0}{\sigma_0} \equiv Q \tag{7.10}$$

An explicit expression for I_D is

$$I_D = \frac{\sigma_0 I_1 + \sigma_1 I_0}{\sigma_0 + \sigma_1} \tag{7.11}$$

When $\sigma_1 = \sigma_0$, I_D becomes

$$I_D = \frac{I_1 + I_0}{2} \tag{7.12}$$

Equation (7.12) corresponds to setting the decision threshold in the middle. This is the situation for most pin photoreceivers whose noise is dominated by thermal noise and is independent of the average current.

The BER with the optimum setting of the decision threshold is obtained by using Eqs. (7.9) and (7.10) and is given by

$$\text{BER} = \frac{1}{2}\text{erfc}\left(\frac{Q}{\sqrt{2}}\right) \approx \frac{\exp(-Q^2/2)}{Q\sqrt{2\pi}} \tag{7.13}$$

where Q is obtained by using Eqs. (7.10) and (7.11) and is given by

$$Q = \frac{I_1 - I_0}{\sigma_1 + \sigma_0} \tag{7.14}$$

Q is called the "Q factor" (quality factor) and is defined as in Eq. (7.14) for optimum setting. The approximate form of BER in Eq. (7.13) is obtained by using the asymptotic expansion of $\text{erfc}(Q/\sqrt{2})$ and is reasonably accurate for $Q > 3$. Figure 7.17 shows how the BER varies with the Q factor. The BER improves as Q increases and becomes lower than 1×10^{-12} for $Q > 7$. The BER of 1×10^{-9} can be obtained at $Q \approx 6$.

The performance of the optical receiver depends on the signal-to-noise ratio (SNR). The SNR of electrical signal is defined as (Agrawal, 1997)

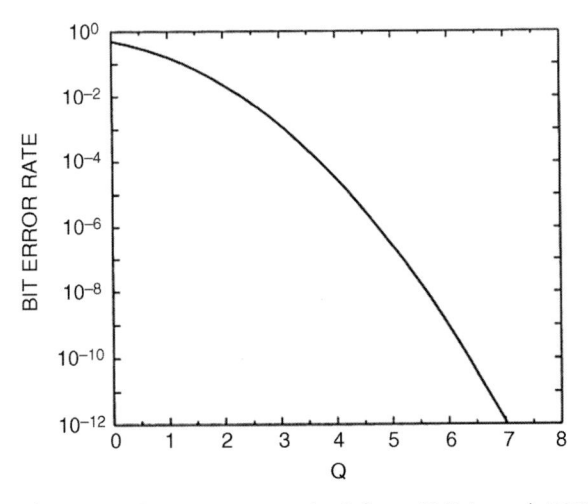

Figure 7.17 Bit error rate *versus* the Q factor. (G.P. Agrawal, (1997). © 1997 Wiley.)

$$\text{SNR} = \frac{\text{average signal power}}{\text{noise power}} = \frac{I_1^2}{\sigma_1^2} \tag{7.15}$$

where the fact that electrical power varies as the square of the current is used. Under the symmetric conditions of $I_0 \approx -I_1$ and $\sigma_0 \approx \sigma_1$, the Q factor of Eq. (7.14) is simplified as

$$Q = \frac{2I_1}{2\sigma_1} = \frac{I_1}{\sigma_1} \tag{7.16}$$

Therefore, the Q factor and the SNR have the following relationship from Eqs. (7.15) and (7.16)

$$Q = \sqrt{\text{SNR}} \tag{7.17}$$

Substituting Eq. (7.17) into Eq. (7.13), and the BER is given by the function of SNR,

$$\text{BER} = \frac{1}{2}\text{erfc}\left(\sqrt{\frac{\text{SNR}}{2}}\right) \approx \frac{\exp(-\text{SNR}/2)}{\sqrt{2\pi \cdot \text{SNR}}} \tag{7.18}$$

In fact, SNR is proportional to the number of received photons N_p

$$\text{SNR} \approx \eta N_p \tag{7.19}$$

where η is the quantum efficiency of the photoreceiver. Substituting Eq. (7.19) into Eq. (7.18), and the BER is given by the function of the number of received photons as,

$$\text{BER} = \frac{1}{2}\text{erfc}\left(\sqrt{\frac{\eta N_p}{2}}\right) \approx \frac{\exp(-\eta N_p/2)}{\sqrt{2\pi \eta N_p}} \tag{7.20}$$

The BER curves for conventional optical communication systems are generally measured as a function of the received optical power ηN_p.

For optical chaos communication systems, the BER is degraded due to several reasons: imperfection of synchronization of chaos, desynchronization burst, and channel noise. The errors due to the synchronization can be improved when the amplitude of the message signal is enhanced compared with the chaotic carrier signal. Therefore, the BER curves are often measured as a function of the message amplitude, instead of the received optical power, for optical chaos communication systems (see Section 8.3 in detail).

7.5.2
Modulation Format and Eye Diagram

In digital optical communication, the output of optical source such as a semiconductor laser is modulated by applying the electrical signal to an external modulator. There are two main choices for the digital modulation format of the resulting optical bit stream (Agrawal, 1997). These are shown in Figure 7.18 and are known as the return-to-zero (RZ) and nonreturn-to-zero (NRZ) formats. In the RZ format

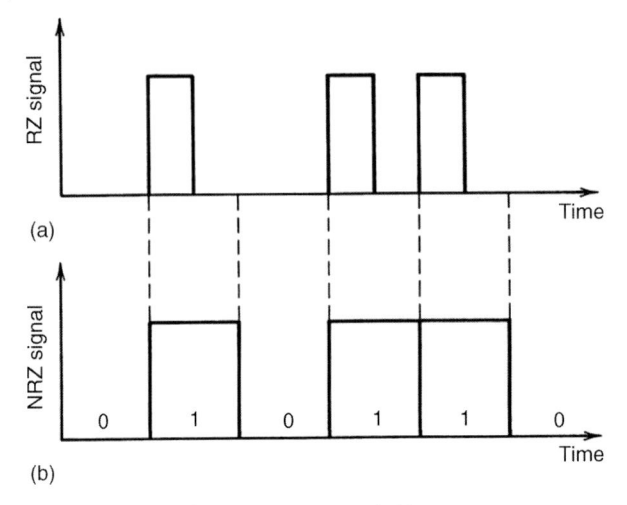

Figure 7.18 Digital bit stream 010110 coded by using (a) return-to-zero (RZ) and (b) nonreturn-to-zero (NRZ) formats. (G.P. Agrawal, (1997). © 1997 Wiley.)

(Figure 7.18a), each optical pulse representing bit 1 is shorter than the bit slot, and its amplitude returns to zero before the bit duration is over. In the NRZ format (Figure 7.18b), the optical pulse remains on throughout the bit slot and its amplitude does not drop to zero between two or more successive 1 bits. As a result, pulse width varies depending on the bit pattern, whereas it remains the same in the case of the RZ format. An advantage of the NRZ format is that the bandwidth associated with the bit stream is smaller than that of the RZ format by about a factor of 2 simply because on-off transitions occur fewer times. However, the NRZ format requires tighter control of the pulse width and may lead to bit-pattern-dependent effects if the optical pulse spreads during transmission. The NRZ format is often used in practice because of a smaller signal bandwidth associated with it.

An important issue is related to the choice of the physical variable that is modulated to encode the data on the optical carrier (Agrawal, 1997). The optical carrier wave before modulation is of the form

$$\mathbf{E}(t) = \hat{\mathbf{e}} \, A\cos(2\pi f_0 t + \phi) \tag{7.21}$$

where $\mathbf{E}$ is the electric field vector, $\hat{\mathbf{e}}$ is the polarization unit vector, A is the amplitude, f_0 is the carrier frequency, and ϕ is the phase. One may choose to modulate the amplitude A, the frequency f_0, or the phase ϕ. In the case of analog modulation, the three modulation schemes are known as amplitude modulation (AM), frequency modulation (FM) and phase modulation (PM). The same modulation techniques can be applied in the digital case and are called amplitude-shift keying (ASK), frequency-shift keying (FSK), and phase-shift keying (PSK), depending on whether the amplitude, frequency, or phase of the carrier wave is shifted between the two levels of a binary digital signal. The simplest technique consists of simply changing the signal intensity between two levels, one of which is set to zero, and is often called

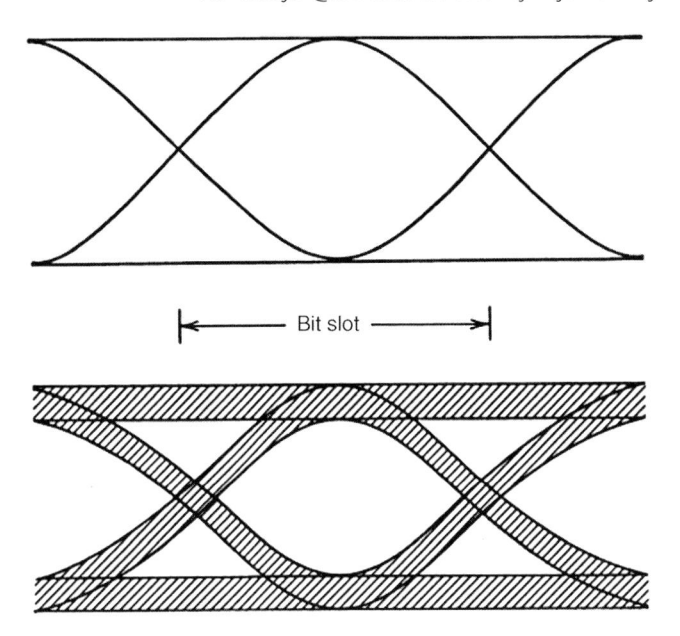

Figure 7.19 Ideal and degraded eye diagrams for the NRZ format. (G.P. Agrawal, (1997). © 1997 Wiley.)

on-off keying (OOK) to reflect the on-off nature of the resulting optical signal. Most chaos communication systems, as well as the conventional optical communication systems, use the ASK format imposed on the optical intensity of the chaotic carrier (see Sections 8.2 and 8.3).

For the signal recovery at the receiver, the decision circuit compares the received NRZ signal to a threshold level, at sampling times determined by the clock-recovery circuit, and decides whether the signal corresponds to bit 1 or 0. The best sampling time corresponds to the situation in which the signal level difference between 1 and 0 bits is maximum. It can be determined from the bit pattern formed by superposing 2- or 3-bit-long electrical sequences in the NRZ bit stream on top of each other. The resulting pattern is called an "eye diagram" because of its appearance. Figure 7.19 shows an ideal eye diagram together with a degraded one in which the noise and the timing jitter lead to a partial closing of the eye. The best sampling time corresponds to maximum opening of the eye.

Because of noise inherent in any receiver, there is always a finite probability that a bit would be identified incorrectly by the decision circuit. Digital receivers are designed to operate in such a way that the error probability is quite small (typically less than 1×10^{-9}). The eye diagram provides a visual way of monitoring the receiver performance: closing of the eye is an indication that the receiver is not performing properly.

The Q factor can be obtained from the eye diagram. The Q factor is one of the quantitative measures of the performance of digital communication systems. The means and variances of the received signals for bit 1 and 0 are used for Q factors as (rewriting Eq. (7.14)),

$$Q = \frac{I_1 - I_0}{\sigma_1 + \sigma_0} \tag{7.22}$$

where I_1 and I_0 are the means of the sampled values for 1 and 0 bits, σ_1 are σ_0 are the standard deviations of the sampled values for 1 and 0 bits, respectively. The Q factor can be measured from the eye diagram in Figure 7.19.

The measurement tools described in this section, such as BER, SNR, Q factor, and eye diagram, have been used for the quantitative evaluation of the performance of optical chaos communication systems, which will be described in detail in Sections 8.2 and 8.3.

8
Implementation of Optical Communication with Chaotic Lasers

In this chapter, practical implementation of optical chaos communication is described and discussed. This chapter describes recent advance of the implementation of optical chaos communication systems, including the demonstration of chaos communication in commercial optical-fiber networks at 10 Gb/s. Review articles and books for optical chaos communication have been published in the literature (Donati and Mirasso, 2002; Larger and Goedgebuer, 2004; Kane and Shore, 2005; Uchida *et al.*, 2005; Larson *et al.*, 2006; Ohtsubo, 2008).

8.1
History of Chaos Communication

8.1.1
Chaos Communication in Electronic Circuits

The history of chaos communication started from the early 1990s. The original idea of chaos communications came from its demonstration in chaotic electronic circuits. The application of synchronization of chaos to secret communication systems was suggested in 1990 (Pecora and Carroll, 1990, 1991). A chaotic transmitter could consist of an electronic circuit that simulated the dynamics of the Lorenz model (see Section 2.2.2), as shown in Figure 8.1 (Ditto and Pecora, 1993). A message to be concealed, assumed small in magnitude, is added to chaotic fluctuations, assumed to be much larger, of one of the variables (let us choose the z variable of the Lorenz model for this purpose) and transmitted to the receiver, while another chaotic variable (let us choose x variable) is separately transmitted. The receiver consists of a subsystem of the circuits in the transmitter that generates the dynamics of the y and z variables, and is driven by the signal from the x variable of the transmitter. The receiver synchronizes to the chaos of the transmitter if the conditional Lyapunov exponents for the response systems are negative for the given operating parameters. The message can be recovered from the chaos through a subtraction at the receiver.

Optical Communication with Chaotic Lasers: Applications of Nonlinear Dynamics and Synchronization, First Edition. Atsushi Uchida.
© 2012 Wiley-VCH Verlag GmbH & Co. KGaA. Published 2012 by Wiley-VCH Verlag GmbH & Co. KGaA.

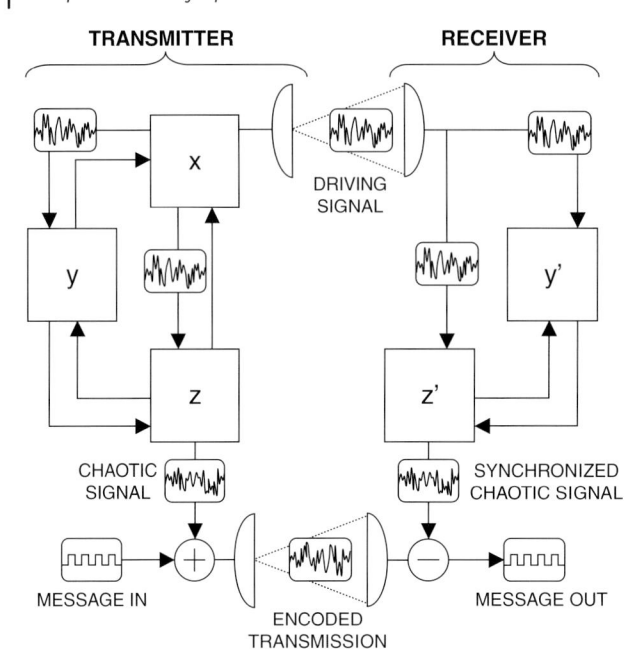

Figure 8.1 Schematics of private communications using chaos. (Adapted from W. L. Ditto and L. M. Pecora (1993), © 1993 Scientific American.)

An elegant variation of the above-described method that does not require the separate transmission of a driving signal to the receiver has been proposed, as shown in Figure 8.2 (Cuomo and Oppenheim, 1993). It is shown that the receiver can actually synchronize to the chaotic dynamics of the transmitter even when a message is added to the chaotic driving signal from the transmitter. The synchronized output from the receiver is then used to subtract out the information from the transmitted signal. The synchronization is not perfect, and the message, treated as a perturbation of the chaotic signal, has to be small compared to the chaos (i.e., chaos masking, see Section 7.4.1). The chaos shift keying scheme has also been demonstrated by switching one of the system parameter values as a binary message signal.

The techniques in which the message actually drives the chaotic transmitter system (i.e., chaos modulation) have been also investigated in chaotic electronic circuits (Halle et al., 1993; Volkovskii and Rul'kov, 1993). The synchronization between receiver and transmitter can be exact, so message recovery can be very accurate in principle (see Section 7.4.2).

8.1.2
Chaos Communication in Optical Systems

Following the original implementation of chaos communication in electronic circuits, the first numerical proposal of optical communication with synchronized chaotic lasers was carried out in a solid-state Nd:YAG laser model (Colet and Roy, 1994).

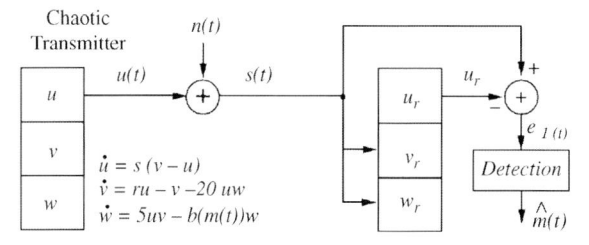

Figure 8.2 Schematics of chaos communication systems with electronic circuits of the Lorenz model. (K. M. Cuomo and A. V. Oppenheim, (1993), © 1993 APS.)

Equations in figure:
$$\dot{u} = s(v - u)$$
$$\dot{v} = ru - v - 20\,uw$$
$$\dot{w} = 5uv - b(m(t))w$$

In that pioneering work, a spiky chaotic carrier is generated from a loss-modulated Nd:YAG laser. To encode a digital message on the chaotic carrier, bits are encoded in the pulses, with only a single bit per pulse. Intensity maxima of the chaotic pulses are either increased or decreased according to a message bit of 1 or 0, as shown in Figure 8.3. The encoded chaotic waveform is transmitted to the receiver laser through a communication channel. The receiver laser synchronizes to the original chaotic signal rather than to the transmitted signal, so that the message can be recovered by subtraction of the original signal from the transmitted signal. The message signal has to be small enough to achieve synchronization between the receiver laser and the transmitted signal in the presence of a perturbation, which corresponds in this case to encoding a bit (i.e., chaos masking).

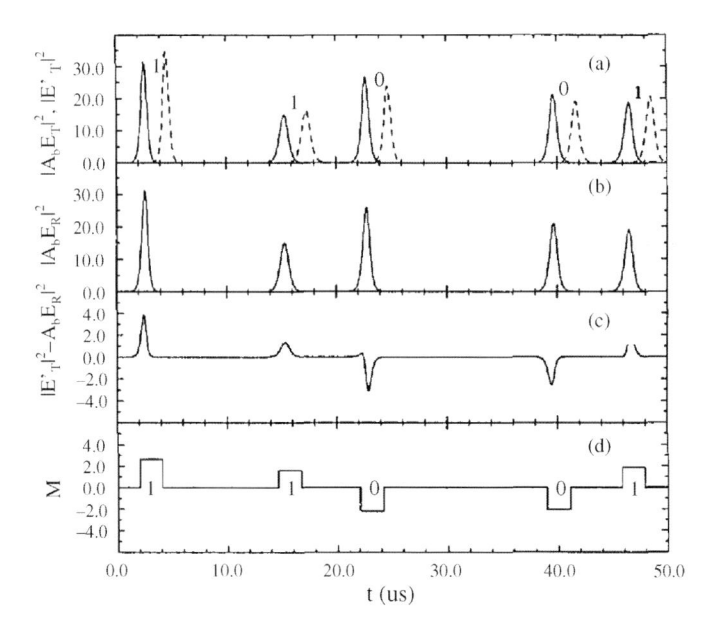

Figure 8.3 Illustration of encoding and decoding for 11 001 bit sequence: (a) original chaotic pulses (solid curve) and chaos with encoded message (dashed curve) displaced from the original pulses by 2 μs for clarity, (b) receiver output, (c) detected difference, (d) value of the message. (P. Colet and R. Roy, (1994), © 1994 OSA.)

8 Implementation of Optical Communication with Chaotic Lasers

One of the major motivations for using lasers is that optical chaotic systems offer the possibility of high-speed data transfer, as shown in simulations of numerical models that include realistic operational characteristics of the transmitter and receiver using semiconductor lasers and fiber communication channels. Some numerical simulations on chaos communication with synchronized chaotic semiconductor lasers have been reported (Mirasso et al., 1996; Annovazzi-Lodi et al., 1996; Sánchez-Díaz et al., 1999). Figure 8.4 shows the numerical results of chaos communication in two separated semiconductor lasers with optical feedback over a 50-km fiber transmission (Mirasso et al., 1996). Encoding of a message signal is done by variable modulation of the transmitter output, as shown in Figures 8.4a and b (i.e.,

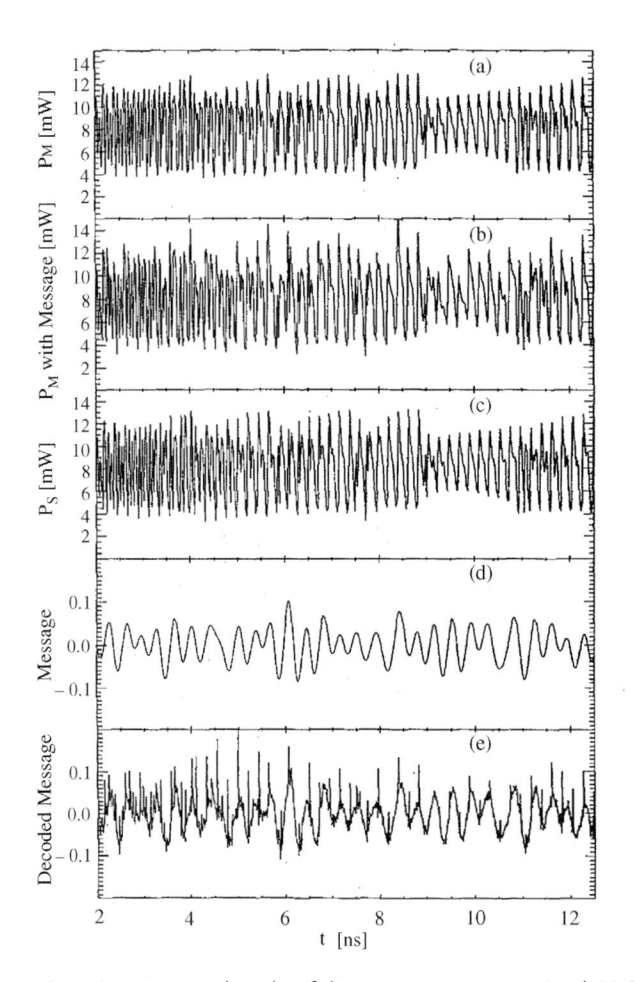

Figure 8.4 Numerical results of chaos communication using semiconductor lasers. (a) Typical laser output power from transmitter. (b) Transmitter output after encoding the message signal. (c) Output of the synchronized receiver. (d) Encoded original message. (e) Decoded message. (C. R. Mirasso, P. Colet, and P. García-Fernández, (1996), © 1996 IEEE.)

chaos masking). The message has to be small, not only to ensure privacy but also to avoid large distortions of the transmitter output that could prevent the receiver from synchronization. After 50 km of transmission, the output of the fiber is amplified to compensate the initial attenuation and losses in the fiber. The amplified output of the fiber is injected as an external signal in the receiver laser for synchronization. The decoding process is based on the synchronization of the receiver output to the chaotic transmitter carrier field, rather than to the transmission signal (i.e., chaos + message), so that the message can be recovered by comparing the transmission signal with synchronized chaotic signal. Figure 8.4c shows the output of the synchronized receiver laser. In Figure 8.4e the decoded message is shown, obtained by the division of the transmission signal by the synchronized receiver signal. The original analog message (Figure 8.4d) and the decoded one (Figure 8.4e) look similar, and the differences are originated by the lack of perfect synchronization between the transmitter and the receiver. Another scheme of chaos communication (chaos shift keying) has been also investigated numerically by using optically injected chaotic semiconductor lasers (Annovazzi-Lodi et al., 1996). These results indicate the strong potential of experimental realization of chaos-based optical communication with semiconductor lasers.

After the numerical analysis of chaos communication, two notable experimental demonstrations of chaos communication were reported in 1998 in fiber lasers (VanWiggeren and Roy, 1998a, 1998b) and electro-optic semiconductor laser systems (Goedgebuer et al., 1998; Larger et al., 1998a). A scheme for communication with chaotic waveforms was experimentally demonstrated in which the intensity fluctuations generated by an erbium-doped fiber-ring laser (EDFRL) were used either to mask or carry a message as shown in Figure 8.5 (left) (VanWiggeren and Roy, 1998a). A small-amplitude, 10-MHz square-wave message is introduced into the transmitter's cavity *via* an output coupler. Under these conditions synchronization between the receiver's output signal and the transmitter's unperturbed signal (i.e., without the message) is preserved, so that on subtracting the signals from photodiodes A and B, the message is recovered. This is shown in Figure 8.5a (right), which indicates the result of such a subtraction. The low-pass filtered signal difference reproduces the message intensity, since the typical frequency of the chaotic carrier fluctuations (hundreds of MHz) is much faster than the message frequency (10 MHz). Figure 8.5b (right) shows the result of such low-pass filtered signal (solid curve), and compares it with the original message (dashed curve). The good quality of the message recovery is evident. When a similar filtering is applied to the raw transmitted signal (chaos + message), the result shows no trace of the original message, as shown in Figure 8.5c (right), indicating that the message is efficiently masked by the chaos.

Another experimental demonstration of optical chaos communication was carried out by using the chaotic wavelength fluctuations output from an electro-optic system consisting of a semiconductor laser with an optoelectronic feedback loop (Goedgebuer et al., 1998). The experimental setup is already shown in Figure 5.22 (right) of Section 5.5.1. The transmitter is formed by a distributed Bragg reflector (DBR), a two-section tunable semiconductor laser with a feedback loop containing a

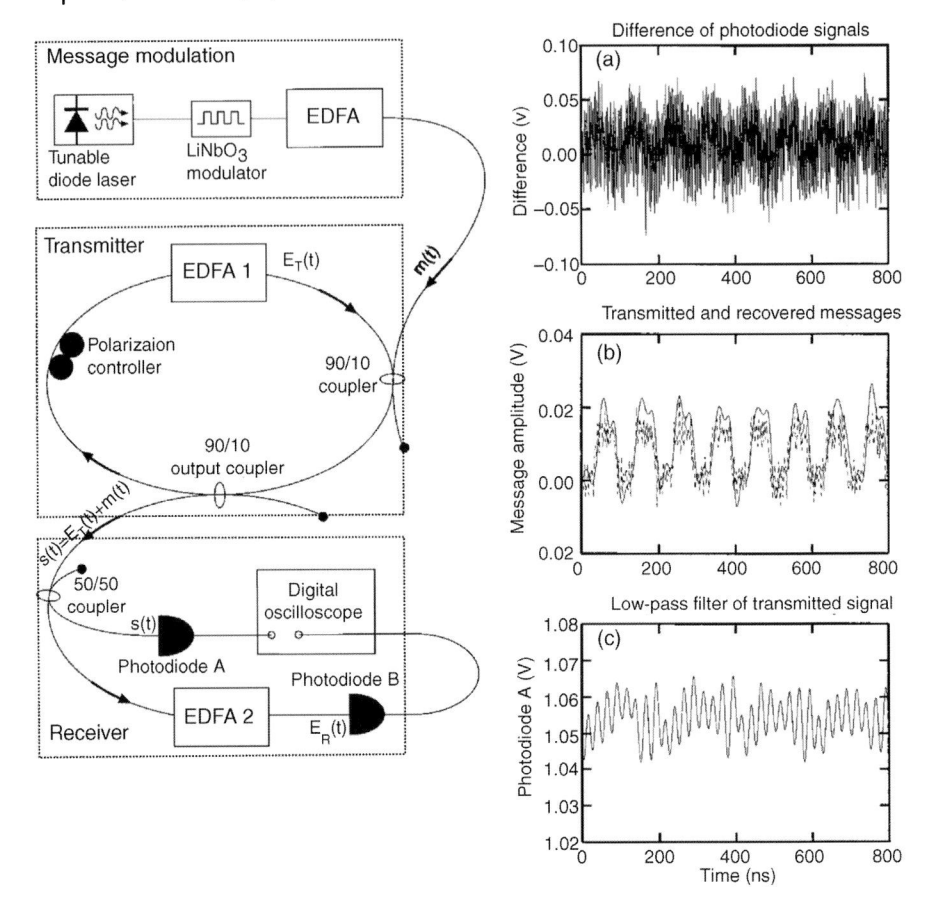

Figure 8.5 (Left) Experimental system for optical communication using chaotic erbium-doped fiber-ring lasers (EDFRL) (Right). The difference of the receiver input recorded from photodetector A and receiver output recorded from photodetector B, after the transmitter signal has been shifted by the appropriate time delay $\tau = 51$ ns. (b) The solid curve is the recovered square-wave message after low-pass filtering of the difference signal in (a). The dashed curve shows the original message. (c) Low-pass filtered version of the transmitted signal showing no trace of the square-wave message. (G. D. VanWiggeren and R. Roy, (1998a). © 1998 AAAS.)

nonlinear birefringent plate (BP) with an optical path difference, a photodetector (PD), and a delay line loop. The spectral transmission curve of BP is a function of $\sin^2 x$. The wavelength can be varied continuously around a center wavelength of 1550 nm by the injection current of the tunable semiconductor laser. The message to be encrypted is introduced in the feedback loop (i.e., chaos modulation) in such a way that the modulation current is formed by the addition of the message and the time-delayed signal provided by the PD. Chaotic fluctuations occur in the wavelength emitted by the laser diode. The receiver is a replica of the transmitter, except that the feedback loop is open and fed directly with the transmitted light beam that features the chaotic wavelength fluctuations. This beam is directed into the nonlinear birefringent plate in the receiver, which exhibits the same spectral transfer function as the transmitter. The receiver semiconductor laser emits light with synchronized chaotic fluctuations of its wavelength. The original message is recovered by subtraction of the output signal in the receiver from the transmission signal. Encoding and decoding of a 2-kHz-sine message signal have been experimentally demonstrated through free-space propagation with a distance of 2 m between the transmitter and the receiver.

Optical chaos communication was also demonstrated experimentally by using chaotic semiconductor lasers with time-delayed optical feedback (Sivaprakasam and Shore, 1999b; Fischer *et al.*, 2000; Kusumoto and Ohtsubo, 2002; Argyris *et al.*, 2004). At the same time, optical chaos communication in semiconductor lasers with optoelectronic feedback has been intensively investigated (Tang and Liu, 2001c; Liu *et al.*, 2002a) as a part of MURI (Multidisciplinary University Research Initiative) project in USA (Larson *et al.*, 2006). Many researchers have been involved in developing new experimental techniques and numerical models and the literature in this field has grown rapidly over a decade. The numerical evaluation of chaos communication schemes have been performed for achieving a bit rate of 10 Gb/s and higher (Liu *et al.*, 2002a; Kanakidis *et al.*, 2003). Research activities of optical chaos communication in the early stage (between the late 1990s and early 2000s) in the literature are summarized in Table 8.1.

8.1.3
European Project for Chaos Communication

Research activities for optical chaos communication had led to an international research project in Europe. One of the remarkable projects is called the OCCULT (Optical Chaotic Communication Using Laser-diode Transmitters) project, which was funded by the European Union from 2001 to 2004 (OCCULT, 2001; Annovazzi-Lodi *et al.*, 2008). The project aimed to demonstrate the basic principles of chaos-based transmission. Researchers from universities, research centers, and industry in many European countries (France, Germany, Greece, Italy, Spain, Switzerland, and United Kingdom) participated in the project.

The OCCULT project had led to the demonstration of several different transmission schemes, some based on the all-optical approach and others that included an optoelectronic feedback loop. Based on the main project outcomes, researchers had succeeded in performing several experiments of digital data transmission on a chaotic carrier over the commercial metropolitan fiber network of Athens in Greece (Argyris *et al.*, 2005) (see Section 8.2.1.1).

The OCCULT project has addressed the effect of chromatic dispersion in fibers, polarization-mode dispersion, nonlinear effects, optimum coding, and proposed phase-modulation encoding for increased privacy. The project has performed trans-

Table 8.1 Research activities of optical chaos communication in the early stage (between late 1990s and early 2000s).

Laser medium	Chaos Generation Method	Simulation or Experiment	Encoding Method (Message)	Speed [Gb/s]	Reference
Semiconductor lasers	Optical feedback	Simulation	CMA	~0.3	(Mirasso et al., 1996)
	Optical injection	Simulation	CMA, CSK (Square)	5×10^{-3}	(Annovazzi-Lodi et al., 1996)
	Optical feedback	Experiment	CMA (Sine)	2.5×10^{-6}	(Sivaprakasam and Shore, 1999b)
	Optical feedback	Experiment	CMA (Sine)	0.58	(Fischer et al., 2000)
	Optical feedback	Experiment	CMA (Sine)	1.5	(Kusumoto and Ohtsubo, 2002)
	Optical feedback	Simulation	COOK(PRBS)	0.064	(Heil et al., 2002)
	Optical feedback	Experiment	CMA (Sine)	~5	(Uchida et al., 2003d)
	Optical feedback	Experiment	CMA (Sine)	~6	(Paul et al., 2004)
	Optical feedback	Simulation	CMA, CMO, CSK (PRBS)	1–20	(Kanakidis et al., 2003)
	Incoherent optical feedback	Simulation	CSK (PRBS)	0.25	(Rogister et al., 2001)
	Optoelectronic feedback, Optical feedback, Optical injection	Simulation	CMA, CMO, CSK (PRBS)	10	(Liu et al., 2002a)
	Optoelectronic feedback	Experiment	CMO (PRBS)	2.5	(Tang and Liu, 2001c)
Fiber lasers	EDFA with self-feedback	Experiment	CMO (PRBS)	0.25	(VanWiggeren and Roy, 1998a, 1998b)
	EDFA with self-feedback	Experiment	CMO (Sine, PRBS)	1.0	(Luo et al., 2000a)
Electro-optic systems	EOM with self-feedback	Experiment	CMO (Sine)	0.14	
	EOM with self-feedback	Experiment	CMO (PRBS)	2×10^{-6}	(Goedgebuer et al., 1998)
		Experiment		3.0	(Larger et al., 2004)
Solid-state lasers	External modulation	Simulation	CMA (PRBS)	1×10^{-4}	(Colet and Roy, 1994)
	External modulation	Experiment	COOK (Square)	1×10^{-4}	(Uchida et al., 2001b)

CMA, chaos masking; CMO, chaos modulation; CSK, chaos shift keying; COOK, chaos on-off keying; PRBS, pseudorandom bit sequence; Sine, sinusoidal waveform; Square, square waveform; EDFA, erbium-doped fiber amplifier; EOM, electro-optic modulator.

mission of analog radio and video signals in fibers (Annovazzi-Lodi et al., 2005) (see Section 8.2.1.2).

After the OCCULT project succeeded, a second project started from 2006, called PICASSO (Photonic Integrated Components Applied to Secure chaoS encoded Optical communications systems) project, and continued to 2009 (PICASSO, 2006; Annovazzi-Lodi et al., 2008). The aim of this project is to design and build integrated modules (both monolithic and hybrid systems) for chaotic transmission. To achieve this target, new researchers from Germany, Ireland, and United Kingdom have joined the project to produce optical integrated devices for chaos-based communications.

The goal of the PICASSO project is to develop fully functional transmitter and receiver modules at four adjacent wavelengths of the ITU grid (100 GHz spacing) in the C band of the telecommunications window operating at 2.4 and 10 Gb/s. The PICASSO researchers investigated compatibility with the existing transmission infrastructure. Different types of photonic integrated circuits have been developed (Argyris et al., 2008), to enable both all-optical and electro-optic approaches. The development of integrated optics or hybrid modules will offer better mechanical stability, less temperature sensitivity, and more compact and rugged systems at lower volume cost (Argyris et al., 2010c).

8.2
Examples of Communication Systems with Various Chaotic Lasers

There have been many investigations of optical communication with various chaotic laser systems. In Section 8.2 several examples of optical chaos communication systems using different chaotic lasers are described: semiconductor lasers with optical feedback, semiconductor lasers with optoelectronic feedback, electro-optic systems, and fiber lasers. These research activities on optical chaos communication shown in Section 8.2 are summarized in Table 8.2.

8.2.1
Semiconductor Lasers with Optical Feedback

8.2.1.1 Field Experiment of Chaos Communication

The performance of all-optical-fiber communication systems has been experimentally investigated by using chaotic semiconductor lasers with optical feedback at high bit rate in the OCCULT European project (Argyris et al., 2005). The experimental system is shown in Figure 8.6 (top). Two distributed-feedback (DFB) lasers from the same wafer with almost identical characteristics and operating at 1552.5 nm are selected as the transmitter (emitter) and the receiver lasers. The pair of lasers is selected to exhibit parameter mismatches that are constrained below 3% for accurate synchronization of chaos. The chaotic carrier is generated by means of time-delayed optical feedback in a 6-m optical external cavity formed between the transmitter laser and a digital variable reflector. 2% of the laser's output power is fed back to the

Table 8.2 Examples of experiments on optical chaos communication by the mid 2000s (~2005), described in Section 8.2.

Laser system	Technique for Performance Enhancement	Distance [km] (Type)	Speed [Gb/s]	Bit-Error Rate for Legitimate Users	Bit-Error Rate for Eavesdroppers	Encoding Method (Message Type)	Reference, Section
Semiconductor lasers with optical feedback (open loop)	Dispersion compensation, Erbium-doped fiber amplifier	120 (Field)	1.0 2.4	10^{-7} 10^{-3}	N/A	CMA (PRBS)	(Argyris et al., 2005) Section 8.2.1.1
Semiconductor lasers with optical feedback (closed loop)	Fiber-optic components	1.2 (Lab)	2.4	N/A (S/N: 16–18 dB)	N/A	CMA (TV signal)	(Annovazzi-Lodi et al., 2005) Section 8.2.1.2
Semiconductor lasers with optoelectronic feedback	N/A	N/A (Lab)	0.5 2.5	1.7×10^{-4} N/A	N/A	CMO (PRBS)	(Tang and Liu, 2001c; Liu et al., 2002a Section 8.2.2
Electro-optic system with intensity modulator	Intensity chaos	N/A (Lab)	3.0	7×10^{-9}	N/A	CMO (PRBS)	(Larger et al., 2004) Section 8.2.3
Fiber lasers	Double loop	35 (Lab)	0.25	10^{-5}	N/A	CMO (PRBS)	(VanWiggeren and Roy, 1998b, 1999) Section 8.2.4

CMA, chaos masking; CMO, chaos modulation; CSK, chaos shift keying; COOK, chaos on-off keying; PRBS, pseudorandom bit sequence; Lab, laboratory experiment; Field, field experiment.

transmitter's laser cavity to generate chaos. A polarization controller inside the external cavity is used to adjust the polarization state of the light reflected back from the reflector. A nonreturn-to-zero (NRZ) pseudorandom message with small amplitude is encrypted into the chaotic carrier of the laser output by externally modulating a Mach–Zehnder lithium niobate ($LiNbO_3$) modulator (i.e., chaos masking). The chaotic carrier output with the message is unidirectionally injected into the receiver laser in order to reproduce the chaotic waveform. An erbium-doped fiber amplifier (EDFA) and an optical bandpass filter are used to amplify the transmission signal and to eliminate the amplified spontaneous emission (ASE) noise of the EDFA, respectively. The polarization controller in the injection path is critical, since the most efficient reproduction of the chaotic carrier by the receiver can be achieved only for an appropriate polarization state. The chaotic waveform of the transmitter (including the message) and the synchronized waveform of the receiver are sent through an optical coupler to two fast photodetectors that convert the optical input into an electronic signal. The photodetector for the receiver signal adds a 180-degree phase shift to the electrical output related to the optical one. By combining the transmitter's output and the inverted receiver's output with a microwave coupler, an efficient subtraction is carried out for message recovery. The adjustment of the equal optical power and the time jitter between the two electric outputs is mandatory for high-quality message recovery. The subtraction signal is amplified and low-pass filtered for the measurement of the decoded message signal.

This chaos-based all-optical communication system is implemented in a real-world optical network infrastructure of single-mode fiber belonging to the metropolitan area network of Athens in Greece (Argyris *et al.*, 2005). The network has a total length of 120 km and is provided by Attika Telecom SA. The topology consists of three fiber rings, linked together at specific cross-connect point as shown in Figure 8.6a (bottom). Through three cross-connect points, the transmission path follows the Ring-1, Ring-2, and Ring-3 routes. A dispersion compensation fiber module with 6 km, set at the beginning of the link (precompensation technique), cancels the chromatic dispersion that would be induced by the single-mode fiber transmission. Two amplification units that consist of EDFA and optical filters are used within the optical link for compensation of the optical losses and ASE noise filtering, respectively.

The original message signal of a NRZ pseudorandom bit sequence applied to the $LiNbO_3$ external modulator in the transmitter is shown in the upper panel of Figure 8.6b (bottom). The message amplitude is attenuated 14 dB with respect to the chaotic carrier, so the message is sufficiently masked in the chaotic carrier of the transmitted signal (the middle panel of Figure 8.6b (bottom)). The bit-error rate (BER) of the transmitted signal after filtering without the appropriate decoding is always larger than 6×10^{-2}. A good synchronization performance between the transmitter and the receiver leads to an efficient cancelation of the chaotic carrier and a satisfactory decoding process (the lower panel of Figure 8.6b (bottom)).

The performance of the chaotic transmission system is studied in terms of BER for different message bit rates up to 2.4 Gb/s and for different code lengths of 2^7-1 and $2^{23}-1$. All BER values are measured after filtering the subtracted electric signal by

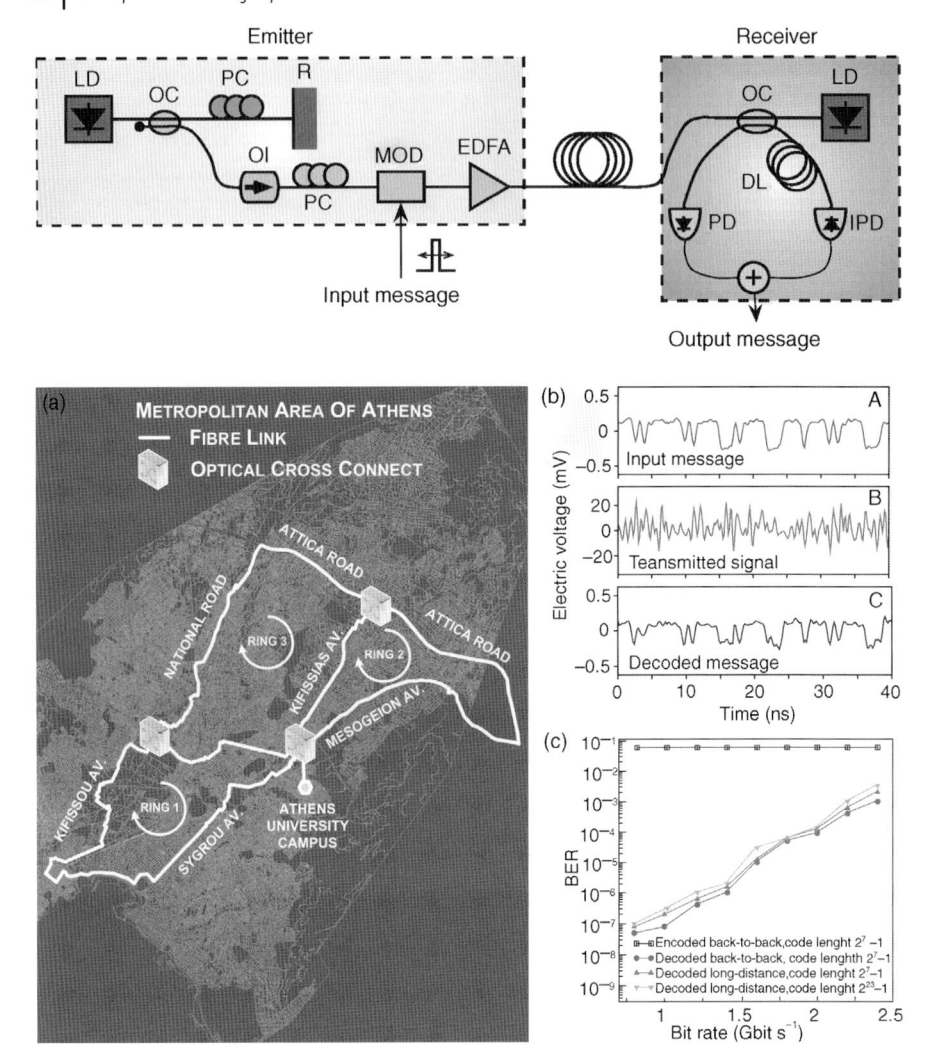

Figure 8.6 (Top) Schematic setup for optical chaos communication using semiconductor lasers with optical feedback. DL, delay line; EDFA, erbium-doped fiber amplifier; IPD, sign-inverting amplified photodiode; LD, laser diode; MOD, modulator; OC, optical fiber couplers; OI, optical isolator; PC, polarization controller; PD, amplified photodiode; R, reflector. (Bottom) Field experiment of fiber transmission. (a) Implementation of chaos-encoded communications in the optical communication network of Athens, Greece. (b) Time traces of a 1.0 Gbit/s applied message (trace A; BER $< 10^{-12}$), carrier with the encoded message (trace B; BER $\approx 6 \times 10^{-2}$) and recovered message after 120-km transmission (trace C; BER $\approx 10^{-7}$). (c) The BER performance of the encoded signal (squares), back-to-back decoded message (circles) and decoded message after transmission for two different code lengths (triangles). (A. Argyris, D. Syvridis, L. Larger, V. Annovazzi-Lodi, P. Colet, I. Fischer, J. García-Ojalvo, C. R. Mirasso, L. Pesquera, and K. A. Shore, (2005). © 2005 Nature.)

using low-pass filters with appropriate bandwidth adjusted each time to the message bit rate. The result of BER measurement is summarized in Figure 8.6c (bottom). The BERs of 10^{-7} and 10^{-3} are obtained at the bit rate of 1.0 and 2.4 Gb/s, respectively. The BER increases exponentially with increase of the bit rate. This deterioration of the BER is due to the fact that the synchronization is not perfect, because of parameter differences between the two lasers. Moreover, signal extraction is less efficient at bit rates comparable to the relaxation oscillation frequencies of the lasers around 3 GHz for chaos masking (Uchida et al., 2003d; Paul et al., 2004) (see Section 7.4.1). When the code length increases from 2^7-1 to $2^{23}-1$ bits, only a minor increase in BER is observed. Also, relatively small differences in the BER values exist between the back-to-back and the long-distance transmission setups, revealing only a slight degradation of the system performance due to the transmission link. Communication bit rates are mainly limited by the bandwidth of the chaotic carrier, which depends on the transmitter components such as relaxation oscillation frequencies of the lasers. The bandwidth of chaos can be enhanced by optical injection of semiconductor lasers (Simpson et al., 1995; Takiguchi et al., 2003; Uchida et al., 2003b) (see Section 3.2.4.2). The results presented here provide a convincing proof of a practical concept for optical chaos communication technology.

8.2.1.2 Transmission of TV Video Signal

Another demonstration of the fiber transmission of video signals with chaotic semiconductor lasers with optical feedback has been reported (Annovazzi-Lodi et al., 2005). A transmission experiment of a composite TV signal, generated by a standard TV camera, has been reported along a 1.2-km fiber span including splitters, joints, couplers, and an optical amplifier. The experiments have been performed on the chaos masking setup of Figure 8.7 (left). The drive (master) laser at the transmitter is driven to chaos by backreflection from the fiber tip positioned in front of its launching lens, which defines an external cavity of about 3 cm. This short-cavity scheme is compact and suitable for future integration, and offers a continuous and flat chaotic spectrum, where the message can be easily hidden. The characteristics of the chaotic regime of the laser depend on working conditions, such as supply current, cavity length, and backinjected power, which can be varied by changing the alignment. The response (slave) laser at the receiver is also routed to chaos by building an external cavity identical to that of the drive laser, which is the closed-loop configuration (see Section 5.3.3.2). The fiber path between the transmitter and the receiver is of about 1.2 km and, besides splitters, couplers, and joints, it also includes a semiconductor optical amplifier to increase the maximum injection level from the drive into the response laser. To prevent undesired backreflection into the drive laser, photodiode PD1 is slightly tilted, and the fiber facing is cleaved at an angle of 10 degree; this also prevents injection from the message laser into the drive laser. The optical isolator ensures unidirectional injection. Polarizers in front of the lasers select the same polarization of the laser emission for both reflection and injection. Moreover, the polarizer on the response laser is used, together with the polarization controller, to trim the injection level.

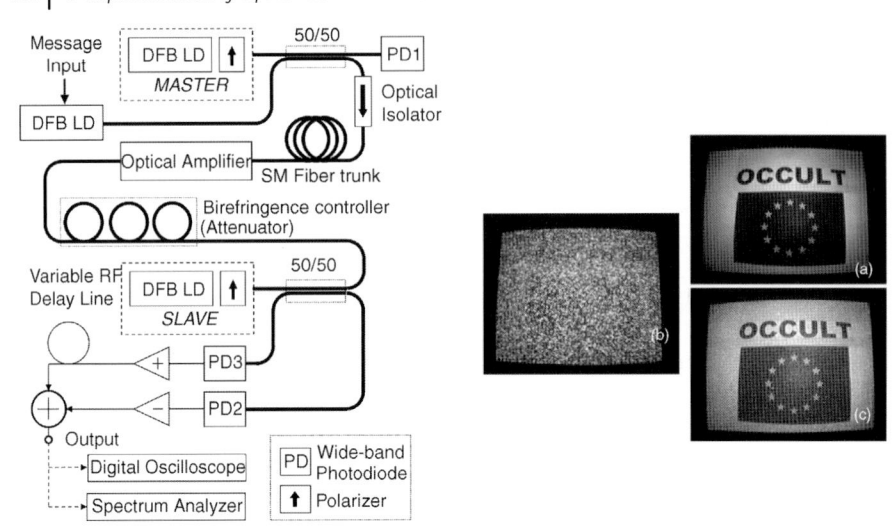

Figure 8.7 (Left) Setup for transmission with chaos-masking cryptographic scheme. (Right) TV frames of a still image transmitted with the setup of Figure 8.7 (left): (a) no encryption; (b) encrypted; (c) decrypted. (V. Annovazzi-Lodi, M. Benedetti, S. Merlo, M. Norgia, and B. Provinzano, (2005). © 2005 IEEE.)

The laser pair consists of standard DFB telecommunication lasers, which were selected between first neighbors on the same wafer for parameter matching. Their difference of threshold and of differential efficiency is less than 1%. Both external cavities are also trimmed in length with a resolution of a fraction of wavelength. Synchronization of chaos is obtained by adjusting the working point, the alignment, and the temperature of both lasers, as well as the injection level. The regimes of the two lasers can be compared from PD1, PD2 with a fast oscilloscope or with a radio-frequency (RF) spectrum analyzer. Synchronization is better checked by observing the spectrum of the difference between the drive and response lasers, obtained by amplifying the outputs of PD2 and PD3 by one inverting and one noninverting amplifier (Figure 8.7 (left)), and then passively summing their outputs. The difference should be minimized for optimum synchronization. To compensate the differential propagation delay of drive and response laser intensities to PD3, PD2, respectively, an RF delay line is introduced.

The communication setup has been inserted in a transmission link connecting a standard TV camera and its receiver. The camera output is a composite TV signal at the carrier frequency of 2.4 GHz and connected to the communication system input. The output of the system is sent to the TV receiver to be displayed on a monitor. In Figure 8.7 (right), three photographs of the monitor screen are shown, taken while the camera was aimed at a still picture. Figure 8.7a (right) is relative to transmission with no added chaos. Figure 8.7b (right) shows the picture hidden within chaos, and represents the message as it would be recovered by an eavesdropper. Figure 8.7c (right) shows the extracted message after synchronization. The signal level is adjusted as a trade-off between sufficient image masking by chaos and acceptable image quality after chaos cancelation. Figure 8.7b (right) is obtained by setting the AM sideband level at about 4 dB over chaos (a rapid deterioration of the image quality is observed below 5 dB, probably due to threshold in AM detection and to loss of TV synchronism). In these conditions, a signal-to-noise ratio (SNR) of 16–18 dB is obtained for the decoded message of Figure 8.7c (right).

8.2.2
Semiconductor Lasers with Optoelectronic Feedback

Optical chaos communication through synchronization of semiconductor lasers with optoelectronic feedback has been experimentally demonstrated (Tang and Liu, 2001c; Liu et al., 2002a). The schematic experimental setup of the chaos communication system that uses chaotic pulsing semiconductor lasers is shown in Figure 8.8 (left) (Tang and Liu, 2001c). The two lasers are identical InGaAsP/InP single-mode DFB semiconductor lasers at 1.299-μm wavelength. The two lasers are parameter matched by careful choice of a pair of lasers from the same batch with the closest characteristics and then fine tuning of their operating conditions. In the experiment, both lasers are temperature stabilized. The optical outputs from the lasers are detected by InGaAs photodetectors and observed with a digital oscilloscope.

In this setup, the transmitter laser has an optoelectronic feedback loop that drives the laser into chaotic pulsing when the feedback delay time τ is carefully adjusted. The receiver laser operates in the open-loop configuration (see Section 5.3.3.1) and is driven by the signal from the transmitter. A message is encoded by means of incoherent addition onto a chaotic carrier of the transmitter laser. When the message is encoded onto the chaotic carrier and the combined signal is sent to the

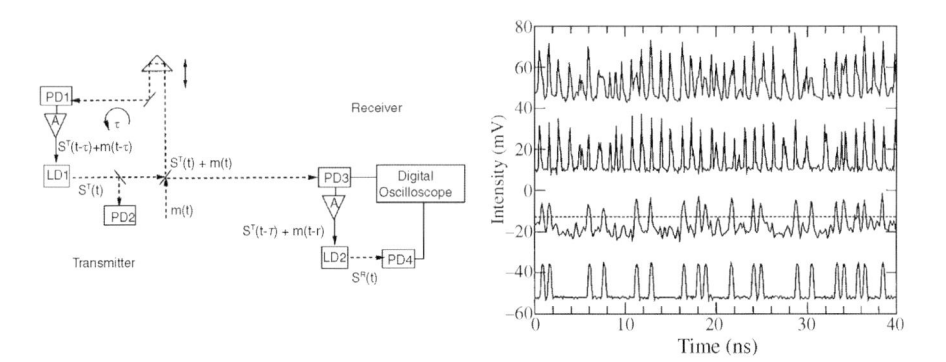

Figure 8.8 (Left) Schematic of the chaotic communication system with chaotic pulsing semiconductor lasers for encoding and decoding messages through chaos modulation. LDs, laser diodes; PDs, photodetectors; As, amplifiers. (Right) Transmission of a pseudorandom nonreturn-to-zero (NRZ) bit sequence at 2.5 Gb/s. Time series of received signal (top), receiver laser output (second from top), recovered signal (third), and encoded signal (bottom). (S. Tang and J. M. Liu, (2001c). © 2001 OSA.)

receiver laser, it is also fed back to the transmitter laser (i.e., chaos modulation). Therefore, the transmitter laser is driven by the combined signal shifted by τ because of the time delay in the feedback loop, and the receiver laser is driven by the combined signal shifted by T because of the time delay in transmission, where T indicates the transmission time. With parameters matched between the transmitter and the receiver lasers, the receiver laser output is synchronized with the original chaotic carrier with the time shift $T-\tau$ (see Section 7.4.2). The message can be recovered by subtraction of the synchronized receiver laser output from the received signal.

To show the potential communication capacity of this chaotic pulsing system, a pseudorandom sequence at a 2.5 Gb/s bit rate is transmitted, that fits the standard of the commercial digital OC-48 system. Figure 8.8 (right) shows the results of message recovery; time series of received signal (top), synchronized receiver laser output (second from top), recovered signal (third), and encoded signal (bottom). A 2.5 Gb/s NRZ message is generated by a pseudorandom pattern generator. Even though there are some fluctuations in the recovered time series, when a decision threshold is set at the position of the dashed line in Figure 8.8 (right), it is clear that the message can be successfully recovered at the bit rate of 2.5 Gb/s. The significance of this experiment is that it proves that chaos communication can be implemented with commercial standards of the digital OC-48 system at 2.5 Gb/s.

8.2.3
Electro-Optic Systems

Another experimental chaos communication system has been demonstrated based on nonlinear electro-optic dynamics and involving standard 1550-nm telecommunication components (Larger et al., 2004; Gastaud et al., 2004). A block diagram of the electro-optic system is shown in Figure 8.9. The nonlinear component is a 10-Gb/s $LiNbO_3$ integrated electro-optic Mach–Zehnder interferometer, acting as a nonlinear intensity modulator (see Section 3.3.3). The primary optical source is a 5-mW DFB semiconductor laser at 1550 nm, modulated electrically by a voltage V, providing a transfer function $F(x) = \cos^2(x+\Phi)$, where $\Phi = \pi V_b/2V_\pi$ is the static phase shift, determined by the bias direct-current (DC) voltage of V_b and the DC half-wave voltage of V_π ($V_\pi = 4.5$ V). The modulator output is fed into a fiber delay line of 5 m, then into a fiber coupler. One of the coupler output signals is sent into the transmission line, while the other output is fed back to drive the modulator for generating chaos, after an optical-to-electrical conversion and amplification. The peak optical power is $P_0 = 0.5$ mW. The actual electro-optic voltage swing ensures an intensity modulation in a highly nonlinear regime, that is, a Mach–Zehnder transmission curve with multiple extrema, thus allowing for a high complexity chaos and wideband nonlinear spectrum spreading of the optical chaotic carrier (see Section 3.3.1).

For the encoding process, chaotic oscillation is significantly perturbed by the seeded message inside the feedback loop (i.e., chaos modulation). Another fast

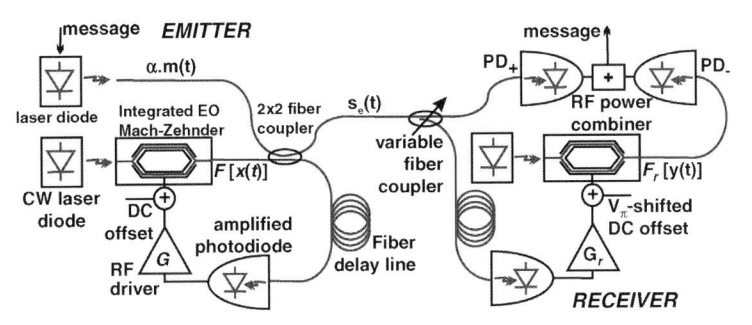

Figure 8.9 Multi-Gbits/s intensity chaos encoder and decoder. (L. Larger, J.-P. Goedgebuer, and V. Udaltsov, (2004). © 2004 Elsevier.)

direct-modulated semiconductor laser seeds an NRZ pseudorandom bit sequence (PRBS) (i.e., $m(t) = 0$ or 1) inside the feedback loop, with an adjustable weight determined by the peak binary optical power αP_0. The parameter α determines the masking efficiency of the message within the chaotic carrier. If the message is not sufficiently masked (i.e., too large), message recovery is possible through a direct detection by eavesdroppers. For an authorized receiver, the aim is to cancel the chaotic signal through knowledge of the chaos-synchronization process. The receiver is hence composed of the same components (matched pairs are used), with operating conditions adjusted carefully according to those in the transmitter. The receiver is used in an open-loop architecture, which consists of two branches for synchronization of chaos and a direct detection of a transmission signal, as shown in Figure 8.9. Under perfect transmitter–receiver matching conditions, synchronization of chaos can be achieved. The synchronized chaos is subtracted from the direct detection signal, thus performing message decoding. For experimental convenience, the subtraction uses an electronic RF combiner, together with a receiver Mach–Zehnder bias adjusted so that the actual interference function is inverted (out of phase) with respect to the transmitter one by adding a V_π-shifted DC offset.

Typical experimental encoding and decoding results at 3.0 Gb/s are depicted in Figure 8.10 (left) as eye diagrams (Larger et al., 2004). The message is obtained through a direct semiconductor laser modulation with a $2^7 - 1$ binary pseudorandom sequence. The masking efficiency is determined by adjusting the relative message to chaos optical power, thus varying the parameter α. For $\alpha > 1.7$, it is found that the chaotic carrier is not strong enough to prevent eavesdropping from direct detection of the transmitted signal, leading to a measurable BER in the order of 10^{-2}. Figure 8.10 (left) was obtained with $\alpha = 1.4$, thus preventing bit recovery from direct detection, but also leading to an acceptable BER for the authorized receiver of 7×10^{-9}.

Figure 8.10 (right) shows the experimentally recorded BER as a function of SNR at the standard bit rates of OC-24 (1.24 416 Gb/s) and OC-48 (2.48 832 Gb/s) (Kouomou et al., 2005). Here, SNR is defined as a signal-to-chaotic-carrier ratio, since the chaotic carrier is considered as noise to mask the message signal for privacy. The analytically

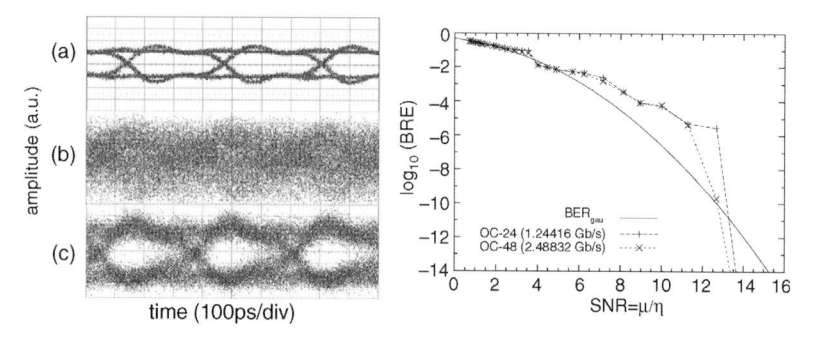

Figure 8.10 (Left) Bit-error rate (BER) test with binary pseudorandom sequence (length $2^7 - 1$) at 3.0 Gb/s: (a) eye diagram for the direct transmission without chaos encryption (BER $< 10^{-12}$); (b) eye diagram for the direct detection by an eavesdropper of the chaos encoded message (BER $> 10^{-2}$); (c) eye diagram of the recovered bits corresponding to BER of 7×10^{-9}. (L. Larger, J.-P. Goedgebuer, and V. Udaltsov, (2004). © 2004 Elsevier.) (Right) Experimental variations of the BER at the standard bit rates of OC-24 (dashed curve, 1.244 16 Gb/s) and OC-48 (dotted curve, 2.488 32 Gb/s). The theoretical BER curve is also shown (solid curve). (Y. C. Kouomou, P. Colet, L. Larger, and N. Gastaud, (2005). © 2005 IEEE.)

predicted BER curve is also plotted (solid curve). It should be emphasized that up to SNR values as high as 11, the experimental curves at 1.2 and 2.5 Gb/s (dashed and dotted curves) are well superimposed, in agreement with the theoretical BER curve, the bit rate has no influence on the BER variation law. A different feature is experimentally observed around SNR = 14, the BER drops sharply to zero, or at least, to values that are far below 10^{-15}. From a more global perspective, one can note that above SNR = 12, a small increase of the SNR leads to a very strong improvement of the BER. This limit, therefore, seems to be the target to reach for the fulfillment of the BER requirements in standard optical communication systems. However, for such high SNR values, the message does noticeably deform the statistical property of the optical carrier, and thereby induce a faster decay to 0 for the BER.

8.2.4
Fiber Lasers

Experimental demonstrations of chaos communication with fiber lasers have been reported (VanWiggeren and Roy, 1998a, 1998b, 1999). Compared with their previous setup shown in Figure 8.5 (left), several qualitatively new developments were implemented. The first one is the fact that the encoding method of chaos modulation is used. The message modulation drives the chaotic dynamics of the laser and the laser dynamics incorporate the digital message into the chaotic waveform. The second is the introduction of double time-delayed feedback loop. Both transmitter and receiver have configurations that involve two separate time delays, thereby enhancing the privacy of the transmission: Successful recovery of the message depends on multiple parameter settings and a matched geometric configuration in the receiver.

The experimental setup is shown in Figure 8.11 (left) (VanWiggeren and Roy, 1998b). The inner ring of the transmitter includes EDFA 1 and a polarizing LiNbO$_3$ intensity modulator that encodes the message onto the chaotic light. The inner ring is approximately 40 m in length. The outer loop contains EDFA 2, to adjust the amplitude of the light field, and a polarization controller. The polarization controller consists of three wave plates ($\lambda/4$, $\lambda/2$, and $\lambda/4$, respectively, where λ is the wavelength) which allow control over the relative phase and polarization of the light fields in the inner ring and outer loop. The outer loop itself is approximately 36 m long and provides a time delay between these light fields. Light coupled out of the

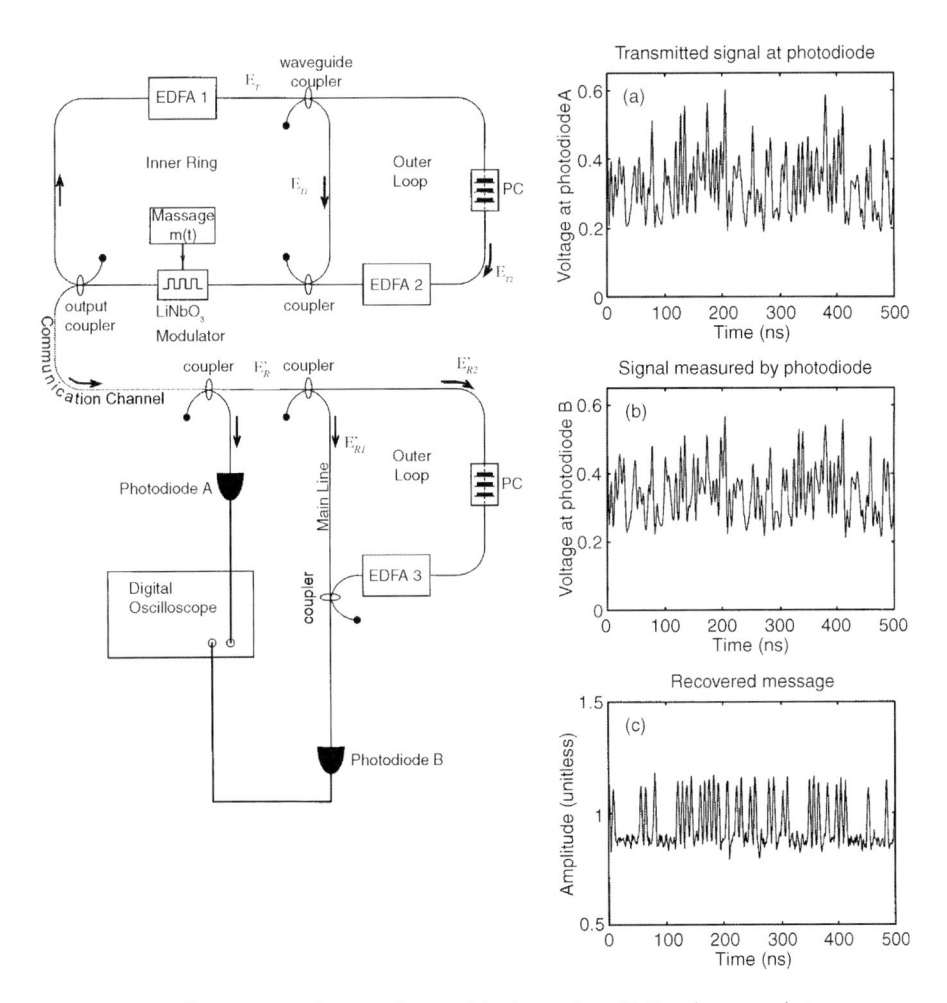

Figure 8.11 (Left) Experimental system for optical chaotic communication using erbium-doped fiber-ring lasers (EDFRL). PC: polarization controller. (Right) (a) The transmitter output as recorded by photodiode A in the receiver. (b) Signal measured at photodiode B after passing through the matched receiver. (c) A division of (A) by (B) results in recovery of the message. (G. D. VanWiggeren and R. Roy, (1998b). © 1998 APS.)

transmitter propagates through a fiber-optic communication channel to the receiver. 10% of the transmitted light is sent through an attenuator to photodiode A. The length of the fiber in the outer loop of the receiver is matched to the length of the transmitter's outer loop. The time delay between reception of a signal at photodiode A and reception at photodiode B is matched to the round-trip time in the inner ring of the transmitter. The signals detected by the photodiodes are recorded by a digital oscilloscope.

The transmitter in this system actually comprises two coupled ring lasers. The inner ring is one of these lasers; the other is formed partly by the inner ring and partly by the outer loop. A solitary erbium-doped fiber-ring laser (EDFRL) naturally produces chaotic fluctuations of intensity at frequencies ranging to many gigahertz (see Section 3.4.3). The chaotic dynamics are the result of the combination of nonlinearities in the amplifier and a round-trip time delay. While the nonlinear effects in the amplifier of a solitary EDFRL are sufficient to induce chaotic intensity fluctuations, they do not significantly change a waveform from one round-trip to the next. However, introducing the outer loop and its additional time delay into the transmitter causes the chaotic waveforms to change significantly after one round-trip around the inner ring. Thus, any signals encoded onto the chaotic light by the $LiNbO_3$ modulator very quickly become incorporated into the chaotic dynamics of the transmitter. In this way, the information signal both modulates the chaotic lightwave and drives the chaotic dynamics of the transmitter, ensuring high dimensionality.

Experimental results of this communication system are shown in Figure 8.11 (right). For this experiment, a repeating sequence of 100 000 pseudorandom bits was communicated at 126 Mb/s. Figure 8.11a (right) shows the transmitted temporal waveform measured at photodiode A. Figure 8.11b (right) shows the temporal waveform measured by photodiode B. Dividing the signal shown in Figure 8.11a (right) by the signal in Figure 8.11b (right) gives the decoded message shown in Figure 8.11c (right). The bits are clearly recovered from the chaos. The data analysis indicates that recovery is very consistent. The system is able to achieve BER less than 10^{-5}, except for occasional bursting of the transmitter ($\sim$1 ms in duration).

The recovery of the message requires the conditions where the receiver's configuration, time delays, and relative amplitudes must be matched to those in the transmitter. Figure 8.12 shows the robustness of the message recovery against parameter mismatch between the transmitter and the receiver (VanWiggeren and Roy, 1999). Figure 8.12a shows the temporal waveforms of the properly recovered message from the matched receiver laser, and other figures indicate incorrect message recovery due to the parameter mismatch. The message signal is degraded when the geometrical configuration of the receiver is mismatched to that of the transmitter, as shown in Figures 8.12b and c. The lengths of the fiber in the outer loop and the time-delay between photodiode A and B must be matched fairly precisely, otherwise the message recovery is failed, as shown in Figures 8.12d and e. The relative power levels in the receiver must also be properly matched to the

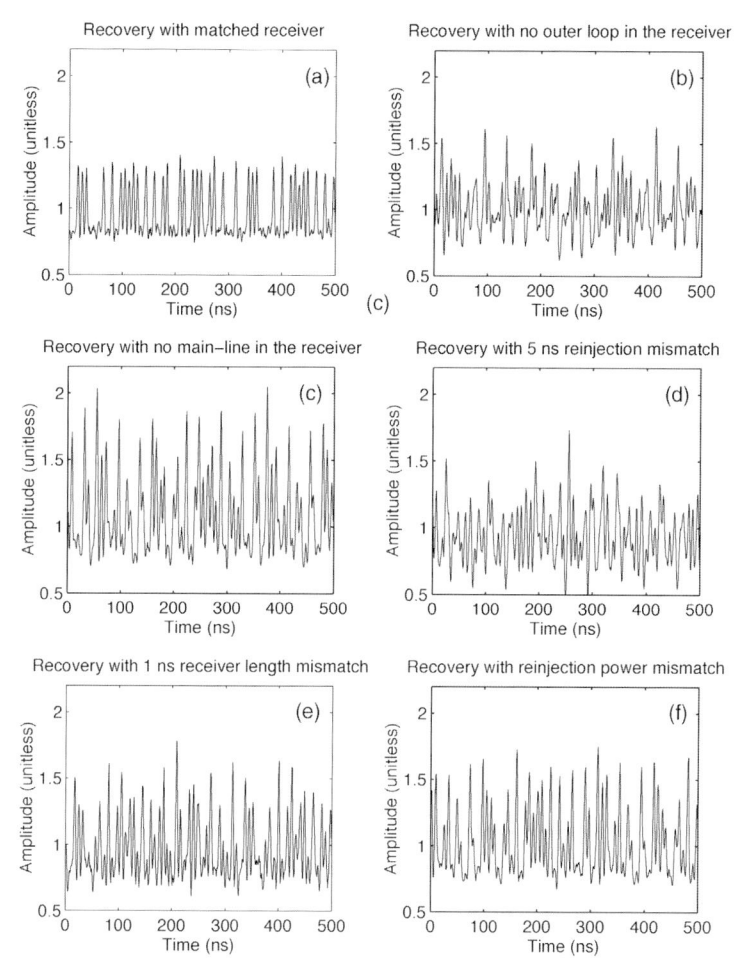

Figure 8.12 Recovery of the message requires that certain parameters in the transmitter be matched in the receiver. (a) Recovery of the message with all appropriate parameters matched. The other panels show attempted recovery with just one mismatched parameter. (b) Attempted recovery with a geometrical configuration in the receiver that lacks the outer loop. (c) Recovery is attempted without the main-line part of the receiver. An extra meter of optical fiber is used in the outer loop for the attempted recovery shown in (d). (e) Attempted recovery when the time delay between photodiodes A and B is mismatched by just 1 ns (± 20 cm of optical fiber). (f) Recovery when the amplitude of the receiver output power is too large when it is recombined with the receiver input power. (G.D. VanWiggeren and R. Roy, (1999). © 1999 World Scientific.)

power levels in the transmitter (Figure 8.12f). The transmitter parameters that must be known in order to recover the message form a multidimensional space in which to hide the key (i.e., a set of parameter values) for recovery. Conceptually, it seems possible to construct even more geometrically complicated transmitters (and thus receivers) so that the parameter space can become even higher dimensional.

Bit rates of 125 and 250 Mb/s (for a NRZ waveform) have been demonstrated in this experimental setup of Figure 8.11 (left). The method works well even over long communication channels (~35 km), and in both ordinary and dispersion-shifted fibers.

Another demonstration of chaos communication with fiber lasers has been also demonstrated (Luo *et al.*, 2000a). A sinusoidal signal of 1 GHz or a random bit stream at 140 Mb/s was encoded on a chaotic waveform and successfully recovered in the receiver laser.

8.3
Performance Evaluation of Optical Communication with Chaotic Lasers

The performance of optical chaos communication has been improved and enhanced in recent years. Transmission speeds up to 10 Gb/s have been achieved experimentally, and the error-free operation (i.e., BER less than 10^{-12}) has been obtained. The transmission encoded on a chaotic carrier in commercial optical-fiber networks at a distance over 100 km has also been demonstrated by using dispersion compensation and optical amplification techniques. The detailed techniques for the improvement and evaluation of communication performance are described in this section. Examples of state-of-the-art performance of optical chaos communication experiments shown in this section are summarized in Table 8.3.

8.3.1
Subcarrier Modulation

The RF spectral distribution of the chaotic carrier is of great importance for chaos data encryption, since it determines the microwave spectral region in which an efficient message encryption can be applied. This distribution is not uniform and depends on the operating parameters that lead the transmitter to generate chaotic dynamics such as the injection current, the feedback strength, and external cavity length. The initial dynamics always arise around the relaxation oscillation frequency of the laser and spectrally expand further when increasing the feedback strength. However, the power deficiency of chaotic spectral components in the low-frequency regime (less than 1 GHz) is the normal case for such chaotic carriers. Thus, any baseband message encryption requires its amplitude to be very small, resulting in a reduced SNR that will finally introduce errors in the decoding process.

In order to increase the encryption efficiency, one could embed the message within the spectral region with the highest power density at around the relaxation oscillation frequency of a few GHz. This can be realized by modulating the message over a tone of frequency comparable with the relaxation oscillation frequency of the laser. This radio-frequency tone is the so-called "subcarrier frequency". The subcarrier frequency is set at values where the chaotic carrier has its maximum power density,

Table 8.3 State-of-the-art performance of experiments on optical chaos communication (~2010), described in Section 8.3.

Laser system	Technique for Performance Enhancement	Distance [km] (Type)	Speed [Gb/s]	Bit-Error Rate for Legitimate Users	Bit-Error Rate for Eavesdroppers	Encoding Method (Message Type)	Reference, Section
Semiconductor lasers with optical feedback (open loop)	Subcarrier modulation	N/A (Lab)	1.0–2.4	10^{-11} 10^{-4}	10^{-1} 10^{-1}	CMA (PRBS)	(Argyris et al., 2010a) Section 8.3.1
Photonic integrated circuits for semiconductor lasers with optical feedback (closed loop)	Forward error correction (FEC)	100, (Lab)	1.25 2.5	10^{-3} (native) 10^{-12} (FEC)	0.5 0.5	CMA (PRBS)	(Argyris et al., 2010c; Bogris et al., 2010) Section 8.3.2
Electro-optic system with phase modulator	Phase chaos	120 (Field) 70 (Lab)	10.0 10.0	10^{-6} 10^{-12}	N/A N/A	CMO (PRBS)	(Lavrov et al., 2010) Section 8.3.3

CMA, chaos masking; CMO, chaos modulation; PRBS, pseudorandom bit sequence; Lab, laboratory experiment; Field, field experiment; FEC, forward error correction.

allowing increased encryption efficiency for data while having larger modulation amplitudes and thus improved initial SNR. This technique results in an error-free performance of the final system for data bit rates of 1 Gb/s, significantly improving the results achieved by using the chaos-based communication configuration (Bogris et al., 2007).

The chaos shift keying (CSK) technique combined with subcarrier modulation is mathematically described by the formula,

$$M(t) = m(t)\cos(2\pi f_s t) \tag{8.1}$$

where $M(t)$ is the modulation signal, $m(t)$ is the binary message that equals to "0" or "1," and f_s is the microwave subcarrier frequency. When $f_s = 0$ Hz, the typical CSK method is the case. At the receiver side, the message is decoded utilizing a local oscillator, followed by a low-pass filter.

The subcarrier modulation technique has been implemented experimentally and evaluated in an all-optical chaotic communication system that incorporates a hybrid integrated chaotic emitter (Argyris et al., 2010a). The experimental setup shown in Figure 8.13 (top) is based on the common configuration of the all-optical chaotic systems with semiconductor lasers in an open-loop receiver architecture. Two DFB semiconductor lasers, selected from the same fabrication wafer, emit a single longitudinal mode at 1553 nm with identical characteristics and are used as the transmitter (emitter) and receiver active elements in a one-way coupling configuration. The transmitter's laser is biased at 10 mA, or 1.3 times its current threshold, with a relaxation oscillation frequency around 3 GHz. The transmitter consists of the semiconductor laser and a 30-cm-long integrated fiber external cavity that includes a 90/10 coupler, a thermally tunable on-fiber variable optical attenuator and a high-reflective fiber end with a reflectivity greater than 95%. By controlling the loss within the cavity through the variable optical attenuator voltage adjustment, the desirable optical feedback is sent back to the transmitter laser, generating a broadband optical signal.

The data streams that are applied for encryption are pseudorandom bit sequences (PRBS) of $2^{31} - 1$ code length. In order to implement the subcarrier modulation technique and to shift data spectral components around a subcarrier frequency, the electric data stream signal is mixed by a RF mixer, with a sinusoidal signal produced by a local oscillator, and the mixing product externally modulates the optical carrier through a $LiNbO_3$ modulator. A fraction of the transmitter's optical power is injected, after optical amplification and filtering, into the receiver's laser to drive it to a synchronized response. At the receiver, an optical amplifier and filter are used in the first optical output branch to provide equal optical power between the two outputs, while a variable optical delay line in the second optical branch determines temporal alignment of both signal waveforms. Both the transmitter's and the receiver's chaotic optical waveforms are detected by two photoreceivers. The photoreceiver used to collect the receiver's waveform is an inverting diode; so the additive combination of the two electrical chaotic signals will lead to the subtraction signal. This signal is the subcarrier decrypted message, which is spectrally located at the subcarrier frequency, and is downgraded mainly by the synchronization error induced by the chaotic carriers' cancelation. Finally, a RF mixer identical to the one used at the transmitter is

8.3 Performance Evaluation of Optical Communication with Chaotic Lasers

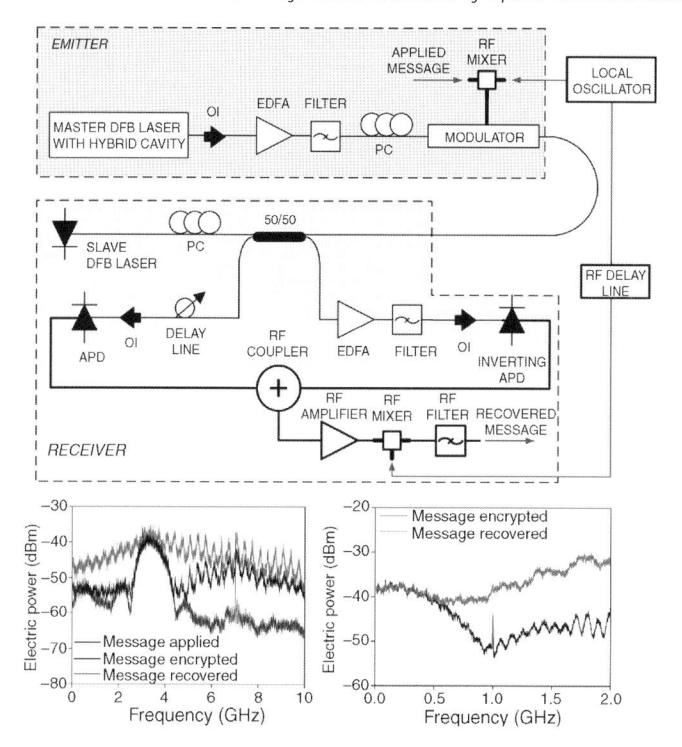

Figure 8.13 (Top) Experimental setup for an chaotic optical communication system using a subcarrier modulation technique. DFB, distributed feedback; OI, optical isolator; 50/50, optical coupler; PC, polarization controller; EDFA, erbium-doped fiber amplifier; APD, avalanche photoreceiver. (Bottom) Spectral distribution of (left) the applied (BER $\sim 10^{-12}$), encrypted (BER $\sim 10^{-1}$), and recovered (BER $\sim 10^{-11}$) 1 Gb/s message, using subcarrier modulation at 3.5 GHz and (right) of the partially encrypted (BER $\sim 10^{-4}$) and recovered (BER $\sim 10^{-12}$) 1 Gb/s message using baseband modulation. (A. Argyris, A. Bogris, M. Hamacher, and D. Syvridis, (2010a). © 2010 OSA.)

used to convert this subcarrier message, after the appropriate low-pass filtering, into the baseband recovered message.

Figure 8.13 (bottom, left) shows the case of a 1 Gb/s pseudorandom bit sequence centered on a 3.5-GHz subcarrier frequency. The amplitude of the applied message (Figure 8.13 (bottom) left, lowest trace) is adjusted in order to be encrypted adequately into the available chaotic carrier (Figure 8.13 (bottom, left) top trace); on the other hand, this amplitude cannot be extremely small, since a sufficient SNR of the initially applied message must be preserved for a successful message recovery at the receiver, after the subtraction of the chaotic carrier (Figure 8.13 (bottom, left) center trace).

For privacy reasons, the message amplitudes with a SNR as high as 20 dB are considered hidden within the chaotic signal and an eavesdropper can extract the data with BER $\sim 10^{-1}$ even after linear filtering. Under this constraint, the decoding result of BER $\sim 10^{-11}$ is obtained after the chaotic carrier cancelation at

1 Gb/s transmission rate for the subcarrier modulation scheme. By contrast, significantly worse decoding performance of BER $\sim 10^{-6}$ is achieved for the baseband modulation scheme because of less-powerful chaotic components in that spectral region of 1 GHz (Figure 8.13 (bottom, right)). Without this constraint (i.e., larger message amplitudes are allowed), error-free operation of BER $\sim 10^{-12}$ can be achieved for both the subcarrier and baseband modulation schemes. However, the eavesdropper that performs direct detection and linear filtering can extract the data with BER $\sim 10^{-2}$ and $\sim 10^{-4}$ for the subcarrier and baseband modulation schemes, respectively, indicating better masking performance for the subcarrier modulation scheme.

Higher data rates of 2.4 Gb/s have also been tested by using the same experimental configuration. However, because subcarrier modulation utilizes the double bandwidth compared with the baseband modulation, the degradation of the system performance in terms of data bit rate is abrupt. For encrypted data with BER value of 10^{-1}, the corresponding BER value of the recovered data is 10^{-4}. A small improvement in these values should be expected by using a single-sideband spectral product of data, thus halving the recovering data bandwidth.

8.3.2
Photonic Integrated Circuit and Forward Error Correction for High-Performance Chaos Communication

8.3.2.1 Photonic Integrated Circuit

Photonic integrated circuits (PIC) appear very attractive, since they could be easily included in the existing infrastructure of optical-fiber networks (see Section 8.5 in detail). The first experimental demonstration of chaos communication with PICs has been reported (Argyris et al., 2010c; Bogris et al., 2010). Figure 8.14 (top) shows the PIC used for the communication experiment (Argyris et al., 2010c). The PIC has been fabricated using selective-area epitaxial growth and incorporates the fundamental principles of the time-delayed all-optical feedback effect, using different sections that provide the capability to control the chaotic properties of the optical emission accurately and reproducibly (Argyris et al., 2008). The criterion for the selection of the short external cavity length is the ability of the device to produce chaotic attractors with high complexity. The selected length of 1 cm corresponds to a feedback round-trip time of approximately 280 ps, a delay long enough to provide a fully chaotic behavior. Depending on the biasing current of the DFB laser and the feedback strength, the bandwidth of the chaotic carrier may be increased from several GHz up to 20 GHz. The integration of an amplifying/absorbing section (SOA/VOA in Figure 8.14a (top)) offers control on the feedback strength; no biasing of this section leads to a specific optical power feedback ratio equal to 1.6%, which is predetermined by the inherent losses of the external cavity. Even though the value of 1.6% proves to be high enough to generate chaotic dynamics, the feedback ratio can be increased up to 5% by positively biasing the semiconductor optical amplifier (SOA) section for increased complexity. The phase section (PM in Figure 8.14a (top)) accurately tunes

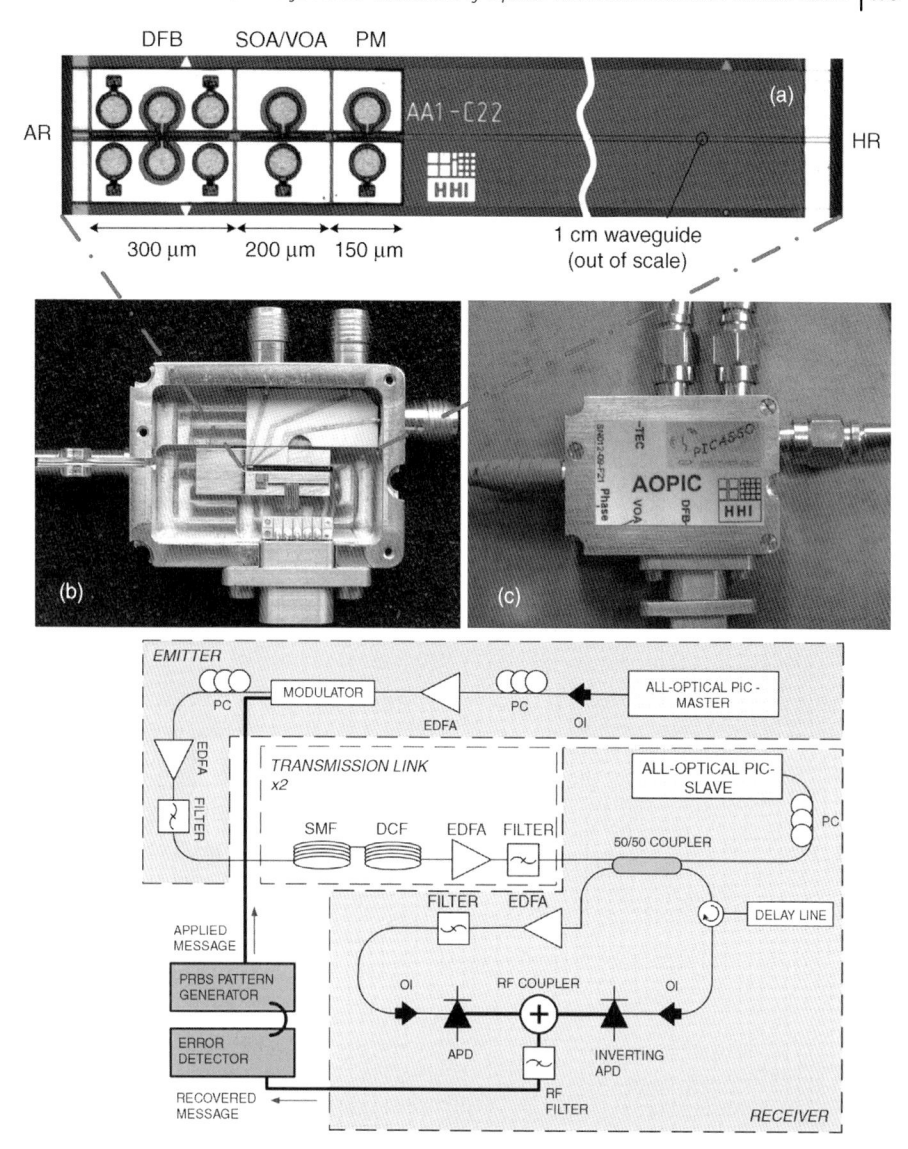

Figure 8.14 (Top) Photonic monolithic integrated chaos generator: (a) Device structure. AR: antireflective coated, DFB: distributed feedback semiconductor laser, HR: highly reflective coated, PM: phase modulator, SOA: semiconductor optical amplifier, VOA: variable optical attenuator. (b) Internal structure of the packaging module. (c) Packaged module. (Bottom) Topology of the chaos-secured data transmission experiment using PIC chaos generators. Thin lines: optical fibers, thick lines: electrical cables, OI: optical isolator, PC: polarization controller, EDFA: erbium-doped fiber amplifier, SMF: single-mode fiber, DCF: dispersion compensating fiber, 50/50 COUPLER optical coupler, APD: avalanche photoreceiver. (A. Argyris, E. Grivas, M. Hamacher, A. Bogris, and D. Syvridis, (2010c), © 2010 OSA.)

the round-trip time of the cavity with subwavelength resolution; this operation is extremely crucial for synchronization with other matched devices that might exhibit a small mismatch in the cavity length and optical feedback phase. Precise thermo-electric cooling of the devices, by using the appropriate packaging for the PICs (see Figures 8.14b and c), provides controllability and operating long-term stability, not only in terms of wavelength and optical power, but also in terms of the spectral distribution of the chaotic carrier and the feedback phase-matching conditions. All the above are vital parameters for the accomplishment of high-performance synchronization of chaos in the closed-loop configuration.

The PIC modules support closed-loop receiver configurations, which presume a PIC replica of the transmitter at the receiver. In the case of comprehensive device identity, enhanced synchronization performance can be obtained, compared with open-loop receiver architectures that incorporate only an identical laser section and exclude the existence of any external cavity. This discrepancy accredits increased privacy in closed-loop communication systems, since an eavesdropping receiver becomes more demanding in its specifications (Soriano et al., 2009) (see Section 8.4). The standardized fabrication process of the designed PICs guarantees a tolerable matching of the internal parameters, the operating characteristics, and the external cavity length for all devices; however, even from the same fabrication wafer, only a few matched-pairs out of dozens of devices prove to be identical in order to provide efficient synchronization of chaos in the closed-loop configuration.

8.3.2.2 Chaos Communication Experiment

The structure of a 100-km transmission communication system based on chaotic carriers generated from PICs has been implemented, as illustrated in Figure 8.14 (bottom). Both PICs, selected at the transmitter (emitter) and the receiver, emit optical carriers exactly at 1556.111 nm. The encoded data is a PRBS with tunable amplitude and $2^{31}-1$ length and is applied externally using a $LiNbO_3$ electro-optic modulator (EOM). The optical signal is amplified by an EDFA, followed by a bandpass optical filter in order to suppress amplified spontaneous emission, and then launched into the transmission link. Two identical dispersion-compensated transmission links in series, each one consisting of 50 km single-mode fiber (SMF), 6 km dispersion compensation fiber (DCF) and the prerequisite amplification stages, channel the chaos-encrypted data to the receiver. The optical power of a broadband signal that is sent for transmission in fiber links of tens or hundreds of km is a crucial parameter. Optical signals with low power are vulnerable to the noise coming from the amplifiers, while high-power optical signals trigger nonlinear effects that alter the transmitted chaotic waveforms. Both extreme cases result in a worse synchronization performance at the receiver. The input optical power of 4 mW is used in order to maintain the best synchronization performance and to minimize the effects of the transmission impairments. The receiver synchronizes only to the fluctuations of the chaotic signal after injecting a small fraction of the transmitted signal into the receiver's PIC. The temporal alignment through an optical delay line and power equalization through amplification of the two detected signals lead to chaotic carrier cancelation and message data recovery. The subtraction is practically performed by adding the electrically converted signals, after inverting the transmitter's one by using an inverting photoreceiver. The appropriate microwave filters have been used at the recovery stage to suppress the undesirable residual spectral components coming from the subtraction and always the bandwidth is matched to the data bit rate: 1.1 GHz for 1.25 Gb/s data, and 2 GHz for 2.5 Gb/s data, respectively.

8.3.2.3 Forward Error Correction Technique

The receiver has the task to cancel the chaotic carrier and recover data with a native bit-error rate as low as needed in order to acquire bit-error rates below 10^{-12} (i.e., error-free operation) using coding techniques. The BER improvement is achieved by an additional processing unit that employs a fast transceiver that enables forward error-collection (FEC) methods (Wilson, 1996). The FEC method used for improving the inherent recovered data employed the Reed–Solomon (RS) code (255 223), where 32 check symbols were added to every 223 symbols of useful information, leading to an overhead of 14.35%. The size of the Galois field used is 2^8 (256) with 8-bit long symbols. This configuration allows the correction of up to 16 symbols per codeword (255 symbols), meaning that an error-free operation can be achieved while the input BER is as high as 6.3×10^{-2}. This theoretical threshold value applies to the ideal situation where there are exactly 16 erroneous symbols in every codeword. For the specific system studied the threshold value was found to be $R = 1.8 \times 10^{-3}$ due to burst errors. The effective data rates for the plain link rates of 1.25 Gb/s and 2.5 Gb/s are 1.09375 Gb/s and 2.1875 Gb/s, respectively. The FEC scheme has been implemented along with the BER tester in a field programmable gate array (FPGA) device.

Small message amplitudes that externally modulate the chaotic carrier can be completely hidden within the chaotic signal. The message amplitude is of high importance in this encryption technique, providing a two-fold benefit: low message amplitudes cannot be distinguishable by tapping the transmission line and are not susceptible to filtering, while at the same time they favor successful recovery only by an authorized user with an identical receiver. Small-amplitude data sequences with SNR as low as 12 dB may lead to partial recovery, which can be significantly improved to error rates below 10^{-12} using FEC techniques.

8.3.2.4 Bit-Error-Rate (BER) Performance

The evaluation of this chaotic-PIC-based communication system (Figure 8.14 (bottom)) in terms of data encryption and recovery is performed through BER *versus* the applied message amplitude, as presented in Figure 8.15. This analysis does not follow the typical form of BER curves that characterize conventional optical communication systems that is expressed *versus* the optical power at the receiver. In the present work, the major investigation is to simultaneously achieve data encryption along the transmission line and efficient data recovery (postprocessed to BER values lower than 10^{-12}) by an authorized receiver.

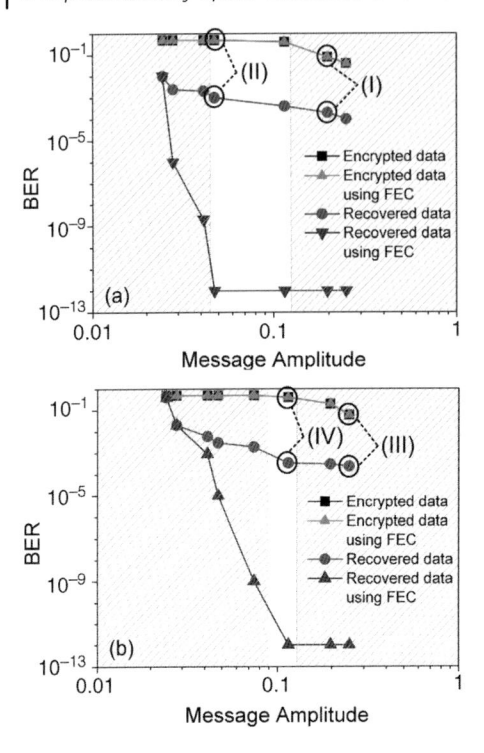

Figure 8.15 System performance in terms of data encryption and recovery: Bit-error-rate (BER) measurements vs. data amplitude for (a) 1.25 Gb/s and (b) 2.5 Gb/s PRBS. Cases II and IV ensures complete encryption with BER ~0.5, while the recovered data have an inherent BER value below 10^{-3} that is enhanced by FEC methods to error-free operation. For larger message amplitudes (cases I and III) the probability of error for the encrypted data is below 0.5 providing reduced security, however maintaining an error-free data recovery. (A. Argyris, E. Grivas, M. Hamacher, A. Bogris, and D. Syvridis, (2010c), © 2010 OSA.)

In Figure 8.15a, a 1.25 Gb/s data stream is characterized *versus* its modulation amplitude, signifying three different operating regions. The first one involves large message modulation amplitudes (over 12%) of the chaotic carrier's mean optical power. Such amplitude values provide efficient data recovery, with inherent BER values below $R = 1.8 \times 10^{-3}$ (case I in Figure 8.15a). The error threshold value R is determined under the condition where a BER value of at least 10^{-12} will finally emerge by applying the FEC method using the appropriate transceiver. However, such large amplitudes act at the expense of the quality of message encryption that is below 0.5, indicating that an eavesdropper could extract some bits with direct detection. The second operating region with message amplitudes from 12% down to 4.5% ensures what is desired: an encryption level that corresponds to a tapped BER value equal to 0.5, recovered data at the receiver with an inherent BER below the R

value and an after-FEC BER performance of 10^{-12} at maximum (case II in Figure 8.15a). Finally, even smaller amplitudes (below 4.5%) that also preserve secure data encryption cannot lead to efficient recovery, since the inherent BER values are above the value R, providing errors at the decoding process even for the authorized receiver that uses the FEC technique.

An analogous study is performed after doubling the data rate to 2.5 Gb/s, as presented in Figure 8.15b. The extended bandwidth of the message induces a different encryption efficiency compared to the previous case; however, as previously, large message amplitudes cannot be sufficiently encrypted (case III in Figure 8.15b). A secure transmission can be obtained in a much narrower operating window of amplitude values around 10% with an encryption level of the BER value of 0.5 and an after-FEC satisfactory performance of the BER value of 10^{-12} (case IV in Figure 8.15b).

8.3.2.5 Analysis for Unauthorized Users

The above performance analysis is the main reason for claiming an additional layer of security in optical communication systems, which are summarized in Table 8.4. For security reasons the message amplitude at the transmitter is set to low values that allow the authorized receiver to recover a native BER just below $R (= 1.8 \times 10^{-3})$ threshold. This BER value is the "digital threshold" set for the security of the system, is associated to the operating characteristics of the FEC method and when applied in the system pushes down the BER values to 10^{-12}. The message amplitude should be set to 4.5% for 1.25 Gb/s data series, and 10% for 2.5 Gb/s data series, as shown in Figures 8.15a and b. In this condition, the authorized user can obtain the native BER less than $R (= 1.8 \times 10^{-3})$ and the BER less than 10^{-12} after FEC. For unauthorized users, it is assumed that an eavesdropper has exactly the same laser with the transmitter, but does not have any information about the external cavity. This situation can be realized that the authorized PIC of the receiver is operated it in an open-loop configuration. In this situation, the best values for the recovered BER are 2×10^{-2} and 8×10^{-2}, which are higher than R threshold, for bit rates of 1.25 Gb/s and 2,5 Gb/s, respectively, The eavesdropping receiver fails to exceed the FEC threshold limit R and thus fails to improve the decoding performance. Finally, unauthorized users may gain access to the transmission line by tapping the physical medium. It is found that they should only be capable (after linear or nonlinear filtering) of recovering data with BER values of ~ 0.5. This value practically means total randomness of the extracted bits.

An essential issue for exploiting such systems in real-world conditions is their operational stability and sustainability over long periods of time. Chaotic carriers should preserve their characteristics that ensure the initial encryption, while synchronization between chaotic carriers after including transmission links should be precisely controlled. A stability analysis performed in the presented system shows that the incorporated chaotic PICs prove to be a reliable solution in terms of securing fiber transmission links, after operating continuously for hundreds of hours, thus guaranteeing a stable encryption and recovery performance.

Table 8.4 Security allocation of different types of users *via* their decoding performance.

Type of User	Action Performed	Decoding Performance	
		Native BER	BER After FEC
Authorized	Identical hardware receiver	$<R$	$<10^{-12}$
Unauthorized	Nonidentical hardware receiver	$>R$	$>R$
Unauthorized	Carrier filtering	~ 0.5	~ 0.5

The threshold value of BER, which guarantees the BER less than 10^{-12} by using the forward error correction (FEC) technique, is $R = 1.8 \times 10^{-3}$. (A. Argyris, E. Grivas, M. Hamacher, A. Bogris, and D. Syvridis, (2010c), © 2010 OSA.)

8.3.3
Optical Phase Chaos for 10-Gb/s Chaos Communication

Successful 10 Gb/s transmission of a message hidden in a chaotic optical phase over more than 100 km of an installed fiber-optic network has been reported (Lavrov et al., 2010). Phase modulators are used for encoding and decoding the message signal, instead of intensity modulators. The setup depicted in Figure 8.16 (top) shows the transmitter–receiver architecture of the phase chaos communication system (Lavrov et al., 2010) (see also Section 3.3.4). It consists of a standard communication link based on differential phase shift keying (DPSK) modulation. Note that any wavelength-division multiplexing (WDM) channel can thus be selected through the use of the proper external laser source, independently of the subsequent chaos communication processing. A message phase modulator (MΦM in Figure 8.16 (top)) performs the binary DPSK phase modulation φ_m corresponding to the message to be transmitted. It is worth noting here that MΦM could be equivalently placed before seeding the chaotic oscillator, or inside this oscillator.

The encoding process consists of the superposition of the message phase modulation and the chaos phase modulation, the latter being partly determined also by the message phase modulation. The chaos-encoding technique used in this scheme is thus of the same kind of in-loop addition approaches (i.e., chaos modulation). The phase chaos generator (ΦCG in Figure 8.16 (top)) performs the modulation of the DPSK message. At the receiver side, phase chaos cancelation (ΦCD in Figure 8.16 (top)) is processed from the input light beam, after which a standard DPSK demodulation MΦD recovers the original binary message. Note that if ΦCG and ΦCD are deactivated, the communication link becomes a standard optical DPSK transmission (chaos modulation can be switched on and off).

A brief description of the novel chaos-generation process at the transmitter is as follows. The dynamics belongs to the class of nonlinear delay differential equations (see Section 3.3.4). A nonlinear transformation is performed while modulating the phase condition in an interferometer. Instead of implementing this interferometer with an electro-optic Mach–Zehnder, the interferometer function and the phase

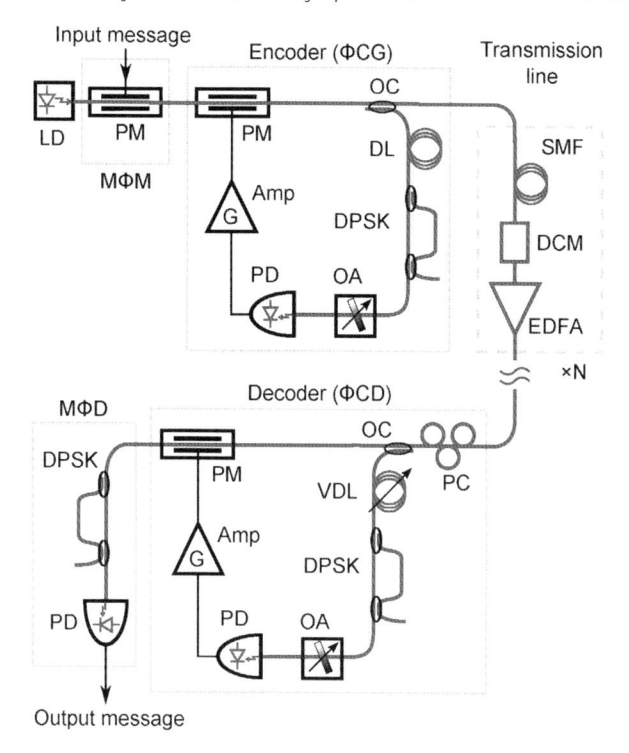

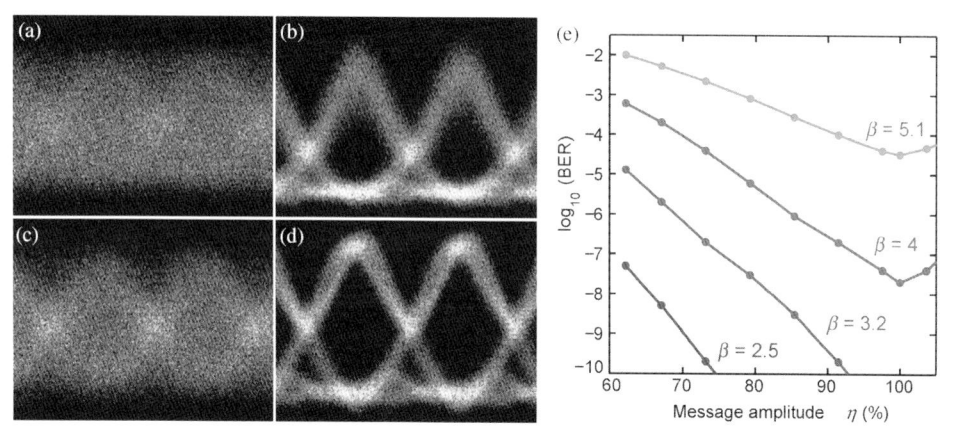

Figure 8.16 (Top) Point-to-point transmission setup using electro-optic phase chaos. $M\Phi M$: message phase modulation, ΦCG: phase chaos generation and masking, ΦCD: phase chaos cancellation and demodulation, LD: laser diode, PM: phase modulator, DPSK: differential phase shift keying demodulator, OA: optical attenuator, PD: photodiode, Amp: telecom driver, OC: optical coupler, SMF: single-mode fiber channel, EDFA: erbium doped fiber amplifier, DCM: dispersion compensation module, PC: polarization controller, VDL: variable delay line). (Bottom) 10 Gbps back-to-back experimental results. Eye diagrams for (a), (c) unauthorized receiver (direct detection) and (b), (d) authorized receiver for different message amplitudes: (a), (c) $\eta = 60\%$ and (b), (d) $\eta = 100\%$ at $\beta = 4$. (e) BER dependence on message amplitude for an authorized receiver for different β values. (R. Lavrov, M. Jacquot, and L. Larger, (2010). © 2010 IEEE.)

modulation operation are split (see ΦCG in Figure 8.16 (top)). The actual interferometer (DPSK in Figure 8.16 (top)) is imbalanced with a delay δT, leading to an interference condition ruled by the phase difference between times t and $t-\delta T$. As soon as the electro-optic phase modulation is performed faster (response time τ) than the time imbalancing ($\delta T \gg \tau$), a dynamic scanning of the nonlinear interference modulation transfer function is achieved. This modified Ikeda dynamics is inspired by conventional DPSK modulation and demodulation techniques. Note, however, that *a priori* any multiple wave imbalanced interferometer can be used as a hardware key determining the chaos-generation process.

At the receiver side, phase chaos cancelation is achieved using the open-loop receiver scheme: the receiver is aimed at replicating the transmitter nonlinear delay processing path (with a minus sign). At the end of this path, the receiver electro-optic phase modulation is applied onto the incoming light beam, with the aim of suppressing the chaos phase modulation performed at the transmitter. Due to the antiphase replicated nonlinear delayed processing, chaos cancelation at the output of the receiver phase modulator is successfully achieved, when all receiver parameters are matched as closely as possible with those set at the transmitter. The stability of the synchronization manifold can be derived, and the solution for this synchronization manifold is found to be stable, resulting in a synchronized receiver, and thus in a canceled chaotic phase modulation.

The standard DPSK light beam after the receiver phase modulator, is practically processed by a data-rate-matched DPSK demodulator, resulting in a binary intensity modulated light beam. The latter is finally detected by a photodiode and electronically amplified. When recovering the demodulated binary data stream in the electronic domain for BER measurements, a 7.73-GHz low-pass filter, whose cut-off is matched with the spectrum of 10 Gb/s data, is applied for both the authorized and nonauthorized receiver. This signal is the one used for the eye diagrams in Figures 8.16a–d (bottom). Chaos modulation is sufficient to strongly perturb the eye diagram as shown in Figures 8.16a and c (bottom), whereas chaos cancelation allows for nicely recovered eyes in Figures 8.16b and d (bottom).

When adjusting the operating parameters of the chaos encoding and decoding, one has to deal with a trade-off in practical chaos communications: masking (or embedding) efficiency *versus* decoding quality. Both are tuned by the relative weight of the message phase-modulation amplitude ($\eta\pi$ at MΦM, where $\eta = 1$ is the optimum for standard DPSK transmission), and the phase chaos amplitude (tuned by β). Figure 8.16e (bottom) shows the BER performance on message amplitude η for different chaos amplitude β. As illustrated in Figure 8.16e, the decoding quality rises significantly on increasing the message amplitude η (see also Figure 8.16b and d (bottom)). However, one can note an impact on the security: the shape of the bits becomes more visible, as shown in Figures 8.16a and c (bottom). For $\beta = 4$ and message amplitude $\eta = 60\%$, only 10^{-3} BER is obtained, and one would need to push the message amplitude to $\eta = 100\%$ so that it can be properly extracted from the chaotic carrier (BER is then close to 10^{-8}, as seen in Figure 8.16e (bottom)).

For field experiments, dispersion issues are critical, and successful data recovery is only possible with accurately tuned dispersion compensation modules (in addition to an EDFA for loss compensation). The polarization control (using standard loop controllers) also had to be used appropriately. No other particular fiber channel disturbance is noticed. More specifically, since the transmission is based on fast differential phase modulation, no influence of slowly varying phase perturbations is found while the optical signal is traveling through the long fiber channel. Out of channel disturbances, long-term stable operation of the encoding and decoding requires the use of control systems for the operating point of the DPSK demodulators.

The field experiment has been performed on an optical network in Athens, Greece (see Figure 8.6a (bottom)). In this case the fiber loop path is close to 120 km and involved two EDFAs and two dispersion compensation units. All the fibers need to be fixed properly before the polarization state is stable enough and do not require any adjustment during several tens of minutes. Under these conditions a BER of 10^{-6} can be achieved at 10 Gb/s for $\beta = 2$. These 10 Gb/s field experiments, however, do not involve all the currently available optical signal processing techniques for 10 Gb/s link optimization. Improved transmission conditions in the frame of laboratory experiments have achieved error-free (BER of 10^{-12}) 10-Gb/s phase-chaos communications with a 70-km fiber spool. This investigation shows that the performances of phase chaos communication can be even further improved.

8.3.4
Comparison of the Encoding and Decoding Schemes for Chaos Communication

The comparison of the performance of the three encoding and decoding methods (chaos masking, chaos modulation, and chaos shift keying) has been reported (Liu et al., 2002a; Larson et al., 2006). Figure 8.17 represents the three encoding and decoding schemes applied to the semiconductor laser system with optoelectronic feedback. The same configuration can be used and compared for the three encoding and decoding schemes except for the place where the message is encoded, as shown in Figure 8.17.

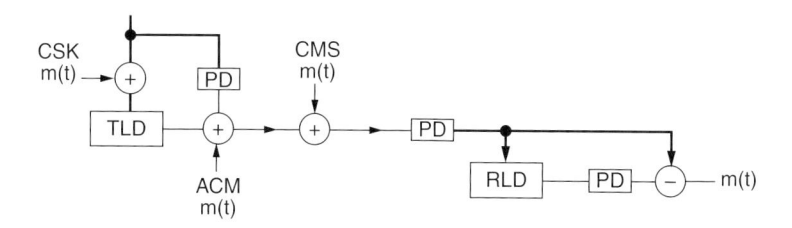

Figure 8.17 Schematic diagrams of three encoding and decoding methods in synchronized chaotic optical communication systems using semiconductor lasers with optoelectronic feedback. Thick lines indicate the electrical paths. Thin lines indicate the optical paths. CSK: chaos shift keying. CMS: chaos masking. ACM: additive chaos modulation. PD: photodetector–amplifier combination. TLD: transmitter laser diode. RLD: receiver laser diode. $m(t)$: encoding messages. (J. M. Liu, H. F. Chen, and S. Tang, (2002a), © 2002 IEEE.)

The system performance of the three encoding and decoding schemes has been numerically investigated at the OC-192 standard bit rate of 10 Gb/s in the presence of channel noise and internal laser noise. Figure 8.18 (left) shows the time series of the original message and the decoded messages for different encryption methods (Liu et al., 2002a). The recovery of the high-bit-rate message for chaos shift keying (CSK in Figure 8.18 (left)) is not possible, because the resynchronization time after a desynchronization burst has to be shorter than the bit duration for a following bit to be recoverable. For chaos masking (CMS in Figure 8.18 (left)), the errors in message recovery are primarily generated by the timing errors that are caused by the breaking of the symmetry between the transmitter and the receiver due to the presence of the message. The recovered message is thus contaminated by frequent spikes. The performance of chaos modulation (ACM in Figure 8.18 (left)) is the best among the three encryption schemes. The error bits in the recovered message for chaos modulation are caused by synchronization error due to the channel noise and the laser noise. No desynchronization bursts are observed in the recovered message for chaos modulation.

The communication system performance measured numerically by BER as a function of SNR is shown in Figure 8.18 (right). A BER lower than 10^{-5} can be obtained for chaos modulation (ACM) when the SNR is larger than 38 dB, whereas BER is always higher than 10^{-1} for chaos masking (CMS) and chaos shift keying (CSK). Among the three encryption schemes, it is found that only the performance of chaos modulation is acceptable at 10 Gb/s because chaos modulation allows

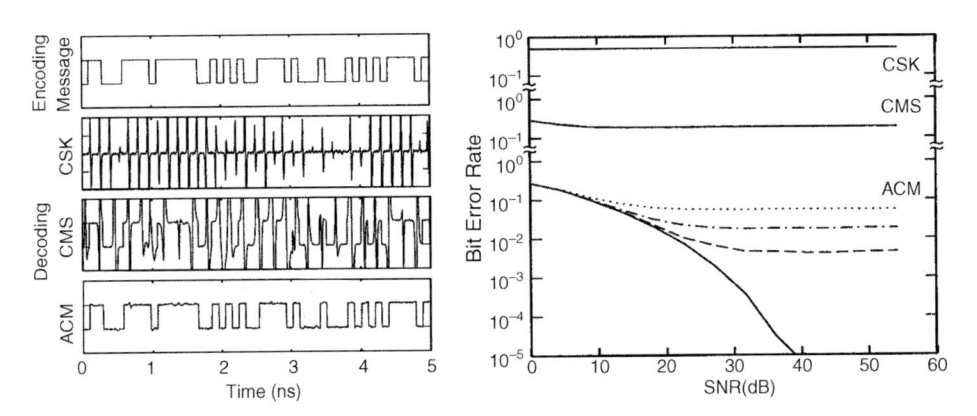

Figure 8.18 (Left) Time series of the decoded messages of the three encoding and decoding methods in the optoelectronic feedback system. CMS: chaos masking. ACM: chaos modulation. CSK: chaos shift keying. (Right) Bit-error rate (BER) versus signal-to-noise ratio (SNR) for the three encoding and decoding methods in the optoelectronic feedback system. CMS: chaos masking. ACM: chaos modulation. CSK: chaos shift keying. Solid line: obtained when the laser noise is absent. Dashed line: obtained at the laser noise level of 100 kHz (represented as the equivalent laser linewidth) for both the transmitter and receiver lasers. Dot-dashed line: obtained for the laser noise level of 1 MHz. Dotted line: obtained for the laser noise level of 10 MHz. (J. M. Liu, H. F. Chen, and S. Tang, (2002a), © 2002 IEEE.)

high-quality synchronization in the process of message encoding by maintaining the mathematical identity between the transmitter and the receiver (see Section 7.4.2). Both chaos shift keying and chaos masking cause significant desynchronization bursts or synchronization deviation in the systems because they break the identity between the transmitter and the receiver in the process of message encoding. The best performance is thus obtained for chaos modulation in the three encoding and decoding methods.

8.4
Privacy Issues in Optical Communication with Chaotic Lasers

8.4.1
Introduction

The achievement of privacy is an important motivation for chaos communication research. In his pioneering paper, "Communication Theory of Secrecy Systems", Claude Shannon discussed three aspects of secret communications systems: "concealment", "privacy", and "encryption" (Shannon, 1949; Hellman, 1977; Welsh, 1988; VanWiggeren and Roy, 1999). These aspects apply to systems that use chaotic waveforms for communication and can be interpreted in that context. Concealment of the information occurs because the chaotic carrier or masking waveform is irregular and aperiodic; it is not obvious to an eavesdropper that an encoded message is being transmitted at all. Privacy in chaos communication systems results from the fact that an eavesdropper must have the proper hardware and parameter settings in order to decode and recover the message. In conventional encryption techniques, a key is used to alter the symbols used for conveying information. The transmitter and receiver share the key so that the information can be recovered. In chaos communication systems, a transmitter generates a time-evolving chaotic waveform to mask information symbols. The information can be recovered with a receiver possessing the nearly identical hardware key, which is the knowledge of its configuration and parameter settings of the transmitter. The key in chaos communication is a set of parameter values as well as similar hardware structure that are required for chaos synchronization and cancelation. In this sense, the key is static even though the time-varying chaotic carrier is used in chaos communication.

Two factors that are important to privacy considerations in chaos communication systems are the dimensionality of the chaos and the effort required to obtain the necessary parameters for a matched receiver. Earlier work has shown that for certain chaos communication techniques, particularly those involving addition of a message to the chaotic carrier, the message can be recovered from the transmitted signal by mathematically reconstructing the transmitter's chaotic attractor if the chaos is low dimensional (Short, 1994; Pérez and Cerdeira, 1995; Short, 1996). Higher-dimensional signals, especially those involving hyperchaotic dynamics, are likely to provide improved privacy. The number of parameters that have to be matched for information recovery and the precision with which they must be matched are important aspects of receiver design (Yoshimura, 2004) (see Section 8.4.3). In fact, the privacy of chaos communication techniques is a complex and involved issue. Several attempts at the measurement and evaluation of privacy are described in Section 8.4.

8.4.2
Reconstruction of Model and Parameter Settings by Time-Series Analysis

An interesting analysis of data from a chaotic fiber-laser system has been perfored (Geddes *et al.*, 1999). Encoded waveforms and decoded data are provided, together with a detailed description of the configuration and parameters for the chaotic fiber-laser system shown in Figure 8.11 (left) (VanWiggeren and Roy, 1999). It has been shown that in the parameter regime when the double-loop geometry resulted in stable steady-state operation of the laser output without message modulation, one can construct an accurate computational model and recover the message encoded in the transmitted signal. In this case, the laser dynamics is governed almost entirely by the modulation signal that is echoed in the two loops, and nonlinear effects can be neglected to a good approximation (Geddes *et al.*, 1999). It is, however, presumably difficult to estimate all the parameter values from the chaotic temporal waveforms with message modulation in the nonlinear dynamical regime.

8.4.3
Parameter Estimation by Using Similar Hardware

Conventional cryptologists have a formal definition for security that is difficult to apply directly to chaos communication schemes, since chaos communication is more likely steganography (concealing the "existence" of a message) rather than cryptography (scrambling the "meaning" of a message), as described in Section 7.1. However, an analogy between an attack for software-based cryptography and that for chaos communication can be introduced as a measure of privacy. One of the simplest cryptanalytic attacks is called the "ciphertext-only attack" for software-based cryptography (van der Lubbe, 1998), where the cryptanalyst only has access to the encrypted signal (i.e., chaos plus message signals for chaos communication). The privacy level of chaos communication can be estimated by the analogy of the ciphertext-only attack as an example.

Suppose that an eavesdropper has a similar laser device but no information about all the parameter settings (i.e., hardware key) for chaos communication systems. This is a similar situation to software-based cryptography, where the algorithm is open and the key is hidden. The eavesdropper may search all the combination of parameter settings in order to obtain synchronization of chaos for message extraction. In this situation, the privacy of chaos communication relies on the two following factors:

(a) the number of adjustable laser parameters required for synchronization of chaos, and (b) the ratio of the parameter range where synchronization can be achieved, to the whole parameter range.

Let us consider the situation where there are n independent variable parameters to be adjusted for synchronization, and the tolerance of parameter values for synchronization against parameter mismatch is defined as p ($0 \leq p \leq 1$). p is the ratio of the synchronizable parameter range to the whole parameter range for one parameter value. An eavesdropper can repeat searching one parameter value at $1/p$ times to find the correct parameter setting where synchronization is achieved. The number of all the combination of the parameter search for synchronization is thus estimated as $(1/p)^n$ for n independent variable parameters. This indicates that the eavesdropper is able to find all the correct parameter settings after the $(1/p)^n$th attempts of the parameter search for the worst case. For example, when a chaotic laser system has the values of $p = 0.1$ and $n = 10$, the number of all the combination of the parameter search can be estimated as $(1/p)^n = 10^{10} \approx 2^{33}$. Therefore, this chaos communication system has the security level that is comparable to the 33-bit software encryption key, in terms of the attack of parameter search (i.e., ciphertext-only attack). Note that it takes much more time to search all the possible parameter values for hardware attack than to scan the encryption key in a computer for software attack.

This is the least favorable situation for an attacker and indicates the worst scenario for the legitimate users, since the chaos communication systems based on synchronization of chaos would be broken after $(1/p)^n$ attempts of the parameter search by attackers. However, there may be some internal parameters that can be only matched during fabrication process, but cannot be matched after the device was made, as a "hidden" hardware key. The existence of such internal parameters makes the attack more difficult. This assumption, however, contradicts the fact that all the system configurations need to be open, such as software-based cryptography. A special situation may be considered where there exist some specific internal parameters that are hidden and unadjustable, and the system presumably has much more privacy, although the quantitative privacy estimation becomes more difficult. It would be an interesting situation by using such a hardware key for implementing extremely secure communication for a specific purpose.

To increase the level of privacy in the proposed optical chaos communication systems, two methods can be considered: increase of the number of the independent variable parameters (large n), and decrease of the parameter region for synchronization against parameter mismatch (small p). As an example, the first situation (large n) can be achieved by using cascaded laser systems, where many chaotic optical devices are introduced and sequentially coupled (Yoshimura, 2004). The second situation (small p) can also be obtained by using a closed-loop feedback semiconductor laser system, instead of the open-loop configuration as an example (Soriano et al., 2009). The matching of the optical feedback phase between the transmitter and the receiver is very critical to achieve synchronization in the closed-loop configuration, and the parameter range of the optical feedback phase where synchronization is achieved can be narrow. The closed-loop feedback system may require the use of stable external cavities, which can be attained by using photonic integrated circuits as described in Section 8.3.2 (Argyris et al., 2010c).

8.4.4
Parameter Estimation by Time-Series Analysis

Another method for the attack on a chaos communication system is the estimation of system-parameter values by using time-series analysis. Several attempts of time-series analysis have been demonstrated to speculate on some of the laser parameter values in chaos communication schemes.

A new type of modular neural network has been used to obtain the transmitter nonlinear dynamics in a DFB laser with a time-delayed electro-optic feedback loop (Ortín et al., 2009). To estimate the delay time of the feedback loop, a neural network is introduced that consists of two modules: one for the nonfeedback part with input data delayed by the sampling time, and a second one for the feedback part with input data delayed by the feedback time. The neural network is trained and a training error around 5% is achieved for all the available values of the nonlinear strength. The nonlinear function of the type $\cos^2 x$ (see Section 8.2.3) is extracted from the neural network, although the dimension of the chaotic attractor is very high (>100). The message can be recovered by using the neural-network model as an unauthorized receiver. This result suggests that chaotic cryptosystems based on electro-optic feedback with one fixed time delay are not safe from the attack of time-series analysis.

Another technique of extracting a message from chaos is the use of partial differential equations. A message signal encoded on chaotic carrier intensity by chaos masking has been extracted by using partial differential equations (Jacobo et al., 2010). A characteristic is used for message detection that the message is encoded in chaotic carrier intensity such that the mean power of a bit 1 differs from that of bit 0. The Ginzburg–Landau equation in one dimension with an external forcing is used as a nonlinear filter to find changes in the mean value and to recover the message. This succeeds in extracting the open eye diagram of the recovered message signal from the transmission signal. This result suggests that the mean value (optical power) of the transmission signal needs to be fixed in the presence of the message component. The modulation to optical phase may be more preferable than optical intensity for message encoding to protect against this type of attack.

An estimation of the delay time is crucial for chaos communication systems with time-delayed feedback. The delay time can be estimated from autocorrelation and/or mutual information of chaotic time series (Rontani et al., 2007). A novel approach to identify delay phenomena in noisy time series has been investigated. It is possible to perform a reliable time-delay identification by using quantifiers derived from information theory, more precisely, permutation entropy and permutation statistical complexity (Zunino et al., 2010). These quantifiers show clear extrema when the embedding delay time matches the characteristic delay time of the system. The identification of the time delay is even more efficient in a noise environment.

To avoid the delay-time identification and enhance the privacy, the introduction of double time-delayed feedback loops in semiconductor lasers may be effective, instead of a single delay (Lee *et al.*, 2005; Wu *et al.*, 2009). A scheme of delay-time modulation has been also proposed for privacy enhancement (Kye *et al.*, 2004).

8.4.5
Direct Detection of Presence of Message from Time-Series Analysis

One of the measures for privacy in chaos communications is the direct detection of the presence of a message signal in a chaotic carrier. Although it would be desirable to directly extract the message it is always important to know in advance whether any information is hidden or not into the chaotic carrier, especially in those cases where multiple chaotic signals are being transmitted simultaneously. This initial test would also prevent the extra effort required to extract the message in case the message is not present in the carrier. So, it is important to evaluate the encryption efficiency that is referred to as the ability of the scheme to hide the presence of the message. In this sense, this work aims to provide an upper bound for the message amplitude in terms of encryption purposes.

One of the simplest approaches for the direct detection of the message embedded on the chaotic carrier is filtering. A low-pass or bandpass filter can be applied for the chaotic carrier with message and the bit-error rate is measured from the filtered signal. This attempt has been done to determine the proper message amplitude and to avoid an attack with direct detection of chaotic carrier in chaos communication systems, as already described in Section 8.3 (Argyris *et al.*, 2010a; Argyris *et al.*, 2010c; Lavrov *et al.*, 2010). It has also been shown that smaller message amplitude can be used in the closed-loop feedback system than the open-loop feedback system to avoid an attack of the direct observation of the transmission signal (Soriano *et al.*, 2009). This result indicates that higher privacy can be achieved for the closed-loop system than for the open-loop system.

More sophisticated attempts have been carried out to estimate the presence of a message signal embedded on a chaotic carrier by using a multifractal approach (Zunino *et al.*, 2009) and an information theory approach (Rosso *et al.*, 2008). For the multifractal approach, it has been shown that the multifractal indicators are powerful tools for quantifying the efficiency of encryption schemes (Zunino *et al.*, 2009). Under some circumstances, the proposed measures are useful approaches to distinguish the presence of a hidden message within a chaotic carrier. In particular, the "Hurst exponent" seems to be a powerful tool for detecting and quantifying the presence of this hidden message. It allows a very clear discrimination of the signals with message amplitudes of the order of 5–10% or larger when compared to pure chaotic carriers without message (0% amplitude). From the multifractal approach it is found that the chaos modulation encoding technique is better that the chaos shift keying. For the same amplitude, messages encoded with the former technique are more difficult to detect. The multifractal approach appears to be a suitable and versatile way of assessing the encryption efficiency for the optical chaos communication schemes.

For the information theory approach, the computation of two measures based on information theory grounds has been carried out to assess the performance of chaos modulation and chaos shift keying schemes (Rosso et al., 2008). In particular, a disorder and complexity quantifiers (normalized Shannon entropy and intensive MPR-statistical complexity) are evaluated for different time series generated by a chaotic laser in which a message is encrypted. One-way analysis of variance has proved that both measures successfully detect the presence of a message, provided that a message amplitude larger than 10% and the proper sampling time are used. By contrast, message amplitudes smaller than 10% are almost undetectable for both chaos modulation and chaos shift keying schemes although chaos modulation appears to be more secure. The statistical measures offer criteria to decide for optimum encoding techniques.

8.4.6
Summary of Privacy Issues

In Section 8.4 several attempts for privacy evaluation of chaos communication systems are described and discussed. It has, however, still been difficult to assess the privacy or security offered by any specific schemes for chaos communication. In particular, a comprehensive comparison is still lacking between the security levels offered by chaotic encryption and those of standard software-based cryptography. The main reason for this is that the security measures introduced to quantify the efficiency of software-based encoding are too much tailored to the key-distribution process, which is something that chaos communication schemes have not addressed so far. Moreover, chaotic waveforms provide an additional layer or level of privacy beyond any conventional key distribution or encryption scheme that can be simultaneously part of the communication protocol. It is hoped that eventually there will be comprehensive research on the security of methods of communication that use chaotic dynamical systems.

8.5
Photonic Integrated Circuit for Optical Communication with Chaotic Lasers

The trend in the optical communication systems is towards robust and sophisticated photonic integrated circuits (PIC). Monolithic PICs that are capable of producing broadband high-dimensional chaotic dynamics need to be designed, fabricated, and investigated for advanced chaos communication systems (Argyris et al., 2008). Dynamics can be easily controlled experimentally *via* the phase current and feedback strength, therefore establishing this device as a compact integrated chaos emitter. Since the dynamics are well identified, the advantages of the PICs are fully exploited to the benefit for applications to chaos-based optical communications. Some examples of PICs are described in Section 8.5.

8.5.1
Photonic Integrated Circuit for a Semiconductor Laser with all-Optical Feedback

The PIC proposed in this section incorporates the fundamental principles of the all-optical feedback-induced chaos, using elements that provide the capability to control the chaotic properties of the optical emission accurately and reproducibly (Argyris et al., 2008). It consists of four successive sections as shown in Figure 8.19 (left): A DFB InGaAsP semiconductor laser operating at 1561 nm, followed by a gain-absorption section (GAS), a phase section (PH) and a 1-cm-long passive waveguide.

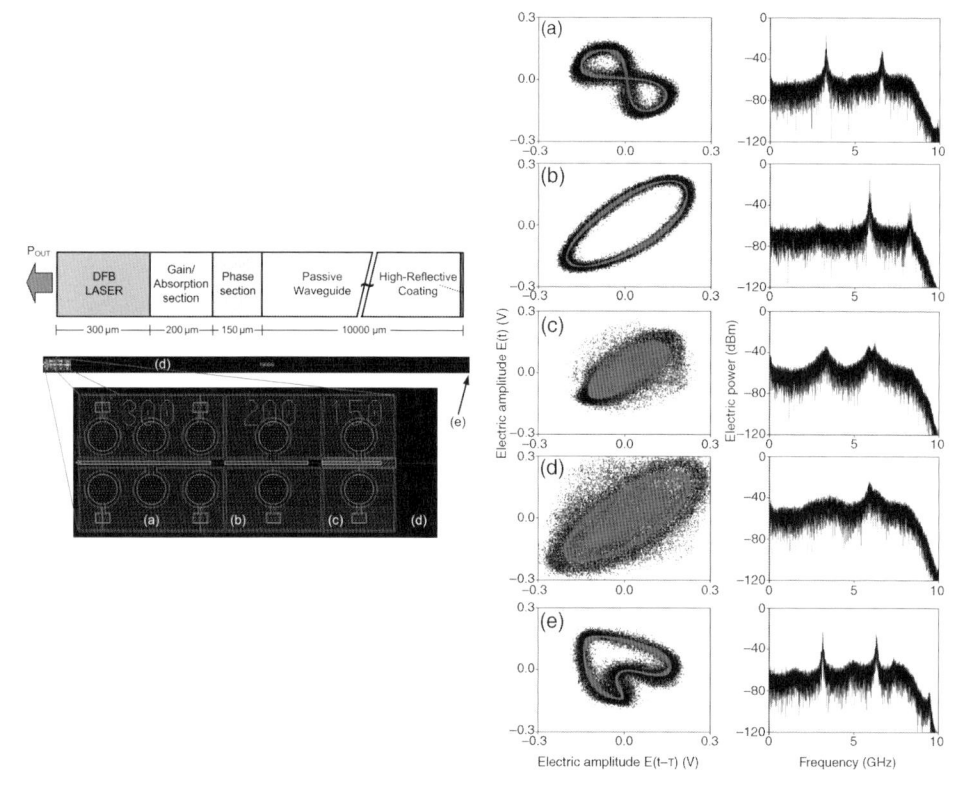

Figure 8.19 (Left) (Top) Schematic diagram of the photonic integrated device. (Bottom) Detail of the mask design of the integrated device: (a) InGaAsP DFB laser; (b) gain absorption section; (c) phase section; (d) passive waveguide. While the laser facet is antireflection coated, the end of the waveguide is highly reflective coated. (e) The lengths of the different sections are shown (Right) Feedback phase dependence: (left column) Experimental attractor plots in the phase space and (right column) corresponding RF spectra of the device output for $I = 3I_{TH}$ and $I_{GAS} = 0$ mA. From (a) to (e), $I_{PH} = 3.3$, 4.8, 5.5, 5.9, and 6.9 mA. The grayscale trace denotes the filtered attractor. (A. Argyris, M. Hamacher, K. E. Chlouverakis, A. Bogris, and D. Syvridis, (2008), © 2008 APS.)

The passive waveguide is grown by selective-area epitaxial growth. The overall resonator length is defined by the internal laser facet and the chip facet of the waveguide that is highly reflective coated (HRC) ($R = 95\%$) and includes the GAS and PH. A criterion of the selection of the cavity length is the ability of the device to produce chaotic attractors with high complexity. The selected length of 1 cm provides an increased effective feedback round-trip time, therefore enhancing the possibility to encounter fully chaotic behavior. The integration of a GAS emerges from the requirement to control the optical feedback strength. Taking into account that the PICs optical feedback is determined by its total internal losses and the passive waveguide facet reflectivity, the GAS provides a two-fold operation: In cases where the desirable feedback values surpass the provided feedback of the passive cavity (GAS and PH unbiased), positive biasing of the GAS, thus acting as a semiconductor optical amplifier (SOA), provides the ability to amplify the optical field emitted from the inner facet of the laser and thus increase the feedback strength. In cases where lower values of optical feedback are needed, the GAS can be reverse biased and acts as a variable optical attenuator (VOA). Consequently, a very wide range of optical feedback values can be set and thus various types of dynamics can be generated. The integration of a PH emerges from the requirement to adjust the phase of the optical feedback field. While the GAS affects both intensity and the phase of the feedback signal, the PH allows for fine tuning that can almost continuously go over beyond 2π, even several multiples of it. This is crucial for controlling the dynamics performance of lasers with external short-cavity optical feedback, since the response of the laser is triggered by a coherent delayed optical field.

The laser biasing current is not crucial for achieving enhanced nonlinear dynamics when it is kept well above threshold. When the DFB laser is biased at $3I_{th}$ ($I_{th} = 10.6$ mA) different types of dynamics can be observed depending on GAS and PH operation. When the GAS is reverse biased, the total attenuation that the emitted power experiences after a round-trip within the cavity is enhanced. Only limit-cycle dynamics or stable solutions (CW operation) are observed excluding any chaos dynamics. When the bias current for the GAS is $I_{GAS} = 0$ mA, the dynamics of the output signal differs considerably when altering the cavity phase through the PH current. By increasing the PH current the observed dynamic states are periodically repeated, indicating a cycle of 2π phase change. The different dynamic states, which include limit cycles and chaotic states, are depicted in Figure 8.19 (right) and emerge for different phase values. In some narrow phase current regions two intense peaks are observed at 3.3 and 6.6 GHz, while the rest of the spectral components lay just above the noise floor, as shown in Figures 8.19a and e (right). These peaks correspond to the external cavity modes; however, their frequencies shift slightly with phase tuning. In other phase current regions, various limit-cycle conditions arise: the first peak may be suppressed entirely or additional peaks of moderate power may arise (Figure 8.19b (right)). There is also a range for the phase section current values in which the device operates in the coherence collapse regime providing a chaotic output, as revealed from Figures 8.19c and d (right). In this operating region the radio-frequency spectrum of the output signal becomes broadband and extends up to the cut-off

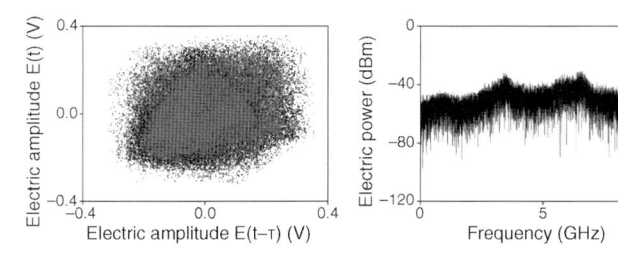

Figure 8.20 (Left) Experimental attractor plots in the phase space and (Right) corresponding RF spectra of the device output for $I = 3I_{TH}$, $I_{GAS} = 10$ mA, and 2.9 mA. The grayscale trace denotes the filtered attractor. (A. Argyris, M. Hamacher, K.E. Chlouverakis, A. Bogris, and D. Syvridis, (2008), © 2008 APS.)

frequency of the optical receiver (around 8 GHz). The frequency peaks are now much less intense, however, still distinguishable from the rest strengthened spectral components.

When the GAS is slightly positively biased (e.g., $I_{GAS} = 5$ mA), the provided gain in the cavity compensates the internal losses, increasing the optical power feedback ratio to 2.8%. In this case, there is only a single phase condition in which an intense peak at 6.3 GHz emerges in the microwave spectrum. The rest phase region was found to provide only broad-spectrum chaotic dynamics, therefore deteriorating the phase dependence. When the GAS is moderately biased (e.g., $I_{GAS} = 10$ mA), the optical power feedback ratio rises to 3.3% and the output signal now is fully chaotic and independent of any phase condition, as shown in Figure 8.20 (Argyris et al., 2008). Any further increase in the GAS current does not provide any alteration in the characteristics of the chaotic spectrum. For the filtered time series in Figure 8.20, the correlation dimension is measured ($D_2 = 4.8 \pm 0.3$), and the Kolmogorov–Sinai entropy is calculated ($K_2 = 10\, \text{ns}^{-1}$).

Similar PICs designed specifically for high-speed random number generators have been proposed (Harayama et al., 2011), which will be described in Section 10.3.6.

8.5.2
Photonic Integrated Circuit for two Mutually Coupled Semiconductor Lasers

Another configuration for generating chaos in photonic integrated device has been proposed, known as integrated tandem device (ITD) (Wünsche et al., 2005). In this configuration, two DFB semiconductor lasers are integrated at a distance of less than 1 mm with mutual optical coupling on the same chip as shown in Figure 8.21 (left). The time scale governing the dynamics of a solitary semiconductor laser is given by the period τ_{RO} of the relaxation oscillations of the carrier-photon dynamics. This period is typically in the range of a few 100 ps. The coupling delay τ results from the propagation of the optical fields between the spatially separated lasers. In the long-delay limit $\tau \gg \tau_{RO}$, a complex behavior comprising low-frequency fluctuations and synchronization of chaos in conjunc-

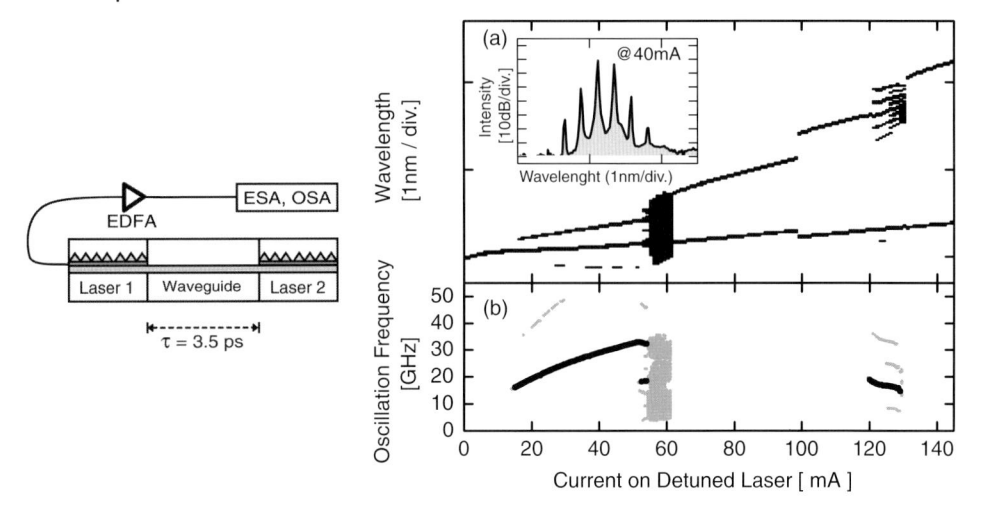

Figure 8.21 (Left) Schemes of experimental setup for the integrated tandem device. ESA: electrical spectrum analyzer; OSA: optical spectrum analyzer; EDFA: erbium-doped fiber amplifier. (Right) Detuning scenario of the integrated tandem device (ITD). (a) Dominant spectral lines (20 dB down of the maximum peak) in the optical spectrum and their shift as a function of the pump current on the detuned laser. Inset: Optical spectrum at 40 mA. (b) Spectral lines in the RF spectrum. Black: 25 dB above noise floor; gray: side peaks more than 5 dB above noise floor. The pump current on the stationary laser is always 45 mA. (H.-J. Wünsche, S. Bauer, J. Kreissl, O. Ushakov, N. Korneyev, F. Henneberger, E. Wille, H. Erzgräber, M. Peil, W. Elsäßer, and I. Fischer, (2005), © 2005 APS.)

tion with symmetry breaking has been observed (Heil et al., 2001a). The case of ultrashort delay $\tau \ll \tau_{RO}$ is realized by an ITD where both lasers are arranged on a single chip.

The ITD consists of two DFB lasers, each 220 μm long, separated by a 300 μm long passive waveguide section, as shown in Figure 8.21 (left) (Wünsche et al., 2005). The coupling in the ITD is distinctly stronger, as about 50% of the field intensity passes the passive section. Irrespective of the stronger interaction, owing to a specific grating design, both lasers of the tandem always operate in single mode. Current-induced heating changes the refractive index in the device sections. Thus, the emission wavelengths of the lasers can be selectively detuned by asymmetric current pumping. The length of the passive section defines a delay of only $\tau \approx 3.5$ ps.

On the stationary laser section of the ITD, the pump current is fixed, while the bias on the detuned laser is increased in steps of 0.5 mA. Except for small current ranges, two sharp peaks dominate the optical spectrum, as the two lines shown in Figure 8.21a (right). The second peak appears already at about 18 mA, that is, below the 25 mA solitary threshold, owing to optical pumping by the other laser. Whether the synchronization at smaller biases is frequency locking or caused by switching off one laser could not be distinguished experimentally. The overall shift of 2.2 nm seen for the detuned laser is consistent with that of solitary lasers from the same wafer. The peak associated with the stationary laser also moves slightly due to thermal cross-talk. Both emission lines have comparable strength although the spectrum is recorded from only one side of the tandem. Thus, the field of each laser propagates to the opposite facet without significant attenuation. The resultant intensity oscillations are hence pronounced. Their frequencies monitored by the RF spectra and the line separations in the optical spectra agree within the 50-GHz bandwidth of the electrical spectrum analyzer (Figure 8.21b (right)). The step separation in the optical spectra are again close to the round-trip frequency of $1/2\tau \approx 140\,\text{GHz}$ defined by the passive section. These discontinuities are a common feature of mutually interacting lasers with short delay.

8.5.3
Photonic Integrated Circuit for Colliding-Pulse Mode-Locked Lasers

A photonic integrated device for colliding-pulse mode-locked lasers has been reported (Yousefi et al., 2007). The device is part of a series of colliding-pulse mode-locked lasers (CPMLL) that were designed and fabricated for 40 GHz colliding-pulse passively mode-locked laser operation at an emission wavelength of 1.52 μm as shown in Figure 8.22 (left). Each chip consists of a 2-μm-wide active ridge waveguide SOA terminated by cleaved facets ($R = 0.35$) at the two ends. The small region in the center is electrically isolated and can be contacted independently. By reverse biasing this section a saturable absorber can be achieved. Such a configuration, consisting of a long amplifier section with a short absorber in the center, is designed to produce a train of picosecond optical pulses, but in practice the intended dynamics are hard to achieve. Instead, a variety of other dynamics may be observed, depending on the precise settings of the external control parameters, in this case the bias current and the absorber voltage.

The measurement setup is also depicted in Figure 8.22 (left) together with a picture of an array of five CPMLLs. The chip is glued on a temperature-controlled copper mount and the current and voltage are provided through two probes. Light from the laser output waveguides is collected using a lensed fiber. The signal is amplified using a SOA between two optical isolators in order to avoid feedback into the laser. The light is analyzed using a photodiode connected to a RF spectrum analyzer, and a photodiode connected to an oscilloscope with 1 GHz bandwidth. The measured RF spectrum of the device suggests operation on many dynamical attractors as a function of the operation parameters. Strong dynamics in the frequency range up to 1 GHz are observed at a reverse bias voltage of -2.4 V on the absorber. The absorber voltage is fixed at this value and the pump current of the SOA sections used as the bifurcation parameter.

Figure 8.22 (right) depicts a sequence of dynamics as the pump current is varied from 133.5 mA to 139.5 mA in steps of 1 mA. The left column shows the phase portrait in the $(P(t), P(t+\tau))$ plane, where $P(t)$ and $P(t+\tau)$ indicate the time series of the laser power and the time-advanced laser power, respectively. The middle column shows the phase portrait in the $(P(t), Q(t))$ plane, where $Q(t) = (P(t) - P(t-h))/h$ indicate the time derivative of the laser power. The right

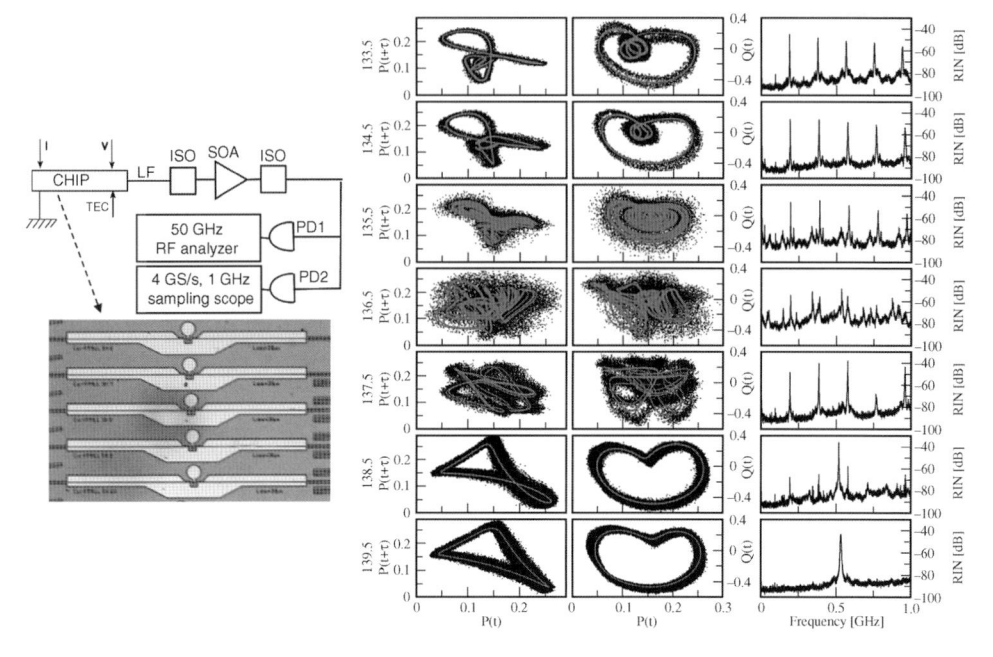

Figure 8.22 (Left) The measurement setup with a picture of an array of five 40-GHz CPMLLs. The 40-μm-long absorber is shifted gradually from the center guaranteeing at least one laser with a perfectly cantered saturable absorber for mode locking. The results presented here are from the most asymmetric device (absorber 40 μm off from the center, lowest in the array). In the setup: I and V are the current and voltage sources, respectively, TEC is the temperature controller, ISO denotes the optical isolator, LF is the lensed fiber, and PD1 (50 GHz) and PD2 (1.25 GHz) are the two photodiodes. (Right) Phase plane portraits and rf spectra of the transition in and out of chaos of the colliding-pulse mode-locked lasers (CPMLL). The left column shows the phase portraits in the $(P(t), P(t+\tau))$ plane, the middle column in the $(P(t), Q(t))$ plane, where $P(t)$ and $P(t+\tau)$ indicate the time series of the laser power and the time advanced laser power, and $Q(t)$ indicate the time derivative of the laser power. The right column is the rf spectrum up to 1 GHz. The pump current is shown in the left corner of the plot in units of mA. (M. Yousefi, Y. Barbarin, S. Beri, E. A. J. M. Bente, M. K. Smit, R. Nötzel, and D. Lenstra, (2007), © 2007 APS.)

column shows the RF spectra of the output power up to 1 GHz. In the interval from 133.5 to 134.5 mA (the first and second rows of Figure 8.22 (right)), the laser is operating on a period-4 limit cycle with a fundamental period of 1.31 ns and a full period of 5.22 ns. When the current is increased to 135.5 mA (the third row), the limit cycle has lost its stability and the laser shows intermittent type of dynamics, which can be a precursor to chaos. Indeed, further increase of the pump current to 136.5 mA (the fourth row) results in chaotic complex motion on a very long period, as can be concluded from both the RF spectra and the reconstructed deterministic trajectory. The chaotic dynamics are excited through a period-3 limit cycle, which appears when the current is further increased to 137.5 mA (the fifth row). The fundamental period of the limit cycle is 5 ns and a third of the period is 1.67 ns (600 MHz). At 138.5 mA (the sixth row), the laser shows single-period limit-cycle oscillations at a new frequency of 530 MHz, while the corresponding spectrum shows a weak intermittency with the previous periodic dynamics. Note that the 530-MHz oscillations are probably related to the fast oscillations underlying the period-3 limit cycle at 137.5 mA (the fifth row). This period-1 limit cycle at 530 MHz stabilizes as the current is increased to 139.5 mA (the seventh row).

Many similar nonlinear dynamics have been observed in the 1-GHz range in the output of the CPMLLs for various pump currents of the SOA and voltages and positions of the absorbers. The physical explanation for one of the fundamental oscillation frequencies of the limit cycles in Figure 8.22 (right) is the beating between the two resonant frequencies of the compound cavity of the CPMLL. This beat frequency is a direct measure for the asymmetric positioning of the absorber region.

8.6
Other Encoding and Decoding Techniques

8.6.1
Spatiotemporal Encoding

The chaos communication systems discussed so far involve serial transmission of data through a single communication channel, using temporal chaotic signals as information carriers. A generalization of this approach to systems with spatial degrees of freedom would enable the use of spatiotemporal chaos for the parallel transfer of information, which would yield a substantial increase in channel capacity. However, multichannel optical chaos communication systems have not been implemented so far. Implementing a parallel chaos communication scheme in electronic systems would be a complex task due to the need of a comprehensive extended coupling between a transmitter and a receiver. Optical systems, on the other hand, provide a natural arena for the parallel transfer of information. Spatiotemporal optical communication utilizes the inherent large-scale parallelism of information transfer that is possible with broad-area optical wavefronts.

Spatiotemporal chaos communications require the existence of synchronization between a transmitter and a receiver (Amengual *et al.*, 1997; Kocarev and Parlitz, 1996). A communication system has been proposed and demonstrated numerically by using the synchronization of the spatiotemporal chaos generated by a broad-area nonlinear optical cavity (García-Ojalvo and Roy, 2001a, 2001b). The setup is shown in Figure 8.23 (top). Two optical ring cavities are unidirectionally coupled by a light beam extracted from the left ring (the transmitter) and partially injected into the right one (the receiver). Each cavity contains a broad-area nonlinear absorbing medium, and is subject to a continuously injected plane wave. Light diffraction is

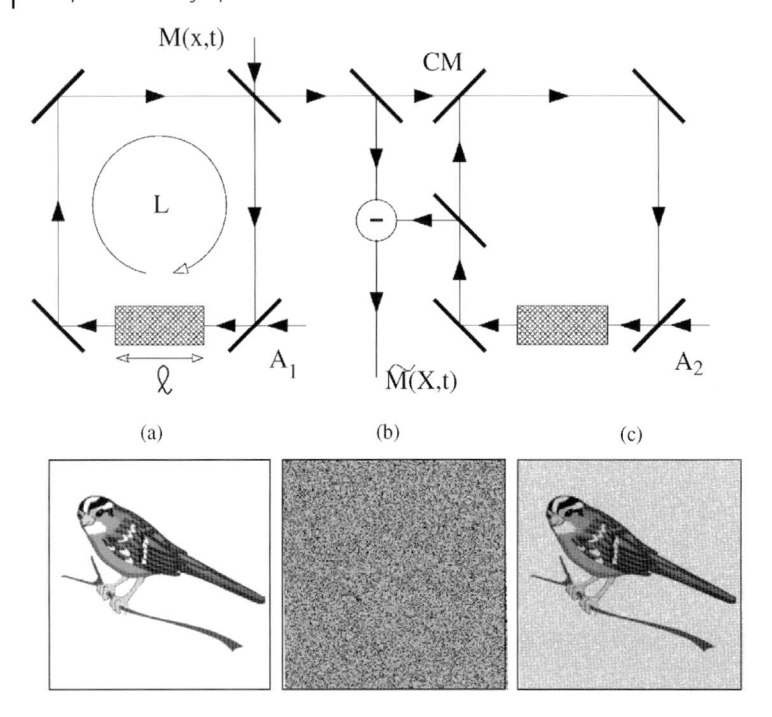

Figure 8.23 (Top) Scheme for communicating spatiotemporal information using optical chaos. CM is a coupling mirror. (Bottom) Transmission of a 2D spatiotemporal static image. (a) Input image, (b) real part of the transmitted signal at a certain time, and (c) recovered data. (J. García-Ojalvo and R. Roy, (2001a). © 2001 APS.)

taken into account during propagation through the medium, in such a way that a nonuniform distribution of light in the plane transverse to the propagation direction appears. In fact, an infinite number of transverse modes can oscillate within the cavity.

In the absence of a message, the transmitter is a standard nonlinear ring cavity, well known to exhibit temporal optical chaos (Ikeda, 1979). When transverse effects due to light diffraction are taken into account, a rich variety of spatiotemporal instabilities appear, including solitary waves and spatiotemporal chaos. This latter behavior can be used for chaotic waveforms as information carriers.

Numerical simulations have been performed on a square array with 256 by 256 pixels of width. Figure 8.23 (bottom) shows an example of data encoded and decoded using the scheme of spatiotemporal chaos communication, where a static two-dimensional (2D) image has been transmitted in space and time. Figure 8.23a (bottom) depicts the input image, Figure 8.23b (bottom) plots the real part of the transmitted signal (spatiotemporal chaos with message), and Figure 8.23c (bottom) plots the recovered data. The message amplitude is set to be small enough. The image of a bird is clearly recognizable, as shown in Figure 8.23c (bottom).

8.6.2
Polarization Encoding

Information for communication can be encoded on the amplitude, frequency (phase) or polarization of a light wave. While the first two options have been explored extensively for decades (Agrawal, 1997), techniques to encode information on the polarization state of light have been developed only recently. Techniques in which the state-of-polarization of light is used to carry information include multiplexing and polarization shift keying.

Some methods of polarization encoding with optical chaos have been developed in fiber lasers (VanWiggeren and Roy, 2002) and vertical-cavity surface-emitting lasers (Sciré *et al.*, 2003). In the scheme of the fiber lasers, there is no one-to-one correspondence between the state-of-polarization of the lightwave detected in the receiver and the value of the binary message that it carries. Instead, the binary message is used to modulate parameters of an erbium-doped fiber-ring laser (i.e., chaos shift keying). The modulation generates output light from the laser with fast, irregular polarization fluctuations. This light propagates through a communication channel to a suitable receiver, which is able to detect changes in the transmitter's polarization dynamics caused by the message signal and ultimately recover the message from the irregular polarization fluctuations of the transmitted light.

The experimental apparatus used to demonstrate optical communication with dynamically fluctuating polarization states is shown in Figure 8.24 (left) (VanWiggeren and Roy, 2002). The transmitter consists of a unidirectional erbium-doped fiber-ring laser (EDFRL) with a polarization controller (PC) and a phase modulator within the ring. The polarization controller consists of loops of fiber that can be twisted to alter their net birefringence. The phase modulator comprises a titanium indiffused $LiNbO_3$ strip waveguide. Electrodes on either side of the waveguide induce a difference in the index of refraction between the transverse-electric (TE) and transverse-magnetic (TM) polarization modes of the waveguide. In this way, it can also be used to alter the net birefringence in the ring. The length of the ring is about 50 m, which corresponds to a round-trip time for light in the ring of roughly 240 ns. This time delay in circulation of the light makes the dynamics observed in this type of laser quite different from those of more conventional cavities. A 70/30 output coupler directs roughly 30% of the light in the coupler into a fiber-optic communication channel, while the remaining 70% continues circulating around the ring.

The communication channel, consisting of several meters of standard single-mode fiber, transports the light to a receiver comprising two branches. Such fiber does not maintain the polarization of the input light. Instead, due to random changes in birefringence along the length of the fiber, the polarization state of the input light evolves during its propagation. The receiver is designed to divide the transmitted light into two branches. Light in the first branch of the receiver passes through a polarizer before being detected by photodiode A. Light in the other branch passes through a polarization controller before it is incident on a polarizer.

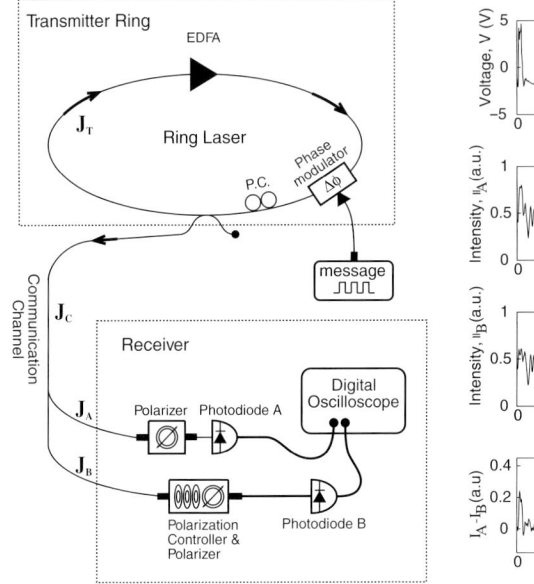

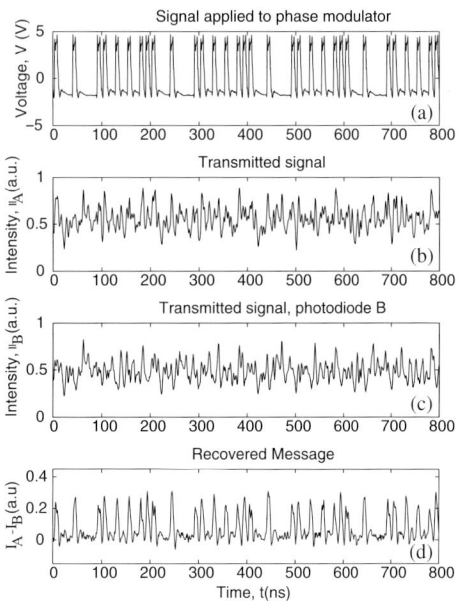

Figure 8.24 (Left) The experimental apparatus for the communication technique is shown. The transmitter consists of an erbium-doped fiber amplifier (EDFA), a polarization controller, and a phase modulator. A time-delay is present between the two photodiodes of the receiver. (Right) Experimental demonstration of message recovery using the irregular polarization fluctuations to carry messages is demonstrated. (a) The message signal applied to the phase modulator in the transmitter. (b) The transmitted signal as measured by photodiode A. (c) The signal as measured by photodiode B. (d) The result of a subtraction of the data in (c) from those in (b). Loss of synchronization (nonzero values of this difference) is interpreted as "1" bits, whereas synchronization (roughly zero values) is interpreted as "0" bits. (G.D. VanWiggeren and R. Roy, (2002). © 2002 APS.)

After passing through the polarizer, the light is measured by photodiode B. Signals from these photodiodes are recorded by a digital oscilloscope. A crucial element of receiver operation is the time delay between the signals arriving at the photodiodes, which is equal to the round-trip time of the light in the transmitter laser. It is thus dual detection with time delay that allows differential detection of the dynamical changes that occur in the transmitter ring and recovery of the sequence of perturbations that constitute the digital message (i.e. differential chaos shift keying).

Figure 8.24 (right) shows experimental results demonstrating the communication technique. The message, the binary voltage signal shown in Figure 8.24a (right), drives the phase modulator as described above. It consists of a repeating sequence of 16 bits transmitted at 80 Mb/s. At this rate, roughly 19 bits can be transmitted during the time light takes to complete one circuit around the ring. A completely random and nonrepeating sequence of bits can also be transmitted successfully, but the repetitive message signal provides some additional insights into the dynamics of the transmitter. Figure 8.24b (right) shows the signal measured by photodiode A in the receiver. Despite the repeating nature of the message, the measured signal does not show the same periodicity. Figure 8.24c (right) shows the signal measured simultaneously by photodiode B. As mentioned above, photodiode B measures the intensity, with a time delay, of a different polarization component of the same signal measured by photodiode A. A subtraction of the two detected signals is shown in Figure 8.24d (right). A "0" bit is interpreted when the subtracted signal is roughly zero because this value corresponds to synchronization of the signals measured by the two photodiodes. A "1" bit is interpreted when the difference signal is nonzero. It is found that the transmitted message bits are accurately recovered by the receiver, as shown in Figure 8.24d (right).

8.6.3
Multiplexing Communications

Despite the many schemes of chaos communication, most of the studies on chaos communication are limited to one-to-one communication using one pair of chaotic lasers. A scheme for multiuser communication using chaos may be important for more sophisticated communication applications (Yoshimura, 1999; White and Moloney, 1999).

Chaotic waveforms could be useful as subcarriers in wavelength-division multiplexing (WDM) in multiuser optical communications. The WDM using chaos as a subcarrier (called chaotic wavelength-division multiplexing, CWDM) has been demonstrated experimentally and numerically in two pairs of one-way coupled Nd:YVO$_4$ microchip lasers (Matsuura et al., 2004). The experimental setup for the CWDM scheme is shown in Figure 8.25 (left). Four Nd:YVO$_4$ microchip lasers are used, where two of the microchip lasers play the role of transmitters (referred to as T1 and T2) and the other two lasers are used as receivers (R1 and R2). Chaotic pulsations can be obtained by sinusoidally modulating the injection current of the laser diodes pumping the transmitters. Two optical isolators are used to achieve one-way coupling. Two individual digital messages are encoded on the chaotic laser outputs of T1 and T2 by using two acousto-optic modulators (AOM) in front of the transmitters (i.e., chaos masking). The two laser beams of T1 and T2 with the two messages are then mixed at a beam splitter and are propagated through one transmission channel in free space. This combined signal is injected into the two laser cavities of R1 and R2 with precise temperature control. Synchronization can be achieved individually in two pairs of lasers by using injection locking, which is referred to as "dual synchronization" (Liu and Davis, 2000b; Uchida et al., 2003c). Two wavelength filters (WF) are used to separate the two different wavelengths of the T1 and T2 for detection. The individual transmission signal consisting of T1 or T2 with the corresponding message is detected by photodiodes and a digital oscilloscope, together with the synchronized chaotic waveform from the corresponding receiver. The message signal can be recovered by subtracting the

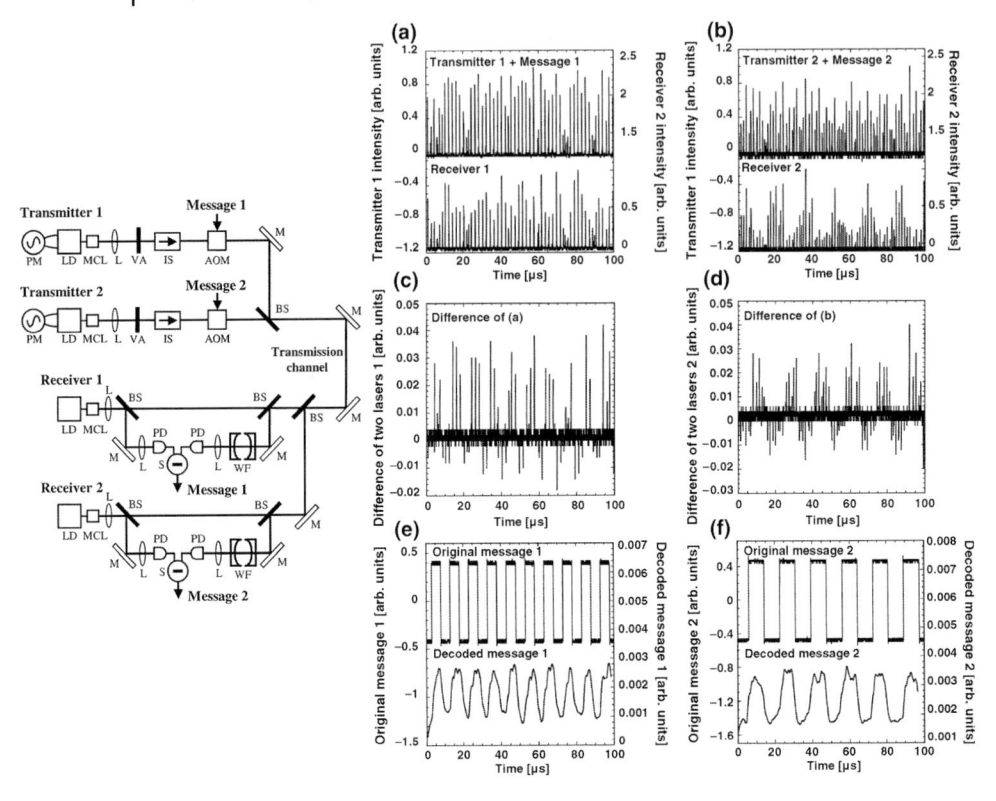

Figure 8.25 (Left) Experimental setup for chaotic wavelength-division multiplexing (CWDM) with microchip lasers: AOM, acousto-optic modulator; BS, beam splitter; IS, optical isolator; L, lens; LD, laser diode; M, mirror; MCL, Nd:YVO$_4$ microchip laser; PD, photodiode; PM, pump modulation; S, subtractor; VA, variable attenuator; WF, wavelength filter. (Right) Experimental results showing (a), (b) temporal waveforms of the transmitter outputs with the digital message and the synchronized receiver outputs. (c), (d) Temporal waveforms of the difference between the transmitter and receiver outputs shown in (a) and (b). (e), (f) Temporal waveforms of the original messages and the decoded messages obtained by filtering the signals in (c) and (d) with a low-pass filter. (T. Matsuura, A. Uchida, and S. Yoshimori, (2004). © 2004 OSA.)

synchronized chaotic signal from the individual transmission signal and by applying a low-pass filter.

The experimental temporal waveforms of the chaotic laser outputs and decoded message signals are shown in Figure 8.25 (right). The chaotic outputs of the transmitters (including messages) and the synchronized outputs of the receivers are displayed in Figures 8.25a and b (right). The message components of the square waveforms cannot be seen in these temporal waveforms. When the two laser outputs are normalized and one is subtracted from the other, the messages can be obtained as an envelope of chaotic oscillations (Figures 8.25c and d (right)). The original messages can be recovered by filtering the difference of the two outputs with a low-pass filter, as shown in the lower traces of Figures 8.25e and f (right). Compared with the original square waveforms shown in the upper traces of Figures 8.25e and f (right), the digital sequences are successfully recovered and the bits can be detected by introducing a certain threshold value.

Another investigation of multiplexed encryption has been performed numerically (Rontani et al., 2009, 2011). Successful data transmission and recovery between multiple users have been achieved at several Gb/s on a single communication channel.

For another application of chaos communication, a scheme for message authentication and integrity protection based on optically generated chaos has been proposed (Rizomiliotis et al., 2010). The introduced protocol relies physically on the dependence of synchronization of chaos on the phase experienced by the delayed portion of the optical field inside the external cavity. The protocol's security is analytically evaluated, considering data modification and impersonation attack scenarios, and issues related to the practical implementation of the scheme are investigated.

8.7
New Perspective of Optical Communication with Chaotic Lasers

8.7.1
Analogy to Biological Communication Systems

There is a longer-term perspective regarding chaos communication that needs to be emphasized (Uchida et al., 2005). In the development of traditional communication techniques, researchers have learned how to use sinusoidal waveforms as carriers of information, through the use of amplitude-, frequency-, or phase-modulation techniques. Receivers for such communication use a single parameter to discriminate between different transmitters; this is the frequency or wavelength of the sine-wave carrier. The receiver typically possesses a tuned circuit that synchronizes to the carrier frequency, and the modifications of the carrier sine waveform that are the information content of the message are then extracted by a suitable demodulation technique. This basic approach has been developed for a century with great success.

It is also very clear that nature does not use such methods for communication, particularly in biological systems. If one asks how information is encoded in the electrical signals that propagate in biological systems, it becomes clear that non-sinusoidal, spiky waveforms are often used for conveying information. The need for highly precise, synchronized clocks that have been used in modern communication systems is also avoided in biological systems.

The basic question then becomes: Can one discover alternate methods of communication that use irregular waveforms as carriers of information and explore such schemes to investigate new methods for communication that do not rely exclusively on sine waves and precise clocks, but that utilize more adaptive and versatile approaches to encoding and decoding information?

This question has already been answered positively (albeit to a very limited extent). Synchronization of chaotic dynamical systems offers such possibilities. A multidimensional parameter space for transmitter–receiver synchronization with complex waveforms may allow the channel capacity for communications to be more fully utilized without the need for high-precision synchronization. New ways of multiplexing information based on the dynamical properties or characteristics of the carrier waveforms may emerge, and a fuller utilization of the degrees of freedom available in electromagnetic waves could be developed in the future.

From the studies of synchronization of chaos, a receiver can synchronize to a chaotic temporal waveform, given appropriate matching of parameters to the transmitter, and that there may be two or more parameters that need to be tuned for successful decoding of a message. Thus, complex temporal carrier waveforms can be used to transmit and receive information, and synchronization is a natural process for extraction of the information. This process of synchronization could be extended for optical systems to spatiotemporal waveforms and include vector dynamical variables such as polarization of light waves (see Sections 8.6.1 and 8.6.2). These studies indicate that the degrees of freedom available for communication with electromagnetic waves may be more fully used, and exploring the dynamics of coupled systems is an important area for investigation for such applications.

It is very natural to ask at all stages of such research: What is the possible advantage of such methods of communication over the more traditional methods that have been developed over the past century and found to be so effective and useful? While this is a very important question, one must recognize the need for research that attempts to discover qualitatively new ways of communicating information. Such attempts will hopefully lead to an understanding of basic concepts that underlie communication in biological systems, and eventually may lead to schemes that have useful applications in other specific contexts.

8.7.2
Towards the World of Scientific Fiction

The concept of encoding and decoding messages with chaos may lead to a conceptually novel technique for information security. Here, is an interesting idea of the use of chaotic complex dynamics in the most complicated biological systems (i.e., human brain) for message encoding and decoding, introduced by Haruki Murakami, the world-famous Japanese writer, in his scientific fiction (Murakami, 2003). The hero in the scientific fiction is a technician to encode and decode confidential data by using his brain with a dynamical hardware key that is never known even by himself. This type of dynamical information hiding technique may be promising in the future. A quote from the scientific fiction is the following (Murakami, 2003).

"I (the hero) input the data-as-given into my right brain, then after converting it *via* a totally unrelated sign-pattern, I transfer it to my left brain, which I then output as completely recoded numbers and type up on paper. This is what is called 'laundering.' Grossly simplified, of course. The conversion code varies with each technician. This code differs entirely from a random number table in its being diagrammatic. In other words, the way in which right brain and left brain are split (which is a convenient fiction; left and right are never actually divided) holds the key. Significantly, the way the jagged edges do not precisely match up means that it is impossible to reconvert data back into its original form."

9
Secure Key Distribution Based on Information-Theoretic Security with Chaotic Lasers

9.1
Introduction

9.1.1
Secure Key Distribution

As seen in Chapters 7 and 8, the chaos communication schemes with unidirectionally coupled chaotic lasers are based on a private-key procedure, where two communicating parties have a common secret key prior to the communication process and they use it to encrypt the message they wish to transmit. The unidirectionally coupled lasers are synchronized in a drive–response configuration, and the secret key consists of the system's parameter values. The system parameters provide a private key because the two communicating lasers must have nearly identical parameter values, otherwise synchronization is failed.

The distribution of a private key is most probably the main weakness of any secure communication system. To establish a completely secure information transfer, it is necessary for two users to share a secret key, known only to them, before the communication can take place. In many practical scenarios, especially when the two users are separated by a large distance, this requirement is difficult to realize because secure transmission of the key requires a previously shared key. This is known as a "secure key distribution" problem. Public channel cryptography such as RSA cipher is one of the major approaches for algorithmic-based secure key distribution (van der Lubbe, 1998).

As opposed to algorithmically, physically secure key-distribution schemes based on the fundamental properties of quantum mechanics have been intensively developed, known as quantum key distribution (QKD) (Bouwmeester et al., 2000; Gisin et al., 2002). Although ideally such communication protocols are perfectly secure, their practical implementation is not straightforward. Noise and attenuation in the quantum channel significantly reduce their efficiency, especially from the range and data rate aspects. Theoretical and experimental studies show that channel attenuation, noise, and detector dark counts limit the key establishing rates and the operational ranges of QKD systems (Bouwmeester et al., 2000; Gisin et al., 2002). In addition, specially designed transmitters and photoreceivers are required for QKD, and the optical transmission signal cannot be amplified without losing the quantum state.

Instead of using quantum mechanics, chaotic lasers could be useful for the physical implementation of secure key distribution. The architecture based on chaotic lasers offers potential key establishing rates that are larger by several orders of magnitude than those of the QKD systems, especially at long communication ranges. In addition, optical transmitters and receivers used for conventional optical communication systems, including EDFA and dispersion compensation fibers, can be used for chaos-based secure key distribution without using specially designed hardware. The chaos-based secure key distribution systems would be simpler and may have potential to provide a new way of secure key distribution.

9.1.2
Computational Security and Information-Theoretic Security

A new concept of security is required in order to utilize chaotic lasers for secure key distribution. This is known as "information-theoretic security" based on the probability theory (Maurer, 1993). There are two main security paradigms, namely computational security and information-theoretic security. Computational security is based on the assumed hardness of computational problems such as the integer-factoring or discrete logarithm problems, and has been already implemented in most computer network systems (van der Lubbe, 1998). Information-theoretic security, on the other hand, is based on probability theory and on the fact that an adversary's information is limited. Such a limitation can come from classical uncertainty in communication channels. Information-theoretic security avoids the reliance on unproven assumptions about the complexity of computations, and is "future-proof" in the sense that the security of keys generated today will not be compromised by improvements in computing technology, including quantum computing, in the future.

One approach to information-theoretic security assumes that there is a public source of randomness and that all parties have limited storage so that they cannot record all the randomness from the source, which is known as the bounded storage model (Chachin, and Maurer, 1997; Aumann et al., 2002). Some theoretical articles envisioned the random source as a high-rate stream of perfectly random bits being broadcast from some natural or artificial source of randomness (Maurer, 1993). These theoretical works relaxed conditions on the randomness of the source required to prove security (Chachin and Maurer, 1997; Muramatsu et al., 2006). The concept of information-theoretic security can be applied for chaos-based secure key distribution, which will be described in Section 9.3.

9.2
Concept of Information-Theoretic Security

9.2.1
History of Information-Theoretic Security and Maurer's Satellite Scenario

Historically, a concept of information-theoretic security has been introduced, known as the "wire-tap channel" (Wyner, 1975). A scenario is considered in which an eavesdropper (Eve) is assumed to receive messages transmitted by a sender (Alice) over a channel that is noisier than a legitimate receiver's (Bob's) channel. Secure keys can be generated between Alice and Bob when it is assumed that Eve's channel is worse than the main channel between Alice and Bob. A concept of "keyless cryptography" based on information-theoretic security has also been proposed (Alpern and Schneider, 1983). In that scheme, Alice and Bob anonymously post their uncorrelated choices of binary strings on a public blackboard, and only they can recognize the generator of individual messages.

A theoretical framework of information-theoretic security has been proposed (Maurer, 1993). This situation could be implemented by a satellite-broadcast scenario as shown in Figure 9.1 (Muramatsu et al., 2008). In this scenario, a satellite broadcasts messages U that are unknown to Alice, Bob, and Eve. They have antennas that enable them to receive the messages. Inevitably, there is noise between the satellite and each antenna: these noises are independent. Thus, Alice, Bob, and Eve have access to correlated sources, which are the respective outputs of mutually independent channels with the input corresponding to the broadcast message. Alice, Bob, and Eve can receive these bits over independent binary symmetric channels with error probabilities ε_A, ε_B, and ε_E, respectively, where $\varepsilon_A \neq 0$, $\varepsilon_B \neq 0$, and $\varepsilon_E \neq 0$ are satisfied. The condition is required where the information that is common between Alice and Bob is larger than the information that is common among Alice, Bob, and Eve. This fact guarantees the secrecy of the system based on the information theory (Maurer, 1993).

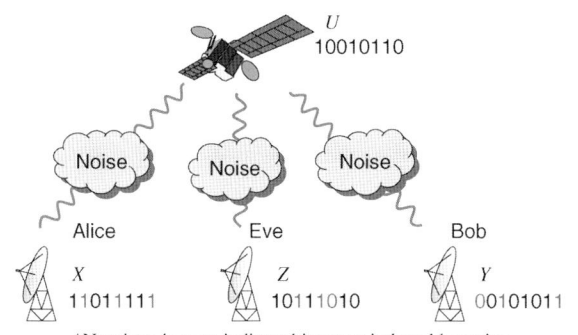

*Numbers in gray indicate bit errors induced by noise.

Figure 9.1 Satellite scenario for secret key agreement. (J. Muramatsu, K. Yoshimura, K. Arai, and P. Davis, (2008), © 2008 NTT.)

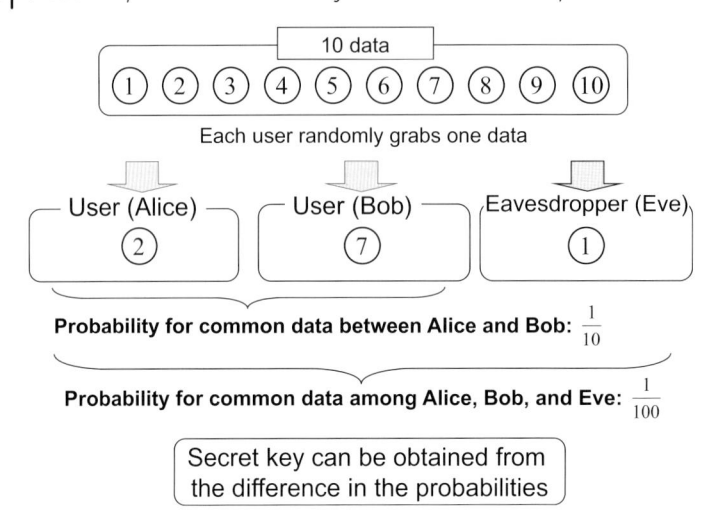

Figure 9.2 Example of the scheme for information-theoretic security based on bounded observability.

9.2.2
Bounded Observability

Another example of information-theoretic security is based on "bounded observability". Figure 9.2 shows the situation of the bounded observability model. A situation is considered where Alice and Bob can grab a part of the data from a common information source, but not all of them. For example, let us assume that Alice and Bob randomly grab one data from 10 whole data. The probability of having the common data between Alice and Bob is estimated as $1/10$. It is thus likely that Alice and Bob can share one common data for 10 trials on average. Let us also assume that Eve randomly grabs one data from the 10 whole data. Eve's probability of having the common data that are also common between Alice and Bob is estimated as $(1/10) \times (1/10) = 1/100$. Therefore, Eve can share only one data that is also common between Alice and Bob for 100 trials on average. The gap of the probability of having the common data between Alice and Bob (i.e., $1/10$), and that among the three users (i.e., $1/100$) can be used to generate a secret key. Alice and Bob may need to exchange partial information to identify their common data as a tag.

Figure 9.3 shows the probability for sharing the common information for Alice, Bob, and Eve by using a Venn diagram. The important assumption is that any user cannot possess the whole data, which is the prerequisite condition of bounded observability (Maurer and Wolf, 1999). Alice, Bob, and Eve have limited data and some of them are common with some probabilities. Alice and Bob can use the common information that is not common for Eve to create secret keys (gray part in Figure 9.3). The superiority of the amount of information between the two users (Alice, Bob) than that for the three users (Alice, Bob, Eve) can be used for a secret key

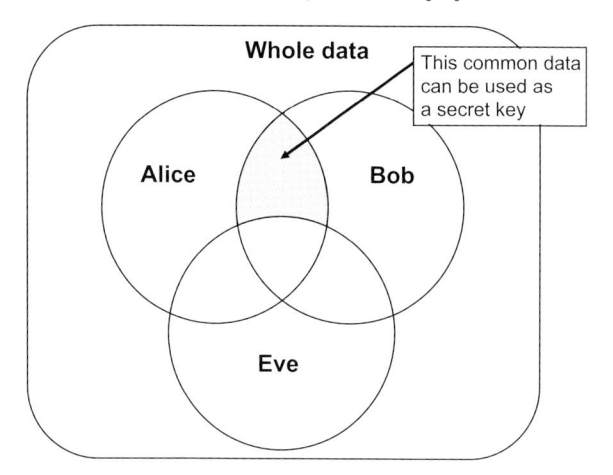

Figure 9.3 Venn diagram for the scheme of information-theoretic security based on bounded observability.

for the implementation of information-theoretic security. If Eve has a superior amount of data so that her probability is larger, then her chances for eavesdropping improve. It is very important to design the secure system where Eve's probability is sufficiently smaller than that for the legitimate users.

9.3
Implementation of Information-Theoretic Security with Chaotic Lasers

9.3.1
Bounded Observability with Chaotic Semiconductor Lasers

Information-theoretic security based on bounded observability described in Section 9.2.2 has been implemented experimentally (Uchida et al., 2003a). The secure key distribution scheme used in this experiment is based on the idea of uncertainty due to the physical difficulty of obtaining a complete record of a fast random signal. Figure 9.4 shows the schematics of the concept of this scheme. Two legitimate users, Alice and Bob, wish to share keys that are secure with respect to an eavesdropper, Eve. Each of Alice and Bob has a detector and independently samples the signal from a random signal source, and stores the random data signal with tag information in a computer. The detected random analog signal can be converted into a digital bit by using a threshold. It is important to satisfy the condition where all the data cannot be obtained by any users and eavesdroppers. To generate a secure key, they use an open communication channel to exchange information about which samples they acquired, by using the tag information. If there is a correspondence tag, Alice and Bob can generate a bit from the random signal that are common between them. Otherwise they discard the random signal.

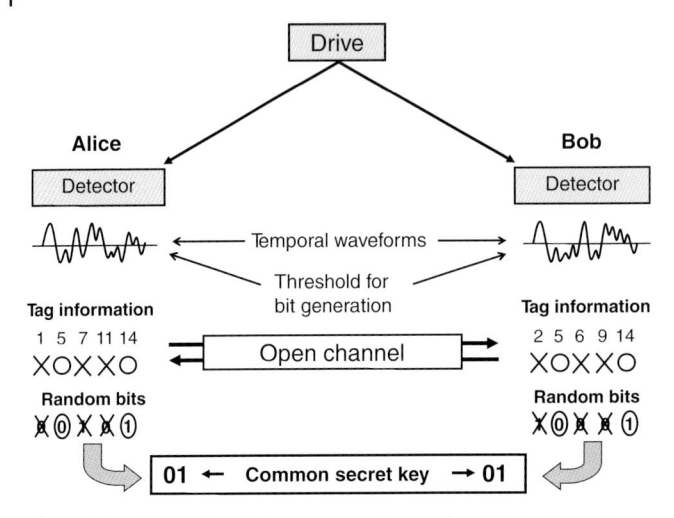

Figure 9.4 Schematics of the concept of secure key distribution scheme.

From listening to the exchange between Alice and Bob on the open channel, Eve also learns which samples Alice and Bob have recorded. However, since the content of the samples is not revealed on the open channel, and the exchange on the open channel only takes place after the transmission from the random signal source is finished, Eve cannot know the keys constructed by Alice and Bob unless she had already acquired the corresponding samples before Alice and Bob revealed which samples they have. Such samples can be used to make a secure key if no one can record the whole random signal and the probability of Eve coincidentally acquiring the same samples as both Alice and Bob is sufficiently small (see Figure 9.3).

A key component to implement this scheme is a signal source that is so fast and random that an adversary cannot record or reproduce the whole signal. A physically chaotic semiconductor laser operating in the gigahertz regime is used in this scheme. Fast chaotic oscillations in semiconductor lasers can provide a strong experimental basis for the implementation of information-theoretic security.

Figure 9.5 shows the schematic of the data acquisition in this scheme. The essential limiting feature of the acquisition system is that it requires a certain acquisition time T_{acq} to record the waveform and prepare for the next grab. One frame time of the data (0.4×10^{-6} seconds) and the trigger period T_{trg} need to be much shorter than the acquisition time T_{acq} (0.74 seconds on average) in the experiment. Moreover, this acquisition time has a random jitter due to intrinsic system fluctuations, so Alice and Bob randomly sample the data.

Figure 9.6 shows the experimental setup for secure key generation (Uchida et al., 2003a). A single chaotic semiconductor laser emitting at 1538 nm is used as a source of large bandwidth random signal. High-frequency chaotic intensity oscillations are induced in the laser by reinjection of delayed optical feedback from an external mirror. The broad radio-frequency (RF) spectrum of the chaotic intensity oscillations extends to around the relaxation oscillation frequency of the laser near 2.4 GHz.

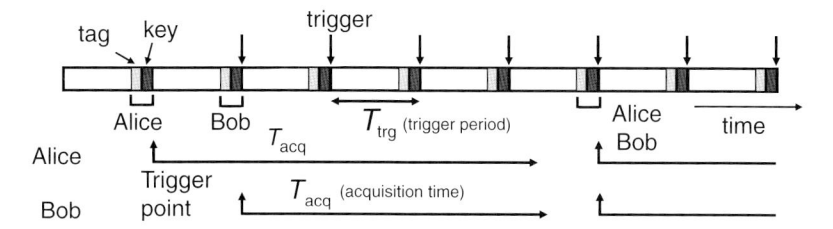

Figure 9.5 Schematic of the data acquisition in this secure key distribution scheme. One frame time of the data and the trigger period need to be much shorter than the acquisition time to satisfy the condition of incompleteness of the whole data sampling. This acquisition time has a random jitter due to intrinsic system fluctuations.

A periodic modulation with adjustable period is applied to the semiconductor laser to induce periodic pulse features in the chaotic waveform that can be used as trigger features to assist sampling. The laser output is coupled into two optical fibers for transmission to two separate detection circuits (Alice and Bob), which include high-speed digital oscilloscopes, programmed to automatically trigger, grab a sample frame, and download to a large volume memory. Sections of chaotic waveform are independently grabbed by Alice and Bob and recorded as sequences of digital data.

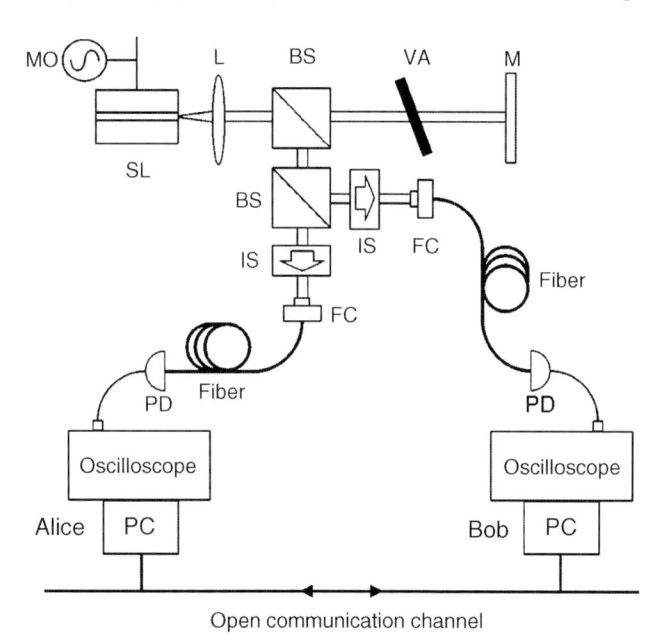

Figure 9.6 Experimental setup for secure key generation. The random signal source is a semiconductor laser with optical feedback. The output from the laser is sent to two separate receivers, which record independent random samples of the signal. BS: beam splitter, FC: fiber collimator, IS: isolator, L: lens, M: mirror, MO: modulation for trigger, PC: personal computer, PD: photodetector, SL: semiconductor laser, and VA: variable attenuator. (A. Uchida, P. Davis, and S. Itaya, (2003a), © 2003 AIP.)

After taking a large number of samples, Alice and Bob start to process the samples. They dissect each waveform section into tag, key and trigger parts. Specifically, 50% of the whole temporal waveform is used as a tag, and 45% is used as key part, and the remaining 5% is used as trigger part. To avoid transients immediately following the trigger pulses the trigger point is set to a position corresponding to 95% of the acquisition window. Since the actual length of the key part in the experiment is about 180 ns and the correlation time of the chaotic laser signal is at most of the order of a few nanoseconds, random bits that have no correlation with the tag part can be extracted from the key part of each waveform sample (see Chapter 10).

Next, Alice and Bob exchange their acquired tag parts. They can do this exchange on an open communication channel. By comparing each pairs of tags, they can identify all the matching tags. Alice and Bob use an error function to compare tags. Figure 9.7 (left) shows experimental results of the distribution of minimum tag errors and the corresponding key errors for 100 waveforms acquired by Alice and Bob in one session. An important point to notice is that the distributions are clearly bimodal, so that Alice and Bob can distinguish matching tag from mismatching tag by the threshold of −10 dB, and see that a match of the tag part also means a good match between the key parts. This procedure is done by using analog temporal waveforms, however, digitized random signals obtained from chaotic waveforms can also be used.

Let us consider the chances of Alice and Bob acquiring the same samples. The chances of acquiring the same waveform is defined as the hit rate, $R_{hit} = N_{hit}/N_{data}$, where N_{hit} is the number of coincident waveforms and N_{data} is the total number of

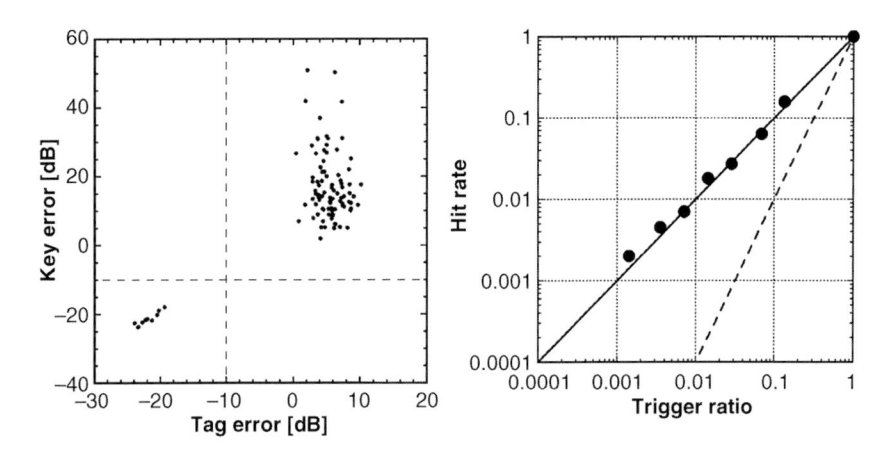

Figure 9.7 (Left) Matching common samples. Experimental results of the distribution of minimum tag errors and the corresponding key errors for 100 waveform samples. (Right) Data hit rate (R_{hit}) of the tag part as a function of the trigger-acquisition time ratio ($R_{ta} = T_{trg}/T_{acq}$). The solid dots indicate the experimental data of the hit rate between Alice and Bob. The solid line indicates the expected theoretical line for the number of Alice–Bob hits ($R_{hit} = R_{ta}$). The dashed line indicates the theoretical line for the number of Alice–Bob hits that are also expected for Eve ($R_{hit} = (R_{ta})^2$). (A. Uchida, P. Davis, and S. Itaya, (2003a), © 2003 AIP.)

waveforms acquired in one session. As a way of controlling the hit rate, the chaotic laser is externally modulated so that the chaotic waveforms include pulse features that can be used as acquisition triggers. The trigger period T_{trg} is set to be shorter than the acquisition time T_{acq} to guarantee that sampling is incomplete, as shown in Figure 9.5. Furthermore, if the trigger period is less than the jitter range, then subsequent samples by Alice and Bob become statistically independent, and R_{hit} is thus expected to be proportional to the trigger-acquisition time ratio $R_{ta} = T_{trg}/T_{acq}$ (see Figure 9.5).

Figure 9.7 (right) shows how the hit rate changes as the trigger-acquisition time ratio R_{ta} is changed from 1 to 10^{-3}, at 10^3 samples per session. The solid dots indicating the experimental data of the hit rate between Alice and Bob agree well with the anticipated theoretical line $R_{hit} = R_{ta}$. The dashed line in Figure 9.7 (right) indicates the theoretical line for the number of Alice–Bob hits that are also expected for Eve, which is $R_{hit} = (R_{ta})^2$. As R_{ta} is decreased, the difference between the solid and dashed lines increases, showing the smaller ratio of key parts expected to be acquired by Eve. For example, in the case of $R_{ta} = 10^{-3}$ at 10^3 samples per session, Alice and Bob typically get only one common sample per session, and one can expect Eve would get this same sample only once every 10^3 sessions.

Alice and Bob can increase the security of keys by combining all the N_{hit} common samples found in one session into a single common key. This is one of the methods of "privacy amplification" which can be used if Eve can accidentally acquire some but not all of the keys (Bennett et al., 1995). The chance of Eve of getting all the N_{hit} common Alice–Bob samples needed to make this combined key is $(R_{ta})^{N_{hit}}$. For the session length of the experiment, this chance for Eve is only 10^{-20} in the case of $R_{ta} = 10^{-2}$ and $N_{hit} = 10$, or only 10^{-100} in the case of $R_{ta} = 10^{-1}$ and $N_{hit} = 100$. If Eve has superior equipment so that her acquisition rate R_{ta} is greater, then her chances improve. The security of the system relies on this probability being sufficiently small.

9.3.2
Public-Channel Cryptography with Coupled Chaotic Semiconductor Lasers

Another scheme based on information-theoretic security has been implemented as "public channel cryptography" (Klein et al., 2006b). An all-optical public channel cryptographic system is proposed, in which there is no need to conceal any of the system's parameters or to exchange private information prior to the public-channel communication process. A regime in the coupling parameter space is used for which the mutually coupled lasers synchronize very well, yet a unidirectionally coupled laser does not. In an application in which synchronization is desirable, synchronization is more easily achieved for mutually coupled lasers than for unidirectionally coupled lasers. A secret communication scheme over a public channel is proposed based on this advantage of mutual coupling over unidirectional coupling.

Figure 9.8 (top) shows the schematic of this key distribution scheme. In the proposed system, the two communicating lasers (Alice and Bob) are mutually coupled in such a way that they exhibit identical synchronization, in which there

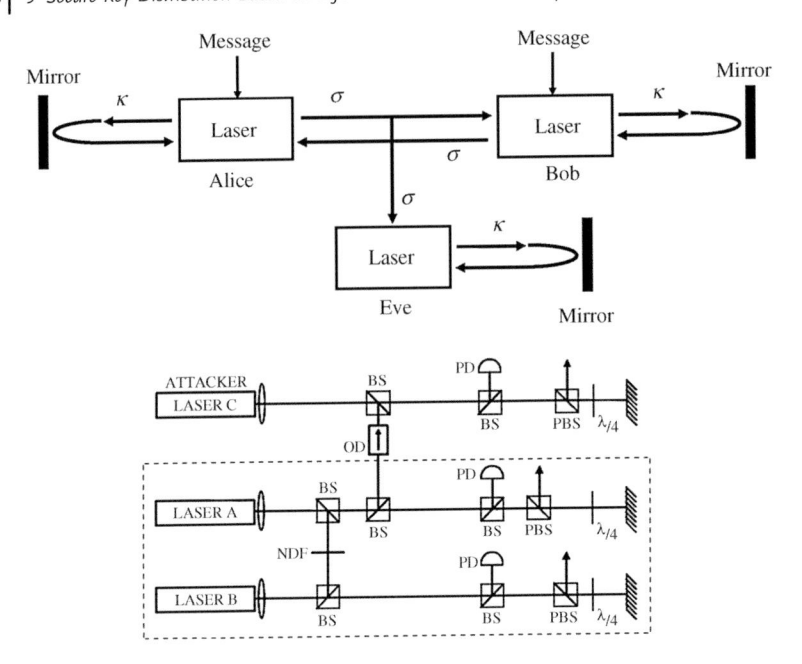

Figure 9.8 (Top) Schematic of the secure key distribution scheme based on superiority of synchronization of chaos in mutually coupled lasers than unidirectionally coupled lasers. All the lasers have external mirrors and optical self-feedback. σ is the coupling strength and κ is the self-feedback strength. (Bottom) Schematic diagram of the experimental setup. Lasers A and B are mutually coupled, and C, the attacker, is unidirectionally coupled. BS, beam splitters; PBS, polarization beam splitters; OD, optical isolator; PD, photodetectors. (E. Klein, N. Gross, E. Kopelowitz, M. Rosenbluh, L. Khaykovich, W. Kinzel, and I. Kanter, (2006b), © 2006 APS.)

is no delay in their synchronized signals (see Section 5.3.1). Stable identical synchronization can be achieved due to the self-feedback of each laser that forms the symmetry of the coupled laser systems (Klein et al., 2006b). Message encryption is accomplished by adding a low-amplitude binary message to the chaotic laser fluctuations so that the transmitted signal still appears to be chaotic and random. Both lasers are allowed to send, simultaneously, independent messages to each other and the messages are independently recovered at both ends of the communication line, ensuring bidirectional information flow. At the receiving end, both lasers use the chaos pass filtering procedure (Fischer et al., 2000) (see Section 7.4.1) to decode the message from the received chaotic signal, which is called a "mutual chaos pass filtering" (MCPF) procedure (Klein et al., 2006b).

Communication security is based on the fact that a third laser (Eve), who tries to synchronize herself to the transmitted signals, is disadvantaged compared to the mutually coupled lasers, and although she can manage to partly recover the message, she has considerably more error bits in her recovered message, and so her eavesdropping attack can be considered unsuccessful. The use of the MCPF procedure provides two distinct advantages: it is public key, that is, it does not require a secret key prior to communication, and it allows for simultaneous bidirectional communication.

The experimental setup is schematically depicted in Figure 9.8 (bottom), where two mutually coupled external feedback lasers A and B represent the communicating pair and the third laser C that is identically configured but coupled unidirectionally to one of the pair represents the attacker. The low-frequency fluctuations (LFFs) regime is used for convenience (see Section 3.2.1.4). In the experiments three single-mode lasers A, B, and C are used, emitting at 660 nm and operating close to their threshold. The temperature of each laser is stabilized and all the lasers are subjected to a similar optical feedback. The length of the external cavity is equal for all lasers and is set to 1.8 m (round-trip time of 12 ns). The feedback strength of each laser is adjusted using a $\lambda/4$-wave plate and a polarizing beam splitter. The two lasers (A and B) are mutually coupled by injecting a fraction of each one's output power to the other. Coupling powers are adjusted using a neutral-density filter. The attacker laser (C) is coupled unidirectionally to one of the mutually coupled lasers, with unidirectionality ensured by an optical isolator. All the coupling optical paths are set to be equal to the self-feedback round-trip path. Three fast photodetectors are used to monitor the laser intensities that are simultaneously recorded with a digital oscilloscope.

Mutual *versus* unidirectional coupling is compared over a range of coupling strength σ and self-feedback strength κ (see Figure 9.8 (top)). In the experiment, the total feedback intensity is set such that it results in the reduction of the solitary laser's threshold current by approximately 5%. While keeping the total feedback of all lasers equal, the values of σ and κ are varied over the entire parameter space. The degree of synchronization between the lasers is evaluated *via* the cross-correlation function (Eq. (5.19) in Section 5.4). The two-dimensional (2D) phase space is defined by the parameters σ and κ of lasers A and B and the attacker to either A or B. This 2D phase space is characterized by the following three regimes as depicted in Figure 9.9 (left): the dark gray regime where σ is strong enough in comparison to κ and all lasers

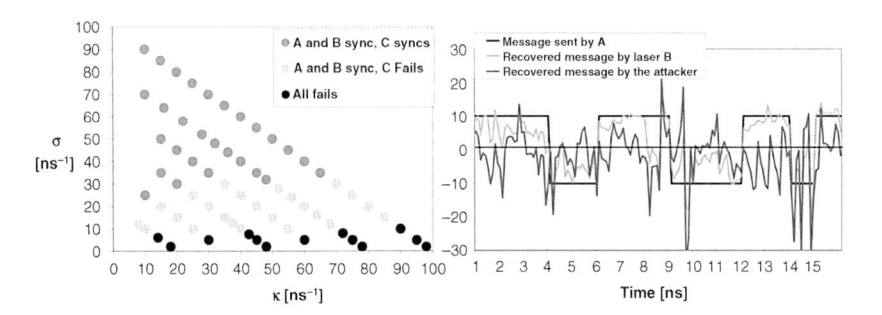

Figure 9.9 (Left) Success or failure of synchronization for the parties and attacker for a range of parameter values: κ, feedback strength, and σ, coupling strength. (Right) (black) A trace of the message sequence sent from laser A to B, which consists of 111001110001101, (light gray) the recovered message by B, and (dark gray) the recovered message by the attacker. The attacker amplifies the coupling signal to a maximum value. (E. Klein, N. Gross, E. Kopelowitz, M. Rosenbluh, L. Khaykovich, W. Kinzel, and I. Kanter, (2006b), © 2006 APS.)

are synchronized, and the black regime where the coupling is negligible and there is a lack of synchronization between any of the lasers. Most interesting is the window of the light gray regime where A and B are synchronized, but C fails to synchronize (the failure is defined when the average correlation is less than 0.7). It is in this regime that mutual coupling is superior to unidirectional coupling; thus, even if laser C uses the exact same parameters as lasers A and B, the fact that laser C is not mutually coupled to A or B but only listening, affects his synchronization ability. This phenomenon is at the center of the proposed cryptographic system.

Note that if the attacker amplifies the coupling signal and uses a stronger σ (outside the light-gray regime) such as $\sigma_{\text{attacker}} = \kappa + \sigma$ of A and B, he manages to synchronize very well. However, in the cryptographic system, such synchronization does not allow the attacker to reliably decode the message. The reason for the decoding failure is that the coupling signal consists both of a fraction of the laser's signal and the message, and so the attacker amplifies the message as well as the laser's signal.

In the proposed public-channel cryptography system, each laser transmits a signal to the other laser that consists of the original chaotic laser signal with some added low-amplitude message signal. Synchronization can be achieved even in the presence of the added message. For a wide range of parameter values of σ and κ, the two lasers achieve stable identical synchronization, despite the fact that each laser is receiving an additional and different time-dependent message. This message, sent from A to B, is recovered by laser B *via* the chaos pass filter procedure by subtracting the output from the received input and then averaged over 1-ns windows, giving the recovered message. The same method is used by the other laser.

Figure 9.9 (right) displays the traces of the message sent by A, the message recovered by B, and the message recovered by the attacker, for an attacker using a maximal coupling constant σ. One can observe that the attacker's recovered message, although generally following the original message, has several mistakes even in this short sequence. When sending compressed data, even several mistakes can corrupt the entire message. An accepted measure of the ability to recover a message successfully is the bit error rate (BER).

The BER of the communicating lasers is measured for many different points in the "light-gray regime" in Figure 9.9 (left), for which there is a strong self-feedback. The best possible κ and σ values are selected for the attacker, in the entire space (κ, σ), which gives the minimal BER value. For all the parameters, the BER of lasers A and B is considerably smaller than the BER of the best attacker. For example, the BER of lasers A and B is $\sim 10^{-4}$, whereas the BER of the best attacker is $\sim 10^{-1}$.

The added message is changed randomly every 1 ns, giving a rate of 1 Gb/s. It is clear that the BER of the attacker is a few orders of magnitude higher than the BER of the parties. Hence, while a compressed block (for instance, of 10 kbits) is recovered properly with high probability by the parties, the attacker has many $O(10^3)$ error bits. By reducing the transmission rate, for instance, and transmitting a bit every 3 ns (0.33 Gb/s), the BER of the parties can be reduced and the gap between the BER of lasers A and B and the BER of the best attacker is enhanced. The security may be also enhanced by introducing a pair of instance private commutative filters in front of the coupled lasers (Kanter et al., 2008).

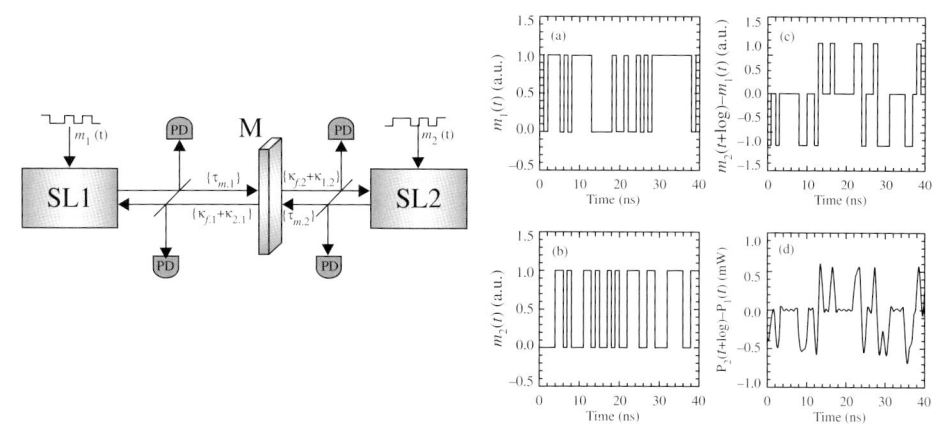

Figure 9.10 (Left) Scheme of two semiconductor lasers coupled through a partially transparent mirror: PD, photodiode; $m_{1,2}$, encoding messages. (Right) Illustration of the message decryption process. (a) and (b) Original messages encoded by SL1 and SL2, respectively. (c) Difference between the two original messages $[m_1(t) - m_2(t + \Delta t)]$ with a given time lag. (d) Difference between the two recovered messages. This signal is filtered with a fifth-order Butterworth filter with a cut-off frequency of 0.8 GHz. (R. Vicente, C. R. Mirasso, and I. Fischer, (2007), © 2007 OSA.)

9.3.3
Bidirectional Message Transmission with Mutually Coupled Semiconductor Lasers

A scheme that allows for simultaneous bidirectional transmission of information encoded into a chaotic carrier generated by coupled lasers has been proposed and demonstrated numerically (Vicente et al., 2007). The scheme is depicted in Figure 9.10 (left). Two semiconductor lasers (SL1 and SL2) are mutually coupled through a partially transparent mirror (M) placed in the pathway connecting both lasers. Due to the mirror the light injected into each laser is the sum of its delayed feedback from the mirror M and the light coming from the other laser. The coupling coefficients and the feedback strengths have been chosen such that the lasers operate in a chaotic regime. It is assumed that both the feedback and coupling times are larger than the typical time scale of the lasers given by the relaxation oscillation period. With this configuration identical synchronization between the dynamics of both lasers can be obtained for arbitrary distances between the lasers. Even for strongly asymmetric positioning of the mirror one can still obtain identical synchronization, with temporal offset given by the difference of the corresponding delay times (see Eq. (5.9) of Section 5.3.3.1 for instance). The reflection and transmission characteristics of the mirror turn out not to be critical for synchronization, provided that the transmission coefficient is above a threshold value that guarantees synchronization occurs. Due to the condition $T_i + R_i = 1$, where T_i and R_i are the transmission and reflection coefficients of the mirror, which is satisfied for the two branches of Figure 9.10 (left), it is guaranteed that both lasers receive the same levels of light injection.

The proposed scheme is introduced to simultaneously exchange information between SL1 and SL2 by using a single communication channel. The information is encoded by simultaneously modulating the bias currents of both lasers with two independent pseudorandom digital messages of amplitude 0.12 I_{th} at 1 Gb/s. The two transmitted messages (m_1 and m_2) are shown in Figures 9.10a and b (right), respectively. Since the amplitude of the messages is kept small, the information is well hidden within the chaotic carriers. The procedure to decipher the messages starts by subtracting the optical power of both lasers. The synchronization error between the two lasers' powers allows recovery of the difference between the messages that have been sent (Figure 9.10d (right)), which reproduces the difference between the original messages (Figure 9.10c (right)) after the appropriate lag has been compensated by a variable RF delay line. After digitalizing this difference, only the sender of m_2 can completely recover the content of m_1 and *vice versa*. The maximum encoding rate depends on the inverse of the resynchronization time after a bit arrives at one of the lasers. Under these conditions, it turns out to be ~0.3 ns. Consequently, a maximum bit rate of ~3 Gb/s can be achieved.

The security aspects of this scheme are discussed as follows. Since both output powers (P_1 and P_2) are accessible from the same communication channel (a simple beam splitter easily allows for separating the signals coming from SL1 and SL2), an eavesdropper could easily monitor the difference $P_1 - P_2$, and consequently the difference of the messages being transmitted. Thus, a level of 1 in the message difference would clearly indicate that at the proper time the bit associated with SL1 was a "1" while the one sent by SL2 was a "0", as seen in Table 9.1. A similar argument holds when the message difference is -1. Only when the message difference is zero (i.e., both lasers are coding the same bit), the eavesdropper has no clue as to which are the bits that are being sent (see Table 9.1). Based on this result, this type of mutually synchronized configuration could be used for secure key distribution. Both sides of the link can agree to discard those bits that are different from each other while accepting that the key that is formed by the first N bits that coincide with each other. The main advantage of this approach resides in the fact that both sender and receiver now can share a secret key through a public channel.

Table 9.1 Concept of secret key distribution shown in Section 9.3.3.

Alice (SL1)	Bob (SL2)	Difference (SL1 − SL2)
0	0	0
0	1	−1
1	0	1
1	1	0

When the bit difference is 0 (i.e., the same bits are selected), Alice and Bob can share the bit that is sent, whereas there is no clue from the bit difference whether Alice and Bob have a 0 or 1 bit for an eavesdropper. The bits are discarded for other cases (i.e., different bits are selected).

9.4
Information-Theoretic Security with Optical Noise

9.4.1
Ultralong Fiber-Laser System

Instead of using chaotic lasers, optical noise signals can be used to implement information-theoretic security. A concept for classical key distribution based on establishing a laser oscillation between the sender and receiver using an ultralong fiber laser (UFL) system has been proposed (Scheuer and Yariv, 2006; Zadok et al., 2008). The suggested architecture offers large key establishing rates at long communication ranges.

The system consists of a fiber link with a terminal at each end, one controlled by Alice and the other by Bob. Each terminal includes an erbium-doped fiber amplifier (EDFA) and a set of two spectrally selective mirrors. The peak reflectivity frequencies of the two mirrors in the set are f_0 (mirror "0"), and f_1 (mirror "1"). In each bit cycle, both Alice and Bob randomly choose one of the mirrors ("0" or "1") as an end mirror of the UFL. The combination of mirror choices is identified through measurements of the UFL spectrum, and represents a single bit. Here, mirror choices are denoted as (Alice's choice, Bob's choice). Mirror choice (0, 0) or (1, 1) leads to oscillations near f_0 or f_1, respectively, as shown in Table 9.2. An eavesdropper (Eve) measuring peak frequencies f_0 or f_1 can thus easily infer the corresponding mirror choices. These data are thus discarded. Both of the choices (1, 0) and (0, 1) lead to oscillation close to $f_c = (f_0 + f_1)/2$, as shown in Table 9.2. If Eve measures f_c, she cannot easily determine which arrangement, (1, 0) or (0, 1), was used. Alice, knowing her own mirror choice, can determine the complementary choice of Bob, and *vice versa*. The two of them can therefore assign, for example, a logical "1" to the choice of (1, 0), and a logical "0" to (0, 1). This scheme is thus based on a similar (but complementary) concept of the scheme shown in Section 9.3.3 (see Table 9.1 for comparison).

The experimental setup used in the UFL system demonstration is shown in Figure 9.11a. The spectrally selective mirrors are implemented by fiber Bragg

Table 9.2 Concept of secret key distribution shown in Section 9.4.1.

Alice	Bob	Frequency
0	0	f_0
0	1	f_c
1	0	f_c
1	1	f_1

When the frequency is $f_c = (f_0 + f_1)/2$ (i.e., the different bits are selected), Alice and Bob can share the bit ((1, 0) is assigned as "1" and (0, 1) is assigned as "0"), whereas there is no clue from the frequency f_c whether Alice and Bob have 0 or 1 bit for an eavesdropper. The bits are discarded for other cases (i.e., the same bits are selected). This bit-generation scheme is complementary to the scheme shown in Table 9.1.

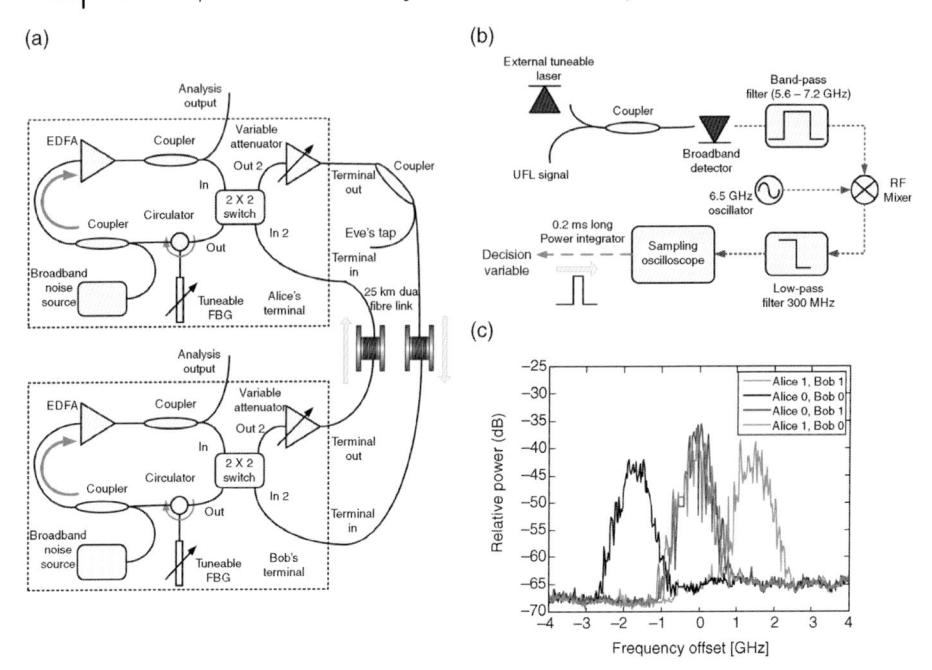

Figure 9.11 (a) Experimental setup. (b) Detection scheme for calculating Alice, Bob, and Eve's decision variables. Solid lines indicate optical signals. Dotted lines indicate RF electrical signals. The dashed line indicated off-line software processing of the sampled data. (c) Measured steady-state spectra of the UFL subject to all four combinations of mirror choices. (A. Zadok, J. Scheuer, J. Sendowski, and A. Yariv, (2008), © 2008 OSA.)

gratings (FBGs). During each bit-exchange cycle, the peak reflectivity frequencies of Alice's and Bob's FBGs is tension tuned to either f_0 or f_1. The frequency separation $f_1 - f_0$ is 3 GHz. The terminals are connected by two 25-km long spans of standard single-mode fiber. Eve's tapping coupler is placed at the very beginning of the fiber span connected to Alice's terminal output port. Each terminal is buffered from the fiber spans by a 2×2 voltage controlled optical switch. When the switches are set to reflection mode, the UFL is effectively split into two local loops at the terminals, with no light transmitted outside the terminals. This mode of operation is used for individually tuning the peak reflectivity frequencies of the FBGs to f_0 or f_1, while literally leaving Eve "in the dark". Once the tuning is completed, the two switches are simultaneously set to transmission mode and the UFL is re-established. Light from a 30-nm wide, external noise source is coupled to the input of each EDFA, and the UFL is set to operate close to the lasing threshold. The peak reflectivity frequencies of both FBGs are randomly varied in between bits, within a range of ± 500 MHz around either f_0 or f_1.

Figure 9.11b shows the detection scheme used to generate the time-dependent decision variables $V_{AB}(t)$ for Alice and Bob, and $V_E(t)$ for Eve. The UFL signal,

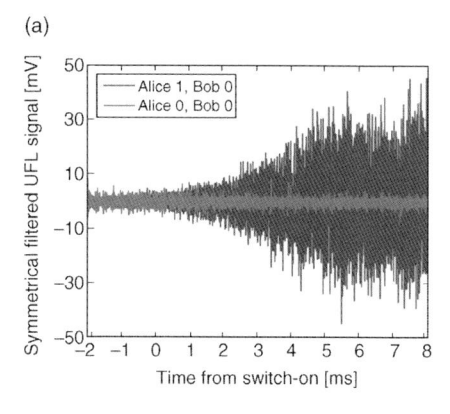

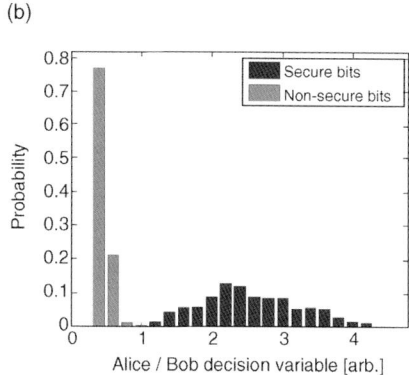

Figure 9.12 (a) Alice and Bob's decision variable $V_{AB}(t)$, *versus* time following switch-on. Significant signal power is observed when Alice and Bob share a secure key bit (dark gray, complementary mirror choices), no signal is observed when information represented by mirror choices is nonsecure (light gray, identical mirror choices). (b) Histogram of the RMS value of $V_{AB}(t)$, taken 3 ms after the switch-on of the UFL. Dark gray: the decision variable distribution for secure bits, (1, 0) and (0, 1) mirror choices. Light gray: the distribution for nonsecure bits, (1, 1) and (0, 0) choices. Setting a threshold value for $V_{AB}(t)$, 994 out of 1000 bits are properly categorized. (A. Zadok, J. Scheuer, J. Sendowski, and A. Yariv, (2008), © 2008 OSA.)

emerging from either the analysis output ports of the terminals or from the eavesdropping coupler, is initially downconverted to the RF domain through heterodyne beating with an external tunable laser of optical frequency f_{lo}. The difference in frequencies $f_{lo} - f_c$ is set to fall within the bandwidth of a broadband detector. The detected photocurrent is observed using an electrical RF spectrum analyzer, or processed further.

Figure 9.11c shows the measured spectra of the UFL at all four possible mirror choices. The spectra for (1, 0) and (0, 1) choices are overlapped and indistinguishable at the center frequency offset of 0 GHz, which can be used to generate a secure key. Such a spectral reconstruction, however, requires many seconds. Alice, Bob and Eve have to identify the key bit within several round-trip propagation cycles. By tuning f_{lo}, different portions of the UFL spectrum are analyzed separately.

Figure 9.12a shows $V_{AB}(t)$ for two different key bits, one with complementary mirror choices by Alice and Bob, and the other with identical choices. When the mirror choices of Alice and Bob are complementary, the UFL central frequency is close to f_c and the magnitude of $V_{AB}(t)$ increases following the UFL switch-on. This build-up of the signal power is an indication of the secure generation of a single key bit. On the other hand, when Alice and Bob choose identical mirrors, the lasing frequency of either f_0 or f_1 is detuned from the frequency offset by approximately 1.5 GHz, and no build-up is observed in $V_{AB}(t)$.

Figure 9.12b shows the histograms of the root-mean-square (RMS) values of $V_{AB}(t = 3\,\text{ms})$, for 1000 random bits. As seen in the figure, a clear distinction between securely generated bits and those who should be discarded is established. The probability of Alice or Bob making a wrong decision (i.e., BER) is 0.006.

The experimental key generation rate of 167 b/s is obtained. While the UFL system is at a disadvantage over the relatively short distance of 25 km, its key generation rate decays only linearly with the link length, rather than exponentially. In order to maintain a given security performance over a longer link, it is anticipated that Alice and Bob would need to employ more elaborate intermediate filters in their terminals, or cascade a larger number of such filters. While this requirement may raise the system cost and complexity, the solution paths are nevertheless feasible. Each terminal may also be equipped with a larger set of mirrors (more than two), enabling the generation of a multilevel key rather than a binary one.

10
Random Number Generation with Chaotic Lasers

10.1
Introduction

10.1.1
Needs for Random Number Generation

High-bandwidth chaotic lasers have been used to demonstrate transmission of messages hidden in complex optical waveforms, as described in Chapters 7 and 8. The use of fast optical chaos for secure key distribution has also been proposed in Chapter 9. In this chapter, the application of physical random number generation using fast chaotic lasers is introduced and discussed. The research field of random number generation with chaotic semiconductor lasers has attracted increasing interest since the first demonstration was reported in 2008 (Uchida *et al.*, 2008a). The characteristics of random number generation based on chaotic lasers are described in detail in Chapter 10.

Random numbers play a crucial role in guaranteeing the secrecy of the system in most cryptographic systems. The performance and reliability of the digital networked society relies on the ability to generate large quantities of randomness. Random numbers are widely used in communication and computing. For example, random numbers are used in information security schemes as codes, keys and challenges to ensure confidentiality (encryption), authenticity (challenge–response protocols) and data integrity (digital signatures) (Eastlake *et al.*, 2005). Quantum cryptography systems require the generation of trusted random numbers to ensure unpredictable detection parameters (Bouwmeester *et al.*, 2000; Gisin *et al.*, 2002). Random numbers are also used for sampling in numerical computations to solve problems in many fields, including nuclear medicine, computer graphics, finance, biophysics, computational chemistry, and materials science. This chapter describes the use of chaotic semiconductor lasers as a physical entropy source to generate random bits at high rates that are suitable for random number generation for the applications of communication and computing.

A true random number generator is mathematically defined as a number generator that can generate any one of a set of N numbers with the probability of generating each of the numbers is $1/N$. Coin-toss, dice or roulette are common examples of random number generators. Computing or communication applications use bit generators that generate just two numbers "1" or "0" with probability $1/2$ (called random bit sequences). Common statistical tests of randomness (see Section 10.9) are designed to test random bit sequences. They test the statistical likelihood that a sequence of bits could have been generated by a true random bit generator.

Random number generators can be classified into two categories: pseudorandom number generators (PRNG) and physical random number generators (RNG or physical RNG). Pseudorandom numbers are generated from a single random seed using deterministic algorithms, and these are used in modern digital electronic information systems. Finite sequences of numbers generated by PRNGs can appear sufficiently random to pass statistical tests of randomness. However, sequences of pseudorandom numbers generated deterministically from the same seed will be identical, and this may cause serious problems for applications in security or parallel computation systems. Truly random numbers should be nonreproducible as well as statistically unbiased, where "bias" indicates the deviation from the equal probability for each random number. For this reason, physically random processes are often used as entropy sources in random number generators (Kelsey, 2004).

Random phenomena such as photon noise, thermal noise in resistors and frequency jitter of oscillators have been used as physical entropy sources for RNGs in combination with PRNGs. However, RNGs have been limited to much slower rates than PRNGs due to limitations of the rate and power of the mechanisms for extracting bits from physical noise. For example, typical rates are tens of megabit per second (Mb/s) using thermal noise and quantum noise (see Section 10.6).

Another terminology has been used for the research field of information-communication technologies, such as deterministic random number generators (DRNG) and nondeterministic random number generators (NRNG), which correspond to PRNG and RNG, respectively. The terms of digital and physical random number generators are also used. A PRNG (and DRNG) generates bits using digital logic operations (a dedicated circuit, FPGA or an algorithm executed on a central processing unit (CPU)) to generate a sequence of bits from a given seed. A physical RNG (and NRNG) generates bits by digital sampling of a physical random process as a source of physical entropy.

A promising approach for fast physical RNGs is the use of chaotic semiconductor lasers. It has been shown that chaotic lasers could be exploited to achieve efficient and stable generation of good-quality random bits at high frequencies (Uchida et al., 2008a). The output of chaotic devices can be both unpredictable as well as statistically random because they generate large-amplitude random signals from microscopic noise by nonlinear amplification and mixing mechanisms (Bracikowski et al., 1992) (see Section 10.5).

10.1.2
Extraction of Randomness from Chaotic Lasers

The output of chaotic lasers provides fast temporal dynamics of chaos with large spectral bandwidth. The typical bandwidth of semiconductor lasers is a few GHz, which is determined by the relaxation oscillation frequency (see Section 2.3.4). The speed of lasers is advantageous for the applications of physical RNGs. The combination of the characteristics of the complexity in chaos and large bandwidth in lasers has opened up a new research field of fast physical random number generation.

A sequence of bits obtained by sampling a theoretically ideal deterministic chaotic system is, in principle, completely predictable if the initial states of the system are known with infinite precision. However, in any real physical device, there is always thermal noise in the device, and this will be amplified by the dynamical instability in a chaotic device so that the macroscopic state of the device after some transient time cannot be determined. This effect is used for example in dice and roulette to produce unpredictable results (Galton, 1890). Thus, the output of chaotic lasers can be both unpredictable as well as statistically random because they amplify the effect of microscopic noise (Bracikowski et al., 1992). It is very important to guarantee theoretically the fact that nondeterministic random bits can be generated from chaotic lasers, which will be described in Section 10.5.

10.2
Types of Random Number Generators

10.2.1
What are Random Numbers?

What are random numbers? Figure 10.1 shows an example of binary random numbers (0 or 1) obtained from coin tossing when one side of the coin is assigned as "0" and the other side is "1". Bits 0 and 1 are randomly distributed and it seems difficult to determine the next bit from the past sequences. However, the length of the sequence is not enough to determine whether it is random or not from the statistics point of view. Integer random numbers can be generated by combining the binary sequences. For example, integer random numbers ranged from 0 to 2^n-1 can be obtained by combining n-bit binary sequences (non-overlapping).

```
0001111000000110101110010001001011011011011101101
1000001000010100101010101011111110010010110010100
0101101000110100011110011010110101011111001110101
1100000000110110110011000011000010100101000011001
1001000100011100100111011011010000110100100001011101
```

Figure 10.1 Binary random bit sequences (250 bits).

Figure 10.2 Example of random bit sequence plotted in two-dimensional plane. Bits "1" and "0" are converted into black and white dots, respectively, and placed from left to right (and from top to bottom). 500 by 500 bits are shown. The data is obtained from the experiment in Section 10.3.1

Figure 10.2 shows a random bit sequence with longer length (250 kilobits) obtained from chaotic lasers, plotted on a two-dimensional (2D) plane. Bits "1" and "0" are converted into white and black dots, respectively, and placed from left to right (and from top to bottom). 500×500 bits are shown. There is no obvious periodic pattern on the 2D plane, so the sequence looks random. However, the randomness of the bit sequences in both Figures 10.1 and 10.2 is not easily determined without statistical tests of randomness.

True random numbers must have the characteristics of "independence" and "unpredictability" (Rukhin *et al.*, 2010). These two important features are described in the following sections.

10.2.1.1 Independence

A random bit sequence could be interpreted as the result of the flips of an unbiased fair coin with sides that are labeled "0" and "1", with each flip having a probability of exactly 1/2 of producing a 0 or 1. Furthermore, the flips are "independent" of each other: the result of any previous coin flip does not affect future coin flips. The unbiased fair coin is thus the perfect random bit stream generator, since the 0 and 1 values will be randomly and uniformly distributed. All elements of the sequence are generated independently of each other, and the value of the next element in the sequence cannot be predicted, regardless of how many elements have already been produced.

Obviously, the use of unbiased coins for cryptographic purposes is impractical. Nonetheless, the hypothetical output of such an idealized generator of a true random sequence serves as a benchmark for the evaluation of RNGs.

10.2.1.2 Unpredictability

Random numbers generated for cryptographic applications should be "unpredictable". In the case of PRNGs, if the seed is unknown, the next output number in the sequence should be unpredictable in spite of any knowledge of previous random numbers in the sequence. This property is known as forward unpredictability. It should also not be feasible to determine the seed from knowledge of any generated values (i.e., backward unpredictability is also required). No correlation between a seed and any value generated from that seed should be evident; each element of the sequence should appear to be the outcome of an independent random event whose probability is $1/2$.

To ensure forward unpredictability, care must be exercised in obtaining seeds for PRNGs. The values produced by PRNGs are completely predictable if the seed and generation algorithm are known. Since in many cases the generation algorithm is publicly available, the seed must be kept secret and should not be derivable from the pseudorandom sequence that it produces. In addition, the seed itself must be unpredictable.

10.2.2
Two Types of Random Number Generators

There are two basic types of generators used to produce random sequences: physical random number generators (RNG) and pseudorandom number generators (PRNG). For cryptographic applications, both of these generator types produce a stream of zeros and ones that may be divided into substreams or blocks of random numbers. The schemes for RNGs and PRNGs and related techniques treated in Chapter 10 are summarized in Figure 10.3.

10.2.2.1 Physical Random Number Generators (RNG)

The first type of sequence generator is a physical random number generator (RNG). The term of true (or truly) random number generators (TRNG) has often been used in the literature, however, the ambiguity of the meaning of "true" should be avoided (No "true" RNG is available in practice) and the term of RNG is preferably used. An RNG uses a nondeterministic source (i.e., the entropy source (Kelsey, 2004)), along with some processing function (i.e., the entropy distillation process) to produce randomness. The use of a distillation process is needed to overcome any weakness in the entropy source that results in the production of nonrandom numbers (e.g., the occurrence of long strings of zeros or ones). The entropy source typically consists of some physical quantity, such as the noise in an electronic circuit, the timing of user processes (e.g., key strokes or mouse movements), or the quantum effects in a semiconductor. Various combinations of these inputs may be used.

450 | *10 Random Number Generation with Chaotic Lasers*

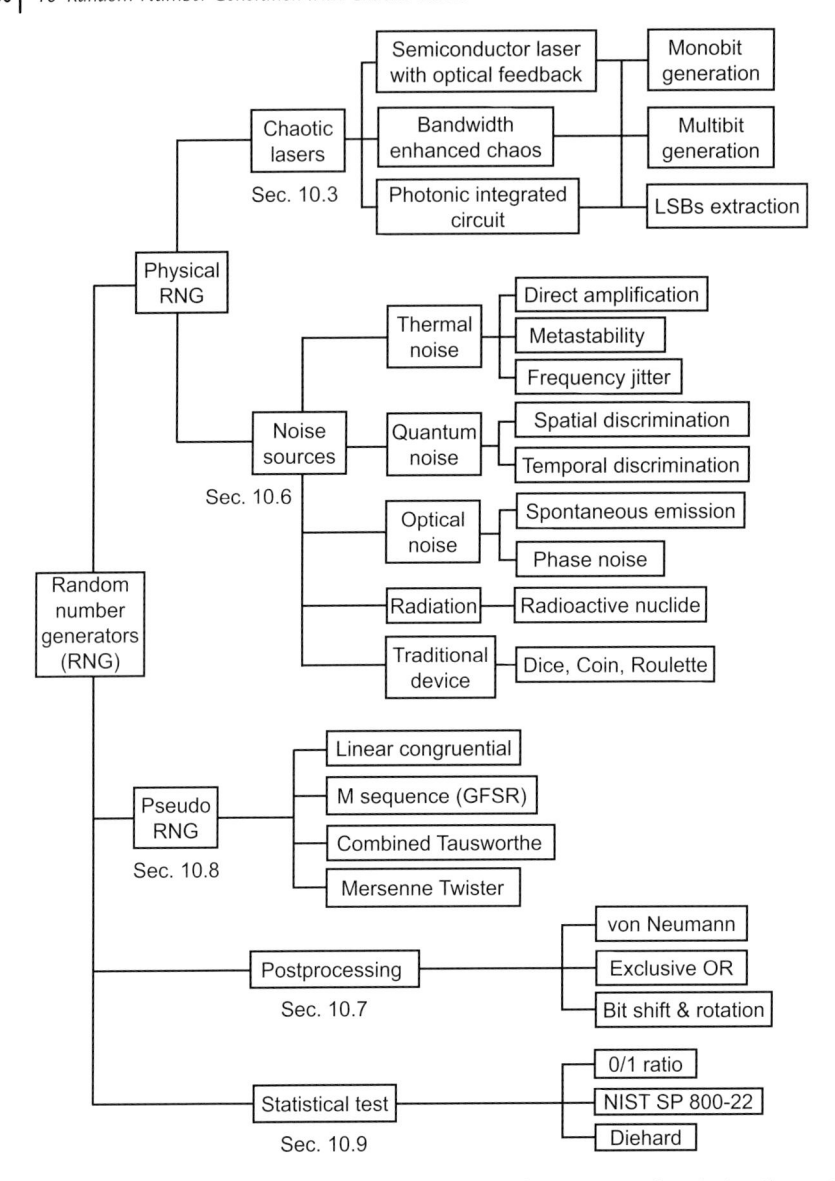

Figure 10.3 Summary of the techniques for random number generation described in Chapter 10.

The outputs of a RNG may be used directly as a random number or may be fed into a PRNG. Postprocessing for the output of RNGs may be required to satisfy strict randomness criteria as measured by statistical tests that determine that the physical sources of the RNG inputs appear random. The production of high-quality random numbers from the RNG may be time consuming when a large quantity of random numbers is needed. The detailed schemes for RNGs are described in Sections 10.3 and 10.6.

10.2.2.2 Pseudorandom Number Generators (PRNG)

The second generator type is a pseudorandom number generator (PRNG). A PRNG uses one or more inputs and generates multiple "pseudorandom" numbers by using deterministic mathematical equations. Initial inputs to PRNGs are called "seeds". In contexts in which unpredictability is needed, the seed itself must be random and unpredictable.

The outputs of a PRNG are typically deterministic functions of the seed; that is, all true randomness is confined to seed generation. The deterministic nature of the process leads to the term "pseudorandom." Since each element of a pseudorandom sequence is reproducible from its seed, only the seed needs to be saved if reproduction or validation of the pseudorandom sequence is required. The detail schemes for PRNGs are described in Section 10.8.

10.2.3
Issues of Conventional Random Number Generators

PRNGs have often been used for producing the keys used to securely store and transmit data in computer and communication systems. The advantage of PRNGs is that they are inexpensive and can operate at rates limited only by the processor speed (Murphy and Roy, 2008). On the other hand, the disadvantage is that because they are not truly random, it is vulnerable for security applications if an attacker can acquire partial knowledge about the algorithm or its seeds.

Physical RNGs rely on inherently random or unpredictable processes in the physical world. Such events could be either fundamentally random, as for example with quantum-mechanical uncertainty or thermal noise, or deterministic but difficult to predict phenomena, such as rolling dice, coin tossing or the spin of a roulette wheel, which exhibit a sensitive dependence on initial conditions (Galton, 1890). Although such physical RNGs are now commercially available, they are restricted to rates below around tens of Mb/s for a single device. Multiple sources of randomness have been integrated and multibits have been extracted from a single device to increase the bit-generation speed.

The use of chaotic lasers for physical RNGs can significantly improve the bit-generation rate over 1 gigabits per second (Gb/s) with a single device, because chaotic fluctuations can be easily obtained at frequencies of several to tens of GHz. Moreover, the integration of laser arrays on a chip and a multibit extraction technique could potentially result in bit-generation rates of tens or hundreds of Gb/s. The use of chaotic lasers for RNGs becomes a breakthrough to improve the bit-generation rate for physical RNGs.

In fact, a random number generator in a modern CPU consists of a combination of physical RNG and PRNG, with various user options for supporting both nondeterministic and deterministic generation of random bits. For example, the RNG module inside a general-purpose computation device, such as a modern CPU chip, has several components: (nondeterministic RNG) + (digital postprocessing) + (buffer) + (digital PRNG). This can be used to generate bits in a number of different ways. Bits can be extracted from the buffer, or from the output of the digital PRNG. One can choose to use bits from the physical RNG output buffer to seed the PRNG, or one can specify a particular seed.

There are two reasons for using a PRNG. One is that users ideally would like to use a pseudorandom output, so that they are able to reproduce it. The other is that they ideally would like to use a true random output but the available physical RNG device is not fast enough, so they use a slow physical RNG to seed a fast PRNG. The latter case can benefit from a faster physical RNG such as a fast chaotic laser. The aim of the use of fast chaotic laser devices could be to improve the quality of the output of the physical RNG by increasing the rate of nondeterministic bits (i.e., the entropy rate).

10.3
Examples of Random Number Generators with Chaotic Lasers

The random number generators with chaotic lasers have been intensively investigated because of the advantage of fast complex signal generation since the first demonstration was published in 2008 (Uchida et al., 2008a). Several schemes have been proposed and demonstrated, which are summarized in Table 10.1. Random-bit-generation rates from 1.7 to 400 Gb/s have been reported by 2011. The details of these schemes are described in Section 10.3.

10.3.1
Monobit Generation with Two Lasers

10.3.1.1 Scheme
The first experimental demonstration of random number generators with chaotic semiconductor lasers was reported at a generation speed of 1.7 Gb/s in real time (Uchida et al., 2008a). The scheme for generating random bit sequences using chaotic lasers is shown in Figure 10.4a. The scheme uses two semiconductor lasers with chaotic intensity oscillations. The output intensity of each laser is converted to an alternating-current (AC) electrical signal by photodetectors, amplified and converted to a binary signal using a 1-bit analog-to-digital converter (ADC) driven by a fast clock. The ADC first converts the input analog signal into a binary signal by comparing with a threshold voltage, and then samples the binary signal at the rising edge of the clock. The binary bit signals obtained from the two lasers are combined by a logical Exclusive-OR (XOR) operation (see Section 10.7.2) to generate a single random bit sequence. No other digital postprocessing is required.

An implementation of the laser scheme is shown in Figure 10.4b. The lasers are the type used in optical-fiber communications. They are single-mode distributed-feedback (DFB) semiconductor lasers, operating at 1.5 μm wavelength and generating around 1.5 mW of optical power. The lasers are prepared without standard optical isolators, to allow optical feedback from an external fiber reflector that reflects a fraction of the light back into the laser, inducing high-frequency chaotic oscillations of the optical intensity in the gigahertz regime. The amplitude of the optical feedback can be adjusted by the variable fiber reflector. Polarization-maintaining fibers are

Table 10.1 Several schemes on laser-chaos-based physical random number generators (2008–2011).

Optical sources for random number generator (configuration)	Bit generation method	Postprocessing method	Maximum bit-generation rate [Gb/s]	Real time or offline generation	Reference/Section
Two semiconductor lasers with optical feedback (fiber optics)	1-bit ADC (Comparator)	XOR	1.7	Real time (NRZ format)	(Uchida et al., 2008a) Section. 10.3.1
One semiconductor laser with optical feedback (space optics)	8-bit ADC	Differential signal and LSBs extraction	12.5 (2.5 GS/s × 5 LSBs)	Offline	(Reidler et al., 2009) Section 10.3.3
One semiconductor laser with optical feedback (space optics)	8-bit ADC	High derivatives and LSBs extraction	300 (20 GS/s × 15 LSBs)	Offline	(Kanter et al., 2010) Section 10.3.4
Bandwidth-enhanced two semiconductor lasers with optical feedback and injection (fiber optics)	8-bit ADC	Bitwise XOR and LSBs extraction	75 (12.5 GS/s × 6 LSBs)	Offline	(Hirano et al., 2010) Section 10.3.5
Photonic integrated circuit of a semiconductor laser with optical feedback	8-bit ADC (16-bit enhancement)	LSBs extraction	140 (10 GS/s × 14 LSBs)	Offline	(Argyris et al., 2010b) Section 10.3.6
Photonic integrated circuit of a semiconductor laser with optical feedback	1-bit ADC	XOR	2.08	Offline	(Harayama et al., 2011) Section 10.3.6
Bandwidth-enhanced two semiconductor lasers with optical feedback and injection (fiber optics)	8-bit ADC	Bit-order reverse and bit-wise XOR	400 (50 GS/s × 8 bits)	Offline	(Akizawa et al., 2011) Section 10.3.5

ADC, analog-to-digital converter; XOR, exclusive OR; LSB, least-significant bit; NRZ, nonreturn-to-zero.

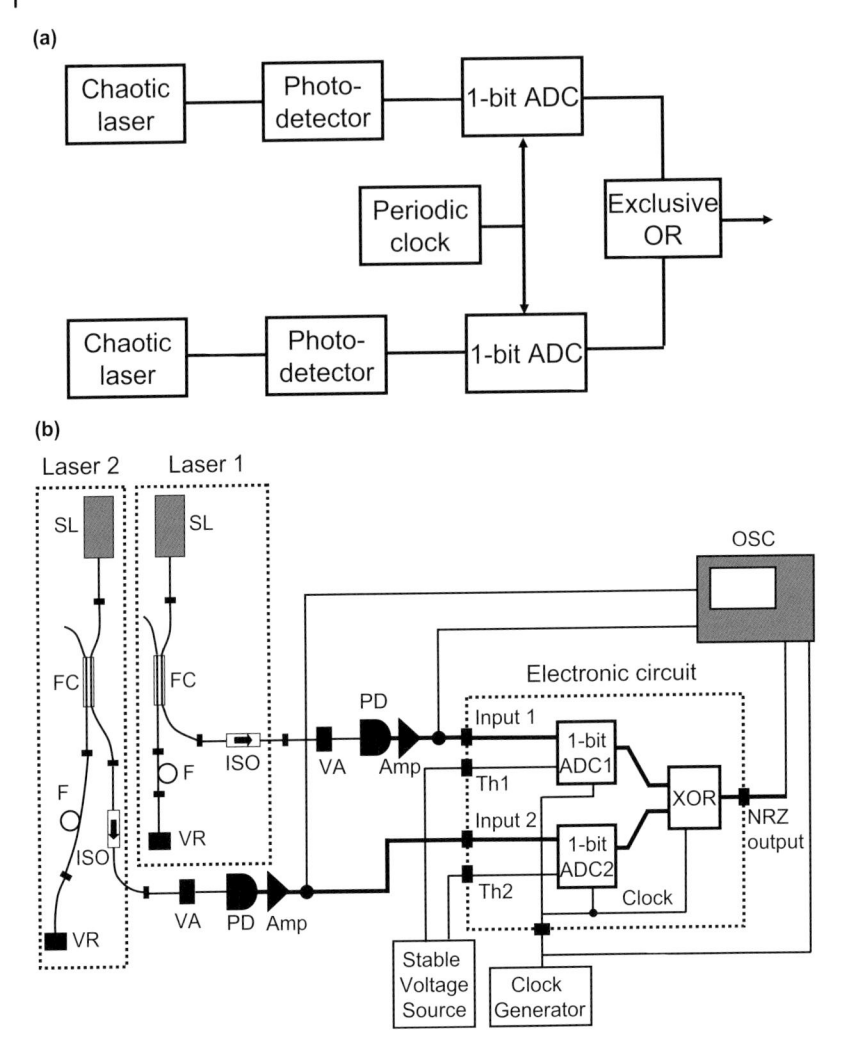

Figure 10.4 (a) Schematic diagram of random bit sequence generator using two chaotic lasers. ADC, 1-bit analog-to-digital converter. (b) Experimental setup for the random bit generator with two chaotic semiconductor lasers. Amp, electronic amplifier; F, optical fiber; FC, fiber coupler; ISO, fiber isolator; OSC, digital oscilloscope; PD, photodetector; SL, semiconductor laser; Th1,2, threshold voltages; VA, variable fiber attenuator; VR, variable fiber reflector; XOR, exclusive OR. (A. Uchida, K. Amano, M. Inoue, K. Hirano, S. Naito, H. Someya, I. Oowada, T. Kurashige, M. Shiki, S. Yoshimori, K. Yoshimura, and P. Davis, (2008a), © 2008 Nature.)

used for the optical-fiber components. The intensity of the laser light output is converted to an electrical signal by a photodetector with AC coupling that removes the direct-current (DC) component, the signal is amplified by an electronic amplifier and then converted to a binary digital signal by a 1-bit ADC. The ADC consists of a comparator and a D flip-flop, which first converts the input analog signal into a binary level by comparing with the threshold voltage, and then samples the binary level at the rising edge of an external clock. The two independent binary digital signals obtained from the two lasers are then combined by an XOR operation. The output signal from the XOR operation is a stream of bits generated in real time with the format of nonreturn-to-zero (NRZ) that is suitable for high-speed data communications.

An example of random bit generation at the maximum rate achievable with the lasers in this setup is shown in Figure 10.5. The rate is 1.7 Gb/s, corresponding to a clock with frequency of 1.7 GHz. Figure 10.5a shows the temporal waveforms of the

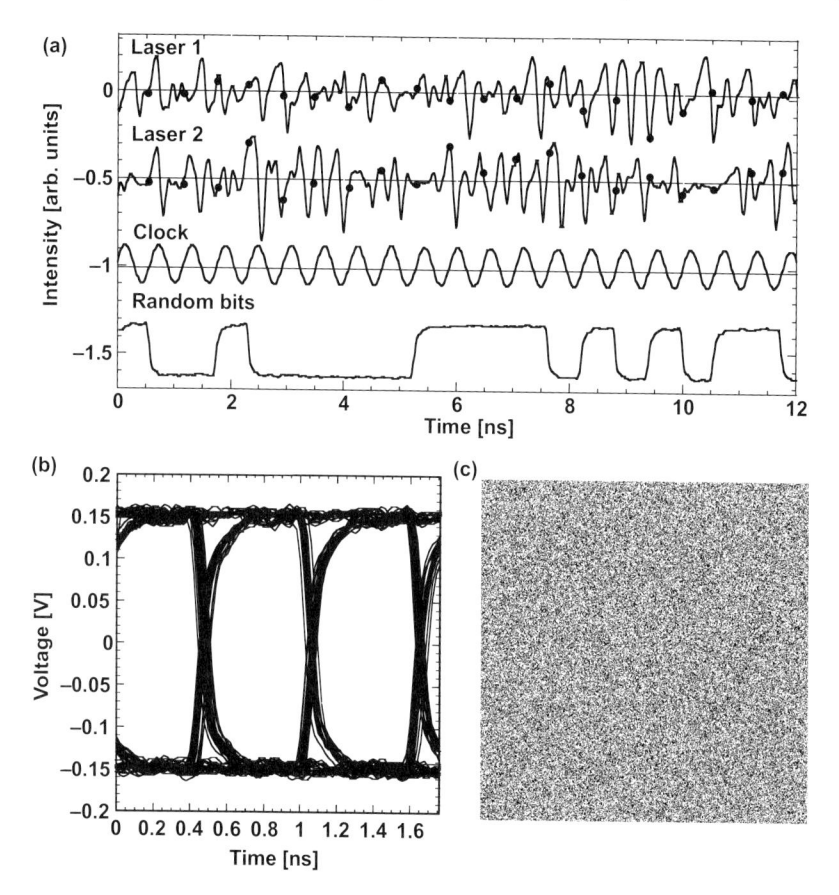

Figure 10.5 Typical output signals from an experimental system. (a) Temporal waveforms of (top and second) laser output signals, (third) external clock and (bottom) corresponding random bit sequence. The threshold values for the ADCs are shown as solid lines. Solid dots mark points sampled at the rising edge of the clock. (b) Eye diagram of the random bit signal. (c) Random bit patterns in a two-dimensional plane. Bits 1 and 0 are converted into black and white dots, respectively, and placed from left to right (and from top to bottom); 500 by 500 bits are shown. (A. Uchida, K. Amano, M. Inoue, K. Hirano, S. Naito, H. Someya, I. Oowada, T. Kurashige, M. Shiki, S. Yoshimori, K. Yoshimura, and P. Davis, (2008), © 2008 Nature.)

two chaotic laser outputs (the top and second plots), the clock (the third), and the corresponding random bits sequences (the bottom). The signal at the bottom of Figure 10.5a is the sequence of random bits output from the XOR operation, in the format of NRZ, generated in real time at 1.7 Gb/s. The eye diagram of the output NRZ signal is shown in Figure 10.5b. A visualization of the randomness of the bits is shown in Figure 10.5c by plotting a single bit sequence as a 500 by 500 pattern of black and white dots.

To evaluate the statistical randomness of digital bit sequences, the standard statistical test suite for random number generators provided by the National Institute of Standards and Technology in USA (NIST Special Publication 800-22) and the Diehard test suite are used (Rukhin et al., 2010; Marsaglia, 1995) (see Section 10.9). The tests are performed using 1000 instances of 1-Mbit sequences for the NIST SP 800-22 tests and using 74-Mbit sequences for the Diehard tests. Bit sequences obtained from the experiment pass all of the NIST and Diehard tests. Typical results of the NIST and Diehard tests are shown in Tables 10.2 and 10.3, respectively. These results confirm that random bit sequences obtained from the two chaotic semiconductor lasers can be sufficiently random that they pass the statistical tests of NIST SP 800-22 and Diehard, as described in the table captions for the pass criteria of the tests (see also Section 10.9 for details).

The use of optical-fiber components ensures stable oscillation conditions. Both the chaotic oscillations and sampling operations are stable with respect to mechanical and thermal perturbations, to the extent that the statistical properties of the

Table 10.2 Results of NIST Special Publication 800-22 statistical tests.

Statistical Test	P-Value	Proportion	Result
frequency	0.366918	0.9920	SUCCESS
block-frequency	0.639202	0.9900	SUCCESS
cumulative-sums	0.101311	0.9920	SUCCESS
runs	0.223648	0.9920	SUCCESS
longest-run	0.603841	0.9890	SUCCESS
rank	0.031012	0.9900	SUCCESS
fft	0.274341	0.9910	SUCCESS
nonperiodic templates	0.013760	0.9810	SUCCESS
overlapping templates	0.893482	0.9910	SUCCESS
universal	0.903338	0.9920	SUCCESS
approximate entropy	0.880145	0.9920	SUCCESS
random excursions	0.142248	0.9836	SUCCESS
random excursions variant	0.068964	0.9869	SUCCESS
serial	0.440975	0.9860	SUCCESS
linear complexity	0.291091	0.9970	SUCCESS
Total			15

For "success" using 1000 samples of 1 Mbit data and significance level of 0.01, the P-value (uniformity of p-values) should be larger than 0.0001 and the proportion should be in the range of 0.99 ± 0.0094392. For the tests that produce multiple P-values and proportions, the worst case is shown.

Table 10.3 Results of Diehard statistical tests.

Statistical Test	P-Value	Result	
birthday spacing	0.493353	SUCCESS	KS
overlapping 5-permutation	0.860647	SUCCESS	
binary rank for 31 × 31 matrices	0.905862	SUCCESS	
binary rank for 32 × 32 matrices	0.324578	SUCCESS	
binary rank for 6 × 8 matrices	0.026400	SUCCESS	KS
bitstream	0.019300	SUCCESS	
Overlapping-Pairs-Sparse-Occupancy	0.035000	SUCCESS	
Overlapping-Quadruples-Sparse-Occupancy	0.096900	SUCCESS	
DNA	0.013200	SUCCESS	
count-the-1s on a stream of bytes	0.650262	SUCCESS	
count-the-1s for specific bytes	0.079061	SUCCESS	
parking lot	0.231274	SUCCESS	KS
minimum distance	0.770625	SUCCESS	KS
3D spheres	0.883907	SUCCESS	KS
sqeeze	0.717981	SUCCESS	
overlapping sums	0.702774	SUCCESS	KS
runs	0.252852	SUCCESS	KS
craps	0.150910	SUCCESS	
Total		18	

For "success" for significance level of 0.01, the P-value (uniformity of p-values) should be larger than 0.0001.

"KS" indicates that single P-value is obtained by the Kolmogorov–Smirnov (KS) test. For the tests that produce multiple P-values without KS test, the worst case is shown.

sequences are maintained over many hours of continual operation. The stability of statistics could be improved further by adaptive control of the detection threshold while monitoring the 1/0 ratio. The susceptibility to fluctuations would be less of a problem in practical implementations using photonic integrated circuits (see Section 10.3.6).

10.3.1.2 Parameter Dependence

The dependence of randomness of generated bit sequences on the laser parameters has been intensively investigated in this scheme (Hirano et al., 2009). The external cavity length, the injection current, and the feedback strength are three crucial parameters that need to be adjusted in order to generate chaos suitable for extracting random sequences. In the bit-extraction stage, the clock frequency and the threshold voltage are the key parameters.

One of the most important parameters is the threshold for ADC and the adjustment of the threshold values is crucial to obtain good random sequences for 1-bit extraction by using ADC. Figure 10.6 shows the number of passed NIST SP800-22 tests and the frequency (occurrence) of bits "0" as a function of the threshold value of one of the lasers, with the value for the other laser fixed. A value of "15" for the number of NIST SP800-22 tests passed means that all the tests are passed. It is found that in this experimental system, the frequency of 1/0 is a good indicator of the

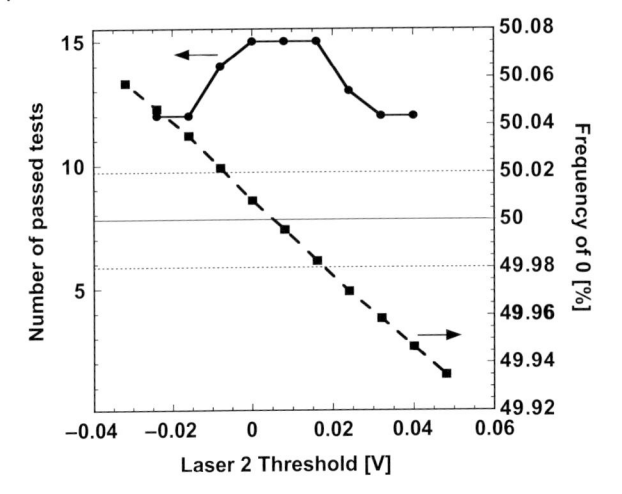

Figure 10.6 The number of passed NIST SP 800-22 tests (solid curve with circles) and the frequency (occurrence) of "0" (dashed curve with squares) as a function of the threshold value of one of the lasers, with the value for the other laser fixed. "15" indicates that all the tests are passed. The dotted lines show the range 50 ±0.02%, and it is found empirically that the 1-Gbit sequences generated within this range can pass the frequency test of the NIST tests. The 0/1 ratio is a good indicator of the randomness of the bit sequence. (K. Hirano, K. Amano, A. Uchida, S. Naito, M. Inoue, S. Yoshimori, K. Yoshimura, and P. Davis, (2009), © 2009 IEEE.)

randomness of the bit sequence. This is useful since the 1/0 ratio is easy to monitor in real time and could be used for real-time adaptive control of system parameters to maintain the quality of the randomness.

Next, the dependence of randomness on the external cavity lengths of the two lasers is investigated. The cavity lengths are varied over the range 0.6–3.1 m. Table 10.4 shows the summary of the number of passed tests for NIST SP 800-22 at different combinations of the two external cavity lengths for the two lasers. In each case, the ratio of 0/1 is optimized by adjusting the threshold values. In Table 10.4, "15" indicates that all the tests are passed. All of the 15 tests are passed for many

Table 10.4 Summary of the number of passed tests for NIST SP 800-22 at different combinations of the two external cavity lengths for the two lasers.

	0.6 [m]	0.7 [m]	1.4 [m]	2.0 [m]	2.4 [m]	2.8 [m]	3.1 [m]
0.6 [m]	5	15	15	15	15	15	15
0.7 [m]	15	10	15	15	15	15	15
1.4 [m]	15	15	12	15	15	14	15
2.0 [m]	15	15	15	14	15	15	15
2.4 [m]	15	15	15	15	13	15	15
2.8 [m]	15	15	14	15	15	13	15
3.1 [m]	15	15	15	15	15	15	15

combinations of the external cavity lengths, but not when the external cavity lengths are the same (diagonal components in Table 10.4) or when the lengths have the particular 1:2 combination of 1.4 m and 2.8 m. This demonstrates that the two external-cavity round-trip times need to be an incommensurable ratio (or at least a high-order commensurable ratio) to avoid periodic recurrences in the extracted bit sequences. It is also necessary to ensure that the sampling (clock) time of the ADC and the external-cavity round-trip times are incommensurable. The requirement for random bit generation is summarized as follows: The lasers should have incommensurable external cavity lengths, such that $l\tau_{ext.1} \neq m\tau_{ext.2} \neq n\tau_s$ for any low-order integers l, m, n (where $\tau_{ext.1}$ and $\tau_{ext.2}$ correspond to the external-cavity round-trip times of the two lasers and τ_s is the sampling time).

The effects of the injection currents of the two lasers are also investigated. It is found that the number of passed NIST tests increases as the injection currents are increased, because the center frequency of the radio-frequency (RF) spectra also increases. Since the clock frequency is fixed, faster chaotic waveforms can result in more random bit sequences. Higher relaxation oscillation frequencies of the two lasers are preferable for random bit generation with a fixed clock frequency.

Finally, the effect of the strength of the feedback light in the external cavity is investigated. It is found that there is an overall tendency for the difference in heights of spectral peaks to decrease, that is, for the RF spectrum envelope to become flatter, as intermediate feedback strength is used (see Figures 4.5 and 4.6 in Section 4.1.3). The RF spectra with less difference (more flatness) in the peak heights are suitable for random bit generation (e.g., Figures 4.5d and 4.6d); that is, the RF spectra need to be as flat as possible to obtain high randomness of bit sequences generated from the chaotic waveforms. Note that it is important to adjust both the injection current and the feedback strength together, since the optimal feedback strength depends on the laser power.

The use of two lasers enhances the randomness of the bit sequences and also makes the generation more robust. A very simple method could be used for extracting bits from the optical waveform, namely 1-bit sampling of optical intensity at fixed clock rate and fixed threshold level. Bit sequences generated from just one laser in this way tend to exhibit periodic recurrences associated with the harmonic relations between the relaxation oscillation frequency and the clock frequency. When two lasers are used, the additional characteristic frequencies result in increasing recurrence times in the combined state of two waveforms, to the extent that, combined with the effects of oscillation instability, recurrences are no longer detectable in the output of the XOR device. In addition, the 0/1 ratio is more finely controllable and robust against variation of the detection threshold in the system with two laser outputs and two threshold values.

The procedure of preparing for high-quality random bit generation in the two-laser scheme with 1-bit ADC and XOR operation is summarized in the following way (Hirano et al., 2009). Initially, the injection current, the length of the external cavity and the external feedback strength are roughly adjusted to put the lasers in a regime of high-bandwidth chaos. The parameters of the two lasers are finely adjusted to "detune" their chaotic oscillations so their external cavity frequencies and the clock frequency are incommensurable, and correlations are small. The injection current and the feedback strength need to be adjusted in order to obtain a flat RF spectrum with less difference in the peak heights. The threshold values of the detectors are then adjusted to optimize the ratio of 0/1 close to 50% in the output of the final XOR stage. The detailed explanation of this procedure is summarized in Appendix 10.A.1.

10.3.2
Monobit Generation with One Laser

The two-laser method described in Section 10.3.1 is very effective to generate random numbers since two independent chaotic waveforms and two controllable threshold values are used. However, it is costly to have two lasers and two photoreceivers to be implemented in a package or in an integrated chip. In this section, a modified method with the output from one chaotic laser and its time-delayed signal is proposed.

The block diagram of the monobit generation method with one laser is shown in Figure 10.7. The output of a chaotic semiconductor laser is detected by using a photoreceiver to convert the optical signal to the electronic signal. The detected electronic signal is divided into two signals by a power divider; one of them is time delayed to the other. The two signals (the chaotic semiconductor laser output and its time-delayed signal) are converted into digital bit sequences by comparing with a threshold value for each chaotic signal. The two-bit sequences are combined with a logical XOR operation and the resulting bit sequences are considered as random numbers.

The delay time between the two chaotic waveforms needs to be carefully adjusted. The three time scales (i.e., the delay time between the two chaotic waveforms, the external-cavity round-trip time of the laser, and the sampling time) need to be incommensurable in order to avoid the recurrence of the periodicity. Bit sequences obtained from this scheme with appropriate parameter settings are tested with the statistical tests of the NIST SP 800-22 and Diehard tests. The results confirm that random bit sequences obtained from this scheme can be sufficiently random that they pass the NIST SP 800-22 and Diehard tests (Hirano et al., 2010). This method is used in the scheme described in Section 10.3.5.

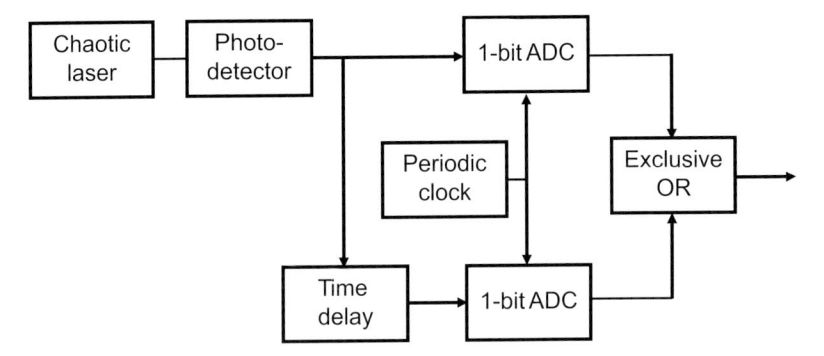

Figure 10.7 Schematic diagram of random bit sequence generator using one chaotic laser and its time-delayed signal. ADC, 1-bit analog-to-digital converter.

10.3.3
Multibit Generation with One Laser

In Sections 10.3.1 and 10.3.2, a scheme for bit extraction that requires just a single binary threshold operation at a fixed clock rate is considered. This scheme requires accurate control of the threshold voltages for 1-bit ADCs and precise parameter tuning for external cavity length and clock period to achieve robust practical generation of random bit streams. However, it has been shown that a higher effective bit rate can be achieved by extracting multiple bits from the difference between successive 8-bit samples of laser intensity in offline processing (Reidler et al., 2009). A rate equivalent to 12.5 Gb/s has been achieved by extracting 5 random bits from the difference in 8-bit samples acquired at a 2.5-GHz sampling rate.

Figure 10.8 shows a schematic diagram of the RNG setup. The laser diode wavelength is near 656 nm and partial feedback is obtained from a reflector placed at a photon round-trip time of 12.22 ns. The laser is operated moderately above the threshold current, $I = 1.55 I_{th}$, and the optical feedback strength is a few percent of the output intensity, though these parameters can be varied without affecting the RNG. The AC component of the detected signal is digitized by an 8-bit ADC triggered by a 2.5-GHz clock. The digital signal is stored and the difference between consecutive digital values is obtained using a software implementation. The m least significant bits (LSBs) of the difference value are stored as the next m bits in a final RNG string ($m \leq 8$). The rate of random number generation is therefore m times the ADC clock rate, and m can be varied up to a maximum value that depends on the resolution of the ADC.

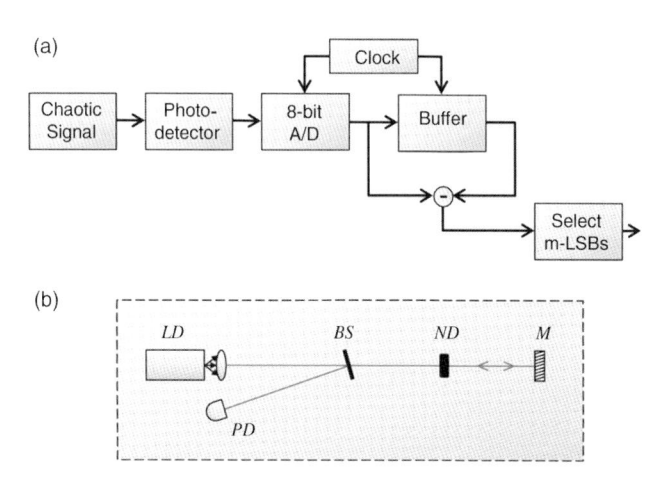

Figure 10.8 (a) A schematic diagram of the RNG. (b) Laser implementation, laser diode (LD), beam splitter (BS), neutral density filter (ND), mirror (M), high-speed photodetector (PD). (I. Reidler, Y. Aviad, M. Rosenbluh, and I. Kanter, (2009). © 2009 APS.)

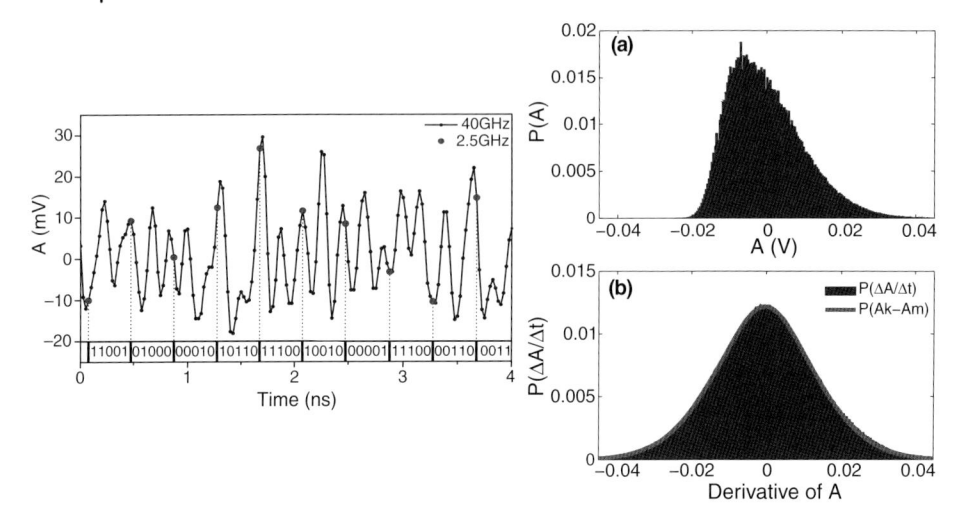

Figure 10.9 (Left) A 4-ns trace of laser intensity digitized at 40 GHz (small dots connected by a line to guide the eye), and the sampling points at 2.5 GHz (large circles). At each sampling point, m-LSBs obtained from the difference between the current and the previous sampled point are generated and attached to the random bit stream. The generated bit stream for $m = 5$ is depicted in the strip below the signal trace. (Right) (a) Histogram of the laser intensity obtained via an 8-bit ADC. (b) Histogram of the derivative of the laser intensity as obtained from the time series of laser intensities. The solid curve represents a histogram generated from the histogram shown in Figure 10.28(a), whereby the temporal position of the values has been eliminated. (I. Reidler, Y. Aviad, M. Rosenbluh, and I. Kanter, (2009). © 2009 APS.)

Figure 10.9 (left) shows the AC component of the chaotic signal as recorded at a 40-GHz digitizing rate by the 8-bit ADC (Reidler et al., 2009). The characteristic fluctuation time of the intensity is clearly shorter than the 2.5-GHz sampling rate used for the RNG, indicated by the large circles in the figure.

The raw data is converted into a time series consisting of the derivative of the ADC signal amplitudes A_t, that is, $\Delta_t = A_t - A_{t-1}$. The histogram of Δ_t exhibits a very high degree of symmetry and is unbiased, compared with that of A_t, as shown in Figure 10.9 (right). Since adjacent Δ_t values are temporally correlated even if the original amplitudes are random, (for instance, it is very unlikely or impossible to have two successive large positive Δ_t, since the amplitudes are bounded from above), one might expect that such a series is not a good candidate for a random sequence. This difficulty is solved, however, by taking into account only the m LSBs of each Δ_t and relying on the chaotic nature of the time-varying laser intensity. The values of the discrete derivatives form a highly symmetric histogram, centered at zero, and allows for the unbiased, even division of the bins into even or odd bins based on the LSB of the bin.

The bit sequences obtained from the differentiated chaotic laser intensity fluctuations using the 5 LSBs passed all of the NIST SP 800-22 and Diehard tests, thus allowing effective generation of random bits at an equivalent rate of 12.5 Gb/s (= 2.5 gigasamples per second (GS/s) × 5 LSBs). As expected, random sequences with verified randomness can be generated also with a lower m, as was verified, for instance, in the tests for $m = 4$, but not for $m \geq 6$ for the 8-bit ADC. The maximum value of m_{max} to pass all of the tests depends on the shape of the distribution of intensity derivatives. When this distribution becomes narrow m_{max} decreases since the distribution of all m_{max} bits becomes biased (see Section 10.3.5).

10.3.4
Postprocessing for High-Speed Random Number Generation

Postprocessing is an effective technique to increase the generation speed of random number generators as well as to enhance the quality of random numbers (see Section 10.7 for details). A random bit generator at up to 300 Gb/s generation rate has been reported (Kanter et al., 2010). In this method high derivatives of a chaotic signal, which eliminates the necessity of using incommensurable timing, are used to generate random numbers. The problem of bias is also eliminated, because the distribution of results for the derivative function is highly symmetric, smooth and can be evenly divided into equal numbers of ones and zeroes by a mapping of even/odd bins to logical 0/1 values. Two statistical tests (NIST SP 800-22 and Diehard) have become accepted as the standard by which random sequences are certified and are used in this work to confirm the random nature of the generated bit strings.

The experimental system consists of a semiconductor laser emitting near a wavelength of 660 nm, operated at ~1.5 times the threshold current. The laser is subject to delayed self-feedback with a propagation loop time ($\tau = 10$ ns) as shown in Figure 10.10a (Kanter et al., 2010). A neutral density filter is used to control the feedback strength and a beam splitter is used to couple a small portion of the beam into a fast detector biased with a bias tee. The AC current from the bias tee is sampled by an 8-bit ADC capable of 40 GHz sampling rate with 12 GHz analog bandwidth.

The intensity fluctuations of the chaotic laser in the experiments are digitized at a 20-GHz sampling rate with an 8-bit resolution as illustrated by the large circles in Figure 10.10b. At this digitization rate, each intensity spike may be sampled at a number of sample points, all lying on the same spike, as shown in Figure 10.10b. The generation of the random bit stream consists of the following two steps. First, the nth derivative is calculated (this uses $n + 1$ successive values of the recorded waveform). Secondly, the m least significant bits (LSBs) of the results of the nth derivative are appended to the bit sequence. A schematic of the algorithm is illustrated in Figure 10.10c, and the example in Figure 10.10d shows the calculated decimal value of the 5-LSBs obtained from the fourth derivative. In the particular example shown in the figure (4th derivative, 20 GS/s, 5-LSBs) the random bit-generation rate is 100 Gb/s. The random bit sequence passes the NIST SP 800-22 and the Diehard tests. The maximum generation rate of 300 Gb/s (16th derivative, 20 GS/s, 15-LSBs) is obtained in the experiments.

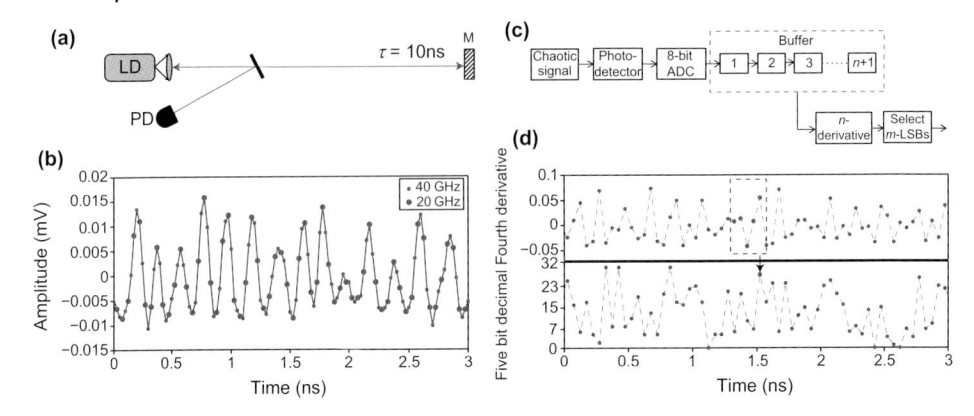

Figure 10.10 Schematic of the random bit generator and examples of digitized data. (a) Schematic of the optical setup (LD, laser diode; M, mirror, PD, high-speed photodetector. (b) A 3-ns recording of the fluctuating laser output at 40 and 20 GHz sampling rates. (c) Operational schematic of the random bit generator. (d) Fourth derivative of the digitized laser signal, at a sampling rate of 20 GHz (upper panel). Decimal representation of the 5-LSBs of the fourth derivative (lower panel). The box in the upper panel indicates the five successive sampling data points required for the calculation of the fourth derivative. (I. Kanter, Y. Aviad, I. Reidler, E. Cohen, and M. Rosenbluh, (2010). © 2010 Nature.)

For a given order of derivative of the digitized signal, two parameters control the maximum generation rate: the sampling rate of the signal and the number of retained LSBs. Use of higher-order derivatives enables an increase in the generation rate by allowing the retention of more LSBs and, simultaneously, allowing the use of a higher sampling rate. To demonstrate the dependence of the generation rate on the order of derivative, the maximum possible generation rate for each derivative order is determined. The number of retained LSBs is allowed to vary up to the maximum number of bits available for a given order of derivative, and the sampling rate is allowed to vary up to 20 GHz, the maximum sampling rate used in the data acquisition. For derivatives higher than or equal to two, it is possible to use the 20-GHz sampling rate, and the increase in maximal generation rate results solely from the increase in the number of retained LSBs. The maximum generation rate increases linearly with the order of derivative up to 300 Gb/s (= 15 LSBs × 20 GS/s) in this experiment.

The use of higher derivatives may enable the use of an increasing sampling rate and an increasing number of retained LSBs. However, extracting more bits from high derivatives could be more susceptible to the effects of physical noise in the ADC – a potential additional source of randomness that is separate from the characteristics of the chaotic lasers. A strong motivation for using direct sampling of optical chaos for random number generation is to reduce the dependence on digital electronic operations, which may be difficult to implement at high frequencies, and that in principle cannot increase the rate for generation of nondeterministic bits. From these points of view, it is very important to increases the bandwidth of the chaotic lasers used for random number generation, as described in the next section.

10.3.5
Bandwidth Enhancement of Chaotic Lasers for High-Speed Random Number Generation

One promising approach to increasing the generation speed of random bit generators is a technique for bandwidth enhancement of optical chaos in semiconductor lasers (Simpson et al., 1995; Uchida et al., 2003b; Someya et al., 2009). The bit rate of generated random sequences is limited by the dominant oscillation frequency of chaotic waveforms, which is around the relaxation oscillation frequency of a several GHz. To achieve significantly higher rates it is necessary to largely enhance the bandwidth of chaotic waveforms. Random bit generation by using bandwidth-enhanced chaos in semiconductor lasers has been experimentally demonstrated (Hirano et al., 2010). Chaotic fluctuation of laser output is generated in a semiconductor laser with optical feedback and the chaotic output is injected into a second semiconductor laser for bandwidth enhancement. The enhancement of the chaos bandwidth up to 16 GHz has been achieved, and random bit generation at rates equivalent to 75 Gb/s has been realized with multibit sampling at 12.5 GS/s with 6 LSBs.

Figure 10.11 (top) shows the experimental setup for the scheme of fast physical random bit generation with bandwidth enhancement of chaos (Hirano et al., 2010). Two DFB semiconductor lasers mounted in butterfly packages with optical-fiber pigtails are used (optical wavelength of 1547 nm), developed for optical-fiber communications. One laser (referred to as Laser 1) is used for the generation of chaotic intensity fluctuations induced by optical feedback. The other laser (referred to as Laser 2) is used for the bandwidth enhancement of chaotic waveforms. The relaxation oscillation frequencies are set to be 6.5 GHz for both Laser 1 and 2 by adjusting the injection current of the lasers. Both Laser 1 and 2 are prepared without standard optical isolators, to allow optical feedback and injection. Laser 1 is connected to a fiber coupler and a variable fiber reflector that reflects a fraction of the light back into the laser, inducing high-frequency chaotic oscillations of the optical intensity. The fiber length between Laser 1 and the variable fiber reflector is 4.55 m, corresponding to a feedback delay time (round-trip) of 43.8 ns. On the other hand, there is no optical feedback for Laser 2.

A portion of the chaotic Laser 1 beam is injected into Laser 2 for bandwidth enhancement of chaos. Two optical isolators are used to achieve one-way coupling from Laser 1 to Laser 2. A portion of Laser 2 output is extracted by a fiber coupler, and divided into two beams by another fiber coupler. An extra optical fiber (1-m length) is inserted into one of the optical paths after the two beams are divided, so that a chaotic waveform and its time-delayed signal (5.0 ns delay) can be detected by two 25-GHz photodetectors.

To enhance the bandwidth of chaos, the optical wavelength of Laser 2 is detuned to the positive direction with respect to that of Laser 1, that is, the optical wavelength is set to be 1547.585 nm for Laser 1 and 1547.718 nm for Laser 2 by precisely controlling the temperature of the two lasers. The optical wavelength detuning ($\Delta\lambda = \lambda_2 - \lambda_1$, λ is the wavelength) is thus set to 0.133 nm (16.6 GHz in frequency), which corresponds to the positive detuning condition (see Sections 6.1 and 6.3). No injection locking (no matching of the optical wavelengths) is achieved between Laser 1 and 2 at this

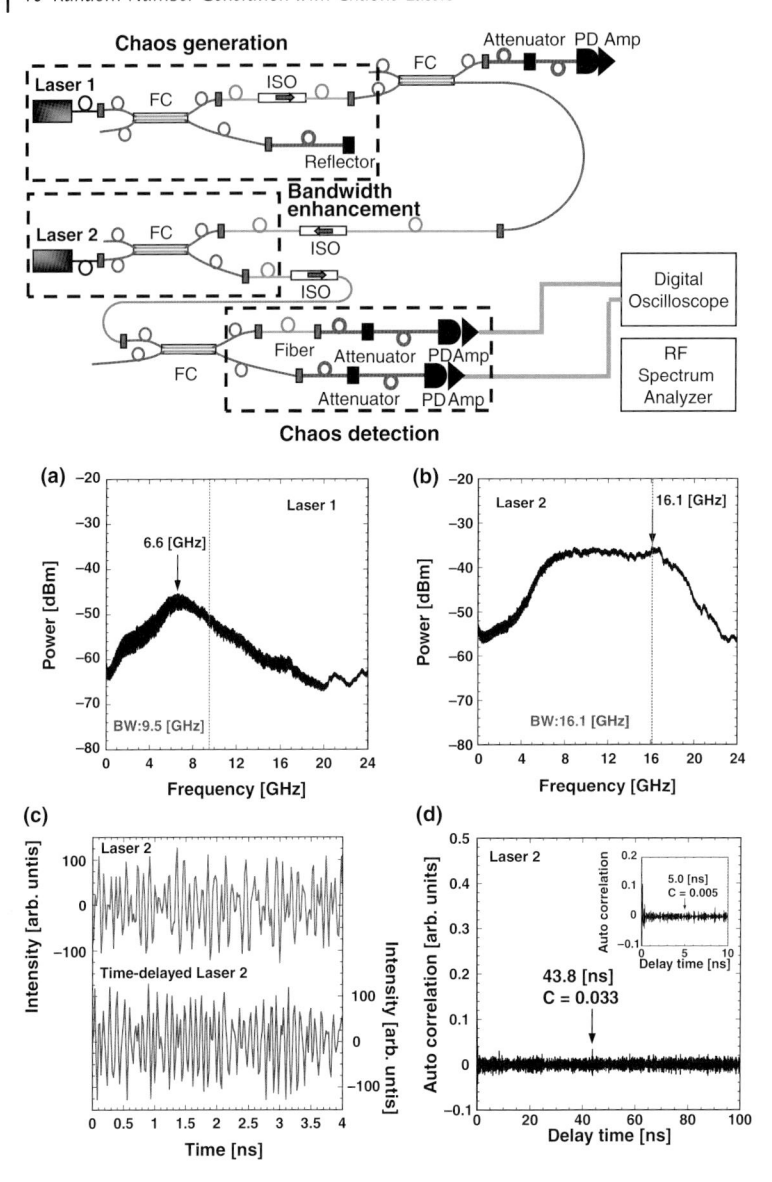

Figure 10.11 (Top) Experimental setup for random bit generation with chaotic lasers. Amp, electronic amplifier; FC, fiber coupler; ISO, optical isolator; PD, photodetector. (Bottom) (a) RF spectrum of Laser 1, (b) RF spectrum of Laser 2, (c) temporal waveforms of Laser 2 output and the same output optically delayed in time (5.0 ns delay), and (d) autocorrelation function of the temporal waveform of Laser 2 output. The inset is the enlargement of (d). (a), (b) BW: bandwidth. (K. Hirano, T. Yamazaki, S. Morikatsu, H. Okumura, H. Aida, A. Uchida, S. Yoshimori, K. Yoshimura, T. Harayama, and P. Davis, (2010), © 2010 OSA.)

condition. The existence of the frequency component corresponding to the optical wavelength detuning $\Delta\lambda$ is crucial for the bandwidth enhancement of the laser chaos; it results in nonlinear frequency mixing between the optical wavelength detuning and the relaxation oscillation frequency of the laser.

Figures 10.11a and b (bottom) show the RF spectra of Laser 1 and 2. Bandwidth enhancement of Laser 2 by the optical injection of the Laser 1 output is achieved, where the center frequency is changed from 6.6 to 16.1 GHz. The bandwidths of Laser 1 and 2, defined as the frequency band starting at zero frequency and containing 80% of the spectrum power (Someya *et al.*, 2009), are 9.5 and 16.1 GHz, respectively. It can also be seen that the RF spectrum of the output of Laser 2 is much flatter than that of Laser 1, where the flatness of the RF spectrum is advantageous for the generation of random bit sequences. The enhancement of chaos bandwidth up to 16 GHz can result in generating random bit sequences by sampling at the rate of 12.5 GS/s. Figure 10.11c (bottom) shows the 8-bit-resolution temporal waveform of the Laser 2 output and the same output optically delayed by 5.0 ns, used for the generation of random bit sequences. The autocorrelation function of the Laser 2 output is also shown in Figure 10.11d (bottom). The peak value of the autocorrelation function at 43.8 ns (i.e., the round-trip time of the external cavity for Laser 1) is only 0.033, showing that periodicity due to the external cavity of the Laser 1 is suppressed in the Laser 2 output. The correlation value at the time of 5.0 ns (the delay time between the two detected signals) is only 0.005, as shown in the inset of Figure 10.11d (bottom).

Random bit sequences are generated by using the output of Laser 2 and its time-delayed signal. The two chaotic optical signals are detected by AC coupled photodetectors, amplified and converted to digital 8-bit signals by a dual-channel oscilloscope sampling at 12.5 GS/s per channel. Corresponding pairs of bits in the two 8-bit digital signals are combined by bitwise XOR operation, giving a single 8-bit digital signal, as illustrated in Figure 10.12 (top). A subset of m LSBs from each sample are then selected and interleaved to generate a single bit sequence (Reidler *et al.*, 2009). In the experimental system, the XOR operation and bit interleave are done offline after acquisition by the oscilloscope.

The randomness of generated bit sequences is tested using a standard statistical test suites, NIST SP 800-22 and Diehard test suites. It is confirmed that sequences obtained by interleaving up to 6 consecutive LSBs at each sample passed the statistical tests of randomness. These results correspond to random bit-generation rates from 12.5 to 75 Gb/s ($= 12.5$ GS/s $\times$ 6 LSBs).

To investigate the dependence of the number of LSBs on the randomness of bit sequences, the probability density function of multibit states obtained from various LSBs are measured. Figure 10.12a (bottom) shows the distribution for 8-bit samples from one chaotic waveform. Figure 10.12b (bottom) shows the distribution of 8-bit XOR samples after bitwise XOR of 8-bit samples from a chaotic waveform and the delayed waveform. The distribution becomes more uniform, but there are some discontinuities in the distribution of Figure 10.12b (bottom). When only 7 LSBs of the 8-bit XOR data are selected (Figure 10.12c (bottom)), the distribution becomes flatter. In the case of 6 LSBs of the 8-bit XOR data (Figure 10.12d (bottom)), the distribution appears almost uniform. This result is consistent with the fact that the random bit

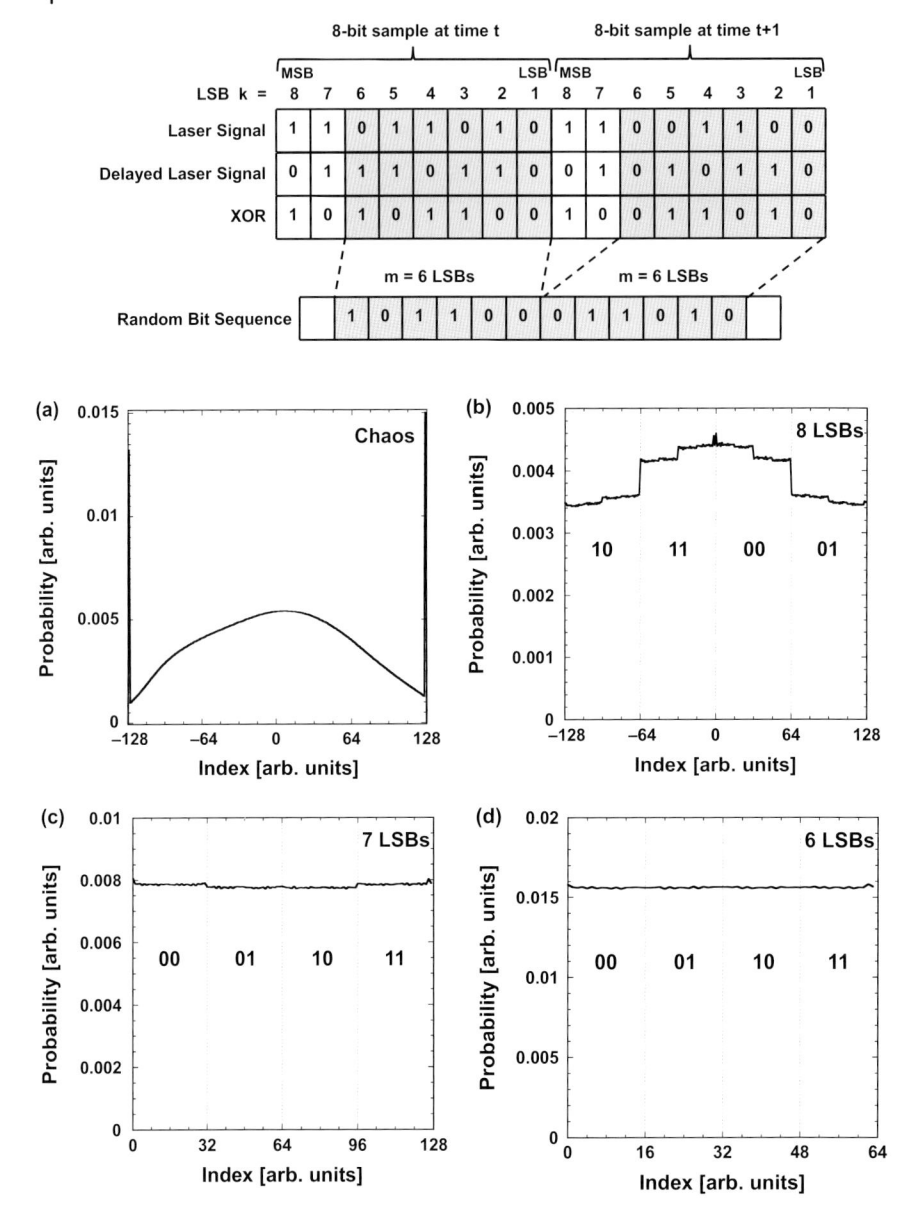

468 | 10 Random Number Generation with Chaotic Lasers

Figure 10.12 (Top) Method for generating a random bit sequence using multiple ($m = 1, \ldots, 8$) significant bits. Example for $m = 6$. LSB, least significant bit; MSB, most significant bit; XOR, exclusive-OR operation. (Bottom) Probability density functions for (a) 8-bit digitized chaotic waveform, (b) 8-bit random bits after bitwise XOR operation is applied to the two 8-bit digitized chaotic waveforms, (c) 7 LSBs selected from the 8-bit XOR data, and (d) 6 LSBs selected from the 8-bit XOR data. The two-bit labels shown in (b)–(d) indicate the first two most significant bits (MSBs) of the data generated from m LSBs. (K. Hirano, T. Yamazaki, S. Morikatsu, H. Okumura, H. Aida, A. Uchida, S. Yoshimori, K. Yoshimura, T. Harayama, and P. Davis, (2010), © 2010 OSA.)

sequences generated from up to 6 LSBs can pass the statistical tests of randomness. The uniformity of the probability density function of multibit states is crucial to achieve high-quality random bit sequences in the multibit generation scheme.

A scheme for fast random bit generation with a post-processing method has been demonstrated at a rate of 400 Gb/s with the same experimental setup for the bandwidth-enhanced chaotic semiconductor lasers (Akizawa et al., 2011). Chaotic laser output and its time-delayed signal are sampled at 50 GS/s and converted into 8 bit values. The order of the 8 bit sequences of the time-delayed signal is reversed and bitwise XOR operation is executed between the bit-order-reversed signal and the original chaotic signal, resulting in random bit sequences. For this method no elimination of some of the MSBs is required to obtain good-quality random bit sequences. Experiments with off-line post-processing indicate that the equivalent generation rate of random bit sequences can be achieved up to 400 Gb/s ($= 8$ bit $\times$ 50 GS/s). The randomness of the generated bit sequences are tested by using statistical tests of randomness. This method can succeed in generating uniform distribution of the bias of each significant bit (Akizawa et al., 2011).

10.3.6
Photonic Integrated Circuit for Random Bit Generators

A photonic integrated circuit (PIC) has been applied for a compact physical random bit generator (Argyris et al., 2010b). The proposed generator is simple and consists of the PIC, a photodetector and an oscilloscope. Depending on the operating conditions of the PIC and by using the LSB extraction postprocessing, random bit sequences extracted from the oscilloscope Ethernet output port, with verified randomness and rates as high as 140 Gb/s are obtained.

The PIC employed in this work has been specifically designed and fabricated for providing fully controllable and stable chaotic optical signals at 1556 nm with broadband microwave spectral characteristics over 10 GHz (Argyris et al., 2008). As shown in Figure 10.13a (top), it consists of a DFB laser followed by an integrated 10-mm passive waveguide cavity with a highly reflective coated (HRC) end; an active gain-absorption section (GAS) that controls the optical feedback of the cavity and a passive phase section (PH) that controls the phase of the generated optical field are also included within the cavity (Argyris et al., 2010b) (see also Figure 8.19 (left) in Section 8.5.1). The chaotic signal generation is based on the delayed optical feedback effect. The prerequisite for a potential analog signal to seed successfully an ultrafast random bit generation is a broad and flat spectrum without periodicities. Under these specifications, the most prominent operating condition of the PIC is selected at high injection current values of 50 mA, producing the bandwidth of chaotic laser output over 10 GHz.

The proposed random bit-generation scheme is shown in Figure 10.13b (top). The chaotic optical signal from the PIC is fed through its fiber pigtail to a 15-GHz photodetector (PD) after optical isolation in order to eliminate any residual reflectivity that would disturb the PIC operation. The converted electrical signal is sent finally to an oscilloscope with 40 GS/s sampling rate and 12 GHz bandwidth. The oscilloscope's 8-bit ADC, along with the internal 16-bit digital-to-analog converter

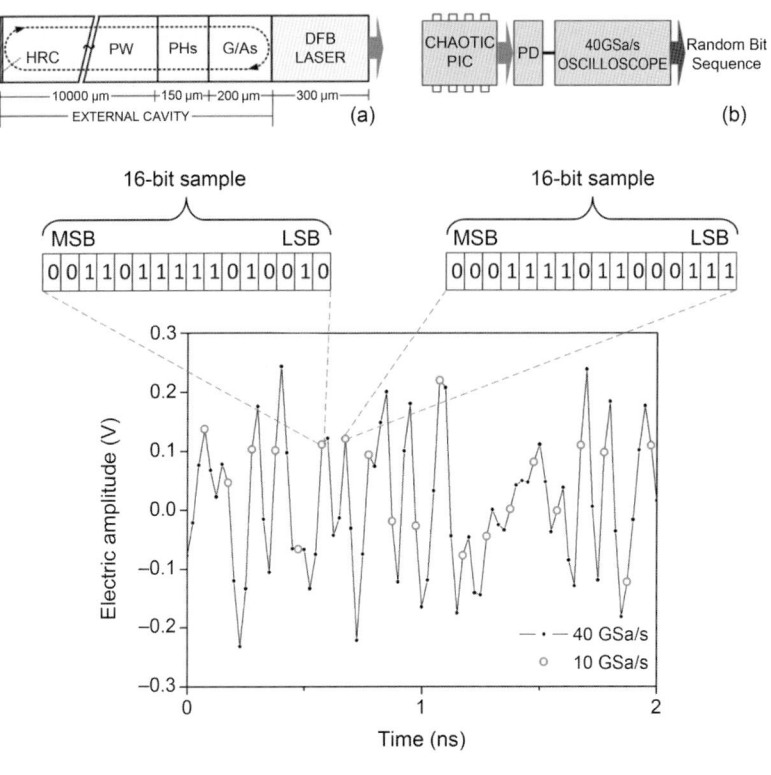

Figure 10.13 (Top) (a) Schematic of the photonic integrated circuit (PIC). (b) A physical random bit generator based on the emitted chaotic signal from the PIC. (Bottom) Time trace of the chaotic analog signal. The sampling rate is set to 40 GS/s (solid dots) while only 10 GS/s (open circles) are considered for the output of the random bit generator. Only k LSBs out of the 16-bit digitized representation of each sample is included in the final output sequence. (A. Argyris, S. Deligiannidis, E. Pikasis, A. Bogris, and D. Syvridis, (2010b). © 2010 OSA.)

(DAC) and the rest processing units, provides a "word-type", noise-enhanced, 16-bit output binary sequence for each sample. An external downsampling to 10 GS/s is applied (i.e., one out of each four samples is considered) in order to eliminate any effect of interpolation samples of the oscilloscope. In such a way the initial bandwidth of the signal is preserved. The downsampling operation in the chaotic time trace is shown in Figure 10.13 (bottom). The number k of the LSBs retained in each sample of the output sequence determines the final bit rate generation.

The randomness of the PIC-based random bit generator is verified when all the 15 statistical tests of the NIST SP 800-22 test suite are passed. The characterization is investigated as the biasing laser current and the number of LSBs (from $k=1$ to $k=16$) of each sample are changed. The absence of the external cavity modes and the powerful broadband spectral distribution of the chaotic signal allows the favorable performance. All NIST SP 800-22 tests are passed, even when including 14 LSBs, which is translated into a bit-generation rate of 140 Gb/s (= 10 GS/s × 14 LSBs).

Similar PICs designed specifically for high-speed random number generators have been proposed (Harayama et al., 2011). Figure 10.14 shows the random bit-

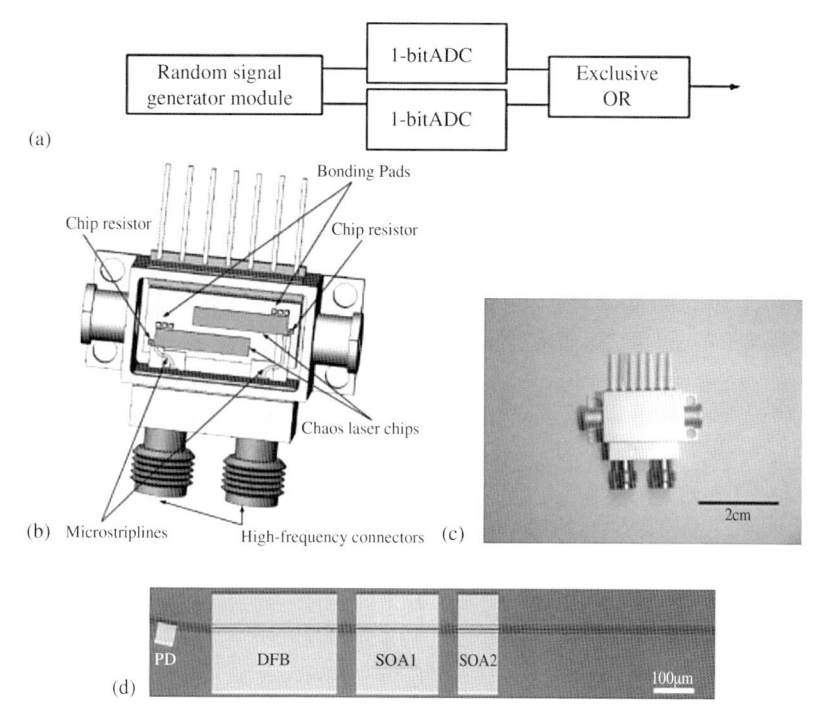

Figure 10.14 Random bit-generation scheme and PIC device structures. (a) Schematic diagram of random bit generation. ADC: analog-to-digital converter. (b) Schematic of the random signal generator module consisting of two chaos laser chips. Chip resistors are used for impedance matching and current is injected to the distributed feedback (DFB) laser and semiconductor optical amplifier (SOA) contacts of the chaos laser chips via bonding pads (wires not shown here). Electrical signal is output from the photodiodes (PDs) in the chaos laser chips via microstriplines to high-frequency connectors. (c) Photo of a random signal generator module. (d) Optical image showing the monolithically integrated optical components. A small section of the 10-mm-long passive waveguide for optical feedback can be seen. DFB laser, semiconductor distributed feedback laser; SOA1 and SOA2, semiconductor optical amplifiers, The width of the waveguide is 2 μm. The lengths of the PD, DFB laser, SOA1, SOA2, and passive waveguide are 50, 300, 200, 100, and 10 000 μm, respectively. (T. Harayama, S. Sunada, K. Yoshimura, P. Davis, K. Tsuzuki, and A. Uchida, (2011), © 2011 APS.)

generation scheme and the PIC used in the experiment. Two DFB laser chips are contained in a single module with two high-frequency connectors to output the electrical signals from the integrated photodiodes, as shown in Figures 10.14b and d. For the generation of high-frequency chaotic laser output, high-reflective coating at the edge of the passive waveguide reflects the light back into each DFB laser. The feedback delay length is set at 10 mm. The strength and phase of the optical feedback is controlled with the current of semiconductor optical amplifiers.

The AC components of the electrical signals from the photodiodes are digitized at a 2.08-GHz sampling rate. The AC signals are converted to binary signals by comparing with a threshold voltage (i.e., 1-bit ADC), and finally the binary bit signals are combined by a logical XOR operation to generate a single random bit sequence, as shown in Figure 10.14a (see also Section 10.3.1). Bit sequences obtained from the experimental device for sampling rate up to 2.08 Gb/s passed all of the statistical tests of randomness (Harayama *et al.*, 2011). This achievement opens the door to reduction of size and cost of fast physical random bit generators operating at rates beyond Gb/s.

Another PIC with a passive ring waveguide has been proposed for random number generation (Sunada *et al.*, 2011). The PIC consists of a DFB semiconductor laser, two semiconductor optical amplifiers (SOAs), a photodiode, and a passive ring waveguide. The ring-type structure with the two separate SOAs achieves stronger delayed optical feedback, which makes possible the generation of strong broadband chaotic signals with flat RF spectrum up to 10 GHz. High-quality random bit sequences that passed statistical tests of randomness can be generated at a rate up to 1.56 Gb/s (Sunada *et al.*, 2011).

10.4
Application of Chaotic-Laser-Based Random Number Generators to High-Speed Quantum Key Distribution

One of the potential applications of fast physical random number generators is quantum key distribution (QKD) (Bouwmeester *et al.*, 2000; Gisin *et al.*, 2002). Many QKD experiments have been successfully demonstrated, and the performance of the QKD system has been improved. In particular, GHz-clocked differential phase-shift QKD (DPS-QKD) experiments have been demonstrated (Inoue *et al.*, 2003; Honjo *et al.*, 2004). In order to properly implement such a high-speed QKD system, a high-speed physical random number generator is indispensable.

The first GHz-clocked DPS-QKD experiment that employs a fast physical random bit generator has been reported (Honjo *et al.*, 2009). A 1-Gb/s random bit signal that is generated by a physical random bit generator with chaotic semiconductor lasers (described in Section 10.3.1) is used to perform random phase modulation in DPS-QKD system. Stable operation for over one hour has been successfully demonstrated and a sifted key generation rate of 9.0 kb/s with a quantum bit-error rate (QBER) of 3.2% has been achieved after 25-km fiber transmission (Honjo *et al.*, 2009).

In the DPS-QKD scheme, a sender (called Alice) randomly phase modulates an optical pulse train of weak coherent states by $\{0, \pi\}$ for each pulse and sends it to a receiver (called Bob) with an average photon number of less than one per pulse. Bob measures the phase difference between two sequential pulses using a 1-bit delay Mach–Zehnder interferometer and photon detectors, and records the photon arrival time and the detector that observed the photon. After transmission of the optical pulse train, Bob tells Alice the time instances at which a photon was counted. From this time information and her modulation data, Alice knows which detector observed the photon at Bob's site. Under an agreement that an observation by one detector denotes "0" and an observation by the other detector denotes "1", Alice and Bob can obtain an identical bit string.

The DPS-QKD scheme has certain advantageous features including a simple configuration, efficient time-domain use, and robustness against photon number splitting attack. In particular, a high repetition frequency is possible through the use

10.4 Application of Chaotic-Laser-Based Random Number Generators

of one-way transmission and a pulse train. Long transmission distance and a high key generation rate in QKD systems were demonstrated experimentally in this scheme (Takesue et al., 2007).

A 1-GHz clocked DPS-QKD system using the physical random bit generator with chaotic semiconductor lasers is implemented. For the random bit generator the same experimental system shown in Section 10.3.1 (i.e., monobit generation with two lasers) is used with the clock rate reduced to generate a random bit stream at a repetition rate of 1.0 Gb/s with the NRZ format. The laser parameters are adjusted to achieve randomness of bit sequences at the required clock rate. It is confirmed that bit sequences generated at 1.0 Gb/s can pass all of the statistical tests of randomness in the NIST SP 800-22 test suite.

Figure 10.15 shows the experimental setup for the 1-GHz clocked DPS-QKD system using the physical random bit generator (Honjo et al., 2009). At Alice's site the

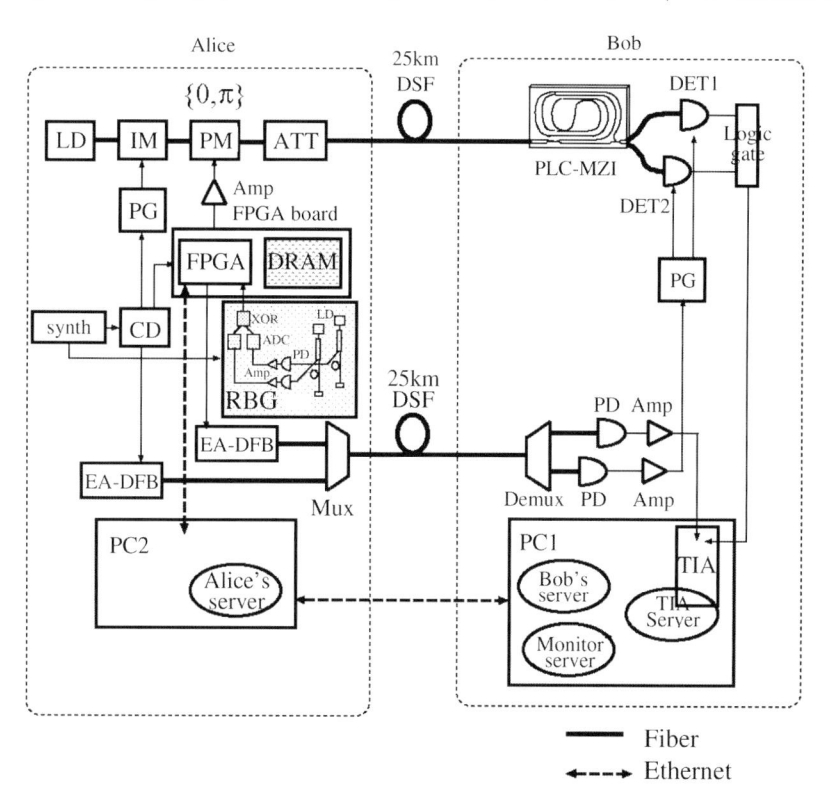

Figure 10.15 Experimental setup for the 1-GHz clocked differential-phase-shift quantum-key-distribution system using the physical random bit generator.: DET, single-photon detector; EA-DFB, distributed-feedback semiconductor laser; LD, laser diode; IM, intensity modulator; PG, pulse generator; PM, phase modulator; ATT, attenuator; PLC-MZI, planar lightwave circuit Mach–Zehnder interferometer; Synth, synthesizer; CD, clock divider; RBG, random bit generator; PD, photodiode; TIA, time interval analyzer; PC, personal computer. (T. Honjo, A. Uchida, K. Amano, K. Hirano, H. Someya, H. Okumura, K. Yoshimura, P. Davis, and Y. Tokura, (2009). © 2009 OSA.)

DPS-QKD system and the physical random bit generator are synchronized with a 1-GHz clocked synthesizer. The random bit signal from the random bit generator is captured by the field programmable gate array (FPGA) board. The FPGA board saves the random bit signal to the 512-MB dynamic random access memory (DRAM), and simultaneously generates a 1-GHz NRZ random signal to drive the phase modulator.

A quantum channel is organized as follows. A 1551-nm continuous light from an external cavity semiconductor laser is modulated into a pulse stream with a 1-GHz clock frequency using a $LiNbO_3$ intensity modulator. The intensity modulator is driven by the pulse generator synchronized with the 1-GHz clock. Each pulse is randomly phase modulated by $\{0, \pi\}$ with a $LiNbO_3$ phase modulator that is driven by the random bit signal from the FPGA board. The optical pulse is attenuated to 0.2 photons per pulse and then transmitted to Bob's site over a 25-km dispersion shifted fiber (DSF). After the transmission, the 1-GHz pulse stream is input into a Mach–Zehnder interferometer based on planar lightwave circuit technology. The path length difference is 0.20 m, corresponding to the optical pulse interval. The phase difference between the two paths in the Mach–Zehnder interferometer can be stably adjusted by controlling the temperature of the waveguide chip. The output ports of the Mach–Zehnder interferometer are connected to single-photon detectors (DET) based on InGaAs avalanche photodiodes.

The detected signals are input into a time-interval analyzer (TIA) by way of a logic gate to record the photon detection events. The TIA server continuously retrieves detection event packets from the TIA device, and sends them to Bob's server. From these packets, Bob's server generates his sifted key and sends the time information to Alice's server through the Ethernet. Bob's server also sends his key to the monitor server. On the other hand, Alice's server generates her key from the phase modulation information and the time information received from Bob's server. Her key is sent to the monitor server. The monitor server receives the keys from Alice's and Bob's servers, and estimates the key generation and QBER.

Figure 10.16 (left) shows the sifted key generation rate and quantum bit-error rate as a function of time. It is successfully demonstrated the continuous operation over an hour and generated sifted keys at a rate of 9.0 kb/s with an average QBER of 3.2% after 25-km fiber transmission. Note that the sifted key generation rate is not an estimated value but an actually obtained value including data processing, which means the sifted key is continuously generated at a rate of 9.0 kb/s at Alice's and Bob's servers. Although only sifted keys are generated in this experiment, the observed error rate is good enough to distill a secure key using error correction and privacy amplification. Based on an analysis of the security against a general individual attack (Inoue et al., 2003), the secure key generation rate is estimated to be 0.98 kb/s.

Figure 10.16 (right) shows the occurrence of bits "1" in the physical random bit generator as a function of time. The occurrence of bits 1 of 1-Mbit samples is measured to monitor the quality of the random bit signal. This confirms the stability of the operation of the physical random bit generator throughout the QKD experiment. The fast physical RNGs with chaotic semiconductor lasers are promising candidates for RNG components in high-speed QKD systems.

10.5 Numerical Evaluation of Random Number Generator as Entropy Source

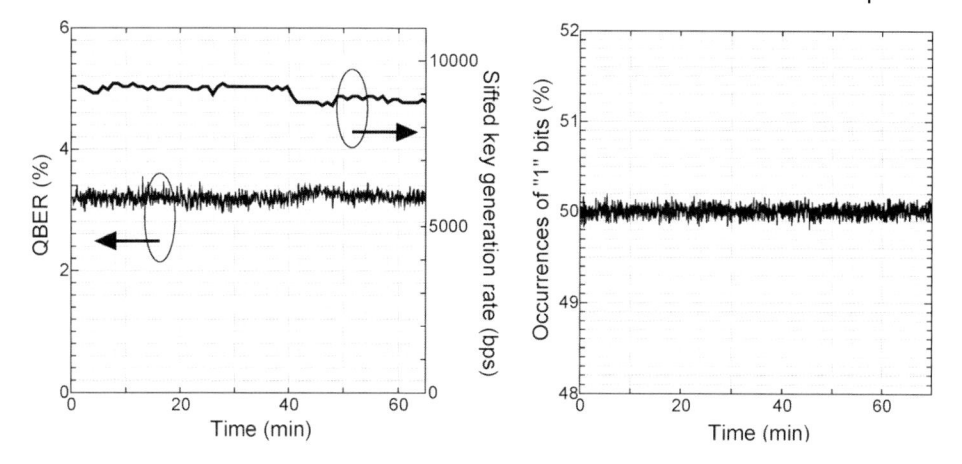

Figure 10.16 (Left) Experimental results of quantum bit-error rate and sifted key generation rate. (Right) Experimental result of occurrences of bits "1" for the fast physical random bit generator used in the quantum-key-distribution experiment. (T. Honjo, A. Uchida, K. Amano, K. Hirano, H. Someya, H. Okumura, K. Yoshimura, P. Davis, and Y. Tokura, (2009). © 2009 OSA.)

10.5
Numerical Evaluation of Random Number Generator as Entropy Source

10.5.1
Entropy Generation from Internal Noise by Chaotic Dynamics

It is considered that the output of chaotic devices can be unpredictable because they amplify intrinsic noise fluctuations. In any real physical devices, there is always some microscopic noise in the device, and the noise is amplified by the dynamical instability in a chaotic device so that the macroscopic state of the device cannot be determined after some transient time.

It is very important to estimate theoretically the rate at which nondeterministic random bits can be generated, that can be defined as "entropy rate". In Section 10.5 a numerical analysis is presented that shows how unpredictable random bits can be generated at fast rates in the chaotic laser system, and estimate the entropy rate for the generation of nondeterministic bits (Mikami et al., 2011). It is difficult to achieve this proof experimentally, because of the need to repeatedly prepare the laser in the same initial state.

A set of rate equations for a semiconductor laser with time-delayed optical feedback (i.e., the Lang–Kobayashi equations, see Chapter 4) is used in this numerical simulation. The gain-saturation effect is included in the rate equations, as described in Section 4.6. An intrinsic noise is added in the rate equations and the noise-amplification effect by chaotic dynamics is investigated. The noise is added to the same chaotic trajectory at $t = 0$. Figure 10.17 shows an example of two temporal waveforms starting from the same initial conditions with different intrinsic noise sequences. The two trajectories are separated due to the amplification of the small noise by chaotic dynamics. On the scale of this plot, the two trajectories appear to start diverging after

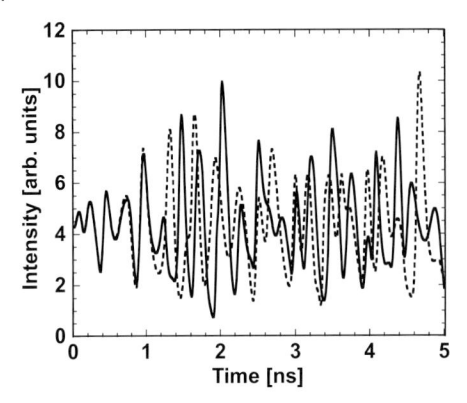

Figure 10.17 Example of two temporal waveforms starting from the same initial states with different small noise sequences. The two trajectories are separated due to the amplification of the small noise by chaotic dynamics.

1–2 ns. The difference in the trajectories indicates the loss of the memory of the initial conditions due to the internal noise amplification by the chaotic dynamics.

A simulation is executed for 1000 chaotic temporal waveforms that are generated from the same initial conditions, but added by different microscopic noise sequences. The probability density function (PDF) $p(t)$ (i.e., histogram) of the laser-intensity distribution for the ensemble of the 1000 temporal waveforms is calculated at time t. The time-dependent histogram $p(t)$ is obtained as shown in Figure 10.18. For $t = 0$,

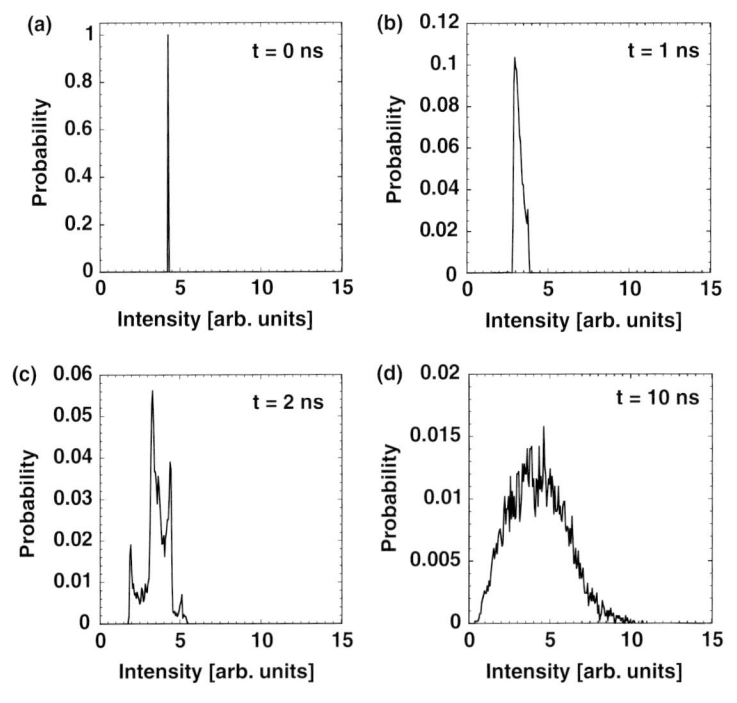

Figure 10.18 Time-dependent histogram $p(t)$ for the temporal waveforms starting from the same initial conditions with different small noise sequences. (a) $t = 0$ ns, (b) 1 ns, (c) 2 ns, and (d) 10 ns.

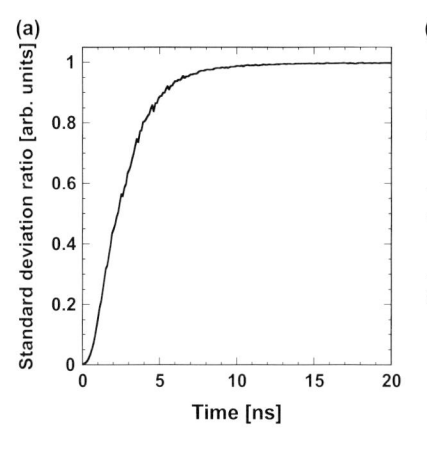

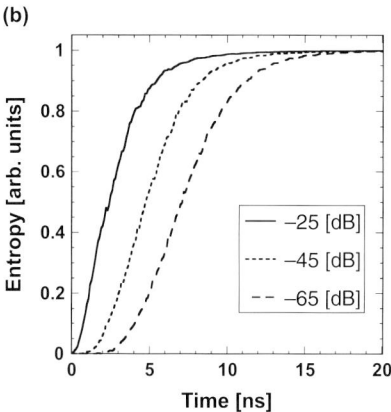

Figure 10.19 (a) Time-dependent standard deviation ratio $\sigma(t)/\sigma(\infty)$ of the histogram $p(t)$ as a function of time. (b) Bit entropy against time for an ensemble of trajectories starting at exactly the same state at $t=0$, but with different noise instances. The noise levels of -25, -45, and -65 dB are used.

the histogram has only one peak because all of the 1000 chaotic temporal waveforms start from the same initial intensity value (Figure 10.18a). As t is increased, the added microscopic noise is amplified by the chaotic dynamics and the standard deviation of $p(t)$ increases, as shown in Figures 10.18b and c. After a sufficient transient, $p(t)$ converges into a certain distribution, as shown in Figure 10.18d. The change in $p(t)$ indicates the macroscopic effect of entropy (uncertainty) generation from microscopic intrinsic noise by chaotic nonlinear dynamics.

The standard deviation $\sigma(t)$ of the histogram $p(t)$ is calculated and plotted as a function of time, as shown in Figure 10.19a. $\sigma(t)$ is normalized by the standard deviation $\sigma(\infty)$ at the invariant distribution $p(\infty)$. $\sigma(t)$ is also averaged over 1000 different initial conditions. Thus, the standard deviation of $p(t)$ is estimated by using 10^6 temporal waveforms (1000 different noise instances for each initial condition $\times$ 1000 different initial conditions). For Figure 10.19a, the ratio of $\sigma(t)$ to $\sigma(\infty)$ increases monotonically and reaches more than 0.99 ($\sigma(t)/\sigma(\infty) > 0.99$) after 10.7 ns. This indicates that the memory of the initial conditions for the chaotic temporal waveforms is lost after $\sim$10.7 ns.

10.5.2
Estimation of Entropy

Instead of the evaluation by using the standard deviation of analog temporal waveforms, time-dependent entropy is computed to evaluate the entropy rate of generated bits. A threshold level is set to detect whether the optical intensity is above or below a specified level at fixed time intervals for extracting random bits (see Section 10.3.1). The threshold level is predefined for digitizing temporal waveforms, and is adjusted to equalize the long-term average 0/1 ratio. Each point on the temporal waveform is compared with the threshold level, and a bit 1 is generated when the sampled point is larger than the threshold level, and a bit 0 is generated otherwise. The temporal waveform is thus converted into a bit sequence in time. This procedure is executed for 1000 temporal waveforms that are generated from the same initial conditions, but added by different microscopic noise sequences. The probabilities of bits 0 and 1 are calculated at time t from the 1000 temporal waveforms, and time-dependent entropy $H(t)$ is estimated by the following equation (Mikami et al., 2011).

$$H(t) = -\sum_{i=0}^{1} P_i(t)\log_2 P_i(t) \tag{10.1}$$

where $P_i(t)$ is the probability of the occurrence of bits i ($i=0$ or 1) at the time t for an ensemble of bits generated from the temporal waveforms with different additive noise instances. The time-dependent entropy is also averaged over 1000 different initial conditions. The time-dependent entropy is thus estimated by using 10^6 temporal waveforms as well.

Figure 10.19b shows examples of plots of bit entropy against time for an ensemble of trajectories starting at exactly the same state at $t=0$, but with the different noise sequences. The results for different noise strengths of -25, -45, and -65 dB are plotted. The noise strength is defined as a ratio of the peak values of the RF spectra (FFT) between the chaotic and additive noise signals. It can be seen that the entropy reaches more than 0.99 after 10.9 ns for the noise strength of -25 dB. This result indicates that even though the state of the laser in the dynamical model is known to high precision at some time, it is impossible to predict whether the waveform will correspond to bit 0 or 1 after 10.9 ns, and there is no information about the initial conditions in the bit after this time. For other noise strengths, the entropy reaches more than 0.99 after 13.8 ns and 16.3 ns for the noise strengths of -45 dB and -65 dB, respectively. The convergence time of the entropy is thus strongly dependent of the intrinsic noise strength: The time for converging to $H(t) \geq 0.99$ increases as the noise strength is decreased.

The inverse of the convergence time of the entropy may be considered as one definition of the entropy rate, for example, the entropy rate is $1/10.9 = 0.092$ ns^{-1} for the noise strength of -25 dB. This value indicates the lower limit of a rate of the loss of all the information about the initial conditions. However, this definition of the entropy rate is strongly dependent on the noise strengths in the laser. Therefore, the intrinsic entropy rate that is independent of the noise strength is preferable and considered in the next section.

10.5.3
Entropy Rate and Nondeterministic Bit Generation

Let us define the "memory time" of the initial conditions T_m as the time when the entropy reaches more than 0.99 in this calculation. Figure 10.20 shows the result of a systematic evaluation of the dependence of memory time on noise strength, obtained from the results of Figure 10.19b. The plot is an almost straight line on the semi-logarithmic plot and it is found that the memory time decreases as the noise strength is increased.

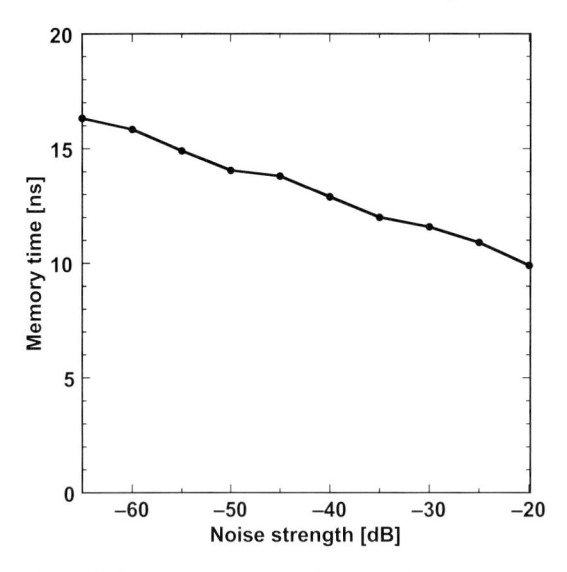

Figure 10.20 Memory time as a function of noise strength.

The line of Figure 10.20 can be fitted with the empirical relation as,

$$T_m = T_0 + L(10\log_{10} S_n) \tag{10.2}$$

where T_m is the memory time, T_0 is the offset time, S_n is the noise strength, and L is the slope of the line in Figure 10.20. The value of L obtained from the plot in Figure 10.20 is $L = -0.139$ ns/dB.

The entropy rate can be defined as the ratio of the change in the logarithm of the noise strength to the corresponding change in the memory time,

$$R = -\Delta \log_e(S_n)/\Delta T_m \tag{10.3}$$

where ΔX is the change in X.

The value of the entropy rate R can be obtained from the slope L by using Eqs. (10.2) and (10.3),

$$R = \frac{1}{10\log_{10} e} \cdot \frac{1}{|L|} = 1.7 \text{ ns}^{-1} \tag{10.4}$$

This value R indicates the amplification rate of microscopic noise by chaotic dynamics in lasers (Mikami et al., 2011).

The entropy rate of random bits generated in a chaotic semiconductor laser is evaluated and it is shown that bits are nondeterministic if they are extracted at rates slower than the entropy rate due to the amplification of microscopic noise by the chaotic dynamics. It is shown numerically that nondeterministic bits can be generated at a rate more than Gb/s in a chaotic semiconductor laser. This is because of the persistent uncertainty in the state of the laser, due to the property that the rate of the generation of entropy (due to the amplification of noise by the chaotic dynamics) is large compared to the bit-generation rate.

Another method for the estimation of the rate of nondeterministic bit generation has been proposed by using an autocorrelation function of chaotic waveforms (Harayama et al., 2011). These theoretical and numerical studies for the estimation of nondeterministic bit generation are very important for chaos-based RNGs to avoid the misunderstanding that chaos is completely deterministic and unsuitable for RNGs. The existence of internal noise in a chaotic system plays a crucial role to guarantee nondeterministic bit generation in chaos-based RNGs. These chaos-based RNGs have several advantages: controllability of the dynamics in RNGs and robustness against external perturbations, which are beneficial for engineering implementation.

10.6
Conventional Methods for Physical Random Number Generators

In this section, conventional techniques for physical random number generators (RNG), such as thermal noise, quantum noise, optical noise, and radioactive nuclide, are introduced and described in detail. Other methods such as chaos in electronic circuits and traditional physical device are also discussed. Some examples of commercially available physical random number generators are introduced.

10.6.1
Thermal Noise

10.6.1.1 Direct Amplification of Thermal Noise

The use of thermal noise is one of the most traditional methods as an entropy source for physical RNG. Direct amplification of thermal noise is obtained with a high-gain, high-bandwidth amplifier to process voltage changes produced by a noise source. Thermal noise is present inherently in a resistor due to random electron and material behavior and has electrically measurable characteristics as a voltage or current.

An example of a RNG based on direct amplification of thermal noise is shown in Figure 10.21 (Holman et al., 1997). Voltage fluctuation is generated across a large resistor due to thermal noise. A white noise source with uniform noise spectral density exhibits a Gaussian distribution in its output voltage value. The noise voltage is amplified by an electronic amplifier with large bandwidth. The output of the white noise voltage is sampled at a constant clock rate and compared to a threshold value using a comparator. A binary sequence of 0 or 1 is obtained when the measured voltage value is below or above the threshold. The value of the noise voltage at any given time has an equal probability of being above or below the threshold value in the comparator, resulting in random binary output sequence. The comparator could be replaced to a multibit ADC to obtain multiple bits from one sampling voltage value.

10.6 Conventional Methods for Physical Random Number Generators

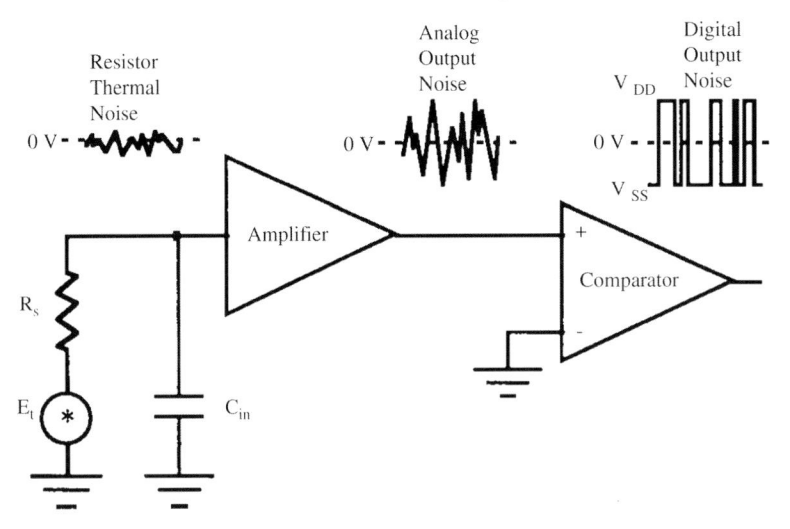

Figure 10.21 Schematics of direct amplification of thermal noise for a physical random number generator. (W. T. Holman, J. A. Connelly, and A. B. Dowlatabadi, (1997). © 1997 IEEE.)

The setting of the threshold value in the comparator strongly affects the occurrence of 0 or 1 bit. In addition, thermal noise is typically coupled with local characteristics such as substrate noise and power-supply fluctuations. If the circuit is not properly shielded, those factors will dominate and affect the ability to capture the randomness of the thermal-noise source. To avoid these problems, differential measurement of two noise sources is often used. A pair of resistors can be used to transform the thermal noise of the resistor pairs into a voltage-variation signal. One can overcome these problems by sampling an adjacent pair of resistors and using the differential to minimize the effect of the other sources.

The integration of this type of RNG has been reported. A compact physical RNG can selectively extract the high-frequency noise signal generated from stochastic physical phenomenon of electrons trapped in the silicon nitride (SiN) layer of a transistor (Toshiba, 2008). The RNG achieves a random number generation rate of 2.0 Mb/s. The high-frequency noise signals generated by the noise source device allow reduction of the size of the ADC, and of the entire RNG circuit area to 1200 square micrometers, including the noise-source device.

Different types of noise sources for physical RNG have been proposed, such as a Zener diode (Tamura et al., 2006) and a CMOS semiconductor device (Holman et al., 1997).

10.6.1.2 Metastability

For another method of physical RNGs, metastability based RNGs have been proposed (Tokunaga et al., 2008; Srinivasan et al., 2009). A metastable cross-coupled inverters are used to generate individual bits that result from the effect of thermal noise. The inverters hold the bistable close to metastability, and then allow to resolve, producing

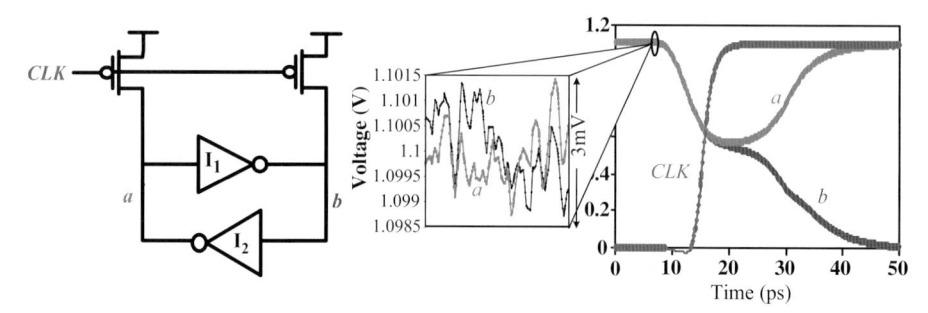

Figure 10.22 (Left) Schematics of metastability based physical random number generator with a pair of inverters. (Right) Bistable element behavior with thermal noise. (S. Srinivasan, S. Mathew, V. Erraguntla, and R. Krishnamurthy, (2009). © 2009 IEEE.)

a high or a low level according to the polarity of the thermal noise at the time of release.

A cross-coupled inverter can be driven towards a metastable state by forcing input/output nodes "a" and "b" to identical values, as shown in Figure 10.22 (left) (Srinivasan et al., 2009). This is achieved by using a pair of precharge devices to initialize both nodes to "1" when the clock signal is set to be zero (CLK = 0). At the rising edge of CLK, the bistable element enters a metastable state, where nodes a and b both settle to a value V_{meta}, which represents the intersection point of the voltages of inverters I_1 and I_2. Resolution to either of the stable states of ($a = 0$, $b = 1$) or ($a = 1$, $b = 0$) depends on the magnitude of differential noise at a and b during the metastable period. A random noise source such as thermal noise at nodes "a" and "b" results in a random resolution state during each evaluation phase of the clock. Thus, a random bit is generated every cycle. Figure 10.22 (right) shows the behavior of the bistable element in the presence of noise. Thermal noise magnitudes up to 3 mV on nodes a and b quickly push the system out of metastability. As a result, the element settles into the stable state of $a = 1$ and $b = 0$, as shown in Figure 10.22 (right).

10.6.1.3 Two Oscillators with Frequency Jitter

Another practical approach in RNG design is the oscillator-sampling method with frequency jitter due to thermal noise (Jun and Kocher, 1999; Bucci et al., 2003). This design exploits the relationship between two independent free-running oscillators, as shown in Figure 10.23 (top) (Bucci et al., 2003). One centered at a low frequency and another centered at a high frequency, to capture a nondeterministic noise source. The basic concept is that a low-frequency oscillator with high jitter is used to sample the high-frequency oscillator to produce the random number sequences. In the context of a digital circuit implementation, a low-frequency square-wave source would be used to clock a positive-edge-triggered D flip-flop, and the high-frequency square-wave source is applied to the flip-flop data input and is sampled on the rising edge of the clock.

The key components behind the creation of random values in this system are that the low-frequency oscillator is designed to have some frequency instability

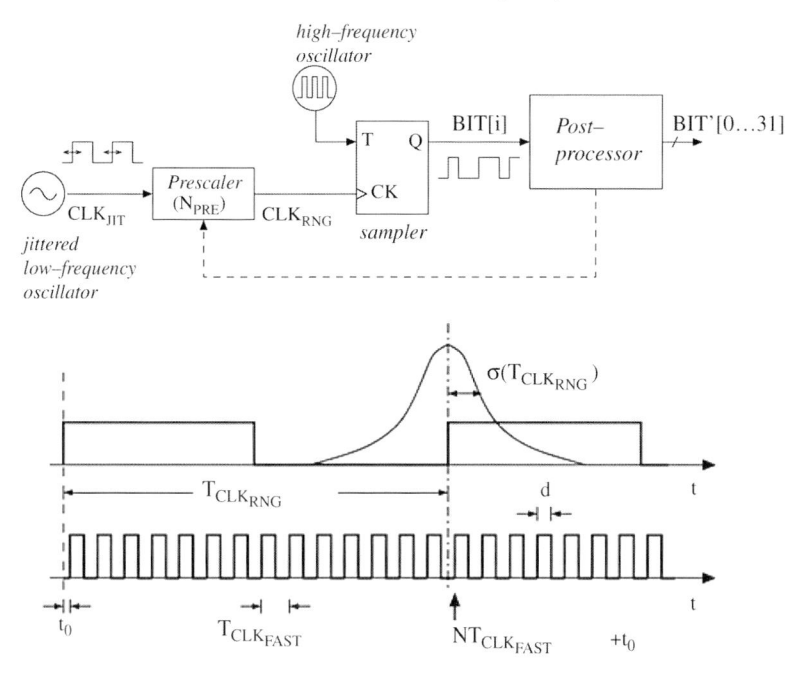

Figure 10.23 (Top) Schematics of physical random number generator based on two oscillators with frequency jitter. (Bottom) Oscillator ourput signals. (M. Bucci, L. Germani, R. Luzzi, A. Trifiletti, and M. Varanonuovo, (2003). © 2003 IEEE.)

(i.e., jitter), as shown in Figure 10.23 (bottom), and the ratio of the low and high frequencies is carefully selected to meet certain conditions. The key design component is the amount of jitter in the low-frequency oscillator: This jitter is the source of randomness. This frequency instability can be generated as a function of the type of oscillator or can be seeded directly by another nondeterministic noise source. Thus, it is this variation in the sampling clock phase with respect to the high-frequency data input that provides the mechanism to capture a random bit stream.

If the two oscillators are run without drift, the sampled bits exhibit predictable periodicity with respect to the frequency ratio, commonly known as beat frequencies. There exists a significant bit-to-bit correlation, and an individual bit will become more predictable than preceding bits. Additionally, the ratio of the two oscillator frequencies has a critical effect upon the resulting bit stream.

A prototype chip has been fabricated in a standard digital n-well CMOS process, which features a 10 Mb/s throughput, and fulfills the NIST Federal Information Processing Standards Publication (FIPS) tests for randomness (Bucci et al., 2003). Small and low-power implementation of this type of RNG has been reported, where the macrocell area is 16 000 square micrometers and 2.3-mW power consumption is measured.

10.6.2
Quantum Noise

A useful advance in realizing high-quality RNGs has been reported to utilize quantum-mechanical uncertainty in the generation of random numbers, so-called quantum RNGs (QRNGs). The QRNGs based on single-photon detection in conjunction with spatial discrimination have been demonstrated (Jennewein et al., 2000). Spatial discrimination is simply realized by use of a 50:50 beam splitter with two single-photon detectors; each detector monitors one of the beam splitter output ports, as shown in Figure 10.24 (Stipčević and Medved Rogina, 2007). A source of photons, conveniently derived from a laser attenuated down to the single-photon level, impinges upon the input port of the beam splitter. There is a 1/2 probability that one of the beam splitter output port detectors will measure an incident photon. Thus, a random binary bit stream can be realized based on quantum-mechanical uncertainty.

From a practical point of view the random bits generated will not be completely free from bias as the two detectors will almost certainly have unequal detection efficiencies and it is very difficult to ensure the beam splitter will have exactly 50:50 transmission reflection probability. The first problem can be somewhat circumvented by using one detector. However, the second problem is very difficult to solve. Therefore, bit correction postprocessing is inevitable to unbias the raw random bits.

A QRNG based on temporal discrimination of photon arrival times has been demonstrated, achieving a random bit rate of ~1 Mb/s (Stipčević and Medved Rogina, 2007). Timing information of detected photons is used to generate binary random bits. A bit extraction method eliminates both bias and autocorrelation, while

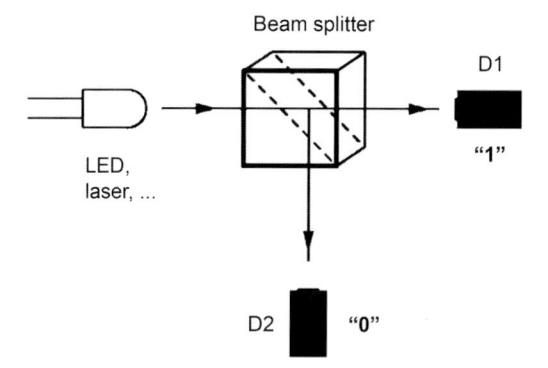

Figure 10.24 Schematics of quantum physical random number. A beam splitter is a frequently used component for quantum random number generators. The two photon detectors D1 and D2 are used to detect two possible outcomes corresponding to one of the two possible paths a photon can take. Thus, each photon entering the beam splitter generates one random binary digit-bit. (M. Stipčević and B. Medved Rogina, (2007). © 2007 AIP.)

reaching an efficiency of almost 0.5 bits per random event. However, similar to many of the spatial discrimination schemes, some postprocessing of the raw bits is needed to produce random bits of high enough quality.

Another QRNG has been proposed to generate random bits (Dynes et al., 2008). It is a simple device requiring only one source of photons (a cw laser attenuated to the single-photon level) and one single-photon detector. The photons are emitted at random intervals determined by the quantum mechanics of photonic emission. They also have very long ($\sim$1 μs) coherence times. This long photon coherence time means the photon wave function extends over many gating cycles of the detector. The collapse of the photon wave function on a random detector gating cycle signals a detection event and these random arrival events are converted into sequences of random bits. Furthermore, the QRNG satisfies the following two key requirements: it outputs high-quality randomness and at a high rate up to 4 Mb/s without any need for classical postprocessing either in software or hardware. Other methods for QRNG based on measurement of vacuum states have been also reported (Gabriel et al., 2010; Shen et al., 2010).

The randomness of QRNG is in principle guaranteed by the laws of quantum mechanics though one still has to be very careful not to introduce any experimental artifact that could correlate adjacent bits (Gisin et al., 2002). One particular problem is the dead time of the detectors, which may introduce a strong anticorrelation between neighboring bits. Similarly, afterpulses may provoke a correlation. These detector-related effects increase with higher pulse rates, limiting the bit rate of QRNGs.

10.6.3
Optical Noise (Spontaneous Emission Noise)

Spontaneous emission noise in incoherent optical systems can be used for random number generators. Spontaneous emission noise is inherently quantum mechanical in origin and cannot be described by deterministic equations. The use of amplified spontaneous emission (ASE) and phase noise of lasers has been reported for high-speed random number generators.

A RNG based on spectrally sliced ASE noise produced by a fiber amplifier has been demonstrated experimentally (Williams et al., 2010). The experimental setup is shown in Figure 10.25. The erbium/ytterbium (Er:Yb) codoped fiber amplifier generates broadband, incoherent, and unpolarized optical noise through ASE. The broadband optical noise from the amplifier is filtered by an optical bandpass filter (bandwidth of 14.5 GHz), comprised of a fiber Bragg grating and optical circulator. The resulting filtered noise signal is amplified in an erbium-doped fiber amplifier. A fiber polarization splitter divides the noise into independent, identically distributed, orthogonally polarized noise signals that are separately detected in a pair of photoreceivers.

To generate random bits, the two independent noise signals are connected to the differential logic input of a bit error rate tester (BERT in Figure 10.25). The BERT performs a clocked comparison of the two input signals, producing a logical bit 1 when one of the noise signals is larger than the other, and a logical bit 0 otherwise.

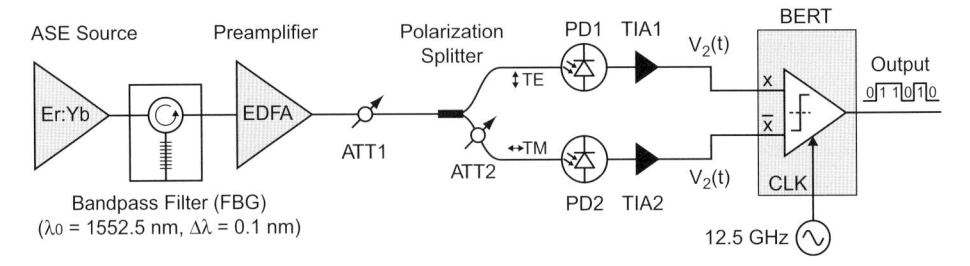

Figure 10.25 System used to generate random bits at 12.5 Gb/s. Amplified spontaneous emission (ASE) is generated in an Er/Yb-doped fiber that is continuously pumped by a 1 W, fiber-coupled 915 nm semiconductor laser diode. The resulting broadband ASE spectrum is bandpass-filtered using a 14.5-GHz (0.1 nm) fiber Bragg grating and optical circulator. The filtered noise is amplified in a conventional Er-doped fiber amplifier (EDFA). A fiber polarization splitter is used to produce two independent, identically distributed optical noise signals that are separately detected in a pair of matched 11 GHz photoreceivers, each comprised of a photodiode (PD) and transimpedance amplifier (TIA). A 12.5-Gb/s bit-error rate tester (BERT) is used to perform a clocked comparison of the two received signals, producing a random string of bits. Two variable attenuators (ATT1, ATT2) are used to control the power of the noise signal, and compensate for loss mismatch between the two arms. (C. R. S. Williams, J. C. Salevan, X. Li, R. Roy, and T. E. Murphy, (2010), © 2010 OSA.)

A bit-generation rate of 12.5 Gb/s has been achieved with verified randomness of statistical tests (Williams et al., 2010). Amplified spontaneous emission noise generated from a superluminescent light emitting diode has been also used to generate random bit sequences at a rate of 20 Gb/s (Li et al., 2011).

Another method has been proposed based on measuring the phase noise of a single-mode semiconductor laser (Qi et al., 2010). The phase noise of a laser originates from spontaneous emission: each spontaneous emitted photon has a random phase, which in turn contributes a random phase fluctuation to the total electric field and results in a linewidth broadening. The spontaneous emission and the corresponding phase noise are quantum mechanical in origin. A practical laser source also exhibits additional classical noises. The phase noise (manifested as the fundamental laser linewidth) is inversely proportional to the laser output power. By operating the laser at a low intensity level near the lasing threshold, a 32-fold broadening of its emission spectrum is measured. This ensures that the main contribution to the phase noise is from spontaneous emission. One significant advantage of this scheme is the high random number generation rate up to 500 Mb/s. Another RNG based on measurement of phase noise of a VCSEL has been reported at a generation rate of 20 Mb/s (Guo et al., 2010).

10.6.4
Radiation from Radioactive Nuclide

A method of generating random bits using time intervals of radiation from radioactive nuclide has been proposed (Inoue et al., 1983). This method, which utilizes gamma rays emitted from a radioactive nuclide, is based on the following characteristics: Nuclei decay independently, the energy of gamma rays is high enough to permit reliable identification from background noise, and the contamination by high-energy radiations such as cosmic rays is unimportant since gamma rays and cosmic rays both obey Poisson's distribution law.

The generator is made up of a pulse generator, a Ge (Li) gamma-ray detector, a multichannel analyzer used as a multiscalar and a computer. Clock pulses generated by the pulse generator are sent to the multiscalar. The gamma-ray detector yields random pulses that are triggered by gamma-rays emitted from a radioactive source. The consecutive addressing of memory locations of the multiscalar is advanced by the random pulses received from the gamma-ray detector. The counts of clock pulses stored in each memory represent the time interval between two successive random pulses. The counts are transferred to the computer where the last digit of the counts in each memory location is taken as a random digit in the range 0 to 9. The counting rate of about 2 kHz and clock pulse frequency of 500 kHz are used, giving a value of random bits.

10.6.5
Chaotic Dynamics in Electronic Circuits

A RNG based on chaotic dynamics in electronic circuits has been proposed (Yalçın et al., 2004). The output of chaotic systems is unpredictable because of the sensitive dependence on initial conditions for a long term. The RNG includes a chaotic oscillator and thresholds circuit. The chaotic oscillator is a continuous-time third-order autonomous system and exhibits a double-scroll attractor, which has been observed in Chua's circuit (Chua et al., 1986). The chaotic oscillator based on the Chua's circuit is used as a source for the RNG.

The phase space of the double-scroll attractor is partitioned into three subspaces with two thresholds, as shown in Figure 10.26. Two of the subspaces are related with the location of the scrolls in the phase space. The third subspace is a region where the transition between the scrolls occurs. The voltage signal from the chaotic oscillator is discretized in the two subspaces of the scrolls and converted into 0 or 1 bit, depending on the subspace V_0 or V_1. Sampling the chaotic signal in space gives an irregular sampling of the signal in time. Statistical properties of the RNG can be adjusted by the threshold values. The RNG successfully passes all statistical tests of randomness of FIPS 140-1 and Diehard.

10.6.6
Traditional Physical Devices

Traditional physical randomizing devices, such as rolling dice, tossing coin, shuffling playing cards, and roulette wheels, have long been used for games and gambling as well as scientific purposes for physical random number generators (Galton, 1890). These are based on "chaotic" dynamical phenomena, exhibiting a sensitive dependence on initial conditions. Small intrinsic errors of the initial conditions are expanded in time and randomized macroscopic states can be obtained.

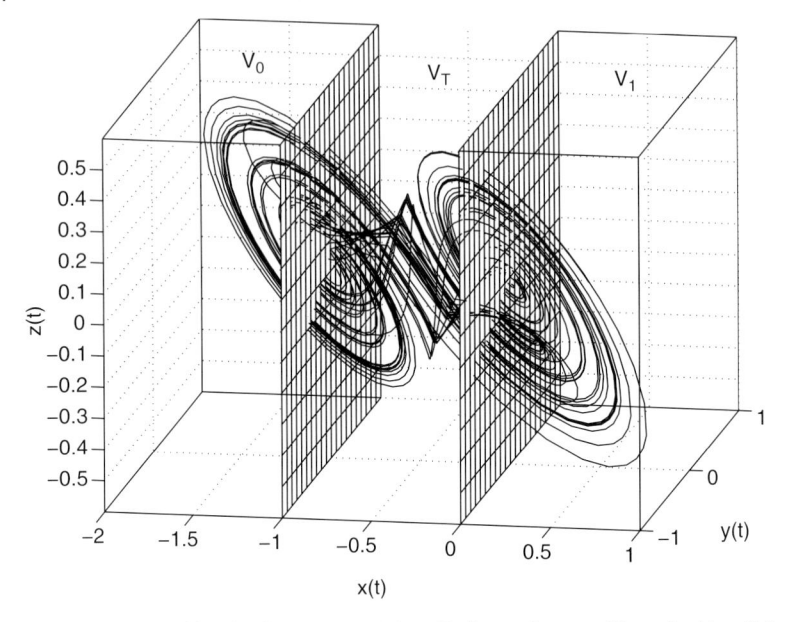

Figure 10.26 Double-schroll attractor partitioned in three subspaces. The region V_0 or V_1 is used to generate 0 or 1 bit by sampling the chaotic trajectory in the phase space. (M. E. Yalçın, J. A. K. Suykens, and J. Vandewalle, (2004). © 2004 IEEE.)

10.6.7
Other Methods

Another method of RNG based on chemical materials has been proposed, in which a two-dimensional crystal of protein molecules is used (Ikezoe et al., 2009). Binary self-assemblies of protein molecules have different inner structures, forming two-dimensional monomolecular-layer crystals. Statistical analysis shows a random molecular distribution in the crystal. This molecular pattern is readily prepared, but it is neither reproducible nor predictable and hence could be used as a nanometer-scale cryptographic device or an identification tag.

Other methods of RNGs have been proposed based on bistability of magnetic spin (Fukushima et al., 2011) and turbulent electroconvection (Gleeson, 2002).

10.6.8
Commercial Physical Random Number Generators

There have been many commercially available physical random number generators. Some examples of them are introduced in this section.

10.6.8.1 Intel Chip (Intel)
The RNG developed by Intel Corporation in USA uses a random source that is derived from two free-running oscillators, one fast and one much slower (Jun and Kocher,

1999) (see Section 10.6.1.3). The thermal noise source is used to modulate the frequency of the slower clock. The variable, noise-modulated slower clock is used to trigger measurements of the fast clock. Drift between the two clocks thus provides the source of random binary digits. The slow oscillator frequency must be significantly perturbed by the noise source, in addition to any pseudorandom environmental, electrical, or manufacturing conditions. Recorded histograms of modulated frequency resemble a normal distribution. The modulated frequency has standard deviation that spans approximately 10–20 high-frequency clock periods; indicating that the sampling process is significantly varied by the random source.

10.6.8.2 Random Master (Toshiba)

Toshiba Corporation in Japan has produced physical RNGs (called Random Master) (Onodera, 2006; Tamura *et al.*, 2006). The RNG has been implemented on a peripheral-component-interconnect (PCI) board as shown in Figure 10.27. The generator uses Zener diode noise as the random number source, and changes the noise signal into the digital time-series data using high-speed ADC, and generates the uniform random numbers by the digital signal processing technique. The random numbers of 16 MByte/s can be generated from the 1 corner random number source. The equipment, which has the plural random number sources on a substrate, can provide the physical random numbers above 100 MByte/s. The maximum generation speed of the commercial Random Masters is 133 Mbyte/s, limited by the transmission rate of the PCI bus.

Figure 10.27 Random Master (Onodera, 2006, © Toshiba Corporation).

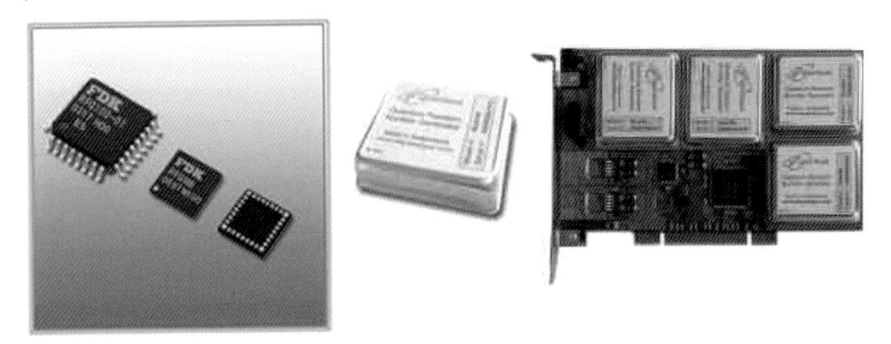

Figure 10.28 (Left) Random Streamer (FDK, 2011, © FDK Corporation). (center and Right) The quantis random number generator (Quantis), available for example as an OEM component (center) and as a PCI-card (right), offers high-quality random numbers at a speed of up to 16 Mb/s. (ID Quantique, 2011, © ID Quantique).

10.6.8.3 Random Streamer (FDK)

FDK Corporation in Japan has produced physical RNGs (called Random Streamer, RPG100) as small-packaged integrated-circuit (IC) generators, as shown in Figure 10.28 (left) (FDK, 2011). The generator has two random number generating circuit parts and two amplifier parts that make use of the thermal noise produced inside the semiconductor. Random Streamer generates high-quality random numbers at high speed, which meets the statistical tests of randomness of FIPS 140-2, NIST SP 800-22, and Diehard. The generation speed of a single generator is 250 kb/s, however, parallel implementation of the generators (16 bits × 32) results in the total generation speed of up to 128 Mb/s. Random Streamers with universal serial bus (USB) output port are also available.

10.6.8.4 Quantis (ID Quantique)

ID Quantique in Switzerland has produced quantum RNGs (called Quantis), exploiting an elementary quantum optics process (see Section 10.6.2), as shown in Figure 10.28 (center and right) (ID Quantique, 2011). Photons are sent one by one onto a semitransparent mirror and detected. The exclusive events (reflection–transmission) are associated to "0" – "1" bit values. The operation of Quantis is continuously monitored to ensure immediate detection of a failure and disabling of the random bit stream. The product produces a random stream through a USB device at the generation rate of 4 Mb/s. Four RNGs are also launched on PCI board (see Figure 10.28 (right)) to increase the generation speed up to 16 Mb/s.

10.7
Postprocessing Techniques for Improvement of Randomness

Physical random number generators may show some deviation from the mathematical ideal of statistically independent and uniformly distributed bits (Dichtl, 2007). Algorithmic postprocessing can be used to eliminate or reduce the imperfections of the output. The per bit entropy of the output can only be increased if the postprocessing algorithm is compressing, that is, more than one bit of input is used to get one bit of output. For the examples in Section 10.7, one bit is generated from two input bits by using logical execution for postprocessing.

A deviation of the probability of the occurrence of bits "1" from 1/2, which is defined as "bias", is a very frequent problem. There are various ways to deal with it, when one assumes that bias is the only problem, that is, the bits are statistically independent. In Section 10.7, several methods to eliminate the statistical bias are introduced and discussed as postprocessing of physical random number generators.

10.7.1
von Neumann Method

Probably the oldest method of postprocessing biased random numbers was invented by John von Neumann (von Neumann, 1963). The bits generated from a random number generator are partitioned in adjacent, non-overlapping pairs. The results of a "01" pair is converted into a "1" output bit. A "10" pair results in a "0" output bit. Pairs "00" and "11" are just discarded. This procedure is summarized in Table 10.5. Since the bits are assumed to be independent and the bias is assumed to be constant, the pairs "01" and "10" have exactly the same probability. The output is therefore completely unbiased.

Let us assume the probability of the occurrence of bits "1" as p_1 and that of bits "0" as p_0 for the input bits. The statistical bias for bits "1" is denoted as b ($|b| \leq 0.5$). The probabilities are written as,

$$p_1 = 0.5 + b \tag{10.5}$$

$$p_0 = 0.5 - b \tag{10.6}$$

For a pair of bits, the probability of $p_{ij}(i,j = 0, 1)$ becomes,

$$p_{00} = (0.5 - b)^2 \tag{10.7}$$

$$p_{01} = (0.5 - b)(0.5 + b) \tag{10.8}$$

$$P_{10} = (0.5 + b)(0.5 - b) \tag{10.9}$$

Table 10.5 von Neumann postprocessing method.

Input bits	Output
0, 0	None
0, 1	1
1, 0	0
1, 1	None

$$p_{11} = (0.5 + b)^2 \tag{10.10}$$

Therefore, $p_{01} = p_{10}$ is satisfied and the output bits of "1" and "0", generated from the input bits of "01" and "10", have the same probability.

The von Neumann postprocessing method is very simple and easy to implement. However, this method has two drawbacks: first, even for a perfect random number generator it results in an average output data rate 4 times slower than the input data rate, due to the combination of 2 bits into 1 bit as an output data and the elimination of 50% of the input data of "00" and "11". Secondly, the waiting times until output bias are available can become arbitrarily large. Although pairs "01" and "10" statistically turn up quite often, it is not certain that they will do so within any fixed number of pairs considered. This is an unsatisfactory situation for the software developer who has to use the random numbers. In many protocols, time-outs occur when reactions to messages take too long. Although the probability for this event can be made arbitrarily low, it cannot be reduced to zero when using the von Neumann postprocessing method.

10.7.2
Exclusive-OR (XOR) Method

One of the simplest postprocessing methods is to operate exclusive-OR (XOR) of a pair of bits from the input data in order to get one bit of output. The bits from the random number generator are partitioned in adjacent, non-overlapping pairs. The results of a "00" or "11" pair is converted into a "0" output bit. A "01" or "10" pair results in a "1" output bit. This procedure is summarized in Table 10.6. No input bits are discarded in this method, unlike the von Neumann method.

When the bias b exists for the occurrence of bits "1", the probability of a pair of the bits are written as Eqs. (10.7)–(10.10). The probability of the output bit p'_i ($i = 0, 1$) after XOR operation is thus calculated as,

$$p'_0 = p_{00} + p_{11} = (0.5 - b)^2 + (0.5 + b)^2 = 0.5 + 2b^2 \tag{10.11}$$

$$p'_1 = p_{01} + p_{10} = (0.5-b)(0.5+b) + (0.5+b)(0.5-b) = 0.5 - 2b^2 \tag{10.12}$$

Therefore, the original bias b of the input is reduced to $2b^2$ after XOR operation, where $b \geq 2b^2$ under the condition of $|b| \leq 0.5$. As the bias is close to 0, the XOR operation is more effective to reduce the original bias b to $2b^2$.

Table 10.6 Exclusive OR (XOR) postprocessing method.

Input bits	Output
0, 0	0
0, 1	1
1, 0	1
1, 1	0

The XOR postprocessing results in an average output data rate 2 times slower than the input data rate, due to the combination of 2 bits into 1 bit as an output data, but no elimination of the input bits. Two independent bits may be used as an input of the XOR operation to maintain the original bit-generation rate. In addition, no waiting times are available in the XOR postprocessing, and easy hardware implementation can be expected with a constant clock signal, as described in Section 10.3.1. The XOR postprocessing is often used for physical random number generators to reduce the bias (see Sections 10.3 and 10.6).

This concept can be expanded to XOR n bits from the generator in order to get one bit of output where n is a fixed integer greater than 1. The bias is more reduced for larger n, however, the bit-generation rate is decreased to n times slower than that for the input data.

A powerful technique based on bit shift and rotation has been proposed to reduce the bias of random bits (Dichtl, 2007). This method eliminates the powers of the bias b up to the fourth order b^4 and obtains the bias of b^5 for the output bits. It is assumed that a stationary source of random bits exists that produces statistically independent, but biased output bits. The postprocessing algorithm uses 16 bits from the physical source of randomness in order to produce 8 bits of output. The aim is to choose an algorithm such that the entropy of the output byte is high. The detail procedure of this method is described in Appendix 10.A.2.

10.8
Pseudorandom Number Generators

Pseudorandom number generators (PRNG) rely on numerical algorithms to produce irregular sequences of numbers. Numbers generated in this way are actually only pseudorandom: two systems that begin in the same initial state will produce identical sequences. The majority of computer telecommunications, online commerce and data encryption systems rely on such PRNGs for producing the keys used to securely store and transmit data. Several examples of uniformly distributed PRNGs are described in this section: the linear congruential method, generalized feedback shift register, combined tausworthe, and mersenne twister. The methods of PRNGs have been standardized by the International Standard Organization in 2010, which has been published as ISO 28640:2010 (ISO, 2010).

10.8.1
Linear Congruential Method

One of the most traditional pseudorandom number generators is called the linear congruential method (Knuth, 1998). Random numbers of integer X_n between zero and some number m are generated from the following linear recurrence,

$$X_{n+1} = aX_n + c \, (\text{mod } m), \, n \geq 0 \tag{10.13}$$

where a is the multiplier ($0 \leq a \leq m$), c is the increment ($0 \leq c \leq m$), and m is the modulus (mod), indicating the residue from dividing integers by m. The starting integer value X_0 ($0 \leq X_0 \leq m$), called a "seed", is used to obtain X_n. Taking the remainder mod m is somewhat like determining where a ball will land in a spinning roulette wheel labeled from zero to m. The terms of multiplicative and mixed congruential methods are used to denote linear congruential sequences with $c = 0$ and $c \neq 0$, respectively. Dividing X_n by m, real random numbers are obtained in [0,1].

For example, the sequence obtained when $m = 10$ and $X_0 = a = c = 7$ is

$$7, 6, 9, 0, 7, 6, 9, 0, \ldots \tag{10.14}$$

As the example shows, the sequence is not always random for all choice of m, a, c, and X_0; the appropriate choice of the magic numbers is very important.

The example of Eq. (10.14) illustrates the fact that the congruential sequences always get into a loop; that is, there is ultimately a cycle of numbers that is repeated endlessly. This property is common to all sequences having the general form $X_{n+1} = f(X_n)$. The repeating cycle is called the "period": the sequence (10.14) has a period of length 4. A useful sequence will have a relatively long period.

The linear congruential method has been used for the standard function of "rand()" in ANSI C libraries. For the rand() function, the parameters are selected as $a = 1103515245$, $c = 12345$, and $m = 2^{32}$, thus the period of generated random numbers is $2^{32} - 1$. It has been known that the rand() function is not satisfactory to pass statistical tests of randomness. It has also been known that the linear congruential method has the disadvantage that it is not free from sequential correlation on successive calls, even though it has the advantage of being very fast, requiring only a few operations per call (Press et al., 1992). The linear congruential method is thus not recommended as a practical PRNG, and is only included in the annex in ISO 28640:2010 (ISO, 2010).

10.8.2
M sequence and Generalized Feedback Shift Register (GFSR)

The generalization of the linear congruential method has been introduced (Knuth, 1998). One technique is to make X_{n+1} dependent on both X_n and X_{n-1}, instead of just on X_n; then the period length can be as high as m^2, since the sequence will not begin to repeat until the condition of $(X_{n+\lambda}, X_{n+\lambda+1}) = (X_n, X_{n+1})$ is satisfied. Let us consider generators of the general from,

$$X_{n+1} = X_n + X_{n-k} \pmod{m} \tag{10.15}$$

where k is a comparatively large value.

Useful random number generators can be constructed by taking general linear combinations of $X_{n-1}, \ldots, X_{n-k}$ for small k. In this case the best results occur when the modulus m is a large prime number; for example, m can be chosen to be the largest prime number that fits in a single computer word. When $m = p$ is prime, the theory of finite fields tells us that it is possible to find multipliers $a_1, \ldots, a_k$ such that the sequence defined by

$$X_n = a_1 X_{n-1} + a_2 X_{n-2} + \cdots + a_k X_{n-k} \pmod{p} \tag{10.16}$$

has period length $p^k - 1$; here the seeds $X_0, \ldots, X_{k-1}$ may be chosen arbitrarily but not all zero. The special case $k = 1$ corresponds to a multiplicative congruential sequence with prime modulus. The constants $a_1, \ldots, a_k$ in Eq. (10.16) have the desired property if and only if the polynomial

$$f(x) = x^k - a_1 x^{k-1} - \cdots - a_{k-1} x - a_k \tag{10.17}$$

is a "primitive polynomial modulo p", that is, if and only if this polynomial has a root that is a primitive element of the field with p^k elements (Knuth, 1998). The sequence generated from Eqs. (10.16) and (10.17) is called Maximum-length linearly recurring sequence, namely the "M sequence".

Base on the theory of the M sequence, a generalized feedback shift register (GFSR) method has been proposed. This method uses a characteristic polynomial and it generates "binary" integer sequences X_n of w bits by the following recurrence formula. A trinomial (3-term) GFSR method is described as,

$$X_{n+p} = X_{n+q} \oplus X_n \tag{10.18}$$

where $\oplus$ is the bitwise XOR of binary integers (equivalent to mod 2). The period of this sequence is $2^p - 1$ for an appropriate choice of p and q (e.g., $p = 1279$ and $q = 418$ for 32-bit binary integers).

A pentanomial (5-term) GFSR is a more practical GFSR method and is described as,

$$X_{n+p} = X_{n+q1} \oplus X_{n+q2} \oplus X_{n+q3} \oplus X_n \tag{10.19}$$

The period of this sequence is also $2^p - 1$ for an appropriate choice of p, q_1, q_2, and q_3. The values of $p = 521$, $q_1 = 86$, $q_2 = 197$, and $q_3 = 447$ can be used as an example for generating 32-bit ($w = 32$) binary-integer pseudorandom sequences (ISO, 2010).

The GFSR methods can be easily implemented in electrical circuits by using a set of feedback shift registers and logical XOR devices, as shown in Figure 10.29.

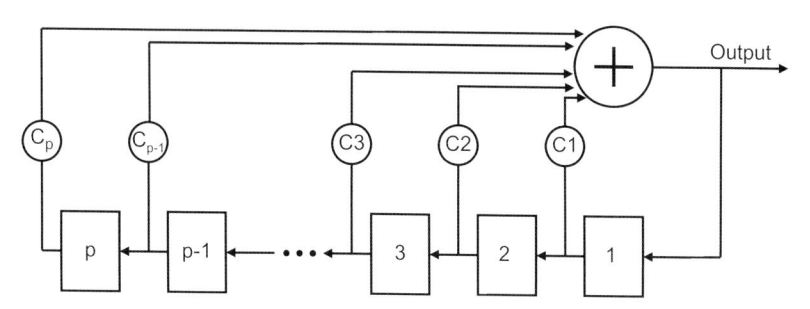

Figure 10.29 Schematic of the implementation of generalized feedback shift register (GFSR) method as a pseudorandom number generator.

Therefore, the GFSR methods have been widely used to generate pseudorandom number sequences in communication systems.

10.8.3
Combined Tausworthe Method

The combined Tausworthe method has been also commonly used as a pseudorandom number generator. Let $x_0, x_1, x_2, \ldots$ be an M-sequence generated by the recurrence relationship:

$$x_{n+p} = x_{n+q} + x_n \pmod{2} \quad (n = 0, 1, 2 \ldots) \tag{10.20}$$

Using this M-sequence, a w-bit integer sequence called a simple Tausworthe sequence with parameters (p, q, t) is obtained as follows:

$$X_n = x_{nt} x_{nt+1} \ldots x_{nt+w-1} \quad (n = 0, 1, 2 \ldots) \tag{10.21}$$

where t is a natural number that is coprime to the period $2^p - 1$ of the M-sequence, and w is the word length that does not exceed p. Note that two integers are said to be coprime, or relatively prime, when they have no common divisors other than unity. The period of this sequence is also $2^p - 1$.

Let us show an example when a primitive polynomial $t^4 + t + 1$ is chosen, set $p = 4$, and $q = 1$ in the above recurrence relationship. If the seeds $(x_0, x_1, x_2, x_3) = (1, 1, 1, 1)$ are given to the recurrence, then the M-sequence obtained by the recurrence will be 1, 1, 1, 1, 0, 0, 0, 1 0, 0, 1, 1 0, 1, 0, 1 1, 1, 1, 0 $\ldots$, and the period of the sequence is $2^4 - 1 = 15$. Taking, for example, $t = 4$ which is coprime to 15, and $w = 4$, the simple Tausworthe sequence $\{X_n\}$ with parameters $(4, 1, 4)$ is obtained as follows:

$$\begin{aligned}
X_0 &= x_0 x_1 x_2 x_3 = 1111 (= 15) \\
X_1 &= x_4 x_5 x_6 x_7 = 0001 (= 1) \\
X_2 &= x_8 x_9 x_{10} x_{11} = 0011 (= 3) \\
X_3 &= x_{12} x_{13} x_{14} x_0 = 0101 (= 5) \\
X_4 &= x_1 x_2 x_3 x_4 = 1110 (= 14) \\
X_5 &= x_5 x_6 x_7 x_8 = 0010 (= 2) \\
&\ldots
\end{aligned} \tag{10.22}$$

The simple Tausworthe sequence obtained in this way will be, in decimal notation, 15, 1, 3, 5, 14, 2, 6, 11, 12, 4, 13, 7, 8, 9, 10, 15, 1, 3, $\ldots$, and its period is $2^4 - 1 = 15$.

10.8.4
Mersenne Twister

The Mersenne Twister (MT) is one of the most practical pseudorandom number generating algorithms (Matsumoto and Nishimura, 1998; Mersenne Twister, 2011). MT is a type of GFSR that has a matrix called Twister for the mixture of bit sequences. MT has a far longer period and a far higher order of equidistribution than any other implemented generators. It has been proven that the period is $2^{19937} - 1$, and

623-dimensional equidistribution property is assured. In addition, it is reported that MT is sometimes faster than the standard ANSI-C library in a system with pipeline and cache memory, and efficient use of the memory is attainable.

MT is designed to pass the k-distribution test. For example, for 32-bit accuracy, the 623-tuples from the output in a whole period are equidistributed in the 623-dimensional unit cube. MT passes many stringent tests, including the NIST SP 800-22 and Diehard test suites. The generator is implemented to generate the output only by fastest arithmetic operations: no division, no multiplication. By generating an array at one time, it takes the full advantage of cache memory and pipeline processing. The MT algorithm is described in Appendix 10.A.3. The implementation of MT in C source codes is also available (Mersenne Twister, 2011).

10.9
Statistical Evaluation of Random Numbers with NIST Special Publication 800-22 Test Suite

10.9.1
Strategies for Statistical Analysis of Random Number Generators

To evaluate the randomness of bit sequences, a *de-facto* standard statistical test suite for pseudorandom number generators has been widely used, which has been provided by National Institute of Standards and Technology (NIST) in USA. The test suite is known as the NIST Special Publication 800-22 (NIST SP 800-22) (Rukhin et al., 2010; Kim et al., 2004). The NIST SP 800-22 test suite consists of 15 statistical tests, summarized in Table 10.7. The detail explanation for each test is described in Appendix 10.A.4. The NIST SP 800-22 has been updated several times since 2001, and the latest version of NIST SP 800-22 is the version 2.1.1 (Revision 1a, August 2010) (Rukhin et al., 2010).

In practice, there are many distinct strategies employed in the statistical analysis of a random number generator. NIST SP 800-22 provides the following stages involved in the statistical testing of a random number generator.

Stage 1: Selection of a generator
Select a hardware- or software-based generator for evaluation. The generator should produce a sequence of bits "0" and "1" of a given length n.

Stage 2: Binary sequence generation
For a fixed sequence of length n and the preselected generator, construct a set of m binary sequences ("0" and "1") and save the sequences to a file.

Stage 3: Execution of the statistical test suite
Invoke the NIST SP 800-22 using the file produced in Stage 2 and the desired sequence length. Select the statistical tests and relevant input parameters (e.g., block length) to be applied. Typical parameter values used for NIST SP 800-22 are summarized in Table 10.8.

Stage 4: Examination of the p-values

Table 10.7 NIST Special Publication 800-22 test suite (NIST SP 800-22).

Test Type	Test Focus	Defect Detected
Frequency	Probability of occurrence of zeroes (0) and ones (1)	Too many zeroes or ones
Frequency within a block	Probability of occurrence of ones within M-bit blocks	Too many zeroes or ones in specific block sizes
Cumulative sums	Maximal excursion (from zero) of the random walk defined by the cumulative sum of adjusted digits	Random walk excursions away from zero too large
Runs	Total number of runs (uninterrupted sequence of identical bits)	Too many (or too few) runs of zeroes or ones
Longest run of ones in a block	Longest run of ones within M-bit blocks	Too many long runs of ones in specific block sizes
Binary matrix rank	Rank of disjoint submatrices	Linear dependence among fixed-length substrings of original
Discrete Fourier transform	Peak heights in the discrete Fourier transform	Periodic features in the bitstream
Non-overlapping template matching	Number of occurrences of prespecified target strings	Too many occurrences of nonperiodic templates
Overlapping template matching	Number of occurrences of prespecified target strings	Too many or too few occurrences of runs of ones
Maurer's universal statistical	Number of bits between matching patterns	Too easy to compress bitsteam without loss of information
Approximate entropy	Frequency of all possible overlapping m-bit patterns	Nonuniform distribution of specific length words
Random excursions	Number of cycles having exactly K visits in a cumulative sum random walk	Too many visits of a random walk to a certain state
Random excursions variant	Total number of times that a particular state is visited in a cumulative sum random walk	Too many total visits (across many random walks) to a certain state
Serial	Frequency (occurrence) of all possible overlapping m-bit patterns	Nonuniform distribution of specific length words
Linear complexity	Length of a linear feedback shift register	Sequence not complex enough to be considered random

Table 10.8 Typical parameters for NIST SP 800-22 test suite.

Test Type	Parameter	Value
Block frequency	Block length	128
Overlapping templates	Template length	9
Non-overlapping templates	Template length	9
Serial	Block length	16
Approximate entropy	Block length	10
Linear complexity	Block length	500
Universal	Block length	7
Universal	Number of initialization step	1280

An output file will be generated by the test suite with relevant intermediate values, such as test statistics, and p-values (see Section 10.9.2) for each statistical test. Based on these p-values, a conclusion regarding the quality of the sequences can be made.

Stage 5: Pass/fail assignment

For each statistical test, a set of p-values (corresponding to the set of sequences) is produced. For a fixed significance level, a certain percentage of p-values are expected to indicate failure. For example, if the level of significance α is chosen to be 0.01 (i.e., $\alpha = 0.01$), then about 1% of the sequences are expected to fail. A sequence passes a statistical test whenever the p-value $\geq \alpha$ and fails otherwise. For each statistical test, the "proportion" of sequences that pass is computed and analyzed accordingly. The "uniformity" of p-values is also evaluated. This analysis should be performed using additional statistical procedures, that is, the proportion and the uniformity of p-values, which are the two main criteria to determine randomness of bit sequences for each test.

10.9.2
Evaluation of p-Values

Various statistical tests can be applied to a sequence to attempt to compare and evaluate the sequence to an ideal random sequence. Randomness is a probabilistic property; that is, the properties of a random sequence can be characterized and described in terms of "probability" (Rukhin et al., 2010). The likely outcome of statistical tests, when applied to a truly random sequence, is known *a priori* and can be described in probabilistic terms. There are an infinite number of possible statistical tests, each assessing the presence or absence of a "pattern" that, if detected, would indicate that the sequence is nonrandom. Because there are so many tests for judging whether a sequence is random or not, no specific finite set of tests is deemed "complete." In addition, the results of statistical testing must be interpreted with some care and caution to avoid incorrect conclusions about a specific generator.

A statistical test is formulated to test a specific null hypothesis. The null hypothesis under test is that the sequence being tested is random. For each test, a relevant randomness statistic must be chosen and used to determine the acceptance or rejection of the null hypothesis. If the data is, in truth, random, then a conclusion to reject the null hypothesis (i.e., conclude that the data is nonrandom) will occur a small percentage of the time. The probability of this type of error is often called "the level of significance" of the test. This probability can be set prior to a test and is denoted as α. For the test, α is the probability that the test will indicate that the sequence is not random when it really is random. That is, a sequence appears to have nonrandom properties even when a "good" generator produced the sequence. Common values of α in cryptography are about 0.01.

The test statistic is used to calculate a "p-value" that summarizes the strength of the evidence against the null hypothesis. For these tests, each p-value is the probability that a perfect random number generator would have produced a sequence less random than the sequence that was tested, given the kind of nonrandomness assessed by the test. If a p-value for a test is determined to be equal to 1, then the sequence appears to have perfect randomness. A p-value of zero indicates that the sequence appears to be completely nonrandom. A level of significance α can be chosen for the tests. If the p-value $\geq \alpha$, then the null hypothesis is accepted; that is, the sequence appears to be random. If p-value $< \alpha$, then the null hypothesis is rejected; that is, the sequence appears to be nonrandom. The parameter α denotes the probability of the error that indicates the sequence is not random when it really is random.

An α of 0.01 indicates that one would expect 1 sequence in 100 sequences to be rejected. A p-value ≥ 0.01 would mean that the sequence would be considered to be random with a confidence of 99%. A p-value < 0.01 would mean that the conclusion was that the sequence is nonrandom with a confidence of 99%.

For the recommendation of the NIST SP 800-22 test suite, one p-value is produced from a bit stream with the length of 1 M bits ($n = 10^6$). This evaluation is repeated for 1000 times with different bit sequences, that is, the number of different bit streams is 10^3 ($m = 10^3$). These parameter values for the data are summarized in Table 10.9. Thus, the total of a 1-Gbit ($m \cdot n = 10^9$) binary sequence is required for NIST SP 800-22 tests (1 Mbits × 1000 sequences). As a result, 1000 p-values are produced and are used for the statistical evaluation of the proportion and uniformity of the p-values (see Section 10.9.3). The pass criteria are chosen based on the length of the sequence and a specified significance level. The significance level of $\alpha = 0.01$ is commonly used, since α should be larger than $1/m$, in order to have more than one p-value in the range of p-values $\leq \alpha$ (e.g., 10 p-values would be expected in the range of p-values ≤ 0.01 for

Table 10.9 Data lengths and significance level for NIST SP 800-22 test suite.

Parameters	Symbols	Values
Number of different bit streams	m	1000
Length of one bit stream	n	1000000
Significance level	α	0.01

1000 p-values of an ideal random number sequence where p-values are uniformly distributed in [0,1]).

10.9.3
Interpretation of Empirical Results

The interpretation of empirical results can be conducted in any number of ways. Two approaches NIST has adopted include (1) the examination of the "proportion" of sequences that pass a statistical test and (2) the "uniformity" of distribution of p-values (Rukhin et al., 2010). In the event that either of these approaches fails (i.e., the corresponding null hypothesis must be rejected), additional evaluation should be conducted on different samples of the generator to determine whether the phenomenon was a statistical anomaly or clear evidence of nonrandomness.

10.9.3.1 Proportion of p-Values
Given the empirical results for a particular statistical test, the proportion of sequences that pass the test is computed. For example, if 1000 binary sequences of 1 Mbit data are tested (i.e., $m = 1000$) with $\alpha = 0.01$ and 996 binary sequences have p-values ≥ 0.01, then the proportion is $996/1000 = 0.9960$.

The range of acceptable proportions is determined using the confidence interval defined as,

$$(1-\alpha) \pm 3\sqrt{(1-\alpha)\alpha/m} \tag{10.23}$$

where m is the sample size. If the proportion falls outside of this interval, then there is evidence that the data is nonrandom. Note that other standard deviation values could be used. For the example above, the confidence interval is

$$(1-0.01) \pm 3\sqrt{(1-0.01)0.01/1000} = 0.99 \pm 0.0094392 \tag{10.24}$$

The proportion should lie above 0.9805608 and be less than 0.9994392. The confidence interval is calculated using a normal distribution as an approximation to the binomial distribution, which is reasonably accurate for large sample sizes (e.g., $m \geq 1000$).

10.9.3.2 Uniformity of Distribution of p-Values
The distribution of p-values is examined to ensure uniformity. The interval between 0 and 1 is divided into 10 subintervals (i.e., [0, 0.1], [0.1,0.2], ..., [0.9, 1.0]), and the p-values that lie within each subinterval are counted.

Uniformity is determined *via* an application of a χ^2 test and the determination of a "P-value" corresponding to the goodness-of-fit distributional test on the p-values obtained for an arbitrary statistical test (i.e., a "P-value" of the p-values). This is accomplished by computing

$$\chi^2 = \sum_{i=1}^{10} \frac{(F_i - s/10)^2}{s/10} \tag{10.25}$$

where F_i is the number of p-values in subinterval i of the 10 divided subintervals, and, s is the sample size. A P-value is calculated such that

$$\text{P-value} = \text{igamc}(9/2, \chi^2/2) \tag{10.26}$$

where igamc( ) is the incomplete gamma function. The sequences can be considered to be uniformly distributed if the condition

$$\text{P-value} \geq 0.0001 \tag{10.27}$$

is satisfied.

The "proportion" and the "P-value" of the uniformity (P-value of the p-values) obtained from 1000 p-values are used for pass/fail assignment in NIST SP 800-22. A typical example is already shown in Table 10.2 in Section 10.3.1. For the tests that produce multiple proportions and P-values, the worst case is shown in Table 10.2. For the pass criteria of Eqs. (10.24) and (10.27), the proportion of sequences should be in the range of 0.99 ±0.0094392 and the P-value of the uniformity obtained for each test should be larger than 0.0001. All the tests satisfy these criteria and are passed in Table 10.2. These criteria are used to check the randomness of the bit sequences generated from random number generators, as described in Section 10.3.

10.9.4
Tendency of Passed/Failed NIST SP 800-22 Tests in Laser-Chaos-Based Random Number Generators

NIST SP 800-22 had been mainly used for the test of pseudorandom number generators. This test suite has also been used for the sequences generated from physical random number generators. In this section, empirical results of the tendency of test failure in the NIST SP 800-22 are described for random bit sequences generated chaotic laser systems, as described in Section 10.3. The tendency of test failure depends on the methods for random number generators, however, the general tendency for physical random number generators with chaotic lasers is summarized.

One of the most difficult tests to pass is the "block frequency" test in the NIST SP 800-22 for random bit sequences generated from chaotic lasers (Hirano et al., 2009). This indicates that the frequency of 1 in the block length (128 bits are used) does not satisfy the statistical criteria for an ideal random number sequence. It is also found that the "runs" test and "nonperiodic-templates" test tend to be failed. The tendency for deviation from randomness on the short term can be expected to appear as a manifestation of the structure in the chaotic oscillations. In fact, these results are related to the fact that physical random number generators utilize physical oscillatory behaviors which have short-term autocorrelation. To pass the block frequency, runs, and nonperiodic-template tests, it is important to eliminate short-term autocorrelation of physical random sources (e.g., by using postprocessing).

10.9.5
Other Statistical Tests of Randomness

Another standard statistical test suite for random number generators has also been widely used, called "Diehard" tests (Marsaglia, 1995). The test consists of 18 statistical tests, as already shown in Table 10.3 in Section 10.3.1. The tests are performed using 74-Mbit sequences and significance level $\alpha = 0.01$ for all the Diehard tests. For "success" at significance level $\alpha = 0.01$, the P-value (uniformity of p-values) should be larger than 0.0001. "KS" indicates that a single P-value is obtained by the Kolmogorov–Smirnov (KS) test. For the tests that produce multiple P-values without the KS test, the worst case is shown in Table 10.3. All the tests are passed in Table 10.3.

Appendix

10.A.1
Recipe for High-Quality Random Number Generators

The detail procedure of preparing for high-quality random bit generation in the two-semiconductor-laser scheme with 1-bit (mono-bit) ADC and XOR operation (see Section 10.3.1) is summarized as follows (Hirano et al., 2009).

1) The external cavity length should be long so that the external cavity frequency is small compared with the clock frequency.
2) External cavity lengths should be adjusted so that the ratio of external cavity frequencies of the two lasers, and the ratio of each external cavity frequency and clock frequency are incommensurable, in particular avoiding low-order commensurable ratios.
3) The injection current should be large enough that relaxation oscillation frequency and dominant chaotic oscillation frequency are fast compared to the sampling clock frequency. (There is no need to adjust the relaxation oscillation frequencies of the two lasers to be incommensurable if the relaxation oscillation frequencies are much higher than the clock frequency.)
4) The strength of the feedback light should be adjusted so that the RF spectra of the chaotic waveform is as flat as possible – in particular, minimize the amplitude of the modulation of spectrum corresponding to the external cavity frequency. This procedure corresponds to minimizing the autocorrelation peak of the round-trip time of the external cavity in the time domain.
5) The detection thresholds for the two lasers are initially set to the mean value of the chaotic waveforms. Then, the detection threshold of one laser is adjusted so that the output of the XOR has 0/1 ratio as close as possible to 50%. Then, the threshold of the second laser is adjusted so that the output of the XOR has a 0/1 ratio as close as possible to 50%. (Note that one should not adjust the threshold values for the outputs of each laser separately before combining the sequences at the XOR. In each case one needs to monitor the output of the XOR and adjust the

threshold values. Independent adjustment of each threshold value does not result in the optimal threshold settings.)

6) Confirm that the output of the XOR has 0/1 ratio within a certain range of 50%. If not, repeat and improve the above adjustments from (1) to (5). Specifically, it is found empirically that for this system to pass all the NIST tests, the 0/1 ratio for 1-Gbit data should satisfy the condition that deviation from 50% is less than 0.02%.

10.A.2
Dichtl method for Postprocessing of Random Number Generators

A technique to reduce the bias has been proposed as postprocessing, in which the powers of the bias b up to the fourth order b^4 are eliminated and the bias of b^5 is obtained for the output bits, as described in section 10.7.2 (Dichtl, 2007). The procedure of this method is shown in Table 10.A.1. Let $a_0, a_1, \ldots, a_{15}$ be the input 16 bits ($a_i = 0$ or 1) for postprocessing. The first-half 8 bits is written as,

$$A = a_0, a_1, a_2, a_3, a_4, a_5, a_6, a_7. \ (A_i = a_i) \tag{10.A.1}$$

The 1-bit left-shifted (and left-rotated) bits of A are generated,

$$B = a_1, a_2, a_3, a_4, a_5, a_6, a_7, a_0. \ (B_i = a_{(i+1) \bmod 8}) \tag{10.A.2}$$

The 2-bit left-rotated bits of A are generated,

$$C = a_2, a_3, a_4, a_5, a_6, a_7, a_0, a_1. \ (C_i = a_{(i+2) \bmod 8}) \tag{10.A.3}$$

The 4-bit left-rotated bits of A are written as,

$$D = a_4, a_5, a_6, a_7, a_0, a_1, a_2, a_3. \ (D_i = a_{(i+4) \bmod 8}) \tag{10.A.4}$$

The second-half 8 bits of the 16-bit input is written as,

$$E = a_8, a_9, a_{10}, a_{11}, a_{12}, a_{13}, a_{14}, a_{15}. \ (E_i = a_{(i+8)}) \tag{10.A.5}$$

Finally, the bit-wise XOR is executed from A to E to obtain the 8-bit output F as,

$$F = A \oplus B \oplus C \oplus D \oplus E \tag{10.A.6}$$

Table 10.A.1 Dichtl postprocessing method (Dichtl, 2007).

Input bits	Bit sequences
First-half 8-bit sequence	$a_0, a_1, a_2, a_3, a_4, a_5, a_6, a_7$
1-bit left-rotated sequence	$a_1, a_2, a_3, a_4, a_5, a_6, a_7, a_0$
2-bit left-rotated sequence	$a_2, a_3, a_4, a_5, a_6, a_7, a_0, a_1$
4-bit left-rotated sequence	$a_4, a_5, a_6, a_7, a_0, a_1, a_2, a_3$
Second-half 8-bit sequence	$a_8, a_9, a_{10}, a_{11}, a_{12}, a_{13}, a_{14}, a_{15}$

The 16-bit input $a_0, a_1, \ldots, a_{15}$ are used ($a_i = 0$ or 1) to obtain 8-bit output. Bit-wise XOR is applied to the five 8-bit sequences shown in Table 10.A.1 to obtain the final output.

where $\oplus$ denotes the bit-wise XOR operation. The general form of the output bits can be described as,

$$F_i = a_i \oplus a_{(i+1) \bmod 8} \oplus a_{(i+2) \bmod 8} \oplus a_{(i+4) \bmod 8} \oplus a_{(i+8)} \quad (10.A.7)$$

where $i = 0, 1, \ldots, 7$. For example, the first bit of the 8-bit output is obtained as $F_0 = a_0 \oplus a_1 \oplus a_2 \oplus a_4 \oplus a_8$.

The Dichtl method has been mathematically proven that the bias b is eliminated to the order of b^5 (Dichtl, 2007). For example, when the bias of the input bits is 0.1 (10%), the bias of the postprocessed output bits is 10^{-5} (0.001%), which can easily pass the frequency test of NIST SP 800-22 for 1 Gbit data. The Dichtl method is powerful to eliminate bias of 8-bit output from 16-bit input data by postprocessing.

10.A.3
Algorithm of Mersenne Twister Pseudorandom Number Generator

The algorithm of Mersenne Twister described in Section 10.8.4 is based on the linear recurrence as follows (Matsumoto and Nishimura, 1998; ISO 28640:2010, 2010; Mersenne Twister, 2011). Let X_n be a binary integer of w bits. Then, the Mersenne Twister method generates a sequence of binay integer pseudorandom numbers of w bits according to the following recurrence formula with integer constants p, q, r and a binary integer a of w bits.

$$X_{n+p} = X_{n+q} \oplus \left(X_n^f | X_{n+1}^l\right)^{(r)} A, \quad (n = 1, 2, 3 \ldots) \quad (10.A.8)$$

where $m \oplus k$ represents bitwise exclusive OR (disjunction) of binary integers m and k, and $\left(X_n^f | X_{n+1}^l\right)^{(r)}$ represents a binary integer that is obtained by a concatenation of X_n^f and X_{n+1}^l, the first $w-r$ bits of X_n and the last r bits of X_{n+1} in this order. A is a $w \times w$ 0–1 matrix, which is determined by a, and the product XA is given by the following formula.

$$X \gg 1 \text{ (when the last bit of } X = 0) \quad (10.A.9)$$

$$XA = (X \gg 1) \oplus a \text{ (when the last bit of } X = 1) \quad (10.A.10)$$

Here, X is regarded as a w-dimensional 0–1 vector, $m \gg k$ represents k-bit right shift of binary integer m.

The necessary amount of memory for this computation is p words, the period becomes $2^{pw-r}-1$, and the efficiency is better than that of the GFSR methods described in Section 10.8.2. To improve the randomness of the first $w-r$ bits, the following series of conversions can be applied to X_n.

$$y := X_n \quad (10.A.11)$$

$$y := y \oplus (y \gg u) \quad (10.A.12)$$

$$y := y \oplus [(y \ll s) \wedge b] \qquad (10.\text{A}.13)$$

$$y := y \oplus [(y \ll t) \wedge c] \qquad (10.\text{A}.14)$$

$$y := y \oplus [(y \gg l)] \qquad (10.\text{A}.15)$$

where $m := k$ represents replaces value m by k, $m \ll k$ represents k-bit left shift of binary integer m, and b, c are constant bits masks to improve the randomness of the first $w-r$ bits. The parameters of this algorithm are $(p, q, r, w, a, u, s, t, l, b, c)$. The seeds are $X_2, \ldots, X_{q+1}$ and the first $w-r$ bits of X_1. The final value of y is the output pseudorandom number.

10.A.4
Detailed Description of NIST Special Publication 800-22

The NIST test suite is a statistical package consisting of 15 tests that were developed to test the randomness of (arbitrarily long) binary sequences produced by either hardware- or software-based cryptographic random or pseudorandom number generators, as described in Section 10.9 (Rukhin et al., 2010). These tests focus on a variety of different types of nonrandomness that could exist in a sequence. Some tests are decomposable into a variety of subtests. The 15 tests are described as follows (see also Table 10.7).

10.A.4.1 Frequency (Monobit) Test
The focus of the test is the proportion of zeroes and ones for the entire sequence. The purpose of this test is to determine whether the numbers of ones and zeros in a sequence are approximately the same as would be expected for a truly random sequence. The test assesses the closeness of the fraction of ones to 1/2, that is, the number of ones and zeroes in a sequence should be about the same. All subsequent tests depend on the passing of this test.

10.A.4.2 Frequency Test within a Block
The focus of the test is the proportion of ones within M-bit blocks. The purpose of this test is to determine whether the frequency of ones in an M-bit block is approximately $M/2$, as would be expected under an assumption of randomness. For block size $M = 1$, this test degenerates to test 1, the frequency (monobit) test.

10.A.4.3 Cumulative Sums (Cusum) Test
The focus of this test is the maximal excursion (from zero) of the random walk defined by the cumulative sum of adjusted (–1, +1) digits in the sequence. The purpose of the test is to determine whether the cumulative sum of the partial sequences occurring in the tested sequence is too large or too small relative to the expected behavior of that cumulative sum for random sequences. This cumulative sum may be considered as a random walk. For a random sequence, the excursions of

the random walk should be near zero. For certain types of nonrandom sequences, the excursions of this random walk from zero will be large.

10.A.4.4 Runs Test

The focus of this test is the total number of runs in the sequence, where a run is an uninterrupted sequence of identical bits. A run of length k consists of exactly k identical bits and is bounded before and after with a bit of the opposite value. The purpose of the runs test is to determine whether the number of runs of ones and zeros of various lengths is as expected for a random sequence. In particular, this test determines whether the oscillation between such zeros and ones is too fast or too slow.

10.A.4.5 Tests for the Longest Run of Ones in a Block

The focus of the test is the longest run of ones within M-bit blocks. The purpose of this test is to determine whether the length of the longest run of ones within the tested sequence is consistent with the length of the longest run of ones that would be expected in a random sequence. Note that an irregularity in the expected length of the longest run of ones implies that there is also an irregularity in the expected length of the longest run of zeroes. Therefore, only a test for ones is necessary.

10.A.4.6 Binary Matrix Rank Test

The focus of the test is the rank of disjoint submatrices of the entire sequence. The purpose of this test is to check for linear dependence among fixed length substrings of the original sequence. Note that this test also appears in the Diehard battery of tests.

10.A.4.7 Discrete Fourier Transform (Spectral) Test

The focus of this test is the peak heights in the discrete Fourier transform of the sequence. The purpose of this test is to detect periodic features (i.e., repetitive patterns that are near each other) in the tested sequence that would indicate a deviation from the assumption of randomness. The intention is to detect whether the number of peaks exceeding the 95% threshold is significantly different from 5%.

10.A.4.8 Non-overlapping Template Matching Test

The focus of this test is the number of occurrences of prespecified target strings. The purpose of this test is to detect generators that produce too many occurrences of a given nonperiodic (aperiodic) pattern. For this test and for the overlapping template matching test, an m-bit window is used to search for a specific m-bit pattern. If the pattern is not found, the window slides one bit position. If the pattern is found, the window is reset to the bit after the found pattern, and the search resumes.

10.A.4.9 Overlapping Template Matching Test

The focus of the overlapping template matching test is the number of occurrences of prespecified target strings. Both this test and the non-overlapping template matching test use an m-bit window to search for a specific m-bit pattern. As with the non-overlapping template matching test, if the pattern is not found, the window slides one bit position. The difference between this test and the non-overlapping template matching test is that when the pattern is found, the window slides only one bit before resuming the search.

10.A.4.10 Maurer's "Universal Statistical" Test
The focus of this test is the number of bits between matching patterns (a measure that is related to the length of a compressed sequence). The purpose of the test is to detect whether or not the sequence can be significantly compressed without loss of information. A significantly compressible sequence is considered to be nonrandom.

10.A.4.11 Approximate Entropy Test
The focus of this test is the frequency of all possible overlapping m-bit patterns across the entire sequence. The purpose of the test is to compare the frequency of overlapping blocks of two consecutive/adjacent lengths (m and $m+1$) against the expected result for a random sequence.

10.A.4.12 Random Excursions Test
The focus of this test is the number of cycles having exactly K visits in a cumulative sum random walk. The cumulative sum random walk is derived from partial sums after the (0, 1) sequence is transferred to the appropriate (-1, $+1$) sequence. A cycle of a random walk consists of a sequence of steps of unit length taken at random that begin at and return to the origin. The purpose of this test is to determine if the number of visits to a particular state within a cycle deviates from what one would expect for a random sequence. This test is actually a series of eight tests (and conclusions), one test and conclusion for each of the states: $-4, -3, -2, -1$ and $+1, +2, +3, +4$.

10.A.4.13 Random Excursions Variant Test
The focus of this test is the total number of times that a particular state is visited (i.e., occurs) in a cumulative sum random walk. The purpose of this test is to detect deviations from the expected number of visits to various states in the random walk. This test is actually a series of eighteen tests (and conclusions), one test and conclusion for each of the states: $-9, -8, \ldots, -1$ and $+1, +2, \ldots, +9$.

10.A.4.14 Serial Test
The focus of this test is the frequency of all possible overlapping m-bit patterns across the entire sequence. The purpose of this test is to determine whether the number of occurrences of the $2m$ m-bit overlapping patterns is approximately the same as would be expected for a random sequence. Random sequences have uniformity; that is, every m-bit pattern has the same chance of appearing as every other m-bit pattern. Note that for $m=1$, the serial test is equivalent to the frequency test.

10.A.4.15 Linear Complexity Test
The focus of this test is the length of a linear feedback shift register (LFSR). The purpose of this test is to determine whether or not the sequence is complex

enough to be considered random. Random sequences are characterized by longer LFSRs. An LFSR that is too short implies nonrandomness.

The order of the application of the above-described 15 tests is arbitrary. However, it is recommended that the frequency test be run first, since this supplies the most basic evidence for the existence of nonrandomness in a sequence, specifically, nonuniformity. If this test fails, the likelihood of other tests failing is high.

11
Controlling Chaos in Lasers

In this chapter, the concept and techniques of controlling chaos are introduced and described. The term "controlling chaos" indicates that a chaotic temporal waveform of laser output is stabilized into a steady-state output or a periodic temporal waveform by applying a self-feedback signal or an external perturbation. Since chaos is generally considered as an undesirable behavior in many engineering applications, the techniques for controlling chaos have been applied to many types of nonlinear dynamical systems. The principle of controlling chaos is explained and examples of controlling chaos in laser systems are described from the literature in Chapter 11. Some possible applications using controlling chaos are also discussed.

Comprehensive reviews on controlling chaos in nonlinear dynamical systems including laser systems have been reported in the literature (Kapitaniak, 1996; Schuster, 1999; Gauthier, 2003; González-Miranda, 2004; Illing *et al.*, 2006).

11.1
Classification of Controlling Chaos

11.1.1
Feedback Control Method

11.1.1.1 OGY Method

The concept of controlling chaos was originally proposed in 1990 (Ott *et al.*, 1990). This technique is known as the OGY method, derived from the inventors' names (Ott, Grebogi, and Yorke). The concept of this method is the following. There are infinite numbers of unstable periodic orbits (UPOs) in a chaotic attractor, which is an ensemble of chaotic trajectories in the phase space. A chaotic trajectory can be stabilized on one of the UPOs by perturbing a system parameter, as shown in Figure 11.1. It is possible to stabilize a chaotic trajectory on an UPO by applying a small perturbation to one of the system parameters. When the chaotic trajectory starts deviating from the target UPO, a perturbation is applied to the opposite direction of the movement of the trajectory so that the chaotic trajectory can approach the target UPO in the phase space. This technique is similar to an example of standing a stick

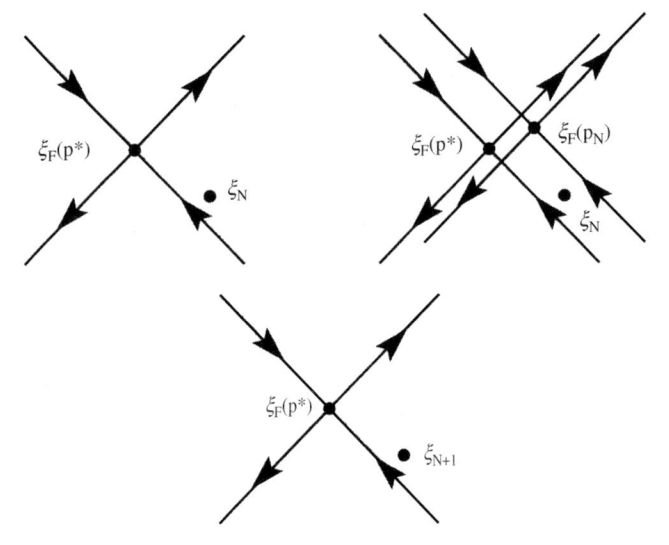

Figure 11.1 Idea of OGY method in the phase space. (T. Kapitaniak, (1996), © 1996 Academic Press.)

(or cane) on a hand by slightly moving the position of the hand, so that the stick (or cane) can be kept standing on the hand. The mathematical formula of the OGY method is described in Appendix 11.A.1.

The validity of the OGY method has been proven mathematically, and some examples of the OGY method have been reported. The OGY method was considered as an important milestone work since the idea of controlling chaos is revolutionary at that time. However, there are some difficulties for the implementation of the OGY method in real-world dynamical systems. It is required to calculate a target UPO beforehand and the deviation of the chaotic trajectory from the target UPO needs to be monitored continuously. In addition, the amount of the perturbation for the control signal cannot be estimated for the dynamical systems that cannot be described by a mathematical model (or equations).

11.1.1.2 Occasional Proportional Feedback (OPF) Method

A modification of the OGY method has been applied, known as the occasional proportional feedback (OPF) method (Hunt, 1991). In the OPF method, one of the system variables is monitored continuously and the perturbation, whose amount is proportional to the deviation of the chaotic trajectory from a predetermined value of the selected variable, is applied to one of the system parameters. The perturbation is applied to the parameter only if the chaotic trajectory approaches the predetermined value, otherwise no perturbation is applied. The amount of the perturbation and the predetermined value are selected empirically, and different types of periodic oscillations can be produced, depending on these values.

The OPF method has been implemented in an electric circuit experiment (Hunt, 1991). The advantages of the OPF method are the following. There is no need to detect the position of the target UPO in the phase space beforehand. In addition, the OPF method can be applied to the dynamical systems that cannot be described by a mathematical model. On the other hand, it takes some time to monitor and calculate the amount of the perturbation during the control, and the OPF method cannot be directly applied to fast chaotic oscillations. Instead, the transient-time pulsewidth feedback method has been proposed for controlling fast chaotic oscillations (Myneni et al., 1999), where the time for applying the perturbation is changed depending on the deviation from the target, while the amount of the perturbation is kept constant.

11.1.1.3 Continuous Feedback Control Method

Both the OGY and OPF methods are classified into the discrete-time feedback control method, where a control signal is applied to a system parameter at discrete time only when a chaotic trajectory approaches a target UPO very closely. A continuous-time feedback control method has been proposed, known as the continuous feedback control method (Pyragas, 1992). This method is shown in Figure 11.2. An external periodic signal is prepared and the difference between the chaotic waveform and the external periodic signal is fed back to one of the system parameters, as shown in Figure 11.2a. Another version of this method is that the difference between the chaotic waveform and its time-delayed signal is fed back to the system parameter, without using the external periodic signal, as shown in Figure 11.2b. The latter case is called the continuous time-delayed feedback method. For both schemes, the negative feedback that is proportional to the deviation of the chaotic waveform from the periodic signal is applied to the system parameter when the control is executed, while the feedback signal vanishes under the control because the difference between the chaotic signal and the target signal becomes close to zero. Different types of

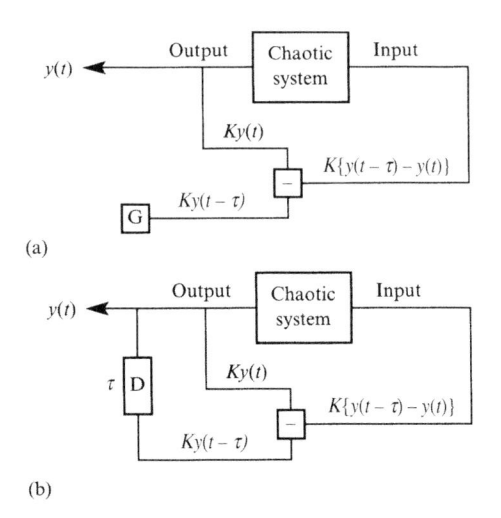

Figure 11.2 Block diagram of (a) external force control, and (b) delayed feedback control. G is a special external periodic oscillator and D is a delay line. (K. Pyragas, (1992), © 1992 Elsevier.)

periodic waveforms can be obtained by changing the external periodic signal or the delay time. The continuous feedback control method is simple and useful for the implementation to real-world dynamical systems.

11.1.2
Nonfeedback Control Method

The feedback control methods described in Section 11.1.1 require continuous detection of the chaotic signal. On the other hand, the detection is not required for nonfeedback control methods. Nonfeedback control methods has been proposed and demonstrated, where a constant or periodic modulation is continuously applied to one of the system parameters, regardless of the position of the chaotic trajectory in the phase space (Braiman and Goldhirsch, 1991). The nonfeedback control methods can be considered as a simple external modulation to a chaotic dynamical system.

The mechanism of the nonfeedback control methods has been explained in the literature (Braiman and Goldhirsch, 1991; Breeden et al., 1990), however, the common understanding of this method is still lacking since the effectiveness of the nonfeedback control methods is strongly dependent on the dynamical systems themselves. The modulation amplitude and frequency are selected empirically so that chaotic oscillations can be stabilized into a steady state or a periodic oscillation. The nonfeedback control methods slightly change the controlled dynamical system by a small permanent shift of control parameter, changing the system behavior from chaos to a periodic oscillation that is close (but not identical) to an UPO, while the feedback control methods do not change the controlled system and stabilize an UPO on an original chaotic attractor. The mechanism of the control is thus completely different between the nonfeedback and feedback control methods. The nonfeedback control methods can be interpreted as the change in the bifurcation of the dynamical system by adding an external modulation, where a chaotic oscillation can be changed into a steady state or a periodic oscillation.

The characteristics of the nonfeedback control methods are in the following. The configuration for the control is simple, and fast chaotic oscillations can be stabilized by this method since no detection and monitoring are required. On the other hand, the types of periodic oscillations and the corresponding control parameter values are strongly dependent on the structures of bifurcation and chaotic attractors, and cannot be determined *a priori*. It is difficult to predict the types of periodic oscillations and the control parameter values before the control is applied. In addition, the number of obtained periodic oscillations is limited since the controlled periodic oscillations are different from the UPOs in the original chaotic attractor.

To solve this problem, a nonfeedback control method has been proposed, where the period of the stabilized oscillation can be predicted (Li and Chern, 1996). When a small sinusoidal modulation whose frequency equals the $(1/n)$-th subharmonic of the fundamental frequency of chaotic oscillation is applied to a system parameter, a period-n oscillation can be obtained from the chaotic oscillation by this method. This method utilizes the resonance of subharmonics of chaotic oscillations. There is however, a limitation of n: the control is not effective for too large n. Another nonfeedback control method is an injection of a prerecorded UPO into the dynamical system to obtain a periodic oscillation corresponding to the UPO (Pyragas, 1993). This method can also be applied to a technique for synchronization of chaos when a prerecorded chaotic signal is used as a control signal (see Section 5.2.1.2).

11.2
Examples of Controlling Chaos in Lasers

11.2.1
Feedback Control Method for Controlling Chaos in Lasers

11.2.1.1 Occasional Proportional Feedback (OPF) Method

The first experimental demonstration of controlling chaos in lasers was reported in 1994 with a laser-diode-pumped Nd:YAG solid-state laser with an intracavity KTP doubling crystal (Roy et al., 1992). The OPF method was applied to the laser system. The basic technique for achieving dynamical control is as follows. The total intensity of the laser output is sampled within a window of selected offset and width. The sampling frequency is related to the relaxation oscillation frequency of the laser. A signal proportional to the deviation of the sampled intensity from the center of the window is generated and applied to perturb a laser parameter from its ambient value. This control signal repeatedly attempts to bring the system closer to an UPO that is embedded in the chaotic attractor, resulting in a realization of the periodic orbit with accuracy limited by the frequency and extent of feedback.

A block diagram of the laser system and controller is shown in Figure 11.3 (Roy et al., 1992). The fundamental 1.06-μm radiation is monitored by a photodiode, the output from which serves as the input to the control circuit. A stable oscillator is used to generate the synchronizing frequency at which the output from the chaotic laser is sampled. A variable offset is added to the laser signal to bring it within a window of

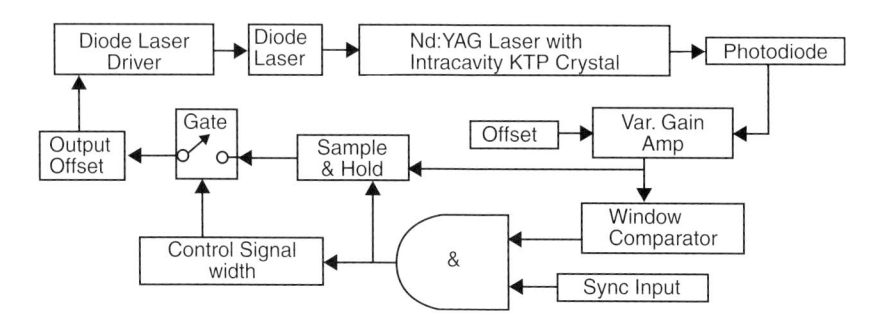

Figure 11.3 Schematic of the laser system and occational proportional feedback (OPF) controller. The perturbation of the laser-diode injection current is proportional to the deviation of a sampled waveform value from the window center. The synchronizing oscillator frequency, waveform offset, and control signal width are varied to optimize stability of the periodic waveform. (R. Roy, T. W. Murphy, Jr., T. D. Maier, Z. Gills, and E. R. Hunt, (1992), © 1992 APS.)

adjustable width. The window comparator is activated when the waveform makes a transit through the window. When the synchronizing input is coincident with this event, the sample and hold acquires the waveform voltage. The sampled signal is output through the gate only for time periods short compared to the period of the synchronizing oscillator. A typical time period for application of the correction signal is less than 10 µs. An inverting amplifier with variable offset and gain delivers the control signal to the laser-diode driver, and the injection current of the laser diode for pumping is perturbed.

It is very easy to make these adjustments and to obtain many higher-order periodic waveforms of the laser intensity. Some examples are shown in Figure 11.4 (Roy et al., 1992). The waveform of the chaotic laser without any control signal is shown in Figure 11.4a. The FFT of this waveform shows a broad peak centered at the relaxation oscillation frequency of 118 kHz. When the OPF control is applied to the laser system, period-1, -4, and -9 waveforms are obtained with different control signals, as shown in Figures 11.4b–d. A rich variety of waveforms can be obtained and maintained in stable operation by the OPF control method. For the low-period waveforms, control can be established with small perturbations applied near the relaxation oscillation frequency or its subharmonics (Figures 11.4b and c). For higher-period waveforms, the nature of the control signal grows progressively more complex (Figure 11.4d). The perturbations applied to the drive current of the laser diode are only a few percent of the ambient-bias current for the low-order periodic waveforms. Even for the higher-order periodic waveforms, the maximum perturbations observed are less than 10%, and the original chaotic attractor may be modified to some extent by the feedback control.

The OPF method has also been applied to a semiconductor laser interferometer system with a delayed optoelectronic feedback (Liu and Ohtsubo, 1994). A schematic of the control circuit together with the delayed-feedback system is shown in Figure 11.5. The laser-diode output intensity is detected by a built-in photodiode and is converted into a time-dependent electric signal. A variable offset is added to the detected electric signal to bring it within a window of adjustable width. When the waveform transits within the window, the window comparator outputs a pulse, with the pulse width coincident with the length of transition. Meanwhile, a synchronizing signal is generated by a microcomputer. The frequency of the synchronizing signal (referred to as a sampling frequency) is related to the delay and relaxation times of the system. When the synchronizing signal is coincident with the pulse from the window comparator the sample-and-hold circuit is activated and acquires the waveform voltage. A gate is employed to select only a part of the sampled signal. The gate width is adjustable within a range from 1 µs to several milliseconds. An amplifier with variable gain, offset, and polarity delivers the control signal to the drive current of the second laser diode (LD2) and, in turn, perturbs the injection current of the light source (LD1) from its ambient value.

The dynamical control of chaos is performed. A large number of periodic orbits is extracted and stabilized. For the delay time of 0.12 ms the free-running state in the absence of the control signal is adjusted to be a weak chaotic state, as shown in

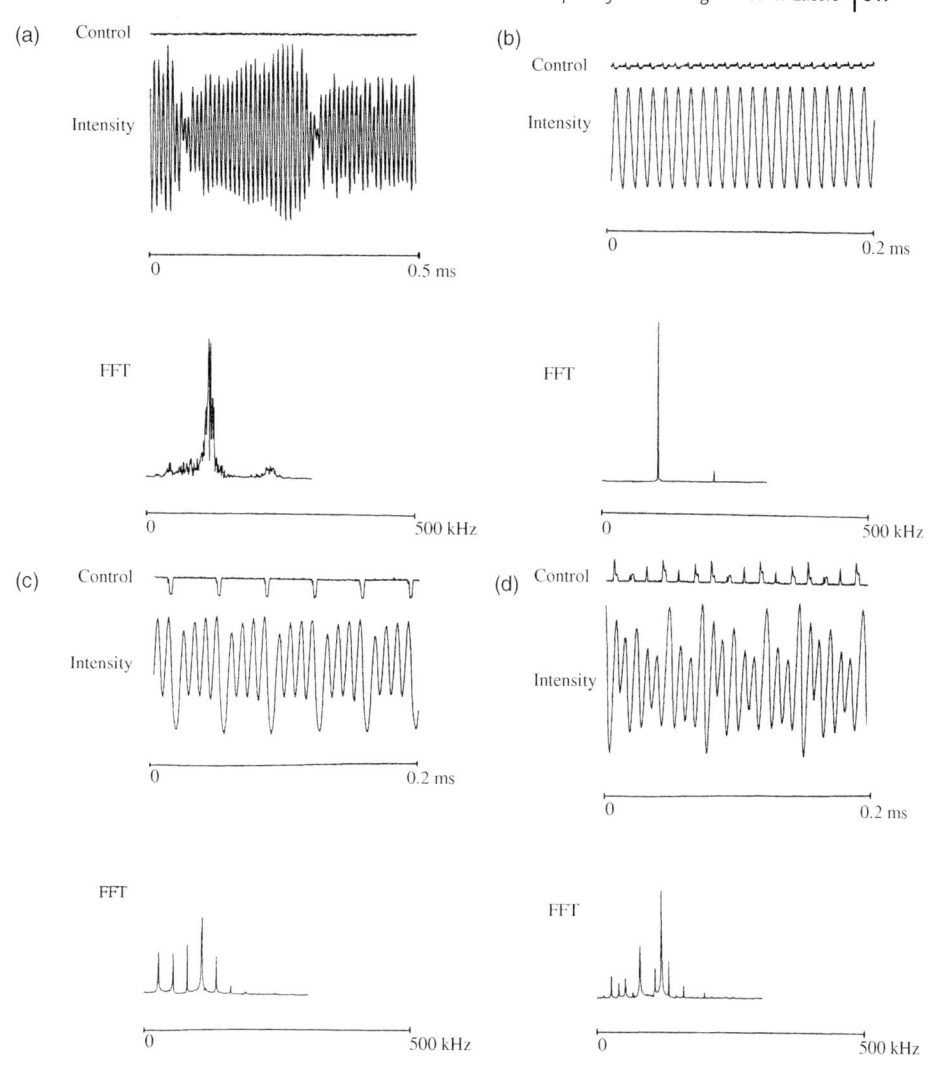

Figure 11.4 Time traces of the laser intensity and the corresponding FFTs. The control signal is shown at the top, and then the intensity time trace and FFT. (a) The chaotic intensity fluctuations of the laser output without any applied control signal. The FFT below the waveform shows the broad peak corresponding to the relaxation oscillation frequency. (b) A period-1 oscillation obtained by adjusting the synchronizing oscillator frequency to approximately the relaxation oscillation frequency. (c) A period-4 oscillation obtained by adjusting the synchronizing oscillator frequency to approximately 1/4 of the relaxation oscillation frequency. (d) A period-9 oscillation obtained by adjusting the synchronizing oscillator frequency at 6/9 of the dominant frequency shown in the FFT. The complex nature of the control signal is clearly visible. (R. Roy, T. W. Murphy, Jr., T. D. Maier, Z. Gills, and E. R. Hunt, (1992), © 1992 APS.)

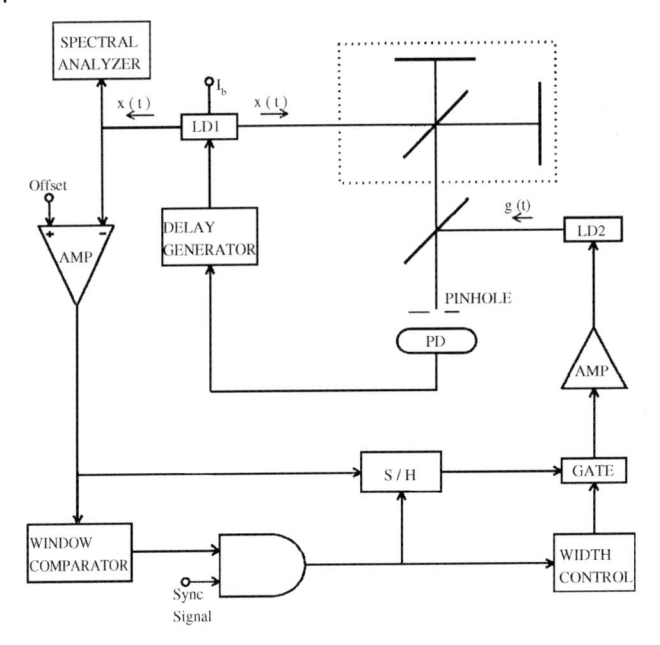

Figure 11.5 Schematic of the delayed optical bistable system and the OPF controller: PD, photodiode; S/H, sample-and-hold circuit; AMPs, amplifiers. (Y. Liu and J. Ohtsubo (1994), © 1994 OSA.)

Figure 11.6a (Liu and Ohtsubo, 1994). The frequency can be obtained from the peak position of the spectral distribution corresponding to the waveform in Figure 11.6a and the obtained value is 2.45 kHz. The dynamical control can be realized by adjusting the synchronizing frequency, the offset, the window width, and the gate width. In the experiment, subharmonic periodic orbits can be successfully stabilized to as high as the tenth order by setting the synchronizing frequency to be rational fractions of the fundamental frequency. Figures 11.6b and c display the stable waveforms of the fundamental and fourth subharmonic periodic orbits, respectively. Their corresponding synchronizing frequencies are read as 2.45 and 0.61 kHz, respectively. A more complex control signal can be observed for the stabilization to a higher periodic oscillation.

11.2.1.2 Continuous Feedback Control Method

The continuous feedback control methods have been widely used to stabilize chaos due to its simplicity and effectiveness. Fast chaotic oscillations can be stabilized in various nonlinear dynamical systems by using these techniques (Blakely et al., 2004).

The continuous feedback control method has been applied to a chaotic CO_2 gas laser (Bielawski et al., 1994). The applied control signal is proportional to the difference between a chaotic laser intensity and its time-delayed signal as,

$$k(t) = a[I(t) - I(t-T)] \tag{11.1}$$

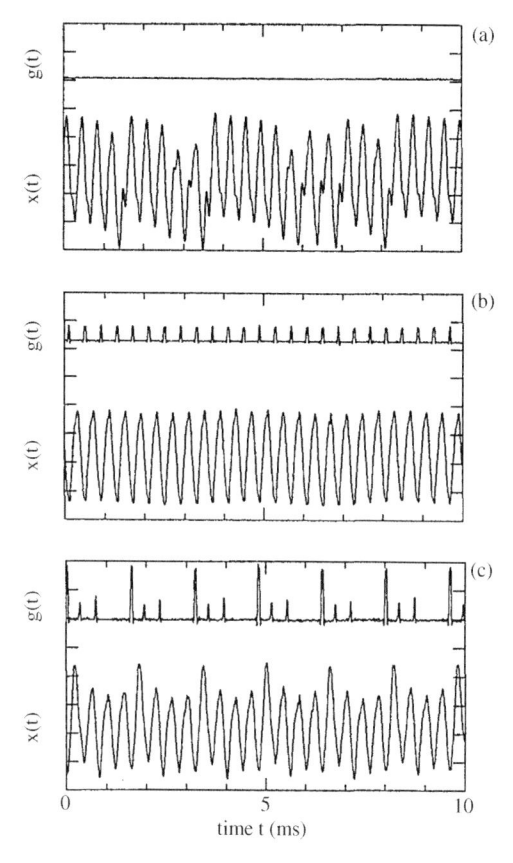

Figure 11.6 Experimental results of the temporal waveforms of the control signal $g(t)$ and the laser intensity $x(t)$. (a) Without the control, and (b), (c) with the control. The synchronizing frequencies are (b) 2.45 kHz and (c) 0.61 kHz, respectively. (Y. Liu and J. Ohtsubo, (1994), © 1994 OSA.)

where $k(t)$ is the additional loss of the electro-optic modulator in the CO_2 laser for chaos control, $I(t)$ is the chaotic laser intensity, T is the delay time, and α is the gain parameter of the delayed continuous feedback.

Stabilization of the unstable T-periodic orbit has been achieved with the delayed continuous feedback method. Figure 11.7 shows the bifurcation diagram without and with stabilization of the unstable T-periodic orbit as a function of the modulation index (Bielawski et al., 1994). The period-doubling route to chaos is observed in the laser output of the original CO_2 laser system without stabilization (Figure 11.7a). When the control is applied to the laser, a period-1 oscillation is obtained for all the parameter region of the modulation index shown in Figure 11.7b. Note that the same unstable period-1 orbit is stabilized for different modulation parameters. The correction applied to the system remains small, showing that the stabilized orbit is identical to the UPO existing in the original chaotic attractor without stabilization.

Another continuous time-delay feedback control method has been applied to a NH_3 laser experimentally (Dykstra et al., 1998). The laser pump power is varied in

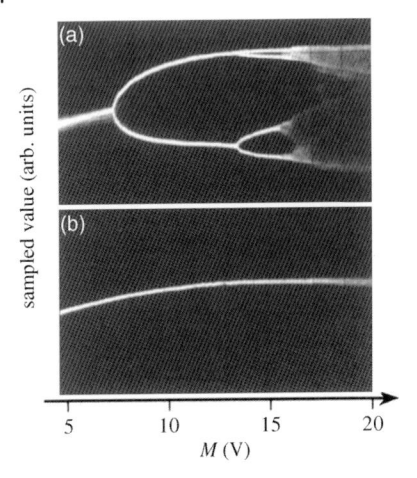

Figure 11.7 Bifurcation diagram (a) without and (b) with stabilization of the unstable period-1 orbit. The stable period-1 orbit can be observed for a wide range of the modulation index M with stabilization. (S. Bielawski, D. Derozier, and P. Glorieux, (1994), © 1994 APS.)

response to a continuous error signal generated by comparing the laser output with its value at an earlier time. The experimental setup is shown schematically in Figure 11.8. The most significant parts of the setup include the CO_2 laser pump and the method whereby its output is modulated. The rest of the setup including the NH_3 laser operating at the 153-μm wavelength and Schottky barrier diode detector. The pump power is controlled by the use of an acousto-optic modulator (AOM), where amplitude modulating the radio-frequency (RF) drive gives a simple means of producing a time-varying pump power.

The continuous time delay error signal is generated in the following way. Approximately 90% of the output signal from the Schottky diode detector is fed back into a simple power splitter–recombiner arrangement. The remaining 10% is used to sample the output intensity of the laser. The power splitter–recombiner resistors have values such that the reflection losses are minimized and 50% of the incident electrical power is injected into a delay line. That part of the signal that enters the delay line is reflected from the short-circuited end and therefore inverted. Upon returning to the junction this signal recombines with that part that arrives at the junction the delay time later and does not enter the delay line. The recombined signal forms an error signal that is then amplified, put through a switch that allows it to be turned on and off, then attenuated if necessary, and finally used to modulate the drive power for the AOM.

Figure 11.9 shows the laser displaying Lorenz chaos (see Section 3.6.1) without and with stabilization (Dykstra *et al.*, 1998). Once the feedback loop is closed at 260 μs, the laser quickly moves onto the fixed point, after which the error signal becomes very small. The maximum change in pump power in this record is only about 3%. The simple form of the transient as the loop gains control is not the most general or usual one, there is a period of metastable chaotic behavior. The control signal is possibly visible and of magnitude 0.1% of the total pump power in Figure 11.9. This control is therefore clearly a perturbative method that stabilizes the fixed points. This could be a

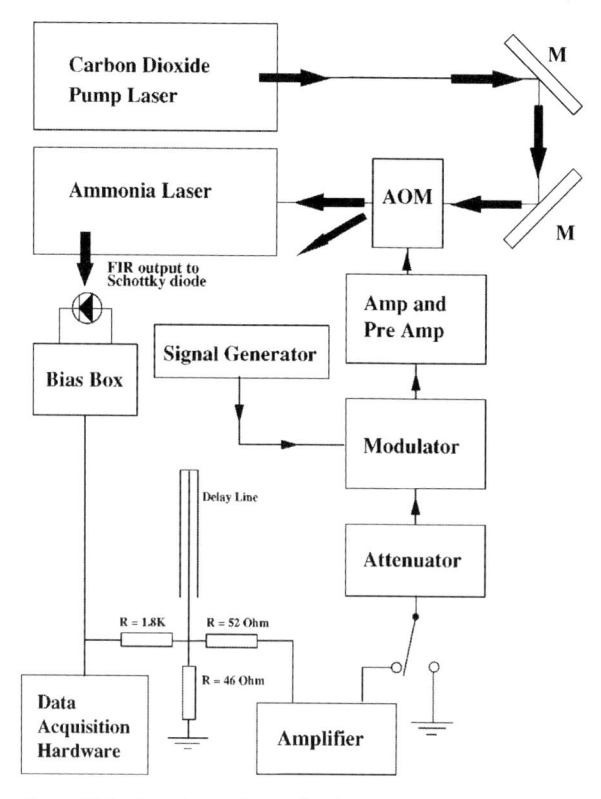

Figure 11.8 Experimental setup for the continuous time-delay feedback control method in a NH_3 laser. (R. Dykstra, D. Y. Tang, and N. R. Heckenberg, (1998), © 1998 APS.)

robust method for obtaining greater continuous-wave (CW) powers out of the single-mode laser. The control demonstrated in Figure 11.9 is obtained only for limited parameter ranges and for fixed delay-line lengths.

The most dominant behavior is the control to a period-one pulsation as shown in Figure 11.10 (Dykstra *et al.*, 1998). In the Lorenz chaos there exist three unstable fixed points in the phase space, one at the origin and two from which the trajectories always spiral away. These spirals increase in period as their distance from the fixed point increases. The control mechanism with the delays incorporated as they are embodies a specific resonance frequency because of the signal's phase. When the delay time is set to be equal to the natural period of the laser's output intensity, the error signal is minimized due to the resonance.

11.2.2
Nonfeedback Control Method for Controlling Chaos in Lasers

11.2.2.1 Loss Modulation
A nonfeedback control method has been applied to a single-mode CO_2 laser with an intracavity electro-optic modulator (EOM) (Meucci *et al.*, 1994). The stabilization of

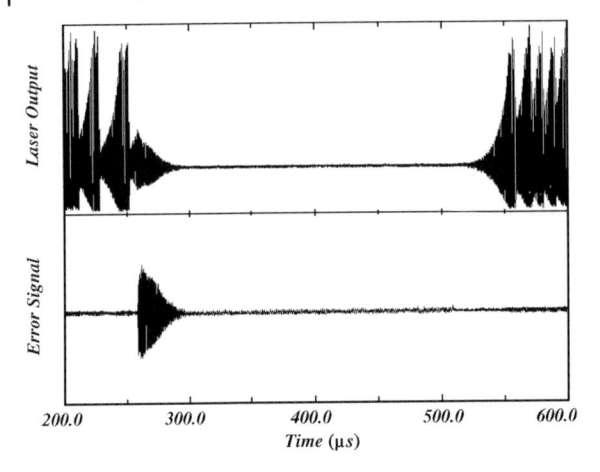

Figure 11.9 Control to the fixed point when the laser demonstrates Lorenz chaos with the lower trace showing the error signal and the upper the laser output. The feedback loop is closed at 260 μs and opened at 520 μs. (R. Dykstra, D. Y. Tang, and N. R. Heckenberg, (1998), © 1998 APS.)

periodic orbits within the chaotic region has been obtained by slight modulations of the control parameter of the EOM voltage as,

$$V(t) = V_1[1 + \varepsilon \sin(2\pi f\, t + m\pi)] \tag{11.2}$$

where f is the modulation frequency, ε is the modulation amplitude, $m\pi$ is the adjustable phase offset, and V_1 is the bias voltage.

The experimental results are shown in Figure 11.11 for the modulation parameters of $f = 50$ kHz and $\varepsilon = 0.025$ (Meucci et al., 1994). In Figure 11.11, a stroboscopic

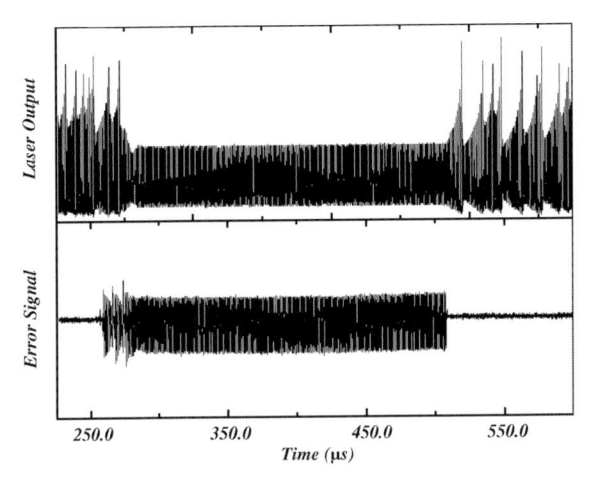

Figure 11.10 Control to the period-1 oscillation when the laser demonstrates Lorenz chaos with the lower trace showing the error signal and the upper the laser output. The feedback loop is closed at 260 μs and opened at 520 μs. (R. Dykstra, D. Y. Tang, and N. R. Heckenberg, (1998), © 1998 APS.)

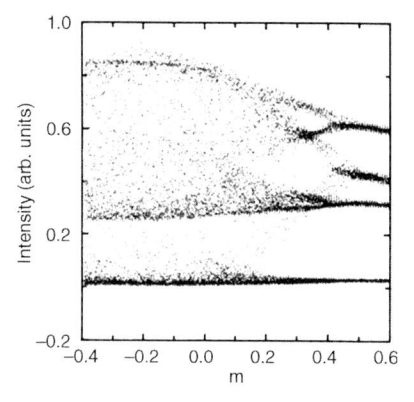

Figure 11.11 Experimental result of controlling chaos in the CO_2 laser with sinusoidal loss modulation. Stroboscopic recording of the laser intensity *versus* the relative phase offset m of the perturbation. (R. Meucci, W. Gadomski, M. Ciofini, and F. T. Arecchi, (1994), © 1994 APS.)

recording of the laser intensity is plotted as a function of the relative phase offset m. It is clear that it is possible to stabilize period-4 orbits embedded in the chaotic attractor by choosing a suitable phase offset. Once the appropriate phase is selected, the stability is of the order of several minutes, until uncontrolled drifts spoil the resonance condition between the cavity mode and the gain line. This experimental result agrees well with the numerical result (Meucci *et al.*, 1994).

11.2.2.2 High-Frequency Injection (HFI) Method for Semiconductor Lasers

The study of the stabilization of the output intensity in a semiconductor laser has been of great importance to the applications to optical communications and optical data recording systems. In most practical environments, optical feedback to the semiconductor laser is inevitably induced and a feedback level as small as 0.1% may cause intensity noise enhancement that dramatically deteriorates the performance. One of the methods to avoid such feedback-induced instabilities that is used in optical data recording systems is known as the "high-frequency injection" (HFI) technique, which modulates the injection current of a semiconductor laser with frequencies much higher than the data rate at hundreds of MHz (Yamada and Higashi, 1991; Gray *et al.*, 1993). The experimental results show that the relative intensity noise (RIN) enhancement does not occur if the modulation frequency is appropriately chosen and if the modulation amplitude is large enough (typically tens of percent of the bias injection current).

The HFI method has been analyzed from the viewpoint of controlling chaos (Liu *et al.*, 1995; Kikuchi *et al.*, 1997). The optimization of the modulation frequency for the HFI method is converted to a problem of how to select the frequency by which chaos can be most effectively stabilized. The linear stability analysis (see Section 4.3.4) is performed to yield the mode distributions whose frequencies are selected as the modulation frequencies for controlling chaos. Effective and small modulation amplitudes of the order of a few percent can be attained for chaos stabilization by using the knowledge of mode distributions and the corresponding modulation frequencies.

The Lang–Kobayashi equations describing the dynamics of a semiconductor laser with optical feedback are used and the linearized equations around the steady-state solutions are obtained (see Section 4.3.4). The characteristic equation for the eigenvalues of the linearized equations is derived, and the real and imaginary parts of the eigenvalues are numerically calculated, as already shown in Figure 4.12 of Section 4.3.4 (Liu et al., 1995). The real and imaginary parts of the eigenvalues are plotted in Figure 4.12, which are the indicators of the stability and the oscillation frequency, respectively.

The stabilization has been performed by modulating the injection current of the semiconductor laser with frequencies equal to the imaginary parts of those modes indicated with arrows in Figure 4.12. Figure 11.12 shows the results of the stabilization (Liu et al., 1995). Figures 11.12a and c are the time series of the laser intensity and the corresponding attractor for the chaotic output without the modulation, and Figure. 11.12b and d represent those after the modulation. The modulation frequency is chosen to be the same value as the largest stable mode. The chaotic output is reduced to a limit cycle under such modulation frequency, although the modulation

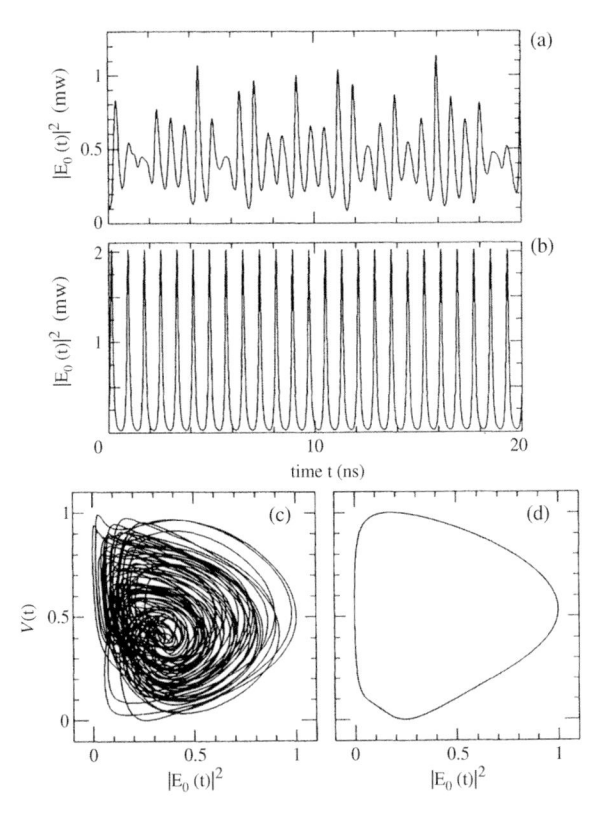

Figure 11.12 Numerical result of controlling chaos in the semiconductor laser with optical feedback by using the high frequency injection (HFI) method. (a) Time series of chaotic output in the absence of the modulation, (b) periodic oscillations with the modulation amplitude of 0.021 and the modulation frequency of 1.251 GHz. (c), (d) Attractors of (a) and (b), respectively, in the phase space. (Y. Liu, N. Kikuchi, and J. Ohtsubo, (1995), © 1995 APS.)

amplitude is as small as 2.1%. Note that the amplitude of the periodic oscillations is about twice that of the original chaotic oscillations. This implies that the system is in a resonant state with respect to the frequency of the linear mode.

11.2.2.3 Stabilization to High-Periodic Oscillations

The nonfeedback control method that can stabilize chaotic oscillations into high-periodic orbits has been proposed and applied to a laser-diode-pumped Nd:YVO$_4$ microchip solid-state laser by using an internal frequency resonance among the relaxation oscillation frequencies and the modulation frequencies (Uchida et al., 1998b). A frequency-shifted feedback light-injection scheme is utilized as a loss-modulation system (see Section 3.5.2). The laser light is injected upon a rotating circular paper sheet, and weak scattered light whose center frequency is shifted because of the Doppler effect returns to the laser cavity. The laser intensity is then modulated as a result of self-mixing between the two light components in the cavity, referred to as a loss modulation. At the same time the injection current of the laser diode for pumping is sinusoidally modulated, and this acts as a pump-modulation system. Chaotic instabilities appear in the laser output with only one of the two modulations when the Doppler-shifted feedback light power or the pump-modulation amplitude is increased.

When only the loss modulation is applied to the microchip laser, a chaotic temporal waveform of the laser output can be observed, as shown in Figure 11.13a (Uchida et al., 1998b). The pump modulation is then applied to the microchip laser to stabilize the chaotic oscillation caused by the loss modulation. The pump-modulation parameters are selected so that a new periodic orbit is generated without the loss modulation. When the frequency and amplitude of the pump modulation are set to 857 kHz and 0.080, respectively, a periodic laser output in a period-6 orbit is obtained without the loss modulation, as shown in Figure 11.13b. When the pump modulation that creates the period-6 orbit is applied in addition to the loss modulation, the temporal waveform of the laser output is stabilized to a period-18 regular oscillation, as shown in Figure 11.13c, which is different from the periodic orbit caused by the pump modulation. These results show that a high-period oscillation whose orbit does not exist in the original chaotic attractor and its bifurcation can be obtained. Various periodic orbits, which are not restricted to those in the original chaotic attractor, can be generated from the chaotic oscillation by changing the pump-modulation parameters in this control scheme. This control technique could be useful for the implementation of high-periodic pattern generators.

11.3
Applications of Controlling Chaos in Lasers

11.3.1
Suppression of Relative Intensity Noise (RIN)

The techniques of controlling chaos can be applied for the stabilization and noise reduction of the laser output intensity. The relative intensity noise (RIN) can be

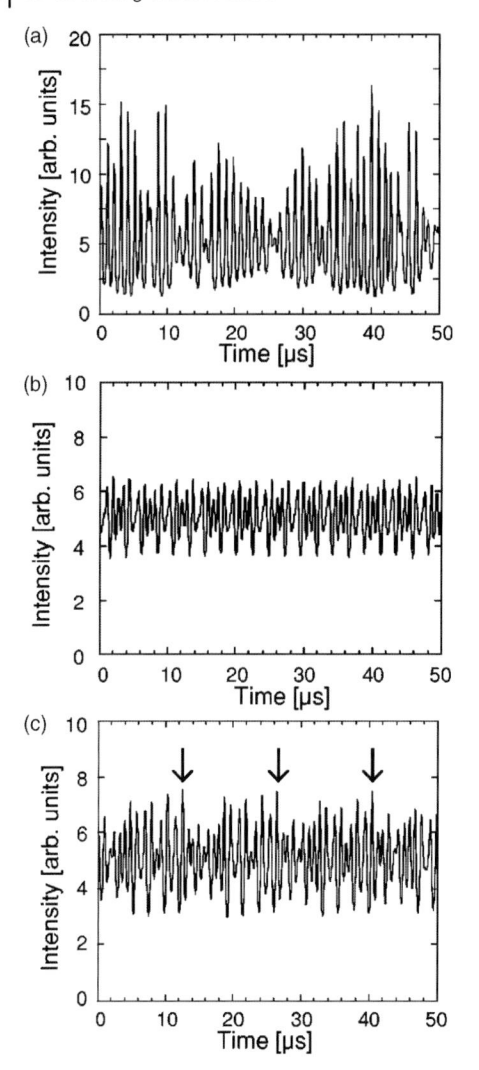

Figure 11.13 Temporal waveforms obtained from the experiment in the Nd:YVO$_4$ microchip solid-state laser with loss and pump modulations. (a) Chaotic oscillation with only loss modulation. (b) Period-6 oscillation with only pump modulation. (c) Period-18 oscillation with both loss and pump modulations. Arrows indicate the period. (A. Uchida, T. Sato, T. Ogawa, and F. Kannari, (1998b), © 1998 APS.)

suppressed in a semiconductor laser by using the HFI method for controlling chaos, as described in Section 11.2.2.2. This research direction of controlling chaos is straightforward from the engineering point of view.

Figure 11.14 shows the experimental result of the RIN *versus* optical feedback ratio (OFB) for different modulation depths of the injection current using the modulation frequency of 480 MHz applied to a semiconductor laser (Gray *et al.*, 1993). The bias

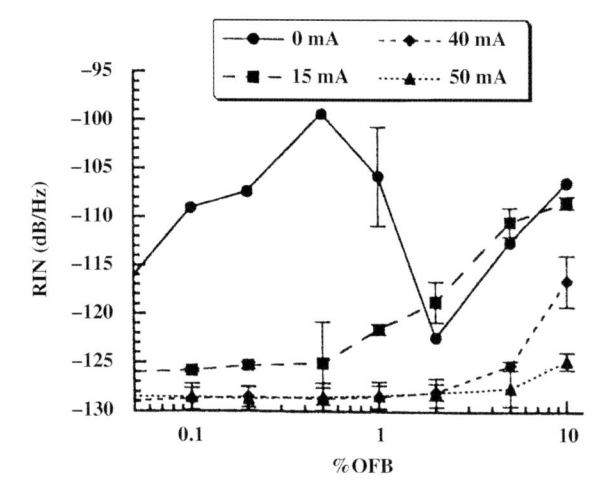

Figure 11.14 Experimental measurement of relative intensity noise (RIN) *versus* optical feedback level (OFB) for different modulation depths in the semiconductor laser with optical feedback. The high frequency injection (HFI) method is applied for suppressing chaos. (G. R. Gray, A. T. Ryan, G. P. Agrawal, and E. C. Gage, (1993), © 1993 SPIE.)

injection current is set to 65 mA. Without modulation (0 mA), the experimental RIN has already increased to high values for a feedback of 0.1%. At high feedback levels, the RIN actually decreases to low values, which correspond to a frequency-locked state. The high-feedback regime is very unstable, however, the RIN fluctuates between high and low values. A typical error bar shown on the unmodulated RIN curve is 10 dB. The frequency-locked states get narrower and closer together as OFB increases, so that variations in OFB switch the laser between chaotic states and frequency locking. When the HFI control is turned on, the transition of the RIN to high values is delayed and suppressed as OFB increases. It is found that larger modulation depths are more effective in suppressing the RIN and delaying the onset of chaos for all OFB.

11.3.2
Chaotic Search and Adaptive Mode Selection

Chaotic search and adaptive mode selection based on chaos control has been proposed as a novel application of controlling chaos. The basic idea of adaptive mode (wavelength) selection using chaotic search (Aida and Davis, 1994; Davis, 1990) was proposed to adaptively select a dominant lasing mode that fits the environment by switching a control parameter such as the injection current between multistability and chaotic mode-transition states (Liu and Davis, 1998). A multimode Fabry–Perot laser diode with optical feedback is used for the adaptive wavelength selection.

The chaotic search consists of two steps: mode evaluation and parameter adjustment (Liu and Davis, 1998). In the first step the output of each mode is detected, and dominant-mode ratio (DMR) that shows how frequently a specific mode dominates

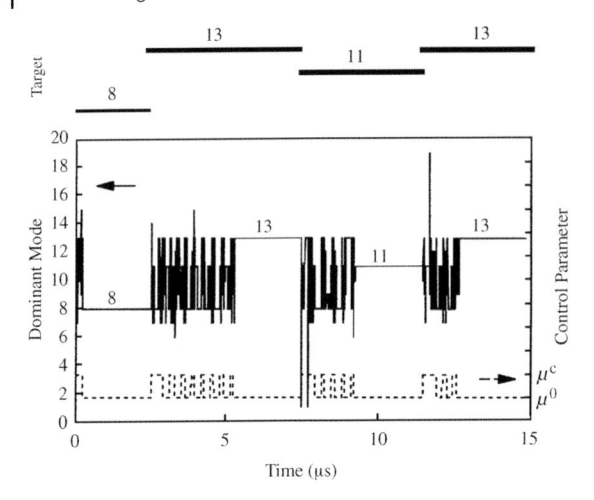

Figure 11.15 Mode search process. Target modes are mode 8, 13, 11, and 13. The control parameter (bias injection current) is switched between $\mu^c = 1.15 I_{th}$ and $\mu^0 = 1.35 I_{th}$. Upper trace: dominant mode (DMR). Lower trace: the control signal. (Y. Liu and P. Davis, (1998), © 1998 World Scientific.)

the laser output is evaluated. The mode evaluation time is chosen as several times the relaxation oscillation period of the laser. In the second step, the control parameter (the injection current) is chosen as a two-level variable for simplicity, with parameter value μ^C corresponding to the chaotic mode-transition regime and parameter value μ^0 to the multistability regime. When a mode gets a good external response, the control parameter is set to μ^0. If the laser output falls into a different mode that gets a bad response, the control parameter is reset to μ^C and the search loop is repeated.

Figure 11.15 shows an example of adaptive mode selection (Liu and Davis, 1998). The injection current is taken as a control parameter and is switched between the chaotic mode-transition regime at $\mu^C = 1.15 \, I_{th}$ and multistability regime at $\mu^0 = 1.35 \, I_{th}$. Three modes (mode 8, 11, and 13) are alternatively set as targets. During each search loop, there are several switch-on and -off processes. Such behavior is typical in this adaptive mode selection method due to mismatch between the external classification of output states and the basins of attraction. When the target mode appears, the laser is set in the multistability regime ($\mu = \mu^0$) and the output mode converges to one of the multiple stable modes. If the output converges to the target mode, the good response will keep the control parameter at μ^0. If the output falls into some other mode, the change in the external response will automatically reset the control parameter to be μ^C. This chaotic search technique could be applied to various temporal-pattern generators (Aida and Davis, 1994; Davis 1990).

11.3.3
Dynamical Memory

Another novel application based on chaos control has been proposed, such as dynamical memory (Aida and Davis, 1992). Temporal waveforms of binary codes can be generated and maintained as a memory device. The dynamical memory has been demonstrated in an electro-optic system with large delay, as shown in Figure 3.29 (left) of Section 3.3.3. The method utilizes multistable oscillation modes for the memory state with binary coding and the method to select a mode corresponding to a particular binary code.

For the writing process, a seeded bifurcation switch is used for selective excitation of oscillation modes corresponding to binary-coded input data. The parameter μ, which is the height of the nonlinear function in the electro-optic system, is changed by increasing input optical power (laser diode (LD) power). The seed signal is injected by modulation of the bias voltage V_0 of the electro-optic modulator (EOM) in Figure 3.29 (left). It is necessary to determine two voltage levels, corresponding to μ_{off} and μ_{on}, to be used in driving the LD, where the seed signal is memorized in the electro-optic system during the input optical power of μ_{on}. A code sequence is converted to a V_0 seed signal of length T_r and modulates the EOM as the seed signal.

Figure 11.16 shows an example of writing process of 21-bit data (Aida and Davis, 1992). The input optical power is changed from μ_{off} to μ_{on} for writing. The pattern of the seed signal V_0 determines a particular mode that is excited. It can be seen that the stable asymptotic oscillation is similar to the initially excited oscillation, in which 21 bits of binary data are successfully encoded and memorized as the output of the electro-optic system after the seed signal disappears.

The read operation is done by thresholding peak levels of the oscillation at the output detector. Erasing of the data is done simply by switching the input power from μ_{on} to μ_{off}.

Another method has been proposed by controlling the antiphase dynamics in multimode lasers with spatial hole burning (Otsuka and Chern, 1992). Coexisting dynamical spatial or temporal patterns can be selectively excited by applying a seed signal to the modulated multimode lasers. The memory capacity can be estimated as $\log(N-1)!/\log 2$, where N is the number of lasing modes.

11.3.4
Communication with Chaos by Controlling Chaos

The use of chaos to transmit information has been proposed in a different way described in Chapters 7 and 8. A small control perturbation to chaos can cause a signal to follow an orbit whose binary sequence represents the information for communication (Hayes *et al.*, 1993). Positive or negative peaks of a chaotic signal can be associated with bits 1 or 0, respectively.

The basic strategy is as follows. A chaotic oscillator is examined and a symbolic dynamics is extracted from it to assign symbol sequences to the orbits on the chaotic attractor. The next step is to choose a code whereby any message that can be emitted by the information source can be encoded using symbol sequences that satisfy suitable constraints imposed by the dynamics in the presence of control. Once the code is selected, the next step is to specify a control method whereby the orbit can be made to follow the symbol sequence of the information to be transmitted. Finally, the transmitted signal must be detected and decoded.

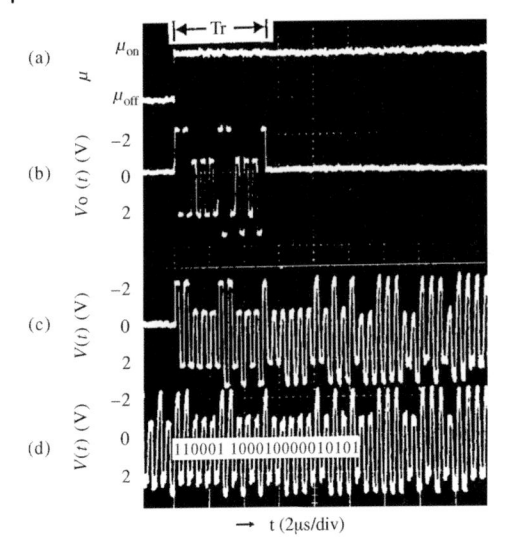

Figure 11.16 An example of 21-bit data WRITE to dynamical memory. Bit sequence "110001100010000010101". (a) The change of μ from μ_{off} to μ_{on} corresponds to change of input optical power. (b) $V_0(t)$ modulated by the seed signal. (c) the excited oscillation $V(t)$, one of the (21, 2) modes. (d) The stable asymptotic state of the oscillation $V(t)$, observed a long time ($\geq 10^6 T_r$) after WRITE. (T. Aida and P. Davis, (1992), © 1992 IEEE.)

Figure 11.17 shows the encoded waveform for the double-scroll electronic circuit system (Hayes et al., 1993). This waveform corresponds to the voltage across the passive conductance. Each letter of the Roman alphabet is represented by the five-bit binary number for its location in the alphabet, and the extra bits added to satisfy the no-signal-oscillation constraint to encode the word "chaos". Small perturbations are applied to the chaotic electronic circuit to generate the coding sequence of "chaos". The amplitude of the control signal over the writing of the word is of order 10^{-3} in the normalized units. The signal produced by the chaos control can be transmitted through the channel for communication purpose if the conductance of the electronic circuit is replaced by a transmission channel of the same impedance.

Appendix

11.A.1
OGY Method for Controlling Chaos

Ott, Grebogi and Yorke have proposed and developed a method by which chaos can always be suppressed by shadowing one of the infinitely many unstable periodic orbits (or steady states) embedded in the chaotic attractor, as described in Section 11.1.1.1 (Ott et al., 1990). The OGY algorithm is schematically explained in Figure 11.1 of Section 11.1.1.1. The basic assumptions of this method are described as follows (Kapitaniak, 1996).

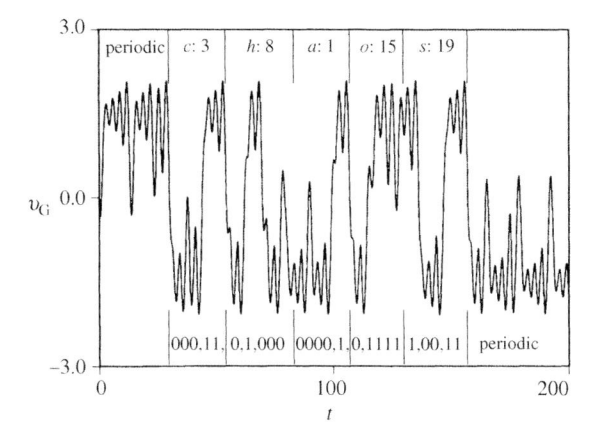

Figure 11.17 Controlled signal for the double-scroll chaotic system encoded with "chaos". Each letter is shown at the top of the figure, along with its numerical position in the alphabet. Shown at the bottom are the corresponding binary code words. Extra bits (indicated by commas) are added to satisfy the constraints imposed by the grammar. (S. Hayes, C. Grebogi, and E. Ott, (1993), © 1993 APS.)

a) The dynamics of the system can be described by an n-dimensional map of the form.

$$\varsigma_{n+1} = f(\varsigma_n, p) \tag{11.A.1}$$

This map, in the case of continuous-time systems, can be constructed, for example, by introducing a transversal surface of section for system trajectories (Poincaré map).

b) p is some accessible system parameter that can be changed in some small neighborhood of its nominal value p^*.

c) For this value p^* there is a periodic orbit within the attractor around which one would like to stabilize the system.

d) The position of this orbit changes smoothly with changes in p, and there are small changes in the local system behavior for small variations of p.

Let ς_F be a chosen fixed point of the map f of the system existing for the parameter value p^*. In the close vicinity of this fixed point with good accuracy it is assumed that the dynamics are linear and can be expressed approximately by

$$\varsigma_{n+1} - \varsigma_F = \mathbf{M}(\varsigma_n - \varsigma_F) \tag{11.A.2}$$

The elements of the matrix $\mathbf{M}$ can be calculated using the measured chaotic time series and analyzing its behavior in the neighborhood of the fixed point. Further, the eigenvalues λ_s, λ_u and eigenvectors e_s, e_u of this matrix can be found. These eigenvectors determine the stable and unstable directions in the small neighborhood of the fixed point.

Denoting by f_s, f_u the contravariant eigenvectors ($f_s e_s = f_u e_u = 1$, $f_s e_u = f_u e_s = 0$) one can find the linear approximation valid for small $|p_n - p^*|$:

$$\varsigma_{n+1} = p_n g + (\lambda_u e_u f_u + \lambda_s e_s f_s)(\varsigma_n - p_n g) \tag{11.A.3}$$

where

$$g = \left.\frac{\partial \varsigma_F(p)}{\partial p}\right|_{p+p^*} \tag{11.A.4}$$

Because ς_{n+1} should fall on the stable manifold of ς_F, choose p_n, such that $f_u \varsigma_{n+1} = 0$:

$$p_n = \frac{\lambda_u \varsigma_n f_u}{(\lambda_u - 1) g f_u} \tag{11.A.5}$$

p_n is the perturbation to the parameter for the OGY method.

12
Other Applications with Chaotic Lasers

In this chapter, other promising applications using chaotic lasers are introduced and discussed. First, remote-sensing applications with chaotic lasers are described. Blind source separation using independent component analysis is also discussed and applied to chaotic temporal waveforms of laser output. Finally, fractal optics are introduced and described for wireless optical communication applications.

12.1
Remote Sensing with Chaotic Lasers

12.1.1
Chaotic Lidar

Remote sensing using LIDAR (light detection and ranging) has been pursued for applications in target detection, imaging, and range finding. To precisely resolve the range, either high-peak-power short-pulse lasers or modulated continuous-wave (CW) laser diode have been used as the light sources.

In the short-pulse technique, the range resolution, which is determined by the pulse-width, is typically in the range of meters. In the modulated CW technique, target detection and localization are accomplished either by correlating the signal waveform reflected or backscattered from the target with the time-delayed reference waveform or by interfering them optically with a Michelson interferometer, where the range resolution is determined by the bandwidth of the modulated waveform. However, limited by the code rate and the modulation speed, the resolution of a pseudorandom-number (PN) code-modulated CW lidar is in the range of several tens of meters. Moreover, the range of unambiguous detection is limited due to the finite length of the PN code, and expensive external modulator is necessary for high-speed modulation in order to achieve better resolution.

The chaotic lidar (CLIDAR) is a lidar system utilizing the nonlinear dynamics of semiconductor lasers (Lin and Liu, 2004a). By perturbing a semiconductor laser with either optical injection, optical feedback, or optoelectronic feedback, the laser can be operated in various nonlinear dynamical states. Different chaotic oscillation states

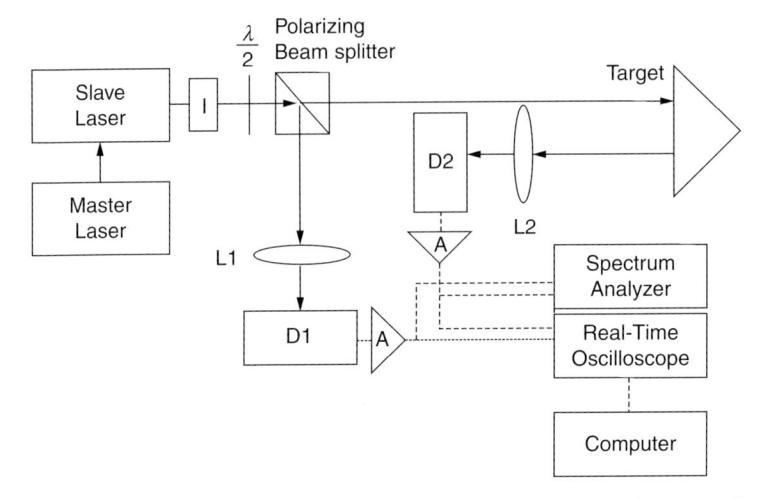

Figure 12.1 Schematic setup of the chaotic lidar (CLIDAR) system with an optically injected semiconductor laser. I: isolator, L1, L2: lens, D1, D2: detectors, A: amplifiers, $\lambda/2$: half-wave plate. (F.-Y. Lin and J.-M. Liu, (2004a), © 2004 IEEE.)

can be found at proper operating conditions by adjusting the operational parameters of a laser system. Using the chaotic output as the light source, detection can be realized by correlating the signal waveform reflected or backscattered from the target with a delayed reference waveform. Compared with both short-pulse and modulated CW lidars, CLIDAR has a much higher range resolution, benefiting from the broad bandwidth of the optical chaos, which can be easily generated and amplified. With a bandwidth broader than 15 GHz, a centimeter resolution is readily achieved. Moreover, while CLIDAR has all the advantages that a modulated CW lidar has by using a laser diode as the light source, the needs of expensive high-speed code generation and modulation electronics no longer exist. The ambiguity caused by the limited length of PN codes or a repeated waveform are also eliminated because a chaotic waveform never repeats itself. Ranging using a chaotic laser pulse train generated from a semiconductor laser subject to optical feedback has also been proposed (Myneni et al., 2001).

The schematic setup of the proposed CLIDAR system is shown in Figure 12.1 (Lin and Liu, 2004a). An optical injection scheme is used to generate chaotic waveforms. The dynamics of the optically injected response (slave) semiconductor laser is controlled by adjusting the controllable operational parameters, which are the injection current of the response laser, the frequency detuning between the drive (master) and response semiconductor lasers, and the optical injection strength from the drive to the response laser. With proper adjustments, the response laser can be operated in various chaotic states. The chaotic optical waveforms are noise-like, and the spectra are broadband. An optical isolator is placed right after the chaotic laser to prevent unwanted optical feedback. The chaotic output is split by a polarizing beam splitter into two beams, one serving as the probe beam and the other as the reference. By rotating the angle of the half-wave plate relative to the polarizing beam splitter, the power ratio between these two beams can be adjusted. The probe beam is directed to the target, and the signal light that is backscattered or reflected from the target is collected and detected by a combination of lens and detector. Together with the waveform received from the reference beam, these two waveforms are amplified and simultaneously recorded with a radio-frequency (RF) spectrum analyzer and an oscilloscope. Target detection and localization are accomplished by correlating the signal waveform with the time-delayed reference waveform. This procedure of correlating the signal and the reference can be carried out either electronically or optically. This flexibility is an advantage of CLIDAR in practical applications.

The lasers used in the experiment are single-mode distributed-feedback (DFB) InGaAsP/InP semiconductor lasers. The frequency detuning between the two lasers is around 2.7 GHz, and the relaxation oscillation frequency of the response laser is 3 GHz. The signal and reference waveforms are detected by high-speed InGaAs photodetectors. These electronic signals are amplified by RF amplifiers. Waveforms are displayed and recorded on a digitizing real-time oscilloscope, while the corresponding power spectra are simultaneously measured with a RF spectrum analyzer. The correlation and data processing are carried out with a personal computer.

CLIDAR uses optical chaos as the light source. Its performance is mainly determined by the chaotic state chosen. To have a delta-function-like correlation trace that has the highest possible resolution and lowest possible detection ambiguity, CLIDAR should be operated in a state that its chaotic waveform has similar properties to those of white noise, such as a flat, smooth, and broad spectrum, a complex phase portrait, and a noise-like time series.

To show the feasibility of the CLIDAR system, experiments of range finding using a mirror as the target are conducted. The CLIDAR system is operated in the chaotic state with flat RF spectrum. A mirror is arranged at about 2 m away from the CLIDAR system on a translation stage. A set of signal and reference waveforms are first obtained, and the cross-correlation trace of them is plotted in Figure 12.2 (left)

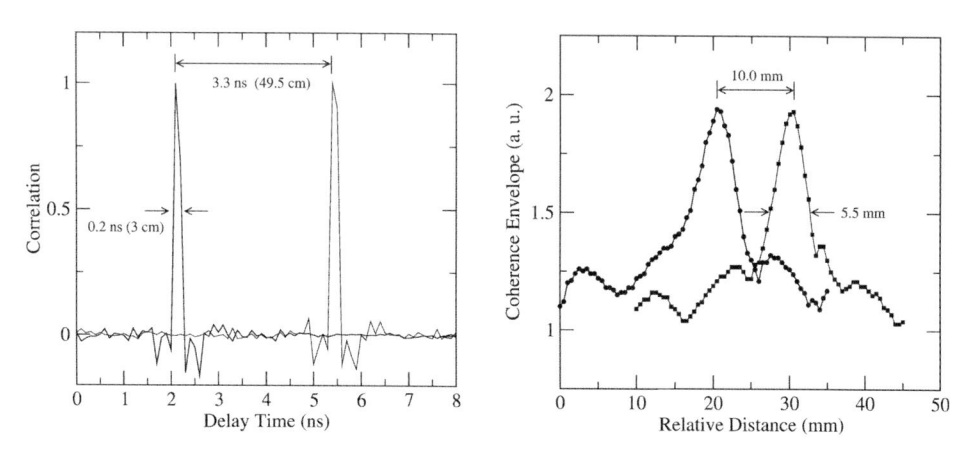

Figure 12.2 (Left) Cross-correlation traces of a target moving 50.0 cm in the line of sight. (Right) Coherence envelopes of a target moving 10.0 mm in the line of sight. (F.-Y. Lin and J.-M. Liu, (2004a), © 2004 IEEE.)

(Lin and Liu, 2004a). By translating the mirror 50.0 cm away in the line of sight, a second set of signal and reference waveforms are obtained, and their cross-correlation trace is also plotted in Figure 12.2 (left). In both cases, the correlation lengths are 2 μs. From the separation between the correlation peaks shown in Figure 12.2 (left), the relative range difference is measured to be 49.5 cm showing a subcentimeter accuracy in ranging. Moreover, with a 0.2-ns full-width half-maximum (FWHM) of the cross-correlation peak, a 3-cm range resolution is achieved and demonstrated experimentally, which is currently limited by the bandwidth of the oscilloscope used (3 GHz). The CLIDAR system clearly shows an excellent performance in correlation such that target detection can be done unambiguously with a very high resolution. To examine the signal-to-noise ratio (SNR) of the CLIDAR system, five consecutive sets of data are recorded, and the average and the variance of the values of the cross-correlation peaks representing the signal and the noise, respectively, are calculated. The SNR of 27.5 dB is obtained in experiment.

To fully utilize the broad bandwidths of chaotic waveforms and to eliminate the bandwidth limitation from the electronics, a CLIDAR system utilizing a Michelson interferometer that interferes the signal light with the reference light optically can be adopted. The relative target distance can be determined by measuring the coherence envelope (fringe contrast function) that modulates the interference fringes. Experiments using a mirror as the target located at about 1 m away from the CLIDAR system are carried out. By translating the mirror for 10.0 mm in the line of sight, two coherence envelopes are obtained and plotted in Figure 12.2 (right) with the two curves. As can be seen, a distance of 10.0 mm is measured from the separation of the peaks that has a submillimeter accuracy, while a 5.5-mm range resolution is achieved deriving from the FWHM of the peaks.

12.1.2
Chaotic Radar

Instead of using chaotic lidars as light sources, chaotic semiconductor lasers can be used as microwave sources for chaotic RADAR (radio detection and ranging) systems (Lin and Liu, 2004b). Random signal radars (RSRs) using random noise have been studied extensively (Taylor, 2000). Compared with conventional radars, RSRs have some advantages, including good accuracy and resolution in range and velocity, high immunity to noise, and low probability of intercept. However, the main difficulty for a practical RSR system is the generation of the broadband microwave random signal. The typical power spectral density of either the random noise or the chaotic signal that is generated by a microwave circuit or device has a bandwidth of less than 1 GHz. To meet the demands of high-resolution radar imaging and target identification, a source that generates noise-like waveforms with broader bandwidths is desired. This can be done by using an optically injected semiconductor laser, where a chaotic state that has a bandwidth greater than 10 GHz can be easily obtained.

A chaotic radar (CRADAR) system using laser chaos has been proposed (Lin and Liu, 2004b). The correlation property and performances are investigated, and the

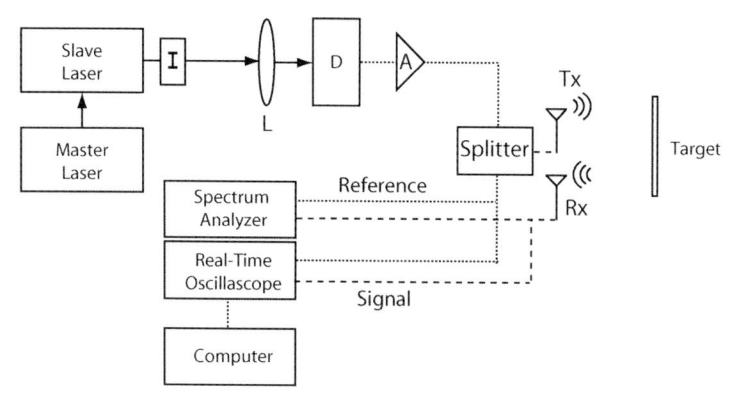

Figure 12.3 Schematic setup of the chaotic radar (CRADAR) system. I: isolator, L: lens, D: photodetector, A: amplifier, Tx, Rx: transmitter and receiver antennas. (F.-Y. Lin and J.-M. Liu, (2004b), © 2004 IEEE.)

immunity of noise and jamming signals are presented. Using a pair of planar antennas that have a transmission band from 1.5 to 3 GHz, a range resolution of 9 cm is demonstrated. An experiment is carried out, through which the feasibility of the proposed CRADAR system is shown.

The schematic setup of the proposed CRADAR system is shown in Figure 12.3 (Lin and Liu, 2004b). An optically injected semiconductor laser is used to generate chaotic waveforms. Similar to the CLIDAR system shown in Figure 12.1, the dynamics of the injected response (slave) semiconductor laser is controlled by adjusting the controllable operational parameters: the injection current, the detuning frequency between the drive (master) and response semiconductor lasers, and the optical injection strength from the drive to the response laser. With proper adjustments, the response laser can be operated in a chaotic state that has a noise-like waveform and a broadband spectrum. An optical isolator is placed after the chaotic laser to prevent unwanted optical feedback. The chaotic light is collected and detected by a combination of lens and detector. The chaotic waveform is then amplified and split into two channels, the probe and the reference. The probing signal is transmitted and received by a pair of antennas. Together with the reference waveform, these waveforms are recorded simultaneously with a RF spectrum analyzer and an oscilloscope, where target detection and localization are accomplished through cross-correlation.

Single-mode DFB InGaAsP–InP semiconductor lasers are used as the two lasers. The laser output power is approximately 4 mW, the relaxation oscillation frequency is around 10 GHz, the laser wavelength is 1313.3 nm, and the detuning frequency is 4.84 GHz. The chaotic light is detected by a high-speed InGaAs photodetector. This electronic signal is amplified by RF amplifiers. Using a microwave splitter, the chaotic waveform is split into the reference and the probe channels. Two identical planar antennas are used, respectively, as the transmitter and the receiver, both of which have a bandwidth from 1.5 to 3 GHz. The signal waveform received by the receiver is

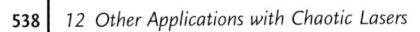

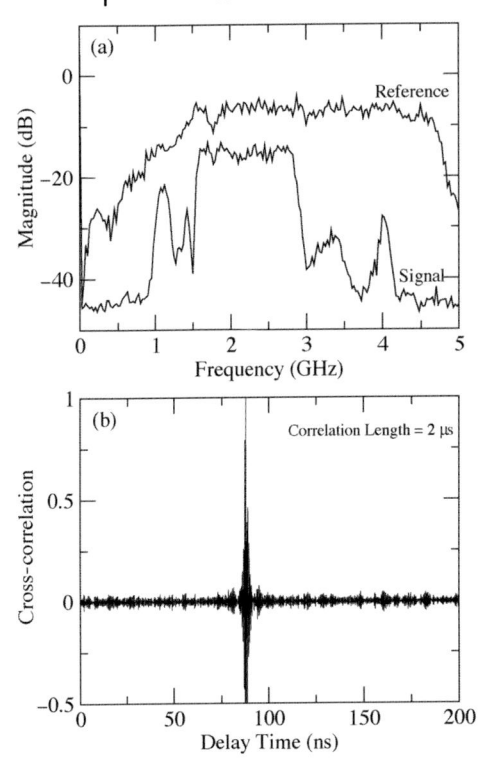

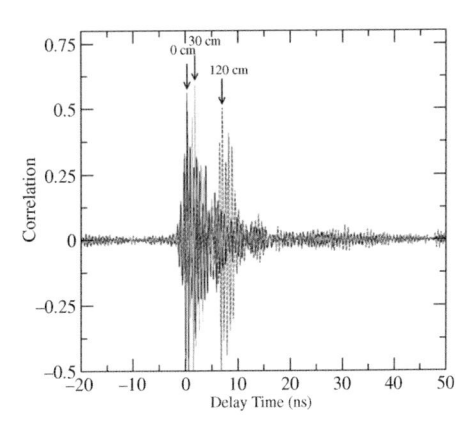

Figure 12.4 (Left) (a) RF spectra and (b) the cross-correlation trace of the signal and reference waveforms. The full-width half-maximum (FWHM) of the correlation peak is measured to be 0.6 ns. (Right) Cross-correlation traces with the target 0, 30, and 120 cm away from the CRADAR, respectively. (F.-Y. Lin and J.-M. Liu, (2004b), © 2004 IEEE.)

recorded simultaneously with the reference waveform on the digitizing real-time oscilloscope, and the corresponding power spectra are measured with the RF spectrum analyzer. The data processing is carried out on a personal computer.

Figure 12.4a (left) shows the RF spectra of the reference and signal waveforms. (Lin and Liu, 2004b) To minimize the effect of loss and noise, the signal waveform is obtained by pointing the transmitter antenna to the receiver antenna directly without any separation. As can be seen in Figure 12.4a (left), both the reference and signal waveforms have a flat and smooth power spectral density in the range of 1.5–3 GHz. By correlating for 2 µs, the FWHM of the correlation peak shown in Figure 12.4b (left) is measured to be 0.6 ns. Deriving from this result, a high range resolution of 9 cm is achieved and demonstrated experimentally, which outperforms the conventional radars. Even filtered by the limited bandwidths of the antennas and the amplifiers, the chaotic and broadband properties are well preserved.

The experiment demonstrating the ability of CRADAR in range finding is carried out. The cross-correlation traces of the reference and signal waveforms are plotted. The range finding is demonstrated with a 1 m × 1 m metal plate as the target. In this case, the antennas are facing the same direction and are separated by 20 cm perpendicular to the field of view to eliminate the crosstalk. Figure 12.4 (right) shows the cross-correlation traces with the target 0, 30, and 120 cm away from the CRADAR, respectively. As can be seen in Figure 12.4 (right), the peaks indicating the location of the target can still be identified unambiguously, while the side lobes are relatively high, which is mainly due to the low reflection of the target and the omnidirectional antennas used. Better resolution can be achieved by using horn antennas and by amplifying the received signal further.

12.1.3
Chaotic Correlation Optical Time-Domain Reflectometer

Another application with chaotic lasers for remote sensing is the optical time-domain reflectometer (OTDR), which is an important diagnostic tool for the testing of fiber transmission systems and components (Wang et al., 2008b). Many techniques have been proposed and successfully applied in OTDR for dynamic range enlargement, SNR improvement, and spatial resolution enhancement (Tateda and Horiguchi, 1989). In order to overcome the trade-off between SNR and resolution, the correlation optical time-domain reflectometer (COTDR) was developed (Zoboli and Bassi, 1983). In COTDR, PN code sequences are used to modulate the optical pulse, and the reflection point is located by correlating the backscattered light with a delayed code sequence. However, limited by the bandwidth of the electrical codes, the spatial resolution of the current COTDR is only of the order of tens of meters. Moreover, it is difficult and costly to generate large-amplitude broadband electrical PN or random codes, and the dynamic range of the COTDR is also barely enlarged.

By perturbing a semiconductor laser with either optical injection, optical feedback, or optoelectronic feedback, the laser can generate ultrawideband chaotic light. By utilizing the broadband chaotic laser, a chaotic correlation optical time-domain reflectometer (CCOTDR) has been proposed and demonstrated experimentally (Wang et al., 2008b). Compared with the COTDR, the CCOTDR has some significant advantages: The CCOTDR has high spatial resolution since the bandwidths of chaotic waveforms generated by the semiconductor laser are over 10 GHz. The high spatial resolution can meet the needs of small-size network diagnoses. In addition, the ambiguity caused by the limited length of PN codes or repeated waveform in COTDR is eliminated because a noise-like chaotic waveform never repeats itself. Expensive high-speed electronic devices are no longer needed for code generation, amplification, and modulation, while the chaotic waveforms can be generated easily with ordinary semiconductor lasers.

The experimental setup of the CCOTDR is shown in Figure 12.5 (Wang et al., 2008b). The chaotic laser system is framed in dashed lines. The chaotic laser is a DFB laser diode (DFB-LD) operating at 1550.6 nm with optical feedback from an external optical fiber ring cavity. The DFB laser is biased at 33.0 mA, and the temperature and the wavelength are stabilized by a temperature controller. An optical circulator (OC1) is used to construct a 6-m external optical fiber ring cavity,

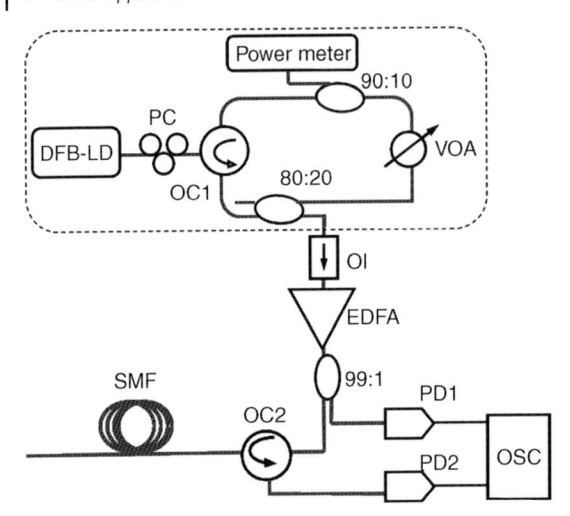

Figure 12.5 Experimental setup of the chaotic correlation optical time-domain reflectometer (CCOTDR). (Y. Wang, B. Wang, and A. Wang, (2008b), © 2008 IEEE.)

which is composed of an 80:20 fiber coupler, a variable optical attenuator (VOA), and a 90:10 fiber coupler. The 80:20 fiber coupler provides 80% for the output and 20% for the feedback light power. The feedback strength is adjusted by the VOA and monitored by an optical power meter. The polarization controller (PC) between the DFB-LD and OC1 is used to adjust the polarization of the light feedback into the DFB-LD. The chaotic laser is amplified by an erbium-doped fiber amplifier (EDFA), and an optical isolator (OI) is used to prevent unwanted optical feedback into the chaotic laser transmitter. The amplified chaotic laser is then split by a 99:1 fiber coupler, which provides 1% of the chaotic laser output for the reference light and 99% for the signal light. The signal light is launched into the fiber under test through another optical circulator (OC2). The fiber under test is a single-mode fiber with loss coefficient 0.2 dB/km. The backscattered light from the fiber end is detected by a photodetector (PD2) and recorded by a real-time oscilloscope. The cross-correlation between the reference light and the backscattered light can be calculated by using a computer.

The laser's chaotic temporal waveform is shown in Figure 12.6a (left) (Wang et al., 2008b). The RF spectrum is measured by a RF spectrum analyzer with the photodetector and is plotted in Figure 12.6b (left). Obviously, this noise-like chaotic waveform has a broadband spectrum over 5 GHz and can be readily obtained by a laser diode with optical feedback.

To demonstrate the feasibility of the CCOTDR for locating the fiber reflection point, several open-ended fibers with different lengths are measured. The average input power launched into the fibers under test is 11.2 dBm. From the cross-correlation traces shown in Figure 12.6a (right), it can be seen clearly that four peaks are located at a distance of 0, 46.1, 90.3, and 140.2 m, respectively. The peak at the distance of 0 m is used to scale and is obtained by correlating the reference light

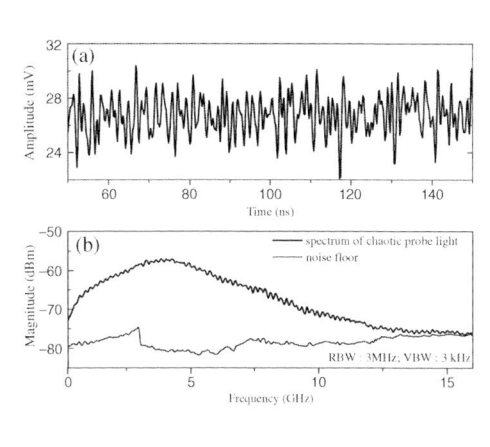

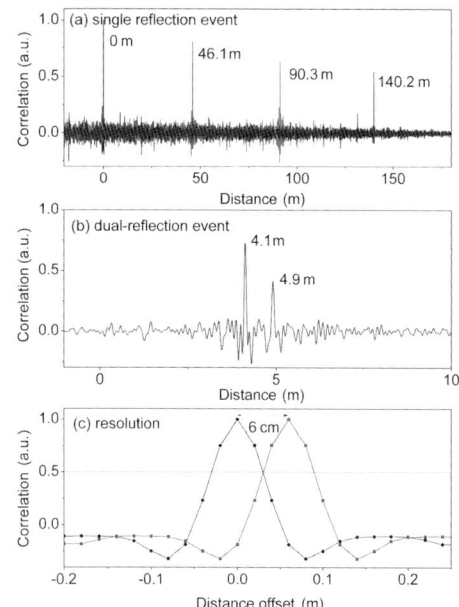

Figure 12.6 (Left) (a) Noise-like waveform and (b) broadband power spectrum of chaotic light employed as probe light for CCOTDR. (Right) Experimental correlation detection of single reflection events located at 46.1, 90.3, and 140.2 m away from the CCOTDR, respectively. (b) Detection of two reflection events. (c) A 6-cm spatial resolution. (Y. Wang, B. Wang, and A. Wang, (2008b), © 2008 IEEE.)

with the backscattered light from the OC2. The peak amplitude degrades with the tested fiber length increasing, which is caused by the fiber loss.

Detection of dual-reflection events is demonstrated in Figure 12.6b (right). The signal light is injected simultaneously into two different-lengths open-ended fibers *via* a 50/50 fiber coupler. Consequently, the location of two open ends and relative distance can be measured by the cross-correlation trace between this signal light and the reference light, as clearly shown in Figure 12.6b (right).

The spatial resolution of the CCOTDR scheme is examined by a tunable fiber delay-line with a 0.05-mm step. The resolution is identified when two cross-correlation traces are just apart from its original position according to a 3-dB criterion. A 6-cm spatial resolution is obtained, which is shown in Figure 12.6c (right). Note that the 6-cm resolution is actually limited by the bandwidth of the oscilloscope used in the experiment (500 MHz). The resolution determined by the bandwidth of chaotic waveform in the experiment is only 4 mm, which is characterized by interference from the signal light and the backscattered light with a Michelson interferometer. Moreover, the resolution is range independent because it is only determined by the chaotic waveform's bandwidth.

In the experiment, the detection of the fiber fault location is demonstrated by using the proposed CCOTDR. The CCOTDR can also measure the fiber loss and Rayleigh scatters just like the conventional COTDR does. The background noise and the side lobes can be reduced or erased by developing a method (like the Gray code) which will improve significantly its SNR.

12.2
Blind Source Separation of Chaotic Signals by Using Independent Component Analysis

12.2.1
Motivation for Blind Source Separation

Many complex signals obtained from high-dimensional nonlinear dynamical systems are not purely simple chaotic signals. They may be contaminated by different components of individual chaotic signals or internal noise signals. The extraction of each chaotic component from mixed signals is a very attractive and important research area for chaos-based communication systems. Correlation detection techniques for the separation of a chaotic component from mixed chaotic signals are typically used for multiplexing communications using chaos such as code-division multiple-access (CDMA) (Kennedy et al., 2000). The technique of dual synchronization of chaos is another method to separate mixed chaotic signals by using the dynamical stability of synchronization (Liu and Davis, 2000b; Uchida et al., 2003c). In this method, parameter sensitivity of synchronization limits the number of mixed chaotic signals that can be separated. A different approach for the separation of linearly superimposed uncorrelated signals has been proposed by using time-delayed correlation functions (Molgedey and Schuster, 1994). The adaptation process for the inhibitory interactions in a neural network is used to solve the source separation problem.

Blind source separation by using independent component analysis (ICA) has been rapidly developed and applied for many complex signals (Hyvärinen et al., 2001). ICA is a statistical estimation technique to separate mixed complex signals based on the non-Gaussianity of the probability density function (PDF) of signals. The PDF of the sum of multiple source signals approaches the Gaussian distribution when compared with that of each source signal, because of the central limit theorem. Independent components can be extracted from the observed mixed signals by maximizing the non-Gaussianity of the PDF of estimated signals. ICA has been successfully applied for auditory signal separation, neural networks without teaching data, analysis of electroencephalogram (EEG) data in the brain, signal processing in medical engineering, and image processing (Hyvärinen et al., 2001). ICA may be used for the auditory signal separation process in the brain, known as the cocktail party effect, in which a native language can be easily recognized from mixed conversations at an international party.

For communication applications, ICA for CDMA system with chaotic codes has been investigated recently (Umeno, 2007). The use of chaos and ICA for a new type of CDMA application with multiple-input multiple-output (MIMO) architecture is very promising since the PDF of many chaotic signals displays non-Gaussian distribution that allows successful signal separation by ICA. It is expected that not only the data rate of such communication systems could be greatly enhanced but also a novel signal-receiving technique without CDMA spreading codes could be implemented (Umeno, 2007). Next-generation intelligent transport systems (ITS) are also promising candidates where car-to-car communications are realized among more than one thousand cars in a traffic jam. Such supermultiplexing communications could be implemented by using chaotic codes and ICA algorithms for blind source separation.

Blind source separation of mixed optical chaotic signals by using ICA has been demonstrated experimentally (Kuraya *et al.*, 2008). Neodymium-doped yttrium orthovanadate (Nd:YVO$_4$) microchip solid-state lasers with external modulation are used as optical chaotic sources. Chaotic signals are mixed together with randomly selected mixing ratio and ICA is applied to the mixed chaotic signals for blind source separation. Many chaotic signals are also used and the ability of ICA is investigated to separate a large number of mixed signals (see Section 12.2.3).

12.2.2
Principle of Independent Component Analysis

ICA is a signal-processing technique whose goal is to express a set of random variables as linear combinations of statistically independent component variables (Umeno, 2007). In the simplest form of ICA, particularly for blind source separation, m scalar random variables $x_1(t), x_2(t),\ldots, x_m(t)$ are observed at each time instant t, which are assumed to be linear combination of n unknown mutually independent components $s_1(t), s_2(t),\ldots, s_n(t)$, where $n \leq m$. In the vector notation it is written as,

$$\mathbf{x}(t) = \mathbf{A}\mathbf{s}(t) \tag{12.1}$$

where $\mathbf{x} = (x_1, x_2,\ldots, x_m)^T$, $\mathbf{s} = (s_1, s_2,\ldots, s_n)^T$, and $\mathbf{A}$ is an unknown $m \times n$ matrix of full rank called the mixing matrix. T denotes the transpose. The goal is to find $\mathbf{A}$ or $\mathbf{s}$ from the observed random vector $\mathbf{x}$ under the assumptions that $\mathbf{s}$ are independent variables and their probability distributions are non-Gaussian. Equivalently, the inverse model of Eq. (12.1) can be formulated as,

$$\mathbf{y}(t) = \mathbf{W}\mathbf{x}(t) \tag{12.2}$$

where $\mathbf{W}$ is a demixing matrix to be estimated and $\mathbf{y}$ is to be compared with the source signals $\mathbf{s}$. Neither the variances nor the signs of the independent components can be determined as both $\mathbf{s}$ and $\mathbf{A}$ are unknown in Eq. (12.1). Also, no order is defined between the independent components.

The problem of estimating the mixing matrix can be somewhat simplified by preprocessing: centering $E\{\mathbf{x}\} = 0$ and whitening $E\{\mathbf{x}\mathbf{x}^T\} = \mathbf{I}$, where $E\{\cdot\}$ represents the ensemble average and $\mathbf{I}$ is an identity matrix. It is assumed that the data $\mathbf{x}$ is preproceeded by centering and whitening.

Among algorithms for ICA, FastICA is simple and highly efficient (Hyvärinen et al., 2001). The FastICA algorithm is based on a fixed point scheme for finding a maximum of the non-Gaussianity of $\mathbf{w}^T \mathbf{x}$ in which $\mathbf{w}^T$ corresponds to each row vector of $\mathbf{W}$. The one-unit FastICA algorithm finding the unit vector $\mathbf{w}$ has the following form,

$$\mathbf{w}(k+1) = E[g(\mathbf{w}(k)^T \mathbf{x})\mathbf{x}] - E[g'(\mathbf{w}(k)^T \mathbf{x})]\mathbf{w}(k) \qquad (12.3)$$

where $\mathbf{w}$ is also normalized after every iteration, and the function g is directly related to the measure of non-Gaussianity. The function $g(y) = \tanh(y)$ is often used, which provides good performance for chaotic signal separation ($g(y) = y^3$ is also used). The one-unit algorithm can be easily generalized to estimate several independent components. The details of the FastICA algorithm are provided in (Hyvärinen et al., 2001).

12.2.3
Examples of Blind-Source Separation with Chaotic Lasers

To test blind source separation of mixed optical chaotic signals by using ICA, two Nd:YVO$_4$ microchip lasers with external modulations are used, as shown in Figure 12.7 (left) (Kuraya et al., 2008). Two independent chaotic laser beams are combined at a beam splitter (BS). The two outcoming beams from BS contain a portion of the two laser outputs with different mixing ratio: one beam contains $R\%$ of the laser 1 output and $T\%$ of the laser 2 output, and *vice versa* for the other beam,

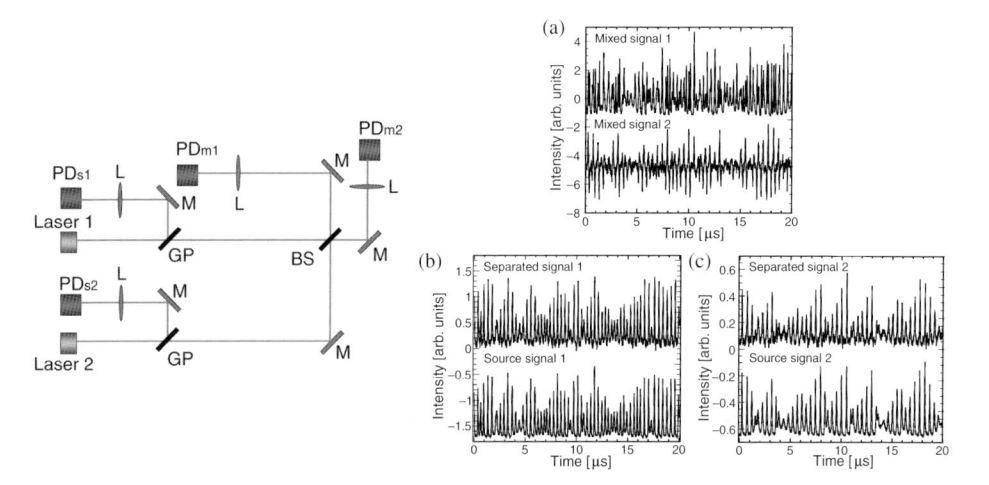

Figure 12.7 (Left) Experimental setup of Nd:YVO$_4$ microchip solid-state lasers with external modulation for blind source separation by independent component analysis (ICA). BS, beam splitter; GP, glass plate; L, lens; M, mirror; and PD, photodiode. (Right) Experimental results of blind source separation of chaotic laser signals by using ICA. Temporal waveforms of (a) mixed signals 1 and 2, (b) separated signal 1 and source signal 1, and (c) separated signal 2 and source signal 2. (M. Kuraya, A. Uchida, S. Yoshimori, and K. Umeno, (2008), © 2008 OSA.)

where R and T indicate the reflectivity and the transmittance of the beam splitter (i.e., $R + T = 100\%$). Note that no information on the values of R and T is required for blind source separation by ICA. The two mixed chaotic signals are detected as mixed signal 1 and 2 by using two photodiodes (PD_{m1} and PD_{m2} in Figure 12.7 (left)), so that the incoherent sum of the laser intensities are observed. These two mixed chaotic signals are used for blind source separation by ICA. The chaotic source signals of the two microchip lasers are also detected for reference by using two other photodiodes (PD_{s1} and PD_{s2}).

Figure 12.7 (right) shows the experimental results of the blind source separation by using ICA. Figure 12.7a (right) indicates temporal waveforms of the two mixed chaotic signals that are used for signal separation. The two separated signals are obtained from the two mixed chaotic signals by using ICA, and compared with the source signals, as shown in Figures 12.7b and c (right). The separated signals are almost identical to the source signals. Note that corrections of the signs, amplitudes, and the orders of the independent components are made in the separated signals by postprocessing.

The degree of successful signal separation is quantitatively measured by using the cross-correlation value C between the source and separated temporal waveforms (see Eq. (5.19) in Section 5.4). The best signal separation is obtained at a cross-correlation value of $C = 1$. The corresponding pairs of the source and separated signals show very good correlation. The cross-correlation of 0.980 is obtained between the source signal 1 and the separated signal 1 in Figure 12.7b (right). The cross-correlation of 0.952 is also obtained between the source signal 2 and the separated signal 2 in Figure 12.7c (right). However, the different pairs of the source and separated signals show low correlation close to zero. It is thus shown that blind source separation of two mixed chaotic laser waveforms is succeeded by using ICA.

To test the separation ability of ICA for many chaotic signals, the number of chaotic signals used for the mixed signals is increased. Due to the experimental limitation, many independent chaotic source signals are detected from one microchip laser at different times and stored in a computer. These source signals are examined with a randomly selected mixing matrix and the mixed signals are generated numerically. The FastICA algorithm is applied for these mixed signals to test blind source separation.

Figure 12.8a (left) shows the length of temporal waveform T_L required for successful signal separation as a function of the number of mixed signals N_m at different data-sampling frequencies f_S (Kuraya et al., 2008). Here, the successful signal separation is defined as the case when the average value of the cross-correlations between the source and corresponding separated signals is more than 0.95. Note that blind source separation of 100 mixed chaotic laser signals is succeeded by using ICA, as shown in Figure 12.8a (left). It is found that the longer T_L is required with increase of N_m for all f_S. Moreover, the curves of T_L decrease and saturate as f_S is increased, ($f_S \geq 50$ MHz in Figure 12.8a (left)). Therefore, the minimum T_L required for successful blind source separation by ICA is almost proportional to N_m for large f_S (i.e., oversampled data).

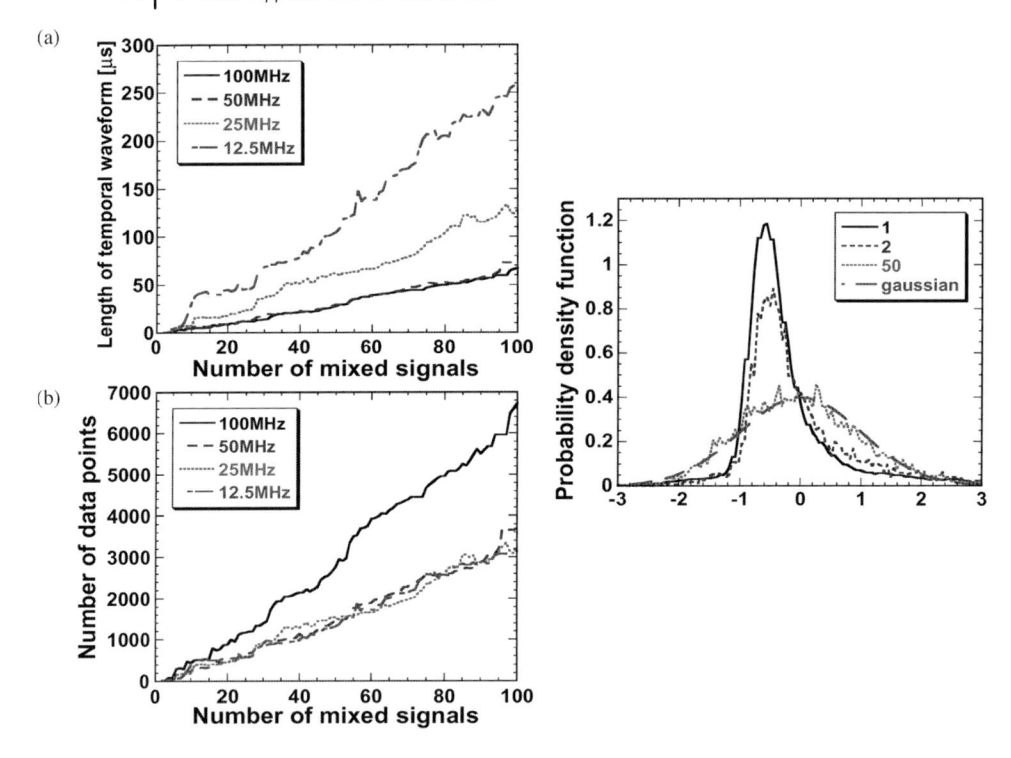

Figure 12.8 (Left) (a) The length of the temporal waveform T_L and (b) the number of data points N_d ($=T_L f_S$) required for successful signal separation as a function of the number of mixed signals N_m at different data-sampling frequencies f_S for the microchip lasers. The curves correspond to the condition at which the average value of cross-correlations between the source and successfully separated signals is 0.95. The fundamental frequency of chaotic laser signals is 3.25 MHz. Solid curve; $f_S = 100$ MHz, dashed curve; $f_S = 50$ MHz, dotted curve; $f_S = 25$ MHz, and dotted-dashed curve; $f_S = 12.5$ MHz. (Right) Probability density function (PDF) of the source, two-mixed, and 50-mixed signals obtained from the microchip lasers. Solid curve; source signal, dashed curve; two mixed signal, dotted curve; 50-mixed signal, and dotted-dashed curve; Gaussian distribution for reference. The kurtosis is 7.858 for the source signal, 4.647 for the two mixed signals, and 0.0841 for 50 mixed signals. (M. Kuraya, A. Uchida, S. Yoshimori, and K. Umeno, (2008), © 2008 OSA.)

Figure 12.8b (left) shows the number of data points N_d required for successful signal separation as a function of N_m at different f_S. Note that N_d equals to $T_L f_S$. It is found that the curves of N_d decrease and saturate as f_S is decreased, ($f_S \leq 50$ MHz in Figure 12.8b (left)). Therefore, the minimum N_d required for successful blind source separation is almost proportional to N_m for small f_S (i.e., undersampled data).

From the results of Figures 12.8a and b (left), the optimal f_S for successful separation with the minimum T_L and N_d is 50 MHz (15.4 times faster than the fundamental frequency of chaos, 3.25 MHz). This condition indicates that roughly 200 periods of chaotic oscillations are required for the separation of 100 mixed chaotic signals at $f_S = 50$ MHz.

Next, the non-Gaussianity of mixed chaotic signals is measured by using the PDFs. Figure 12.8 (right) shows the PDFs of the source, two mixed and 50 mixed signals obtained from the chaotic laser signals. As the number of mixed signals is increased the PDFs approach the Gaussian distribution. The kurtosis of these PDFs is calculated, which is an indicator of non-Gaussianity (i.e., a larger absolute value of kurtosis implies more non-Gaussianity). The kurtosis is 7.858 for the source signal, 4.647 for the two mixed signal, and 0.084 for 50 mixed signals. The kurtosis becomes close to zero, indicating more Gaussianity, as the number of mixed signals is increased. Therefore, non-Gaussianity of chaotic source signals is crucial for successful blind source separation of chaotic laser signals by ICA.

Experimentally observed chaos in lasers does not satisfy the independent and identically distributed condition, and chaos has a time correlation in general. The PDF of chaotic laser signals has a non-Gaussian distribution, unlike quantum noise signals. The degree of non-Gaussianity depends on the type of chaotic laser sources and how to generate chaotic signals. The method of blind source separation by using ICA is applicable to other chaotic laser sources to separate mixed signals when the PDF of the chaotic signals has a non-Gaussian distribution.

12.3
Fractal Optics

The concept of "fractal" has been used for a complex spatial structure with self-similarity. The fractal is a shape made of parts similar to the whole in some way. The fractal is a "spatial" complex pattern generated from a deterministic rule, whereas chaos is a "temporal" dynamics with a deterministic rule. Fractal patterns of light images can be observed in optical complex systems. Fractal optics and their applications are described in this section.

Photonic crystals have been intensively investigated for many years as optical devices to control the localization and propagation characteristics of photons. Photonic fractals have been invented (Takeda *et al.*, 2004), which are three-dimensional (3D) fractal cavities based on the Menger sponge structure fabricated from epoxy resin by stereolithography. Localization of an electromagnetic field at a resonant frequency can be achieved in the photonic fractals. The use of fractal characteristics may have the potential to control and localize photons in a different way from the conventional optical devices, and may lead to new engineering applications of photon localization devices (Takeda *et al.*, 2004).

As an example of fractal optical devices, a 3D optical billiard has been proposed (Sweet *et al.*, 1999), in which the centers of four spherical reflectors are located at the vertices of a regular tetrahedron. Light is injected into the optical billiard and fractal patterns of light scattering with the Wada basin property are found. A similar configuration has been used as a chaos mirror (Harayama and Davis, 1998), which provides free-space optical beam links for optical wireless networks. The chaos mirror has a feature that a one-dimensional spread of input optical rays results in a two-dimensional spread of the rays reflected from the chaos mirror.

12.3.1
Chaos Mirror for Wireless Optical Communications

A novel reflector based on a 3D optical billiard (i.e., spherical reflectors are located at the vertices of a regular tetrahedron) has been proposed for making free-space optical wireless links (Harayama and Davis, 1998). Two-dimensional coverage of the plane can be achieved by a scattered beam from the reflector owing to one-dimensional spread at the transmitter. As this reflector involves multiple reflections that obey chaotic (not temporal, but spatial) ray dynamics, the reflector is called a "chaos mirror". In the case of indoor wireless communication it is possible to link any pairs of points on the floor, corresponding to emitter and receiver, *via* reflection from such a chaos mirror suspended above the floor. A ray entering the mirror from the transmitter enters the chaos mirror, reflects chaotically, and then leaves the mirror for a certain point on the floor. The location of the final reception point on the floor depends in a complicated way on the initial input condition, the position, and the direction of the ray at the transmitter, because of the chaotic behavior of the ray inside the mirror. A key feature of the chaos mirror is that finding a path to a receiver located at an arbitrary point on the two-dimensional floor surface requires only a one-dimensional sweep of the beam at the transmitter. As well as being useful for practical optical links, this feature is a new result from the viewpoint of the physics of chaotic scattering.

The method for constructing the chaos mirror is explained as follows. Consider four spherical reflectors whose radii are identical and a virtual regular tetrahedron whose vertices are labeled T1–T4, as shown in Figure 12.9a (left) (Harayama and Davis, 1998). First, place the centers of the spherical reflectors (O1–O4) at the vertices of the regular tetrahedron, as shown in Figure 12.9b (left). Next, cut the spherical reflectors in a plane that contains vertices T1, T2, and T4, as shown in Figure 12.9c (left). The intersections between the spherical reflectors and the plane are points A, F, and D where the spheres touch. Then, cover the opening surrounded by the intersection points A, F, and D with a plane mirror of the same shape. Finally, cut the spheres in the three other planes that connect the contact points of the spheres similarly and cover two more of the openings with plane mirrors, leaving the opening ABC through which rays can enter and exit the chaos mirror.

It is shown that rays behave chaotically owing to multiple reflections from the chaos mirror. Suppose that the direction of the ray emitted into the mirror from the center of the opening ABC is varied, as shown in Figure 12.9 (right). In Figure 12.9a (right) the incident ray reflects just once at a point DFE on the spherical surface and then exits the chaos mirror. In Figure 12.9b (right) the ray reflects twice, from surfaces DFE and ABF, and in Figure 12.9c (right) it reflects three times, from surfaces DFE, ABF, and BCE, before it exits. The final reception points S1, S2, and S3 in the plane below the chaos mirror differ depending on the itinerary of the reflections on the spherical surfaces. In Figure 12.9d (right) the incident ray finally reaches point S4 after 16 reflections from the mirror. Owing to the positive curvature of the mirror surfaces, small differences between input rays are amplified by reflections.

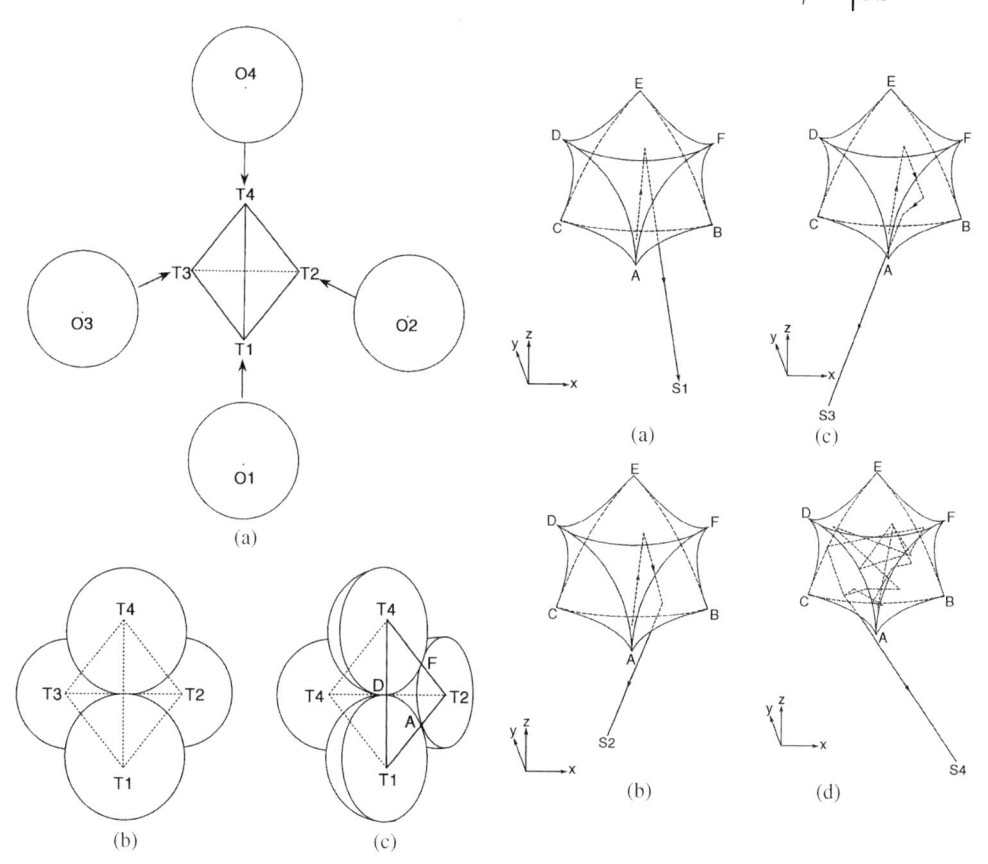

Figure 12.9 (Left) Construction of the chaos mirror. (a) Spherical reflectors and a virtual regular tetrahedron. (b) Spherical reflectors glued together so that they touch and their centers form a regular tetrahedron. (c) Cutting the spherical reflectors: Cuts are made in the planes containing points where the spheres touch. Three of the resulting apertures are covered by planar reflectors. (Right) Reflections on the chaos mirror: Examples of rays exhibiting (a) 1, (b) 2, (c) 3, and (d) 16 reflections are shown. The initial rays all have the same position in the ABC plane but are rotated in direction. The chaos mirror consists of seven reflecting surfaces joined at the edges and facing inward to form a cavity. The dashed curves show obscured joins. Pieces ACD, ABF, DEF, and BCE are spherical reflecting surfaces. ADF, BEF, and CDE are planar reflecting surfaces. ABC is open and acts as the entrance and exit port. (T. Harayama and P. Davis, (1998), © 1998 OSA.)

An example of the dependence of the final reception point on the initial direction of the ray emitted from the transmitter is demonstrated. In this example it is assumed that the chaos mirror constructed with reflecting spheres of unit radius is attached to the ceiling in the center of a rectangular room with dimensions $50 \times 50 \times 8$. The origin of position coordinates (x, y, z) is fixed at a corner of the floor. Rays are emitted from position $(5, 5, 0)$, fixed on the floor and emitted in the direction of vector $(21, v_y, 16)$, where v_y is varied. Ray trajectories can be calculated with 3D ray-tracing algorithms.

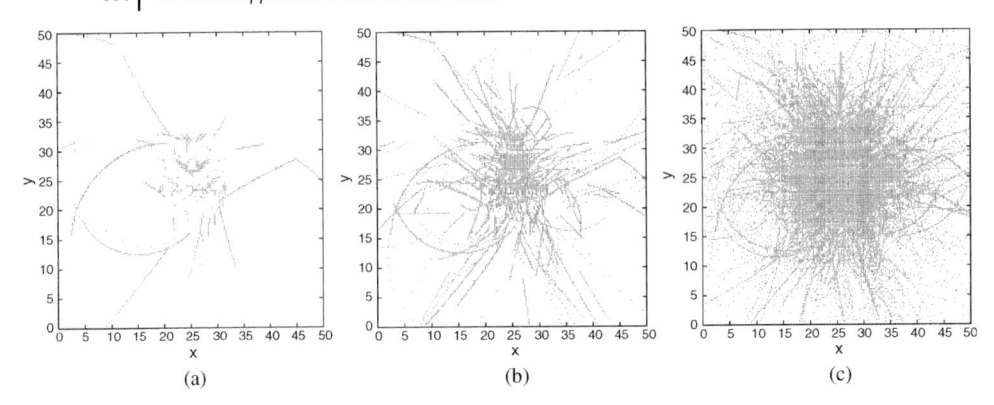

Figure 12.10 Two-dimensional spread of the rays in the reception plane caused by fanning out of the rays at the transmitter. Rays are fanned by variation of the v_y coordinate from the initial value of $v_y = 20$ in the direction vector $(21, v_y, 16)$ for rays emitted from position $(5, 5, 0)$. The range of the fanning angle in radians is (a) 2.0×10^{-4}, (b) 2.0×10^{-3}, and (c) 2.0×10^{-2}. (T. Harayama and P. Davis, (1998), © 1998 OSA.)

Figure 12.10 shows the two-dimensional spread of the rays in the reception plane caused by fanning out of the rays at the transmitter (Harayama and Davis, 1998). The direction of the ray varies in one dimension along the line linking point A with the center of arc BC. The floor is partitioned into discrete reception cells. In Figure 12.10 the dots indicate cells on the floor where the ray is detected. The number of cells covering the floor is 200×200. When the direction of the input rays is fanned out to cover the whole range between A and the center of arc BC, all the floor cells are reached. Fractal structures that are due to multiple reflections can be observed in this chaotic scattering system.

In practice, the contribution of rays undergoing large numbers of reflections will be reduced owing to the decreasing intensity of the rays with each reflection. However, for the geometry of the example in Figure 12.10, more than 80% of the rays reflect from the mirror fewer than 20 times before reaching the floor. When the loss of intensity of the ray at a reflection from the mirror is 1%, this number of reflections causes only 12% loss.

The chaos mirror has the feature that a one-dimensional spread of input rays in direction or position results in a two-dimensional spread of the rays reflected from the chaos mirror, with the spread pattern obeying the rules of deterministic chaos. Either narrow point-to-point links or wide-area broadcast links can be realized, depending on the spread of the transmitted beam. Most significantly, it is possible to reach any site in a two-dimensional reception plane just by variation of the transmitted beam in one dimension.

12.3.2
Fractal Patterns in Regular Polyhedral Mirror-Ball Structures

The characteristics of fractal optical devices need to be well understood to investigate the behavior of photons in the fractal optical devices. The analysis of the fractal dimension is one of the basic characteristics of fractal structures.

Regular polyhedral mirror–ball structures have been introduced (Amano *et al.*, 2007), which are expansion of a 3D optical billiard (Sweet *et al.*, 1999), and fractal patterns of chaotic light scattering have been observed experimentally. The fractal dimension of the basin boundaries of the light-scattering patterns has been analyzed by using the models obtained from the experimental observation. Figure 12.11 shows five examples of the regular polyhedral mirror–ball structures (Amano *et al.*, 2007). The five regular polyhedra correspond to a regular tetrahedron (with 4 faces), hexahedron (6 faces), octahedron (8 faces), dodecahedron (12 faces), and icosahedron

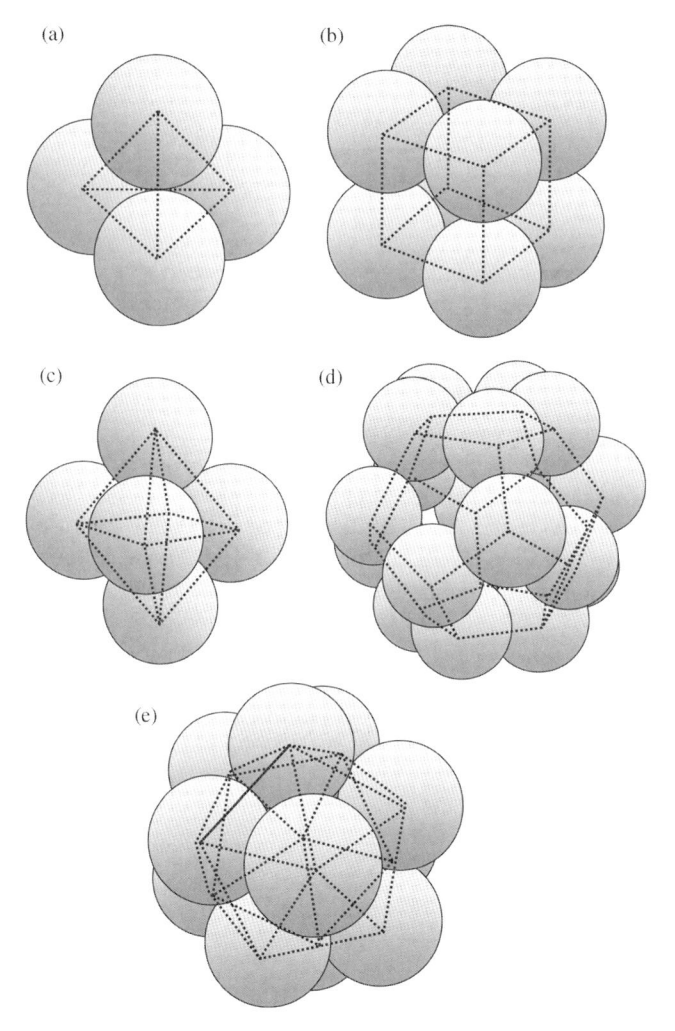

Figure 12.11 Regular polyhedral mirror–ball structures. (a) Regular tetrahedron (4 faces), (b) regular hexahedron (6 faces), (c) regular octahedron (8 faces), (d) regular dodecahedron (12 faces), and (e) regular icosahedron (20 faces). (K. Amano, D. Narimatsu, S. Sotome, S. Tashiro, A. Uchida, and S. Yoshimori, (2007), © 2007 APS.)

(20 faces), respectively. The centers of spherical reflectors are located at the vertices of a polyhedron whose edge length is equal to the diameter of the spherical reflectors. Each of the spheres is in contact with the nearest-neighboring spheres. Mirrored balls with 85-mm diameter are used as spherical reflectors in this experiment. These mirrored balls are known as "Christmas balls" for the ornaments of Christmas trees. These 3D structures can be made of the Christmas balls and bonds.

To observe chaotic light scattering in the regular polyhedral mirror–ball structures, light emitting diodes (LEDs) with different colors are located at the center of the faces of the polyhedra. $N-1$ LEDs with different colors are placed on the faces of the polyhedron, where N is the number of the faces. The light beams from the LEDs are injected and scattered in the mirror–ball structure. Light-scattering patterns are observed in the structures from one face without a LED by using a digital camera.

Figure 12.12 shows the photographs of the light-scattering patterns in the five regular polyhedral mirror–ball structures (see also the cover of this book) (Amano et al., 2007). Self-similarity of the light-scattering patterns is observed, which is an indication of fractal structure. For the tetrahedral structure shown in Figure 12.12a, many triangle patterns with different colors and sizes are shown in the triangle region with some contraction ratios. These triangle patterns indicate the light source coming from one of the faces of the tetrahedron, which can be considered as a set of initial conditions of the light scattering. Therefore, these color patterns are called "basins". The boundaries of these basins are very complex and they are considered as fractal basin boundaries (Sweet et al., 1999). Different fractal patterns are observed in the five regular polyhedral mirror–ball structures, and the basin boundaries look more complex as the number of the faces is increased.

Based on the experimental observation of Figure 12.12, corresponding models are created for the observed fractal patterns in the five regular polyhedral mirror–ball structures, as shown in Figure 12.13 (Amano et al., 2007). The color patterns indicate the basins of light scattering, corresponding to the faces of the polyhedron. The distorted patterns are modified at the edge of the entire region, so that the essence of the fractal patterns from Figure 12.12 can be extracted. Note that the shape of each color pattern is determined by the shape of the faces of the polyhedra, whereas the shape of the entire region is determined by the number of the nearest-neighboring vertices. The fractal patterns of the hexahedron and octahedron are complementary (triangle and square), as well as those of dodecahedron and icosahedron (pentagon and triangle). The basin boundaries are very complicated for the models of all the five structures, as shown in Figure 12.13.

To identify the basic characteristics of the observed fractal patterns, the fractal dimensions of the basin boundaries of the light-scattering patterns are estimated in the five regular polyhedral mirror–ball structures. The dimension of the basin boundaries is referred to as measured in a two-dimensional slice as shown in Figure 12.13. The fractal dimension is thus between 1 and 2. Two methods are introduced to measure the fractal dimension: the self-similar and box-counting dimensions.

Figure 12.12 Photographs of fractal patterns of chaotic light scattering in the five regular polyhedral mirror–ball structures. (a) Regular tetrahedron, (b) regular hexahedron, (c) regular octahedron, (d) regular dodecahedron, and (e) regular icosahedron. See also the cover of this book. (K. Amano, D. Narimatsu, S. Sotome, S. Tashiro, A. Uchida, and S. Yoshimori, (2007), © 2007 APS.)

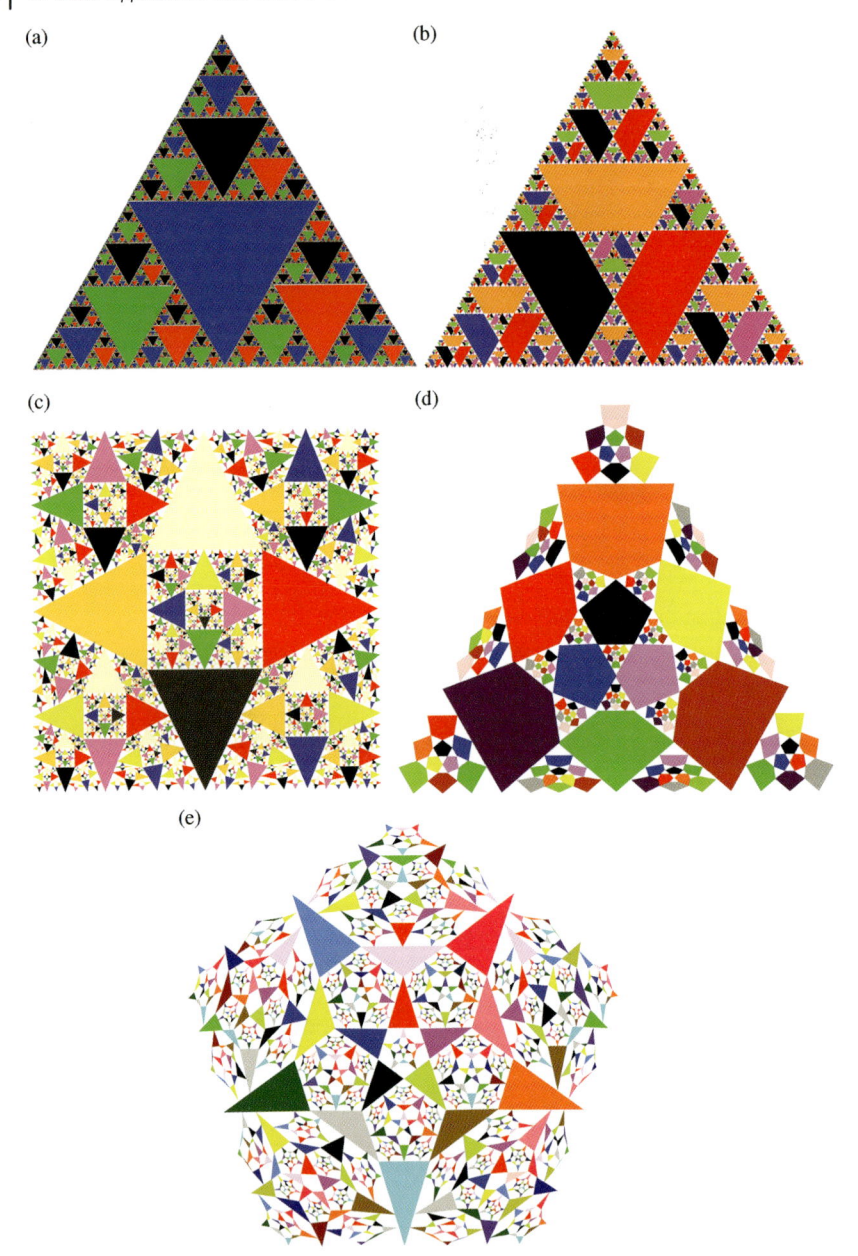

Figure 12.13 Models of fractal patterns of chaotic light scattering in the five regular polyhedral mirror–ball structures. (a) Regular tetrahedron, (b) regular hexahedron, (c) regular octahedron, (d) regular dodecahedron, and (e) regular icosahedron. (K. Amano, D. Narimatsu, S. Sotome, S. Tashiro, A. Uchida, and S. Yoshimori, (2007), © 2007 APS.)

First the self-similar dimension D_s can be analytically calculated,

$$\sum_{i=1}^{M} b_i(a_i)^{D_s} = 1 \qquad (12.4)$$

where a is the contraction ratio ($a \leq 1$) of the size of self-similar patterns and b is the expansion ratio ($b \geq 1$) of the number of increasing self-similar patterns. The size and number of the self-similar patterns are changed with the ratios of a^n and b^n at the nth stage of iteration, respectively, where n indicates the number of reflection of the light on the spheres. M is the total number of the combination of different a_i and b_i. a and b can be different at different regions of the models. The values of a_i and b_i are obtained from the models in Figure 12.13 and the self-similar dimension is calculated analytically.

Next, the box-counting dimension of the basin boundary is estimated from the models shown in Figure 12.13. The box-counting dimension D_c can be calculated numerically,

$$D_c = \lim_{\varepsilon \to 0} \frac{\ln N(\varepsilon)}{\ln(1/\varepsilon)} \qquad (12.5)$$

where ε is the a box size (length) and $N(\varepsilon)$ is the number of boxes needed to cover the basin boundary. The edges of all the fractal patterns are extracted in the models shown in Figure 12.13. The edges are converted with ε-size boxes and counted the number of the boxes to calculate numerically the box-counting dimension of the basin boundaries for the five regular polyhedral mirror-ball structures. A commercially available software is used (e.g., Benoit, TruSoft International Inc.) for the calculation of the box-counting dimension.

Figure 12.14 shows the summary of the self-similar and box-counting dimensions of the basin boundaries of the fractal models for the five regular polyhedral mirror–ball structures. The self-similar and box-counting dimensions match each other with the maximum error of 2%. The fractal dimension of the tetrahedron (4 faces) is identical to that of the Sierpinski gasket ($D = 1.585...$) since the models are exactly the same, as shown in Figure 12.13a. The dimensions of the tetrahedron (4 faces) and hexahedron (6 faces) are very close to each other. The dimension of the icosahedron (20 faces) is the maximum value of the fractal dimension in the five regular polyhedral mirror–ball structures. It is found that the fractal dimension increases as the number of the faces of the polyhedral structures is increased, except for the dodecahedron (12 faces). For the dodecahedral structure shown in Figure 12.13d, the pentagonal color patterns share the large region and the basin boundaries are not as long as the other polyhedral structures.

The fractal dimension obtained from the models may be an important measure to characterize optical scattering devices with fractal structures. The fractal dimension may be related to the Q value of optical confinement devices. The investigation of the relationship between the fractal dimension and the functionality of fractal optical devices would be an interesting topic.

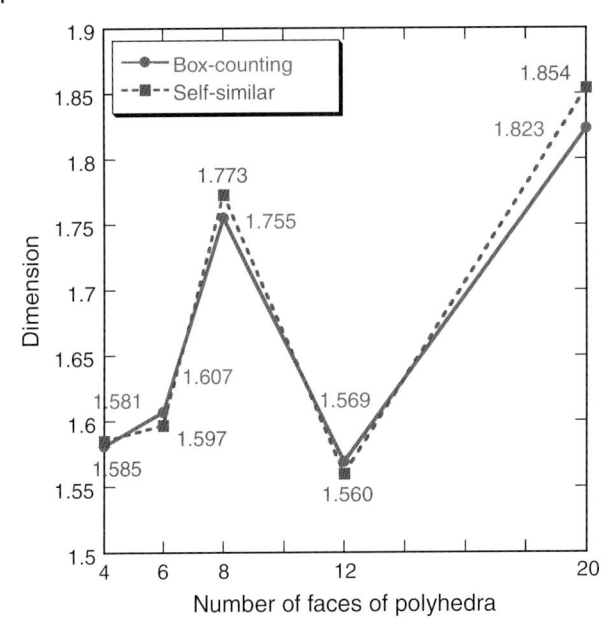

Figure 12.14 Summary of the self-similar and box-counting dimensions (D_s and D_c) of the fractal models for the five regular polyhedral mirror–ball structures. (K. Amano, D. Narimatsu, S. Sotome, S. Tashiro, A. Uchida, and S. Yoshimori, (2007), © 2007 APS.)

For the models in Figure 12.13, ray optics are only taken into account to analyze the fractal patterns since the size of the spherical structures (85-mm diameter) is much larger than the wavelength of visible light (hundreds of nm). The effects of wave optics and quantum optics may become dominant when the size of these structures is decreased in the order of the optical wavelength. It is expected to find more interesting behavior of photons in such nanoscale fractal optical devices.

References

Abarbanel, H.D.I. (1996) *Analysis of Observed Chaotic Data*, Springer-Verlag, New York.

Abarbanel, H.D.I., Rulkov, N.F., and Sushchik, M.M. (1996) Generalized synchronization of chaos: The auxiliary system approach. *Physical Review E*, Vol. **53** (No. 5), pp. 4528–4535.

Abarbanel, H.D.I. and Kennel, M.B. (1998) Synchronizing high-dimensional chaotic optical ring dynamics. *Physical Review Letters*, Vol. **80** (No. 14), pp. 3153–3156.

Abarbanel, H.D.I., Kennel, M.B., Buhl, M., and Lewis, C.T. (1999) Chaotic dynamics in erbium-doped fiber ring lasers. *Physical Review A*, Vol. **60** (No. 3), pp. 2360–2374.

Afraimovich, V.S., Verichev, N.N., and Rabinovich, M.I. (1986) Stochastic synchronization of oscillations in dissipative systems. *Radiofizika*, Vol. **29** (No. 9), pp. 1050–1060.

Agrawal, G.P. and Dutta, N.K. (1993) *Semiconductor Lasers*, 2nd edn, Van Nostrand Reinhold, New York (International Thomson Publishing, London).

Agrawal, G.P. (1997) *Fiber-Optic Communications Systems*, 2nd edn, John Wiley & Sons, New York.

Ahlers, V., Parlitz, U., and Lauterborn, W. (1998) Hyperchaotic dynamics and synchronization of external-cavity semiconductor lasers. *Physical Review E*, Vol. **58** (No. 6), pp. 7208–7213.

Aida, T. and Davis, P. (1992) Oscillation modes of laser diode pumped hybrid bistable system with large delay and application to dynamical memory. *IEEE Journal of Quantum Electronics*, Vol. **28** (No. 3), pp. 686–699.

Aida, T. and Davis, P. (1994) Oscillation mode selection using bifurcation of chaotic mode transitions in a nonlinear ring resonator. *IEEE Journal of Quantum Electronics*, Vol. **30** (No. 12), pp. 2986–2997.

Akizawa, Y., Yamazaki, T., Uchida, A., Harayama, T., Sunada, S., Arai, K., Yoshimura, K., and Davis, P. (2011) Post-processing method for fast random bit generation with semiconductor lasers. *Proceedings of 2011 International Symposium on Nonlinear Theory and its Applications (NOLTA 2011)*.

Alpern, B. and Schneider, F.B. (1983) Key exchange using 'keyless cryptography'. *Information Processing Letters*, Vol. **16** (No. 2), pp. 79–81.

Altés, J.B., Gatare, I., Panajotov, K., Thienpont, H., and Sciamanna, M. (2006) Mapping of the dynamics induced by orthogonal optical injection in vertical-cavity surface-emitting lasers. *IEEE Journal of Quantum Electronics*, Vol. **42** (No. 2), pp. 198–207.

Amano, K., Narimatsu, D., Sotome, S., Tashiro, S., Uchida, A., and Yoshimori, S. (2007) Fractal dimension of chaotic light scattering in regular polyhedral mirror ball structures. *Physical Review E*, Vol. **76** (No. 4), pp. 046213.

Amengual, A., Hernández-García, E., Montagne, R., and San Miguel, M. (1997) Synchronization of spatiotemporal chaos:

the regime of coupled spatiotemporal intermittency. *Physical Review Letters*, Vol. **78** (No. 23), pp. 4379–4382.

Anishchenko, V.S., Vadivasova, T.E., Postnov, D.E., and Safonova, M.A. (1992) Synchronization of chaos. *International Journal of Bifurcation and Chaos*, Vol. **2** (No. 3), pp. 633–644.

Annovazzi-Lodi, V., Donati, S., and Scirè, A. (1996) Synchronization of chaotic injected-laser systems and its application to optical cryptography. *IEEE Journal of Quantum Electronics*, Vol. **32** (No. 6), pp. 953–959.

Annovazzi-Lodi, V., Benedetti, M., Merlo, S., Norgia, M., and Provinzano, B. (2005) Optical chaos masking of video signals. *IEEE Photonics Technology Letters*, Vol. **17** (No. 9), pp. 1995–1997.

Annovazzi-Lodi, V., Argyris, A., Benedetti, M., Hamacher, M., Merlo, S., and Syvridis, D. (2008) A chaos-based approach to secure communications. *Optics and Photonics News*, Vol. **19** (No. 10), pp. 36–41.

Arecchi, F.T., Meucci, R., Puccioni, G., and Tredicce, J. (1982) Experimental evidence of subharmonic bifurcations, multistability, and turbulence in a Q-switched gas laser. *Physical Review Letters*, Vol. **49** (No. 17), pp. 1217–1220.

Arecchi, F.T., Lippi, G.L., Puccioni, G.P., and Tredicce, J.R. (1984) Deterministic chaos in laser with injected signal. *Optics Communications*, Vol. **51** (No. 5), pp. 308–314.

Argyris, A., Kanakidis, D., Bogris, A., and Syvridis, D. (2004) Experimental evaluation of an open-loop all-optical chaotic communication system. *IEEE Journal of Selected Topics in Quantum Electronics*, Vol. **10** (No. 5), pp. 927–935.

Argyris, A., Syvridis, D., Larger, L., Annovazzi-Lodi, V., Colet, P., Fischer, I., García-Ojalvo, J., Mirasso, C.R., Pesquera, L., and Shore, K.A. (2005) Chaos-based communications at high bit rates using commercial fibre-optic links. *Nature*, Vol. **438**, pp. 343–346.

Argyris, A., Hamacher, M., Chlouverakis, K.E., Bogris, A., and Syvridis, D. (2008) Photonic integrated device for chaos applications in communications. *Physical Review Letters*, Vol. **100** (No. 19), pp. 194101.

Argyris, A., Bogris, A., Hamacher, M., and Syvridis, D. (2010a) Experimental evaluation of subcarrier modulation in chaotic optical communication systems. *Optics Letters*, Vol. **35** (No. 2), pp. 199–201.

Argyris, A., Deligiannidis, S., Pikasis, E., Bogris, A., and Syvridis, D. (2010b) Implementation of 140 Gb/s true random bit generator based on a chaotic photonic integrated circuit. *Optics Express*, Vol. **18** (No. 18), pp. 18763–18768.

Argyris, A., Grivas, E., Hamacher, M., Bogris, A., and Syvridis, D. (2010c) Chaos-on-a-chip secures data transmission in optical fiber links. *Optics Express*, Vol. **18** (No. 5), pp. 5188–5198.

Aumann, Y., Ding, Y.Z., and Rabin, M.O. (2002) Everlasting security in the bounded storage model. *IEEE Transactions on Information Theory*, Vol. **48** (No. 6), pp. 1668–1680.

Baer, T. (1986) Large-amplitude fluctuations due to longitudinal mode coupling in diode-pumped intracavity-doubled Nd:YAG lasers. *Journal of Optical Society of America B*, Vol. **3** (No. 9), pp. 1175–1180.

Barsella, A., Lepers, C., Dangoisse, D., Glorieux, P., and Erneux, T. (1999) Synchronization of two strongly pulsating CO_2 lasers. *Optics Communications*, Vol. **165** (No. 4–6), pp. 251–259.

Bennett, C.H., Brassard, G., Crepeau, C., and Maurer, U.M. (1995) Generalized privacy amplification. *IEEE Transactions on Information Theory*, Vol. **41** (No. 6), pp. 1915–1923.

Bennett, S., Snowden, C.M., and Iezekiel, S. (1997) Nonlinear dynamics in directly modulated multiple-quantum-well laser diodes. *IEEE Journal of Quantum Electronics*, Vol. **33** (No. 11), pp. 2076–2083.

Bergé, P., Pomeau, Y., and Vidal, C. (1984) *Order within Chaos: Towards a Deterministic Approach to Turbulence*, John Wiley & Sons, New York.

Bielawski, S., Derozier, D., and Glorieux, P. (1992) Antiphase dynamics and polarization effects in the Nd-doped fiber laser. *Physical Review A*, Vol. **46** (No. 5), pp. 2811–2822.

Bielawski, S., Derozier, D., and Glorieux, P. (1994) Controlling unstable periodic orbits by a delayed continuous feedback. *Physical Review E*, Vol. **49** (No. 2), pp. R971–R974.

Bivalve shells (2011) *Hoshino Koubou*, Yokohama, Japan, URL: http://www.01.246.ne.jp/~hoshi-no/index.html.

Blakely, J.N., Illing, L., and Gauthier, D.J. (2004) Controlling fast chaos in delay dynamical systems. *Physical Review Letters*, Vol. **92** (No. 19), pp. 193901.

Blekhman, I.I. (1971) *Synchronization of Dynamical Systems*, Nauka, Moscow (in Russian).

Boccaletti, S., Kurths, J., Osipov, G., Valladares, D.L., and Zhou, C.S. (2002) The synchronization of chaotic systems. *Physics Reports*, Vol. **366** (No. 1–2), pp. 1–101.

Bogris, A., Chlouverakis, K.E., Argyris, A., and Syvridis, D. (2007) Subcarrier modulation in all-optical chaotic communication systems. *Optics Letters*, Vol. **32** (No. 15), pp. 2134–2136.

Bogris, A., Argyris, A., and Syvridis, D. (2010) Encryption efficiency analysis of chaotic communication systems based on photonic integrated chaotic circuits. *IEEE Journal of Quantum Electronics*, Vol. **46** (No. 10), pp. 1421–1429.

Boulnois, J.L., Van Lerberghe, A., Cottin, P., Arecchi, F.T., and Puccini, G.P. (1986) Self pulsing in a CO_2 ring laser with an injected signal. *Optics Communications*, Vol. **58** (No. 2), pp. 124–129.

Bouwmeester, D., Ekert, A., and Zeilinger F A. (eds) (2000) *The Physics of Quantum Information*, Springer-Verlag, Berlin Heidelberg.

Bracikowski, C. and Roy, R. (1991) Energy sharing in a chaotic multimode laser. *Physical Review A*, Vol. **43** (No. 11), pp. 6455–6457.

Bracikowski, C., Fox, R.F., and Roy, R. (1992) Amplification of intrinsic noise in a chaotic multimode laser system. *Physical Review A*, Vol. **45** (No. 1), pp. 403–408.

Braiman, Y. and Goldhirsch, I. (1991) Taming chaotic dynamics with weak periodic perturbations. *Physical Review Letters*, Vol. **66** (No. 20), pp. 2545–2548.

Breeden, J.L., Dinkelacker, F., and Hübler, A. (1990) Noise in the modeling and control of dynamical systems. *Physical Review A*, Vol. **42** (No. 10), pp. 5827–5836.

Brown, R. (1998) Approximating the mapping between systems exhibiting generalized synchronization. *Physical Review Letters*, Vol. **81** (No. 22), pp. 4835–4838.

Bucci, M., Germani, L., Luzzi, R., Trifiletti, A., and Varanonuovo, M. (2003) A high-speed oscillator-based truly random number source for cryptographic applications on a smart card IC. *IEEE Transactions on Computers*, Vol. **52** (No. 4), pp. 403–409.

Buldú, J.M., García-Ojalvo, J., and Torrent, M.C. (2004) Multimode synchronization and communication using unidirectionally coupled semiconductor lasers. *IEEE Journal of Quantum Electronics*, Vol. **40** (No. 6), pp. 640–650.

Casperson, L. (1978) Spontaneous coherent pulsations in laser oscillators. *IEEE Journal of Quantum Electronics*, Vol. **14** (No. 10), pp. 756–761.

Celka, P. (1995) Chaotic synchronization and modulation of nonlinear time-delayed feedback optical systems. *IEEE Transactions on Circuits and Systems I: Fundamental Theory and Applications*, Vol. **42** (No. 8), pp. 455–463.

Chachin, C. and Maurer, U.M. (1997) Unconditional security against memory-bounded adversaries. *Lecture Notes in Computer Science*, Vol. **1294**, pp. 292–306.

Chen, H.F. and Liu, J.M. (2000) Open-loop chaotic synchronization of injection-locked semiconductor lasers with gigahertz range modulation. *IEEE Journal of Quantum Electronics*, Vol. **36** (No. 1), pp. 27–34.

Chen, H.F. and Liu, J.M. (2004) Unidirectionally coupled synchronization of optically injected semiconductor lasers. *IEEE Journal of Selected Topics in Quantum Electronics*, Vol. **10** (No. 5), pp. 918–926.

Chua, L.O., Komuro, M., and Matsumoto, T. (1986) The double scroll family. *IEEE Transactions on Circuits and Systems*, Vol. **33** (No. 11), pp. 1072–1118.

Chua, L.O., Kocarev, L., Eckert, K., and Itoh, M. (1992) Experimental chaos synchronization in Chua's circuit. *International Journal of Bifurcation and Chaos*, Vol. **2** (No. 3), pp. 705–708.

Colet, P. and Roy, R. (1994) Digital communication with synchronized chaotic lasers. *Optics Letters*, Vol. **19** (No. 24), pp. 2056–2058.

Cuomo, K.M. and Oppenheim, A.V. (1993) Circuit implementation of synchronized chaos with applications to communications. *Physical Review Letters*, Vol. **71** (No. 1), pp. 65–68.

Daido, H. and Nakanishi, K. (2004) Aging transition and universal scaling in oscillator networks. *Physical Review Letters*, Vol. **93** (No. 10), pp. 104101.

Danckaert, J., Nagler, B., Albert, J., Panajotov, K., Veretennicoff, I., and Erneux, T. (2002) Minimal rate equations describing polarization switching in vertical-cavity surface-emitting lasers. *Optics Communications*, Vol. **201** (No. 1–3), pp. 129–137.

Dangoisse, D., Bekkali, A., Papoff, F., and Glorieux, P. (1988) Shilnikov dynamics in a passive Q-switching laser. *Europhysics Letters*, Vol. **6** (No. 4), pp. 335.

Davis, P. (1990) Application of optical chaos to temporal pattern search in a nonlinear optical resonator. *Japanese Journal of Applied Physics*, Vol. **29** (No. 7), pp. L1238–L1240.

DeShazer, D.J., Breban, R., Ott, E., and Roy, R. (2001) Detecting phase synchronization in a chaotic laser array. *Physical Review Letters*, Vol. **87** (No. 4), pp. 044101.

D'Huys, O., Vicente, R., Danckaert, J., and Fischer, I. (2010) Amplitude and phase effects on the synchronization of delay-coupled oscillators. *Chaos*, Vol. **20** (No. 4), pp. 043127.

Dichtl, M. (2007) Bad and good ways of post-processing biased physical random numbers. *Lecture Notes in Computer Science*, Vol. **4593**, pp. 137–152.

Ditto, W.L., and Pecora, L.M. (1993) Mastering chaos. *Scientific American*, Vol. **269** (No. 2), pp. 78–84.

Donati, S. and Mirasso, C.R. (2002) Introduction to the feature section on optical chaos and applications to cryptography. *IEEE Journal of Quantum Electronics*, Vol. **38** (No. 9), pp. 1138–1140.

Dykstra, R., Tang, D.Y., and Heckenberg, N.R. (1998) Experimental control of single-mode laser chaos by using continuous, time-delayed feedback. *Physical Review E*, Vol. **57** (No. 6), pp. 6596–6598.

Dynes, J.F., Yuan, Z.L., Sharpe, A.W., and Shields, A.J. (2008) A high speed, postprocessing free, quantum random number generator. *Applied Physics Letters*, Vol. **93** (No. 3), pp. 031109.

Eastlake, D., Schiller, J., and Crocker, S. (2005) Randomness requirements for security, RFC4086. URL: http://tools.ietf.org/pdf/rfc4086.pdf.

Endo, T. and Chua, L.O. (1991) Synchronization of chaos in phase-locked loops. *IEEE Transactions on Circuits and Systems*, Vol. **38** (No. 12), pp. 1580–1588.

Englert, A., Kinzel, W., Aviad, Y., Butkovski, M., Reidler, I., Zigzag, M., Kanter, I., and Rosenbluh, M. (2010) Zero lag synchronization of chaotic systems with time delayed couplings. *Physical Review Letters*, Vol. **104** (No. 11), pp. 114102.

Fabiny, L., Colet, P., Roy, R., and Lenstra, D. (1993) Coherence and phase dynamics of spatially coupled solid-state lasers. *Physical Review A*, Vol. **47** (No. 5), pp. 4287–4296.

FDK (2011) Random Streamer, FDK Corporation, URL: http://www.fdk.com/.

Fischer, I., van Tartwijk, G.H.M., Levine, A.M., Elsässer, W., Göbel, E., and Lenstra, D. (1996) Fast pulsing and chaotic itinerancy with a drift in the coherence collapse of semiconductor lasers. *Physical Review Letters*, Vol. **76** (No. 2), pp. 220–223.

Fischer, I., Liu, Y., and Davis, P. (2000) Synchronization of chaotic semiconductor laser dynamics on subnanosecond time scales and its potential for chaos communication. *Physical Review A*, Vol. **62** (No. 1), pp. 011801(R).

Fischer, I., Vicente, R., Buldú, J.M., Peil, M., Mirasso, C.R., Torrent, M.C., and García-Ojalvo, J. (2006) Zero-lag long-range synchronization via dynamical relaying. *Physical Review Letters*, Vol. **97** (No. 12), pp. 123902.

Flunkert, V., D'Huys, O., Danckaert, J., Fischer, I., and Schöll, E. (2009) Bubbling in delay-coupled lasers. *Physical Review E*, Vol. **79** (No. 6), pp. 065201(R).

Fujino, H. and Ohtsubo, J. (2000) Experimental synchronization of chaotic oscillations in external-cavity semiconductor lasers. *Optics Letters*, Vol. **25** (No. 9), pp. 625–627.

Fujisaka, H. and Yamada, T. (1983) Stability theory of synchronized motion in coupled-oscillator systems. *Progress of Theoretical Physics*, Vol. **69** (No. 1), pp. 32–47.

Fujiwara, N., Takiguchi, Y., and Ohtsubo, J. (2003) Observation of the synchronization of chaos in mutually injected vertical-cavity surface-emitting semiconductor lasers. *Optics Letters*, Vol. **28** (No. 18), pp. 1677–1679.

Fukushima, A., Seki, T., Yakushiji, K., Kubota, H., Yuasa, S. (2011) Spin based physical random number generator "Spin dice". *The Journal of the Institute of Electronics, Information and Communication Engineers*, Vol. **94** (No. 3), pp. 232–238 (in Japanese).

Gabriel, C., Wittmann, C., Sych, D., Dong, R., Mauerer, W., Andersen, U.L., Marquardt, C., and Leuchs, G. (2010) A generator for unique quantum random numbers based on vacuum states. *Nature Photonics*, Vol. **4** (No. 10), pp. 711–715.

Galton, F. (1890) Dice for statistical experiments. *Nature*, Vol. **42** (No. 1070), pp. 13–14.

García-Ojalvo, J. and Roy, R. (2001a) Spatiotemporal communication with synchronized optical chaos. *Physical Review Letters*, Vol. **86** (No. 22), pp. 5204–5207.

García-Ojalvo, J. and Roy, R. (2001b) Parallel communication with optical spatiotemporal chaos. *IEEE Transactions on Circuits and Systems I: Fundamental Theory and Applications*, Vol. **48** (No. 12), pp. 1491–1497.

Gastaud, N., Poinsot, S., Larger, L., Merolla, J.-M., Hanna, M., Goedgebuer, J.-P., and Malassenet, F. (2004) Electro-optical chaos for multi-10 Gbit/s optical transmissions. *Electronics Letters*, Vol. **40** (No. 14), pp. 898–899.

Gatare, I., Sciamanna, M., Buesa, J., Thienpont, H., and Panajotov, K. (2006) Nonlinear dynamics accompanying polarization switching in vertical-cavity surface-emitting lasers with orthogonal optical injection. *Applied Physics Letters*, Vol. **88** (No. 10), pp. 101106.

Gauthier, D.J. (2003) Resource Letter: CC-1: Controlling chaos. *American Journal of Physics*, Vol. **71** (No. 8), pp. 750–759.

Geddes, J.B., Short, K.M., and Black, K. (1999) Extraction of signals from chaotic laser data. *Physical Review Letters*, Vol. **83** (No. 25), pp. 5389–5392.

Genin, É., Larger, L., Goedgebuer, J.-P., Lee, M.W., Ferriére, R., and Bavard, X. (2004) Chaotic oscillations of the optical phase for multigigahertz-bandwidth secure communications. *IEEE Journal of Quantum Electronics*, Vol. **40** (No. 3), pp. 294–298.

Gioggia, R.S. and Abraham, N.B. (1983) Routes to chaotic output from a single-mode, dc-excited laser. *Physical Review Letters*, Vol. **51** (No. 8), pp. 650–653.

Giraud, L., Langou, J., and Rozloznik, M. (2005) The loss of orthogonality in the Gram-Schmidt orthogonalization process. *Computers and Mathematics with Applications*, Vol. **50** (No. 7), pp. 1069–1075.

Gisin, N., Ribordy, G., Tittel, W., and Zbinden, H. (2002) Quantum cryptography. *Review of Modern Physics*, Vol. **74** (No. 1), pp. 145–195.

Giudici, M., Balle, S., Ackemann, T., Barland, S., and Tredicce, J.R. (1999) Polarization dynamics in vertical-cavity surface-emitting lasers with optical feedback: experiment and model. *Journal of Optical Society of America B*, Vol. **16** (No. 11), pp. 2114–2123.

Gleeson, J.T. (2002) Truly random number generator based on turbulent electroconvection. *Applied Physics Letters*, Vol. **81** (No. 11), pp. 1949–1951.

Goedgebuer, J.-P., Larger, L., and Porte, H. (1998) Optical cryptosystem based on synchronization of hyperchaos generated by a delayed feedback tunable laser diode. *Physical Review Letters*, Vol. **80** (No. 10), pp. 2249–2252.

Goedgebuer, J.-P., Levy, P., Larger, L., Chen, C.-C., and Rhodes, W.T. (2002) Optical communication with synchronized hyperchaos generated electrooptically. *IEEE Journal of Quantum Electronics*, Vol. **38** (No. 9), pp. 1178–1183.

González-Miranda, J.M. (2004) *Synchronization and Control of Chaos: An Introduction for Scientists and Engineers*, Imperial College Press, London.

Grassberger, P. and Procaccia, I. (1983) Characterization of strange attractors. *Physical Review Letters*, Vol. **50** (No. 5), pp. 346–349.

Gray, G.R., Ryan, A.T., Agrawal, G.P., and Gage, E.C. (1993) Control of optical-feedback-induced laser intensity noise in optical data recording. *Optical Engineering*, Vol. **32** (No. 4), pp. 739–745.

Guo, H., Tang, W., Liu, Y., and Wei, W. (2010) Truly random number generation based on measurement of phase noise of a laser. *Physical Review E*, Vol. **81** (No. 5), pp. 051137.

Haken, H. (1975) Analogy between higher instabilities in fluids and lasers. *Physics Letters A*, Vol. **53** (No. 1), pp. 77–78.

Haken, H. (1985) *Light: Volume 2: Laser Light Dynamics*, North-Holland, Amsterdam.

Halle, K.S., Wu, C.W., Itoh, M., and Chua, L.O. (1993) Spread spectrum communication through modulation of chaos. *International Journal of Bifurcation and Chaos*, Vol. **3** (No. 2), pp. 469–477.

Harayama, T. and Davis, P. (1998) Chaos mirror for free-space links. *Optics Letters*, Vol. **23** (No. 18), pp. 1426–1428.

Harayama, T., Sunada, S., Yoshimura, K., Davis, P., Tsuzuki, K., and Uchida, A. (2011) Fast nondeterministic random-bit generation using on-chip chaos lasers. *Physical Review A*, Vol. **83**, pp. 031803(R).

Hayes, S., Grebogi, C., and Ott, E. (1993) Communicating with chaos. *Physical Review Letters*, Vol. **70** (No. 20), pp. 3031–3034.

Hegarty, S.P., Huyet, G., Porta, P., and McInerney, J.G. (1998) Analysis of the fast recovery dynamics of a semiconductor laser with feedback in the low-frequency fluctuation regime. *Optics Letters*, Vol. **23** (No. 15), pp. 1206–1208.

Heil, T., Fischer, I., and Elsäßer, W. (1998) Coexistence of low-frequency fluctuations and stable emission on a single high-gain mode in semiconductor lasers with external optical feedback. *Physical Review A*, Vol. **58** (No. 4), pp. R2672–R2675.

Heil, T., Fischer, I., Elsässer, W., Mulet, J., and Mirasso, C.R. (2001a) Chaos synchronization and spontaneous symmetry-breaking in symmetrically delay-coupled semiconductor lasers. *Physical Review Letters*, Vol. **86** (No. 5), pp. 795–798.

Heil, T., Fischer, I., Elsäßer, W., and Gavrielides, A. (2001b) Dynamics of semiconductor lasers subject to delayed optical feedback: The short cavity regime. *Physical Review Letters*, Vol. **87** (No. 24), pp. 243901.

Heil, T., Mulet, J., Fischer, I., Mirasso, C.R., Peil, M., Colet, P., and Elsäßer, W. (2002) On/off phase shift keying for chaos-encrypted communication using external-cavity semiconductor lasers. *IEEE Journal of Quantum Electronics*, Vol. **38** (No. 9), pp. 1162–1170.

Heil, T., Uchida, A., Davis, P., and Aida, T. (2003) TE-TM dynamics in a semiconductor laser subject to polarization-rotated optical feedback. *Physical Review A*, Vol. **68** (No. 3), pp. 033811.

Hellman, M. (1977) An extension of the Shannon theory approach to cryptography. *IEEE Transactions on Information Theory*, Vol. **23** (No. 3), pp. 289–294.

Hennequin, D., de Tomasi, F., Zambon, B., and Arimondo, E. (1988) Homoclinic orbits and cycles in the instabilities of a laser with a saturable absorber. *Physical Review A*, Vol. **37** (No. 6), pp. 2243–2246.

Herrero, R., Figueras, M., Rius, J., Pi, F., and Orriols, G. (2000) Experimental observation of the amplitude death effect in two coupled nonlinear oscillators. *Physical Review Letters*, Vol. **84** (No. 23), pp. 5312–5315.

Hilborn, R.C. (2000) *Chaos and Nonlinear Dynamics*, 2nd edn, Oxford University Press, Oxford.

Hirano, K., Amano, K., Uchida, A., Naito, S., Inoue, M., Yoshimori, S., Yoshimura, K., and Davis, P. (2009) Characteristics of fast physical random bit generation using chaotic semiconductor lasers. *IEEE Journal of Quantum Electronics*, Vol. **45** (No. 11), pp. 1367–1379.

Hirano, K., Yamazaki, T., Morikatsu, S., Okumura, H., Aida, H., Uchida, A., Yoshimori, S., Yoshimura, K., Harayama, T., and Davis, P. (2010) Fast random bit generation with bandwidth-enhanced chaos in semiconductor lasers. *Optics Express*, Vol. **18** (No. 6), pp. 5512–5524.

Ho, K.-P. (2005) *Phase-Modulated Optical Communication Systems*, Springer, New York.

Holman, W.T., Connelly, J.A., and Dowlatabadi, A.B. (1997) An integrated analog/digital random noise source. *IEEE Transactions on Circuits and Systems I: Fundamental Theory and Applications*, Vol. **44** (No. 6), pp. 521–528.

Hong, Y., Lee, M.W., Spencer, P.S., and Shore, K.A. (2004) Synchronization of chaos in unidirectionally coupled vertical-cavity surface-emitting semiconductor lasers. *Optics Letters*, Vol. **29** (No. 11), pp. 1215–1217.

Honjo, T., Inoue, K., and Takahashi, H. (2004) Differential-phase-shift quantum key distribution experiment with a planar light-wave circuit Mach-Zehnder interferometer. *Optics Letters*, Vol. **29** (No. 23), pp. 2797–2799.

Honjo, T., Uchida, A., Amano, K., Hirano, K., Someya, H., Okumura, H., Yoshimura, K., Davis, P., and Tokura, Y. (2009) Differential-phase-shift quantum key distribution experiment using fast physical random bit generator with chaotic semiconductor lasers. *Optics Express*, Vol. **17** (No. 11), pp. 9053–9061.

Houlihan, J., Huyet, G., and McInerney, J.G. (2001) Dynamics of a semiconductor laser with incoherent optical feedback. *Optics Communications*, Vol. **199** (No. 1–4), pp. 175–179.

Hunt, E.R. (1991) Stabilizing high-period orbits in a chaotic system: The diode resonator. *Physical Review Letters*, Vol. **67** (No. 15), pp. 1953–1955.

Hwang, S.K. and Liu, J.M. (2000) Dynamical characteristics of an optically injected semiconductor laser. *Optics Communications*, Vol. **183** (No. 1–4), pp. 195–205.

Hyvärinen, A., Karhunen, J., and Oja, E. (2001) *Independent Component Analysis*, John Wiley & Sons, New York.

ID Quantique (2011) Quantis quantum random number generator, ID Quantique, URL: http://www.idquantique.com/. Also see the quantis white paper: URL: http://www.idquantique.com/images/stories/PDF/quantis-random-generator/quantis-whitepaper.pdf.

Ikeda, K. (1979) Multiple-valued stationary state and its instability of the transmitted light by a ring cavity system. *Optics Communications*, Vol. **30** (No. 2), pp. 257–261.

Ikeda, K., Kondo, K., and Akimoto, O. (1982) Successive higher-harmonic bifurcations in systems with delayed feedback. *Physical Review Letters*, Vol. **49** (No. 20), pp. 1467–1470.

Ikezoe, Y., Kim, S.-J., Yamashita, I., and Hara, M. (2009) Random number generation by a two-dimensional crystal of protein molecules. *Langmuir*, Vol. **25** (No. 8), pp. 4293–4297.

Illing, L., Gauthier, D.J., and Roy, R. (2006) Controlling optical chaos, spatio-temporal dynamics, and patterns. *Advances in Atomic, Molecular and Optical Physics*, Vol. **54**, pp. 615–697.

Inoue, H., Kumahora, H., Yoshizawa, Y., Ichimura, M., and Miyatake, O. (1983) Random numbers generated by a physical device. *Journal of the Royal Statistical Society. Series C*, Vol. **32** (No. 2), pp. 115–120.

Inoue, K., Waks, E., and Yamamoto, Y. (2003) Differential-phase-shift quantum key distribution using coherent light. *Physical Review A*, Vol. **68** (No. 2), pp. 022317.

Ishiyama, F. (1999) Bistability of quasi periodicity and period doubling in a delay-induced system. *Journal of Optical Society of America B*, Vol. **16** (No. 12), pp. 2202–2206.

ISO (2010) *Random Variate Generation Methods*, International Organization for Standardization, ISO 28640:2010, Geneva, Switzerland.

Jacobo, A., Soriano, M.C., Mirasso, C.R., and Colet, P. (2010) Chaos-based optical communications: Encryption versus nonlinear filtering. *IEEE Journal of Quantum Electronics*, Vol. **46** (No. 4), pp. 499–505.

Jennewein, T., Achleitner, U., Weihs, G., Weinfurter, H., and Zeilinger, A. (2000) A fast and compact quantum random number generator. *Review of Scientific Instruments*, Vol. **71** (No. 4), pp. 1675–1680.

Jun, B. and Kocher, P. (1999) *The Intel Random Number Generator* (White paper prepared for Intel Corporation), Cryptography Research, Inc., URL: http://www.cryptography.com/resources/whitepapers/IntelRNG.pdf.

Kanakidis, D., Argyris, A., and Syvridis, D. (2003) Performance characterization of high-bit-rate optical chaotic communication systems in a back-to-back configuration. *Journal of Lightwave Technology*, Vol. **21** (No. 3), pp. 750–758.

Kane, D.M. and Shore, K.A. (2005) *Unlocking Dynamical Diversity: Optical Feedback Effects on Semiconductor Lasers*, John Wiley & Sons, West Sussex.

Kangou-fu trade, (2011) About Licenses for Trade between Japan and Ming China. Kyoto National Museum, Kyoto, Japan, URL: http://www.kyohaku.go.jp/eng/dictio/data/shoseki/nichimin.htm.

Kanno, K. and Uchida A. (2011) Consistency and complexity in coupled semiconductor lasers with time-delayed optical feedback. *Proceedings of 2011 International Symposium on Nonlinear Theory and its Applications (NOLTA 2011)*.

Kanter, I., Kopelowitz, E., and Kinzel, W. (2008) Public channel cryptography: chaos synchronization and Hilbert's tenth problem. *Physical Review Letters*, Vol. **101** (No. 8), pp. 084102.

Kanter, I., Aviad, Y., Reidler, I., Cohen, E., and Rosenbluh, M. (2010) An optical ultrafast random bit generator. *Nature Photonics*, Vol. **4** (No. 1), pp. 58–61.

Kantz, H. and Schreiber, T. (1997) *Nonlinear Time Series Analysis*, Cambridge University Press, Cambridge.

Kao, Y.H. and Lin, H.T. (1993) Virtual Hopf precursor of period-doubling route in directly modulated semiconductor lasers. *IEEE Journal of Quantum Electronics*, Vol. 29 (No. 6), pp. 1617–1623.

Kapitaniak, T. (1996) *Controlling Chaos: Theoretical and Practical Methods in Non-linear Dynamics*, Academic Press, London.

Kelsey, J. (2004) *Entropy and Entropy Sources in X9.82*, National Institute of Standards and Technology (NIST), URL: http://csrc.nist.gov/groups/ST/toolkit/documents/rng/EntropySources.pdf.

Kennedy, M.P., Rovatti, R. and Setti, G. (2000) *Chaotic Electronics in Telecommunications*, CRC Press, Boca Raton.

Kikuchi, N., Liu, Y., and Ohtsubo, J. (1997) Chaos control and noise suppression in external-cavity semiconductor lasers. *IEEE Journal of Quantum Electronics*, Vol. 33 (No. 1), pp. 56–65.

Kim, S.-J., Umeno, K., and Hasegawa, A. (2004) Corrections of the NIST statistical test suite for randomness, arXiv:nlin.CD/0401040v1.

Kim, M.-Y., Roy, R., Aron, J.L., Carr, T.W., and Schwartz, I.B. (2005) Scaling behavior of laser population dynamics with time-delayed coupling: Theory and experiment. *Physical Review Letters*, Vol. 94 (No. 8), pp. 088101.

Klein, E., Gross, N., Rosenbluh, M., Kinzel, W., Khaykovich, L., and Kanter, I. (2006a) Stable isochronal synchronization of mutually coupled chaotic lasers. *Physical Review E*, Vol. 73 (No. 6), pp. 066214.

Klein, E., Gross, N., Kopelowitz, E., Rosenbluh, M., Khaykovich, L., Kinzel, W., and Kanter, I. (2006b) Public-channel cryptography based on mutual chaos pass filters. *Physical Review E*, Vol. 74 (No. 4), pp. 046201.

Klische, W., Telle, H.R., and Weiss, C.O. (1984) Chaos in a solid-state laser with a periodically modulated pump. *Optics Letters*, Vol. 9 (No. 12), pp. 561–563.

Knuth, D.E. (1998) *The Art of Computer Programming, Volume 2: Seminumerical Algorithms*, 3rd edn, Addison-Wesley Professional.

Kocarev, L. and Parlitz, U. (1996) Synchronizing spatiotemporal chaos in coupled nonlinear oscillators. *Physical Review Letters*, Vol. 77 (No. 11), pp. 2206–2209.

Kouomou, Y.C., Colet, P., Larger, L., and Gastaud, N. (2005) Mismatch-induced bit error rate in optical chaos communications using semiconductor lasers with electrooptical feedback. *IEEE Journal of Quantum Electronics*, Vol. 41 (No. 2), pp. 156–163.

Kuntsevich, B.F. and Pisarchik, A.N. (2001) Synchronization effects in a dual-wavelength class-B laser with modulated losses. *Physical Review E*, Vol. 64 (No. 4), pp. 046221.

Kuraya, M., Uchida, A., Yoshimori, S., and Umeno, K. (2008) Blind source separation of chaotic laser signals by independent component analysis. *Optics Express*, Vol. 16 (No. 2), pp. 725–730.

Kusumoto, K. and Ohtsubo, J. (2002) 1.5-GHz message transmission based on synchronization of chaos in semiconductor lasers. *Optics Letters*, Vol. 27 (No. 12), pp. 989–991.

Kuwashima, F., Kitazima, I., and Iwasawa, H. (1998) The chaotic oscillation of the single-mode He-Ne (6328Å) class A laser. *Japanese Journal of Applied Physics*, Vol. 37, Part 2 (No. 3B), pp. L325–L328.

Kye, W.-H., Choi, M., Kim, M.-W., Lee, S.-Y., Rim, S., Kim, C.-M., and Park, Y.-J. (2004) Synchronization of delayed systems in the presence of delay time modulation. *Physics Letters A*, Vol. 322 (No. 5–6), pp. 338–343.

Lacot, E., Stoeckel, F., and Chenevier, M. (1994) Dynamics of an erbium-doped fiber laser. *Physical Review A*, Vol. 49 (No. 5), pp. 3997–4008.

Lang, R. and Kobayashi, K. (1980) External optical feedback effects on semiconductor injection laser properties. *IEEE Journal of Quantum Electronics*, Vol. 16 (No. 3), pp. 347–355.

Larger, L., Goedgebuer, J.-P., and Delorme, F. (1998a) Optical encryption system using hyperchaos generated by an optoelectronic wavelength oscillator. *Physical Review E*, Vol. 57 (No. 6), pp. 6618–6624.

Larger, L., Goedgebuer, J.-P., and Merolla, J.-M. (1998b) Chaotic oscillator in wavelength: A new setup for investigating differential difference equations describing nonlinear dynamics. *IEEE Journal of Quantum Electronics*, Vol. 34 (No. 4), pp. 594–601.

Larger, L. and Goedgebuer, J.-P. (2004) Encryption using chaotic dynamics for optical telecommunications. *Comptes Rendus Physique*, Vol. **5** (No. 6), pp. 609–611.

Larger, L., Goedgebuer, J.-P., and Udaltsov, V. (2004) Ikeda-based nonlinear delayed dynamics for application to secure optical transmission systems using chaos. *Comptes Rendus Physique*, Vol. **5** (No. 6), pp. 669–681.

Larger, L. and Dudley, J.M. (2010) Optoelectronic chaos. *Nature*, Vol. **465** (No. 7294), pp. 41–42.

Larson, L.E., Liu, J.-M., and Tsimring, L.S. (eds) (2006) *Digital Communications Using Chaos and Nonlinear Dynamics*, Springer, New York.

Lavrov, R., Peil, M., Jacquot, M., Larger, L., Udaltsov, V., and Dudley, J. (2009) Electro-optic delay oscillator with nonlocal nonlinearity: Optical phase dynamics, chaos, and synchronization. *Physical Review E*, Vol. **80** (No. 2), 026207.

Lavrov, R., Jacquot, M., and Larger, L. (2010) Nonlocal nonlinear electro-optic phase dynamics demonstrating 10 Gb/s chaos communications. *IEEE Journal of Quantum Electronics*, Vol. **46** (No. 10), pp. 1430–1435.

Lee, M.W., Rees, P., Shore, K.A., Ortin, S., Pesquera, L., and Valle, A. (2005) Dynamical characterisation of laser diode subject to double optical feedback for chaotic optical communications. *IEE Proceedings Optoelectronics*, Vol. **152** (No. 2), pp. 97–102.

Lewis, C.T., Abarbanel, H.D.I., Kennel, M.B., Buhl, M., and Illing, L. (2000) Synchronization of chaotic oscillations in doped fiber ring lasers. *Physical Review E*, Vol. **63** (No. 1), pp. 016215.

Li, H.-J. and Chern, J.-L. (1996) Goal-oriented scheme for taming chaos with a weak periodic perturbation: Experiment in a diode resonator. *Physical Review E*, Vol. **54** (No. 2), pp. 2118–2121.

Li, X., Cohen, A.B., Murphy, T.E., and Roy, R. (2011) Scalable parallel physical random number generator based on a superluminescent LED. *Optics Letters*, Vol. **36** (No. 6), pp. 1020–1022.

Lin, F.-Y. and Liu, J.-M. (2003) Nonlinear dynamics of a semiconductor laser with delayed negative optoelectronic feedback. *IEEE Journal of Quantum Electronics*, Vol. **39** (No. 4), pp. 562–568.

Lin, F.-Y. and Liu, J.-M. (2004a) Chaotic lidar. *IEEE Journal of Selected Topics in Quantum Electronics*, Vol. **10** (No. 5), pp. 991–997.

Lin, F.-Y. and Liu, J.-M. (2004b) Chaotic radar using nonlinear laser dynamics. *IEEE Journal of Quantum Electronics*, Vol. **40** (No. 6), pp. 815–820.

Lindner, B., García-Ojalvo, J., Neiman, A., and Schimansky-Geier, L. (2004) Effects of noise in excitable systems. *Physics Reports*, Vol. **392** (No. 6), pp. 321–424.

Liu, Y., de Oliveira, P.C., Danailov, M.B., and Rios Leite, J.R. (1994) Chaotic and periodic passive Q switching in coupled CO_2 lasers with a saturable absorber. *Physical Review A*, Vol. **50** (No. 4), pp. 3464–3470.

Liu, Y. and Ohtsubo, J. (1994) Experimental control of chaos in a laser-diode interferometer with delayed feedback. *Optics Letters*, Vol. **19** (No. 7), pp. 448–450.

Liu, Y., Kikuchi, N., and Ohtsubo, J. (1995) Controlling dynamical behavior of a semiconductor laser with external optical feedback. *Physical Review E*, Vol. **51** (No. 4), pp. R2697–R2700.

Liu, Y. and Davis, P. (1998) Adaptive mode selection based on chaotic search in a Fabry-Perot laser diode. *International Journal of Bifurcation and Chaos*, Vol. **8** (No. 8), pp. 1685–1691.

Liu, Y. and Davis, P. (2000a) Synchronization of chaotic mode hopping. *Optics Letters*, Vol. **25** (No. 7), pp. 475–477.

Liu, Y and Davis, P. (2000b) Dual synchronization of chaos. *Physical Review E*, Vol. **61** (No. 3), pp. R2176–R2179.

Liu, J.M., Chen, H.F., and Tang, S. (2002a) Synchronized chaotic optical communications at high bit rates. *IEEE Journal of Quantum Electronics*, Vol. **38** (No. 9), pp. 1184–1196.

Liu, Y., Takiguchi, Y., Davis, P., Aida, T., Saito, S., and Liu, J.M. (2002b) Experimental observation of complete chaos synchronization in semiconductor lasers. *Applied Physics Letters*, Vol. **80** (No. 23), pp. 4306–4308.

Lorenz, E.N. (1963) Deterministic nonperiodic flow. *Journal of the Atmospheric Sciences*, Vol. **20** (No. 2), pp. 130–141.

Lorenz, E.N. (1993) *The Essence of Chaos*, The University of Washington Press, Seattle.

Luo, L.G., Chu, P.L., and Liu, H.F. (2000a) 1-GHz optical communication system using chaos in erbium-doped fiber lasers. *IEEE Photonics Technology Letters*, Vol. **12** (No. 3), pp. 269–271.

Luo, L., Chu, P.L., Whitbread, T., and Peng, R.F. (2000b) Experimental observation of synchronization of chaos in erbium-doped fiber lasers. *Optics Communications*, Vol. **176** (No. 1–3), pp. 213–217.

Maiman, T.H. (1960) Stimulated optical radiation in ruby. *Nature*, Vol. **187** (No. 4736), pp. 493–494.

Maiman, T.H., Hoskins, R.H., D'Haenens, I.J., Asawa, C.K., and Evtuhov, V. (1961) Stimulated optical emission in fluorescent solids. II. Spectroscopy and stimulated emission in ruby. *Physical Review*, Vol. **123** (No. 4), pp. 1151–1157.

Mainen, Z.F. and Sejnowski, T.J. (1995) Reliability of spike timing in neocortical neurons. *Science*, Vol. **268** (No. 5216), pp. 1503–1506.

Mandel, P. (1997) *Theoretical Problems in Cavity Nonlinear Optics*, Cambridge University Press, Cambridge, UK.

Marsaglia, G. (1995) The Marsaglia random number CD-ROM including the Diehard battery of tests of randomness, URL: http://www.stat.fsu.edu/pub/diehard/.

Martin-Regalado, J., Prati, F., San Miguel, M., and Abraham, N.B. (1997) Polarization properties of vertical-cavity surface-emitting lasers. *IEEE Journal of Quantum Electronics*, Vol. **33** (No. 5), pp. 765–783.

Masoller, C. (2001) Anticipation in the synchronization of chaotic semiconductor lasers with optical feedback. *Physical Review Letters*, Vol. **86** (No. 13), pp. 2782–2785.

Matsumoto, M., and Nishimura, T. (1998) Mersenne Twister: A 623-dimensionally equidistributed uniform pseudorandom number generator. *ACM Transactions on Modeling and Computer Simulation*, Vol. **8** (No. 1), pp. 3–30.

Matsuura, T., Uchida, A., and Yoshimori, S. (2004) Chaotic wavelength division multiplexing for optical communication. *Optics Letters*, Vol. **29** (No. 23), pp. 2731–2733.

Maurer, U.M. (1993) Secret key agreement by public discussion from common information. *IEEE Transactions on Information Theory*, Vol. **39** (No. 3), pp. 733–742.

Maurer, U.M. and Wolf, S. (1999) Unconditionally secure key agreement and the intrinsic conditional information. *IEEE Transactions on Information Theory*, Vol. **45** (No. 2), pp. 499–514.

Mersenne Twister (2011) Mersenne Twister, Home Page: A very fast random number generator $2^{19937} - 1$. URL: http://www.math.sci.hiroshima-u.ac.jp/~m-mat/MT/emt.html.

Meucci, R., Gadomski, W., Ciofini, M., and Arecchi, F.T. (1994) Experimental control of chaos by means of weak parametric perturbations. *Physical Review E*, Vol. **49** (No. 4), pp. R2528–R2531.

Mikami, T., Kanno, K., Aoyama, K., Uchida, A., Harayama, T., Sunada, S., Arai, K., Yoshimura, K., and Davis, P. (2011) Estimation of entropy rate for random bit generators with chaotic semiconductor lasers. *Proceedings of 2011 International Symposium on Nonlinear Theory and its Applications (NOLTA 2011)*.

Mirasso, C.R., Colet, P., and García-Fernández, P. (1996) Synchronization of chaotic semiconductor lasers: application to encoded communications. *IEEE Photonics Technology Letters*, Vol. **8** (No. 2), pp. 299–301.

Molgedey, L. and Schuster, H.G. (1994) Separation of a mixture of independent signals using time delayed correlations. *Physical Review Letters*, Vol. **72** (No. 23), pp. 3634–3637.

Moniz, L., Carroll, T., Pecora, L., and Todd, M. (2003) Assessment of damage in an eight-oscillator circuit using dynamical forcing. *Physical Review E*, Vol. **68** (No. 3), pp. 036215.

Mooradian, A. (1985) Laser linewidth. *Physics Today*, Vol. **38** (No. 5), pp. 42–48.

Mørk, J., Tromborg, B., and Christiansen, P.L. (1988) Bistability and low-frequency fluctuations in semiconductor lasers with optical feedback: a theoretical analysis. *IEEE Journal of Quantum Electronics*, Vol. **24** (No. 2), pp. 123–133.

Mørk, J., Mark, J., and Tromborg, B. (1990) Route to chaos and competition between relaxation oscillations for a semiconductor laser with optical feedback. *Physical Review Letters*, Vol. **65** (No. 16), pp. 1999–2002.

Murakami, A. and Ohtsubo, J. (2002) Synchronization of feedback-induced chaos in semiconductor lasers by optical injection. *Physical Review A*, Vol. **65** (No. 3), pp. 033826.

Murakami, H. (2003) *Hard-boiled Wonderland and the End of the World*, Vintage, London (Translated from "Sekai no owari to haado-boirudo wandaa-rando" Shinchosha Tokyo 1985).

Muramatsu, J., Yoshimura, K., Arai, K., and Davis, P. (2006) Secret key capacity for optimally correlated sources under sampling attack. *IEEE Transactions on Information Theory*, Vol. **52** (No. 11), pp. 5140–5151.

Muramatsu, J., Yoshimura, K., Arai, K., and Davis, P. (2008) Some results on secret key agreement using correlated sources. *NTT Technical Review*, Vol. **6** (No. 2), pp. 1–7.

Murphy, T.E. and Roy, R. (2008) Chaotic lasers: The world's fastest dice. *Nature Photonics*, Vol. **2** (No. 12), pp. 714–715.

Myneni, K., Barr, T.A., Corron, N.J., and Pethel, S.D. (1999) New method for the control of fast chaotic oscillations. *Physical Review Letters*, Vol. **83** (No. 11), pp. 2175–2178.

Myneni, K., Barr, T.A., Reed, B.R., Pethel, S.D., and Corron, N.J. (2001) High-precision ranging using a chaotic laser pulse train. *Applied Physics Letters*, Vol. **78** (No. 11), pp. 1496–1498.

Namajūnas, A., Pyragas, K., and Tamaševičius, A. (1995a) An electronic analog of the Mackey-Glass system. *Physics Letters A*, Vol. **201** (No. 1), pp. 42–46.

Namajūnas, A., Pyragas, K., and Tamaševičius, A. (1995b) Stabilization of an unstable steady state in a Mackey-Glass system. *Physics Letters A*, Vol. **204** (No. 3–4), pp. 255–262.

Naumenko, A.V., Loiko, N.A., Sondermann, M., and Ackemann, T. (2003) Description and analysis of low-frequency fluctuations in vertical-cavity surface-emitting lasers with isotropic optical feedback by a distant reflector. *Physical Review A*, Vol. **68** (No. 3), pp. 033805.

Newell, A.C. and Moloney, J.V. (1992) *Nonlinear Optics*, Addison-Wesley, Redwood City.

OCCULT (2001) Optical Chaotic Communication Using Laser-diode Transmitters (OCCULT) project. URL: http://ifisc.uib-csic.es/project/occult/.

Ogawa, T., Uchida, A., Shinozuka, M., Yoshimori, S., and Kannari, F. (2002) Numerical study for secure communications using the chaotic masking method in two microchip lasers. *Japanese Journal of Applied Physics*, Vol. **41**, Part 2 (No. 11B), pp. L1309–L1311.

Ohtsubo, J. (2008) *Semiconductor Lasers: Stability, Instability and Chaos*, 2nd edn, Springer-Verlag, Berlin Heidelberg.

Onodera, T. (2006) Technology of generating hardware random numbers. Technical Meeting on Cyber Security of Nuclear Power Plant Instrumentation, Control and Information Systems, URL: http://entrac.iaea.org/I-and-C/TM_IDAHO_2006/CD/IAEA%20Day%202/14%20Onodera.pdf.

Oowada, I., Ariizumi, H., Li, M., Yoshimori, S., Uchida, A., Yoshimura, K., and Davis, P. (2009) Synchronization by injection of common chaotic signal in semiconductor lasers with optical feedback. *Optics Express*, Vol. **17** (No. 12), pp. 10025–10034.

Ortín, S., Jacquot, M., Pesquera, L., Peil, M., and Larger, L. (2009) Unmasking Chaotic Cryptosystems Based on Delayed Optoelectronic Feedback. Book of Abstracts for CHAOS 2009 conference, Chania, Crete, Greece, URL: http://www.chaos2009.net/proceedings/ABSTRACTS_PDF/Ortin_et_al-Unmasking_chaotic_cryptosystems_based_on_delayed_optoelectronic_feedback_ABSTRACT_CHAOS2009.pdf.

Otsuka, K. and Chern, J.-L. (1991) High-speed picosecond pulse generation in semiconductor lasers with incoherent optical feedback. *Optics Letters*, Vol. **16** (No. 22), pp. 1759–1761.

Otsuka, K. and Chern, J.-L. (1992) Dynamical spatial-pattern memory in globally coupled lasers. *Physical Review A*, Vol. **45** (No. 11), pp. 8288–8291.

Otsuka, K. (1999) *Nonlinear Dynamics in Optical Complex Systems*, KTK Scientific Publishers (Kluwer Academic Publishers), Tokyo.

Otsuka, K., Kawai, R., Hwong, S.-L., Ko, J.-Y., and Chern, J.-L. (2000) Synchronization of mutually coupled self-mixing modulated lasers. *Physical Review Letters*, Vol. **84** (No. 14), pp. 3049–3052.

Ott, E., Grebogi, C., and Yorke, J.A. (1990) Controlling chaos. *Physical Review Letters*, Vol. **64** (No. 11), pp. 1196–1199.

Ott, E. (1993) *Chaos in Dynamical Systems*, Cambridge University Press, Cambridge.

Ozaki, M., Someya, H., Mihara, T., Uchida, A., Yoshimori, S., Panajotov, K. and Sciamanna, M. (2009) Leader-laggard relationship of chaos synchronization in mutually coupled vertical-cavity surface-emitting lasers with time delay. *Physical Review E*, Vol. **79** (No. 2), pp. 026210.

Packard, N.H., Crutchfield, J.P., Farmer, J.D., and Shaw, R.S. (1980) Geometry from a time series. *Physical Review Letters*, Vol. **45** (No. 9), pp. 712–716.

Pappu, R., Recht, B., Taylor, J., and Gershenfeld, N. (2002) Physical one-way functions. *Science*, Vol. **297** (No. 5589), pp. 2026–2030.

Paul, J., Lee, M.W., and Shore, K.A. (2004) Effect of chaos pass filtering on message decoding quality using chaotic external-cavity laser diodes. *Optics Letters*, Vol. **29** (No. 21), pp. 2497–2499.

Pecora, L.M. and Carroll, T.L. (1990) Synchronization in chaotic systems. *Physical Review Letters*, Vol. **64** (No. 8), pp. 821–824.

Pecora, L.M. and Carroll, T.L. (1991) Driving systems with chaotic signals. *Physical Review A*, Vol. **44** (No. 4), pp. 2374–2383.

Peil, M., Heil, T., Fischer, I., and Elsäßer, W. (2002) Synchronization of chaotic semiconductor laser systems: a vectorial coupling-dependent scenario. *Physical Review Letters*, Vol. **88** (No. 17), pp. 174101.

Peil, M., Larger, L., and Fischer, I. (2007) Versatile and robust chaos synchronization phenomena imposed by delayed shared feedback coupling. *Physical Review E*, Vol. **76** (No. 4), pp. 045201(R).

Pérez, G. and Cerdeira, H.A. (1995) Extracting messages masked by chaos. *Physical Review Letters*, Vol. **74** (No. 11), pp. 1970–1973.

Pérez, T. and Uchida, A. (2011) Reliability and synchronization in a delay-coupled neuronal network with synaptic plasticity. *Physical Review E*, Vol. **83** (No. 6), 061915.

Petermann, K. (1988) *Laser Diode Modulation and Noise*, Kluwer Academic, Dordrecht.

PICASSO (2006) Photonic Integrated Components Applied to Secure chaoS encoded Optical communications systems (PICASSO) project. URL: http://picasso.di.uoa.gr/.

Pikovsky, A., Rosenblum, M., and Kurths, J. (2001) *Synchronization: A Universal Concept in Nonlinear Sciences*, Cambridge University Press, Cambridge.

Press, W.H., Teukolsky, S.A., Vetterling, W.T., and Flannery, B.P. (1992) *Numerical Recipes in C: The Art of Scientific Computing*, 2nd edn, Cambridge University Press, Cambridge.

Pyragas, K. (1992) Continuous control of chaos by self-controlling feedback. *Physics Letters A*, Vol. **170** (No. 6), pp. 421–428.

Pyragas, K. (1993) Predictable chaos in slightly perturbed unpredictable chaotic systems. *Physics Letters A*, Vol. **181** (No. 3), pp. 203–210.

Pyragas, K. (1998) Synchronization of coupled time-delay systems: Analytical estimations. *Physical Review E*, Vol. **58** (No. 3), pp. 3067–3071.

Qi, B., Chi, Y.-M., Lo, H.-K., and Qian, L. (2010) High-speed quantum random number generation by measuring phase noise of a single-mode laser. *Optics Letters*, Vol. **35** (No. 3), pp. 312–314.

Reidler, I., Aviad, Y., Rosenbluh, M., and Kanter, I. (2009) Ultrahigh-speed random number generation based on a chaotic semiconductor laser. *Physical Review Letters*, Vol. **103** (No. 2), pp. 024102.

Rizomiliotis, P., Bogris, A., and Syvridis D., (2010) Message origin authentication and integrity protection in chaos-based optical communication. *IEEE Journal of Quantum Electronics*, Vol. **46** (No. 3), pp. 377–383.

Rogers, E.A., Kalra, R., Schroll, R.D., Uchida, A., Lathrop, D.P., and Roy, R. (2004) Generalized synchronization of spatiotemporal chaos in a liquid crystal spatial light modulator. *Physical Review Letters*, Vol. **93** (No. 8), pp. 084101.

Rogister, F., Locquet, A., Pieroux, D., Sciamanna, M., Deparis, O., Mégret, P., and Blondel, M. (2001) Secure communication scheme using chaotic laser diodes subject to incoherent optical feedback and incoherent optical injection. *Optics Letters*, Vol. **26** (No. 19), pp. 1486–1488.

Rogister, F. (2001) Nonlinear dynamics of semiconductor lasers subject to optical feedback, Ph.D. Thesis, Faculté Polytechnique de Mons. URL: http://arxiv.org/ftp/arxiv/papers/0909/0909.4449.pdf.

Rontani, D., Locquet, A., Sciamanna, M., and Citrin, D.S. (2007) Loss of time-delay signature in the chaotic output of a semiconductor laser with optical feedback. *Optics Letters*, Vol. **32** (No. 20), pp. 2960–2962.

Rontani, D., Sciamanna, M., Locquet, A., and Citrin, D.S. (2009) Multiplexed encryption using chaotic systems with multiple stochastic-delayed feedbacks. *Physical Review E*, Vol. **80** (No. 6), pp. 066209.

Rontani, D., Locquet, A., Sciamanna, M., Citrin, D.S., and Uchida, A. (2011) Generation of orthogonal codes with chaotic optical systems. Optics Letters, Vol. **36** (No. 12), pp. 2287–2289.

Rosenblum, M.G., Pikovsky, A.S., and Kurths, J. (1996) Phase synchronization of chaotic oscillators. *Physical Review Letters*, Vol. **76** (No. 11), pp. 1804–1807.

Rosso, O.A., Vicente, R., and Mirasso, C.R. (2008) Encryption test of pseudo-aleatory messages embedded on chaotic laser signals: An information theory approach. *Physics Letters A*, Vol. **372** (No. 7), pp. 1018–1023.

Roy, R., Murphy, T.W. Jr., Maier, T.D., Gills, Z., and Hunt, E.R. (1992) Dynamical control of a chaotic laser: Experimental stabilization of a globally coupled system. *Physical Review Letters*, Vol. **68** (No. 9), pp. 1259–1262.

Roy, R., Gills, Z., and Thornburg, K.S. (1994) Controlling chaotic lasers. *Optics and Photonics News*, Vol. **5** (No. 5), pp. 8–15.

Roy, R. and Thornburg, K.S. Jr. (1994) Experimental synchronization of chaotic lasers. *Physical Review Letters*, Vol. **72** (No. 13), pp. 2009–2012.

Rukhin, A., Soto, J., Nechvatal, J., Smid, M., Barker, E., Leigh, S., Levenson, M., Vangel, M., Banks, D., Heckert, A., Dray, J., Vo, S, and Bassham III, L.E. (2010) *A statistical test suite for random and pseudorandom number generators for cryptographic applications*, National Institute of Standards and Technology (NIST), Special Publication 800-22, Revision 1a, URL: http://csrc.nist.gov/groups/ST/toolkit/rng/documents/SP800-22rev1a.pdf.

Rulkov, N.F., Sushchik, M.M., Tsimring, L.S., and Abarbanel, H.D.I. (1995) Generalized synchronization of chaos in directionally coupled chaotic systems. *Physical Review E*, Vol. **51** (No. 2), pp. 980–994.

Sacher, J., Elsässer, W., and Göbel, E.O. (1989) Intermittency in the coherence collapse of a semiconductor laser with external feedback. *Physical Review Letters*, Vol. **63** (No. 20), pp. 2224–2227.

Sacher, J., Baums, D., Panknin, P., Elsässer, W., and Göbel, E.O. (1992) Intensity instabilities of semiconductor lasers under current modulation, external light injection, and delayed feedback. *Physical Review A*, Vol. **45** (No. 3), pp. 1893–1905.

Sanchez, F., LeFlohic, M., Stephan, G.M., LeBoudec, P., and Francois, P.-L. (1995) Quasi-periodic route to chaos in erbium-doped fiber laser. *IEEE Journal of Quantum Electronics*, Vol. **31** (No. 3), pp. 481–488.

Sánchez-Díaz, A., Mirasso, C.R., Colet, P., and García-Fernández, P. (1999) Encoded Gbit/s digital communications with synchronized chaotic semiconductor lasers. *IEEE Journal of Quantum Electronics*, Vol. **35** (No. 3), pp. 292–297.

Sano, T. (1994) Antimode dynamics and chaotic itinerancy in the coherence collapse of semiconductor lasers with optical feedback. *Physical Review A*, Vol. **50** (No. 3), pp. 2719–2726.

Saucedo Solorio, J.M., Sukow, D.W., Hicks, D.R., and Gavrielides, A. (2002) Bifurcations in a semiconductor laser subject to delayed incoherent feedback. *Optics Communications*, Vol. **214** (No. 1–6), pp. 327–334.

Scheuer, J. and Yariv, A. (2006) Giant fiber lasers: a new paradigm for secure key distribution. *Physical Review Letters*, Vol. **97** (No. 14), pp. 140502.

Schuster, H.G. (ed.) (1999) *Handbook of Chaos Control*, Wiley-VCH, Weinheim.

Sciamanna, M., Masoller, C., Abraham, N.B., Rogister, F., Mégret, P., and Blondel, M. (2003) Different regimes of low-frequency fluctuations in vertical-cavity surface-emitting lasers. *Journal of the Optical Society of America B*, Vol. **20** (No. 1), pp. 37–44.

Sciamanna, M. and Panajotov, K. (2005) Two-mode injection locking in vertical-cavity surface-emitting lasers. *Optics Letters*, Vol. **30** (No. 21), pp. 2903–2905.

Sciamanna, M. and Panajotov, K. (2006) Route to polarization switching induced by optical injection in vertical-cavity surface-emitting lasers. *Physical Review A*, Vol. **73** (No. 2), pp. 023811.

Sciré, A., Mulet, J., Mirasso, C.R., Danckaert, J., and San Miguel, M. (2003) Polarization message encoding through vectorial chaos synchronization in vertical-cavity surface-emitting lasers. *Physical Review Letters*, Vol. 90 (No. 11), pp. 113901.

Shannon, C.E. (1949) Communication theory of secrecy systems. *Bell System Technical Journal*, Vol. 28 (No. 4), pp. 656–715.

Shen, Y., Tian, L., and Zou, H. (2010) Practical quantum random number generator based on measuring the shot noise of vacuum states. *Physical Review A*, Vol. 81 (No. 6), pp. 063814.

Shibasaki, N., Uchida, A., Yoshimori, S., and Davis, P. (2006) Characteristics of chaos synchronization in semiconductor lasers subject to polarization-rotated optical feedback. *IEEE Journal of Quantum Electronics*, Vol. 42 (No. 3), pp. 342–350.

Short, K.M. (1994) Steps toward unmasking secure communications. *International Journal of Bifurcation and Chaos*, Vol. 4 (No. 4), pp. 959–977.

Short, K.M. (1996) Unmasking a modulated chaotic communications scheme. *International Journal of Bifurcation and Chaos*, Vol. 6 (No. 2), pp. 367–375.

Siegman, A.E. (1986) *Lasers*, University Science Books, Sausalito, California.

Simpson, T.B., Liu, J.M., and Gavrielides, A. (1995) Bandwidth enhancement and broadband noise reduction in injection-locked semiconductor lasers. *IEEE Photonics Technology Letters*, Vol. 7 (No. 7), pp. 709–711.

Simpson, T.B., Liu, J.M., Huang, K.F., and Tai, K. (1997) Nonlinear dynamics induced by external optical injection in semiconductor lasers. *Quantum and Semiclassical Optics*, Vol. 9 (No. 5), pp. 765–784.

Singh, S. (2000) *The Code Book: The Science of Secrecy from Ancient Egypt to Quantum Cryptography*, Anchor Books, New York.

Sivaprakasam, S. and Shore, K.A. (1999a) Demonstration of optical synchronization of chaotic external-cavity laser diodes. *Optics Letters*, Vol. 24 (No. 7), pp. 466–468.

Sivaprakasam, S. and Shore, K.A. (1999b) Signal masking for chaotic optical communication using external-cavity diode lasers. *Optics Letters*, Vol. 24 (No. 17), pp. 1200–1202.

Someya, H., Oowada, I., Okumura, H., Kida, T., and Uchida, A. (2009) Synchronization of bandwidth-enhanced chaos in semiconductor lasers with optical feedback and injection. *Optics Express*, Vol. 17 (No. 22), pp. 19536–19543.

Soriano, M.C., Colet, P., and Mirasso, C.R. (2009) Security implications of open- and closed-loop receivers in all-optical chaos-based communications. *IEEE Photonics Technology Letters*, Vol. 21 (No. 7), pp. 426–428.

Spencer, P.S., Mirasso, C.R., Colet, P., and Shore, K.A. (1998) Modeling of optical synchronization of chaotic external-cavity VCSELs. *IEEE Journal of Quantum Electronics*, Vol. 34 (No. 9), pp. 1673–1679.

Srinivasan, S., Mathew, S., Erraguntla, V., and Krishnamurthy, R. (2009) A 4Gbps 0.57pJ/bit process-voltage-temperature variation tolerant all-digital true random number generator in 45nm CMOS. 2009 22nd International Conference on VLSI Design, pp. 301–306.

Stipčević, M. and Medved Rogina, B. (2007) Quantum random number generator based on photonic emission in semiconductors. *Review of Scientific Instruments*, Vol. 78 (No. 4), pp. 045104.

Strogatz, S.H. (1994) *Nonlinear Dynamics and Chaos: With Applications to Physics, Biology, Chemistry, and Engineering*, Westview Press.

Strogatz, S. (2003) *SYNC: The Emerging Science of Spontaneous Order*, Theia books (Hyperion books), New York.

Sugawara, T., Tachikawa, M., Tsukamoto, T., and Shimizu, T. (1994) Observation of synchronization in laser chaos. *Physical Review Letters*, Vol. 72 (No. 22), pp. 3502–3505.

Sukow, D.W., Heil, T., Fischer, I., Gavrielides, A., Hohl-AbiChedid, A., and Elsäßer, W. (1999) Picosecond intensity statistics of semiconductor lasers operating in the low-frequency fluctuation regime. *Physical Review A*, Vol. 60 (No. 1), pp. 667–673.

Sukow, D.W., Blackburn, K.L., Spain, A.R., Babcock, K.J., Bennett, J.V., and Gavrielides, A. (2004) Experimental synchronization of chaos in diode lasers with polarization-rotated feedback and injection. *Optics Letters*, Vol. 29 (No. 20), pp. 2393–2395.

Sunada, S., Harayama, T., Arai, K., Yoshimura, K., Davis, P., Tsuzuki, K., and Uchida, A. (2011) Chaos laser chips with delayed optical feedback using a passive ring waveguide. *Optics Express*, Vol. 19 (No. 7), pp. 5713–5724.

Sweet, D., Ott, E., and Yorke, J.A. (1999) Topology in chaotic scattering. *Nature*, Vol. **399**, pp. 315–316.

Tabaka, A., Peil, M., Sciamanna, M., Fischer, I., Elsäßer, W., Thienpont, H., Veretennicoff, I., and Panajotov, K. (2006) Dynamics of vertical-cavity surface-emitting lasers in the short external cavity regime: Pulse packages and polarization mode competition. *Physical Review A*, Vol. **73** (No. 1), pp. 013810.

Takeda, M.W., Kirihara, S., Miyamoto, Y., Sakoda, K., and Honda, K. (2004) Localization of electromagnetic waves in three-dimensional fractal cavities. *Physical Review Letters*, Vol. **92** (No. 9), pp. 093902.

Takens, F. (1981) Detecting strange attractors in turbulence, in *Dynamical Systems and Turbulence*, Lecture Notes in Mathematics, Vol. **898** (eds D.A. Rand and L.-S. Young) Springer-Verlag, Berlin, pp. 366–381.

Takesue, H., Nam, S.W., Zhang, Q., Hadfield, R.H., Honjo, T., Tamaki, K., and Yamamoto, Y. (2007) Quantum key distribution over a 40-dB channel loss using superconducting single-photon detectors. *Nature Photonics*, Vol. **1**, pp. 343–348.

Takeuchi, Y., Shogenji, R., and Ohtsubo, J. (2008) Chaotic dynamics in semiconductor lasers subjected to polarization-rotated optical feedback. *Applied Physics Letters*, Vol. **93** (No. 18), pp. 181105.

Takiguchi, Y., Fujino, H., and Ohtsubo, J. (1999) Experimental synchronization of chaotic oscillations in externally injected semiconductor lasers in a low-frequency fluctuation regime. *Optics Letters*, Vol. **24** (No. 22), pp. 1570–1572.

Takiguchi, Y., Ohyagi, K., and Ohtsubo, J. (2003) Bandwidth-enhanced chaos synchronization in strongly injection-locked semiconductor lasers with optical feedback. *Optics Letters*, Vol. **28** (No. 5), pp. 319–321.

Tamura, Y., Onodera, T., Nakahata, M., and Shimizu, T. (2006) On physical number generators in Japan. *Journal of the Japan Statistical Society*, Vol. **35** (No. 2), pp. 201–212 (in Japanese).

Tang, C.L., Statz, H., and deMars, G. (1963) Spectral output and spiking behavior of solid-state lasers. *Journal of Applied Physics*, Vol. **34** (No. 8), pp. 2289–2295.

Tang, X., van der Ziel, J.P., Chang, B., Johnson, R., and Tatum, J.A. (1997) Observation of bistability in GaAs quantum-well vertical-cavity surface-emitting lasers. *IEEE Journal of Quantum Electronics*, Vol. **33** (No. 6), pp. 927–932.

Tang, D.Y., Dykstra, R., Hamilton, M.W., and Heckenberg, N.R. (1998) Observation of generalized synchronization of chaos in a driven chaotic system. *Physical Review E*, Vol. **57** (No. 5), pp. 5247–5251.

Tang, S. and Liu, J.M. (2001a) Chaotic pulsing and quasi-periodic route to chaos in a semiconductor laser with delayed opto-electronic feedback. *IEEE Journal of Quantum Electronics*, Vol. **37** (No. 3), pp. 329–336.

Tang, S. and Liu, J.M. (2001b) Synchronization of high-frequency chaotic optical pulses. *Optics Letters*, Vol. **26** (No. 9), pp. 596–598.

Tang, S. and Liu, J.M. (2001c) Message encoding-decoding at 2.5 Gbits/s through synchronization of chaotic pulsing semiconductor lasers. *Optics Letters*, Vol. **26** (No. 23), pp. 1843–1845.

Tang, S. and Liu, J.M. (2003) Experimental verification of anticipated and retarded synchronization in chaotic semiconductor lasers. *Physical Review Letters*, Vol. **90** (No. 19), pp. 194101.

Tateda, M. and Horiguchi, T. (1989) Advances in optical time domain reflectometry. *Journal of Lightwave Technology*, Vol. **7** (No. 8), pp. 1217–1224.

Taylor, J.D. (2000) *Ultra-Wideband Radar Technology*, CRC Press, New York.

Terry, J.R. and VanWiggeren, G.D. (2001) Chaotic communication using generalized synchronization. *Chaos, Solitons and Fractals*, Vol. **12** (No. 1), pp. 145–152.

Tkach, R. and Chraplyvy, A. (1986) Regimes of feedback effects in 1.5-μm distributed feedback lasers. *Journal of Lightwave Technology*, Vol. **4** (No. 11), pp. 1655–1661.

Tokunaga, C., Blaauw, D., and Mudge, T. (2008) True random number generator with a metastability-based quality control. *IEEE Journal of Solid-State Circuits*, Vol. **43** (No. 1), pp. 78–85.

Toshiba (2008) Toshiba Press Release: World's highest performance physical random-number generator circuit. URL: http://www.toshiba.co.jp/about/press/2008_02/pr0702.htm.

Tredicce, J.R., Arecchi, F.T., Lippi, G.L., and Puccioni, G.P. (1985) Instabilities in lasers with an injected signal. *Journal of Optical Society of America B*, Vol. **2** (No. 1), pp. 173–183.

Tredicce, J.R., Arecchi, F.T., Puccioni, G.P., Poggi, A., and Gadomski, W. (1986) Dynamic behavior and onset of low-dimensional chaos in a modulated homogeneously broadened single-mode laser: Experiments and theory. *Physical Review A*, Vol. **34** (No. 3), pp. 2073–2081.

Uchida, A., Ogawa, T., and Kannari, F. (1998a) Synchronization of chaos in two microchip lasers: Comparison between two master-slave types and a mutually-coupled type. *Japanese Journal of Applied Physics*, Vol. **37**, Part 2, pp. L730–L732.

Uchida, A., Sato, T., Ogawa, T., and Kannari, F. (1998b) Nonfeedback control of chaos in a microchip solid-state laser by internal frequency resonance. *Physical Review E*, Vol. **58** (No. 6), pp. 7249–7255.

Uchida, A., Ogawa, T., Shinozuka, M., and Kannari, F. (2000) Accuracy of chaos synchronization in Nd:YVO$_4$ microchip lasers. *Physical Review E*, Vol. **62** (No. 2), pp. 1960–1971.

Uchida, A., Liu, Y., Fischer, I., Davis, P., and Aida, T. (2001a) Chaotic antiphase dynamics and synchronization in multimode semiconductor lasers. *Physical Review A*, Vol. **64** (No. 2), pp. 023801.

Uchida, A., Yoshimori, S., Shinozuka, M., Ogawa, T., and Kannari, F. (2001b) Chaotic on-off keying for secure communications. *Optics Letters*, Vol. **26** (No. 12), pp. 866–868.

Uchida, A., Davis, P., and Itaya, S. (2003a) Generation of information theoretic secure keys using a chaotic semiconductor laser. *Applied Physics Letters*, Vol. **83** (No. 15), pp. 3213–3215.

Uchida, A., Heil, T., Liu, Y., Davis, P., and Aida, T. (2003b) High-frequency broad-band signal generation using a semiconductor laser with a chaotic optical injection. *IEEE Journal of Quantum Electronics*, Vol. **39** (No. 11), pp. 1462–1467.

Uchida, A., Kinugawa, S., Matsuura, T., and Yoshimori, S. (2003c) Dual synchronization of chaos in one-way coupled microchip lasers. *Physical Review E*, Vol. **67** (No. 2), pp. 026220.

Uchida, A., Liu, Y., and Davis, P. (2003d) Characteristics of chaotic masking in synchronized semiconductor lasers. *IEEE Journal of Quantum Electronics*, Vol. **39** (No. 8), pp. 963–970.

Uchida, A., McAllister, R., Meucci, R., and Roy, R. (2003e) Generalized synchronization of chaos in identical systems with hidden degrees of freedom. *Physical Review Letters*, Vol. **91** (No. 17), pp. 174101.

Uchida, A., McAllister, R., and Roy, R. (2004a) Consistency of nonlinear system response to complex drive signals. *Physical Review Letters*, Vol. **93** (No. 24), pp. 244102.

Uchida, A., Shibasaki, N., Nogawa, S., and Yoshimori, S. (2004b) Transient characteristics of chaos synchronization in a semiconductor laser subject to optical feedback. *Physical Review E*, Vol. **69** (No. 5), pp. 056201.

Uchida, A., Rogister, F., García-Ojalvo, J., and Roy, R. (2005) Synchronization and communication with chaotic laser systems, *Progress in Optics*, Vol. **48**, Chapter 5 (ed E. Wolf) Elsevier, The Netherlands, pp. 203–341.

Uchida, A., Amano, K., Inoue, M., Hirano, K., Naito, S., Someya, H., Oowada, I., Kurashige, T., Shiki, M., Yoshimori, S., Yoshimura, K., and Davis, P. (2008a) Fast physical random bit generation with chaotic semiconductor lasers. *Nature Photonics*, Vol. **2** (No. 12), pp. 728–732.

Uchida, A., Yoshimura, K., Davis, P., Yoshimori, S., and Roy, R. (2008b) Local conditional Lyapunov exponent characterization of consistency of dynamical response of the driven Lorenz system. *Physical Review E*, Vol. **78** (No. 3), pp. 036203.

Umeno, K. (2007) Independent component analysis of mixed chaotic signals for communications systems. *Nonlinear Phenomena in Complex Systems*, Vol. **10** (No. 2), pp. 170–175.

van der Lubbe, J.C.A. (1998) *Basic Method of Cryptography*, Cambridge University Press, Cambridge.

van Tartwijk, G.H.M. and Agrawal, G.P. (1998) Laser instabilities: a modern perspective. *Progress in Quantum Electronics*, Vol. **22** (No. 2), pp. 43–122.

VanWiggeren, G.D. and Roy, R. (1998a) Communication with chaotic lasers. *Science*, Vol. **279** (No. 5354), pp. 1198–1200.

VanWiggeren, G.D. and Roy, R. (1998b) Optical communication with chaotic waveforms. *Physical Review Letters*, Vol. **81** (No. 16), pp. 3547–3550.

VanWiggeren, G.D. and Roy, R. (1999) Chaotic communication using time-delayed optical systems. *International Journal of Bifurcation and Chaos*, Vol. **9** (No. 11), pp. 2129–2156.

VanWiggeren, G.D. and Roy, R. (2002) Communication with dynamically fluctuating states of light polarization. *Physical Review Letters*, Vol. **88** (No. 9), pp. 097903.

Vicente, R., Pérez, T., and Mirasso, C.R. (2002) Open-versus closed-loop performance of synchronized chaotic external-cavity semiconductor lasers. *IEEE Journal of Quantum Electronics*, Vol. **38** (No. 9), pp. 1197–1204.

Vicente, R., Daudén, J., Colet, P., and Toral, R. (2005) Analysis and characterization of the hyperchaos generated by a semiconductor laser subject to a delayed feedback loop. *IEEE Journal of Quantum Electronics*, Vol. **41** (No. 4), pp. 541–548.

Vicente, R., Mirasso, C.R., and Fischer, I. (2007) Simultaneous bidirectional message transmission in a chaos-based communication scheme. *Optics Letters*, Vol. **32** (No. 4), pp. 403–405.

Volkovskii, A.R. and Rul'kov, N. (1993) Synchronous chaotic response of a nonlinear oscillator system as a principle for the detection of the information component of chaos. *Technical Physics Letters*, Vol. **19** (No. 2), pp. 97–99.

Volodchenko, K.V., Ivanov, V.N., Gong, S.-H., Choi, M., Park, Y.-J., and Kim, C.-M. (2001) Phase synchronization in coupled Nd:YAG lasers. *Optics Letters*, Vol. **26** (No. 18), pp. 1406–1408.

von Neumann, J. (1963) *Various techniques for use in connection with random digits* (The Collected Works of John von Neumann), Pergamon, London, pp. 768–770.

Voss, H.U. (2000) Anticipating chaotic synchronization. *Physical Review E*, Vol. **61** (No. 5), pp. 5115–5119.

Wang, A., Wang, Y., and He, H. (2008a) Enhancing the bandwidth of the optical chaotic signal generated by a semiconductor laser with optical feedback. *IEEE Photonics Technology Letters*, Vol. **20** (No. 19), pp. 1633-1635.

Wang, Y., Wang, B., and Wang, A. (2008b) Chaotic correlation optical time domain reflectometer utilizing laser diode. *IEEE Photonics Technology Letters*, Vol. **20** (No. 19), pp. 1636–1638.

Wayner, P. (2009) *Disappearing Cryptography: Information Hiding: Steganography & Watermarking*, 3rd edn, Morgan Kaufmann Publishers, Burlington.

Weiss, C.O., Godone, A., and Olafsson, A. (1983) Routes to chaotic emission in a cw He-Ne laser. *Physical Review A*, Vol. **28** (No. 2), pp. 892–895.

Weiss, C.O., Klische, W., Ering, P.S., and Cooper, M. (1985) Instabilities and chaos of a single mode NH_3 ring laser. *Optics Communications*, Vol. **52** (No. 6), pp. 405–408.

Weiss, C.O. and Vilaseca, R. (1991) *Dynamics of Lasers*, VCH Publishers, Weinheim.

Welsh, D. (1988) *Codes and Cryptography*, Oxford Science Publications, Oxford.

White, J.K. and Moloney, J.V. (1999) Multichannel communication using an infinite dimensional spatiotemporal chaotic system. *Physical Review A*, Vol. **59** (No. 3), pp. 2422–2426.

Wieczorek, S., Krauskopf, B., and Lenstra, D. (2000) Mechanisms for multistability in a semiconductor laser with optical injection. *Optics Communications*, Vol. **183** (No. 1–4), pp. 215–226.

Williams, Q.L. and Roy, R. (1996) Fast polarization dynamics of an erbium-doped fiber ring laser. *Optics Letters*, Vol. **21** (No. 18), pp. 1478–1480.

Williams, Q.L., García-Ojalvo, J., and Roy, R. (1997) Fast intracavity polarization dynamics of an erbium-doped fiber ring laser: Inclusion of stochastic effects. *Physical Review A*, Vol. **55** (No. 3), pp. 2376–2386.

Williams, C.R.S., Salevan, J.C., Li, X., Roy, R., and Murphy, T.E. (2010) Fast physical random number generator using amplified spontaneous emission. *Optics Express*, Vol. **18** (No. 23), pp. 23584–23597.

Wilson, S.G. (1996) *Digital Modulation and Coding*, Prentice-Hall, New Jersey.

Winful, H.G. and Rahman, L. (1990) Synchronized chaos and spatiotemporal chaos in arrays of coupled lasers. *Physical Review Letters*, Vol. **65** (No. 13), pp. 1575–1578.

Wu, J.-G., Xia, G.-Q., and Wu, Z.-M. (2009) Suppression of time delay signatures of chaotic output in a semiconductor laser with double optical feedback. *Optics Express*, Vol. **17** (No. 22), pp. 20124–20133.

Wünsche, H.-J., Bauer, S., Kreissl, J., Ushakov, O., Korneyev, N., Henneberger, F., Wille, E., Erzgräber, H., Peil, M., Elsäßer, W., and Fischer, I. (2005) Synchronization of delay-coupled oscillators: A study of semiconductor lasers. *Physical Review Letters*, Vol. **94** (No. 16), pp. 163901.

Wyner, A.D. (1975) The wire-tap channel. *Bell System Technical Journal*, Vol. **54** (No. 8), pp. 1355–1387.

Yalçın, M.E., Suykens, J.A.K., and Vandewalle, J. (2004) True random bit generation from a double-scroll attractor. *IEEE Transactions on Circuits and Systems I: Regular Papers*, Vol. **51** (No. 7), pp. 1395–1404.

Yamada, M. and Higashi, T. (1991) Mechanism of the noise reduction method by superposition of high-frequency current for semiconductor injection lasers. *IEEE Journal of Quantum Electronics*, Vol. **27** (No. 3), pp. 380–388.

Yamamoto, T., Oowada, I., Yip, H., Uchida, A., Yoshimori, S., Yoshimura, K., Muramatsu, J., Goto, S., and Davis, P. (2007) Common-chaotic-signal induced synchronization in semiconductor lasers. *Optics Express*, Vol. **15** (No. 7), pp. 3974–3980.

Yariv, A. (1991) *Optical Electronics*, 4th edn, Saunders College Publishing, Fort Worth.

Ye, J., Li, H., and McInerney, J.G. (1993) Period-doubling route to chaos in a semiconductor laser with weak optical feedback. *Physical Review A*, Vol. **47** (No. 3), pp. 2249–2252.

Yoshimura, K. (1999) Multichannel digital communications by the synchronization of globally coupled chaotic systems. *Physical Review E*, Vol. **60** (No. 2), pp. 1648–1657.

Yoshimura, K. (2004) Secure communications using cascaded chaotic optical rings. *International Journal of Bifurcation and Chaos*, Vol. **14** (No. 3), pp. 1105–1113.

Yousefi, M., Barbarin, Y., Beri, S., Bente, E.A.J.M., Smit, M.K., Nötzel, R., and Lenstra, D. (2007) New role for nonlinear dynamics and chaos in integrated semiconductor laser technology. *Physical Review Letters*, Vol. **98** (No. 4), pp. 044101.

Zadok, A., Scheuer, J., Sendowski, J., and Yariv, A. (2008) Secure key generation using an ultra-long fiber laser: Transient analysis and experiment. *Optics Express*, Vol. **16** (No. 21), pp. 16680–16690.

Zhou, C.S., Kurths, J., Allaria, E., Boccaletti, S., Meucci, R., and Arecchi, F.T. (2003) Constructive effects of noise in homoclinic chaotic systems. *Physical Review E*, Vol. **67** (No. 6), pp. 066220.

Zoboli, M. and Bassi, P. (1983) High spatial resolution OTDR attenuation measurements by a correlation technique. *Applied Optics*, Vol. **22** (No. 23), pp. 3680–3681.

Zunino, L., Soriano, M.C., Figliola, A., Pérez, D.G., Garavaglia, M., Mirasso, C.R., and Rosso, O.A. (2009) Performance of encryption schemes in chaotic optical communication: A multifractal approach. *Optics Communications*, Vol. **282** (No. 23), pp. 4587–4594.

Zunino, L., Soriano, M.C., Fischer, I., Rosso, O.A., and Mirasso, C.R. (2010) Permutation-information-theory approach to unveil delay dynamics from time-series analysis. *Physical Review E*, Vol. **82** (No. 4), pp. 046212.

Glossary

G.1 List of Acronyms

G.1.1 Acronyms of Technical Terms

Each frequently used acronym is defined the first time it appears in a chapter. As a further help, all acronyms are listed here in alphabetical order.

AC	alternating current
ADC	analog-to-digital converter
AM	amplitude modulation
AOM	acousto-optic modulator
APD	avalanche photodiode
AR	antireflection
Ar^+	argon ion
ASE	amplified spontaneous emission
ASK	amplitude shift keying
BAL	broad-area semiconductor laser
BD	blu-ray disk
BER	bit error rate
BS	beam splitter
CCOTDR	chaotic correlation optical time-domain reflectometer
CD	compact disk
CDMA	code division multiple access
CH_3I	iodomethane (methyl iodide)
CLIDAR	chaotic light detection and ranging
CO_2	carbon dioxide
COTDR	correlation optical time-domain reflectometer
CPMLL	colliding-pulse mode-locked laser
CPU	central processing unit
CRADAR	chaotic radio detection and ranging
CW	continuous wave
CWDM	chaotic wavelength division multiplexing

Optical Communication with Chaotic Lasers: Applications of Nonlinear Dynamics and Synchronization, First Edition. Atsushi Uchida.
© 2012 Wiley-VCH Verlag GmbH & Co. KGaA. Published 2012 by Wiley-VCH Verlag GmbH & Co. KGaA.

DAC	digital-to-analog converter
DBR	distributed Bragg reflector
DC	direct current
DCF	dispersion compensation fiber
DFB	distributed feedback
DPSK	differential phase-shift keying
DPSK-d	differential phase-shift keying demodulator
DPS-QKD	differential phase-shift quantum-key distribution
DRAM	dynamic random access memory
DSF	dispersion shifted fiber
DVD	digital versatile disk
EDFA	erbium-doped fiber amplifier
EDFRL	erbium-doped fiber ring laser
EEG	electroencephalogram
EO	electro-optic
EOM	electro-optic modulator
Er	erbium
FA	fiber attenuator
FBG	fiber Bragg grating
FC	fiber coupler
FEC	forward error collection
FIPS	federal information processing standards publication
FM	frequency modulation
FPGA	field programmable gate array
FSK	frequency shift keying
FSR	free spectral range
FWHM	full-width half-maximum
GaAs	gallium arsenide
GAS	gain absorption section
GFSR	generalized feedback shift register
HFI	high-frequency injection
HRC	highly reflective coated
IC	integrated circuit
ICA	independent component analysis
InGaAs	indium gallium arsenide
InGaAsP	indium gallium arsenide phosphide
InP	indium phosphide
ISO	optical isolator or international standard organization
ITD	integrated tandem device
ITS	intelligent transport system
KTP	potassium titanyl phosphate
KS entropy	Kolmogorov–Sinai entropy
KS test	Kolmogorov–Smirnov test
KY	Kaplan–Yorke
LASER	light amplification by stimulated emission of radiation

LD	laser diode
LED	light emitting diode
LFF	low-frequency fluctuation
LFSR	linear feedback shift register
L-I	light power versus injection current
LIDAR	light detection and ranging
$LiNbO_3$	lithium niobate
$LiNdP_4O_{12}$	lithium neodymium tetraphosphate
LNP	lithium neodymium tetraphosphate
LSB	least significant bit
M	mirror
MCPF	mutual chaos pass filtering
MIMO	multiple-input multiple-output
MPR	Martín, Plastino, and Rosso
MSB	most significant bit
M sequence	maximum-length linearly recurring sequence
MT	Mersenne Twister
MZ	Mach–Zehnder
MZI	Mach–Zehnder interferometer
Nd	neodymium
NdP_5O_{14}	neodymium pentaphosphate
Nd:YAG	neodymium-doped yttrium aluminum garnet
Nd:YVO_4	neodymium-doped yttrium orthovanadate
NDF	neutral density filter
NH_3	ammonia
NIST	National Institute of Standards and Technology in USA
NRZ	nonreturn-to-zero
OC	optical circulator
OGY	Ott, Grebogi, and Yorke
OFB	optical feedback
OOK	on-off keying
OPF	occasional proportional feedback
OTDR	optical time-domain reflectometer
PC	polarization controller or personal computer
PCI	peripheral component interconnect
PD	photodetector or photodiode
PDF	probability density function
PH	phase section
PIC	photonic integrated circuit
PM	phase modulator or phase modulation or polarization maintaining
PN	pseudorandom number
PRBS	pseudorandom bit sequence
PRNG	pseudorandom number generator
PSK	phase shift keying
QBER	quantum bit error rate

Q factor	quality factor
QKD	quantum key distribution
QP	quasiperiodic oscillation
QRNG	quantum random number generator
RADAR	radio detection and ranging
RF	radio frequency
RIN	relative intensity noise
RMS	root mean square
RNG	random number generator
RO	relaxation oscillation
RS	Reed–Solomon
RSA	Rivest, Shamir, and Adleman
RSR	random-signal radio-detection and ranging
RZ	return-to-zero
SMF	single-mode fiber
SHG	second harmonic generation
SL	semiconductor laser
SNR	signal-to-noise ratio
SOA	semiconductor optical amplifier
SP	special publication
SVEA	slowly varying envelope approximation
TE	transverse electric
TIA	time interval analyzer
TM	transverse magnetic
UFL	ultralong fiber laser
UPO	unstable periodic orbit
USB	universal serial bus
VA	variable attenuator
VCSEL	vertical-cavity surface-emitting laser
VFR	variable fiber reflector
VOA	variable optical attenuator
WDM	wavelength division multiplexing
XOR	exclusive OR
YAG	yttrium aluminum garnet
Yb	ytterbium
1D	one-dimensional
2D	two-dimensional
3D	three-dimensional

G.1.2 Acronyms of Units

All acronyms for units used in the text are listed in the alphabetical order.

A	ampere
b/s	bit per second
cm	centimeter (10^{-2} m)

dB	decibel
dBm	decibel referenced to one milliwatt (0 dBm = 1 mW)
dB/km	decibel per kilometer
Gb/s	gigabit per second (10^9 b/s)
GHz	gigahertz (10^9 Hz)
GS/s	gigasample per second (10^9 sample/s)
Hz	hertz (s^{-1})
kb/s	kilobit per second (10^3 b/s)
kHz	kilohertz (10^3 Hz)
m	meter
Mb/s	megabit per second (10^6 b/s)
MByte/s	megabyte per second (10^6 Byte/s)
MHz	megahertz (10^6 Hz)
mm	millimeter (10^{-3} m)
ms	millisecond (10^{-3} s)
mW	milliwatt (10^{-3} W)
μm	micrometer (10^{-6} m)
μs	microsecond (10^{-6} s)
nm	nanometer (10^{-9} m)
ns	nanosecond (10^{-9} s)
ns^{-1}	nanosecond inverse ($10^9 s^{-1}$)
pm	picometer (10^{-12} m)
ps	picosecond (10^{-12} s)
s	second
V	volt

G.2 Source Codes of C Programming Language for Numerical Simulations

Readers may use the following source codes of the C programming language for numerical simulations as a start of numerical works to reproduce the original results and figures shown in this book. It is important to use proper initial conditions as well as proper parameter values to avoid divergence of numerical integration. "//" in the source codes indicates a comment and the sentence after "//" can be ignored. The C source codes listed in this glossary were provided by Shinichiro Morikatsu.

G.2.1 Logistic Map (Chapter 2)

G.2.1.1 C Source Code for Sequence of Logistic Map (Figure 2.5a)

```
#include<stdio.h>

double logistic(double a, double x)
```

```c
{
  return a * x * (1.0 - x);
}

int main()
{
  int i;
  int n = 50;

  double a = 4.0;
  double x = 0.314;

  for(i = 0; i < n; i++){
    printf("%d\t%e\n", i, x);
    x = logistic(a, x);
  }
  return 0;
}
```

G.2.1.2 C Source Code for Bifurcation Diagram of Logistic Map (Figure 2.8)

```c
#include<stdio.h>

double logistic(double a, double x)
{
  return a * x * (1.0 - x);
}

int main()
{
  int i, j;
  int n = 50;
  int transient = 100;

  double a, x;
  double x0 = 0.314;

  // range for parameter change
  double aMin = 0.0;
  double aMax = 4.0;
  int aNum = 400;
  double aDiv = (aMax - aMin) / aNum;
```

```c
  for(i = 0; i < aNum; i++){
    a = aMin + aDiv * i; // parameter change
    x = x0;

    // calculation for transient
    for(j = 0; j < transient; j++){
      x = logistic(a, x);
    }

    for(j = 0; j < n; j++){
      printf("%e\t%e\n", a, x);
      x = logistic(a, x);
    }
  }
  return 0;
}
```

G.2.1.3 C Source Code for Lyapunov Exponent of Logistic Map (Figure 2.9)

```c
#include<stdio.h>
#include<math.h>

double logistic(double a, double x)
{
  return a * x * (1.0 - x);
}

double derivativeLogistic(double a, double x)
{
  return a * (1.0 - 2.0 * x);
}

int main()
{
  int i, j;
  int n = 10000;
  int transient = 10000;

  double a, x, lyapunov;
  double x0 = 0.314;

  // range for parameter change
  double aMin = 0.0;
```

```c
        double aMax = 4.0;
        int aNum = 400;
        double aDiv = (aMax - aMin) / aNum;

        for (i = 0; i < aNum; i++) {
          a = aMin + aDiv * i; // parameter change
          x = x0;

          // calculation for transient
          for (j = 0; j < transient; j++) {
            x = logistic(a, x);
          }

          lyapunov = 0.0;
          for (j = 0; j < n; j++) {
            lyapunov +=
              log(fabs(derivativeLogistic(a, x)));
            x = logistic(a, x);
          }
          printf("%e\t%e\n", a, lyapunov / n);
        }
        return 0;
      }
```

G.2.2 Lorenz Euations (Chapter 2)

G.2.2.1 C Source Code for Time Series of Lorenz Equations (Figure 2.10)

```c
        #include<stdio.h>

        #define M 3 // number of equations

        double sigma = 10.0;
        double gam = 28.0;
        double beta = 8.0 / 3.0;

        void lorenz(double x[], double b[])
        {
          b[0] = sigma * (x[1] - x[0]);
          b[1] = - x[0] * x[2] + gam * x[0] - x[1];
          b[2] = x[0] * x[1] - beta * x[2];
        }
```

```c
void rungeKutta(double a[], double h)
{
 int i, j;
 double x[M], b[4][M];
 for(i = 0; i < 4; i++){
  for(j = 0; j < M; j++){
   if(i == 0) x[j] = a[j];
   if(i == 1) x[j] = a[j] + h * b[0][j] / 2.0;
   if(i == 2) x[j] = a[j] + h * b[1][j] / 2.0;
   if(i == 3) x[j] = a[j] + h * b[2][j];
  }
  lorenz(x, b[i]);
 }
 for(i = 0; i < M; i++){
  a[i] += h * (b[0][i] + 2.0 * b[1][i]
        + 2.0 * b[2][i] + b[3][i]) / 6.0;
 }
}

int main()
{
 int i;
 double a[M];

 double h = 1.0e-3; // calculation step
 double transient = 100.0; // transient time
 double tMax = 50.0; // plot time

 int div = 10; // plot interval

 int trans = (int)(transient / h);
 int n = (int)(tMax / h);

 // Initial conditions
 a[0] = 15.0; // x
 a[1] = 20.0; // y
 a[2] = 30.0; // z

 // calculation for transient
 for(i = 0; i < trans; i++){
  rungeKutta(a, h);
 }

 for(i = 0; i < n; i++){
  if(i % div == 0){
```

```c
      printf("%e\t%e\t%e\t%e\n", h * i,
         a[0], a[1], a[2]);
   }
   rungeKutta(a, h);
  }
  return 0;
}
```

G.2.2.2 C Source Code for Bifurcation Diagram of Lorenz Equations (Figure 2.12a)

```c
#include<stdio.h>

#define M 3  // number of equations

double sigma = 10.0;
double gam = 28.0;
double beta = 8.0 / 3.0;

void lorenz(double x[], double b[])
{
 b[0] = sigma * (x[1] - x[0]);
 b[1] = - x[0] * x[2] + gam * x[0] - x[1];
 b[2] = x[0] * x[1] - beta * x[2];
}

void rungeKutta(double a[], double h)
{
 int i, j;
 double x[M], b[4][M];
 for(i = 0; i < 4; i++){
  for(j = 0; j < M; j++){
   if(i == 0) x[j] = a[j];
   if(i == 1) x[j] = a[j] + h * b[0][j] / 2.0;
   if(i == 2) x[j] = a[j] + h * b[1][j] / 2.0;
   if(i == 3) x[j] = a[j] + h * b[2][j];
  }
  lorenz(x, b[i]);
 }
 for(i = 0; i < M; i++){
  a[i] += h * (b[0][i] + 2.0 * b[1][i]
        + 2.0 * b[2][i] + b[3][i]) / 6.0;
 }
}
```

```c
int main()
{
 int i, p;
 double a[M];

 double h = 1.0e-3; // calculation step
 double transient = 100.0; // transient time
 double tMax = 100.0; // plot time

 int plotCnt, plotMax = 50;

 int n = (int)(tMax / h);
 int trans = (int)(transient / h);

 // range for parameter change
 double paraMin = 0.0;
 double paraMax = 400.0;
 int paraNum = 400;
 double paraDiv = (paraMax - paraMin) / paraNum;

 double prev, now, next;

 // Initial conditions
 a[0] = 15.0; // x
 a[1] = 20.0; // y
 a[2] = 30.0; // z

 for(p = 0; p < paraNum; p++){
  gam = paraMin + paraDiv * p;
    // parameter change

  // calculation for transient
  for(i = 0; i < trans; i++){
   rungeKutta(a, h);
  }

  plotCnt = 0;
  prev = now = next = 0.0;
  for(i = 0; i < n; i++){
   prev = now;
   now = next;
   next = a[0];
   if(prev <= now && now >= next && i > 2){
    printf("%e\t%e\n", gam, now);
    if(++plotCnt > plotMax) break;
```

G.2.2.3 C Source Code for Lyapnov Spectrum (All the Lyapunov Exponents) of Lorenz Equations (Figure 2.12b)

```c
#include<stdio.h>
#include<math.h>

#define M 3 // number of equations
#define dimension 3

double sigma = 10.0;
double gam = 28.0;
double beta = 8.0 / 3.0;

void lorenz(double x[], double b[],
     double dx[][M], double db[][M])
{
  int i;

  // original equations
  b[0] = sigma * (x[1] - x[0]);
  b[1] = - x[0] * x[2] + gam * x[0] - x[1];
  b[2] = x[0] * x[1] - beta * x[2];

  // set of linearized equations
  for(i = 0; i < dimension; i++){
    db[i][0] = - sigma * dx[i][0]
             + sigma * dx[i][1];
    db[i][1] = (gam - x[2]) * dx[i][0] - dx[i][1]
             - x[0] * dx[i][2];
    db[i][2] = x[1] * dx[i][0] + x[0] * dx[i][1]
             - beta * dx[i][2];
  }
}

void rungeKutta(double a[],
     double linearized[][M], double h)
```

```c
{
 int i, j, k;
 double x[M], b[4][M];
 double dx[dimension][M], db[4][dimension][M];
 for(i = 0; i < 4; i++){
  for(j = 0; j < M; j++){
   if(i == 0) x[j] = a[j];
   if(i == 1) x[j] = a[j] + h * b[0][j] / 2.0;
   if(i == 2) x[j] = a[j] + h * b[1][j] / 2.0;
   if(i == 3) x[j] = a[j] + h * b[2][j];
  }
  for(j = 0; j < dimension; j++){
   for(k = 0; k < M; k++){
    if(i == 0) dx[j][k] = linearized[j][k];
    if(i == 1) dx[j][k] = linearized[j][k]
         + h * db[0][j][k] / 2.0;
    if(i == 2) dx[j][k] = linearized[j][k]
         + h * db[1][j][k] / 2.0;
    if(i == 3) dx[j][k] = linearized[j][k]
         + h * db[2][j][k];
   }
  }
  lorenz(x, b[i], dx, db[i]);
 }
 for(i = 0; i < M; i++){
  a[i] += h * (b[0][i] + 2.0 * b[1][i]
      + 2.0 * b[2][i] + b[3][i]) / 6.0;
 }
 for(i = 0; i < dimension; i++){
  for(j = 0; j < M; j++){
   linearized[i][j] += h * (db[0][i][j]
     + 2.0 * db[1][i][j] + 2.0 * db[2][i][j]
     + db[3][i][j]) / 6.0;
  }
 }
}

void modifiedGramSchmidt(double norm[],
    double linearized[][M])
{
 int i, j, k;
 double innerProduct;
 for(i = 0; i < dimension; i++){
  for(j = 0; j < i; j++){
   innerProduct = 0.0;
```

```c
      for(k = 0; k < M; k++) {
        innerProduct += linearized[i][k]
          * linearized[j][k];
      }
      for(k = 0; k < M; k++) {
        linearized[i][k] -= innerProduct
          / norm[j] * linearized[j][k];
      }
    }
    norm[i] = 0.0;
    for(j = 0; j < M; j++) {
      norm[i] += linearized[i][j]
        * linearized[i][j];
    }
  }
}

int main()
{
  int i, j, k, p;
  int lyapunovCount;
  double a[M], linearized[dimension][M],
  lyapunov[dimension], norm[dimension];

  double h = 1.0e-3;  // calculation step
  double transient = 100.0;  // transient time
  double tMax = 100.0;  // plot time

  int n = (int)(tMax / h);
  int trans = (int)(transient / h);
  int lyapunovTrans = (int)(n * 0.1);

  // range for parameter change
  double paraMin = 0.0;
  double paraMax = 400.0;
  int paraNum = 400;
  double paraDiv = (paraMax - paraMin) / paraNum;

  // Initial conditions
  a[0] = 15.0;  // x
  a[1] = 20.0;  // y
  a[2] = 30.0;  // z

  for(p = 0; p < paraNum; p++) {
    gam = paraMin + paraDiv * p;  // parameter change
```

```c
  // calculation for transient
  for(i = 0; i < trans; i++){
    rungeKutta(a, linearized, h);
  }

  // setting initial conditions
  for(i = 0; i < dimension; i++){
    linearized[i][0] = -2.0+ (double) M / 2 - i;
    linearized[i][1] = 0.0 + (double) M / 2 - i;
    linearized[i][2] = 2.0 + (double) M / 2 - i;
    lyapunov[i] = 0.0;
  }

  lyapunovCount = 0;
  for(i = 0; i < n; i++){
    modifiedGramSchmidt(norm, linearized);
    for(j = 0; j < dimension; j++){
      norm[j] = sqrt(norm[j]);
      for(k = 0; k < M; k++){
        linearized[j][k] /= norm[j];
      }
    }
    if(i > lyapunovTrans){
      for(j = 0; j < dimension; j++){
        lyapunov[j] += log(norm[j]);
      }
      ++lyapunovCount;
    }
    rungeKutta(a, linearized, h);
  }
  printf("%e", gam);
  for(i = 0; i < dimension; i++){
    printf("\t%e",
       lyapunov[i] / (h * lyapunovCount));
  }
  printf("\n");
 }
 return 0;
}
```

G.2.2.4 C Source Code for Synchronization of Chaos in Lorenz Equations (Diffusive Coupling, Section 5.2.1.2)

```c
#include<stdio.h>

#define M 6 // number of equations

double sigmaDri = 10.0;
double gammaDri = 28.0;
double betaDri = 8.0 / 3.0;

double sigmaRes = 10.0;
double gammaRes = 28.0;
double betaRes = 8.0 / 3.0;

double kappa = 5.0;

void lorenz(double x[], double b[])
{
 b[0] = sigmaDri * (x[1] - x[0]);
 b[1] = - x[0] * x[2] + gammaDri * x[0] - x[1];
 b[2] = x[0] * x[1] - betaDri * x[2];

 b[3] = sigmaRes * (x[4] - x[3]);
 b[4] = - x[3] * x[5] + gammaRes * x[3] - x[4]
       + kappa * (x[1] - x[4]);
 b[5] = x[3] * x[4] - betaRes * x[5];
}

void rungeKutta(double a[], double h)
{
 int i, j;
 double x[M], b[4][M];
 for(i = 0; i < 4; i++){
  for(j = 0; j < M; j++){
   if(i == 0) x[j] = a[j];
   if(i == 1) x[j] = a[j] + h * b[0][j] / 2.0;
   if(i == 2) x[j] = a[j] + h * b[1][j] / 2.0;
   if(i == 3) x[j] = a[j] + h * b[2][j];
  }
  lorenz(x, b[i]);
 }
 for(i = 0; i < M; i++){
  a[i] += h * (b[0][i] + 2.0 * b[1][i]
```

```c
      + 2.0 * b[2][i] + b[3][i]) / 6.0;
  }
}

int main()
{
  int i;
  double a[M];

  double h = 1.0e-3; // calculation step
  double transient = 0.0; // transient time
  double tmax = 50.0; // plot time

  int div = 10; // plot interval

  int trans = (int)(transient / h);
  int n = (int)(tmax / h);

  // Initial conditions
  a[0] = 15.0;  // x (drive)
  a[1] = 20.0;  // y (drive)
  a[2] = 30.0;  // z (drive)
  a[3] = -15.0; // x (response)
  a[4] = -20.0; // y (response)
  a[5] = -30.0; // z (response)

  // calculation for transient
  for(i = 0; i < trans; i++){
    rungeKutta(a, h);
  }

  for(i = 0; i < n; i++){
    if(i % div == 0){
      printf("%e\t", h * i);
      printf("%e\t%e\t%e\t", a[0], a[1], a[2]);
      printf("%e\t%e\t%e\n", a[3], a[4], a[5]);
    }
    rungeKutta(a, h);
  }
  return 0;
}
```

G.2.3 Lang–Kobayashi Equations for a Semiconductor Laser with Time-Delayed Optical Feedback (Chapter 4)

G.2.3.1 C Source Code for Time Series of Lang–Kobayashi Equations (Figure 4.13e)

```c
#include<stdio.h>
#include<math.h>

#define PI 3.141592653589793238
#define C 2.99792458e8

#define M 3 // number of equations

// variable parameter values
double r3 = 0.01; // reflectivity of external mirror
double JRatio = 1.11;
 // normalized injection current
double L = 0.225;
 // external cavity length (one-way) [m]

// fixed parameter values
double r2 = 0.556;
 // reflectivity of internal mirror
double tauIn = 8.0e-12;
 // internal cavity round-trip time [s]
double lambda = 1.537e-6;
 // optical wavelength [m]
double Gn = 8.4e-13; // gain coefficient
double N0 = 1.4e24;
 // carrier density at transparency
double tauP = 1.927e-12; // photon lifetime [s]
double tauS = 2.04e-9; // carrier lifetime [s]
double alpha = 3.0; // alpha parameter

double Nth, Jth;
double J, kap, tau;
double detun, omega0, phaseShift;

#define DELAY_MAX 100000
 // maximum array size for delay
double eDelay[DELAY_MAX], phiDelay[DELAY_MAX];
int delayNum, delayIndex;
```

```c
// calculation of parameter values
void calcParameter(double h)
{
 Nth = N0 + 1.0 / tauP / Gn;
  // carrier density at threshold
 Jth = Nth / tauS;
  // injection current at threshold
 J = JRatio * Jth;  // injection current
 kap = (1 - r2 * r2) * r3 / r2 / tauIn;
  // injection strength
 tau = 2.0 * L / C;
  // external-cavity round-trip time for drive
 omega0 = 2.0 * PI * C / lambda;
   // optical angular frequency
 phaseShift = fmod(omega0 * tau, 2.0 * PI);
  // initial phase shift
 delayNum = (int)(tau / h);
}

void initializeDelay(double a[])
{
 int i;
 delayIndex = 0;
 for(i = 0; i < DELAY_MAX; i++){
   eDelay[i]   = a[0];
   phiDelay[i] = a[1];
  }
}

void laser(double x[], double b[], double theta)
{
 b[0] = 1.0 / 2.0 * (Gn * (x[2] - N0)
     - 1.0 / tauP) * x[0]
     + kap * eDelay[delayIndex] * cos(theta);
 b[1] = alpha / 2.0 * (Gn * (x[2] - N0)
     - 1.0 / tauP) - kap * eDelay[delayIndex]
     / x[0] * sin(theta);
 b[2] = J - x[2] / tauS - Gn * (x[2] - N0)
     * x[0] * x[0];
}

void rungeKutta(double a[], double h, double t){
 int i, j;
 double x[M], b[4][M];
```

```
      double theta = fmod(phaseShift + a[1]
         - phiDelay[delayIndex], 2.0 * PI);

      for(i = 0; i < 4; i++){
       for(j = 0; j < M; j++){
        if(i == 0) x[j] = a[j];
        if(i == 1) x[j] = a[j] + h * b[0][j] / 2.0;
        if(i == 2) x[j] = a[j] + h * b[1][j] / 2.0;
        if(i == 3) x[j] = a[j] + h * b[2][j];
       }
       laser(x, b[i], theta);
      }

      for(i = 0; i < M; i++){
       a[i] += h * (b[0][i] + 2.0 * b[1][i]
            + 2.0 * b[2][i] + b[3][i]) / 6.0;
      }

      // Renew arrays for delay
      eDelay[delayIndex] = a[0];
      phiDelay[delayIndex] = a[1];

      delayIndex = (delayIndex + 1) % delayNum;
     }

     int main()
     {
      int i;
      double a[M], t;

      double h = 5.0e-12; // calculation step
      double transient = 5000.0e-9; // transient time
      double tMax = 20.0e-9; // plot time

      int div = 10; // plot interval

      int trans = (int)(transient / h);
      int n = (int)(tMax / h);

      // Initial conditions
      a[0] = 1.3e10; // electric field amplitude
      a[1] = 0.0; // electric field phase
      a[2] = 1.90e24; // carrier density
      initializeDelay(a);
```

```
calcParameter(h);
// calculation for transient
for(i = 0; i < trans; i++){
  t = h * i;
  rungeKutta(a, h, t);
}

for(i = 0; i < n; i++){
  t = h * (trans + i);
  if(i % div == 0){
    printf("%e\t", h * i * 1e9); // Time [ns]
    printf("%e\n", a[0] * a[0] * 1e-20);
      // Drive Intensity
  }
  rungeKutta(a, h, t);
}
return 0;
}
```

G.2.3.2 C Source Code for Bifurcation Diagram of Lang–Kobayashi Equations (Figure 4.16)

```
#include<stdio.h>
#include<math.h>

#define PI  3.141592653589793238
#define C  2.99792458e8

#define M 3 // number of equations

  // variable parameter values
double r3 = 0.01;
  // reflectivity of external mirror
double JRatio = 1.11;
  // normalized injection current
double L = 0.225;
  // external cavity length (one-way) [m]

// fixed parameter values
double r2 = 0.556;
  // reflectivity of internal mirror
double tauIn = 8.0e-12;
  // internal cavity round-trip time [s]
```

```c
double lambda = 1.537e-6;
  // optical wavelength [m]
double Gn = 8.4e-13; // gain coefficient
double N0 = 1.4e24;
  // carrier density at transparency
double tauP = 1.927e-12; // photon lifetime [s]
double tauS = 2.04e-9;  // carrier lifetime [s]
double alpha = 3.0;  // alpha parameter

double Nth, Jth;
double J, kap, tau;
double detun, omega0, phaseShift;

#define DELAY_MAX  100000
  // maximum array size for delay
double eDelay[DELAY_MAX], phiDelay[DELAY_MAX];
int delayNum, delayIndex;

// calculation of parameter values
void calcParameter(double h)
{
 Nth = N0 + 1.0 / tauP / Gn;
  // carrier density at threshold
 Jth = Nth / tauS;
  // injection current at threshold
 J = JRatio * Jth;  // injection current
 kap = (1 - r2 * r2) * r3 / r2 / tauIn;
  // injection strength
 tau = 2.0 * L / C;
  // external-cavity round-trip time for drive
 omega0 = 2.0 * PI * C / lambda;
  // optical angular frequency
 phaseShift = fmod(omega0 * tau, 2.0 * PI);
  // initial phase shift
 delayNum = (int)(tau / h);
}

void initializeDelay(double a[])
{
 int i;
 delayIndex = 0;
 for(i = 0; i < DELAY_MAX; i++){
  eDelay[i] = a[0];
  phiDelay[i] = a[1];
 }
```

```
}
void laser(double x[], double b[], double theta)
{
 b[0] = 1.0 / 2.0 * (Gn * (x[2] - N0)
      - 1.0 / tauP) * x[0]
     + kap * eDelay[delayIndex] * cos(theta);
 b[1] = alpha / 2.0 * (Gn * (x[2] - N0)
      - 1.0 / tauP) - kap * eDelay[delayIndex]
     / x[0] * sin(theta);
 b[2] = J - x[2] / tauS - Gn * (x[2] - N0)
      * x[0] * x[0];
}

void rungeKutta(double a[], double h, double t){
 int i, j;
 double x[M], b[4][M];

 double theta = fmod(phaseShift + a[1]
     - phiDelay[delayIndex], 2.0 * PI);

 for(i = 0; i < 4; i++){
   for(j = 0; j < M; j++){
     if(i == 0) x[j] = a[j];
     if(i == 1) x[j] = a[j] + h * b[0][j] / 2.0;
     if(i == 2) x[j] = a[j] + h * b[1][j] / 2.0;
     if(i == 3) x[j] = a[j] + h * b[2][j];
   }
   laser(x, b[i], theta);
 }

 for(i = 0; i < M; i++){
   a[i] += h * (b[0][i] + 2.0 * b[1][i]
        + 2.0 * b[2][i] + b[3][i]) / 6.0;
 }
 eDelay[delayIndex]   = a[0];
 phiDelay[delayIndex] = a[1];

 delayIndex = (delayIndex + 1) % delayNum;
}

int main()
{
 int i, p;
 double a[M], t;
```

```
double h = 5.0e-12; // calculation step
double transient = 5000.0e-9; // transient time
double tMax = 100.0e-9; // plot time

int trans = (int)(transient / h);
int n = (int)(tMax / h);

int plotCnt, plotMax = 50;

// range for parameter change
double paraMin = 0.0;
double paraMax = 0.02;
int paraNum = 200;
double paraDiv = (paraMax - paraMin) / paraNum;

double prev, now, next;

a[0] = 1.3e10; // electric field amplitude
a[1] = 0.0; // electric field phase
a[2] = 1.90e24; // carrier density
initializeDelay(a);

for(p = 0; p < paraNum; p++) {
  r3 = paraMin + paraDiv * p;
     // parameter change
  calcParameter(h);

  // calculation for transient
  for(i = 0; i < trans; i++) {
    t = h * i;
    rungeKutta(a, h, t);
  }

  plotCnt = 0;
  prev = now = next = 0.0;
  for(i = 0; i < n; i++) {
    prev = now;
    now = next;
    next = a[0] * a[0];
    if(prev <= now && now >= next && i > 2) {
      printf("%e\t%e\n", kap * 1e-9,
      now * 1e-20);
      if(++plotCnt > plotMax) break;
    }
    t = h * (trans + i);
```

```
    rungeKutta(a, h, t);
   }
  }
  return 0;
}
```

G.2.3.3 C Source Code for Maximum Lyapunov Exponent of Lang–Kobayashi Equations (Figure 4.19)

```
/* This program may be time-consuming. Smaller values of
transient, tMax, or paraNum result in shorter execution
time. */
#include<stdio.h>
#include<math.h>

#define PI  3.141592653589793238
#define C   2.99792458e8

#define M  6 // number of equations

// variable parameter values
double r3 = 0.01;
    // reflectivity of external mirror
double JRatio = 1.11;
    // normalized injection current
double L = 0.225;
    // external cavity length (one-way) [m]

// fixed parameter values
double r2 = 0.556;
    // reflectivity of internal mirror
double tauIn = 8.0e-12;
    // internal cavity round-trip time [s]
double lambda = 1.537e-6;
    // optical wavelength [m]
double Gn = 8.4e-13; // gain coefficient
double N0 = 1.4e24;
    // carrier density at transparency
double tauP = 1.927e-12; // photon lifetime [s]
double tauS = 2.04e-9; // carrier lifetime [s]
double alpha = 3.0; // alpha parameter
```

```
double Nth, Jth;
double J, kap, tau;
double detun, omega0, phaseShift;

#define DELAY_MAX 1000000
    // maximum array size for delay
double eDelay[DELAY_MAX], phiDelay[DELAY_MAX];
int delayNum, delayIndex;

double eDelayDelta[DELAY_MAX];
double phiDelayDelta[DELAY_MAX];
double nDelayDelta[DELAY_MAX];

// calculation of parameter values
void calcParameter(double h)
{
 Nth = N0 + 1.0 / tauP / Gn;
    // carrier density at threshold
 Jth = Nth / tauS;
    // injection current at threshold
 J = JRatio * Jth;  // injection current
 kap = (1 - r2 * r2) * r3 / r2 / tauIn;
    // injection strength
 tau = 2.0 * L / C;
    // external-cavity round-trip time for drive
 omega0 = 2.0 * PI * C / lambda;
    // optical angular frequency
 phaseShift = fmod(omega0 * tau, 2.0 * PI);
    // initial phase shift
 delayNum = (int)(tau / h);
}

void initializeDelay(double a[])
{
 int i;
 delayIndex = 0;
 for(i = 0; i < DELAY_MAX; i++) {
   eDelay[i] = a[0];
   phiDelay[i] = a[1];
   eDelayDelta[i] = 0.0;
phiDelayDelta[i] = 0.0;
nDelayDelta[i] = 0.0;
  }
}
```

```c
void initializeDelayDelta(double a[])
{
  int i;
  for(i = 0; i < DELAY_MAX; i++){
    eDelayDelta[i] = a[3];
    phiDelayDelta[i] = a[4];
    nDelayDelta[i] = a[5];
  }
}

void laser(double x[], double b[], double theta)
{
  b[0] = 1.0 / 2.0 * (Gn * (x[2] - N0)
       - 1.0 / tauP) * x[0]
       + kap * eDelay[delayIndex] * cos(theta);
  b[1] = alpha / 2.0 * (Gn * (x[2] - N0)
       - 1.0 / tauP) - kap * eDelay[delayIndex]
       / x[0] * sin(theta);
  b[2] = J - x[2] / tauS - Gn * (x[2] - N0)
       * x[0] * x[0];

  b[3] = 1.0 / 2.0 * (Gn * (x[2] - N0)
       - 1.0 / tauP) * x[3] - kap
       * eDelay[delayIndex] * sin(theta) * x[4]
       + 1.0 / 2.0 * Gn * x[0] * x[5]
       + kap * cos(theta)
       * eDelayDelta[delayIndex]
       + kap * eDelay[delayIndex]
       * sin(theta) * phiDelayDelta[delayIndex];
  b[4] = kap * eDelay[delayIndex]
       / (x[0] * x[0]) * sin(theta) * x[3]
       - kap * eDelay[delayIndex] / x[0]
       * cos(theta) * x[4] + alpha / 2.0 * Gn
       * x[5] - kap / x[0] * sin(theta)
       * eDelayDelta[delayIndex]
       + kap * eDelay[delayIndex] / x[0]
       * cos(theta) * phiDelayDelta[delayIndex];
  b[5] = - 2.0 * Gn * (x[2] - N0) * x[0] * x[3]
       - (1.0 / tauS + Gn * x[0] * x[0]) * x[5];
}

void rungeKutta(double a[], double h, double t){
  int i, j;
  double x[M], b[4][M];
```

```c
    double theta = fmod(phaseShift + a[1]
          - phiDelay[delayIndex], 2.0 * PI);
    for(i = 0; i < 4; i++){
     for(j = 0; j < M; j++){
      if(i == 0) x[j] = a[j];
      if(i == 1) x[j] = a[j] + h * b[0][j] / 2.0;
      if(i == 2) x[j] = a[j] + h * b[1][j] / 2.0;
      if(i == 3) x[j] = a[j] + h * b[2][j];
     }
     laser(x, b[i], theta);
    }

    for(i = 0; i < M; i++){
     a[i] += h * (b[0][i] + 2.0 * b[1][i]
           + 2.0 * b[2][i] + b[3][i]) / 6.0;
    }

    eDelay[delayIndex] = a[0];
    phiDelay[delayIndex] = a[1];

    eDelayDelta[delayIndex] = a[3];
    phiDelayDelta[delayIndex] = a[4];
    nDelayDelta[delayIndex] = a[5];

    delayIndex = (delayIndex + 1) % delayNum;
   }

   int main()
   {
    int i, j, p;
    double a[M], t;

    double h = 5.0e-12;  // calculation step
    double transient = 5000.0e-9;  // transient time
    double tMax = 5000.0e-9;  // plot time

    int n = (int)(tMax / h);
    int trans = (int)(transient / h);
    int lyapunovTrans = (int)(n * 0.1);

    // range for parameter change
    double paraMin = 0.000;
    double paraMax = 0.025;
    int paraNum = 200;
    double paraDiv = (paraMax - paraMin) / paraNum;
```

```c
int lyapunovCount;
double lyapunov, norm;

// initial conditions
a[0] = 1.3e10;  // electric field amplitude
a[1] = 0.0;  // electric field phase
a[2] = 1.90e24;  // carrier density
initializeDelay(a);

for(p = 0; p < paraNum; p++) {
 r3 = paraMin + paraDiv * p;
  // parameter change
 calcParameter(h);

 lyapunov = 0.0;
 lyapunovCount = 0;

 // calculation for transient
 for(i = 0; i < trans; i++) {
  t = h * i;
  rungeKutta(a, h, t);
 }

 // initial conditions
 a[3] = 0.01;
  // linearized electric field amplitude
 a[4] = 0.01;
  // linearized electric field phase
 a[5] = 0.01;
  // linearized carrier density
 initializeDelayDelta(a);

 for(i = 0; i < n; i++) {
  t = h * (trans + i);
  rungeKutta(a, h, t);

  if(i % delayNum == 0) {
   norm = 0.0;
   for(j = 0; j < delayNum; j++) {
    norm += eDelayDelta[j] * eDelayDelta[j]
    + phiDelayDelta[j] * phiDelayDelta[j]
    + nDelayDelta[j] * nDelayDelta[j];
   }
   norm = sqrt(norm);
```

```c
        for(j = 0; j < delayNum; j++){
          eDelayDelta[j] /= norm;
          phiDelayDelta[j] /= norm;
          nDelayDelta[j] /= norm;
        }
        a[3] /= norm;
        a[4] /= norm;
        a[5] /= norm;

        if(i > lyapunovTrans){
          lyapunov += log(norm);
          ++lyapunovCount;
        }
      }
    }
    printf("%e\t", kap * 1e-9);
    printf("%e\n", lyapunov
        / (tau * lyapunovCount) * 1e-9);
  }
  return 0;
}
```

G.2.4 Synchronization of Chaos in Coupled Lang–Kobayashi Equations for Unidirectionally Coupled Semiconductor Lasers with Time–Delayed Optical Feedback (Chapter 6)

G.2.4.1 C Source Code for Time Series of Synchronization of Chaos in Coupled Lang–Kobayashi Equations in Open-Loop Configuration (Figure 6.9)

```c
#include<stdio.h>
#include<math.h>

#define PI 3.141592653589793238
#define C 2.99792458e8

#define M 6 // number of equations

// variable parameter values
double r3Dri = 0.008;
    // reflectivity of external mirror of drive
double JRatio = 1.3;
    // normalized injection current
```

```
double LDri = 0.6;
    // external cavity length (one-way) for drive [m]
double LInj = 1.2;
    // distance between drive and response lasers [m]
double detun = 0.0e9;
    // optical frequency detuning [Hz]

double kapInjRatio = 1.0;
    // injection strength ratio to drive feedback
strength

// fixed parameter values
double r2 = 0.556;
      // reflectivity of internal mirror
double tauIn = 8.0e-12;
    // internal cavity round-trip time [s]
double lambda = 1.537e-6;
    // optical wavelength [m]
double Gn = 8.4e-13;  // gain coefficient
double N0 = 1.4e24;
    // carrier density at transparency
double tauP = 1.927e-12;  // photon lifetime [s]
double tauS = 2.04e-9;  // carrier lifetime [s]
double alpha = 3.0;  // alpha parameter

double Nth, Jth;
double J, kapDri, kapInj, tauDri, tauInj;
double omega0, phaseShiftDri, phaseShiftInj;

#define DELAY_MAX 100000
    // maximum array size for delay
double eDelayDri[DELAY_MAX], phiDelayDri[DELAY_MAX];
double eDelayInj[DELAY_MAX], phiDelayInj[DELAY_MAX];
int delayDriNum, delayInjNum;
int delayDriIndex, delayInjIndex;

// calculation of parameter values
void calcParameter(double h)
{
 Nth = N0 + 1.0 / tauP / Gn;
    // carrier density at threshold
 Jth = Nth / tauS;
    // injection current at threshold
 J   = JRatio * Jth;  // injection current
 kapDri = (1 - r2 * r2) * r3Dri / r2 / tauIn;
```

```c
    // feedback strength of drive
  kapInj = kapInjRatio * kapDri;
    // injection strength from drive to response
  tauDri = 2.0 * LDri / C;
    // external-cavity round-trip time for drive
  tauInj = LInj / C;
    // coupling time from drive to response
  omega0 = 2.0 * PI * C / lambda;
    // optical angular frequency
  phaseShiftDri = fmod(omega0 * tauDri, 2.0 * PI);
    // initial phase shift for drive
  phaseShiftInj = fmod(omega0 * tauInj, 2.0 * PI);
    // initial phase shift for coupling
  delayDriNum = (int)(tauDri / h);
  delayInjNum = (int)(tauInj / h);
}

void initializeDelay(double a[])
{
  int i;
  delayDriIndex = delayInjIndex = 0;

  for(i = 0; i < DELAY_MAX; i++){
    eDelayDri[i]   = a[0];
    phiDelayDri[i] = a[1];
  }
  for(i = 0; i < DELAY_MAX; i++){
    eDelayInj[i]   = 0.0;
    phiDelayInj[i] = 0.0;
  }
}

void laser(double x[], double b[], double theta[])
{
  b[0] = 1.0 / 2.0 * (Gn * (x[2] - N0)
       - 1.0 / tauP) * x[0] + kapDri
       * eDelayDri[delayDriIndex]
       * cos(theta[0]);
  b[1] = alpha / 2.0 * (Gn * (x[2] - N0)
       - 1.0 / tauP) - kapDri
       * eDelayDri[delayDriIndex]
       / x[0] * sin(theta[0]);
  b[2] = J - x[2] / tauS - Gn * (x[2] - N0)
       * x[0] * x[0];
```

```c
    b[3] = 1.0 / 2.0 * (Gn * (x[5] - N0)
        - 1.0 / tauP) * x[3] + kapInj
        * eDelayInj[delayInjIndex]
        * cos(theta[1]);
    b[4] = alpha / 2.0 * (Gn * (x[5] - N0)
        - 1.0 / tauP) - kapInj
        * eDelayInj[delayInjIndex]
        / x[3] * sin(theta[1]);
    b[5] = J - x[5] / tauS - Gn * (x[5] - N0)
        * x[3] * x[3];
}

void rungeKutta(double a[], double h, double t){
    int i, j;
    double x[M], b[4][M];

    double theta[2];
    theta[0] = fmod(phaseShiftDri + a[1]
            - phiDelayDri[delayDriIndex],
            2.0 * PI);
    theta[1] = fmod(phaseShiftInj + a[4]
            - phiDelayInj[delayInjIndex]
            - 2.0 * PI * detun * t, 2.0 * PI);

    for(i = 0; i < 4; i++){
        for(j = 0; j < M; j++){
            if(i == 0) x[j] = a[j];
            if(i == 1) x[j] = a[j] + h * b[0][j] / 2.0;
            if(i == 2) x[j] = a[j] + h * b[1][j] / 2.0;
            if(i == 3) x[j] = a[j] + h * b[2][j];
        }
        laser(x, b[i], theta);
    }

    for(i = 0; i < M; i++){
        a[i] += h * (b[0][i] + 2.0 * b[1][i]
            + 2.0 * b[2][i] + b[3][i]) / 6.0;
    }

    // Renew arrays for delay
    eDelayDri[delayDriIndex] = a[0];
    eDelayInj[delayInjIndex] = a[0];
    phiDelayDri[delayDriIndex] = a[1];
    phiDelayInj[delayInjIndex] = a[1];
```

```
    delayDriIndex = (delayDriIndex + 1)
        % delayDriNum;
    delayInjIndex = (delayInjIndex + 1)
        % delayInjNum;
}

int main()
{
    int i;
    double a[M], t;

    double h = 5.0e-12; // calculation step
    double transient = 5000.0e-9; // transient time
    double tMax = 50.0e-9; // plot time

    int trans = (int)(transient / h);
    int n = (int)(tMax / h);

    int div = 10; // plot interval

    // initial conditions
    a[0] = 1.3e10;
        // electric field amplitude for drive
    a[1] = 0.0; // electric field phase for drive
    a[2] = 1.90e24; // carrier density for drive
    a[3] = 1.4e10;
        // electric field amplitude for response
    a[4] = 0.0; // electric field phase for response
    a[5] = 1.85e24; // carrier density for response
    initializeDelay(a);

    calcParameter(h);

    // calculation for transient
    for(i = 0; i < trans; i++){
        t = h * i;
        rungeKutta(a, h, t);
    }

    for(i = 0; i < n; i++){
        t = h * (trans + i);
        if(i % div == 0){
            printf("%e\t", h * i * 1e9); // Time [ns]
            printf("%e\t", a[0] * a[0] * 1e-20);
                // Drive Intensity
```

```c
    printf("%e\t", a[3] * a[3] * 1e-20);
        // Response Intensity
    printf("%e\n", eDelayDri[delayDriIndex]
    * eDelayDri[delayDriIndex] * 1e-20);
        // Delayed Drive Intensity
    }
    rungeKutta(a, h, t);
  }
  return 0;
}
```

G.2.4.2 C Source Code for Time Series of Synchronization of Chaos in Coupled Lang–Kobayashi Equations in Closed-Loop Configuration (Appendix 6.A.1)

```c
#include<stdio.h>
#include<math.h>

#define PI  3.141592653589793238
#define C  2.99792458e8

#define M  6  // number of equations

// variable parameter values
double r3Dri = 0.008;
    // reflectivity of external mirror of drive
double r3Res = 0.008;
    // reflectivity of external mirror of response
double JRatio = 1.3;
    // normalized injection current
double LDri = 0.6;
    // external cavity length (one-way) for drive [m]
double LRes = 0.6;
    // external cavity length (one-way) for response [m]
double LInj = 1.2;
    // distance between drive and response lasers [m]
double detun = -4.0e9;
    // optical frequency detuning [Hz]

double kapInjRatio = 12.5;
    //injection strength ratio to drive feedback strength

    // fixed parameter values
double r2 = 0.556;
    // reflectivity of internal mirror
```

```c
    double tauIn = 8.0e-12;
        // internal cavity round-trip time [s]
    double lambda = 1.537e-6;
        // optical wavelength [m]
    double Gn = 8.4e-13; // gain coefficient
    double N0 = 1.4e24;
        // carrier density at transparency
    double tauP = 1.927e-12; // photon lifetime [s]
    double tauS = 2.04e-9; // carrier lifetime [s]
    double alpha = 3.0; // alpha parameter

    double Nth, Jth;
    double J, kapDri, kapRes, kapInj, tauDri, tauRes,
    tauInj;
    double omega0, phaseShiftDri, phaseShiftRes,
    phaseShiftInj;

    #define DELAY_MAX 100000
        // maximum array size for delay
    double eDelayDri[DELAY_MAX], phiDelayDri[DELAY_MAX];
    double eDelayRes[DELAY_MAX], phiDelayRes[DELAY_MAX];
    double eDelayInj[DELAY_MAX], phiDelayInj[DELAY_MAX];
    int delayDriNum, delayResNum, delayInjNum;
    int delayDriIndex, delayResIndex, delayInjIndex;

        // calculation of parameter values
    void calcParameter(double h)
    {
     Nth = N0 + 1.0 / tauP / Gn;
        // carrier density at threshold
     Jth = Nth / tauS;
        // injection current at threshold
     J  = JRatio * Jth;
        // injection current
     kapDri = (1 - r2 * r2) * r3Dri / r2 / tauIn;
        // feedback strength of drive
     kapRes = (1 - r2 * r2) * r3Res / r2 / tauIn;
        // feedback strength of response
     kapInj = kapInjRatio * kapDri;
        // injection strength from drive to response
     tauDri = 2.0 * LDri / C;
        // external-cavity round-trip time for drive
     tauRes = 2.0 * LRes / C;
        // external-cavity round-trip time for response
     tauInj = LInj / C;
```

```c
    // coupling time from drive to response
  omega0 = 2.0 * PI * C / lambda;
    // optical angular frequency
  phaseShiftDri = fmod( omega0 * tauDri, 2.0 * PI );
    // initial phase shift for drive
  phaseShiftRes = fmod((omega0 - 2.0 * PI * detun)
* tauRes, 2.0 * PI );
    // initial phase shift for response
  phaseShiftInj = fmod( omega0 * tauInj, 2.0 * PI );
    // initial phase shift for injection
  delayDriNum = (int)(tauDri / h);
  delayResNum = (int)(tauRes / h);
  delayInjNum = (int)(tauInj / h);
}

void initializeDelay(double a[])
{
  int i;
  delayDriIndex = 0;
  delayResIndex = 0;
  delayInjIndex = 0;

  for(i = 0; i < DELAY_MAX; i++){
    eDelayDri[i] = a[0];
    phiDelayDri[i] = a[1];
  }
  for(i = 0; i < DELAY_MAX; i++){
    eDelayRes[i] = a[3];
    phiDelayRes[i] = a[4];
  }
  for(i = 0; i < DELAY_MAX; i++){
    eDelayInj[i] = 0.0;
    phiDelayInj[i] = 0.0;
  }
}

void laser(double x[], double b[], double theta[])
{
  b[0] = 1.0 / 2.0 * (Gn * (x[2] - N0)
      - 1.0 / tauP) * x[0] + kapDri
      * eDelayDri[delayDriIndex]
      * cos(theta[0]);
  b[1] = alpha / 2.0 * (Gn * (x[2] - N0)
      - 1.0 / tauP) - kapDri
      * eDelayDri[delayDriIndex]
```

```
        / x[0] * sin(theta[0]);
    b[2] = J - x[2] / tauS - Gn * (x[2] - N0)
        * x[0] * x[0];

    b[3] = 1.0 / 2.0 * (Gn * (x[5] - N0)
        - 1.0 / tauP) * x[3] + kapRes
        * eDelayRes[delayResIndex]
        * cos(theta[1]) + kapInj
        * eDelayInj[delayInjIndex]
        * cos(theta[2]);
    b[4] = alpha / 2.0 * (Gn * (x[5] - N0)
        - 1.0 / tauP) - kapRes
        * eDelayRes[delayResIndex]
        / x[3] * sin(theta[1]) - kapInj
        * eDelayInj[delayInjIndex]
        / x[3] * sin(theta[2]);
    b[5] = J - x[5] / tauS - Gn * (x[5] - N0)
        * x[3] * x[3];
}

void rungeKutta(double a[], double h, double t){
    int i, j;
    double x[M], b[4][M];

    double theta[3];
    theta[0] = fmod(phaseShiftDri + a[1]
        - phiDelayDri[delayDriIndex], 2.0 * PI);
    theta[1] = fmod(phaseShiftRes + a[4]
        - phiDelayRes[delayResIndex], 2.0 * PI);
    theta[2] = fmod(phaseShiftInj + a[4]
        - phiDelayInj[delayInjIndex]
        - 2.0 * PI * detun * t, 2.0 * PI);

    for(i = 0; i < 4; i++){
     for(j = 0; j < M; j++){
      if(i == 0) x[j] = a[j];
      if(i == 1) x[j] = a[j] + h * b[0][j] / 2.0;
      if(i == 2) x[j] = a[j] + h * b[1][j] / 2.0;
      if(i == 3) x[j] = a[j] + h * b[2][j];
     }
     laser(x, b[i], theta);
    }
    for(i = 0; i < M; i++){
     a[i] += h * (b[0][i] + 2.0 * b[1][i]
            + 2.0 * b[2][i] + b[3][i]) / 6.0;
    }
```

```c
    // Renew arrays for delay
    eDelayDri[delayDriIndex] = a[0];
    eDelayRes[delayResIndex] = a[3];
    eDelayInj[delayInjIndex] = a[0];

    phiDelayDri[delayDriIndex] = a[1];
    phiDelayRes[delayResIndex] = a[4];
    phiDelayInj[delayInjIndex] = a[1];

    delayDriIndex = (delayDriIndex + 1)
                % delayDriNum;
    delayResIndex = (delayResIndex + 1)
                % delayResNum;
    delayInjIndex = (delayInjIndex + 1)
                % delayInjNum;
}

int main()
{
    int i;
    double a[M], t;

    double h = 5.0e-12;  // calculation step
    double transient = 5000.0e-9;  // transient time
    double tMax = 50.0e-9;  // plot time

    int trans = (int)(transient / h);
    int n = (int)(tMax / h);

    int div = 10; // plot interval

    // Initial conditions
    a[0] = 1.3e10;
        // electric field amplitude for drive
    a[1] = 0.0;  // electric field phase for drive
    a[2] = 1.90e24;  // carrier density for drive
    a[3] = 1.4e10;
        // electric field amplitude for response
    a[4] = 0.0;  // electric field phase for response
    a[5] = 1.85e24;  // carrier density for response
    initializeDelay(a);
    calcParameter(h);

    // calculation for transient
    for(i = 0; i < trans; i++){
```

```c
    t = h * i;
    rungeKutta(a, h, t);
  }

  for(i = 0; i < n; i++){
    t = h * (trans + i);
    if(i % div == 0){
      printf("%e\t", h * i * 1e9);  // Time [ns]
      printf("%e\t", a[0] * a[0] * 1e-20);
        // Drive Intensity
      printf("%e\t", a[3] * a[3] * 1e-20);
        // Response Intensity
      printf("%e\n", eDelayDri[delayDriIndex]
 * eDelayDri[delayDriIndex] * 1e-20);
        // Delayed Drive Intensity
    }
    rungeKutta(a, h, t);
  }
  return 0;
}
```

G.2.4.3 C Source Code for Cross-Correlation Calculation of Synchronization of Chaos in Coupled Lang–Kobayashi Equations in Open-Loop Configuration (Figures 6.10a and c)

```c
/* This program may be time-consuming. Smaller values of
transient, tMax, or paraNum result in shorter execution
time. */
#include<stdio.h>
#include<math.h>

#define PI  3.141592653589793238
#define C   2.99792458e8

#define M 6 // number of equations

// variable parameter values
double r3Dri = 0.008;
    // reflectivity of external mirror of drive
double JRatio = 1.3;
    // normalized injection current
double LDri = 0.6;
    // external cavity length (one-way) for drive [m]
double LInj = 1.2;
```

```c
    // distance between drive and response lasers [m]
double detun = -4.0e9;
    // optical frequency detuning [Hz]
double kapInjRatio = 12.5;
// injection strength ratio to drive feedback strength

// fixed parameter values
double r2 = 0.556;
    // reflectivity of internal mirror
double tauIn = 8.0e-12;
    // internal cavity round-trip time [s]
double lambda = 1.537e-6;
    // optical wavelength [m]
double Gn = 8.4e-13;  // gain coefficient
double N0 = 1.4e24;
    // carrier density at transparency
double tauP = 1.927e-12;  // photon lifetime [s]
double tauS = 2.04e-9; // carrier lifetime [s]
double alpha = 3.0;  // alpha parameter

double Nth, Jth;
double J, kapDri, kapInj, tauDri, tauInj;
double omega0, phaseShiftDri, phaseShiftInj;

#define DELAY_MAX 100000
    // maximum array size for delay
doubleeDelayDri[DELAY_MAX], phiDelayDri[DELAY_MAX];
double eDelayInj[DELAY_MAX], phiDelayInj[DELAY_MAX];
int delayDriNum, delayInjNum;
int delayDriIndex, delayInjIndex;

#define WAVE_MAX 200000
    // maximum array size to save waveforms
double wave[3][WAVE_MAX];

// calculation of parameter values
void calcParameter(double h)
{
 Nth = N0 + 1.0 / tauP / Gn;
    // carrier density at threshold
 Jth = Nth / tauS;
    // injection current at threshold
 J  = JRatio * Jth;  // injection current
 kapDri = (1 - r2 * r2) * r3Dri / r2 / tauIn;
    // feedback strength of drive
```

```
    kapInj = kapInjRatio * kapDri;
       // injection strength from drive to response
    tauDri = 2.0 * LDri / C;
       // external-cavity round-trip time for drive
    tauInj = LInj / C;
       // coupling time from drive to response
    omega0 = 2.0 * PI * C / lambda;
       // optical angular frequency
    phaseShiftDri = fmod(omega0 * tauDri, 2.0 * PI);
       // initial phase shift for drive
    phaseShiftInj = fmod(omega0 * tauInj, 2.0 * PI);
        // initial phase shift for coupling
    delayDriNum = (int)(tauDri / h);
    delayInjNum = (int)(tauInj / h);
}

void initializeDelay(double a[])
{
 int i;
 delayDriIndex = delayInjIndex = 0;

 for(i = 0; i < DELAY_MAX; i++){
  eDelayDri[i] = a[0];
  phiDelayDri[i] = a[1];
 }
 for(i = 0; i < DELAY_MAX; i++){
  eDelayInj[i] = 0.0;
  phiDelayInj[i] = 0.0;
 }
}

void laser(double x[], double b[], double theta[])
{
 b[0] = 1.0 / 2.0 * (Gn * (x[2] - N0)
     - 1.0 / tauP) * x[0] + kapDri
     * eDelayDri[delayDriIndex]
     * cos(theta[0]);
 b[1] = alpha / 2.0 * (Gn * (x[2] - N0)
     - 1.0 / tauP) - kapDri
     * eDelayDri[delayDriIndex]
     / x[0] * sin(theta[0]);
 b[2] = J - x[2] / tauS - Gn * (x[2] - N0)
     * x[0] * x[0];
```

```c
    b[3] = 1.0 / 2.0 * (Gn * (x[5] - N0)
        - 1.0 / tauP) * x[3] + kapInj
        * eDelayInj[delayInjIndex]
        * cos(theta[1]);
    b[4] = alpha / 2.0 * (Gn * (x[5] - N0)
        - 1.0 / tauP) - kapInj
        * eDelayInj[delayInjIndex]
        / x[3] * sin(theta[1]);
    b[5] = J - x[5] / tauS - Gn * (x[5] - N0)
        * x[3] * x[3];
}

void rungeKutta(double a[], double h, double t){
    int i, j;
    double x[M], b[4][M];

    double theta[2];
    theta[0] = fmod(phaseShiftDri + a[1]
        - phiDelayDri[delayDriIndex], 2.0 * PI);
    theta[1] = fmod(phaseShiftInj + a[4]
        - phiDelayInj[delayInjIndex]
        - 2.0 * PI * detun * t, 2.0 * PI);

    for(i = 0; i < 4; i++){
        for(j = 0; j < M; j++){
            if(i == 0) x[j] = a[j];
            if(i == 1) x[j] = a[j] + h * b[0][j] / 2.0;
            if(i == 2) x[j] = a[j] + h * b[1][j] / 2.0;
            if(i == 3) x[j] = a[j] + h * b[2][j];
        }
        laser(x, b[i], theta);
    }

    for(i = 0; i < M; i++){
        a[i] += h * (b[0][i] + 2.0 * b[1][i]
            + 2.0 * b[2][i] + b[3][i]) / 6.0;
    }

    // Renew arrays for delay
    eDelayDri[delayDriIndex] = a[0];
    eDelayInj[delayInjIndex] = a[0];
    phiDelayDri[delayDriIndex] = a[1];
    phiDelayInj[delayInjIndex] = a[1];
```

```
            delayDriIndex = (delayDriIndex + 1)
                    % delayDriNum;
            delayInjIndex = (delayInjIndex + 1)
                    % delayInjNum;
}

double crossCorrelation(double wave1[],
        double wave2[], int n)
{
 int i;
 double sum1, sum2;
 double ave1, ave2;
 double dev1, dev2;
 double var1, var2;
 double cov;

 sum1 = sum2 = 0.0;
 for(i = 0; i < n; i++){
   sum1 += wave1[i];
   sum2 += wave2[i];
 }
 ave1 = sum1 / n;
 ave2 = sum2 / n;

 var1 = var2 = cov = 0.0;
 for(i = 0; i < n; i++){
   dev1 = wave1[i] - ave1;
   dev2 = wave2[i] - ave2;

   var1 += dev1 * dev1;
   var2 += dev2 * dev2;
   cov  += dev1 * dev2;
 }
 var1 /= n;
 var2 /= n;
 cov  /= n;
 if(var1 == 0.0 || var2 == 0.0) return 0.0;
 return cov / (sqrt(var1) * sqrt(var2));
}
int main()
{
 int i, p;
 double a[M], t;
```

```c
double h = 5.0e-12; // calculation step
double transient = 5000.0e-9; // transient time
double tMax = 1000.0e-9; // plot time

int trans = (int)(transient / h);
int n = (int)(tMax / h);

// range for parameter change
double paraMin = 0.0;
double paraMax = 12.0;
int paraNum = 100;
double paraDiv = (paraMax - paraMin) / paraNum;

// Initial conditions
a[0] = 1.3e10;
  // electric field amplitude for drive
a[1] = 0.0; // electric field phase for drive
a[2] = 1.90e24; // carrier density for drive
a[3] = 1.4e10;
  // electric field amplitude for response
a[4] = 0.0; // electric field phase for response
a[5] = 1.85e24; // carrier density for response
initializeDelay(a);

for(p = 0; p < paraNum; p++){
 kapInjRatio = paraMin + paraDiv * p;
   // parameter change
 calcParameter(h);

 // calculation for transient
 for(i = 0; i < trans; i++){
  t = h * i;
  rungeKutta(a, h, t);
 }

 for(i = 0; i < n; i++){
  t = h * (trans + i);
  wave[0][i] = a[0] * a[0];
    // Drive intensity
  wave[1][i] = a[3] * a[3];
    // Response intensity
  wave[2][i] = eDelayDri[delayDriIndex]
   * eDelayDri[delayDriIndex];
    // Delayed drive intensity
  rungeKutta(a, h, t);
 }
```

```c
    printf("%e\t", kapInj * 1e-9);
    printf("%e\t", crossCorrelation(wave[0],
       wave[1], n));
       // Identical synchronization
    printf("%e\n", crossCorrelation(wave[2],
       wave[1], n));
       // Generalized synchronization
  }
  return 0;
}
```

G.2.4.4 C Source Code for Conditional Lyapunov Exponent of Synchronization of Chaos in Coupled Lang–Kobayashi Equations in Open-Loop Configuration (Figure 6.12)

```c
/* This program may be time-consuming. Smaller values of
transient, tMax, or paraNum result in shorter execution
time. */
#include<stdio.h>
#include<math.h>

#define PI  3.141592653589793238
#define C   2.99792458e8

#define M 9 // number of equations

// variable parameter values
double r3Dri = 0.008;
    // reflectivity of external mirror of drive
double JRatio = 1.3;

// normalized injection current
double LDri = 0.6;
    // external cavity length (one-way) for drive [m]
double LInj = 1.2;
    // distance between drive and response lasers [m]
double detun = 0.0e9;
    // optical frequency detuning [Hz]
double kapInjRatio = 1.0;
// injection strength ratio to drive feedback strength
```

```c
// fixed parameter values
double r2 = 0.556;
    // reflectivity of internal mirror
double tauIn = 8.0e-12;
    // internal cavity round-trip time [s]
double lambda = 1.537e-6;
    // optical wavelength [m]
double Gn = 8.4e-13;  // gain coefficient
double N0 = 1.4e24;
    // carrier density at transparency
double tauP = 1.927e-12;  // photon lifetime [s]
double tauS = 2.04e-9;  // carrier lifetime [s]
double alpha = 3.0;  // alpha parameter

double Nth, Jth;
double J, kapDri, kapInj, tauDri, tauInj;
double omega0, phaseShiftDri, phaseShiftInj;

#define DELAY_MAX 100000
    // maximum array size for delay
double eDelayDri[DELAY_MAX], phiDelayDri[DELAY_MAX];
double eDelayInj[DELAY_MAX], phiDelayInj[DELAY_MAX];
int delayDriNum, delayInjNum;
int delayDriIndex, delayInjIndex;

// calculation of parameter values
void calcParameter(double h)
{
 Nth = N0 + 1.0 / tauP / Gn;
    // carrier density at threshold
 Jth = Nth / tauS;
    // injection current at threshold
 J = JRatio * Jth;  // injection current
 kapDri = (1 - r2 * r2) * r3Dri / r2 / tauIn;
    // feedback strength of drive
 kapInj = kapInjRatio * kapDri;
    // injection strength from drive to response
 tauDri = 2.0 * LDri / C;
    // external-cavity round-trip time for drive
 tauInj = LInj / C;
    // coupling time from drive to response
 omega0 = 2.0 * PI * C / lambda;
    // optical angular frequency
 phaseShiftDri = fmod(omega0 * tauDri, 2.0 * PI);
    // initial phase shift for drive
```

```
  phaseShiftInj = fmod(omega0 * tauInj, 2.0 * PI);
     // initial phase shift for coupling
  delayDriNum = (int)(tauDri / h);
  delayInjNum = (int)(tauInj / h);
}

void initializeDelay(double a[])
{
  int i;
  delayDriIndex = delayInjIndex = 0;

  for(i = 0; i < DELAY_MAX; i++){
    eDelayDri[i] = a[0];
    phiDelayDri[i] = a[1];
  }
  for(i = 0; i < DELAY_MAX; i++){
    eDelayInj[i] = 0.0;
    phiDelayInj[i] = 0.0;
  }
}

void laser(double x[], double b[], double theta[])
{
  b[0] = 1.0 / 2.0 * (Gn * (x[2] - N0)
      - 1.0 / tauP) * x[0] + kapDri
      * eDelayDri[delayDriIndex]
      * cos(theta[0]);
  b[1] = alpha / 2.0 * (Gn * (x[2] - N0)
      - 1.0 / tauP) - kapDri
      * eDelayDri[delayDriIndex]
      / x[0] * sin(theta[0]);
  b[2] = J - x[2] / tauS - Gn * (x[2] - N0)
      * x[0] * x[0];

  b[3] = 1.0 / 2.0 * (Gn * (x[5] - N0)
      - 1.0 / tauP) * x[3] + kapInj
      * eDelayInj[delayInjIndex]
      * cos(theta[1]);
  b[4] = alpha / 2.0 * (Gn * (x[5] - N0)
      - 1.0 / tauP) - kapInj
      * eDelayInj[delayInjIndex]
      / x[3] * sin(theta[1]);
  b[5] = J - x[5] / tauS - Gn * (x[5] - N0)
      * x[3] * x[3];
```

```
   b[6] = 1.0 / 2.0 * (Gn * (x[5] - N0)
       - 1.0 / tauP) * x[6] - kapInj
       * eDelayInj[delayInjIndex]
       * sin(theta[1]) * x[7]
       + 1.0 / 2.0 * Gn * x[3] * x[8];
   b[7] = kapInj * eDelayInj[delayInjIndex]
       / (x[3] * x[3]) * sin(theta[1]) * x[6]
       - kapInj * eDelayInj[delayInjIndex]
       / x[3] * cos(theta[1]) * x[7]
       + alpha / 2.0 * Gn * x[8];
   b[8] = - 2.0 * Gn * x[3] * (x[5] - N0) * x[6]
       - (1.0 / tauS + Gn * x[3] * x[3]) * x[8];
}

void rungeKutta(double a[], double h, double t){
  int i, j;
  double x[M], b[4][M];

  double theta[2];
  theta[0] = fmod(phaseShiftDri + a[1]
     - phiDelayDri[delayDriIndex], 2.0 * PI);
  theta[1] = fmod(phaseShiftInj + a[4]
     - phiDelayInj[delayInjIndex]
     - 2.0 * PI * detun * t, 2.0 * PI);

  for(i = 0; i < 4; i++){
    for(j = 0; j < M; j++){
      if(i == 0) x[j] = a[j];
      if(i == 1) x[j] = a[j] + h * b[0][j] / 2.0;
      if(i == 2) x[j] = a[j] + h * b[1][j] / 2.0;
      if(i == 3) x[j] = a[j] + h * b[2][j];
    }
    laser(x, b[i], theta);
  }

  for(i = 0; i < M; i++){
    a[i] += h * (b[0][i] + 2.0 * b[1][i]
       + 2.0 * b[2][i] + b[3][i]) / 6.0;
  }

  // Renew arrays for delay
  eDelayDri[delayDriIndex] = a[0];
  eDelayInj[delayInjIndex] = a[0];
  phiDelayDri[delayDriIndex] = a[1];
  phiDelayInj[delayInjIndex] = a[1];
```

```c
        delayDriIndex = (delayDriIndex + 1)
            % delayDriNum;
        delayInjIndex = (delayInjIndex + 1)
            % delayInjNum;
}

int main()
{
    int i, p;
    double a[M], t;

    double h = 5.0e-12; // calculation step
    double transient = 5000.0e-9; // transient time
    double tMax = 1000.0e-9; // plot time

    int trans = (int)(transient / h);
    int n = (int)(tMax / h);
    int lyapunovTrans = (int)(n * 0.1);

    // range for parameter change
    double paraMin = 0.00;
    double paraMax = 0.05;
    int paraNum = 100;
    double paraDiv = (paraMax - paraMin) / paraNum;

    int lyapunovCount;
    double lyapunov, norm;

    // initial conditions
    a[0] = 1.3e10;
        // electric field amplitude for drive
    a[1] = 0.0; // electric field phase for drive
    a[2] = 1.90e24; // carrier density for drive
    a[3] = 1.4e10;
        // electric field amplitude for response
    a[4] = 0.0; // electric field phase for response
    a[5] = 1.85e24; // carrier density for response
    initializeDelay(a);

    for(p = 0; p < paraNum; p++) {
      r3Dri = paraMin + paraDiv * p;
        // parameter change
      calcParameter(h);
```

```c
  lyapunovCount = 0;
  lyapunov = 0.0;

  // calculation for transient
  for(i = 0; i < trans; i++){
    t = h * i;
    rungeKutta(a, h, t);
  }

  // initial conditions
  a[6] = 0.01;
    // linearized electric field amplitude
  a[7] = 0.01;
    // linearized electric field phase
  a[8] = 0.01;
    // linearized carrier density

  for(i = 0; i < n; i++){
    t = h * (trans + i);
    norm = sqrt(a[6] * a[6] + a[7] * a[7]
         + a[8] * a[8]);
    a[6] /= norm;
    a[7] /= norm;
    a[8] /= norm;
    if(i > lyapunovTrans){
      lyapunov += log(norm);
      ++lyapunovCount;
    }
    rungeKutta(a, h, t);
  }
  printf("%e\t", kapDri * 1e-9);
  printf("%e\n", lyapunov / (h * lyapunovCount)
  * 1e-9);
 }
 return 0;
}
```

Index

a

acousto-optic modulators (AOMs)   120, 260, 520
adaptive mode selection   527–528
adiabatic elimination   48
Alice/Bob's decision variable   443
Alice–Bob samples   435
alpha parameter   203
ammonia (NH$_3$) ring laser   125
amplified spontaneous emission (ASE)   379, 485
– Er/Yb-doped fiber   486
amplitude death   229
amplitude modulation (AM)   356, 366
– transmission function for   356
amplitude-shift keying (ASK)   366
analog communication   353
analog-to-digital converter (ADC)   12, 452
analytical approach   161–172
angular velocity   118
anticipative synchronization   227, 239
antiphase dynamics   96, 110, 120–122, 244–245
antiphase synchronization   213, 244–245
AOMs. *See* acousto-optic modulators (AOMs)
APDs. *See* avalanche photodiodes (APDs)
applications   5, 12
Ar$^+$ pump laser power   117
atomic polarization   41
atom-photon interaction   45, 46
attractor   32, 176, 177
– butterfly   32
– chaotic   32
– for Lorenz model   32
– strange   32
authentication   348–349
autocorrelation function   153–154
auxiliary system approach   313, 314
auxiliary-system approach for generalized synchronization
– experimental analysis on   313
– experimental setup   314–316
– model for   324–327
– numerical analysis on   327–328
avalanche photodiodes (APDs)   393, 395, 474

b

bandwidth enhancement   91, 92
– for random number generation   465
basic sciences   3
basin boundary   555
beam splitter (BS)   544
BER. *See* bit error rate (BER)
bias   491
biasing laser current   470
bidirectional message transmission   439
bifurcation   36–38, 67, 68
bifurcation diagram   520
– for Logistic map   30
– for Lorenz model   34
– for semiconductor laser   178
biological communication   423, 424
binary voltage signal   420
birefringent plate (BP)   139, 246, 374, 375
bit error rate (BER)   3, 353, 362–365, 397–399, 438
– error-free operation of   394
– test   386
– tester   486
– *vs* Q factor   364
bivalve shells   349, 350
blind source separation   533, 542–547
bluray disk (BD)   21
binary random bit sequences   447
bounded observability   430–435
box-counting dimension   555, 556

BP. *See* birefringent plate (BP)
Bragg grating   485
brain   424, 425
bubbling   230

c

carbon dioxide (CO_2) laser   128–132, 220
carrier density   156
carrier-photon interaction   45–46
C codes   581
CCOTDR   539–542
central limit theorem   542
challenge–response protocols   445
chaos   6–8
– and noise from real experimental data 66–67
– control, CO_2 laser, electro-optic modulator   519
– in electro-optic systems   98–107
– history of research activities of   4
– Ikeda-type nonlinear delay dynamics 98–101
– in electro-optic systems   98
– in fiber lasers   107–115
– in gas lasers   124–135
– in semiconductor lasers   69–98
– in solid-state lasers   115–124
– synchronization of (*See* synchronization of chaos)
chaos communication   346–348, 351, 353, 369–425
– advantages for   352
– analog communication   353
– basic idea of   346–348
– chaos-synchronization-based communication   351–352
– coherent   353
– compatibility   352
– hardware keys   348–349
– hardware-based communication   351
– multiplexing and noise tolerance   353–354
– numerical analysis of   373
– performance of
– – bit-error rate   362–365
– – eye diagram   365–368
– – modulation format   365–368
– – Q factor   362–365
– – signal-to-noise ratio   362–365
– privacy   352
– subcarrier   353
– synchronization for   349–351
– systems   365, 383, 405, 408, 417
chaos in laser
– characteristics of   64–65

– generation techniques   59–60
– – class C lasers, satisfying the condition for   63
– – external modulation   62
– – high-dimensional laser systems   63
– – Ikeda-type passive optical systems   63–64
– – insertion of nonlinear element   62–63
– – multimode lasers   63
– – optical coupling and injection   62
– – optical feedback   60–62
– intensity dynamics   8
chaos masking (CMS)   354–357, 371, 373, 403–405
– block diagram of   355
chaos mirror   548
chaos modulation   357–360, 402–405
chaos on-off keying   360–362
chaos pass filtering effect   355–356
chaos shift keying (CSK)   354, 360–362, 392, 403–405
chaos synchronization   8–9, 285
– experimental results of   287–290
– experimental setup   285–290
– generalized synchronization   285–289
– identical synchronization   301
– in two semiconductor lasers with optical feedback   293–300
– parameter dependence of   290–293
chaos theory, basic   26
– logistic map   26
– – bifurcation diagram   28–30
– – chaotic sequence   27–28
– – Lyapunov exponent   30–31
– – recurrence formula   26–27
– Lorenz model   31
– – bifurcation diagram   33
– – Lorenz equations   31
– – Lyapunov exponent   33–34
– – sensitive dependence on initial conditions   32–33
– – temporal waveform and attractor   32
– reconstruction of attractor   35–36
– – in time-delayed phase space   35–36
– route to chaos   36
– – intermittency   37–38
– – period-doubling   36–37
– – quasiperiodicity   37
chaos-based all-optical communication system   379
chaos-masking cryptographic scheme   382
chaotic carriers   348, 352, 390, 399, 439
chaotic correlation optical time-domain reflectometer (CCOTDR)   539–542
chaotic cryptosystems   408

chaotic dynamics in a solid-state laser
– multimode laser model with spatial hole burning   141–142
– single-mode laser model   140–141
chaotic dynamics in gas laser   142–143
chaotic fluctuations   375
chaotic instabilities   525
chaotic lasers   533
– for engineering applications   3
– for random number generation   452
– for remote sensing   539
– intensity fluctuations of   465
– system   475
– waveform of   516
chaotic lidar (CLIDAR) system   533–537
chaotic light scattering   551–554
chaotic optical waveforms   392
chaotic oscillations   131, 422, 526
chaotic oscillator   529
chaotic pulsing system   384
chaotic radar (CRADAR) system   536–539
chaotic scattering   548
chaotic search   527, 528
chaotic signal, AC component of   462
chaotic source signals, non-Gaussianity of   547
chaotic temporal waveform. *See* temporal waveforms
chaotic transmission system   379
chaotic waveforms   360, 371, 421, 513
– bandwidths of   536
– time-delayed signal   513
chaotic wavelength division multiplexing   421–423
chaotic wavelength fluctuations   373
chaotic wavelength-division multiplexing (CWDM)   421–422
chaotic-laser-based random number generators   452
chaotic-PIC-based communication system   397
characteristic equation   164, 172, 207
Christmas balls   552
Chua's circuit   487
cipher   344
ciphertext-only-attack   406–407
class A laser   49–50
class B laser   48–50
class C laser   48, 50
classification of laser models   48–51
CLIDAR   533–537
closed-loop configuration   227–228, 292–293, 312
CMOS semiconductor device   481

code   344
CO_2 gas laser, with saturable absorber   142
coherence   6
coherence collapse   71
coherent communication   353
coherent coupling   223
coherently coupled lasers   215–216
CO_2 laser   220–223
– controlling chaos   518–520
CO_2 ring laser   132
code-division multiple-access (CDMA)   353, 542
coherence, characteristic of laser   6
coherent optical feedback   274
colliding-pulse mode-locked lasers (CPMLL)   415–417
combined Tausworthe method   496
commercial optical-fiber networks   3
commercial physical random number generators   488
– Intel chip (Intel)   488–489
– quantis (ID quantique)   490
– random master (Toshiba)   489–490
– random streamer (FDK)   490
commercial random number generator   488
communicating spatiotemporal information   418
communication applications   446
communication by controlling chaos   529–532
communication security   436
communication system   388, 404 *See also* chaos communication
communication technique   420
communication theory of secrecy systems   405
compact disk (CD)   21
compatibility   352
complete synchronization   225
concealment   405
conditional Lyapunov exponent   229–230, 273, 309–311, 333–336
consistency   263–265. *See also* generalized synchronization
– application of   269–270
– in laser systems   265–268
continuous feedback control   513, 514, 518–521
continuous time delay error signal   520
continuous time-delayed feedback method   513
continuous-wave (CW)   521, 533
controlling chaos   7, 12, 13, 511–532
controlling chaos in lasers
– applications of   525–530

–– dynamical memory   528–529
–– relative intensity noise (RIN)   525–527
– continuous feedback control method   513–514, 518–521
– occasional proportional feedback (OPF) method   512–513, 515–518
– nonfeedback control method   514–515
– loss modulation   521–523
– semiconductor lasers, high-frequency injection method for   523–525
– stabilization to high-periodic oscillations   525
– OGY method   511–512, 530–532
conventional cryptologists   406
correlation optical time-domain reflectometer (COTDR)   539
coupled electro-optic systems   247
coupled Lang–Kobayashi equations   293, 296
– in auxiliary system approach   325
– for generalized synchronization   325–326 (*See also* auxiliary-system approach)
– for identical synchronization   295–296
coupling schemes   223
coupling strength mutual *vs.* unidirectional coupling   437
C programming language   581
CRADAR   536–539
cross-correlation   230–231, 290–293, 301
cryptography   10, 343–344, 406, 429, 500

d

data acquisition   432
data encryption   398
data recovery   398
decoding performance, security allocation   400
delayed optical bistable system   518
deterministic chaos   6–7, 19, 24–26
deterministic random number generators   446
DFB lasers   394, 414, 471
Dichtl method   504
– postprocessing method   504, 505
Diehard   457, 503
Diehard statistical tests   457
Diehard suite   467
Diehard tests   456, 462, 503
differential phase shift keying (DPSK) modulation   400–403
differential phaseshift QKD (DPS-QKD) experiments   472
diffusive coupling   218
digital communication systems   367, 368
digital versatile disk (DVD)   21

digital-to-analog converter (DAC)   469
dimensionless equations   184–188
dimensionless Lang-Kobayashi equations   184–187
dispersion compensation fibers (DCFs)   11, 396
dispersion-shifted fibers (DSF)   352, 474
distillation process   449
distributed Bragg reflector (DBR)   373
distributed feedback (DFB) lasers   377, 452, 471
dominant-mode ratio (DMR)   527
Doppler effect   118, 525
double-heterostructure   40–41
double-schroll attractor   488
doublescroll chaotic system   531
DPS-QKD scheme   472
DPSK light   402
DPSK modulation   402
drug delivery   270
dual synchronization   421
dynamical control of chaotic green laser   21
dynamical information hiding technique   424
dynamical memory   528–530

e

electrical signals, AC components   471
electric field   41, 156, 158
– amplitude   156, 158, 296
– phase   156, 158, 296
electron-hole pairs   40–41
electronic circuit   217, 369–370, 487
electro-optic Mach–Zehnder interferometer   384, 402
electro-optic modulator (EOM)   521
electro-optic phase chaos   401
electro-optic system   14, 64, 69, 98–107, 139, 384–386
– chaos in   98
– intensity chaos in   103–105
– numerical model for chaotic dynamics in   140
– phase chaos in   105–107
– synchronization of chaos in   245
– wavelength chaos in   101–103
electroencephalogram (EEG) data   542
embedding   35
encoded waveforms   406
encoding/decoding method   354–362, 403–405
encoding/decoding techniques
– chaos masking   354–357
– chaos modulation   357–360
– chaos shift keying   360–362

encryption   405
engineering applications   3
– with chaotic lasers   13
entropy generation, macroscopic effect   477
entropy rate   475
entropy source   475
erbium-doped fiber amplifiers (EDFAs)   11, 248, 352, 379, 414, 420, 441
erbium-doped fiber laser   109–110, 373–375, 386–390
– ion-pair concentration of fiber   112
– laser intensity   110
– nonlinear dynamic behavior   110
erbium-doped fiber-ring laser (EDFRL)   216, 373, 388, 419
erbium-doped optical fiber amplifier (EDFA)   540
Euler's method   208–210
exclusive OR (XOR) postprocessing method   452, 492
experimental analysis
– generalized synchronization   314
– semiconductor laser with optical feedback   145–156
– synchronization of chaos   285–293
external cavity frequency   60, 153
external cavity modes   168–169
eye diagram   367

f

Fabry-Perot cavity   38, 199
Fabry–Perot laser diode   527
Fabry–Perot semiconductor lasers   241
fast Fourier transform (FFT) 175, 176
fast temporal dynamics   114
FastICA algorithm   545
feedback
– control   511–521
– optical   69–79
– optoelectronic   83–87
– polarization-rotated optical   79–83
– system   225–227
feedback-induced instability   22
– in a semiconductor laser   21–23
fiber lasers   109, 112, 386–390
– numerical model for chaotic dynamics   140
fiber-optic communication channel   388
field experiment on chaos communication   377–381, 401–403
field programmable gate array (FPGA)   474
fluctuating signal   362
forward error-collection (FEC) methods   397
fourth-order Runge-Kutta method   209
fractal optical devices   547, 550

fractal optics   547
– regular-polyhedral mirror-ball structures, fractal patterns in   550–556
– wireless optical communications, chaos mirror for   548–550
fractal pattern   550
frequency jitter   482
frequency modulation (FM)   366
frequency shift keying (FSK)   366
frequency-shifted feedback-light injection   117
fringe contrast function   536
full-width half-maximum (FWHM)   536, 538

g

gain-absorption section (GAS)   411, 469
gain saturation   194–199
– effect   475
gas laser   142
– with saturable absorber   128
Gaussian distribution   480, 542
Gaussian noise   363
Gaussian statistics   363
generalized feedback shift register (GFSR) method   494, 495
generalized synchronization   225, 227, 258
– auxiliary system approach   259, 313–314
– dependence of synchronization quality on   336–337
– dependence on optical phase of feedback light   322–324
– with high correlation   225
– numerical analysis   327–328
– linear stability analysis   331–334
– with low correlation   258
– parameter dependence   329–331
– numerical simulation for   302
– two-dimensional (2D) maps   334–335
– in auxiliary-system approach   313–324
generation of chaos   7
GFSR   494
GHz-clocked DPS-QKD experiment   472
Ginzburg–Landau equation   408
Gram–Schmidt orthogonalization   210
green problem   20, 21, 122, 123

h

hardware   9, 351
hardware key   348–350, 407
He-Xe laser   127
HFI method   523–527
high-frequency injection (HFI) technique   22, 523–527
high-quality random number generators

– recipe for 503–504
highly reflective coated (HRC) 411, 412, 469
Hilbert transform 282–283
history of laser and chaos 3–4
homoclinic cycle 129
Hopf bifurcation 91
Hurst exponent 409
hyperchaos 191
hysteresis 95, 111,116–117, 130–131

i

ICA 542–547
identical synchronization 214, 224–225, 233, 239, 295
– rate equations in closed-loop configuration 337–339
– linear stability analysis of 308
– conditional Lyapunov exponent 309–310
– linearized equations 308–309
– open- *vs.* closed-loop configurations 312–313
Ikeda chaos 63–64
Ikeda model 99
imaginary electric field 160
incoherent communication 353
incoherent coupling 223
independence 448
independent component analysis (ICA) 542–547
– algorithms for 544
– blind source separation 542
– optical chaotic signals 544
– separation ability of 545
information-theoretic security 16, 427–431
– bidirectional message transmission 439–440
– bounded observability 430–431
– chaotic lasers, implementation of 431–435
– computational security 428
– concept of 429
– history of 429–430
– ultralong fiber laser system 441–444
– Maurer's satellite scenario 429–430
– public channel cryptography 435–439
– scheme for 430
– secure key distribution 427–428
– theoretical framework of 429
– Venn diagram for 431
information theory 428–431
InGaAs photodetector 537
InGaAsP/InP single-mode DFB semiconductor lasers 383

injection locking 88, 89, 130, 215, 216, 223, 227, 232, 288–293
in situ chaotic optical communication system 393
integrated tandem device (ITD) 413–415
intelligent transport systems (ITS) 543
intensity chaos 103–105
intensity fluctuation 23
intermittency route to chaos 37–38
intrinsic entropy 478
irregular pulsations of laser output 8

j

Jacobian matrix 51, 56, 162–164, 170–172, 180, 181, 229, 272, 273

k

Kaplan–Yorke (KY) dimension 190, 192–194
key agreement 429
key distribution scheme 435, 436
keyless cryptography 429
kink 71, 81–82, 148
Kolmogorov–Sinai (KS) entropy 190, 192–194, 413
Kolmogorov–Smirnov (KS) test 457, 503
KY dimension 190, 192–194

l

Lang–Kobayashi equations 14, 15, 135, 156–161, 166, 185, 194, 293, 338, 475, 524
– bifurcation diagrams 197
– dimensionless 185–188
– with gain saturation 194–199
– histogram 195, 196
– linearized 163, 170
– numerical results 195
– temporal waveforms 196
– synchronization of chaos
– – open loop 293
– – closed loop 338
Langevin equation 24
laser amplification 130
laser biasing current 412
laser cavity 38
laser current 22
laser demonstrates Lorenz chaos 522
laser diode (LD) 116, 529
laser dynamics 6
laser emission, polarization of 381
laser instabilities 19–23
– in ruby laser 19
laser intensity 525
– stroboscopic recording of 522, 523
– tools for measurement of instability 66

– traces of   517
laser medium   38
laser models   48–51
– class A lasers   49–51
– class B lasers   48–49
– class C lasers   48
– comparison with   68–69
laser output
– instabilities of   20
– temporal waveforms of   455
laser propagates   415
laser pump power   519
lasers   6–8
– fiber   109–115
– gas   124–135
– semiconductor   69–98
– solid-state   115–124
laser system, block diagram of   515
laser theory, basic
– light–matter interaction for laser radiation   38
– chaotic instability in lasers, mechanism of   45–46
– elements of laser   38–39
– mechanism of laser oscillation   39–40
– radiative recombination of electron–hole pairs   40–41
– rate equations for laser dynamics   41–44
– relaxation oscillation frequency   44–45
– two-atomic-level description   39–40
laser's chaotic output waveform   540
laser, history of research activities of   4
laser-chaos-based physical random number generators   453, 502
laser-diode-pumped Nd:YAG laser   124
lasers powers   440
laundering   425
leader-laggard relationship   240, 245
least significant bits (LSBs)   461, 463
legitimate user   407
LFFs. See low-frequency fluctuations
L-I characteristics   70–71, 80, 95, 147
L-I curve   147–149
light detection and ranging (LIDAR)   533
light-emitting diodes (LEDs)   552
light-matter interaction   38–39
linear congruential method   493
linear derivation   163
linear stability analysis   14, 15, 51
– conditional Lyapunov exponent   229–230
– for oscillatory trajectory   179–181, 188–190, 196–199
– for steady state solutions with optical feedback   169–172

– for steady-state solutions without optical feedback   162–166
– eigenvalues of Jacobian matrix   162–165
– relaxation oscillation frequency   165–166
– for generalized synchronization   331–335
– for identical synchronization   308–312
– oscillatory trajectory   179–184, 196–199, 308, 331
– steady state solutions   162–173
linearized equations
– for calculation of maximum Lyapunov exponent   53
– for coupled differential equations   271–273
– derivation, in Lorenz model   51–53
– general formula   51, 52
– generalized synchronization   331
– identical synchronization   308
– Jacobian matrix   51
– Lang-Kobayashi equations   163, 170, 197
– Lorenz model   52–53
– Rössler model   55
linearized variables   163, 170, 180
linewidth enhancement factor   157, 203
lithium neodymium tetraphosphate   120
LNP laser array   252
Logistic map   26–30
Lorenz chaos   31
Lorenz equations   31
Lorenz–Haken chaos   34, 124–128
Lorenz-Haken equations   25
Lorenz model   31–36
loss modulation   521, 525–526
low-frequency fluctuations (LFFs)   75–77, 149–151, 155, 241, 413, 437
– mechanism of   137–139
Lyapunov exponent
– general formula   53–54
– Lang-Kobayashi equations   183–185, 190, 191, 193, 195
– Logistic map   30–31
– Lorenz model   33–34
– without time delay, general formula for   53–54
Lyapunov spectrum   185, 190–192

m

Mach–Zehnder interferometer   472, 474
Mach–Zehnder lithium niobate (LiNbO$_3$) modulator   379
Mach–Zehnder modulators   245, 246
Mach–Zehnder transmission curve   384
map   26
Maurer's satellite scenario   429

maximum Lyapunov exponent 7, 181–185, 189, 197–198
Maxwell-Bloch equations 46
memory time 478, 479
Mersenne Twister 505
Mersenne twister 496, 505
message decoding 372
– schematic diagrams of 404
– time series of 404
message signal, encoding of 372
message, recovery of 389
metastability 481
methyl iodide 128
Michelson interferometer 533, 536
microscopic intrinsic noise 477
microscopic noise 475
model
– generalized synchronization 324
– semiconductor laser with optical feedback 156
– synchronization of chaos 293
modular neural network 408
monobit generation method 452, 460
monolithic photonic integrated circuits 410
M sequence 494
multibit generation 461
Multidisciplinary University Research Initiative (MURI) 375
multi-Gbits/s intensity chaos encoder 385
multimode laser model, with spatial hole burning 141–142, 280
multimode solid-state lasers 120
– antiphase dynamics 120–122
multiple-input multiple-output (MIMO) 542
multiplexing 353–354
– communication 421–423
mutual chaos pass filtering (MCPF) 436
mutual coupling 223, 228, 229, 435–439

n

N-mode laser 142
Nd-doped fiber laser 110
Nd:YAG laser 20–21, 220, 221, 370, 371, 515–517
NdP_5O_{14} solid-state laser 118
neodymium-doped yttrium aluminum garnet (Nd:YAG) crystal 123
neodymium-doped yttrium orthovanadate (Nd:YVO_4) microchip solid-state lasers 116, 543–544
neutral density filter (NDF) 241, 463
NH_3 laser 520, 521
NIST Federal Information Processing Standards Publication (FIPS) tests 483
NIST special publication 800–22 456, 497, 506
NIST SP 800–22 statistical test 456–459, 497–500
– approximate entropy test 508
– binary matrix rank test 507
– cumulative sums (Cusums) test 506–507
– discrete fourier transform (Spectral) test 507
– frequency (monobit) test 506
– frequency test within block 506
– linear complexity test 508–509
– longest-run-of-ones in block 507
– maurer universal statistical test 508
– nonoverlapping template matching test 507
– overlapping template matching test 507–508
– random excursions test 508
– random excursions variant test 508
– runs test 507
– serial test 508
noise 8, 23, 24, 66, 67
noise communication 346
nondeterministic bit 475
nondeterministic random number generators (NRNG) 446
nonfeedback control 514–515, 521–525
non-Gaussianity 547
nonlinear equations 8
nonperiodic-templates test 502
nonreturn-to-zero (NRZ) 365, 366, 455
– formats 365
– eye patterns for 367
– pseudorandom message 379
– waveform 390
novel chaos-generation process 402
numerical analysis
– dimensionless equations 184–187
– numerical results of chaotic dynamics of semiconductor laser with optical feedback 172
– linear stability analysis for oscillatory trajectory 179–181
– maximum Lyapunov exponent 181–184
– numerical results of chaotic dynamics 172–179
– on generalized synchronization of chaos in auxiliary system approach 327, 373
numerical model, for a chaotic semiconductor laser
– coherent optical feedback 135
– low-frequency fluctuations 137
– optoelectronic feedback 136–137

– polarization-rotated optical feedback 135–136
– semiconductor laser with optical feedback 172–184
– synchronization of chaos 300, 303

o

occasional proportional feedback (OPF) method 512–513, 515–518
OCCULT project 375–377
OGY algorithm 530
OGY method 511, 512, 530–532, 551
on-off phase shift keying 361
open-loop configuration 399
one-pair device 351
open-loop configuration 225–227, 292–293, 312
optical-carrier frequency 62, 215
optical chaos communication 4, 9–11, 369–425 *See also* chaos communication
– electro-optic systems 384–386
– encoding and decoding schemes 403–405, 417
– multiplexing communications 421–423
– polarization encoding 419–421
– spatiotemporal encoding 417–418
– European project for 375–377
– experimental system for 387
– fiber lasers 386–390
– history of 369
– in electronic circuits 369–370
– in optical systems 370–375
– integrated photonic circuit 410–417
– analogy to biological communication systems 423–424
– world of scientific fiction 424–425
– optical phase chaos 400–403
– bit-error-rate (BER) performance 397–399
– chaos communication experiment 396–397
– forward error-collection (FEC) methods 397
– photonic integrated circuits (PIC) 394–396
– subcarrier modulation 390–394
– unauthorized users, analysis for 399–400
– practical implementation of 369
– privacy issues 405–410
– private communications 370
– research activities of 376
– schematic setup for 380
– semiconductor lasers, with optical feedback 377–383
– semiconductor lasers, with optoelectronic feedback 383–384

– state-of-the-art performance of 391
Optical Chaotic Communications Using Laser Transmitters (OCCULT) project 375
optical chaotic systems 372
optical circulator (OC1) 539
optical communication 3, 9–11, 23, 369–425
– with chaotic semiconductor lasers 21, 375
– with synchronized chaotic lasers 9–11
optical data storage systems 21
optical disk 21–23
optical feedback into semiconductor laser 21
optical feedback phase 407
optical fiber laser, antiphase response of 111
optical frequency detuning 302–303
optical injection 91
– bandwidth enhancement of chaos 91–92
optical intensity in laser cavity 7
optical isolators 465, 540
optical network 379
optical noise
– for random number generation 485
– for secure key distribution 441
optical phase of feedback light 322, 336
optical ray 547
optical signals 396
optical spectra 288
optical systems 417
optical timedomain reflectometer (OTDR) 539
optical wavelength detuning 288–293, 465
optical-confinement devices
– Q-value of 555
optical-feedback-induced chaos in semiconductor lasers 355
optical-fiber communications 465
optical-fiber components 454
optoelectronic feedback loop 383
orthogonal polarizations 111
oscillation frequency 8

p

parameter dependence
– generalized synchronization 320, 329
– synchronization of chaos 305
parameter estimation 406, 408
parameter mismatch 225, 228, 388
passive Q-switching 128
– of CO_2 lasers 253
– pulse 129
pattern generator 525
PDF 542, 546–547
Pecora-Carroll method 217, 218, 270, 271
– for synchronization 270

performance evaluation of chaos communication 390–405
period-doubling route to chaos 36–37, 71, 72, 74, 75
phase chaos 105–107, 400–403
– oscillations 255
phase modulation (PM) 366
phase noise 486
phase shift keying 366
phase synchronization 255
phase-shift keying (PSK) 366
photodetector (PD) 374, 469
photon detectors 472
photonic crystals 547
photonic emission, quantum mechanics of 485
photonic integrated circuit (PIC) 377, 394, 411, 469
– for chaos communication 394–400, 410–417
– different types of 377
– for random number generation 469
photonic integrated device 411
photon lifetime 44, 45
photoreceiver 392
physical random number generators 449, 480–488
– chaotic dynamics, in electronic circuits 487
– optical noise 485–486
– quantum noise 484–485
– radiation, from radioactive nuclide 486–487
– thermal noise
– – direct amplification of 480–481
– – metastability 481–482
– – two oscillators with frequency jitter 482–483
– traditional physical devices 487–488
– with chaotic lasers 452–471
PICASSO project 377
PICs designed 470
p-n junction 40–41
Poincaré sections 36, 112
Poisson's distribution law 487
polarization encoding 419–421
polarizaiton-rotated optical feedback 275–276
polarization 22
polarization controller (PC) 419, 540
polarization-maintaining fibers 452
polarizing beam splitter 534
polyhedral mirror–ball structures 554
population inversion 39, 41
population lifetime 44–45

positive Lyapunov exponent 182, 192
postprocessing 463, 490
potassium titanyl phosphate (KTP) crystal 123
privacy 352, 405–410
– amplification 435
private key 427
PRNG. See pseudorandom number generator (PRNG)
probability density function (PDF) 476, 542, 546–547
probability theory 428
proportion of p-values 501
pseudorandom bit sequence (PRBS) 385
pseudorandom number generator (PRNG) 446, 450, 451, 493
– combined Tausworthe method 496
– linear congruential method 493–494
– M sequence and generalized feedback shift register (GFSR) 494–496
– Mersenne Twister (MT) 496–497
pseudorandom number (PN) 533
public channel cryptography 435–439
pump energy 38–39
pump modulation 117, 525
p-value 499
Pyragas method 218–219

q
Q factor 362–368
Q-switching pulses 8, 128, 253
Quantis 490
quantum bit error rate (QBER) 472, 474
quantum cryptography 445
quantum key distribution (QKD) 427, 428, 472
quantum noise 484
quantum physical random number 484
quantum random number generator 484
quantum RNGs (QRNGs) 484
quasiperiodicity 37
quasiperiodicity route to chaos 37, 74, 75, 81, 82, 84–87, 153, 156, 176

r
radiative recombination 40
radio-frequency (RF) 459, 520, 535
radioactive nuclide 486
random bit generation 463, 465
random bit pattern 448, 455
random bit sequences 448, 454, 460, 467–469
random bit signal, eye diagram of 455
random bit stream, generation of 463
Random Master 489

random molecular distribution 488
random number 447
random number generation 11–13, 445–509
random number generators (RNGs) 11–13, 445–509
– chaotic lasers, bandwidth enhancement of 465–469
– chaotic lasers 447, 451–452
– Chua's circuit 487
– computing/communication applications 446
– conventional random number generators 451–452
– Dichtl method, for postprocessing 504–505
– NIST special publication 800–22 test suite 497–502
– p-values 501–502
– generation speed, postprocessing for 463–464
– Mersenne Twister 505–506
– monobit generation, with one lasers 460
– monobit generation, with two lasers 452–460
– NIST special publication 800–22 506–508
– entropy rate 478–480
– p-values 499–501
– photonic integrated circuit (PIC) 469–472
– postprocessing 450
– postprocessing techniques for randomness improvement 490
–– exclusive-OR (XOR) method 492–493
–– von Neumann method 491–492
– randomness, statistical tests of 503
– strategies for statistical analysis 497–499
– independence 448–449
– physical random number generators 449–450
– pseudorandom number generator 451
– unpredictability 449
– with chaotic lasers, examples 452
random signal radars (RSRs) 536
Random Streamer 490
random bit generation 471
rate equations 41–44, 139, 140, 142, 144, 145
rays 550
real and imaginary electric fields 160, 299
real electric field 160
reconstruction of attractor 35–36
recurrence formula 26
Reed–Solomon (RS) code 397
regular-polyhedral mirror–ball structures 550–556
regular pulse package 78–79
relative intensity noise (RIN) 523, 525–527

relaxation oscillation frequency 44, 45, 55–57, 62, 119, 122, 165, 166, 207, 215
remote sensing
– chaotic lidar 533–536
– chaotic radar 536–539
retarded synchronization 227, 239
return-to-zero (RZ) formats 365, 366
RF amplifiers 535
RF spectrum analyzer 540
RIN 525–527
RNGs. See random number generators (RNGs)
root-mean-square (RMS) values 443
Rössler model 54–55
route to chaos
– intermittency 37–38
– period-doubling 36–37
– quasiperiodicity 37, 112–114
rotational relaxation rate 143
ruby laser 19–20
Runge–Kutta method 173, 187, 208–210
RZ 365–366

s
saddle-node bifurcation (SN) 91
scientific fiction 424–425
secret communication scheme 435
secret communication, history of 343
– cryptography 343–344
– noise communication 346
– steganography 344–346
secret key agreement, satellite scenario for 429
secret key distribution 440, 441
secure communication 343
secure key distribution 427–444
secure key generation 5, 8, 433–434
security 9, 344, 399, 406–410, 427–431
self-mixing laser Doppler velocimetry scheme 120, 121
self-similar dimension 555–556
semiconductor laser with optical feedback, analytical approach of 161
– linear stability analysis for steady-state solutions
–– with optical feedback 169–173
–– without optical feedback 162–166
– steady-state solutions
–– with optical feedback 166–169
–– without optical feedback 161–162
semiconductor laser with optical feedback, model for 156
– Lang–Kobayashi equations 156–158
–– derivation of the electric-field amplitude and phase 158–159

−− derivation of the real and imaginary electric fields of 160–161
−− parameter values used for numerical simulation 157
semiconductor lasers 7, 21, 463
− analysis of chaotic laser dynamics 145–210
− analysis of synchronization of chaos 285–313
− bandwidth of 447
− chaos, use of 446
− controlling chaos 523–527
− derivation of rate equations 199–206
− dynamical behavior of 77, 78
− experimental analysis of 145–156
− for optical disks 21
− injection current of 524
− intensity fluctuation of laser output 22
− interferometer 516–519
− numerical model for chaotic dynamics in 135
− radiative recombination of electron–hole pairs in 40–41
− random number generation 452
− secure key distribution 431–440
− short-cavity regime 78–79
− with injection current modulation 93
− with optical feedback 69–77, 206–208, 372, 377–383
−− under the limits of weak and short feedback 206–208
− with optical injection and coupling 88–92
− with optical self-feedback 7
− with optoelectronic feedback 83–87, 383–384
− with polarization-rotated optical feedback 79–83
semiconductor optical amplifier (SOA) 395, 471
short cavity regime 78
sifted key generation rate 474
signal-to-noise ratio (SNR) 22, 353, 364, 383, 404
signal/reference waveforms 535, 538
single-mode DFB InGaAsP–InP semiconductor lasers 537
single-mode distributed-feedback (DFB) 535
single-mode fiber (SMF) 395
single-mode laser model based on Maxwell–Bloch equations 46–47
sensitive dependence on initial conditions 6, 27–28, 32–33
short cavity regime 78–79
single-mode ring laser equations 47
signal-to-noise ratio (SNR) 362–365

slowly varying envelop 201, 204, 206
slowly varying envelope approximation 47, 201
software-based cryptography 406
solid-state lasers 19
− for second-harmonic generation, green problem in 20–21
− with external modulation 116
− with short cavity length 116
spatiotemporal chaos 417–418
spatiotemporal chaos communications 417
spatiotemporal encoding 417–418
special trading authentication system 349
spontaneous emission 39
spontaneous emission noise 485
static keys 348
statistical evaluation of random numbers 497
steady state solutions 161–162
steganography 10, 344–346, 406
stimulated emission 39
subcarrier communication 353
subcarrier frequency 390
subcarrier modulation 390–394
substitution 343–344
symmetry breaking 240–243
synchronizability 350
synchronization 5, 8, 9, 211, 212
synchronization of chaos 5, 8–9, 211, 213–214, 347, 348, 351, 352
− identical synchronization 214
− cross-correlation values 213
− defined 212
− for communication applications 216–217
− electro-optic systems 245–248
− fiber lasers 248–250
− gas lasers 253–254
− solid-state lasers 250–253
− in electronic circuits 217
−− Pecora–Carroll method 217–218, 270–271
−− Pyragas method 218–219
− in feedback systems 225–230
− in semiconductor lasers 230–245, 285–313, 435–440
−− vertical-cavity surface-emitting lasers 243–245
−− with coherent optical feedback 231–236, 285–313
−− with mutual coupling 240–243
−− with optical injection 239–240
−− with optoelectronic feedback 237–239
−− with polarization-rotated optical feedback 236–237
− numerical model for

– – in unidirectionally coupled electro-optic system   277
– – in unidirectionally coupled fiber lasers   278–279
– – in unidirectionally coupled gas lasers   281–282
– – in unidirectionally coupled semiconductor lasers   274–277
– – in unidirectionally coupled solid-state lasers   279–281
– of periodic oscillations   213
– parameter dependence   305
– generalized synchronization   258–263
– phase synchronization   254–258
– stability of the synchronous solutions   214–215
synchronous solution
– for identical synchronization   295
– for generalized synchronization   327

t

Tang–Statz–deMars equations   116, 141, 142, 280
temporal waveforms
– and attractor   32
– and correlation plots   303
– and radio-frequency (RF) spectra of laser output intensity   149, 151, 315
– chaotic   123, 228, 251, 254, 265
– investigation of delay time of   232
– irregular   21
– length of   546
– numerical results of   174, 303, 330
– of binary codes   528
– of laser intensity without gain saturation   195
– of Nd:YVO$_4$ solid-state microchip laser output with pump modulation   117
– of noise   8
– of the two mixed chaotic signals   545
– of VCSEL   245
– relaxation oscillation of laser intensity   44
– restabilization of   82
– unstable   9
tetrahedron   547–556
thermal noise   480–483
time-delay embedding   35, 126
time-delayed feedback systems   357
time-dependent standard deviation   477
time-interval analyzer (TIA)   474
time series analysis   406, 408, 409
time-varying   349, 350
traditional physical device   487, 488
transistor   481

transmission function   356
transmitted signal   371
transmitter
– chaotic signals of   359
– laser output power   372
– RF spectra of   355
– unperturbed signal   373
transmitter–receiver synchronization   424
transposition   343
transverse Lyapunov exponent.   229, 230 *See also* conditional Lyapunov exponent
transverse-magnetic(TM) polarization modes   419
TV camera   382
TV signal   381
TV synchronism   383
TV video signal   381–383
two-atomic-level description   39, 40
two dimensional map   292
two types of synchronization   305

u

ultralong fiber laser (UFL) system   441–444
ultrastable laser   13
unauthorized user   399, 400
unidirectional coupling   223
uniformity of p-values   501
unpredictability   192, 449, 475
unstable periodic orbits (UPOs)   511–513

v

variable optical attenuator (VOA)   392, 412
variational equations   163
vertical-cavity surface-emitting lasers (VCSELs)   93–98, 139, 243
– numerical model for chaotic dynamics in   139
– optical feedback   96, 97
– optical injection   97, 98
– polarization-resolved temporal dynamics of   97
video signals   381
von Neumann method   491
von Neumann postprocessing method   491, 492

w

Wada basin property   547
wave length division multiplexing (WDM)   400
waveforms   535. *See also* temporal waveforms
wavelength chaos   102, 103
wavelength division multiplexing (WDM)   421–423

wavelength filters (WFs) 421
wavelength-tunable distributed-Bragg-reflector (DBR) semiconductor laser 246
wireless optical communication 548
wire-tap channel 429

x

XOR device 452, 459, 492

z

zero-lag synchronization 242, 243